Biomedical Applications and Toxicity of Polymers, Nanoparticles, Biomaterials, and Metal Ions

Deniz Eren Erişen
https://orcid.org/0000-0003-3416-3732
Institute of Semiconductors, Guangdong Academy of Sciences, China

Faez Iqbal Khan
https://orcid.org/0000-0001-9088-0723
Xi'an Jiaotong-Liverpool University, China

Cheng Chang
https://orcid.org/0000-0001-6623-9223
Institute of New Materials, Guangdong Academy of Sciences, China & University of Birmingham, UK

Kai-Cheng Yan
University of Bath, UK

IGI Global
Scientific Publishing
Publishing Tomorrow's Research Today

Vice President of Editorial	Melissa Wagner
Director of Acquisitions	Mikaela Felty
Director of Book Development	Jocelynn Hessler
Production Manager	Mike Brehm
Cover Design	Jose Rosado

Published in the United States of America by
IGI Global Scientific Publishing
701 East Chocolate Avenue
Hershey, PA, 17033, USA
Tel: 717-533-8845 | Fax: 717-533-7115
Website: https://www.igi-global.com E-mail: cust@igi-global.com

Library of Congress Cataloging-in-Publication Data

Names: Erişen, Deniz Eren, 1990- editor | Khan, Faez Iqbal, 1985- | Chang, Cheng, 1993- | Yan, Kai-Cheng, 1995-
Title: Biomedical applications and toxicity of polymers, nanoparticles, biomaterials, and metal ions / edited by Deniz Eren Erişen, Faez Iqbal Khan, Cheng Chang, Kai-Cheng Yan.
Description: Hershey, PA : IGI Global Scientific Publishing, [2026] | Includes bibliographical references and index. | Summary: "The book aims to provide a comprehensive overview of the biomedical applications and potential toxicity of various materials used in the field. The book intends to cover various materials, including polymers, nanoparticles, biomaterials, and metal ions, and their applications in biomedical fields"-- Provided by publisher.
Identifiers: LCCN 2024055585 (print) | LCCN 2024055586 (ebook) | ISBN 9798337300559 hardcover | ISBN 9798337300566 paperback | ISBN 9798337300573 ebook
Subjects: LCSH: Biomedical materials--Biocompatibility http://id.loc.gov/authorities/subjects/sh2010001388 | Nanoparticles--Toxicology | Metals--Toxicology http://id.loc.gov/authorities/subjects/sh85084255
Classification: LCC R857.M3 B5743 2025 (print) | LCC R857.M3 (ebook) | DDC 610.28/4--dc23/eng/20250408
LC record available at https://lccn.loc.gov/2024055585
LC ebook record available at https://lccn.loc.gov/2024055586
British Cataloguing in Publication Data
A Cataloguing in Publication record for this book is available from the British Library.

Table of Contents

Detailed Table of Contents

Saloua Mabsor-Zgandaoui, Ibn Tofail University, Morocco
Khawla Rachmoune, Université Claude Bernard Lyon, France
Ilham Aftais, Ibn Tofail University, Morocco
Fatima Ezzahra Elamrani, Cadi Ayyad University, Morocco
Imade Amradi, Cadi Ayyad University, Morocco
Adil El Housseini, AQUAMAR Laboratory, Faculty of Sciences, Ibn Zohr University, Morocco
Youssef Ait Hamdan, Cadi Ayyad University, Morocco
Youness Zgandaoui, Ibn Tofail University, Morocco
Abdelghani Iddar, Biotechnology and Biomolecule Engineering Unit, CNESTEN, Morocco
Mohammed El Mzibri, Biotechnology and Biomolecule Engineering Unit, CNESTEN, Morocco
Adnane Moutaouakkil, Biotechnology and Biomolecule Engineering Unit, CNESTEN, Morocco
Aboubaker El Hessni, Ibn Tofail University, Morocco
Abdelhalim Mesfioui, Ibn Tofail University, Morocco

Biomedical polymers have played a transformative role in modern medicine, driving innovation in drug delivery, tissue engineering and medical devices. These materials are classified as natural or synthetic and possess unique properties such as biocompatibility, biodegradability and mechanical versatility. Natural polymers support cellular functions and tissue regeneration, while synthetic polymers allow precise control of degradation rates and mechanical properties. Polymers have facilitated the controlled delivery of drugs through polymer-drug conjugates and stimulant systems, enabling targeted and sustained release. However, challenges such as toxicity, immune responses and regulatory hurdles remain. By exploring their applications, challenges and future prospects, this chapter has shed light on the essential role of biomedical polymers in advancing personalized medicine, regenerative therapies and smart medical devices.

Yusuf Olatunji Waidi, Indian Institute of Science, India

Nanotechnology has revolutionized medicine, offering innovative solutions to longstanding healthcare challenges. This chapter provides a comprehensive overview of nanoparticles, their unique properties, and their diverse medical applications. It explores the historical context of nanoparticle use in therapeutic approaches and delves into various nanoparticle types, including their properties and applications. The chapter also examines the interaction between nanoparticles, cells, and tissues, discussing cellular uptake

The transformative role of polymers, biomaterials, nanomaterials, and metal ions in breast cancer treatment have highlighted their potential to enhance diagnosis and therapeutic outcomes. Traditional cancer therapies have often fallen short due to their "one-size-fits-all" approach, leading to increased patient non-responsiveness. The integration of interdisciplinary research—from biomedical science to nanotechnology—promises more precise and effective treatments. Polymers enable targeted drug delivery with reduced side effects, while biomaterials enhance regeneration and drug delivery systems. Nanomaterials improve drug accumulation in tumors and combat resistance, and metal ions disrupt cancer cell metabolism and DNA. By exploring case studies and applications of these advanced materials, this paper underscores their significant contributions to personalized cancer therapies and improved patient quality of life. Looking forward, innovations in multifunctional nanomaterials and biosensors are anticipated to further revolutionize breast cancer treatment strategies.

PdNPs have received much attention in nanomedicine due to their unique physicochemical values such as high thermal and chemical stability, high surface area, & tuneable optical properties. These characteristics make them suitable for many biomedical applications, including cancer therapy, antibacterial treatments, drug delivery, & biosensing. In this chapter, synthesis methods for PdNPs are reviewed from the environmental friendliness view point, & the properties are described. Enhanced PdNPs' stability, biocompatibility, and efficacy in biological systems are explored through the development of functionalization strategies. PdNPs possess great promise for photothermal therapy in targeted cancer treatment, antibacterial effectiveness against multidrug-resistant pathogens, & drug delivery vehicles. They also offer support for early disease detection in biosensing technologies. Finally, safety and toxicity issues are also examined, emphasizing the need for careful evaluation of safe and effective clinical applications of PdNPs.

Chitosan and its derivatives have garnered increasing interest in oncology due to their unique properties, such as biocompatibility, bioactivity and their ability to interact with molecules. These characteristics make them promising candidates for the development of novel anticancer therapies. Chitosan can be employed as drug delivery systems, enhancing the availability and targeted release of therapeutic agents into cancer cells. Additionally, they have demonstrated the ability to inhibit cell proliferation, induce apoptosis, and modulate immune responses, positioning them as multifunctional tools in the fight against cancer. Recent research has focused on the development of chitosan-based nanoparticles for targeted drug delivery in chemotherapy, reducing side effects on healthy cells. Furthermore, Chitosan systems have been used to deliver siRNA and other therapeutic molecules, inhibiting tumor growth at the genetic level. These advancements highlight chitosan's potential in developing more effective and less toxic cancer treatments.

Ionizing radiation (IR) plays a central role in the biomedical field, both for diagnosis and for the treatment of pathologies such as cancer. However, exposure to IR can cause significant cellular damage, affecting both tumor cells and neighbouring healthy tissue. In this context, the use of effective radioprotective agents is a necessity. Chitosan (CS), a natural biopolymer derived from chitin, stands out for its antioxidant, anti-inflammatory and reparative properties, as well as its biocompatibility and biodegradability. These characteristics make it a promising candidate for radiation protection. This chapter examines the radioprotective mechanisms of chitosan, in particular its ability to trap reactive oxygen species and enhance the activity of antioxidant enzymes. It also highlights its many biomedical applications, particularly in tissue engineering and controlled drug release. Finally, the limits to its clinical use and recent advances aimed at improving its physico-chemical properties are discussed, paving the way for new therapeutic strategies in radiation protection.

Owing to their biodegradability, superior mechanical properties, and excellent in-vitro/in-vivo biological compatibility, magnesium (Mg) alloys have attracted tremendous attention as innovative body-absorbable biomaterials. Mg-based alloys application and classification based on mechanical strength and corrosion resistance for biomedical implants are discussed in this book chapter. In addition, the effect of alloying elements: Zn, Ca, and RE-elements for property enhancement in Mg alloys has been also presented. Mg-Zn-Ca alloys, show improved tensile strength and corrosion behavior appropriate to bone implants whereas Mg-RE alloys offer superior strength and ductility more suited towards demanding higher performance applications. This review will also consider the difficulties in controlling Mg alloy corrosion rate so that it does not proceed at a dangerous and predictable pace within the body. The possibilities of magnesium alloys to change the face of the biomaterial area with degradable implants give rise to many exciting options for future medical advancements.

This chapter provides a comprehensive exploration of the biological interactions of bone implants, focusing on biocompatibility and biodegradability in modern medical applications. It discusses the classification of biomaterials and medical devices, emphasizing the importance of understanding their interactions with biological systems. The chapter also delves into the challenges of Periprosthetic joint infection (PJI), and the clinical significance of biocompatibility. By integrating multidisciplinary knowledge from materials science, biomechanics, and clinical medicine, this chapter aims to provide readers with comprehensive theoretical foundations and practical guidance, thereby facilitating the development of safer and more durable joint replacement solutions.

Metal and metal oxide nanoparticles (NPs) show great promise in biomedical applications, but their toxicity varies. Metallic NPs are stable, while metal oxide and alloy NPs dissolve in biological conditions, releasing metal ions that generate reactive oxygen species (ROS), causing oxidative stress and cellular damage. Their rapid distribution and accumulation in tissues raise toxicity concerns. This chapter reviews

Preface

The field of biomedical materials has undergone a remarkable transformation in recent decades, driven by the convergence of materials science, nanotechnology, and biomedical engineering. This book, titled "Biomedical Applications and Toxicity of Polymers, Nanoparticles, Biomaterials, and Metal Ions," aims to provide a comprehensive and critical overview of the latest developments in the use of advanced materials in biomedical applications, while also addressing the crucial aspect of their potential toxicological impacts. The integration of synthetic and natural materials into medical systems has revolutionized diagnostics, therapeutics, and regenerative medicine. Polymers, nanoparticles, biomaterials, and metal ions are now central to innovations such as drug delivery systems, implantable devices, tissue engineering scaffolds, and biosensors. However, as these materials become more complex and their applications more widespread, concerns regarding their biocompatibility, long-term safety, and environmental impact have grown. This book explores both the promise and the peril of these materials. It delves into their physicochemical properties, mechanisms of interaction with biological systems, and the methodologies used to assess their safety. By presenting a balanced view, we aim to inform researchers, clinicians, and policymakers about the dual nature of these materials—both as enablers of medical progress and as potential sources of risk.

In today's rapidly advancing biomedical landscape, the demand for materials that are not only functional but also safe and sustainable is more pressing than ever. The COVID-19 pandemic, for instance, highlighted the critical role of nanomaterials in vaccine delivery and diagnostics, while also raising questions about their long-term effects. Similarly, the rise of personalized medicine and wearable health technologies has intensified the need for materials that can seamlessly integrate with human physiology. This book is particularly timely as it addresses these emerging challenges and opportunities. It serves as a bridge between innovation and responsibility, offering insights that are essential for the development of next-generation biomedical technologies.

This volume is intended for a diverse audience that includes researchers and scientists in materials science, nanotechnology, toxicology, and biomedical engineering; graduate and postgraduate students seeking a comprehensive introduction to the field; medical professionals and clinicians interested in understanding the materials behind modern medical devices and therapies; regulatory bodies and policymakers involved in the approval and monitoring of biomedical products; and industry professionals engaged in the design, testing, and commercialization of biomedical materials.

CHAPTER OVERVIEW

The book launches with the opening chapter entitled "Biomedical Polymers: Foundations and Future Innovations in Medicine," exploring the pivotal roles of polymers within the medical domain. This chapter spotlights their current applications in tissue engineering and drug delivery. Following this, the second chapter, "Nanomaterials in Medical Applications: An Introduction," offers a comprehensive overview of nanoparticles, detailing their properties and interactions with biological systems.

The third chapter, "Nano-Bioscience: Introduction and Nanoparticles in Medicine," introduces readers to the classification and structural diversity of nanomaterials, emphasizing their applications in diagnostics and therapeutics. In contrast, the fourth chapter, "From Imaging to Drug Delivery and Treatment: Nano-Biomaterials for Nervous System Disorders," focuses on the utilization of nano-biomaterials for treating neurological disorders, with particular attention to their ability to traverse the blood-brain barrier.

Chapters five and six, "Application of Responsive Intelligent Biomaterials in Regenerative Medicine" and "Electroactive Biomaterials for Cell Behavior Regulation and Tissue Regeneration," respectively, delve into the use of smart and electroactive biomaterials within the realm of tissue regeneration. The former chapter reviews responsive biomaterials that react to stimuli, while the latter examines biomaterials that replicate bioelectric signals to influence cell behavior and promote healing.

The seventh chapter, "Transformative Approaches in Breast Cancer Treatment: The Role of Polymers, Biomaterials, Nanomaterials, and Metal Ions," presents an interdisciplinary approach to cancer therapy, integrating advanced materials to enhance both diagnosis and treatment outcomes.

Chapters eight and nine, "Therapeutic Applications of Palladium Nanoparticles (PdNPs)" and "Therapeutic Implications of Chitosan and Its Derivatives – Focus on Cancer Treatment," discuss the biomedical potential of PdNPs and chitosan in cancer therapy, antibacterial treatments, and biosensing. The eighth chapter emphasizes the applications of PdNPs, while the ninth chapter focuses on chitosan's role in enhancing drug delivery and modulating immune responses in oncology.

The tenth and eleventh chapters, "Radioprotective Effects of Chitosan Polysaccharides and Biomedical Applications" and "Chitosan Polysaccharides for Osteoporosis Therapy in Postmenopausal Women," investigate chitosan's antioxidant and anti-inflammatory properties in radiation protection, as well as its potential in bone regeneration and non-hormonal osteoporosis treatments.

The twelfth chapter, "Exploring the Role of Metal Ions in Modern Medicine from Antimicrobial Coatings to Diagnostic Imaging and Targeted Therapies: Metal Ions in Medicine: Coatings, Imaging & Therapies," discusses the diverse roles of metal ions in modern medical applications, including antimicrobial coatings and diagnostic imaging.

Chapters thirteen and fourteen, "Role of Magnesium and Its Composite for Biomedical Application: A Future Demand – Role of Magnesium Alloy for Biomedical Application" and "Biological Interactions of Implants: From Biocompatibility to Biodegradability in Modern Medical Applications," explore the potential of magnesium alloys as biodegradable implants and provide a comprehensive overview of implantable materials and their interactions with biological systems.

Finally, the fifteenth chapter, "Toxicity of Metal and Metal Oxide Nanoparticles; Impact of Physical State, Mechanisms of Action, and Strategies for Remediation in Biomedical Applications," examines the toxicological profiles of metal-based nanoparticles and strategies for risk mitigation. This chapter offers insights into the latest advancements in biomaterials and nanotechnology's transformative impact on modern medicine.

This book offers a holistic and interdisciplinary perspective on the biomedical applications and toxicity of advanced materials. By combining foundational knowledge with cutting-edge research, it serves as a valuable resource for understanding both the potential and the pitfalls of modern biomaterials. Our goal is to foster a more informed and responsible approach to the development and deployment of these materials. We hope that this volume will inspire further research, guide safe innovation, and ultimately contribute to the advancement of healthcare technologies that are not only effective but also safe and sustainable.

Deniz Eren Erisen
Guangdong Academy of Sciences, China

Chapter 1
Biomedical Polymers: Foundations and Future Innovations in Medicine

Saloua Mabsor-Zgandaoui
https://orcid.org/0009-0006-4601-9059
Ibn Tofail University, Morocco

Khawla Rachmoune
Université Claude Bernard Lyon, France

Ilham Aftais
https://orcid.org/0000-0002-9609-7411
Ibn Tofail University, Morocco

Fatima Ezzahra Elamrani
https://orcid.org/0009-0009-0248-1964
Cadi Ayyad University, Morocco

Imade Amradi
https://orcid.org/0009-0003-3063-5058
Cadi Ayyad University, Morocco

Adil El Housseini
https://orcid.org/0009-0007-9294-2685
AQUAMAR Laboratory, Faculty of Sciences, Ibn Zohr University, Morocco

Youssef Ait Hamdan
Cadi Ayyad University, Morocco

Youness Zgandaoui
Ibn Tofail University, Morocco

Abdelghani Iddar
https://orcid.org/0000-0001-6111-0607
Biotechnology and Biomolecule Engineering Unit, CNESTEN, Morocco

Mohammed El Mzibri
Biotechnology and Biomolecule Engineering Unit, CNESTEN, Morocco

Adnane Moutaouakkil
https://orcid.org/0000-0002-1762-881X
Biotechnology and Biomolecule Engineering Unit, CNESTEN, Morocco

Aboubaker El Hessni
Ibn Tofail University, Morocco

Abdelhalim Mesfioui
Ibn Tofail University, Morocco

ABSTRACT

Biomedical polymers have played a transformative role in modern medicine, driving innovation in drug delivery, tissue engineering and medical devices. These materials are classified as natural or synthetic and possess unique properties such as biocompatibility, biodegradability and mechanical versatility. Natural polymers support cellular functions and tissue regeneration, while synthetic polymers allow

DOI: 10.4018/979-8-3373-0055-9.ch001

precise control of degradation rates and mechanical properties. Polymers have facilitated the controlled delivery of drugs through polymer-drug conjugates and stimulant systems, enabling targeted and sustained release. However, challenges such as toxicity, immune responses and regulatory hurdles remain. By exploring their applications, challenges and future prospects, this chapter has shed light on the essential role of biomedical polymers in advancing personalized medicine, regenerative therapies and smart medical devices.

INTRODUCTION

The evolution of biomedical science has been significantly affected by advances in materials science, with polymers becoming an essential element of modern medical applications. Biomedical polymers, through their unique structure diversity, biological compatibility and mechanical properties, have transformed the landscape of medical devices, tissue engineering and drug delivery. Their capability to mimic biological systems, sustain regenerative processes and support cellular interactions has made them indispensable in tackling some of the most compelling medical challenges of our time (Hu, T et al. 2024).

The simplest definition of polymers refers to the large macromolecules made up of repeating sub-units called monomers. In the biomedical field, they are classified into two main categories: natural polymers and synthetic polymers. Natural polymers, such as chitosan, hyaluronic acid and collagen, are derived from biological sources and are reputed for their biocompatibility, biodegradability and low immunogenicity. These properties make natural polymers ideal for applications such as wound healing, tissue scaffolding and drug delivery. Meanwhile, synthetic polymers such as polylactic acid (PLA), polyethylene glycol (PEG) and polycaprolactone (PCL) are engineered to provide precise control over mechanical strength, degradation rates and chemical functionality. This control has allowed tailor-made materials to be designed to meet the specific requirements of various biomedical applications. The integration of biomedical polymers into medical practice has been motivated by their ability to overcome the limitations of conventional materials. Natural materials, such as metals and ceramics, were often inadequate due to their rigidity, potential toxicity or lack of biodegradability. Polymers, by contrast, demonstrated superior flexibility and processability and were potentially bioresorbable, making them suitable for temporary implants, drug carriers and scaffolds for tissue regeneration (Satchanska, Davidova, and Petrov 2024).

Drug delivery systems have emerged as one of the most innovative applications for biomedical polymers. Polymers facilitated the controlled release of therapeutic products, ensuring sustained, localised and targeted drug delivery. For example, polymer-drug combinations and drug-encapsulating nanoparticles offer improved bioavailability, reduced side effects and enhanced therapeutic efficacy. Innovations such as pH-, temperature- and stimuli-responsive polymers have further revolutionised targeted drug delivery, enabling the precise release of therapeutic agents in response to changes in the microenvironment (Trucillo 2024).

The role of polymers in tissue engineering and regenerative medicine has been equally transformative. By serving as scaffolds that mimic the extracellular matrix (ECM), biomedical polymers supported cell adhesion, proliferation and differentiation. These scaffolds created an environment conducive to tissue regeneration, making them essential for bone, skin and organ engineering. Hybrid scaffolds, which combine natural and synthetic polymers, offer a synergistic blend of mechanical strength and biological activity, promoting more effective tissue regeneration. Despite their vast potential, the use of biomedical polymers has not been without its problems. Issues of toxicity, long-term biocompatibility and immune responses remain areas of active research. The degradation of certain polymers can lead to the release of acid by-products, causing local inflammation or cytotoxicity. Efforts to overcome these problems have led to innovations in polymer surface modification, the use of bioactive coatings and the incorporation of anti-inflammatory agents to improve material-tissue integration (Hu, T et al. 2024).

Regulatory approval of biomedical polymers is an additional barrier. The clinical application of polymer-based medical devices has required extensive testing for safety, efficacy and long-term impact on human health. Predictive modelling, based on artificial intelligence (AI) and machine learning, has emerged as a powerful tool for streamlining regulatory processes by simulating polymer behaviour under physiological conditions. This approach accelerates the design, testing and approval of new biomaterials, reducing the time and costs associated with clinical trials (Ornaghi, Monticeli, and Agnol 2023).

The objective of this chapter was to provide a comprehensive understanding of the foundational principles, classification, and key applications of biomedical polymers. It aimed to explore the distinct characteristics of natural and synthetic polymers, by highlighting their roles in drug delivery, tissue engineering, and medical devices, and address the challenges associated with their use in biomedical contexts. By presenting the latest advancements and future prospects, this chapter seeks to equip researchers, students, and professionals with a robust knowledge base to support innovation and informed decision-making in the development and application of biomedical polymers.

HISTORICAL AND MODERN ADVANCEMENT IN POLYMER SCIENCE

The evolution of polymer science has played a key role in the transformation of biomedical applications, from early discoveries to cutting-edge innovations. Historically, the development of biodegradable polymers dates back to the mid-20th century, when materials such as polylactic acid (PLA) and polycaprolactone (PLC) were introduced for medical purposes. In the 1960s and 1970s, research focused on bioresorbable synthetic polymers for sutures and drug delivery systems, laying the foundations for modern biomedical materials. The introduction of PEGylation in drug delivery in the 1990s revolutionized pharmacokinetics, improving drug stability and circulation time (Suk et al. 2016). However, these early polymers had limitations in terms od mechanical strength, degradation control and biocompatibility. By the contrast, modern advances have led to the development of hybrid polymers, intelligent biomaterials and polymer design driven by artificial intelligence. Recent breakthroughs include stimuli-responsive polymers that adapt to physiological conditions and 3-D printed biopolymer scaffolds that enhance tissue regeneration. The incorporation of nanotechnology and bioengineered polymers has further expanded applications in targeted drug delivery and regenerative medicine. In addition, the use of machine learning in polymer synthesis has accelerated materials optimization, enabling precise control of polymer properties. These advances demonstrate how polymer science has evolved from basic biomaterials to regenerative medicine (Morawetz 2014).

TYPES OF BIOMEDICAL POLYMERS

1. Natural Polymers

Both natural and synthetic polymers have aroused the interest of many researchers, notably for various biomedical applications (Feng et al. 2022). Polysaccharides, proteins and polyesters from the plant and animal kingdoms make up the family of natural polymers. According to a study by (Satchanska, Davidova, and Petrov 2024). Polysaccharides, made up of monosaccharide units linked by O-glycosidic bonds, are characterized by their diversity. This variability results from differences in monosaccharide composition, linkage types and patterns, chain configurations and molecular weight. Physical properties are directly influenced by these factors, and include viscosity, solubility, gelation potential, as well as surface and interface behaviours, which are essential for their biomedical applications. For example, hyaluronic acid, as a natural polysaccharide, was widely used in medicine for its ability to promote healing and support cell growth, notably in the treatment of wounds and burns. Similarly, proteins such as collagen and fibrin, because of their structure and biocompatibility, were used in tissue regeneration and repair. What's more, these natural polymers are distinguished by their ability to interact favourably with biological tissues, thereby promoting healing, reducing inflammation and repairing lesions (Hu, P and Chen 2024).

These natural polymers are ubiquitous in nature, they are obtained from renewable resources like plants, animals and micro-organisms (Mabsor-Zgandaoui el al., 2025a). Their use has given rise to the revolutionary applications in tissue engineering and regenerative medicine. It's important to note that today, several polysaccharides are identified as shown the Figure 1, for examples, those from higher plants, (starch, cellulose and exudate gums), seaweed polysaccharides (alginates, galactans, and carrageenans), not forgetting of course polysaccharides of animal origin (chitin, chitosan, glycosaminoglycans (GAGs) and hyaluronic acid (HA)) (Satchanska, Davidova, and Petrov 2024).

Figure 1. Examples of polysaccharides from different sources

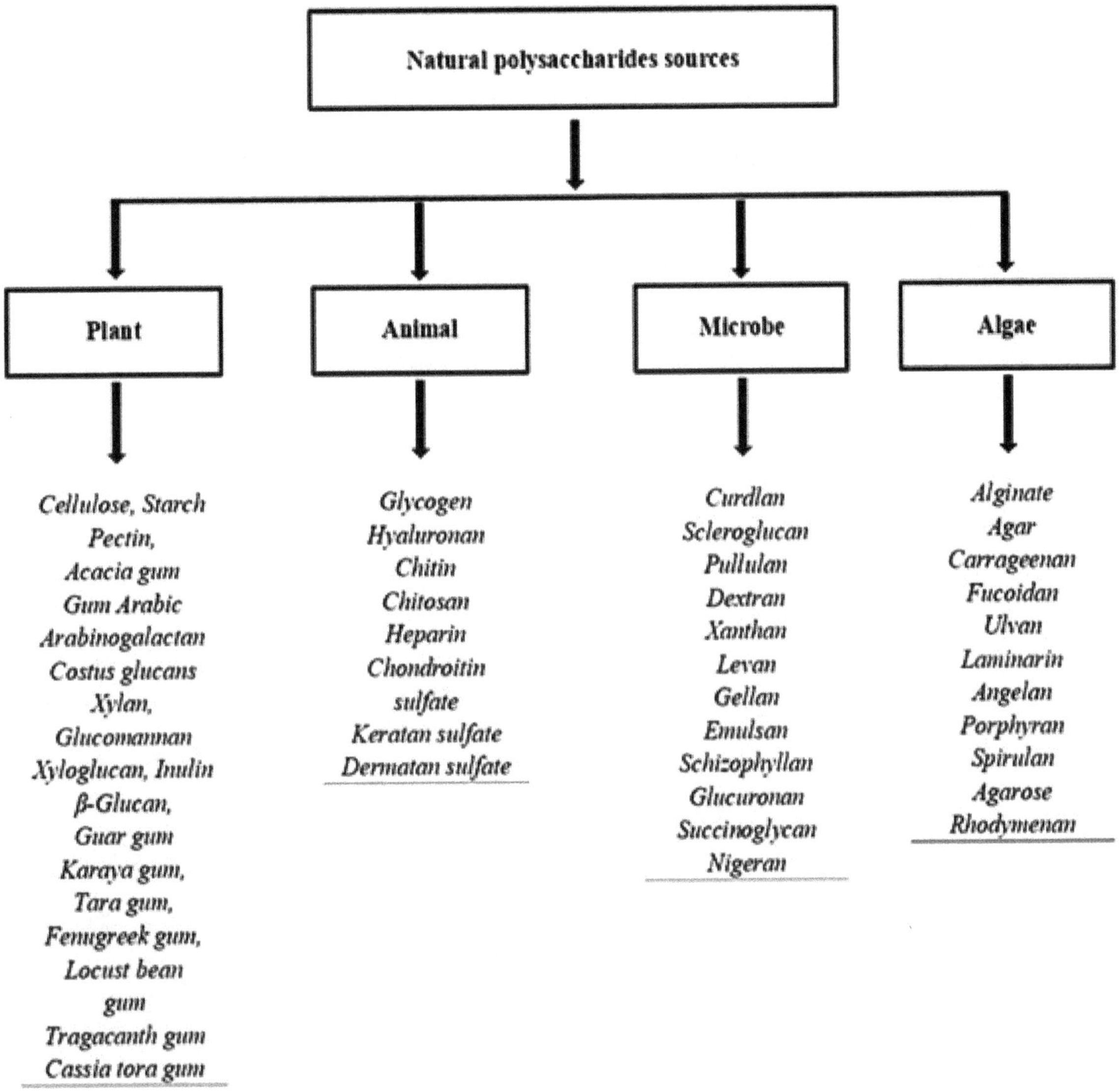

Cellulose

Cellulose, the most abandoned natural polysaccharide on earth, was discovered by French chemist Anselme Payen, who isolated it from plant matter in 1838 and determined its chemical formula $(C_6H_{10}O_5)n$, consisting of a linear chain of several hundred to over ten thousand linked β(1→4) D-glucose units as shown in the Figure 2 (Khopade et al. 2023). This cellulose chain length is expressed in terms of the number of constituents AGUs, known as the degree of polymerization. The degree of linearity and the presence of extended -OH groups throughout the cellulose chain are responsible for the formation of intermolecular and intramolecular hydrogen bonds throughout the polymer chain (Khopade et al. 2023).

Chitin and Chitosan

Chitin and chitosan are two natural biopolymers present in abundance worldwide, and found in a variety of eukaryotic organisms, including crustaceans, insects and fungi (Crini 2019). As illustrated in the Figure 2, chitin is represented as a chain of N-acetyl-β-D-glucosamine monomers linked by a glycosidic (1→4) bond, and nomenclature defines this structure as that of a poly β-(1→4)-2-acetamido-2-deoxy-D-glucopyranose. This biopolymer has a similar composition to cellulose, with a modification at the C2 position. After deacetylation, chitosan was obtained, its chemical structure represented by a chain of ß (1-4)-D-glucosamine monomers. It has three reactive functional groups: an amine in C2 position and two hydroxyl groups in C3 and C6 respectively (Tao et al. 2020).

Figure 2. Chemical structure of chitin chitosan cellulose and hyaluronic acid

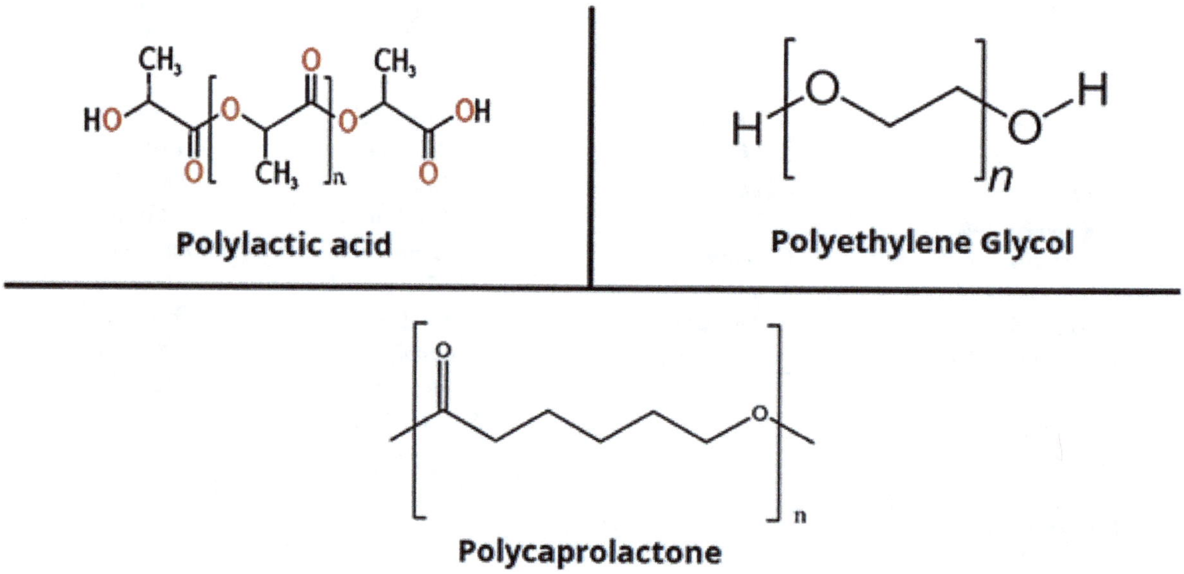

Hyaluronic Acid

Hyaluronic acid (HA), a natural polysaccharide from the animal kingdom, has been widely useful in the biomedical field. known as hyaluronan or hyaluronate, with molecular formula $C_{28}H_{44}N_2O_{23}$, it is an important component of connective tissue, synovial fluid (the fluid that lubricates joints) and the vitreous humor of the eye. Chemically speaking, it is an unsulfated glycosaminoglycan composed of repeating polymeric disaccharides of d-glucuronic acid and N-acetyl-d-glucosamine linked by a glycoside bond in the arrangement of alternating β-(1 → 4) and β-(1 → 3) linkages (Figure 2). The disaccharide units of HA are extended, forming a rigid molecule whose numerous repulsive anionic groups bind cations and water molecules (Bukhari et al. 2018).

Over all, all these polymers have a remarkable property that sets them apart in many contexts. Chitosan has antimicrobial activity against many Gram+ bacteria, such as *Staphylococcus aureus*, and Gram- bacteria, such as *Escherichia coli*, at pH < 6. Its activity was manifested itself on the interaction of positive amino groups with microbial cell membranes, disrupting their barrier properties, causing intracellular leakage or inhibition of DNA/RNA synthesis. It also activated plant tissue defenses and inhibited microbial toxin production by complexing metal ions (Saravanan, Vimalraj, and Anuradha 2018). In addition, its antitumor activity depends on its molecular size, solubility and degree of acetylation. High-molecular-weight chitosan is inactive, underscoring the importance of preparing low-molecular-weight chitosan, which is more soluble in water (Fathy et al. 2018). Many other properties such as biocompatibility, which guarantees safe interaction with living tissue, while biodegradability allowed natural resorption in the body (Crini 2019). The ability to be chemically modified offered flexibility to meet specific needs, e.g. forms: films, hydrogels, sponges, nanoparticles, ... etc (Wang et al. 2021). Chitosan was simply one example among many of natural polymers, which are endowed with enormous properties, having a positive and direct impact on various fields including biomedical, where they are widely exploited for innovative applications.

2. Synthetic Polymers

Moving from natural polymers to synthetic polymers, which also play an important role in this field. Unlike natural polymers, synthetic polymers offer remarkable flexibility in terms of design and properties, making it possible to adapt them precisely to the requirements of biomedical applications. Their importance lied in their ability to overcome certain limitations of natural biomaterials, such as variability, limited availability or sometimes insufficient mechanical properties. In addition, their chemical stability and versatility allow them to be continuously explored to meet emerging needs in medicine, particularly in the fields of tissue regeneration, nanomedicine and intelligent diagnostic and treatment systems. Synthetic polymers are numerous (Satchanska, Davidova, and Petrov 2024). In this study, we mention some of the most frequently mentioned in the literature:

Polylactic Acid

Polylactic acid (PLA) is a thermoplastic aliphatic polyester composed of linear polymer chains. It was generally synthesized from lactic acid monomers, a molecule that exists in different stereoisomeric forms, giving rise to polymers with distinct morphologies: L-PLA, D-PLA, D, L-PLA and meso-PLA. The chemical structure of PLA was shown in the Figure 3, and was marked by unique physical properties such as low processing temperature, thermal stability and good aroma and flavor barrier. These characteristics make PLA a promising material for a variety of applications, especially in the biomedical industry (Ebrahimi and Ramezani Dana 2022).

Figure 3. Chemical structure of Polylactic Acid, Polyethylene Glycol, and Polycaprolactone

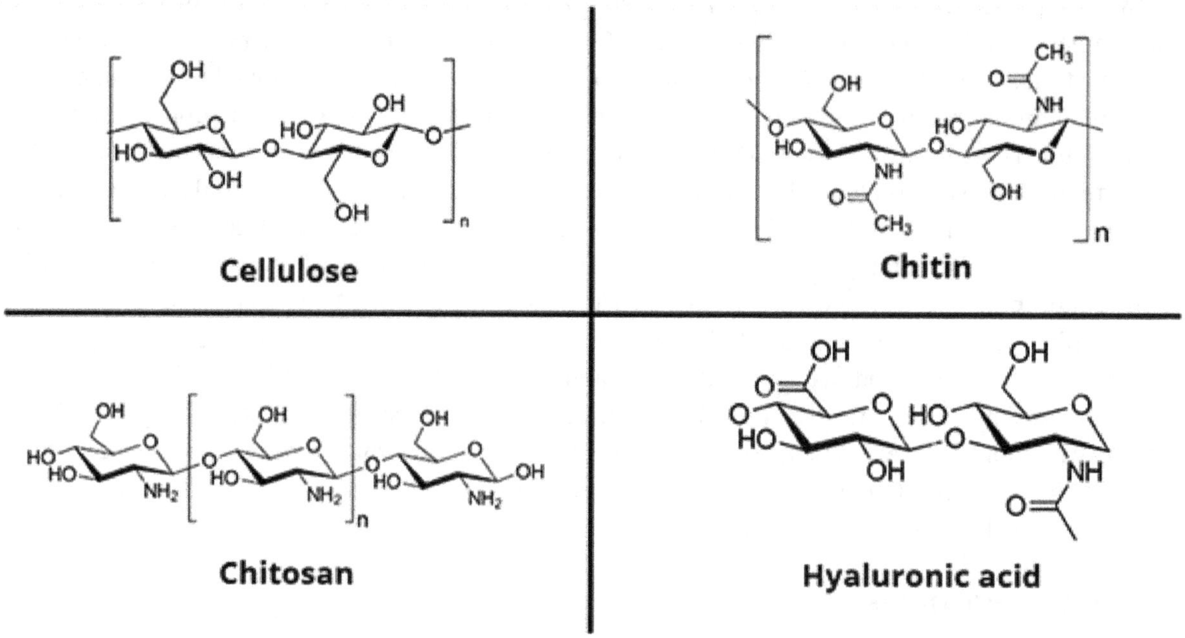

Polyethylene Glycol

Another polymer that was also useful in this context is polyethylene glycol (PEG), which chemically consisted of repeated ethylene glycol units (-CH_2CH_2O-) that form long chains (Figure 3). This structure gives PEG its own unique properties, such as hydrophilicity, biocompatibility and biodegradability, making it an ideal material for biomedical applications, particularly in drug delivery systems. For example, in diabetic people, PEG has attracted particular attention for its ability to boost the efficacy of anti-diabetic drugs. Moreover, by integrating PEG into drug delivery platforms such as micelles, nanoparticles, hydrogels and vesicles, it was possible to create sustainable and controllable delivery systems, thereby improving the therapeutic effects of drugs. However, despite its many advantages, the use of PEG was without risks, as overuse could provoked immune responses, as like the production of anti-PEG antibodies, leading to hypersensitivity reactions or accelerated depletion of PEGylated drugs. (Fu et al. 2021; Yao et al. 2023).

Polycaprolactone

Finally, the polycaprolactone (PCL) is a synthetic polymer also belonging to the aliphatic polyester family. It was derived primarily from the cyclic monomer caprolactone, and was manufactured by opening its ring to create long, linear chains. The chemical structure of PCL consists of repeating units of caprolactone ($C_6H_{10}O_2$), linked by ester bonds (Figure 3). PCL was a semi-crystalline polymer characterized by a low fusion temperature around 60°C, and good flexibility at room temperature. This polymer was particularly appreciated in biomedical applications due to its biocompatibility, biodegradability and ability to blend with other polymers to form higher-performance materials. In the biomedical field, PCL was used to manufacture devices such as implants, dressings, tissue engineering supports and drug delivery systems. (Abrisham et al. 2020).

Synthetic polysaccharides were characterized by unique properties, offering a flexibility that allowed precise control of their structure, including chain length, substitution pattern and degree of modification. This ability to adapt influences their conformation, solubility and tendency to aggregate-essential characteristics for advanced biomedical applications. Imagine synthetic polysaccharides capable of structuring themselves into intelligent nanostructures, mimicking the extracellular environment and thus facilitating cellular communication. Synthetic polysaccharides were constantly evolving, opening up promising new prospects in fields as diverse as targeted drug delivery, tissue engineering and biomaterial design, making them the best candidates for the future of medicine. (Delbianco and Seeberger 2020; Mohammed, Naveed, and Jost 2021).

APPLICATION OF NATURAL AND SYNTHETIC POLYMERS IN THE BIOMEDICAL FIELD

1. Applications of Natural Polymers in Biomedicine

Natural polysaccharides, such as alginate, chitosan and hyaluronan, have been used in biomedicine for a long time, thanks to their biocompatibility, biodegradability and unique functional properties. Alginate, derived from seaweed, is especially appreciated for its ability to form gels, making it useful in applications such as controlled-release drugs and wound-healing dressings. Chitosan, extracted from crustacean shells, was frequently used in drug delivery systems due to its antimicrobial properties and ability to biodegrade naturally in the human body. Hyaluronan, a natural polysaccharide found in the human body, was widely used in joint treatment and skin care products, notably for its ability to promote tissue regeneration and maintain hydration. These materials play a key role in fields ranging from tissue engineering and repair to targeted drug delivery and treatment of joint diseases. (Puertas-Bartolomé, Mora-Boza, and García-Fernåndez 2021; Zhao, L et al. 2023).

2. Applications of Synthetic Polymers in Biomedicine

Synthetic polymers, such as polyethylene glycol (PEG) and polycaprolactone (PCL), offered unique design flexibility, making them ideally suited to a wide range of biomedical applications. These polymers were also used in controlled-release drug delivery systems and medical implants. Polycaprolactone (PCL), meanwhile, is a semi-crystalline polymer particularly suited to tissue engineering and the manufacture of biodegradable medical devices, thanks to its low melting temperature and biocompatibility. These synthetic polymers were therefore essential in the design of tailor-made materials for personalized treatments and advanced medical applications. (Puertas-Bartolomé, Mora-Boza, and García-Fernåndez 2021; Zhao, L et al. 2023). The Table 1 gived a further insight into their applications for bone repair and regeneration (Sheikh et al. 2015).

MECHANISMS OF INTERACTION OF POLYMERS WITH BIOLOGICAL SYSTEMS

1. Polymers Biocompatibility

The biocompatibility was a critical property of materials used in biomedical applications and was defined by their ability to interact with biological systems without inducing harmful immune responses or toxicity. This characteristic was fundamental for ensuring the safety, efficacy, longevity of medical devices, drug delivery systems, and tissue-engineering scaffolds. In addition, to avoiding adverse reactions, biocompatibility facilitated essential biological processed such as cellular adhesion, tissue regeneration, and material integration into the host environment (Chen et al. 2022). The polymers biocompatibility was evaluated by in vitro tests such as cytotoxicity and by in vivo models for systemic responses, also by emerging methods such as organ-on-chip platforms and by computational models. These alternatives improve the precision and ethics of the development of safe and effective biomaterials.

The in vitro studies have evaluated biocompatibility by assessing the cytotoxicity, protein adsorption, and cell adhesion in controlled environments. Cytotoxicity assays, guided by ISO 10993-5 standards, measured the effects of materials on cell viability and morphology, ensuring that they support biological functions without toxicity. Protein adsorption studies have examined the material-protein interactions that influence immune responses and compatibility. Advances in 3D cell models enhance predictive accuracy by simulating complex in vivo-like conditions, providing reliable insights into material performance before in vivo trials (Machla et al. 2024). However, several studies have examined the in vivo performance of biomaterials, focusing on their immune responses, biodegradability, and tissue integration. Polymers such as PLA and PEG demonstrated distinct biocompatibility benefits. PLA's predictable degradation of PLA offered insights into immune response management, whereas PEG reduced macrophage activation and fibrotic encapsulation, enhancing tissue integration. Together, these findings highlighted the effectiveness of PLA and PEG, with minimal adverse effects, supporting their biomedical potential (Ng et al. 2023).

Numerous studies have evaluated the impact of foreign body reactions and inflammation on the integration of biomedical materials, elucidating how these processes impede proper functionality by inducing chronic inflammation or fibrotic encapsulation. Regarding macrophage microRNA-146a, one investigation demonstrated that it is a potential therapeutic target for attenuating inflammation and enhancing material compatibility. In addition, hydrophilic coatings, such as PEG, have been demonstrated to mitigate protein adsorption and immune activation, thereby minimizing the inflammatory cascade. However, the incorporation of anti-inflammatory agents such as dexamethasone into materials facilitated localized, sustained release, which modulates macrophage activity and further attenuates inflammation (Mahanty et al. 2024). Additionally, bio-inspired materials designed to mimic extracellular matrix components were evaluated for their capacity to enhance material integration by promoting cell adhesion and reducing immune recognition. Furthermore, technologies such as nanostructured surfaces and dynamic coatings have been demonstrated to minimize macrophage adhesion and polarize immune responses toward a healing state. Moreover, materials that regulate cytokine production and tissue stiffness exhibit significant potential for mitigating fibrotic encapsulation, thus further improving biocompatibility (Qian et al. 2023).

2. Surface Modification

PEGylation and Coating Techniques

According to Jarad et al. (2024), surface modifications improved biomaterial compatibility by adjusting properties, such as hydrophilicity, charge, and bioactivity, to address immune recognition and integration challenges (Jarad et al. 2024). PEGylation, the attachment of polyethylene glycol (PEG) to surfaces, forms a hydrophilic barrier that reduces protein adsorption and immune cell recognition, enhancing systemic circulation and therapeutic efficacy. Indeed, we also find that advanced coatings such as hydrogels and zwitterionic materials mimic natural cellular environments. Hydrogels with high water content and tunable mechanics have lower immune reactions. Zwitterionic coatings balance the positive and negative charges, resist protein fouling and immune adhesion, and enhance biocompatibility in drug delivery and tissue engineering (Chen et al. 2022). Lastly, the findings about dynamic coatings and stimuli-responsive PEGylation provided precise bioactivity control, adapting to biological conditions, while maintaining efficacy.

Functionalization: Adding Bioactive Molecules

According to Zhao et al., (2022), functionalization integrated bioactive molecules into material surfaces to induced targeted biological effects. Incorporating peptides, growth factors, or antibodies enhances cell adhesion and tissue regeneration (Zhao C. H, Zhang, and Zhang 2022). For example, RGD peptides bind integrin receptors to improve cell-material interactions, which were crucial for scaffold integration in tissue engineering. Many studies were assessed to highlight the effect of anti-inflammatory agents, such as dexamethasone, on macrophage activation and cytokine production, which play a crucial role in creating regenerative environments for better material integration. In the case of the study by Liu et al. (2024), it was confirmed that the localized, sustained release of these agents effectively mitigates inflammation and improves biomaterial performance (Liu et al. 2024). However, advanced methods, such as covalent binding and nanocarrier systems, were further assessed to enable the controlled release of growth factors or drugs, extending their therapeutic effects while minimizing side effects. The study by Agwa et al. (2024), confirmed that functionalized materials, particularly in drug delivery, enhance therapeutic precision by using surface-bound ligands to target specific cells. These techniques have significantly elevated biomaterials, setting new standards for specificity, efficacy, and biocompatibility (Agwa et al. 2024).

Applications: Drug Delivery Systems and Immune Modulation

Studies have shown that surface modifications can enhance drug delivery and immune modulation by improving targeting and reducing immune clearance. PEGylated nanoparticles create hydrophilic barriers, extend circulation, and improve tumor-targeted drug delivery (Hussain et al. 2019). According to Li et al., (2021) Coatings mimicking extracellular matrix properties, such as collagen-based materials, improve biocompatibility, cell adhesion, and proliferation, aid drug delivery, and tissue engineering (Li P. Y. et al. 2021). Functionalized PEG nanoparticles effectively target specific cells while maintaining immune stealth. The findings reveal that the dynamic and stimuli-responsive coatings further enhance the precision and release of drugs based on environmental cues such as pH or temperature (Santhanakrishnan, Koilpillai, and Narayanasamy 2024).

3. Controlled Degradation

Controlled degradation is a critical feature of biomaterials in medical applications, enabling their breakdown into safe byproducts that align with therapeutic objectives (Trucillo 2024).

Types of Degradation for Controlled Biomaterial Breakdown

Hydrolytic degradation was widely utilized for polymers such as PLA and polylactic-co-glycolic acid (PLGA). This process involved water-mediated bond cleavage, producing predictable degradation rates ideal for timed drug delivery. Adjusting the ratio of lactic acid to glycolic acid in PLGA allows for precise control over the degradation timelines, which was crucial for sustained drug release. Recent studies have emphasized the importance of fine-tuning polymer hydrophilicity and crosslinking density to achieve application-specific degradation rates (Ulbricht Aus Großenhain 2017). In other hand, the enzymatic degradation employed biological enzymes, such as collagenase and elastases, to break down polymers. This mechanism enabled tissue-specific degradation, particularly for collagen-based and amino acid-containing polymers used in regenerative medicine. Modern techniques leverage enzymatic activity to improve the responsiveness of biomaterials to specific environments, such as wound sites or tissue scaffolds (Jurišić Dukovski et al. 2023). Finally, the oxidative degradation driven by reactive oxygen species (ROS) in inflammatory environments is valuable for dynamic biological applications (Mabsor-Zgandaoui el al., 2025b). Materials such as poly(ether-urethane) s, often paired with antioxidants, manage oxidative stress while maintaining their functionality. Research has highlighted scaffolds that balance degradation and mechanical integrity, enabling controlled erosion over time (Li, Zhao, and Yi, 2024). These mechanisms underpin biomaterials tailored to diverse therapeutic needs, ensuring precision and minimal adverse effects.

Mechanisms for Regulation of Polymer Degradation

Polymer degradation was carefully controlled by chemical composition and structural modifications to meet therapeutic requirements. Studies have shown that adjusting the monomer ratios in copolymers, such as PLGA, fine-tunes degradation rates; more glycolic acid accelerates degradation, while more lactic acid slows it, ensuring structural integrity for specific applications (Ulbricht Aus Großenhain 2017). Structural modifications, such as cross-linking, enhance polymer stability and reduce susceptibility to hydrolytic or enzymatic attack, making them suitable for inflammatory environments where ROS may degrade materials. Cross-linked polymers maintain their integrity under oxidative stress, thereby supporting wound healing and tissue engineering (Narayanan et al. 2023).

Recent advances have integrated bioresponsive components into polymers that react to environmental stimuli such as pH, temperature, or enzymes. pH-sensitive polymers target acidic tumor microenvironments for localized drug delivery, whereas enzyme-responsive polymers ensure precise therapeutic release (Jurišić Dukovski et al. 2023). Nanoparticles in polymer matrices improve mechanical properties and enable secondary functions such as imaging (Ulbricht Aus Großenhain 2017). Also, the functional groups such as amines or hydroxyls improved biocompatibility, support cell adhesion, and enhance tissue integration. Advanced 3D scaffolds further aid tissue regeneration in vitro and in vivo (Narayanan et al. 2023).

APPLICATIONS OF BIOMEDICAL POLYMERS

1. Drug Delivery Systems

As biomedical polymers were the key tools in contemporary treatment and research due to their unique characteristics, two of their most promising applications were in drug delivery systems and tissue engineering, which take advantage of their structural and functional characteristics to improve therapeutic efficacy and cell regeneration. They can be deployed in 3 paradigms. Firstly, drug modification, including alterations to chemicals and their conjugation, can modulate interactions between the drug and molecules, cells and tissues in the body, and between the drug and its essential target site, thereby facilitating controlled, non-invasive release. Secondly, modification of the microenvironment is a strategy that aims to overcome biological barriers, using modulators to influence biological fluids or mechanical processes that impede drug action. Finally, targeted delivery systems, with their ability to adjust administration frequency, doses and infusion rates, optimize treatment efficacy (Kopeček, 2013).

From this point of view, the conjugation of polymers to therapeutic molecules appears to be a promising alternative for circumventing the problems associated with drug solubility, and for resolving concerns about the systematic release of therapeutic molecules as well as the precision of specific delivery sites. The origins of polymer drug conjugates (PDCs) date back to the 1950s, when Horst Jatzkewitz, improved anti-anxiety drugs.PDCs are synthesised by coupling reactions involving the direct conjugation of biological compounds or drugs with the polymer. They were nanometric in size, stable, flexible and had prolonged half-lives, potential for drug delivery, ability to reduce antigenicity, combination of drugs of different natures (hydrophilic and hydrophobic) which provides better protection against degradation (Kopeček, 2013).

As far as their composition was concerned, they are formed of 3 different units divided between the solubilization zone which facilitated the solubilization of macromolecules without toxicity, a polymer binding zone to the drug molecule and a transport system which was a targeted ligand region located in a hydrophilic region of the polymer. In more detail, the backbone linked by covalent bonds to the therapeutic molecules by a linker, which acts as a spacer for co-delivery and was sometimes specifically cleaved in response to environmental stimuli, not forgetting ligand targeting capable of delivering the PDC to the target site, which was a recognition motif (sugar or antibody) or their important receptors to minimize contact with non-target organs, and drugs constitute the building blocks, not forgetting solubilizing groups to improve the solubility of poorly soluble drugs (Thakor, 2020).

A major advance concern nanoparticle (NPs), promising vectors due to their nanometric size, shape, chemical composition, interconnected macro-pores, tunable porosity and surface that influence the control of therapeutic efficacy, bioavailability and half-life. They are also used for a wide range of pathologies such as cancer (Yao, 2020), cardiovascular disease (Elumalai, 2024) and osteoarthritis (Jin, 2020). To date, the most widely used NPs for the diagnosis and treatment of pathologies are liposomes, micelles, dendrimers, polymer NPs and metal NPs. The release of therapeutic molecules can be further enhanced if these NPs were 'intelligent', i.e. can be stimulated by thermal regulation, pH, oxidoreduction or other factors. Several studies have demonstrated a prolonged and controlled release of drugs using these systems, which can promote the internalization of NPs into the cell (Nasra, 2022).

Moreover, the PLGA nanoparticles have revolutionized drug delivery systems by ensuring consistent and prolonged release of therapeutic agents. By tailoring degradation profiles through copolymer adjustments and surface modifications, these systems enhanced drug bioavailability and minimize systemic side effects. Novel PLGA formulations integrate additional biocompatible agents to improve precision targeting and immune evasion, enhancing their applicability in cancer and autoimmune therapies.

According to Narayanan et al., (2023) biodegradable scaffolds, such as polycaprolactone (PCL), was instrumental in long-term tissue regeneration because of their slow degradation rates and robust mechanical properties. PLA scaffolds were preferred for short-term applications because of their predictability and faster degradation. Innovations in scaffold design have now incorporated bioactive molecules and growth factors to improve cell attachment, proliferation, and tissue integration. These scaffolds were increasingly being employed in soft tissue repair and bone regeneration, offering customizable solutions for diverse medical challenges.

2. Tissue Engineering and Regeneration

The use of polymers in tissue engineering was often in the form of scaffolds, their 3D structure has made it possible to mimic the form and function of the extracellular matrix of native tissue and cause the desirable cell interactions (adhesion, proliferation and differentiation). They were ideal for tissue engineering if they were biocompatible, able to support bio-acting cells and molecules, have structural and dimensional characteristics such as porosity to facilitate the diffusion of nutrients and molecules with biological activity, minimize inflammatory response, biodegradable, cost effective, clinically easy to use. There are different types of biological scaffold (decellularized tissue matrix), synthetics (materials manufactured for desired properties), composites (composed of two or more materials) (Ruiz-Alonso, 2021). An attractive and seemingly effective challenge is the development of scaffolds allowing easy cell seeding and uniform spatial distribution of transplanted cells by combining synthetic polymers with polymers of natural origin and hydroxyapatite (Suhag, 2024).

Tissue regeneration can be achieved using growth factors (GFs) to create a microenvironment favourable to tissue regeneration. These soluble proteins bind to specific receptors, such as tyrosine kinase and serine-threonine kinase receptors, and trigger signalling cascades that influence cell fate and activity. Therapies based on this principle are effective, but very costly and can induce undesirable effects, notably immune reactions. To alleviate these difficulties, bioactive peptides have emerged as promising alternatives. These short amino acid sequences, derived from active protein domains, reproduce the biological activities of their parent proteins. They influence cellular functions, regulate intercellular communication and support tissue regeneration without inducing immunogenicity. Bioactive peptides were cost-effective, easy to synthesize and versatile. Common sources included extracellular matrix proteins (collagen, fibronectin and laminin), soluble growth factors, protein hormones and modified sequences. Many peptides mimic their original proteins, binding to cell receptors to activate signalling pathways. Others were used to functionalize biomaterials, enhancing their regenerative capabilities. Research is demonstrating the significant potential of bioactive peptides in bone formation both in vitro and in vivo. Some, such as PTH1-34 and P-15, have received FDA approval, while others, such as thrombin 508, are in clinical trials (Toosi, 2020)

The application of biomedical polymers in the clinic faces considerable difficulties. Current research is mainly based on in vitro and in vivo studies, but extrapolation of these results to human patients remains an obstacle due to differences between animal models and the human immune system. For example, vascularization remains a major challenge that hinders the clinical application of tissue-engineered products. Several strategies have been developed to promote vascularization in engineered tissues, incorporating key elements of physiological vasculogenic and angiogenesis, such as vascular cells, angiogenic factors, biophysical cues and spatial organization. However, there is no universal approach to vascular network engineering that can be applied to all tissue types (Shakeel, 2023). These discrepancies highlight the need for more human-specific studies to ensure clinical relevance.

Future advances in bioactive materials will depend on the integration of cutting-edge technologies, such as single-cell analysis and computer-aided design, to decipher specific cell behaviours and tailor microenvironments for precise therapeutic outcomes. Innovations in materials design, including adjustments to stiffness, geometry, wettability and surface charge, will improve their effectiveness. By adapting these properties, bioactive materials could significantly improve the speed and quality of wound healing, paving the way for next-generation therapeutic strategies (Amiryaghoubi, 2020) .

3. Medical Devices and Implants

Polymers can be very present in the biomedical field, thanks to their biodegradability, their resistance to degradation, their low toxicity and also thanks to their low immune reaction (Zhao L. et al. 2023).

Polymers in Implants

In 2021, researcher Lam developed a simple method for functionalizing silicone surfaces with sodium poly (styrene sulfonate) (polyNaSS). This approach relied exclusively on UV irradiation at 365 nm, used both to activate the surface by generating hydroxyl groups and to initiate radical polymerization in solution (Lam. M 2021). In the dental field also polymers can be very useful especially in the construction of implant abutments instead of using other materials such as gold, titanium, alumina ... it is thanks to the high performance of polyetheretherketone (PEEK) which is a polymer known for its elasticity comparable to that of bone, as well as its ability to reduce resistance to stress, this is why PEEK was widely used in surgical operations especially in orthopaedic and spinal surgery (AL-Rabab'ah et al., 2019), Also for the purpose of restoring masticatory function, aesthetic satisfaction, polymer-based dental implants play a crucial role in finding solutions to tooth loss and damage (Jiang et al., 2020)

In addition, in eye diseases, intraocular implants were a relative novelty to release drugs in a prolonged, but rather, controlled manner. What makes this method more effective than methods that rely on intravitreal injections or eye drops (García-Estrada et al., 2021). The use of polymers in implants was based on their biocompatibility, which allows improving body functions without disrupting the normal functioning of the body or causing allergies or other adverse effects (Arif et al., 2019)

The Use of Polymers in Medical Devices

Biomaterial polymers played an essential role in modern healthcare technologies, in medical coatings for example, polymers and inorganic materials play an important role in this sense, as well as the combination between the two products improves the biocompatibility and bioactivity of medical devices, offering remarkable potential for biomedical applications (Li et al., 2021), also in additive manufacturing (3D printing) for the production of medical devices, polymers are very present in this use, This technology represents a valuable opportunity to promote better access to healthcare (Culmone et al., 2019). Synthetic polymers can also be widely used in drug delivery devices (Hacker et al., 2019).

4. Case Studies and Real-World Examples of Biomedical Polymers Applications

Cost-Effective of PLGA-Based Drug Delivery Systems in Developing Countries

PLGA has been widely used in drug delivery systems due to its biocompatibility, controlled degradation and FDA approval for various formulations. In developing countries, PLGA-based drug delivery systems offer a cost-effective solution for sustained drug release, particularly for the treatment of chronic diseases such as tuberculosis and cancer (Akpinar Adscheid et al. 2025)(Yang et al. 2024). For example, PLGA microspheres for the treatment of tuberculosis enable prolonged drug release, reducing the need for frequent dosing and improving patient compliance (Bajpai et al. 2008). However, high production costs, regulatory hurdles and the need for cold chain logistics remain challenges to making these systems widely available. Efforts to develop local manufacturing and reduce synthesis costs are being explored to make these systems more accessible.

Affordability and Delivery Issue for 3D-Printed Polymer Scaffolds for Regeneration Medicine

3D-printed polymer scaffolds have evolved into a leading technology in regeneration medicine, allowing patient-specific tissue engineering for bone, cartilage and organ regeneration. Materials such as PCL, gelatin-based hydrogels and polymer-ceramic hybrid composites offer excellent biocompatibility and structural integrity. However, high production costs, lack of standardization and distribution difficulties limit their widespread adoption (Yuan et al. 2024).

INNOVATIVE AND SMART POLYMERS

The 'intelligence' of living cells to adapt to environmental stimuli through specific complex mechanisms has inspired scientists, as this adaptation basically stems from the fact that cells are composed of polymers (proteins, carbohydrates and nucleic acids), and for some decades now, specialists have been discussing the term 'Smart polymers', which refers to stimuli-responsive, shape-memory or bio-inspired polymers with numerous applications as polymeric nanomaterials (NMPs) in the biomedical field, thanks to their biocompatibility, reduced dimensions, high surface area, etc (Aflori, 2021).

1. Stimuli-Responsive Polymers

Smart polymers were sensitive to pathological microenvironment stimuli and helped to improve drug efficacy by targeting release, prolonging blood flow and extending the half-life of the molecule, since the stimuli can be either internal (pH, temperature, redox potential, enzymes) or external (light, electric, magnetic or ultrasonic field). These polymers were the subject of a large number of studies due to their promising effects, for example, the labile acid bonds of pH-responsive polymers are the safest for drug delivery according to (Ofridam, 2021), while redox-responsive polymers are used in cancer therapy (Li D.Z, 2020) et and help overcome multidrug resistance thanks to their deep penetration (Yan, 2021), thermo-responsive polymers were used in tissue engineering (Doberenz, 2020) and magneto-responsive polymers have various applications in targeted prolonged delivery and drug imaging tracking (Liao, 2020). The table 2, illustrated certain characteristics and mechanisms of stimuli-responsive polymers.

The strength of certain polymers such as Near Infra-Red responsive polymers, which were light responsive polymers, because of their favourable tissue penetration force (Xiang, 2018). Hence, their used for brain tumour phototheranostics (Wen, 2020) and on the subject of safe-use polymers, researchers have shown that ultrasound improves the transmission of medicinal substances to regions of interest (Chowdhury, 2017) and assembly in precise regions in cancerology (Das, 2020).

Current studies are moving more towards the use of multi-responsive polymers offering better targeting by combining two or more stimulating signals. Examples of this include the coupling of magnetism and hypothermia, which has been studied *in vivo* and *in vitro* (Thirunavukkarasu, 2018), and the controlled delivery of insulin (Feki, 2020), Multi-stimuli-responsive polymeric nanocarriers are also being used for cancer treatment and are based on pH reactivity, given the significant variations between tumour microenvironments (Chang, 2021).

2. Shape-Memory Polymers

Shape memory polymers (SMPs) and their composites were a class of intelligent materials able to change shape temporarily and regain their permanent shape after being released, depending on the temperature, wavelength, intensity, irradiated position, periodicity and polarization of the stimulating light (Dayyoub T. M., 2022) (Herath, 2020). Their unique structural versatility, light weight, low cost, ease of processing, high elastic deformation (+200%), controllable deformation temperature, biocompatibility and biodegradability gather interest for applications in various fields and were considered practical candidates for intelligent systems that control different conditions and require multiple external stimuli (Dayyoub T. M., 2022). The major disadvantage of these polymers was their structure, and 4D printing technology is currently providing hope of solving this problem (Wang L. Z., 2022).

Shape memory is made up of two initial stages: programming and shape retrieval. This process can be reproduced many times, so SMPs can have different temporary forms but only a single permanent one. (Xia Y. H., 2021) (Figure 4). SMPs have many surgical and vascular applications, for example a biodegradable surgical fibre induced by heat to shrink and tighten the wound after suturing, memory foam to treat aneurysms, a coil coated with a porous foam of SMPs could promote the formation of neointima in the neck of the aneurysm, leading to full recovery, a catheter containing a wire of SMPs, a microgripper with laser jaws activated by SMPs, a dialysis needle adapter made of SMPs to reduce thermodynamic stress, an expandable helix structure of SMPs in capsules to develop a potential gastric device that returns to its original shape in aqueous fluids at 37°C, a biomedical occlusion device printed in 4D, an SMP-based stent that can be inserted into a blood vessel. They have other applications in flexible electronics.

Particular focus should be given to the synthesis and fortification of two-way SMPs, with a potential to exhibit a recovery stress comparable to that of memory material, as well as to the precise control of shape recovery behaviour, which will be advantageous for benefiting from the intermediate form and developing SMPs with a transition phase. Moreover, research into non-reversible and athermal chemical reactions will help to obtain new SMPs. The use of controllable and deformable SMPs would lead to a new generation of stimuli-responsive soft materials to solve scientific challenges (Xia Y. H., 2021).

Figure 4. Shape memory mechanism programming and recovery

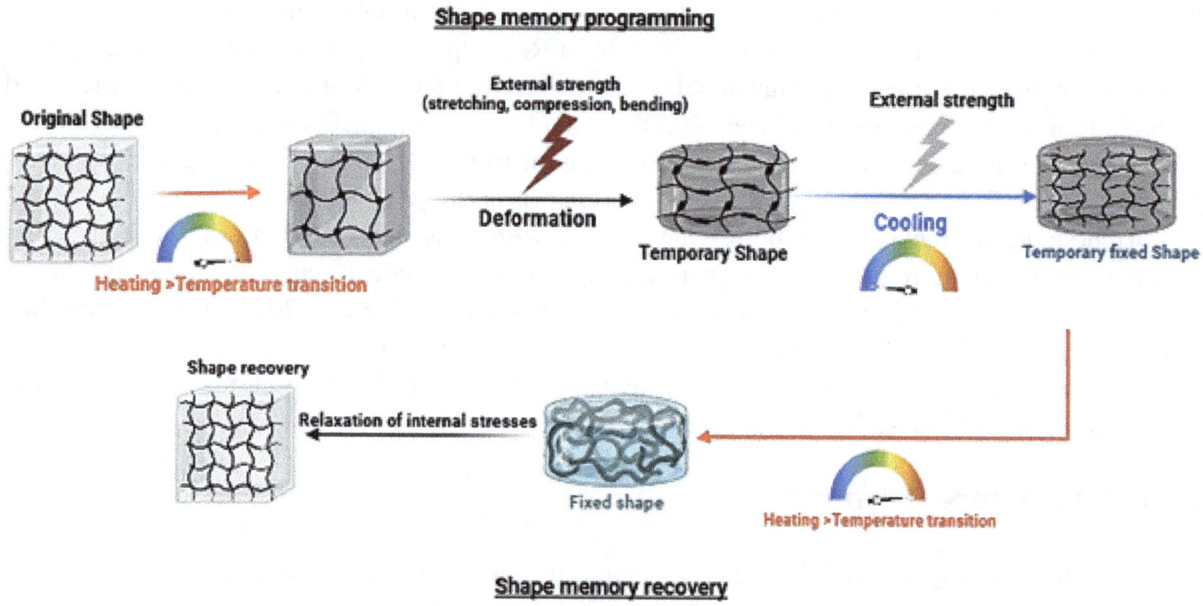

3. Bioinspired and Hybrid Polymers

The inspiration of natural behaviour has made possible to develop intelligent bioinspired polymers capable of interacting specifically with different targeted body systems, offering biocompatibility, low immunogenicity and high stability for drug delivery. Bioinspired polymers are synthesized by functional biomimetic synthesis that mimic the specific properties of substances or by biomimetic process synthesis.

Some promising ideas include yeast microcapsules loaded with biomimetic insulin for oral administration, avoiding the invasive method of administering insulin while bypassing the acidity of the stomach; the generation of antibiotics capable of completely destroying bacteria; anti-inflammatory Nano-capsules, for wound healing using gelatine, collagen or synthetic polymers, prolonged release of single or multiple paracrine factors, cardiovascular therapy (artificial blood vessel that functions like a coronary artery, in addition to unlimited applications in cancer therapy by mimicking the biological configuration of the human body.

Currently, the application of various nanoparticles such as carbon nanotubes, gold nanoparticles, graphene oxide, iron oxide nanoparticles, mesoporous silica nanoparticles and polymeric carriers on their own or by incorporating them into biomimetic scaffolds and hydrogels can mimic the morphology and electrophysiological nature of the heart (Yang, 2020).

The idea of this type of polymer was innovative, except that there were challenges to overcome if they were to be designed effectively, such as the complexity of synthesis and the precise control of reactions. Specific challenges that remain include the development of injectable hydrogels that can be used as artificial cartilage, the design of bioinspired multifunctional hydrogels that combine exceptional mechanical properties with other functionalities, such as electrical conductivity, magnetic or luminescent characteristics (Xu, 2021).

CHALLENGES AND FUTURE DIRECTIONS

1. Toxicity and Biocompatibility

Long-term Safety

Polymers used in biomedical devices, such as implants and drug delivery systems, underwent degradation processes over time. These degradations released products that could affect health in the long term. For example, the degradation of polymers such as poly(L-lactide) (PLLA) in implants led to the release of acidic monomers, triggering inflammatory reactions or local toxicity. Evaluation of the degradation products included study of their systemic and local toxicity, particularly in neighbouring tissues and organs, as well as their impact on cellular metabolism (Ershad-Langroudi et al. 2024).

Immune Responses

Studies show that synthetic materials used in implantable medical devices can induce various immune responses, such as hypersensitivity or rejection reactions. For example, polymers such as polytetrafluoroethylene (PTFE) and polyethylene are often associated with chronic inflammation, compromising their use in long-term applications (Capuani et al. 2022). Particular attention has recently been paid to the phenomenon of biomaterial-induced sterile inflammation (BSI), which plays a critical role in long-term biodegradable implants. This inflammatory response, triggered by damage-associated molecular patterns, can limit the integration of materials and impair their biocompatibility in host tissues (Chen et al. 2022). These observations reveal a negative impact on the safety and efficacy of biodegradable implants, compromising their performance in critical medical applications.

In addition, research on bioresorbable polymers such as poly (D, L-lactide-co-glycolide) (PDLLGA) and poly(L-lactide-co-glycolide) (PLLGA) had demonstrated that their advanced degradation could lead to increased cytotoxicity, particularly for PDLLGA. This toxicity was linked to the release of acidic products that activated receptors such as HCA1, amplifying the inflammatory response (Geddes et al. 2021). These results highlighted the need to design polymers with better-controlled degradation profiles in order to reduce chronic immune reactions. To overcome these challenges, the development of more immunologically compatible polymers was deemed essential. Promising solutions included the use of polymers derived from natural biomaterials such as collagen, or the design of hybrid polymer systems. These innovations are aimed to improve biocompatibility while minimizing the risks associated with the degradation of synthetic materials.

Improvement Strategies

Surface modifications to polymers, such as plasma treatments, grafting of bioactive peptides or the use of functionalized polymers, have played a key role in improving their biocompatibility and reducing the risk of toxicity. For example, the integration of specific proteins or glycans on polymer surfaces has helped minimize inflammatory reactions. In addition, the integration of bioactive thermo-gels, such as those based on modified hyaluronic acid, has demonstrated increased efficacy in limiting inflammatory responses and offering prolonged protection in ocular treatment models (Nguyen, Luo, and Lai 2022). These advances, particularly in hybrid systems combining biological and synthetic polymers, have offered considerable potential for improving biocompatibility while maintaining the material's mechanical stability. Specialized training in polymer surface modification and biocompatibility has thus enhanced their use in tissue engineering and drug delivery systems, paving the way for improved clinical applications (Yoshida et al. 2013). In specific applications, biodegradable polymers used in micro-elastofluidic devices have stood out for their exceptional mechanical elasticity, biocompatibility and degradability into non-toxic products, enhancing their potential for biological interfaces (Tran et al. 2023).

2. Durability and Biodegradability

Studies have revealed that bioplastics offer a promising solution to the global problem of plastic waste accumulation. Thanks to their biodegradable properties, bioplastics represent an environmentally friendly alternative to synthetic plastics, helping to reduce plastic pollution. The development of bioplastic materials, such as polyhydroxyalkanoate (PHA), with characteristics comparable to those of traditional plastics, met the growing demand for environmentally friendly plastics (Thomas et al. 2023). Elkaliny et al. highlighted the growing importance of bioplastics as an ecological alternative to traditional plastics, particularly in the context of the environmental crisis amplified by the COVID-19 pandemic. The use of macroalgae, rich in cellulose and sugars, to produce bioplastics was a promising solution for reducing plastic waste and greenhouse gas emissions (Elkaliny et al. 2024). Despite the challenges associated with large-scale production, including costs and process optimization, these bioplastics have significant potential for a variety of industrial applications and contribute to a more sustainable future, while promoting the creation of green jobs. These materials were evaluated according to criteria such as biocompatibility, biodegradability, mechanical strength and cost, presenting various advantages and disadvantages depending on their specific characteristics (Table 3). This evaluation helped to determine which polymers were most suitable for clinical and therapeutic applications, depending on the specific requirements of medical treatments and devices.

3. Regulatory and Clinical Prospects

Clinical Trials and Safety Testing

To improve the integration of biomaterials into clinical practice, particular attention needed to be paid to understanding their biocompatibility and interactions with the immune system. Challenges such as reduced neovascularization and tissue fibrosis needed to be addressed through in-depth studies of the mechanisms of biomaterial degradation and integration in living tissue. In response to these challenges, clinical regulations had to take these factors into account to ensure the safety and efficacy of biomaterials in various therapeutic applications. Biomaterials were used in fields ranging from implantable medical devices to tissue repair and drug delivery systems, and their regulatory assessment had to be based on a comprehensive analysis of their long-term effects on the human body (Agarwal et al. 2020).

In addition, rigorous clinical trial protocols needed to be put in place to examine not only the mechanical and biological performance of biomaterials, but also their impact on long-term health. Particular attention had to be paid to the safety of biodegradable materials used in complex applications such as vascular grafts, heart valve repair and fracture stabilization, in order to identify and prevent any potential risks to patients. These studies were crucial for the implementation of biomaterials in safe and effective clinical practices. Understanding biomaterial-induced sterile inflammation (BSI) offered insights into the challenges posed by immune responses to synthetic materials, emphasizing the importance of mitigating these reactions to improve tissue-biomaterial integration (Agarwal et al. 2020).

The future of medical materials was looking towards eco-friendly solutions, where biomaterials were designed to minimize their environmental impact. Studies showed that integrating sustainable biomaterials into pharmaceutical and medical applications could revolutionize the healthcare sector while reducing environmental impact. This research highlighted the use of biopolymers, natural extracts and bio-sourced nanoparticles for a variety of applications such as drug delivery, tissue engineering and wound healing. Thanks to their biocompatibility and biodegradability properties, these materials had considerable potential to replace traditional materials and thus contribute to greener, more environmentally responsible medicine (Yingngam 2024).

Emerging Trends: Predictive Modelling for Faster Approvals and Safer Designs

One of the emerging trends in the regulation of biomedical materials is the use of predictive modelling to assess the risks and performance of polymers prior to approval. In the context of emerging trends in biomaterials, the integration of artificial intelligence (AI) was playing a crucial role in transforming the development and design of these materials. The use of AI-powered predictive modelling is improving the speed and accuracy of polymer selection and optimization. By employing technologies such as deep neural networks and machine learning, these predictive models anticipate the behaviour, characteristics and interactions of biomaterials, accelerating their transition from the laboratory to clinical application (Rajitha et al. 2024).

One of the emerging trends in the regulation of biomedical materials was the use of predictive modelling to assess the risks and performance of materials prior to approval. By simulating the behaviour of materials in biological environments, AI made it possible to identify potential risks of toxicity or rejection even before clinical trials. This not only ensured safer designs, but also reduced the time needed for materials approval, speeding up the process of bringing them to market (Rajitha et al. 2024). The studies also highlighted the importance of interdisciplinary collaboration, data standardization and addressing ethical concerns to maximize the impact of AI in this field.

CONCLUSION

This chapter has underlined the essential role of biomedical polymers in enhancing healthcare. By tackling the challenges and exploring future opportunities, it aimed to inspire further research and development in this area. Biomedical polymers are poised to lead the next wave of medical innovation, enabling breakthroughs in personalised medicine, regenerative therapies and smart medical devices. Biomedical polymers have undeniably established themselves as essential tools in modern medical science, revolutionising the fields of drug delivery, tissue engineering and medical device development. The unique properties of these polymers, such as biocompatibility, biodegradability and adjustable mechanical characteristics, allow them to overcome the limitations associated with conventional materials such as metals and ceramics. By providing customisable solutions tailored to specific biomedical needs, biomedical polymers continue to drive innovation in healthcare.

The role of biomedical polymers in drug delivery has been especially noteworthy. By facilitating the controlled, localized and sustained release of therapeutic agents, polymers such as nanoparticles and polymer-drug conjugates have significantly improved drug efficacy and patient outcomes. Innovations in stimuli-reactive polymers that react to pH, temperature or magnetic fields are enabling the creation of next-generation drug delivery systems that precisely target specific sites. Tissue engineering and regenerative medicine also benefit considerably from biomedical polymers. Scaffolds made from these materials mimic the extracellular matrix, enabling cell attachment, proliferation and differentiation. Advances have been made in bone regeneration, skin grafts and even organ engineering. The development of 3D-printed polymer scaffolds tailored to the specific needs of patients highlights the personalized potential of this field. Biomedical devices and implants are another area where polymers play a crucial role. Shape memory polymers and bioinspired materials enable the production of intelligent devices capable of adapting to dynamic physiological conditions. These devices, including minimally invasive surgical tools and self-expanding implants, offer greater flexibility and adaptability in clinical applications.

REFERENCES

Abrisham, M., Noroozi, M., Panahi-Sarmad, M., Arjmand, M., Goodarzi, V., Shakeri, Y., & Golbaten-Mofrad, H.. (2020). The Role of Polycaprolactone-Triol (PCL-T) in Biomedical Applications: A State-of-the-Art Review. *European Polymer Journal, 131*(May), 109701. DOI: 10.1016/j.eurpolymj.2020.109701

Adscheid, A., Selin, M. R.-R., Abdel-Hafez, S. M., Pavone, F. S., Schneider, M., Türeli, A. E., Calamai, M., & Günday-Türeli, N. (2025). Scalable Manufacturing Method for Model Protein-Loaded PLGA Nanoparticles: Biocompatibility, Trafficking and Release Properties. *Pharmaceutics, 17*(1), 87–87. DOI: 10.3390/pharmaceutics17010087 PMID: 39861735

Aflori, M. (2021). Smart nanomaterials for biomedical applications—A review. *Nanomaterials (Basel, Switzerland), 11*(2), 396. PMID: 33557177

Agarwal, K. M., Singh, P., Mohan, U., Mandal, S., & Bhatia, D. (2020). Comprehensive Study Related to Advancement in Biomaterials for Medical Applications. *Sensors International, 1*(January). Advance online publication. DOI: 10.1016/j.sintl.2020.100055

Agwa, M. M., Soliman, H. S. M., Elmotasem, H., & Sabra, S. (2024). Surface Functionalization for Biomaterials. In *Surface Engineering of Biomaterials Synthesis and Processing Techniques* (pp. 344–361). CRC Press., DOI: 10.1201/9781003429920-20

Amiryaghoubi, N. F. (2020). Bioactive polymeric scaffolds for osteogenic repair and bone regenerative medicine. *Medicinal Research Reviews, 40*(5), 1833–1870. PMID: 32301138

Arif, U., Haider, S., Haider, A., Khan, N., Alghyamah, A. A., Jamila, N., & Kang, I. K. (2019). Biocompatible polymers and their potential biomedical applications: A review. *Current Pharmaceutical Design, 25*(34), 3608–3619. DOI: 10.2174/1381612825999191011105148 PMID: 31604409

Bajpai, A. K., Shukla, S. K., Bhanu, S., & Kankane, S. (2008). Responsive polymers in controlled drug delivery. *Progress in Polymer Science, 33*(11), 1088–1118.

Bukhari, S. N. A., Roswandi, N. L., Waqas, M., Habib, H., Hussain, F., Khan, S., Sohail, M., Ramli, N. A., Thu, H. E., & Hussain, Z. (2018). Hyaluronic Acid, a Promising Skin Rejuvenating Biomedicine: A Review of Recent Updates and Pre-Clinical and Clinical Investigations on Cosmetic and Nutricosmetic Effects. *International Journal of Biological Macromolecules*. Advance online publication. DOI: 10.1016/j.ijbiomac.2018.09.188 PMID: 30287361

Capuani, S., Malgir, G., Corrine, Y. X. C., & Grattoni, A. (2022). *"Advanced Strategies to Thwart Foreign Body Response to Implantable Devices." Bioengineering and Translational Medicine*. John Wiley and Sons Inc., DOI: 10.1002/btm2.10300

Chang, D. M. (2021). Stimuli-responsive polymeric nanoplatforms for cancer therapy. *Frontiers in Bioengineering and Biotechnology, 9*, 707319. PMID: 34249894

Chen, W. H., Qi, W. C., Chen, Q., Cui, C., Duan, S., Kang, Y., & Liu, Y.. (2022). *"Biomedical Polymers: Synthesis, Properties, and Applications." Science China Chemistry*. Springer., DOI: 10.1007/s11426-022-1243-5

Chowdhury, S. M. (2017). Ultrasound-guided drug delivery in cancer. *Ultrasonography (Seoul, Korea)*, *36*(3), 171. PMID: 28607323

Crini, G. (2019). Historical Review on Chitin and Chitosan Biopolymers. *Environmental Chemistry Letters*, *17*(4), 1623–1643. DOI: 10.1007/s10311-019-00901-0

Culmone, C., Smit, G., & Breedveld, P. (2019). Additive manufacturing of medical instruments: A state-of-the-art review. *Additive Manufacturing*, *27*, 461–473. DOI: 10.1016/j.addma.2019.03.015

Das, S. S. (2020). Stimuli-responsive polymeric nanocarriers for drug delivery, imaging, and theragnosis. *Polymers*, *12*(6), 1397. PMID: 32580366

Dayyoub, T. M. (2022). Shape memory polymers as smart materials: A review. *Polymers*, *14*(17), 3511. PMID: 36080587

Delbianco, M., & Seeberger, P. H. (2020). Materials Science Based on Synthetic Polysaccharides. *Materials Horizons*, *7*(4), 963–969. DOI: 10.1039/C9MH01936G

Doberenz, F. Z. (2020). Thermoresponsive polymers and their biomedical application in tissue engineering–a review. *Journal of Materials Chemistry. B*, *8*(4), 607–628. PMID: 31939978

Dukovski, J., Bisera, J. L., Kocbek, P., Bočkor, L., Pepić, I., Hafner, A., & Lovrić, J. (2023). Development of a 3D Corneal Epithelial Model for Early Biocompatibility Screening of Topical Ophthalmic Formulations. *Macedonian Pharmaceutical Bulletin*, *69*(03), 247–248. DOI: 10.33320/maced.pharm.bull.2023.69.03.121

Ebrahimi, F., & Dana, H. R. (2022). Poly Lactic Acid (PLA) Polymers: From Properties to Biomedical Applications. *International Journal of Polymeric Materials*, *71*(15), 1117–1130. DOI: 10.1080/00914037.2021.1944140

Elkaliny, N. E., Alzamel, N. M., Moussa, S. H., Elodamy, N. I., Madkor, E. A., Ibrahim, E. M., Elshobary, M. E., & Ismail, G. A. (2024). *"Macroalgae Bioplastics: A Sustainable Shift to Mitigate the Ecological Impact of Petroleum-Based Plastics." Polymers. Multidisciplinary Digital Publishing Institute*. MDPI., DOI: 10.3390/polym16091246

Elumalai, K. S. (2024). Review of the efficacy of nanoparticle-based drug delivery systems for cancer treatment. *Biomedical Technology*, *5*, 109–122.

Ershad-Langroudi, A., Babazadeh, N., Alizadegan, F., Mousaei, S. M., & Moradi, G. (2024). Polymers for Implantable Devices. *Journal of Industrial and Engineering Chemistry*. Advance online publication. DOI: 10.1016/j.jiec.2024.03.030

Fathy, M. M., Mohamed, F. S., Elbialy, N. S., & Elshemey, W. M. (2018). Multifunctional Chitosan-Capped Gold Nanoparticles for Enhanced Cancer Chemo-Radiotherapy: An Invitro Study. *Physica Medica*, *48*(April), 76–83. DOI: 10.1016/j.ejmp.2018.04.002 PMID: 29728233

Feki, A. H. (2020). Conception and characterization of a multi-sensitive composite chitosan-red marine alga-polysaccharide hydrogels for insulin controlled-release. *Carbohydrate Polymers*, *236*, 116046. PMID: 32172860

Feng, Q., Zhang, K., Yang, B., & Yu, Y. (2022). Editorial: Biomedical Applications of Natural Polymers. *Frontiers in Bioengineering and Biotechnology*, *10*(November). Advance online publication. DOI: 10.3389/fbioe.2022.1077823 PMID: 36507260

Fu, Y., Ding, Y., Zhang, L., Zhang, Y., Liu, J., & Yu, P. (2021). Poly Ethylene Glycol (PEG)-Related Controllable and Sustainable Antidiabetic Drug Delivery Systems. *European Journal of Medicinal Chemistry*, *217*(May), 113372. DOI: 10.1016/j.ejmech.2021.113372 PMID: 33744689

García-Estrada, P., García-Bon, M. A., López-Naranjo, E. J., Basaldúa-Pérez, D. N., Santos, A., & Navarro-Partida, J. (2021). Polymeric implants for the treatment of intraocular eye diseases: Trends in biodegradable and non-biodegradable materials. *Pharmaceutics*, *13*(5), 701. DOI: 10.3390/pharmaceutics13050701 PMID: 34065798

Geddes, L., Themistou, E., Burrows, J. F., Buchanan, F. J., & Carson, L. (2021). Evaluation of the in Vitro Cytotoxicity and Modulation of the Inflammatory Response by the Bioresorbable Polymers Poly(D,L-Lactide-Co-Glycolide) and Poly(L-Lactide-Co-Glycolide). *Acta Biomaterialia*, *134*(October), 261–275. DOI: 10.1016/j.actbio.2021.07.049 PMID: 34329786

Hacker, M. C., Krieghoff, J., & Mikos, A. G. (2019). Synthetic polymers. In *Principles of regenerative medicine* (pp. 559–590). Academic press., DOI: 10.1016/B978-0-12-809880-6.00033-3

Herath, M. E. (2020). Light activated shape memory polymers and composites: A review. *European Polymer Journal*, *136*, 109912.

Hu, P., & Chen, G. (2024). Extraction, Structural Characterization, and Physicochemical and Biological Properties of Water-Soluble Polysaccharides from Adlay Bran. *Molecules (Basel, Switzerland)*, *29*(19), 4707. DOI: 10.3390/molecules29194707 PMID: 39407635

Hu, Tao, Jie Fang, Yang Shen, Mingyang Li, Bin Wang, Zushun Xu, and Weikang Hu. 2024. "Advances of Naturally Derived Biomedical Polymers in Tissue Engineering." *Frontiers in Chemistry*. Frontiers Media SA. https://doi.org/.DOI: 10.3389/fchem.2024.1469183

Hussain, Z., Khan, S., Imran, M., Sohail, M., Syed, W. A. S., & de Matas, M. (2019). *"PEGylation: A Promising Strategy to Overcome Challenges to Cancer-Targeted Nanomedicines: A Review of Challenges to Clinical Transition and Promising Resolution." Drug Delivery and Translational Research*. Springer., DOI: 10.1007/s13346-019-00631-4

Jarad, N. A., Chami, A., Weitz, J. I., & Didar, T. F. (2024). Advancements in Surface Modification Strategies of Vascular Grafts to Improve Biocompatibility and Tissue Integration. *Exploration of Biomat-X*, *1*(5), 241–265. DOI: 10.37349/EBMX.2024.00018

Jiang, X., Yao, Y., Tang, W., Han, D., Zhang, L., Zhao, K., & Meng, Y. (2020). Design of dental implants at materials level: An overview. *Journal of Biomedical Materials Research. Part A*, *108*(8), 1634–1661. DOI: 10.1002/jbm.a.36931 PMID: 32196913

Jin, G. Z. (2020). Current nanoparticle-based technologies for osteoarthritis therapy. *Nanomaterials (Basel, Switzerland)*, *10*(12), 2368. PMID: 33260493

Khopade, S., Gomte, S. S., Janrao, C., Bavaskar, A., Agnihotri, T. G., Jain, A., & Khatik, R. (2023). Peptide and Protein Delivery through Cellulose, Hyaluronic Acid, and Heparin. In *Peptide and Protein Drug Delivery Using Polysaccharides* (pp. 75–113). Academic Press., DOI: 10.1016/B978-0-443-18925-8.00003-9

Kopeček, J. (2013). Polymer-drug conjugates: Origins, progress to date and future directions. *Advanced Drug Delivery Reviews*, *65*(1), 49–59. DOI: 10.1016/j.addr.2012.10.014 PMID: 23123294

Lam, M. (2021). Grafting bioactive polymers onto silicone implants to improve biological activity (PhD thesis, Université Paris-Nord-Paris XIII).

Li, D. Z. (2020). Redox-responsive self-assembled nanoparticles for cancer therapy. *Advanced Healthcare Materials*, *9*(20), 2000605. PMID: 32893506

Li, H. Y., Huang, D. N., Ren, K. F., & Ji, J. (2021). Inorganic-polymer composite coatings for biomedical devices. *Smart Materials in Medicine*, *2*, 1–14. DOI: 10.1016/j.smaim.2020.10.002

Li, M., Zhao, Z., & Yi, J. (2024). Biomaterials Designed to Modulate Reactive Oxygen Species for Enhanced Bone Regeneration in Diabetic Conditions. *Journal of Functional Biomaterials*, *15*(8), 220.

Li, P. Y., Bearoff, F., Zhu, P., Fan, Z., Zhu, Y., Fan, M., Cort, L., Kambayashi, T., Blankenhorn, E. P., & Cheng, H. (2021). PEGylation Enables Subcutaneously Administered Nanoparticles to Induce Antigen-Specific Immune Tolerance. *Journal of Controlled Release*, *331*(March), 164–175. DOI: 10.1016/j.jconrel.2021.01.013 PMID: 33450320

Liao, J. (2020). Review on magnetic natural polymer constructed hydrogels as vehicles for drug delivery. *Biomacromolecules*, *21*(7), 2574–2594. PMID: 32543834

Liu, Houqin, Liu, Yingke, Tian, Zhipeng, & Li, Jiaxin, Man Li, and Zhihe Zhao. (2024). Coordinating Macrophage Targeting and Antioxidation by Injectable Nanocomposite Hydrogel for Enhanced Rheumatoid Arthritis Treatment. *ACS Applied Materials & Interfaces*, *16*(29), 37656–37668. DOI: 10.1021/acsami.4c06840 PMID: 38987704

Mabsor-Zgandaoui, S., Moussaif, A., El Hessni, A., Mesfioui, A., El Mzibri, M., Iddar, A., & Moutaouakkil, A. (2025a). Ionising and Ultraviolet Radiation Induce the Secretion of Melanin in Tetrahymena Pyriformis. *The International Journal of Environmental Studies*, *82*(1), 1–14. DOI: 10.1080/00207233.2024.2412455

Mabsor-Zgandaoui, S., Rachmoune, K., El Hessni, A., El Mzibri, M., Iddar, A., & Moutaouakkil, A. (2025b). Effect of Hydrogen Peroxide Pre-Treatment on Enhancing the Cellular Resistance of Tetrahymena Pyriformis Exposed to Ionizing Radiation. *Toxicological and Environmental Chemistry*, *107*(2), 294–309. DOI: 10.1080/02772248.2025.2460804

Machla, F., Bekiari, C., Monou, P. K., Kofidou, E., Theodosaki, A. M., Katsamenis, O. L., & Zisis, V.. (2024). Development of an Oral Epithelial Ex Vivo Organ Culture Model for Biocompatibility and Permeability Assessment of Biomaterials. *Bioengineering (Basel, Switzerland)*, *11*(10), 1035. DOI: 10.3390/bioengineering11101035 PMID: 39451410

Mahanty, M., Dutta, B., Ou, W., Zhu, X., Bromberg, J. S., He, X., & Rahaman, S. O. 2024. "Macrophage MicroRNA-146a Is a Central Regulator of the Foreign Body Response to Biomaterial Implants." BioRxiv *: The Preprint Server for Biology*. Elsevier. https://doi.org/DOI: 10.1016/J.BIOMATERIALS.2024.122855

Mohammed, A. S. A., Naveed, M., & Jost, N. (2021). Polysaccharides; classification, chemical properties, and future perspective applications in fields of pharmacology and biological medicine (a review of current applications and upcoming potentialities). *Journal of Polymers and the Environment*, *29*(8), 2359–2371.

Morawetz, H. (2014). History of Polymer Science. In *Encyclopedia of Polymer Science and Technology* (pp. 1–30). John Wiley & Sons, Ltd., DOI: 10.1002/0471440264.pst631

Narayanan, K. B., Bhaskar, R., Kim, H., & Han, S. S. (2023). In Vitro Cytocompatibility Assessment of Novel 3D Chitin/Glucan- and Cellulose-Based Decellularized Scaffolds for Skin Tissue Engineering. *Sustainability (Switzerland)*, *15*(21). Advance online publication. DOI: 10.3390/su152115618

Nasra, S. B. (2022). Nanoscale advances. *Recent advances in nanoparticle-based drug delivery systems for rheumatoid arthritis treatment*, 4(17), 3479-3494.

Ng, F., Nicoulin, V., Peloso, C., Curia, S., Richard, J., & Lopez-Noriega, A. (2023). In Vitro and In Vivo Hydrolytic Degradation Behaviors of a Drug-Delivery System Based on the Blend of PEG and PLA Copolymers. *ACS Applied Materials & Interfaces*, *15*(48), 55495–55509. DOI: 10.1021/acsami.2c13141 PMID: 38011651

Nguyen, D. D., Luo, L. J., & Lai, J. Y. (2022). Thermogels Containing Sulfated Hyaluronan as Novel Topical Therapeutics for Treatment of Ocular Surface Inflammation. *Materials Today. Bio*, *13*(January). Advance online publication. DOI: 10.1016/j.mtbio.2021.100183 PMID: 34927046

Ofridam, F. T. (2021). pH-sensitive polymers: Classification and some fine potential applications. *Polymers for Advanced Technologies*, *32*(4), 1455–1484.

Ornaghi, H. L., Monticeli, F. M., & Dall Agnol, L. (2023). *"A Review on Polymers for Biomedical Applications on Hard and Soft Tissues and Prosthetic Limbs." Polymers*. Multidisciplinary Digital Publishing Institute., DOI: 10.3390/polym15194034

Puertas-Bartolomé, M., Mora-Boza, A., & García-Fernández, L. (2021). Emerging Biofabrication Techniques: A Review on Natural Polymers for Biomedical Applications. *Polymers*, *13*(8), 1209. DOI: 10.3390/polym13081209 PMID: 33918049

Qian, M., Li, S., Xi, K., Tang, J., Shen, X., Liu, Y., & Guo, R.. (2023). ECM-Engineered Electrospun Fibers with an Immune Cascade Effect for Inhibiting Tissue Fibrosis. *Acta Biomaterialia*, *171*(November), 308–326. DOI: 10.1016/j.actbio.2023.08.058 PMID: 37673231

Rajitha, A., Kansal, L., Raj, G., Kalra, R., Dhamija, K., & Abdul-Zahra, D. S. 2024. "Biomaterials and Artificial Intelligence: Predictive Modeling and Design." In *E3S Web of Conferences*. Vol. 505. EDP Sciences. https://doi.org/DOI: 10.1051/e3sconf/202450501003

Ruiz-Alonso, S. L.-M.-B. (2021). Tendon tissue engineering: Cells, growth factors, scaffolds and production techniques. *Journal of Controlled Release*, *333*, 448–486. PMID: 33811983

Santhanakrishnan, K. R., Koilpillai, J., & Narayanasamy, D. (2024). PEGylation in Pharmaceutical Development: Current Status and Emerging Trends in Macromolecular and Immunotherapeutic Drugs. *Cureus*, (August). Advance online publication. DOI: 10.7759/cureus.66669 PMID: 39262507

Saravanan, S., Vimalraj, S., & Anuradha, D. (2018). Chitosan Based Thermoresponsive Hydrogel Containing Graphene Oxide for Bone Tissue Repair. *Biomedicine and Pharmacotherapy*, *107*(November), 908–917. DOI: 10.1016/j.biopha.2018.08.072 PMID: 30257403

Satchanska, G., Davidova, S., & Petrov, P. D. (2024). *"Natural and Synthetic Polymers for Biomedical and Environmental Applications." Polymers*. Multidisciplinary Digital Publishing Institute., DOI: 10.3390/polym16081159

Shakeel, A. (2023). Mitigating challenges and expanding the future of vascular tissue engineering—Are we there yet? *Frontiers in Physiology*, *13*, 1079421. PMID: 36685187

Sheikh, Z., Najeeb, S., Khurshid, Z., Verma, V., Rashid, H., & Glogauer, M. (2015). Biodegradable Materials for Bone Repair and Tissue Engineering Applications. *Materials (Basel)*, *8*(9), 5744–5794. DOI: 10.3390/ma8095273 PMID: 28793533

Suhag, D. K. (2024). Scaffold Design in Tissue Engineering. In Handbook of Biomaterials for Medical Applications. *Singapore: Springer Nature Singapore.*, Volume 1: Fundamentals. pp. 155-184.

Suk, J. S., Xu, Q., Kim, N., Hanes, J., & Ensign, L. M. (2016). "PEGylation as a Strategy for Improving Nanoparticle-Based Drug and Gene Delivery." *Advanced Drug Delivery Reviews. Advanced Drug Delivery Reviews.* Advance online publication. DOI: 10.1016/j.addr.2015.09.012 PMID: 26456916

Tao, J., Zhang, Y., Shen, A., Yang, Y., Diao, L., Wang, L., Cai, D., & Hu, Y. (2020). Injectable Chitosan-Based Thermosensitive Hydrogel/Nanoparticle-Loaded System for Local Delivery of Vancomycin in the Treatment of Osteomyelitis. *International Journal of Nanomedicine*, *15*, 5855–5871. DOI: 10.2147/IJN.S247088 PMID: 32848394

Thakor, P. B. (2020). Polymer–drug conjugates: Recent advances and future perspectives. *Drug Discovery Today*, *25*(9), 1718–1726. PMID: 32629170

Thirunavukkarasu, G. K. (2018). Magnetic field-inducible drug-eluting nanoparticles for image-guided thermo-chemotherapy. *Biomaterials*, *180*, 240–252. PMID: 30055399

Thomas, A. P., Kasa, V. P., Dubey, B. K., Sen, R., & Sarmah, A. K. (2023). *"Synthesis and Commercialization of Bioplastics: Organic Waste as a Sustainable Feedstock." Science of the Total Environment.* Pergamon., DOI: 10.1016/j.scitotenv.2023.167243

Toosi, S. (2020). Osteogenesis and bone remodeling: A focus on growth factors and bioactive peptides. *BioFactors (Oxford, England)*, *46*(3), 326–340. PMID: 31854489

Tran, D. T., Yadav, A. S., Nguyen, N. K., Singha, P., Ooi, C. H., & Nguyen, N. T. (2023). *"Biodegradable Polymers for Micro Elastofluidics." Small.* John Wiley and Sons Inc., DOI: 10.1002/smll.202303435

Trucillo, P. (2024). *"Biomaterials for Drug Delivery and Human Applications." Materials. Multidisciplinary Digital Publishing Institute.* MDPI., DOI: 10.3390/ma17020456

Ulbricht Aus Großenhain. M Sc Juliane. 2017. "Insights into Polymer Biodegradation Investigations on Oxidative, Hydrolytic and Enzymatic Pathways." https://opus.bibliothek.uni-wuerzburg.de/frontdoor/index/index/docId/15868

Wang, G., Li, R., Parseh, B., & Du, G. (2021). Prospects and Challenges of Anticancer Agents' Delivery via Chitosan-Based Drug Carriers to Combat Breast Cancer: A Review. *Carbohydrate Polymers*, *268*(May). Advance online publication. DOI: 10.1016/j.carbpol.2021.118192 PMID: 34127212

Wang, L. Z. (2022). Shape memory polymer fibers: Materials, structures, and applications. *Advanced Fiber Materials*, *4*(1), 5–23. PMID: 35359822

Wen, G. L. (2020). Effective phototheranostics of brain tumor assisted by near-infrared-II light-responsive semiconducting polymer nanoparticles. *ACS Applied Materials & Interfaces*, *12*(30), 33492–33499. PMID: 32627525

Xia, Y. H. (2021). A review of shape memory polymers and composites: Mechanisms, materials, and applications. *Advanced Materials*, *33*(6), 2000713. PMID: 32969090

Xiang, J. T. (2018). Near-infrared light-triggered drug release from UV-responsive diblock copolymer-coated upconversion nanoparticles with high monodispersity. *Journal of Materials Chemistry*, *6*(21), 3531–354. PMID: 32254448

Xu, X. J. (2021). Bioinspired double network hydrogels: From covalent double network hydrogels via hybrid double network hydrogels to physical double network hydrogels. *Materials Horizons*, *8*(4), 1173–1188. PMID: 34821910

Yan, J. Z. (2021). Redox-responsive polyethyleneimine/tetrahedron DNA/doxorubicin nanocomplexes for deep cell/tissue penetration to overcome multidrug resistance. *Journal of Controlled Release*, *329*, 36–49. PMID: 33259850

Yang, J., Zeng, H., Luo, Y., Chen, Y., Wang, M., Wu, C., & Hu, P. (2024). *"Recent Applications of PLGA in Drug Delivery Systems." Polymers*. Multidisciplinary Digital Publishing Institute., DOI: 10.3390/polym16182606

Yao, X., Qi, C., Sun, C., Huo, F., & Jiang, X. (2023). Poly(Ethylene Glycol) Alternatives in Biomedical Applications. *Nano Today*, *48*(February), 101738. DOI: 10.1016/j.nantod.2022.101738

Yingngam, B. 2024. "Sustainable Biomaterials for Pharmaceutical and Medical Applications." In *Advances in Sustainable Biomaterials*, 8:250–74. https://doi.org/DOI: 10.1201/9781003434313-13

Yoshida, S., Hagiwara, K., Hasebe, T., & Hotta, A. (2013). Surface Modification of Polymers by Plasma Treatments for the Enhancement of Biocompatibility and Controlled Drug Release. *Surface and Coatings Technology*. Advance online publication. DOI: 10.1016/j.surfcoat.2013.02.042

Yuan, X., Zhu, W., Yang, Z., He, N., Chen, F., Han, X., & Zhou, K. (2024). Recent Advances in 3D Printing of Smart Scaffolds for Bone Tissue Engineering and Regeneration. *Advanced Materials*, *36*(34), 2403641. DOI: 10.1002/ADMA.202403641 PMID: 38861754

Zhao, C. H., Zhang, X. P., & Zhang, L. (2022). RGD Peptide Functionalized Graphene Oxide: A Bioactive Surface for Cell-Material Interactions. *Digest Journal of Nanomaterials and Biostructures*, *17*(3), 989–997. DOI: 10.15251/DJNB.2022.173.989

Zhao, L., Zhou, Y., Zhang, J., Liang, H., Chen, X., & Tan, H. (2023). Natural Polymer-Based Hydrogels: From Polymer to Biomedical Applications. *Pharmaceutics*. Advance online publication. DOI: 10.3390/pharmaceutics15102514 PMID: 37896274

Chapter 2
Nanomaterials in Medical Applications:
An Introduction

Yusuf Olatunji Waidi
https://orcid.org/0000-0002-5728-1769
Indian Institute of Science, India

ABSTRACT

Nanotechnology has revolutionized medicine, offering innovative solutions to longstanding healthcare challenges. This chapter provides a comprehensive overview of nanoparticles, their unique properties, and their diverse medical applications. It explores the historical context of nanoparticle use in therapeutic approaches and delves into various nanoparticle types, including their properties and applications. The chapter also examines the interaction between nanoparticles, cells, and tissues, discussing cellular uptake mechanisms, responses, and tissue distribution. Additionally, it addresses the challenges and future prospects of nanoparticle-based therapies, including safety, toxicity concerns, clinical translation barriers, and potential future directions. Overall, the chapter highlights the promising potential of nanoparticles to transform medicine by offering innovative solutions for diagnosis, treatment, and drug delivery.

1. INTRODUCTION

Nanoparticles (NPs), typically ranging from 1 to 100 nanometers in size, exhibit exceptional catalytic and adsorptive properties useful for various applications (Erisen & Uludag, 2024; Waidi & Jain, 2024a; Waidi et al., 2024). These "zero-dimensional" NPs, as they are often referred to, possess unique characteristics due to their nanoscale dimensionss. In contrast to one-dimensional nanomaterials (NMs) like nanowires and nanotubes or two-dimensional NMs like self-assembled monolayer films, NPs have all three dimensions within the nanoscale range. This distinct feature, coupled with their large surface area, imparts unique chemical and physical properties, including size-dependent optical properties that result in distinct colors due to visible light absorption (Jain, Waidi, Vannaladsaysy, Pathote, & Datta, 2025). The unique properties of NPs have led to their incorporation into composite materials, such as the modern rubber tire, which utilizes silica NPs as a reinforcing filler. The early 21st century witnessed a significant surge in nanoparticle usage in various applications, driven by nanocomposite (NCP) technol-

DOI: 10.4018/979-8-3373-0055-9.ch002

ogy advancements (Waidi, 2024a, 2024b). NCPs have found applications in various fields, including the creation of novel materials with tailored dielectric and magnetic properties. Moreover, organic materials incorporating NPs have enabled the development of ultra-thin solar panels that offer a relatively low cost and offer exceptional quantum efficiency and flexibility. Beyond their industrial applications, NPs have also shown promise in synthetic biology. Well-refined NPs of proteins and nucleic acids can self-assemble into highly ordered structures, exhibiting unusual properties that could potentially revolutionize fields like medicine and materials science (Kaltbeitzel & Wich, 2023).

Furthermore, other NPs, such as "quantum dots" or "artificial atoms," are quasi-zero-dimensional (0D) nanostructures with unique properties. Their size-dependent behavior, distinct from bulk materials, arises from quantum mechanical effects (Szczyglewska, Feliczak-Guzik, & Nowak, 2023). They exhibit diverse characteristics based on their composition, structure, and size, leading to a wide range of applications (Mourdikoudis, Pallares, & Thanh, 2018). While traditional methods like physical and chemical synthesis have been employed, recent advancements in green nanotechnology have leveraged biological approaches, particularly plant-mediated synthesis, to produce NPs with controlled properties (Zou et al., 2022). These plant-based NPs offer advantages such as eco-friendliness, biocompatibility, and reduced toxicity compared to conventional methods (O'Dell, Schein, & Erickson, 2016). The applications of NPs span various fields, including medicine, electronics, materials science, and environmental remediation (Waidi, 2024a). Their potential in drug delivery, biosensing, catalysis, and water treatment has garnered significant attention. As research progresses, developing novel NP-based technologies continues to shape advancements in diverse industries.

The evolution of nanoparticles (NPs) in medicine, though termed "nanotechnology" in the 1970s, boasts ancient roots, with Egyptians using colloidal gold for glass coloration and likely tissue penetration, and Ayurvedic practitioners employing NP-containing herbal remedies (Amen et al., 2021; Kagdada, Bhojani, & Singh, 2023). However, modern NP technology truly began to flourish with key scientific advancements: the invention of the scanning electron microscope in the 1930s enabled unprecedented visualization of nanoscale materials (Sudha, Sangeetha, Vijayalakshmi, & Barhoum, 2018), directly impacting our ability to characterize and engineer NPs. The discovery of liposomes by Alec Bangham in the 1960s marked a pivotal moment, as it opened the door to encapsulating drugs and laid the foundation for modern targeted drug delivery systems (Amen et al., 2021). Further revolutionizing diagnostic capabilities, the emergence of quantum dots in the early 1980s transformed fluorescent imaging techniques, paving the way for advanced bioimaging (Amen et al., 2021). These foundational discoveries, coupled with subsequent research in the 21st century, have culminated in the FDA approval of numerous nanomedicine therapies, particularly in cancer treatment, demonstrating the significant impact of NPs in targeted drug delivery and diagnostics (Kagdada et al., 2023). Despite these successes, ongoing research and regulatory oversight remain critical to address persistent challenges in ensuring the safety and efficacy of NP applications in medicine.

Understanding nanoparticle-based therapies is crucial as they offer transformative medical potential through enhanced drug delivery, improved diagnostics, and novel therapeutic approaches. Engineered to target specific diseased cells, NPs significantly improve drug efficacy and minimize side effects (Yetisgin, Cetinel, Zuvin, Kosar, & Kutlu, 2020). Their ability to sustain drug release and penetrate biological barriers, such as the blood-brain barrier, facilitates access to previously unreachable areas (Zubair et al., 2024). Moreover, NPs enable early disease diagnosis through sensitive biomarker detection and enhance imaging modalities, providing valuable insights into disease progression (Umapathy, Natarajan, & Swamikannu, 2024). By delivering therapeutic genes and combining diagnostic and therapeutic functions,

NPs offer potential cures for genetic disorders and pave the way for personalized medicine (Pandey et al., 2024). While the benefits of nanoparticle-based therapies are substantial, addressing concerns regarding their long-term safety and potential toxicity is crucial for their successful integration into clinical practice.

This article attempts to emphasize the significant contributions of NPs to modern medicine. It delves into the multifaceted role of NPs, exploring their diverse types, unique properties, and myriad applications across various medical domains. Elucidating the intricate interactions between NPs and biological systems will help us better understand their potential benefits and limitations. Moreover, the article critically examines the challenges that hinder the widespread adoption of NP-based technologies and proposes promising avenues for future research. Ultimately, this article aims to underscore the immense potential of NPs to revolutionize healthcare and improve patient outcomes.

2. TYPES, PROPERTIES, AND APPLICATIONS OF NANOPARTICLES

2.1 Metallic nanoparticles

Gold NPs (AuNPs) are important metallic NPs to humanity. It has emerged as a transformative force in imaging technology, leveraging its exceptional optical and electronic properties to enhance diagnostic capabilities (Alrahili, 2024). Their unique ability to absorb and scatter light efficiently underpins various imaging modalities, including optical imaging, electron microscopy, and computed tomography. The localized surface plasmon resonance (LSPR) exhibited by AuNPs enables sensitive detection of biomolecules. In contrast, their fluorescence resonance energy transfer (FRET) properties enhance signal-to-noise ratios in biological imaging (Gupta, Sharma, Mishra, Rai, & Verma, 2024). In electron microscopy, AuNPs conjugated to antibodies serve as valuable immunogold labels, facilitating high-resolution imaging of specific cellular targets (Soni & Solanki, 2023). Moreover, AuNPs' high X-ray attenuation properties make them effective contrast agents for computed tomography, improving soft tissue visibility. Additionally, their ability to convert absorbed laser light into heat has led to their application in photoacoustic imaging, enabling non-invasive visualization of deep tissues (Karnwal et al., 2024). Despite these promising advancements, the clinical translation of AuNPs is hindered by concerns regarding potential toxicity and regulatory challenges, necessitating rigorous research to optimize their safety and efficacy for widespread adoption.

AuNPs have also garnered significant attention as promising drug delivery vehicles due to their exceptional properties. Their biocompatibility, surface modifiability, and distinctive optical properties endow them with the ability to enhance drug efficacy while minimizing adverse effects. AuNPs' non-toxic nature and ease of functionalization with various ligands allow for precise targeting of specific cells, including cancer cells, thus minimizing off-target toxicity (Kumar et al., 2024; Shameli et al., 2024). Moreover, their surface chemistry enables the conjugation of diverse therapeutic agents, including drugs, proteins, and nucleic acids, optimizing drug loading and release profiles (Croitoru et al., 2024). The strong LSPR exhibited by AuNPs facilitates their use in imaging and monitoring drug delivery, providing valuable insights into therapeutic efficacy and enabling real-time tracking of drug distribution (Soni & Solanki, 2024; Fatima et al., 2024). Furthermore, AuNPs can be engineered for controlled drug release in response to specific stimuli, such as changes in pH or temperature, ensuring sustained therapeutic effects and improved drug bioavailability (Kumar et al., 2024; Shameli et al., 2024; Croitoru et al., 2024). Despite these advantages, challenges such as potential toxicity from surface coatings and scalability issues must be carefully addressed to fully realize the clinical potential of AuNP-based therapies (Soni & Solanki, 2024; Shameli et al., 2024).

Furthermore, AuNPs have also emerged as a cornerstone of biosensing technology, particularly in medical diagnostics (Sachan et al., 2023). Their unique properties, including tunable size, biocompatibility, and surface chemistry, make them ideal for detecting and diagnosing various diseases (Harzand, 2022). A prime example is their application in the early detection of infectious diseases like HIV. A highly sensitive AuNP-based biosensors capable of detecting HIV at astonishingly low concentrations of 5 femtomolar was developed in a recent study (Farzin, Shamsipur, Samandari, & Sheibani, 2020). This breakthrough enables more accurate and timely diagnosis, improving patient outcomes. Similarly, AuNP biosensors have been employed to identify cancer biomarkers, such as prostate-specific antigen (PSA) and carcinoembryonic antigen (CEA) (Shayesteh & Ghavami, 2020). These sensors exhibit exceptional sensitivity, detecting analytes in the picomolar range (Shah et al., 2022). Furthermore, AuNP-based biosensors have been utilized to detect cardiac biomarkers, including troponin and creatine kinase-MB (CK-MB), aiding in diagnosing heart disorders (R. Gao et al., 2022) (**Fig 1. I**). Their sensitivity, demonstrated by the detection of troponin at concentrations as low as 1.25 picograms per milliliter (pg/mL) (Baryeh, 2019), significantly advances the field of medical diagnostics.

Aside from AuNPs, Silver NPs (AgNPs) are another interesting metallic NPs with long potent antimicrobial properties dating back to ancient civilizations. The unique bactericidal and antiviral activities of AgNPs stem from the release of silver ions, which disrupt the metabolic processes of microorganisms. This antimicrobial efficacy has been harnessed for centuries, as evidenced by using silver filters to purify water and prevent bacterial growth. Recent advances in nanotechnology have enabled the production of AgNPs with exceptional surface area and catalytic activity, further enhancing their antimicrobial potency. AgNPs can effectively penetrate bacterial cell walls and interfere with essential cellular functions, leading to microbial inactivation. Consequently, AgNPs have found widespread applications in various fields, including medicine, textile manufacturing, and healthcare products. Moreover, AgNPs offer a promising alternative to precious metal powders in producing electronic pastes, potentially leading to significant improvements in microelectronic processes (Pitchayyapillai, Seenikannan, Balasundar, & Narayanasamy, 2017).

Furthermore, studies have demonstrated the remarkable ability of AgNPs to eradicate a wide range of microorganisms, including HIV, without causing collateral damage to host cells (Luceri, Francese, Lembo, Ferraris, & Balagna, 2023). Unlike antibiotics, which often exhibit diminishing effectiveness against evolving pathogens, AgNPs selectively target microbial cells, leaving healthy tissues unharmed (Barua & Buragohain, 2024). The unique mechanism of action involves the disruption of microbial cell envelopes, preventing nutrient uptake and ultimately leading to cell death. While AgNPs can effectively combat enveloped viruses, their interaction with human and animal cells is generally benign due to the protective nature of cellular membranes. The potential applications of AgNPs in pharmaceuticals and consumer products are vast and diverse. For instance, AgNP-containing toothpaste, such as "Witch Doctor" by Helios, offers enhanced protection against oral infections (Ansari et al., 2021) (**Fig 1. II**). Additionally, AgNPs are incorporated into various cosmetic products to prevent discoloration and provide anti-inflammatory benefits. Furthermore, AgNPs can be used to modify textiles, imparting self-disinfecting properties and extending their lifespan. AgNP-treated fabrics have found widespread applications in medical settings, children's clothing, and footwear, offering protection against harmful microorganisms. While AgNPs offer numerous advantages, it is crucial to address potential concerns regarding their environmental impact and potential toxicity (Ghose & Kumar, 2022). Ongoing research aims to minimize these risks and ensure the safe and responsible use of AgNPs in various applications.

Figure 1. Examples of metallic NPs in medicine. **I.** *Simultaneous SERS Detection of Cardiac Biomarkers on a Microfluidic Chip (a) This panel depicts a microfluidic chip designed for the simultaneous detection of CK-MB and cTnI cardiac biomarkers using surface-enhanced Raman spectroscopy (SERS). (b) The process of conjugating capture antibodies onto a PNMA substrate is illustrated. (c) The workflow of the SERS detection process is outlined. Mixtures of CK-MB and cTnI antigens, CK-MB/cTnI detection antibody-conjugated SERS nanoprobes, and PBS buffer are injected into inlets I-IV. The CK-MB and cTnI capture antibody-conjugated PNMA substrates are integrated into chambers V and VI, respectively, for biomarker detection. Reproduced from (R. Gao et al., 2022)* **II.** *Interaction of LIV-AgNPs with Microbial Cells. This figure illustrates the interaction between LIV-AgNPs (lipid-encapsulated silver nanoparticles) and various microbial cells: Panels (a, d, g): SEM images of MDR-PA, MRSA, and C. albicans cells without exposure to LIV-AgNPs. Panels (b, c, e-i): SEM images of MDR-PA, MRSA, and C. albicans cells treated with 100 μg/mL of LIV-AgNPs at both low (10-5 μm) and high (5-2 μm) scales, demonstrating cellular damage*

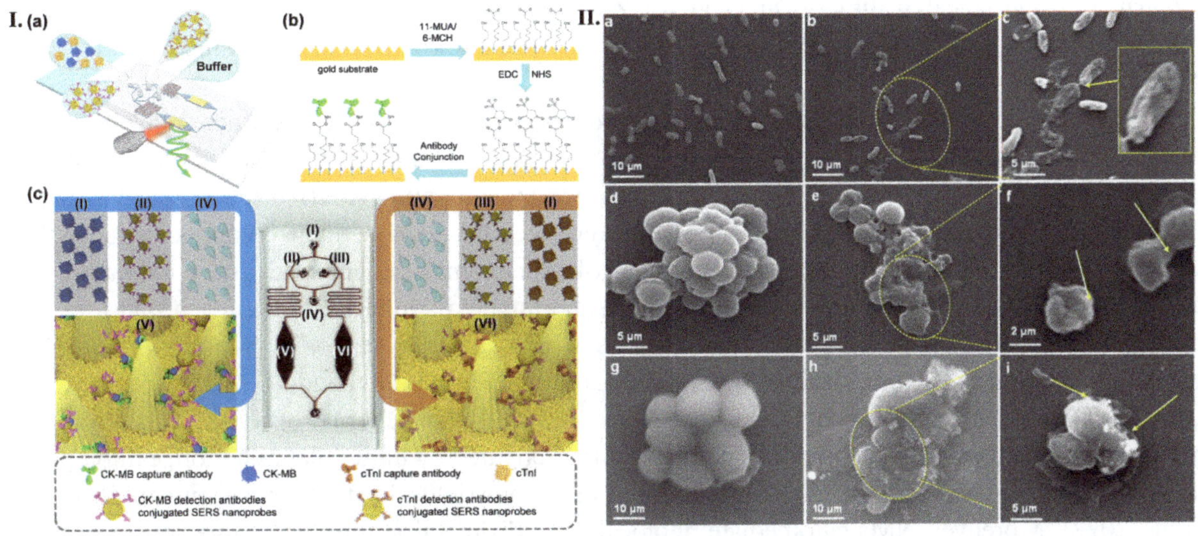

Reproduced from (Ansari et al., 2021).

AgNPs have also garnered considerable attention as promising agents for wound healing due to their potent antimicrobial properties and ability to facilitate tissue regeneration. Their unique physicochemical characteristics enable them to effectively combat various pathogens, including bacteria, fungi, and viruses. AgNPs exert their antimicrobial effects through multiple mechanisms, such as disrupting microbial cell walls, interfering with DNA replication, and inducing oxidative stress, ultimately leading to microbial death (Rodrigues et al., 2024). By mitigating the risk of infections, stimulating cell proliferation, modulating excessive inflammation, and promoting angiogenesis, AgNPs significantly accelerate wound healing and improve patient outcomes (Oselusi, Sibuyi, Meyer, & Madiehe, 2024). AgNPs can be incorporated into topical formulations, wound dressings, or others to ensure effective delivery. While AgNPs offer significant advantages, further research is imperative to elucidate their long-term safety profile and potential cytotoxicity, particularly in vulnerable populations (N. Xie, 2024). By addressing these concerns, AgNPs can be harnessed as valuable tools to enhance wound healing and improve patient care.

2.2 Semiconductor nanoparticles

Beyond graphene, many two-dimensional semiconductor NMs have emerged as promising materials for various biomedical applications. These materials, including transition metal dichalcogenides (TMDCs), graphitic carbon nitride (g-C3N4), transition metal oxides (TMOs), black phosphorus (BP), layered double hydroxides (LDHs), and transition metal carbides and nitrides (MXenes), have demonstrated exceptional antibacterial properties which happens to be one of the major problems in biomedical field (Kurapati, Kostarelos, Prato, & Bianco, 2016; Waidi & Jain, 2024b). Their antibacterial mechanisms are multifaceted, encompassing physical contact damage, oxidative stress, drug-metal ion release, photo-induced effects, and synergistic interactions. One intriguing mechanism involves the direct physical disruption of bacterial cell membranes. For instance, Bi2Se3 nanodiscs have been observed to bind with lipoteichoic acid, a component of Gram-positive bacteria, leading to membrane damage (Ouyang et al., 2019). Additionally, the sharp edges of WS2 nanosheets have been shown to disrupt the structural integrity of bacterial cell walls (X. Liu, Duan, Li, Zhou, & Zhou, 2017). Similar destructive effects have been reported for MoS2, MoO3, RuO2, and MnO2 edges. These findings highlight the potential of 2D NBG materials as novel and effective antibacterial agents.

Furthermore, semiconductor NMs-mediated oxidative stress, a multifaceted process involving both reactive oxygen species (ROS)-dependent and -independent mechanisms, underlies their antibacterial efficacy. ROS-dependent oxidative stress, characterized by generating reactive species such as peroxides, hydroxyl radicals, and superoxides, directly damages bacterial cellular components and DNA. Conversely, ROS-independent oxidative stress involves nanomaterial-induced direct oxidation. Additionally, photodynamic therapy is a non-invasive and targeted antibacterial approach that leverages photoresponsive semiconductors that, when exposed to light, generate heat or ROS through fluorophore excitation, leading to bacterial inactivation (Azizi-Lalabadi, Ehsani, Divband, & Alizadeh-Sani, 2019) (**Fig 2.**). MoS2 nanosheets, renowned for their near-infrared photothermal efficiency, are promising photothermal antibacterial agents. ZnO and TiO2 NPs, activated by ultraviolet radiation, exhibit potent antibacterial properties and have found applications in various fields due to their biocompatibility (Kim et al., 2022). Moreover, NMs can mediate photocatalytic disinfection through charge-separated states and ROS production, facilitated by their large surface area and interaction with oxygen and water molecules (Zeng, Wang, Liu, & Zhang, 2017). Drug-metal ion release, wherein NMs act as nanocarriers delivering antibacterial agents to target sites, represents another antibacterial mechanism. Synergistic approaches, combining multiple mechanisms, enhance antibacterial efficacy. MoS2 polydopamine-Ag nanosheets exhibit efficient antibacterial activity, demonstrating both Ag ion release and photothermal effects (Yuwen et al., 2018). Functionalizing semiconductor NPs with polymers or organic ligands can further augment their antimicrobial properties, as exemplified by the enhanced antibacterial activity of glutamic acid-functionalized ZrO2 nanocomposites (Khan et al., 2020). The electrostatic interactions between positively charged NMs and the more negatively charged gram-negative bacteria contribute to their greater susceptibility to nanoparticle-mediated antibacterial effects.

Figure 2. Disc Diffusion Test of Nanoparticle-Doped 4A Zeolite. This figure presents the results of a disc diffusion test conducted on 4A zeolite doped with various nanoparticles. The following samples were tested: Sample 1: ZnO/4A zeolite. Sample 2: TiO2/4A zeolite. Sample 3: ZnO and TiO2/4A zeolite. Sample 4: Undoped 4A zeolite (control)

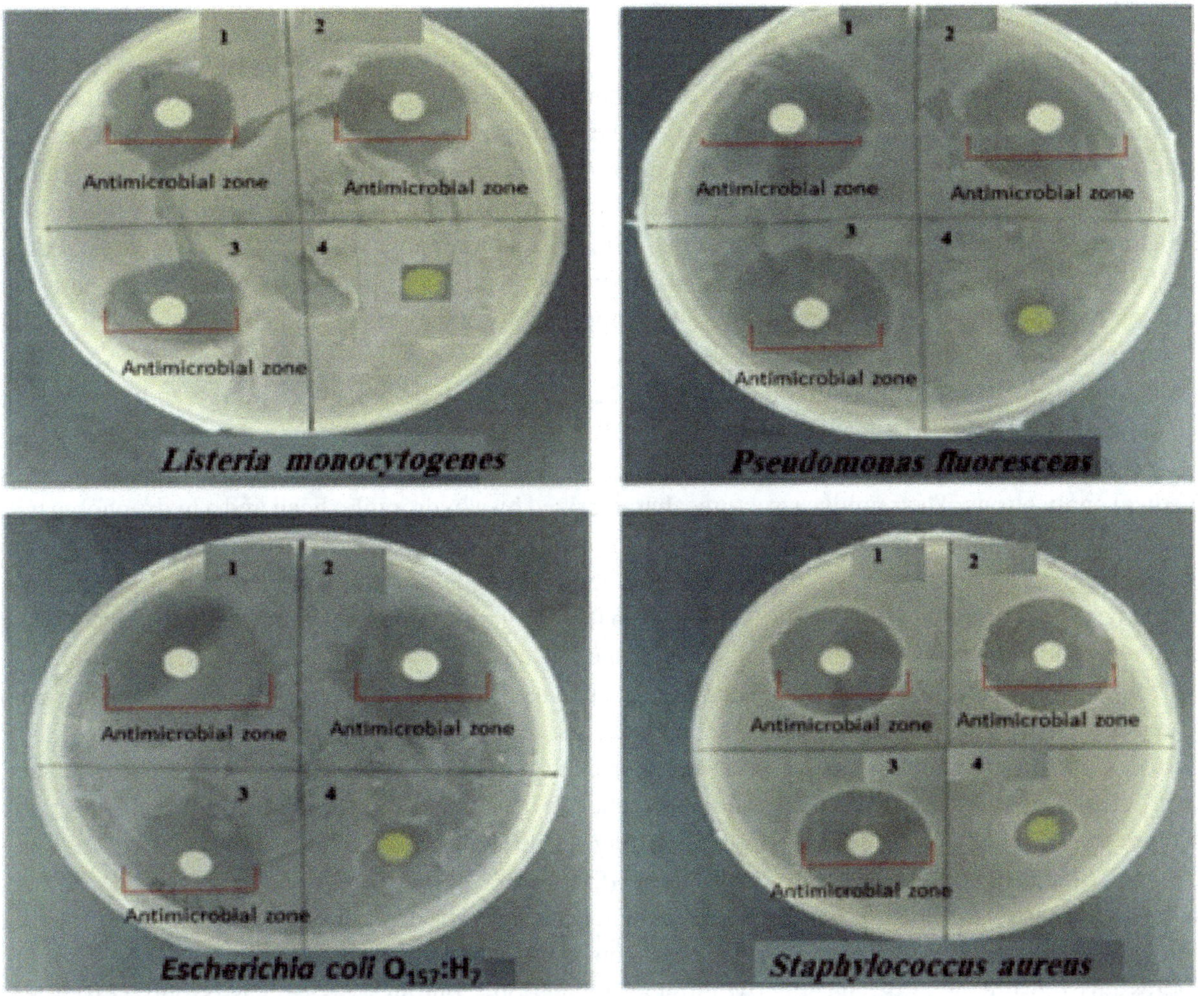

Reproduced from (Azizi-Lalabadi et al., 2019).

2.3 Lipid-based nanoparticles

Lipid-based NPs, particularly solid lipid nanoparticles (SLNs), have garnered significant attention due to their exceptional biocompatibility and low toxicity profiles. Compared to polymeric NPs, such as polyester and poly(alkyl cyanoacrylate) (PACA) NPs, SLNs exhibit significantly reduced cytotoxicity against human granulocytes, often by 10-20 folds (Pawar & Babu, 2010). This relative safety profile makes SLNs promising candidates for drug delivery applications. While the lipid composition and particle size of SLNs play a lesser role in determining toxicity, the choice of surfactant employed exerts a more substantial influence. Notably, SLNs have been shown to avoid macrophage activation and cytokine production, even at varying sizes. The cytotoxic effects of SLNs are primarily attributed to their electrostatic interactions with the negatively charged cellular membrane, emphasizing the importance of surface charge considerations in NP design (Patel, Patel, Patel, Chougule, & Choudhury, 2018).

Liposomes are another interesting lipid-based NPs, also known as vesicles, which are self-assembled structures composed of amphiphilic lipids, such as PEGylated phospholipids like DSPE and DSPC. Their stability in the biological environment is crucial for effective drug delivery. Incorporating proteins or employing strategies like steric stabilization can enhance liposomal stability and prolong circulation time (J. Xie, Lee, & Chen, 2010). Both passive and active targeting strategies are employed to target anticancer agents at tumor sites. Passive targeting, facilitated by enhanced permeability and retention (EPR) effects, involves liposomes evading phagocytosis. PEGylation and maintaining a particle size between 40 and 200 nm are key factors in passive targeting (Onzi, Guterres, Pohlmann, & Frank, 2021). Active targeting, on the other hand, involves surface functionalization of liposomes with targeting ligands like antibodies, peptides, aptamers, or small molecules. This approach enables precise targeting, endocytosis, and therapeutic response, minimizing off-target effects. Immunoliposomes, created by chemically coupling antibodies or their fragments to the liposomal surface, exemplify the potential of targeted liposomes in cancer therapy (Huang, He, Cao, Cai, & Liang, 2012).

Micelles are self-assembling amphiphilic molecules that have emerged as promising tools for various medical applications. Their unique structure, characterized by a hydrophilic head and a hydrophobic tail, enables enhanced solubility, targeted delivery, and controlled release of both hydrophilic and hydrophobic molecules. Polymeric micelles, in particular, have demonstrated exceptional potential in drug delivery, solubilizing hydrophobic drugs and protecting them from degradation, thereby improving bioavailability and therapeutic efficacy (Pathak, Suthar, Nirbhavane, & Jain, 2024; Sharma, Khurana, Patel, Khare, & Kurmi, 2024). Surface modifications with ligands, such as folic acid or hyaluronic acid, allow micelles to target specific cells, enhancing treatment precision and reducing off-target effects (Zheng et al., 2024). Furthermore, smart micelles that respond to external stimuli, like pH or light, can achieve localized drug release, offering improved therapeutic outcomes in cancer therapy (W. Gao et al., 2024). Beyond drug delivery, micelles can be loaded with imaging agents for non-invasive techniques like MRI and CT, facilitating real-time monitoring of biological (Ch, Patel, Biswas, & Ghosh, 2024). Moreover, they serve as biosensors for early disease detection, enhancing diagnostic capabilities. Despite their promising potential, challenges such as stability in biological environments and potential toxicity must be carefully addressed to ensure their safe and effective clinical application. Continued research is essential to harness micelles' potential to revolutionize medicine fully.

2.4 Polymer-based nanoparticles

Polymeric-based NPs are revolutionizing the biomedical field due to their versatility in release mechanisms, including diffusion, erosion, and stimuli-responsive approaches (Herman, Rata, Cadinoiu, Atanase, & Hermenean, 2024; Y. Liu et al., 2024) (**Fig 3.**). The diminutive size of polymeric NPs facilitates efficient targeting of specific tissues, minimizing off-target effects. Applications extend to cancer therapy, where polymeric NPs enhance the delivery of chemotherapeutics like doxorubicin and gene and vaccine delivery, where they protect nucleic acids and antigens. Despite their immense potential, challenges such as stability, toxicity, and scalability persist, necessitating ongoing research to optimize their clinical application.

*Figure 3. In Vivo Antitumor Efficacy of polymer NPs. This figure demonstrates the antitumor efficacy of micellar formulations in 4T1 tumor-bearing mice. A: Relative tumor volume was significantly reduced in mice treated with G1 and G2 micelles compared to the control groups (PBS and DOX). B: Body weight changes were minimal, indicating the formulations had a favorable safety profile. C: Tumor weights at the end of treatment were significantly lower in the G1 and G2 groups. D: Representative tumor images show the reduced tumor size in mice treated with G1 and G2 micelles. E: Histological analysis of major organs revealed no significant toxicity from the micellar formulations. **Data are presented as mean ± standard deviation (n = 5). Statistical significance was determined using a one-way ANOVA followed by Tukey's post-hoc test (*p < 0.05, **p < 0.01, *p < 0.001)*

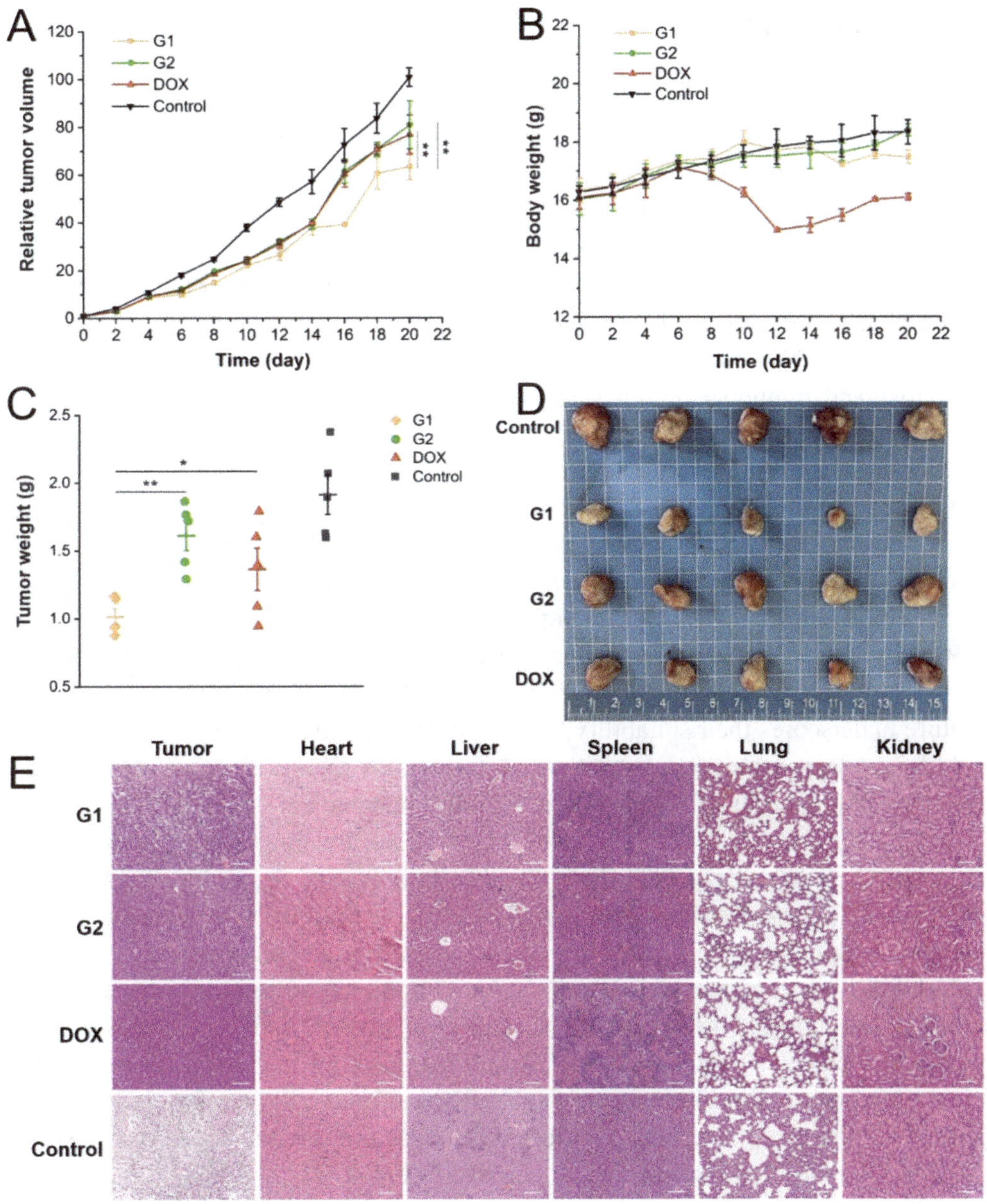

Reproduced from (Y. Liu et al., 2024).

Dendrimers, synthetic polymers with meticulously engineered, highly branched structures, offer unique advantages in various fields, particularly biomedicine. Their precise architecture enables the customization of properties, enhancing their efficacy in drug delivery, gene therapy, and diagnostics. Key attributes include well-defined structures, multivalent interactions, high loading capacities, and generally low toxicity (L. Li et al., 2023). Dendrimers can serve as nanocarriers for drug delivery, protect nucleic acids for gene therapy, act as biosensors for diagnostics, and be used with contrast agents for imaging (Jaisingh & Nebhani, 2024). Despite their immense potential, challenges such as complex synthesis and thorough toxicity evaluation must be addressed to fully harness their capabilities in medicine and materials science.

2.5 Other nanoparticles

Other fantastic NPs have been vital tools in biomedicine research. For instance, Magnetic NPs (MNPs) are transforming biomedical research, particularly in hyperthermia and drug delivery. Their distinctive magnetic properties, amenable to precise manipulation via external fields, enable targeted tumor treatment and enhanced drug delivery. In hyperthermia, MNPs generate heat under alternating magnetic fields, effectively elevating tumor temperatures above 41°C and inducing cell death (R. Singh, Yadav, Ingole, & Ahn, 2024). The acidic tumor microenvironment further amplifies the efficacy of hyperthermia, rendering cancer cells more susceptible to heat-induced damage compared to healthy tissues (Zargar, Bhat, Zargar, & Malik, 2024). Research has revealed that lysosomal localization of MNPs significantly augments apoptosis and immunogenic cell death in cancer cells, stimulating antitumor immune responses (Y. Zhang et al., 2024). Moreover, MNPs can be functionalized to carry therapeutic agents, facilitating their uptake by tumor cells and minimizing systemic toxicity (Mishra & Yadav, 2024). The synergistic combination of MNPs with chemotherapeutic drugs has demonstrated enhanced cytotoxicity against various cancer cell lines, underscoring their potential in combination therapies (García-Soriano et al., 2024). While MNPs hold immense promise, challenges such as toxicity and biocompatibility necessitate ongoing investigation to ensure their safe and effective clinical translation.

Silica NPs (SiNPs), renowned for their biocompatibility and versatility, have emerged as promising materials for various biomedical applications (WAIDI, 2024b). Even at elevated concentrations, their non-toxic nature underscores their suitability for various therapeutic and diagnostic purposes. By strategically modifying their surface chemistry through techniques such as grafting or coupling, researchers can fine-tune the properties of SiNPs to meet specific biomedical demands. This adaptability has paved the way for their integration into drug delivery systems, imaging agents, biosensors, and tissue engineering scaffolds (Khoz et al., 2024; R. P. Singh, Choi, Tiwari, & Pandey, 2014). Functionalized SiNPs can selectively target specific cells or tissues, enhancing drug delivery efficiency and improving imaging contrast. Additionally, their controlled release properties and biocompatible nature make them ideal candidates for tissue engineering applications, providing a supportive environment for cell growth and regeneration. The ability to manipulate the surface chemistry of SiNPs offers many possibilities for addressing critical biomedical challenges, promising significant advancements in healthcare. Overall, NPs offer immense potential in biomedicine, but challenges related to toxicity, biocompatibility, and scalability must be addressed for safe and effective clinical translation. The broad classifications of nanoparticle types used in biomedicine are summarized in **Fig.4**.

Figure 4. Broad classifications of NPs used in biomedicine

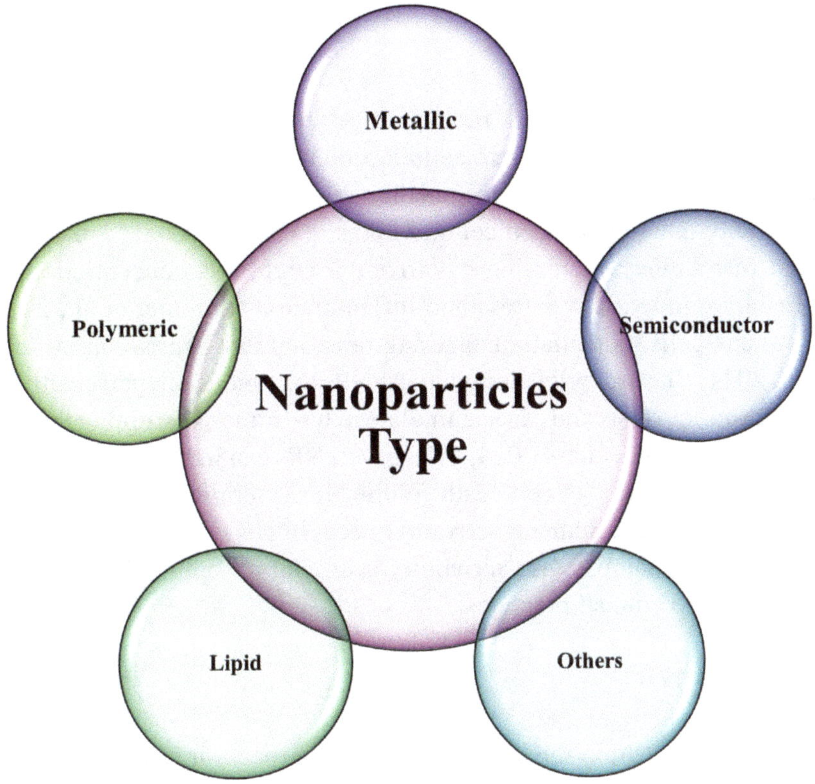

3. INTERACTION WITH CELLS, TISSUES, AND LONG-TERM SAFETY

3.1 Cellular Uptake Mechanisms

Cells have intricate mechanisms to internalize extracellular molecules, primarily relying on endocytosis and passive diffusion. Endocytosis, a multifaceted process, encompasses phagocytosis, pinocytosis, and receptor-mediated endocytosis, each tailored to specific cellular needs (Hivare, Mujmer, Swarup, Gupta, & Bhatia, 2023). Phagocytosis is primarily undertaken by specialized cells like macrophages that engulf large particles, such as bacteria or cellular debris. Conversely, pinocytosis is a more indiscriminate process that facilitates the uptake of fluids and small molecules. Receptor-mediated endocytosis is a highly targeted process driven by binding specific ligands to cell surface receptors, forming clathrin-coated vesicles. In contrast to endocytosis's active and regulated nature, passive diffusion is a simpler process that allows small, uncharged molecules to move across cell membranes down their concentration gradients. This process is essential for exchanging gases like oxygen and carbon dioxide (S. J. Zhang et al., 2023). At the same time, endocytosis and passive diffusion represent distinct mechanisms, both indispensable for cellular function and nutrient acquisition, highlighting cellular uptake strategies' intricate and multifaceted nature.

3.2 Cellular Response

NPs engage in intricate interactions with biological systems, significantly influencing cytotoxicity, biocompatibility, immune responses, and intracellular organelle function. It is paramount to consider these interactions for their safe application in biomedicine. Smaller NPs and those with irregular shapes exhibit higher cytotoxicity due to enhanced cellular uptake (Vanivska et al., 2024). Positively charged NPs can disrupt cell membranes, increasing cytotoxicity. High NPs concentrations may overwhelm cellular defenses, resulting in oxidative stress and inflammation (Summer et al., 2024). Additionally, NPs can activate immune cells, potentially triggering beneficial inflammation or detrimental chronic responses (Sau et al., 2024). In susceptible individuals, NPs may induce hypersensitivity reactions. NPs are internalized through endocytosis and can accumulate in lysosomes, disrupting their function (Fatima, Shehzad, Shahzad, Khan, & Al-Suhaimi, 2024). Moreover, NPs can induce oxidative stress and damage mitochondrial integrity, contributing to cell death. While NPs offer promising therapeutic applications, their potential for cytotoxicity and immune activation necessitates careful design to minimize adverse effects. Understanding individual response variability is crucial for optimizing NP safety and ensuring their effective translation into clinical practice.

3.3 Tissue Distribution

The biodistribution of NPs within the body is a complex process influenced by various factors, including their physicochemical properties, biological interactions, and physiological conditions. Smaller NPs tend to penetrate tissues more effectively than larger ones (Xu et al., 2023), while their shape can also impact interactions with biological components, affecting their distribution (Fu et al., 2023). The surface chemistry of NPs, including charge, hydrophobicity, and functionalization, significantly influences their interactions with proteins and cells, altering their distribution and clearance rates (Chou & Lin, 2024). The route of administration further shapes NP distribution, with intravenous injection and inhalation leading to distinct patterns (Morozova, Isaeva, & Klinov, 2023). Physiological factors, such as age, gender, and organ function, can also modify NPs accumulation, with impaired renal function resulting in increased blood retention. NPs interact with proteins and cells, forming a protein corona that can alter their biological identity and influence tissue targeting. Despite these insights, predicting NPs' in vivo behavior and ensuring effective targeting remains a challenge, particularly when overcoming biological barriers like the blood-brain barrier (J. Li & Wang, 2023).

3.4 Long-Term Safety

The long-term safety of nanoparticle-based therapies is a critical area of research, demanding thorough investigation given their increasing application in drug delivery and various treatment modalities. Despite the significant therapeutic advantages nanoparticles offer, their safety and potential toxicity remain paramount concerns that must be comprehensively addressed. Specifically, nanotoxicity poses a considerable risk, with evidence suggesting that exposure can lead to adverse effects, particularly within reproductive systems where nanoparticles may traverse biological barriers and potentially induce fetal abnormalities

(Ahmad, 2022). The physicochemical properties of nanoparticles, including their size, shape, surface charge, and material composition, are crucial determinants of their safety profile, directly influencing their interactions with biological systems (Kraegeloh, Suarez-Merino, Sluijters, & Micheletti, 2018).

Consequently, assessing the long-term effects of chronic nanoparticle exposure is imperative, as cumulative toxicity may arise, necessitating extensive long-term studies to fully ascertain their impact on human health. To navigate these complexities, a robust regulatory and testing framework is essential. There is a pressing need for comprehensive regulatory guidelines to evaluate nanoparticle safety, encompassing standardized testing protocols for both toxicity and biocompatibility. Furthermore, clinical considerations in safety assessments are vital, requiring evaluations that account for realistic parameters such as dosing, administration routes, and specific patient conditions to ensure accurate assessments of nanoparticle therapies. Moving forward, future directions in this field should emphasize integrated approaches to safety assessment, combining physicochemical analyses with rigorous biological evaluations to enhance the benefit-to-risk ratio of nanoparticle therapies. The development of emerging technologies, such as "smart" nanoparticles designed to respond to environmental triggers, also holds promise for improving safety profiles while maintaining therapeutic efficacy (Dang & Guan, 2020). Ultimately, while nanoparticle-based therapies offer immense potential for enhancing treatment efficacy, their long-term safety remains a complex issue that demands sustained research and vigilant regulatory attention to mitigate associated risks.

4. CHALLENGES AND FUTURE PERSPECTIVES

While NPs offer immense promise in various fields, their potential toxicity and associated safety concerns necessitate careful consideration. NPs can induce cytotoxicity, genotoxicity, inflammation, organ toxicity, and immunological effects. Regulatory challenges, such as the lack of standardized testing protocols, complex risk assessment procedures, and inadequate regulatory frameworks, hinder the safe and responsible use of NPs. To address these issues, future efforts must prioritize the development of advanced risk assessment models, establishing standardized testing protocols, creating robust regulatory frameworks, investing in nanotechnology innovation, and promoting public engagement and education. By proactively addressing these challenges, we can fully realize the benefits of NPs while mitigating their risks, ensuring their safe and sustainable integration into society.

Another major challenge lies in their immunogenicity; nanoparticles can inadvertently trigger the body's immune system, leading to undesirable inflammation or allergic reactions that compromise both their effectiveness and patient safety. To counteract this, researchers are actively developing strategies to manipulate these immune responses, often by engineering nanoparticles with surface modifications like PEGylation. Such modifications help nanoparticles evade immune detection and enhance their uptake by target cells, improving their therapeutic profile.

Simultaneously, bioaccumulation presents another critical concern. The buildup of nanoparticles in tissues can lead to toxicity, as seen with examples like copper nanoparticles, raising questions about their long-term effects within the body. This lack of comprehensive data on their biodegradability and long-term safety creates substantial regulatory hurdles, slowing down their approval and clinical translation. Despite these complexities, ongoing research is dedicated to designing safer and more effective nanoparticle systems that minimize adverse immune reactions and prevent harmful accumulation. This research is crucial for unlocking the full potential of nanoparticles in personalized medicine, although the possibility of adverse effects remains a key consideration in the development pipeline for these innovative therapies.

Regulatory hurdles are another challenge and a complex manufacturing process, biocompatibility concerns, cost-effectiveness, and patient acceptance are among the primary obstacles. To fully realize the benefits of nanoparticle-based therapies, future research must prioritize personalized medicine, combining NPs with other therapeutic modalities, advanced characterization techniques to ensure safety and efficacy, rigorous clinical trials, and international regulatory harmonization. By addressing these challenges and investing in innovative approaches, we can unlock the full potential of NPs to improve human health and revolutionize disease treatment.

5. CONCLUSION

In conclusion, this chapter has comprehensively explored the flourishing field of nanotechnology, highlighting the unique properties and immense potential of NPs in revolutionizing modern medicine. From their historical origins to their diverse types and interactions with biological systems, NPs offer innovative solutions for a wide range of healthcare challenges. While addressing safety, toxicity, and clinical translation concerns remains crucial, ongoing research and advancements in nanoparticle technology are paving the way for their widespread adoption. As the field continues to evolve, NPs are poised to transform medicine, providing novel approaches for diagnosis, treatment, and drug delivery, ultimately improving patient outcomes and enhancing overall healthcare quality.

REFERENCES

Ahmad, A. (2022). Safety and toxicity implications of multifunctional drug delivery nanocarriers on reproductive systems in vitro and in vivo. *Frontiers in Toxicology*, *4*, 895667. PMID: 35785262

Alrahili, M. R. (2024). Insights into Gold Nanoparticle Optical Behavior: Size-Dependent Absorption and Extinction Peaks. *Journal of Nanoelectronics and Optoelectronics*, *19*(8), 783–790.

Amen, R., Mukhtar, A., Saqib, S., Ullah, S., Al-Sehemi, A. G., Mehdi, S. E. H., . . . Bustam, M. A. (2021). History and development of nanomaterials. In *Nanomaterials: Synthesis, Characterization, Hazards and Safety* (pp. 1-14): Elsevier.

Ansari, M. A., Asiri, S. M. M., Alzohairy, M. A., Alomary, M. N., Almatroudi, A., & Khan, F. A. (2021). Biofabricated fatty acids-capped silver nanoparticles as potential antibacterial, antifungal, antibiofilm and anticancer agents. *Pharmaceuticals*, *14*(2), 139. PMID: 33572296

Azizi-Lalabadi, M., Ehsani, A., Divband, B., & Alizadeh-Sani, M. (2019). Antimicrobial activity of Titanium dioxide and Zinc oxide nanoparticles supported in 4A zeolite and evaluation the morphological characteristic. *Scientific Reports*, *9*(1), 17439. PMID: 31767932

Barua, N., & Buragohain, A. K. (2024). Therapeutic Potential of Silver Nanoparticles (AgNPs) as an Antimycobacterial Agent: A Comprehensive Review. *Antibiotics (Basel, Switzerland)*, *13*(11), 1106. PMID: 39596799

Baryeh, K. (2019). *Development of quantitative lateral flow strip biosensors for the detection of cancer biomarkers.* North Dakota State University, Ch, S., Patel, T. K., Biswas, S., & Ghosh, B. (2024). Micelles-based drug delivery for pancreatic cancer. In *Recent Advances in Nanocarriers for Pancreatic Cancer Therapy* (pp. 115-144): Elsevier.

Chou, W.-C., & Lin, Z. (2024). Impact of protein coronas on nanoparticle interactions with tissues and targeted delivery. *Current Opinion in Biotechnology*, *85*, 103046. PMID: 38103519

Dang, Y., & Guan, J. (2020). Nanoparticle-based drug delivery systems for cancer therapy. *Smart Materials in Medicine*, *1*, 10–19. PMID: 34553138

Erisen, D. E., & Uludag, K. (2024). Estimating Biosafety of Biodegradable Biomedical Materials from In Vitro Ion Tolerance Parameters and Toxicity of Nanomaterials in Brain. In *Transformative Approaches to Patient Literacy and Healthcare Innovation* (pp. 201–221). IGI Global.

Farzin, L., Shamsipur, M., Samandari, L., & Sheibani, S. (2020). HIV biosensors for early diagnosis of infection: The intertwine of nanotechnology with sensing strategies. *Talanta*, *206*, 120201. PMID: 31514868

Fatima, A., Shehzad, A., Shahzad, R., Khan, S., & Al-Suhaimi, E. A. (2024). Impact of nanoparticles on structural elements within the cells. In *Molecular impacts of nanoparticles on plants and algae* (pp. 111–141). Elsevier.

Fu, L., Zhang, Y., Farokhzad, R. A., Mendes, B. B., Conde, J., & Shi, J. (2023). 'Passive'nanoparticles for organ-selective systemic delivery: Design, mechanism and perspective. *Chemical Society Reviews*, *52*(21), 7579–7601. PMID: 37817741

Gao, R., Mao, Y., Ma, C., Wang, Y., Jia, H., Chen, X., & Yu, L. (2022). SERS-based immunoassay of myocardial infarction biomarkers on a microfluidic chip with plasmonic nanostripe microcones. *ACS Applied Materials & Interfaces*, *14*(50), 55414–55422. PMID: 36480247

Gao, W., Bigham, A., Ghomi, M., Zarrabi, A., Rabiee, N., Saeb, M. R., & Ashrafizadeh, M. (2024). Micelle-engineered nanoplatforms for precision oncology. *Chemical Engineering Journal*, *495*, 153438.

García-Soriano, D., Milán-Rois, P., Lafuente-Gómez, N., Rodríguez-Díaz, C., Navío, C., Somoza, Á., & Salas, G. (2024). Multicore iron oxide nanoparticles for magnetic hyperthermia and combination therapy against cancer cells. *Journal of Colloid and Interface Science*, *670*, 73–85. PMID: 38759270

Ghose, D., & Kumar, D. (2022). Emerging Nanostructures in Dental Applications. In *Emerging Materials: Design, Characterization and Applications* (pp. 265-313): Springer.

Gupta, P. C., Sharma, N., Mishra, P., Rai, S., & Verma, T. (2024). Role of Gold Nanoparticles for Targeted Drug Delivery. In *Metal and Metal-Oxide Based Nanomaterials: Synthesis, Agricultural, Biomedical and Environmental Interventions* (pp. 243-269): Springer.

Harzand, F. V. (2022). Synthesis of advanced gold nanoparticles for biomedical applications. *Advances in Applied NanoBio-Technologies*, *3*(4), 13–19.

Herman, H., Rata, D. M., Cadinoiu, A. N., Atanase, L. I., & Hermenean, A. (2024). Colloidal and Biological Characterization of Dual Drug-Loaded Smart Micellar Systems. *Polymers*, *16*(9), 1189. PMID: 38732658

Hivare, P., Mujmer, K., Swarup, G., Gupta, S., & Bhatia, D. (2023). Endocytic pathways of pathogenic protein aggregates in neurodegenerative diseases. *Traffic (Copenhagen, Denmark)*, *24*(10), 434–452. PMID: 37392160

Huang, Y., He, S., Cao, W., Cai, K., & Liang, X.-J. (2012). Biomedical nanomaterials for imaging-guided cancer therapy. *Nanoscale*, *4*(20), 6135–6149. PMID: 22929990

Jain, N., Waidi, Y. O., Vannaladsaysy, V., Pathote, D., & Datta, S. (2025). Hydrogels With Nanomaterials Integrated for 3D Bioprinting Technology. In *Design, Fabrication, and Significance of Advanced Nanostructured Materials* (pp. 301-326): IGI Global.

Jaisingh, A., & Nebhani, L. (2024). Dendrimers and Hyperbranched Polymers as Promising and Versatile Additives in Polyolefins. *ACS Applied Engineering Materials*, *2*(2), 236–261.

Kagdada, H. L., Bhojani, A. K., & Singh, D. K. (2023). An overview of nanomaterials: history, fundamentals, and applications. *Nanomaterials: Advances and Applications*, 1-26.

Kaltbeitzel, J., & Wich, P. R. (2023). Protein-based Nanoparticles: From Drug Delivery to Imaging, Nanocatalysis and Protein Therapy. *Angewandte Chemie International Edition*, *62*(44), e202216097. PMID: 36917017

Karnwal, A., Kumar Sachan, R. S., Devgon, I., Devgon, J., Pant, G., Panchpuri, M., & Kumar, G. (2024). Gold Nanoparticles in Nanobiotechnology: From Synthesis to Biosensing Applications. *ACS Omega*, *9*(28), 29966–29982. PMID: 39035946

Khan, M., Shaik, M. R., Khan, S. T., Adil, S. F., Kuniyil, M., Khan, M., & Nawaz Tahir, M. (2020). Enhanced antimicrobial activity of biofunctionalized zirconia nanoparticles. *ACS Omega*, *5*(4), 1987–1996. PMID: 32039336

Khoz, R., Yazdian, F., Pourmadadi, M., Rahdar, A., Fathi-karkan, S., & Pandey, S. (2024). Current Trends in Silica Based Drug Delivery Systems. *European Journal of Medicinal Chemistry Reports*, •••, 100206.

Kim, I., Viswanathan, K., Kasi, G., Thanakkasaranee, S., Sadeghi, K., & Seo, J. (2022). ZnO nanostructures in active antibacterial food packaging: Preparation methods, antimicrobial mechanisms, safety issues, future prospects, and challenges. *Food Reviews International*, *38*(4), 537–565.

Kraegeloh, A., Suarez-Merino, B., Sluijters, T., & Micheletti, C. (2018). Implementation of safe-by-design for nanomaterial development and safe innovation: Why we need a comprehensive approach. *Nanomaterials (Basel, Switzerland)*, *8*(4), 239. PMID: 29661997

Kurapati, R., Kostarelos, K., Prato, M., & Bianco, A. (2016). Biomedical uses for 2D materials beyond graphene: Current advances and challenges ahead. *Advanced Materials*, *28*(29), 6052–6074. PMID: 27105929

Li, J., & Wang, H. (2023). Selective organ targeting nanoparticles: From design to clinical translation. *Nanoscale Horizons*, *8*(9), 1155–1173. PMID: 37427677

Li, L., Deng, Y., Zeng, Y., Yan, B., Deng, Y., Zheng, Z., & Xiao, X. (2023). The application advances of dendrimers in biomedical field. *VIEW*, *4*(6), 20230023.

Liu, X., Duan, G., Li, W., Zhou, Z., & Zhou, R. (2017). Membrane destruction-mediated antibacterial activity of tungsten disulfide (WS 2). *RSC Advances*, *7*(60), 37873–37880.

Liu, Y., Ren, Z., Zhang, X., Zhao, Z., Ma, G., Pei, Y., & Pan, J. (2024). Dual-Triggered Peptide-Based Polymeric Micelles Enhance Doxorubicin Delivery for Targeted Cancer Therapy. *ACS Applied Nano Materials*.

Luceri, A., Francese, R., Lembo, D., Ferraris, M., & Balagna, C. (2023). Silver nanoparticles: Review of antiviral properties, mechanism of action and applications. *Microorganisms*, *11*(3), 629. PMID: 36985203

Mishra, S., & Yadav, M. D. (2024). Magnetic nanoparticles: A comprehensive review from synthesis to biomedical frontiers. *Langmuir*, *40*(33), 17239–17269. PMID: 39132737

Morozova, O., Isaeva, E., & Klinov, D. (2023). Biodistribution of Fluorescent Albumin Nanoparticles among Organs of Laboratory Animals after Intranasal and Peroral Administration. *Current Issues in Molecular Biology*, *45*(10), 8227–8238. PMID: 37886962

Mourdikoudis, S., Pallares, R. M., & Thanh, N. T. (2018). Characterization techniques for nanoparticles: Comparison and complementarity upon studying nanoparticle properties. *Nanoscale*, *10*(27), 12871–12934. PMID: 29926865

O'Dell, D., Schein, P., & Erickson, D. (2016). Simultaneous characterization of nanoparticle size and particle-surface interactions with three-dimensional nanophotonic force microscopy. *Physical Review Applied, 6*(3), 034010. PMID: 30417030

Onzi, G., Guterres, S. S., Pohlmann, A. R., & Frank, L. A. (2021). Passive targeting and the enhanced permeability and retention (EPR) effect. *the ADME encyclopedia: a comprehensive guide on biopharmacy and pharmacokinetics*, 1-13.

Oselusi, S. O., Sibuyi, N. R., Meyer, M., & Madiehe, A. M. (2024). Phytonanotherapeutic Applications of Plant Extract-Synthesized Silver Nanoparticles in Wound Healing—A Prospective Overview. *BioNanoScience*, •••, 1–21.

Ouyang, J., Wen, M., Chen, W., Tan, Y., Liu, Z., Xu, Q., & Liu, Y.-N. (2019). Multifunctional two dimensional Bi 2 Se 3 nanodiscs for combined antibacterial and anti-inflammatory therapy for bacterial infections. *Chemical Communications, 55*(33), 4877–4880. PMID: 30957120

Pandey, N., Singh, K., Kushwaha, S. P., Kumar, A., Khalid, M., Khan, A. A., & Khan, M. (2024). Targeted Drug Delivery to Combat Antibacterial Drug Resistance. In *Contemporary Approaches to Mitigating Antibacterial Drug Resistance* (pp. 250–267). IGI Global.

Patel, S. G., Patel, M. D., Patel, A. J., Chougule, M. B., & Choudhury, H. (2018). Solid lipid nanoparticles for targeted brain drug delivery. In *Nanotechnology-based targeted drug delivery systems for brain tumors* (pp. 191–244). Elsevier.

Pathak, A., Suthar, T., Nirbhavane, G., & Jain, K. (2024). Polymeric micelles in drug delivery and targeting. In *Molecular Pharmaceutics and Nano Drug Delivery* (pp. 161–182). Elsevier.

Pawar, K. R., & Babu, R. J. (2010). Polymeric and lipid-based materials for topical nanoparticle delivery systems. *Critical reviews™ in therapeutic drug carrier systems, 27*(5).

Pitchayyapillai, G., Seenikannan, P., Balasundar, P., & Narayanasamy, P. (2017). Effect of nano-silver on microstructure, mechanical and tribological properties of cast 6061 aluminum alloy. *Transactions of Nonferrous Metals Society of China, 27*(10), 2137–2145.

Rodrigues, A. S., Batista, J. G., Rodrigues, M. Á., Thipe, V. C., Minarini, L. A., Lopes, P. S., & Lugão, A. B. (2024). Advances in silver nanoparticles: A comprehensive review on their potential as antimicrobial agents and their mechanisms of action elucidated by proteomics. *Frontiers in Microbiology, 15*, 1440065. PMID: 39149204

Sachan, R. S. K., Mistry, V., Dholaria, M., Rana, A., Devgon, I., Ali, I., & Bawazeer, S. (2023). Overcoming Mycobacterium tuberculosis drug resistance: Novel medications and repositioning strategies. *ACS Omega, 8*(36), 32244–32257. PMID: 37720746

Sau, S., Dey, A., Pal, P., Das, B., Maity, K. K., Dash, S. K., & Das, B. (2024). Immunomodulatory and immune-toxicological role of nanoparticles: Potential therapeutic applications. *International Immunopharmacology, 135*, 112251. PMID: 38781608

Shah, N. S., Thotathil, V., Zaidi, S. A., Sheikh, H., Mohamed, M., Qureshi, A., & Sadasivuni, K. K. (2022). Picomolar or beyond limit of detection using molecularly imprinted polymer-based electrochemical sensors: A review. *Biosensors (Basel), 12*(12), 1107. PMID: 36551073

Sharma, L., Khurana, D., Patel, P., Khare, S., & Kurmi, B. D. (2024). Advancements in Micellar Formulation: Drug Delivery Vehicle for Water-Insoluble Drugs. *Recent Advances in Drug Delivery and Formulation: Formerly Recent Patents on Drug Delivery & Formulation, 18*(3), 188–207.

Shayesteh, O. H., & Ghavami, R. (2020). A novel label-free colorimetric aptasensor for sensitive determination of PSA biomarker using gold nanoparticles and a cationic polymer in human serum. *Spectrochimica Acta. Part A: Molecular and Biomolecular Spectroscopy, 226*, 117644. PMID: 31614271

Singh, R., Yadav, D., Ingole, P. G., & Ahn, Y.-H. (2024). Magnetic engineering nanoparticles: Versatile tools revolutionizing biomedical applications. *Biomaterials Advances*, •••, 213948. PMID: 38959651

Singh, R. P., Choi, J. W., Tiwari, A., & Pandey, A. C. (2014). Functional nanomaterials for multifarious nanomedicine. *Biosensors nanotechnology*, 141-197.

Soni, S. K., & Solanki, M. K. (2023). CHEMICAL POTENTIAL AND BIOMEDICAL ACTIVITY OF GOLD NANOPARTICLES IN CANCER AND DRUG DELIVERY SYSTEMS: AN UPDATE. *Vietnam Journal of Science and Technology*.

Sudha, P. N., Sangeetha, K., Vijayalakshmi, K., & Barhoum, A. (2018). Nanomaterials history, classification, unique properties, production and market. In *Emerging applications of nanoparticles and architecture nanostructures* (pp. 341–384). Elsevier.

Summer, M., Ashraf, R., Ali, S., Bach, H., Noor, S., Noor, Q., & Khan, R. R. M. (2024). Inflammatory response of nanoparticles: Mechanisms, Consequences, and Strategies for Mitigation. *Chemosphere*, •••, 142826. PMID: 39002651

Szczyglewska, P., Feliczak-Guzik, A., & Nowak, I. (2023). Nanotechnology–general aspects: A chemical reduction approach to the synthesis of nanoparticles. *Molecules (Basel, Switzerland), 28*(13), 4932. PMID: 37446593

Umapathy, V. R., Natarajan, P. M., & Swamikannu, B. (2024). Therapeutic Potential of Nanomaterial-based Drug Delivery. *International Journal of Drug Delivery Technology, 14*(1), 556–563.

Vanivska, K., Dianová, L., Halo, M., Štefunková, N., Lenický, M., Slanina, T., & Massányi, P. (2024). Toxicity of Nanoparticles on Animal and Human Organism: Cell Response. *Journal of Microbiology, Biotechnology and Food Sciences, 14*, e10844.

WAIDI, Y. O. (2024b). pH Stimuli Responsive Dextran Gated Mesoporous Silica as a Viable Delivery Vehicle.

Waidi, Y. O. (2024a). Carbon-Based Technologies for Water Purification. In *Carbon-Based Materials and Environmental Remediation: Graphene, Biochar, and More* (pp. 113-136): IGI Global Scientific Publishing.

Waidi, Y. O., & Jain, N. (2024a). Unlocking new possibilities: Application of MXenes in 3D bioprinting for advanced therapy. *Nanoscale, 16*(43), 20037–20047. PMID: 39405122

Waidi, Y. O., & Jain, N. (2024b). Unlocking New Possibilities: Application of MXenes in 3D Bioprinting for Advanced Therapy. *Nanoscale*.

Waidi, Y. O., Jain, N., Barua, R., Das, S., Vannaladsaysy, V., Prasad, A., . . . Datta, S. (2024). 3 Cellulose: biomedical and engineering applications. *Sustainable Bio-Based Composites: Biomedical and Engineering applications, 20*, 43.

Xie, J., Lee, S., & Chen, X. (2010). Nanoparticle-based theranostic agents. *Advanced Drug Delivery Reviews, 62*(11), 1064–1079. PMID: 20691229

Xie, N. (2024). Synthesis and antibacterial effects of silver nanoparticles (AgNPs) against multi-drug resistant bacteria. [Preprint]. *Bio-Medical Materials and Engineering*, •••, 1–13. PMID: 38995765

Xu, M., Qi, Y., Liu, G., Song, Y., Jiang, X., & Du, B. (2023). Size-dependent in vivo transport of nanoparticles: Implications for delivery, targeting, and clearance. *ACS Nano, 17*(21), 20825–20849. PMID: 37921488

Yetisgin, A. A., Cetinel, S., Zuvin, M., Kosar, A., & Kutlu, O. (2020). Therapeutic nanoparticles and their targeted delivery applications. *Molecules (Basel, Switzerland), 25*(9), 2193. PMID: 32397080

Yuwen, L., Sun, Y., Tan, G., Xiu, W., Zhang, Y., Weng, L., & Wang, L. (2018). MoS 2@ polydopamine-Ag nanosheets with enhanced antibacterial activity for effective treatment of Staphylococcus aureus biofilms and wound infection. *Nanoscale, 10*(35), 16711–16720. PMID: 30156245

Zargar, F., Bhat, H. A., Zargar, M. A., & Malik, S. (2024). Magnetic Nanoparticles in Cancer Thermotherapy: A Mathematical Approach to Optimal Treatment Design.

Zeng, X., Wang, G., Liu, Y., & Zhang, X. (2017). Graphene-based antimicrobial nanomaterials: Rational design and applications for water disinfection and microbial control. *Environmental Science. Nano, 4*(12), 2248–2266.

Zhang, S. J., Lowe, L. A., Anees, P., Krishnan, Y., Fai, T. G., Szostak, J. W., & Wang, A. (2023). Passive endocytosis in model protocells. *Proceedings of the National Academy of Sciences of the United States of America, 120*(24), e2221064120. PMID: 37276401

Zhang, Y., Zhang, Y., Li, J., Liang, C., Shi, K., Wang, S., & Liu, X. (2024). Impact of Nanoheater Subcellular Localization on the Antitumor Immune Efficacy of Magnetic Hyperthermia. *Nano Today, 56*, 102226.

Zheng, Y., Oz, Y., Gu, Y., Ahamad, N., Shariati, K., Chevalier, J., & Annabi, N. (2024). Rational design of polymeric micelles for targeted therapeutic delivery. *Nano Today, 55*, 102147.

Zou, Y., Goei, R., Ong, S.-A., Ong, A. J., Huang, J., & Tok, A. I. Y. (2022). Development of Core-Shell Rh@ Pt and Rh@ Ir Nanoparticle Thin Film Using Atomic Layer Deposition for HER Electrocatalysis Applications. *Processes (Basel, Switzerland), 10*(5), 1008.

Zubair, M., Riaz, M., Kiani, M. N., Aslam, H. M., Fatima, A. E. Z., Sultan, H. M., & Rehman, A. U. (2024). Application of Nanotechnology for Targeted Drug Delivery and Nontoxicity. *International Journal of General Practice Nursing, 2*(2), 57–67.

Chapter 3
Nano-Bioscience:
Introduction and Nanoparticles in Medicine

Ajinkya Nene
https://orcid.org/0000-0003-0302-8650
SIAT CAS, China

Sorour Sadeghzade
https://orcid.org/0000-0002-7034-4808
Westlake University, China

Wenjie Yang
SIAT CAS, China

Prakash Somani
https://orcid.org/0009-0003-7560-2444
Center for Grand Challenges and Green Technologies, Applied Science Innovations Pvt. Ltd., Pune, India

ABSTRACT

Nanotechnology stands at the forefront of materials science, offering opportunities and applications that exploit the unique properties of nanomaterials. These materials exhibit characteristics that differ significantly from their bulk counterparts, enabling innovations across various sectors. This chapter delves into the classification of nanomaterials as outlined by Siegel, categorizing them into zero, one, two, and three-dimensional structures, each with distinct properties and potential applications.

INTRODUCTION

Nanotechnology is at the forefront of materials research, providing opportunities and applications that take advantage of the unique features of nanoparticles. These materials have properties that differ greatly from their bulk equivalents and atomic forms, allowing for advances in a variety of sectors. This chapter looks into Siegel's taxonomy of nanomaterials, which includes zero, one, two, and three-dimensional structures, each having unique features and potential applications. Zero-dimensional (0D) nanomaterials, such fullerenes and quantum dots, are especially significant. Fullerenes, which are made

DOI: 10.4018/979-8-3373-0055-9.ch003

solely of carbon atoms organized in a hollow sphere, have excellent electrical, thermal, and mechanical properties, making them ideal for applications such as drug delivery systems. Quantum dots, semiconductor nanoparticles with size-tunable optical characteristics, are extremely useful in medical imaging and display technology. Magnetic Fe_3O_4 nanoparticles, in addition to 0D nanomaterials, have garnered popularity for their distinct magnetic properties. These nanoparticles improve data storage capacity in computer hard drives and are critical in emerging medical technologies such as magnetic fluid hyperthermia, which targets and heats cancer cells, and targeted drug delivery systems, which enable precise therapeutic delivery. Fe_3O_4 nanoparticles are used to remove pollutants from water and seal rotary shafts in mechanical systems. Nanomaterials have far-reaching ramifications in sectors as diverse as renewable energy, information technology, environmental research, and biology. Their unique features enable advances in solar cell efficiency, smaller data storage, and enhanced diagnostic tools, demonstrating nanotechnology's disruptive promise across numerous industries. As research advances, nanotechnology's uses are projected to grow, tackling some of society's most urgent concerns.

NANOTECHNOLOGY AND NANOBIOTECHNOLOGY

Nanotechnology and Biotechnology are critical scientific fields, and their combination, known as Nanobiotechnology, merges nanotechnology principles with biotechnology applications, making it highly interdisciplinary(Payal et al., 2021). This emerging field offers vast potential for the 21st century with numerous applications. For example, metal oxides like titanium oxide and zinc oxide are already used in skincare products, while nanoparticles have applications in biosensing (e.g., glucose and gas sensing) and drug delivery(Dhara & Mahapatra, 2019; R. Liu et al., 2020; Raveendran et al., 2017). In the food sector, technologies like nutrient delivery systems, biosensors, and edible films are being developed to address issues such as food preservation and microbial contamination(Durán & Marcat, 2013). Nanosensors, for instance, are being designed to monitor proper storage conditions like temperature. In the biomedical field, nanobiotechnology is showing promise in targeted drug delivery, where nanoparticles enhance drug concentration at disease sites. Gold nanoparticles loaded with complementary DNA strands enable highly sensitive pathogen detection, including applications in detecting anthrax. Semiconductor nanoparticles integrated with DNA, enzymes, or antibodies are also being used to create stable optical-electrical biosensors. These advancements are paving the way for significant innovations across various industries(Fakruddin et al., 2012; A. Nene et al., 2023).

NANOMATERIALS

Nanomaterials have at least a dimension in the nanometer range, so they exhibit properties distinct from their bulk counterparts. Nanotechnology is an emerging field dedicated to the generation, characterization and manipulation of nanomaterials classified by Siegel as zero, one, two, or three-dimensional. Current applications of nanomaterials include solar cells, fuel cells, audio/video tapes, computer hard drives, dental bonding agents, catalysis, water purifiers, and skincare products. In biomedical science, they are utilized for targeted drug delivery, diagnostics, and sensing applications(Patel et al., 2021).

Zero-dimensional nanomaterials include fullerenes and quantum dots. Fullerenes (C_{60}) are spherical molecules composed entirely of carbon atoms, with unique electrical and physical properties. Quantum dots, ranging from 1-10 nm, are nanocrystals known as artificial atoms, with examples like cadmium telluride (CdTe) and indium phosphide (InP). One-dimensional nanomaterials include nanotubes, nanowires, and nanorods, such as carbon nanotubes, silicon nanotubes, and boron nitride nanotubes. These nanostructures are important in various technological and scientific applications (Carrow & Gaharwar, 2015)..

Figure 1. Graphical representation on examples of 0-dimensional, 1-dimensional, 2-dimensional and 3-dimensional nanomaterials

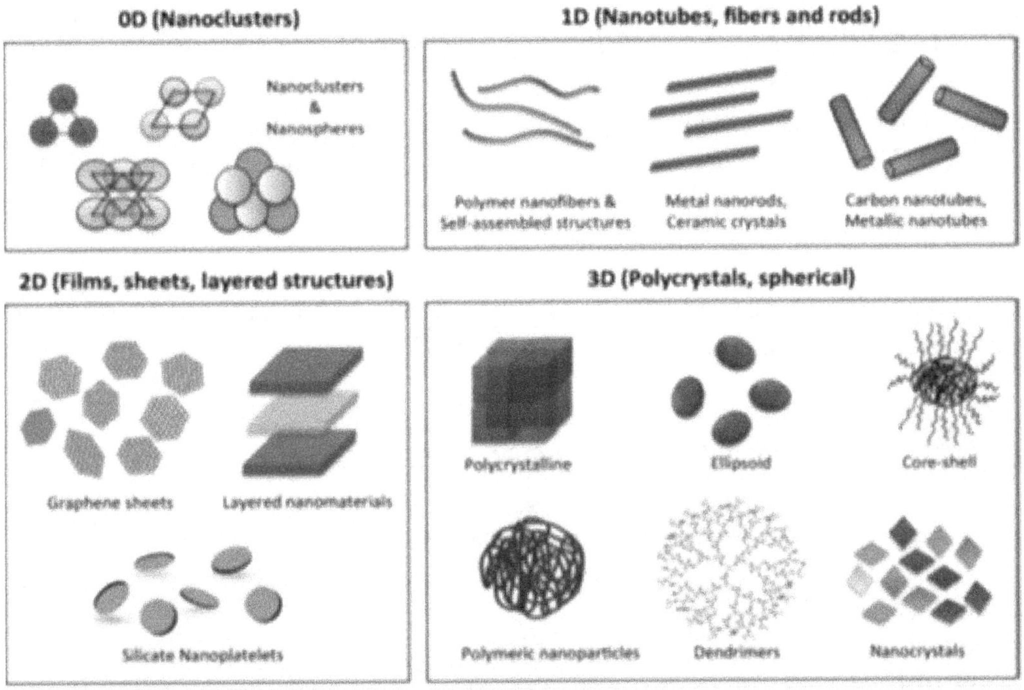

(Carrow & Gaharwar, 2015)

Two-dimensional nanomaterials include thin films, nanosheets, and nanoplates. Three-dimensional nanomaterials, such as nanoballs (dendritic structures), nanocones, polymeric nanoparticles, and nanopillars, are known for their high surface area, offering numerous absorption sites(Mekuye & Abera, 2023; Saleh, 2020).

NANOPARTICLES

Nanoparticles are generally defined as particles with sizes ranging from 1 to 100 nm. They exhibit distinct chemical and physical properties compared to bulk materials(A. G. Nene et al., 2016). Nanoparticles' unique features have made them a primary subject of research across a wide range of scientific areas. Their potential applications cover numerous sectors, with significant attention in biological, electrical, and optical technology.(A. Nene et al., 2021, 2023; Singh Tuli et al., 2019).

APPLICATIONS OF NANOPARTICLES

Nanoparticles have several applications in research and daily life. Their application benefits analytical systems by increasing sensitivity, selectivity, speed, and portability.

- Catalysis: Ag and Pt nanoparticles serve as catalysts in chemical reactions(Astruc, 2020).
- Electronics: Metal nanoparticle pastes (Au, Cu) are used for circuit pattern formation in printed circuit boards. Quantum dots are employed in LEDs, while CdS, CdSe, GaAs, and Si nanoparticles are used in fast optical switches(Payal et al., 2021).
- Magnetism: Magnetic nanoparticles have applications in high-density magnetic recording and MRI, while Fe_3O_4 nanoparticles are used in cancer hyperthermia treatment(Ajinkya et al., 2020).
- Environmental: Nanoparticles (e.g., TiO_2, Al_2O_3, ZrO_2) aid in environmental pollutant separation and wastewater treatment(Demir et al., 2013).
- Biomedical: Nanoparticles are used in targeted drug delivery, biosensing, medical diagnostics, and antimicrobial applications. Noble metal nanoparticles (Au, Ag) are key in molecular diagnostics, drug delivery, and imaging(Mabrouk et al., 2021).

Their unique properties make nanoparticles valuable across various industries, particularly in electronics, healthcare, catalysis, and environmental applications.

MAGNETITE/FE_3O_4 NANOPARTICLES PROPERTIES

Magnetite (Fe_3O_4) is one of several iron oxides and exists alongside other forms such as:

- Magnetite (Fe_3O_4)
- Maghemite (γ-Fe_2O_3)
- Hematite (α-Fe_2O_3)
- Goethite (α-FeOOH)
- Ferrous oxide (FeO)

Magnetite (Fe_3O_4) is distinguished by its composition, which includes both ferrous (Fe^{2+}, reduced) and ferric (Fe^{3+}, oxidized) ions. This is why it is also known as iron oxide (II, III). Figure 1.6.A depicts a simplified magnetite process, including chemical formulae for its several phases.

Figure 2. A Simplified magnetite synthesis reaction

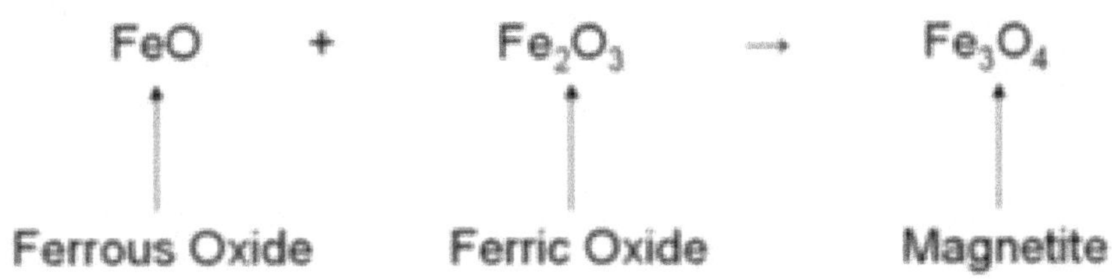

(Preserve et al., 2007)

Fe_3O_4 nanoparticles, often known as magnetite, have unique features that make them useful for a variety of applications. Magnetite's crystal structure is an inverse spin arrangement, with alternating tetrahedral and octahedral layers. In this arrangement, half of the octahedral lattice sites are occupied by Fe^{2+} ions, which have a higher crystal field stabilization energy (CFSE), while the remaining octahedral and tetrahedral sites are filled by Fe^{3+} ions. Magnetite's unit cell is face-centered cubic (FCC), with a lattice parameter of 0.8396 nm. The production process and circumstances affect the effective surface area and particle size of Fe_3O_4 nanoparticles. Furthermore, magnetite has a melting point of 1590°C.

Magnetite's unique magnetic and electrical properties are attributed to the presence of both Fe^{2+} and Fe^{3+} ions. This makes it a popular material in sectors such as magnetism, electronics, and biological applications.

Figure 3. Structure of Unit cell of magnetite (Fe3O4)

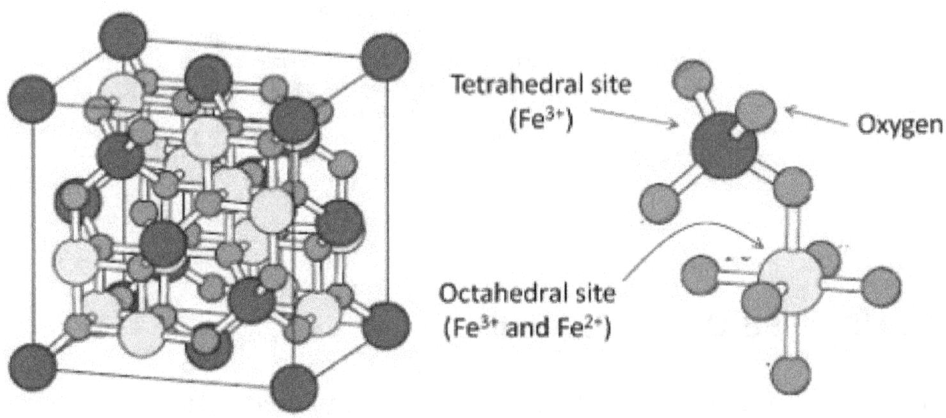

(Paula et al., 2012)

Magnetite's Curie temperature is 850 K, which represents a critical point in its magnetic characteristics. In this structure, the magnetic moments of ferric ions on tetrahedral sites align ferromagnetically, whereas those of ferrous ions on octahedral sites align antiferromagnetically. This cancels out the magnetic moments, resulting in ferrimagnetic behavior. Ferrimagnetism influences traits like coercivity and remanence (or retentivity). As nanoparticles' sizes decrease, their magnetic characteristics move toward paramagnetic or superparamagnetic magnetization. As a result, smaller nanoparticles exhibit increased supermagnetic behavior and decreased ferromagnetic characteristics. Furthermore, the reduction in nanoparticle size results in a decreased relative oxygen content, which can marginally diminish the iron valence state. This increase in ferrous ion content is expected to cause an increase in magnetism. Table 1.6 summarizes the many features of magnetite nanoparticles.(Bárcena et al., 2009; Preserve et al., 2007).

Table 1. Different properties of Fe3O4 nanoparticles, their value and details

Property	Value	Details	Application
Band gap	0.1 eV	almost semi-conductor	Photocatalysis, catalyst
color	black	suspended solutions exhibit jet black color	Optical applications
conductivity	10^2-10^3 /Ω/cm	almost metallic	electromagnetic
crystal structure	inverse spinel	alternating octahedra and tetrahedra-octahedral layers	Semiconductor and superconductor
Curie temperature (nano-scale)	738 K	higher proportion of surface spins in nano-scale particles enhances the dipolar anisotropy	Hyperthermia, spintronics
electrical properties	between metal and semiconductor	almost metallic good conductivity due to closeness of Fe^{2+} and Fe^{3+} ions on octahedral sites	Spintronics & Sensors
Magnetic properties (nano-scale)	superparamagnetic	-	Hypethermia, commercial magnets, drug delivery
morphology	Octahedral crystals	bounded by {111} planes and rhombo-dodecahedra	-
saturation magnetization (nano-scale)	6.4 nm: 1.1 μB/fu 10.8 nm: 2.6 μB/fu 37.8 nm: 2.3 μB/fu	relative oxygen concentrations decline causing valence states of iron cations to be slightly reduced, generating greater ferrous ion content; since the ultimate magnetic moment depends upon ferrous species, subsequent increase in magnetization; below 10 nm surface effects cause saturation magnetization to decrease	hyperthermia

(Preserve et al., 2007)

APPLICATIONS OF FE_3O_4 NANOPARTICLES

Fe3O4 nanoparticles have a variety of applications, including biosensing, targeted drug delivery, photocatalysis, and others. However, their magnetic properties make them suitable for magnetic separation and cancer hyperthermia treatment.Figure 1.7 shows a typical example of magnetic separation process in which magnetically tagged cells (•) can be separated from the mixture of unwanted biomaterial (o) present in a fluid.

Furthermore, if Fe3O4 nanoparticles are exposed to an external alternating magnetic field, they produce heat due to losses during the magnetic reversal process during magnetization. This feature has potential applications in areas such as adhesive hardening and cancer treatment. Hyperthermia is a cancer treatment that generates heat within malignant tissue. High temperatures can degrade, burn, or destroy cancer cells. However, such heat can cause damage to normal tissue, which is a serious issue with hyperthermia therapy. There are three types of hyperthermia: local, regional, and whole-body.

When an external magnetic field is applied to Fe3O4 nanoparticles, heat is created, as are magnetic losses caused by hysteresis, Neel or Brown relaxation, and frictional losses. Under the effect of heat energy, Fe3O4 nanoparticles exhibit spontaneous magnetization change. Consequently, magnetization oscillates or rotates between two equilibrium positions. The typical time between two states can be calculated using Neel relaxation time. If a magnetic field is supplied to Fe3O4 nanoparticles, they align with the magnetic field and rotate. Brownian motion is the rotation of a particle, while Neel relaxation is the rotation of a particle's moment. This is due to the interaction of the magnetic field and magnetization, which causes torque.

Brown relaxation time describes the time it takes for Fe3O4 nanoparticles to align with an external magnetic field. Hysteresis is caused by a delay between magnetic field rotation and magnetization rotation. The hysteresis loop provides thermal energy/heat that can be used for cancer hyperthermia treatment. (Ajinkya et al., 2020; G. Nene et al., 2016; A. G. Nene et al., 2016).

Figure 4. Typical example of magnetic separation process

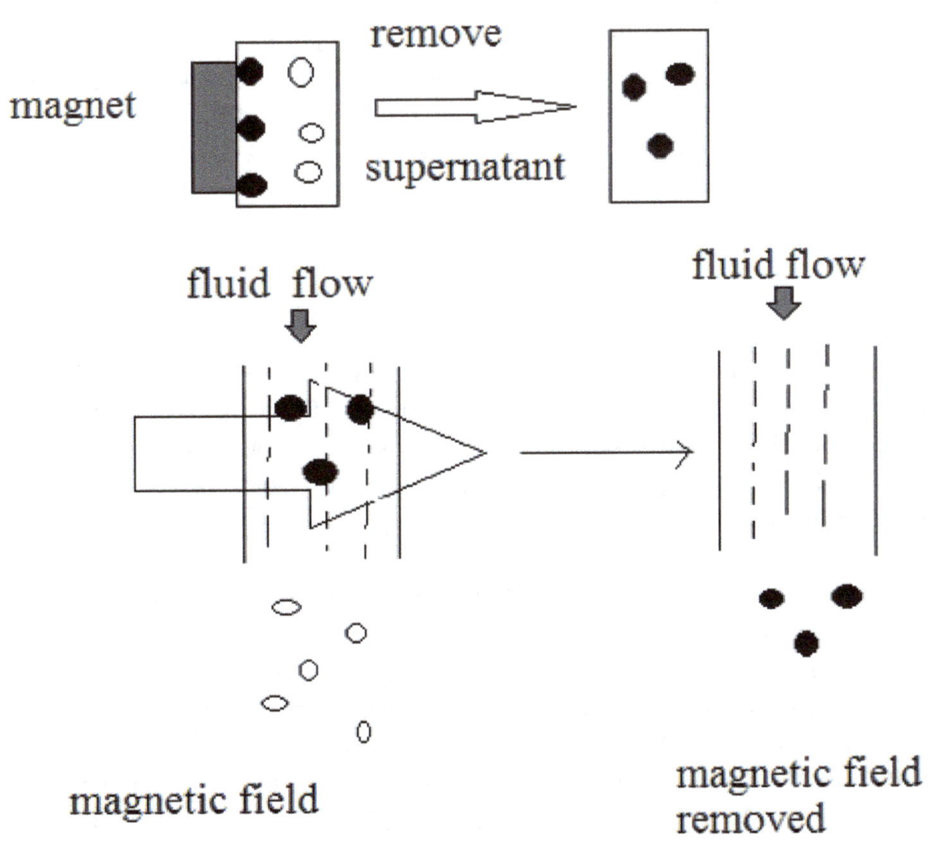

(Bárcena et al., 2009)

Magnetic resonance imaging (MRI) is a potent technology for discriminating between healthy and diseased tissues, noted for its noninvasive nature, tomographic capabilities, and high spatial resolution. Iron oxide nanoparticles (IONPs) have recently been identified as possible contrast agents for MRI applications. Magnetite and maghemite are particularly useful due to their biocompatibility, low toxicity, and ability to disintegrate in vivo.

Superparamagnetic iron oxide nanoparticles are especially important in this context, having been approved for clinical usage as MRI contrast agents. Their excellent sensitivity to MR detection and non-toxicity make them ideal for usage as blood pool agents. Advances in MRI imaging technology have enabled near-microscopic resolution in vivo, offering up new research opportunities, particularly in molecular imaging.

One fascinating breakthrough is the combination of IONPs with antibodies or ligands that target specific receptors on tumor cells. When these nanoparticles enter the human body, they can congregate in tumor tissues, improving the ability to identify and diagnose tiny lesions at an early stage. This novel strategy to incorporate IONPs into MRI demonstrates their potential to dramatically increase imaging accuracy and effectiveness in cancer detection.(Schoepf et al., 1998; Soleymani et al. 2020; Yeh et al., 1993).

Lobel et al. found that iron oxide nanoparticles (IONPs) can aid in tissue repair by coating damaged tissue surfaces with these nanoparticles at specific locations. Lobel et al. demonstrated the effectiveness of IONPs in facilitating tissue repair by placing protein or synthetic polymer-coated nanoparticles between two tissue surfaces and applying a temperature above 500°C—known

Furthermore, nanoparticles with effective light absorption properties, such as gold/silica-coated IONPs, have showed potential in improving tissue repair processes. Beyond just replacing tissue, synthetic tissues must mimic the shape, cellular organization, and behavior of healthy native cells. Tissue engineers continue to face substantial challenges in achieving effective control of biological processes, especially during stem cell differentiation.(Bull et al., 2014; Emadi et al., 2023; Pinaud et al., 2004).

CONCLUSION

Iron oxide nanoparticles (IONPs) have emerged as useful medicinal tools, particularly for detecting nucleic acids and identifying pathogenic bacteria and viruses. Rapid and ultrasensitive detection of disease-causing agents at the point of care (POC) is crucial, as demonstrated by the continuing COVID-19 pandemic, which emphasizes the urgent need for user-friendly, low-cost, and widely available detection technology. Nanoparticle-based detection technologies are particularly well adapted to producing rapid results.

Innovative approaches include combining IONPs into diagnostic tools like RT-PCR and surface-enhanced Raman spectroscopy (SERS). Recent research indicates that magnetic IONP-based SERS probes can greatly improve POC diagnostics. Fe_2O_3 and Fe_3O_4 nanoparticles, especially when coupled with gold, provide synergistic magnetic and optical properties, opening up new opportunities for SERS, cancer hyperthermia, MRI, and photothermal treatments. Furthermore, IONP-based metal-organic frameworks (MOFs) are promising in biomedical applications due to their biocompatibility and ease of implantation and removal after use. Their interconnected porosity allows for efficient targeting of infections, viruses, and macromolecules, while their high surface-to-volume ratio enhances adsorption capacities (Ajinkya et al., 2020; Herawati et al., 2026).

In addition to these novel uses, there is an urgent need to create commercial solutions based on IONPs in a variety of sectors, including biomedicine, electronics, and catalysts.

REFERENCES

Ajinkya, N., Yu, X., Kaithal, P., Luo, H., Somani, P., & Ramakrishna, S. (2020). Magnetic iron oxide nanoparticle (Ionp) synthesis to applications: Present and future. In Materials (Vol. 13, Issue 20, pp. 1–35). MDPI AG. doi: DOI: 10.3390/ma13204644

Astruc, D. (2020). Introduction: Nanoparticles in Catalysis. In Chemical Reviews (Vol. 120, Issue 2, pp. 461–463). American Chemical Society. doi: DOI: 10.1021/acs.chemrev.8b00696

Bárcena, C., Sra, A. K., & Gao, J. (2009). Applications of Magnetic Nanoparticles in Biomedicine. In Liu, J. P., Fullerton, E., Gutfleisch, O., & Sellmyer, D. J. (Eds.), *Nanoscale Magnetic Materials and Applications* (pp. 591–626). Springer US., DOI: 10.1007/978-0-387-85600-1_20

Bull, E., Madani, S. Y., Sheth, R., Seifalian, A., Green, M., & Seifalian, A. M. (2014). Stem cell tracking using iron oxide nanoparticles. In International Journal of Nanomedicine (Vol. 9, Issue 1, pp. 1641–1653). Dove Medical Press Ltd. doi: DOI: 10.2147/IJN.S48979

Carrow, J. K., & Gaharwar, A. K. (2015). Bioinspired polymeric nanocomposites for regenerative medicine. *Macromolecular Chemistry and Physics*, *216*(3), 248–264. DOI: 10.1002/macp.201400427

Demir, E., Burgucu, D., Turna, F., Aksakal, S., & Kaya, B. (2013). Determination of TiO2, ZrO2, and Al2O 3 nanoparticles on genotoxic responses in human peripheral blood lymphocytes and cultured embyronic kidney cells. *Journal of Toxicology and Environmental Health - Part A: Current Issues*, *76*(16), 990–1002. DOI: 10.1080/15287394.2013.830584 PMID: 24156722

Dhara, K., & Mahapatra, D. R. (2019). Recent advances in electrochemical nonenzymatic hydrogen peroxide sensors based on nanomaterials: A review. *Journal of Materials Science*, *54*(19), 12319–12357. DOI: 10.1007/s10853-019-03750-y

Durán, N., & Marcato, P. D. (2013). Nanobiotechnology perspectives. Role of nanotechnology in the food industry: A review. In International Journal of Food Science and Technology (Vol. 48, Issue 6, pp. 1127–1134). doi: DOI: 10.1111/ijfs.12027

Emadi, H., Karevan, M., Masoudi Rad, M., Sadeghzade, S., Pahlevanzadeh, F., Khodaei, M., Khayatzadeh, S., & Lotfian, S. (2023). Bioactive and Biodegradable Polycaprolactone-Based Nanocomposite for Bone Repair Applications. *Polymers*, *15*, 3617. DOI: 10.3390/polym15173617

Fakruddin, M., Hossain, Z., & Afroz, H. (2012). Prospects and applications of nanobiotechnology: A medical perspective. In Journal of Nanobiotechnology (Vol. 10). doi: DOI: 10.1186/1477-3155-10-31

Herawati, A., & Antarnusa, K. Nene, Fe3O4/chitosan nanocomposites for Congo red removal: Adsorption surpasses photodegradation, Materials Chemistry and Physics, Volume 347, 2026, 131431, ISSN 0254-0584, https://doi.org/DOI: 10.1016/j.matchemphys.2025.131431

Liu, R., Ye, X., & Cui, T. (2020). Recent progress of biomarker detection sensors. *Yosetsu Gakkai Ronbunshu/Quarterly Journal of the Japan Welding Society, 38*(3). doi: DOI: 10.2207/QJJWS.38.193

Mabrouk, M., Das, D. B., Salem, Z. A., & Beherei, H. H. (2021). Nanomaterials for biomedical applications: Production, characterisations, recent trends and difficulties. In Molecules (Vol. 26, Issue 4). MDPI AG. doi: DOI: 10.3390/molecules26041077

Mekuye, B., & Abera, B. (2023). Nanomaterials: An overview of synthesis, classification, characterization, and applications. *Nano Select*, *4*(8), 486–501. DOI: 10.1002/nano.202300038

Nene, A., Galluzzi, M., Hongrong, L., Somani, P., Ramakrishna, S., & Yu, X. F. (2021). Synthetic preparations and atomic scale engineering of silver nanoparticles for biomedical applications. *Nanoscale*, *13*(33), 13923–13942. DOI: 10.1039/D1NR01851E PMID: 34477675

Nene, A., Phanthong, C., Surareungchai, W., & Somasundrum, M. (2023). Electrochemical detection of creatinine using Au–Ag bimetallic nanoparticles. *Journal of Solid State Electrochemistry*, *27*(10), 2869–2875. DOI: 10.1007/s10008-023-05571-1

Nene, A. G., Takahashi, M., Wakita, K., & Umeno, M. (2016). Size controlled synthesis of Fe3O4 nanoparticles by ascorbic acid mediated reduction of Fe(acac)3 without using capping agent. *Journal of Nano Research*, *40*, 8–19. DOI: 10.4028/www.scientific.net/JNanoR.40.8

Nene, , A., Takahashi, M., & R. Somani, P. (2016). Fe3O4 and Fe Nanoparticles by Chemical Reduction of Fe(acac)3 by Ascorbic Acid: Role of Water. *World Journal of Nano Science and Engineering*, *06*(01), 20–28. DOI: 10.4236/wjnse.2016.61002

Patel, J. K., Patel, A., & Bhatia, D. (2021). Introduction to Nanomaterials and Nanotechnology. In Patel, J. K., & Pathak, Y. V. (Eds.), *Emerging Technologies for Nanoparticle Manufacturing* (pp. 3–23). Springer International Publishing., DOI: 10.1007/978-3-030-50703-9_1

Paula, A., Teixeira, C., Tristão, J. C., Araujo, M. H., Oliveira, L. C. A., Moura, F. C. C., Ardisson, J. D., Amorim, C. C., & Lago, R. M. (2012). Iron: A Versatile Element to Produce Materials for Environmental Applications. *Journal of the Brazilian Chemical Society*, *23*(9), •••.

Payal, & Pandey, P. (2021). Role of Nanotechnology in Electronics: A Review of Recent Developments and Patents. *Recent Patents on Nanotechnology, 16*(1), 45–66. doi: DOI: 10.2174/1872210515666210120114504

Pinaud, F., King, D., Moore, H. P., & Weiss, S. (2004). Bioactivation and Cell Targeting of Semiconductor CdSe/ZnS Nanocrystals with Phytochelatin-Related Peptides. *Journal of the American Chemical Society*, *126*(19), 6115–6123. DOI: 10.1021/ja031691c PMID: 15137777

Preserve, L., & Blaney, L. (2007). *Magnetite (Fe3O4): Properties, Synthesis, and Applications*. Retrieved from https://preserve.lehigh.edu/cas-lehighreview-vol-15http://preserve.lehigh.edu/cas-lehighreview-vol-15/5

Raveendran, J., Resmi, P. E., Ramachandran, T., Nair, B. G., & Satheesh Babu, T. G. (2017). Fabrication of a disposable non-enzymatic electrochemical creatinine sensor. *Sensors and Actuators. B, Chemical*, *243*, 589–595. DOI: 10.1016/j.snb.2016.11.158

Saleh, T. A. (2020). Nanomaterials: Classification, properties, and environmental toxicities. *Environmental Technology & Innovation*, *20*, 101067. DOI: 10.1016/j.eti.2020.101067

Schoepf, U., Marecos, E. M., Melder, R. J., Jain, R. K., & Weissleder, R. (1998). Intracellular magnetic labeling of lymphocytes for in vivo trafficking studies. *BioTechniques*, *24*(4), 642–651. DOI: 10.2144/98244rr01 PMID: 9564539

Singh Tuli, H., Kaur, P., Nene, A. G., Sharma, D., & Somani, P. R. (2019). Synergistic effect of copper nanoparticles and antibiotics to enhance antibacterial potential Bio-Materials and Technology Synergistic effect of copper nanoparticles and antibiotics to enhance antibacterial potential. In Bio- []. Retrieved from www.bio-mat-tech.com]. *Materials Technology*, *1*(1)

Soleymani, F., Emadi, R., Sadeghzade, S., & Tavangarian, F. (2020). Bioactivity Behavior Evaluation of PCL-Chitosan-Nanobaghdadite Coating on AZ91 Magnesium Alloy in Simulated Body Fluid. *Coatings*, *10*, 231. DOI: 10.3390/coatings10030231

Yeh, T.-C., Zhang, W., Ildstad, S. T., & Ho, C. (1993). Intracellular labeling of T-cells with superparamagnetic contrast agents. *Magnetic Resonance in Medicine*, *30*(5), 617–625. DOI: 10.1002/mrm.1910300513 PMID: 8259062

Chapter 4
From Imaging to Drug Delivery and Treatment:
Nano-Biomaterials for the Nervous System Disorders

Seyyed Mohammad Amin Mousavi-Sagharchi
Shahr-e-Qods Branch, Islamic Azad University, Tehran, Iran

Mahdieh Ranjbar-Jamalabadi
Isfahan University of Technology, Isfahan, Iran

Sama Yavari
Tabriz University, Iran

Elina Afrazeh
Khorramshahr University of Marine Science and Technology, Khorramshahr, Iran

Naresh Poondla
Icahn School of Medicine at Mount Sinai, New York, USA

Mohsen Sheykhhasan
https://orcid.org/0000-0002-4610-2278
Cellular and Molecular Research Center, Qom University of Medical Sciences, Qom, Iran

ABSTRACT

Nano-sized particles, ranging from 1 to 100 nm, uniquely traverse blood vessels and cell membranes, making nano-biomaterials highly promising for treating brain and nervous system disorders. Their small size facilitates efficient drug delivery, enabling them to penetrate the blood-brain barrier as effective nanocarriers. Additionally, nanoparticles play a crucial role in diagnosing various conditions, especially neurological disorders, allowing for non-invasive imaging of the brain and nervous system. Nano-biomaterials can potentially mitigate the irreversible effects of conditions like Alzheimer's and Parkinson's diseases. Various types of nanomaterials, including metal-based nanoparticles (e.g., gold nanoparticles), carbon nanotubes, silicon nanoparticles, nanocrystals, nanodiscs, liposomes, micelles, and polymeric nanoparticles, are applicable to neurological disorders. This chapter aims to explore the applications of nano-biomaterials in the diagnosis and treatment of nervous system disorders, highlighting their innovative contributions to modern medicine.

DOI: 10.4018/979-8-3373-0055-9.ch004

1. INTRODUCTION

The nervous system is a complex network responsible for controlling and coordinating body functions and is particularly vulnerable to a variety of debilitating conditions, such as neurodegenerative disorders, traumatic injuries, and congenital abnormalities (García-González et al., 2024; Haikal & Weissert; Ng & Lee, 2019; Palanisamy et al., 2023; Shi et al., 2024). These neurological disorders often cause consequential morbidity and mortality, leading to a major challenge to healthcare systems around the globe (Hubbers, 2022; Lee, 2009; Steinmetz et al., 2024). Although there is continuous advancement in medical science, drug delivery into brain for treat these disorders remains challenging due to the presence of the blood-brain barrier (BBB) (Sheykhhasan et al., 2022). BBB causes limitations in delivering therapeutic agents to the brain whereas the nervous system has limited capacity for self-healing (Furtado et al., 2018; Hersh et al., 2022). To overcome these limitations, nanotechnology has shown promising advances. Nanotechnology, an interdisciplinary field, involves manipulating materials at the nanoscale, typically ranging from 1 to 100 nanometers. At this scale, materials display unique properties that are significantly different from their bulk forms, including enhancement in their electrical, optical, and mechanical characteristics (Rezghi Rami et al., 2024). The convergence of nanotechnology with biology has led to the development of nano-biomaterials, that are designed for specific biomedical purposes, particularly within the nervous system. These materials are engineered to interact with biological systems at the molecular level. This makes them highly potential to promote neural repair and regeneration (Erisen & Uludag, 2024).

Nano-biomaterials have brought significant improvement in nervous system applications by their potential in delivery of different materials in addition to creating a supportive environment for cell growth, differentiation, and imitation of the natural extracellular matrix (ECM). Due to their minute size, these materials can cross biological barriers, making them ideal for drug delivery that aims to target the central nervous system (CNS). Additionally, it is possible to enhance them with bioactive molecules to reduce probable side effects and increase their therapeutic effectiveness. In the case of repairing the nervous system, nano-biomaterials can be utilized to make scaffolds that boost neural tissue regeneration or to deliver neuroprotective agents directly to the injury site. These applications are vital in the treatment of neurodegenerative diseases and injuries that conventional therapies often fall short (Haleem et al., 2023; Logothetidis, 2011; Dur E Nayab et al., 2023). Biomaterials are specially designed substances for the purpose of interacting with biological systems for therapeutic or diagnostic purposes. Unlike biological materials produced by living organisms, biomaterials are synthesized to replace, enhance, or interact with living tissue, making them essential in medical advancements (e.g., drug delivery systems, tissue engineering, and implants). The field of biomaterials has progressed significantly, drawing on disciplines such as physics, biology, chemistry, and engineering to develop biocompatible materials. Biomaterials can be categorized into structural and functional types. Structural biomaterials such as titanium is used in joint replacements to provides physical support (Cao & Ding, 2022; Chen et al., 2020). As technology advances, there is a growing emphasize in developing materials that not only replace damaged tissues but also promote healing and regeneration, offering significant potential specially in the field of nervous system repair. The potential of nano-biomaterials in revolutionizing nervous system disorder treatments continues to expand (De et al., 2022) (Figure 1).

Figure 1. Nanoparticles cross the BBB. Nanoparticles are able to cross from barriers and enter brain tissue that are suitable material for drug delivery and imaging purposes

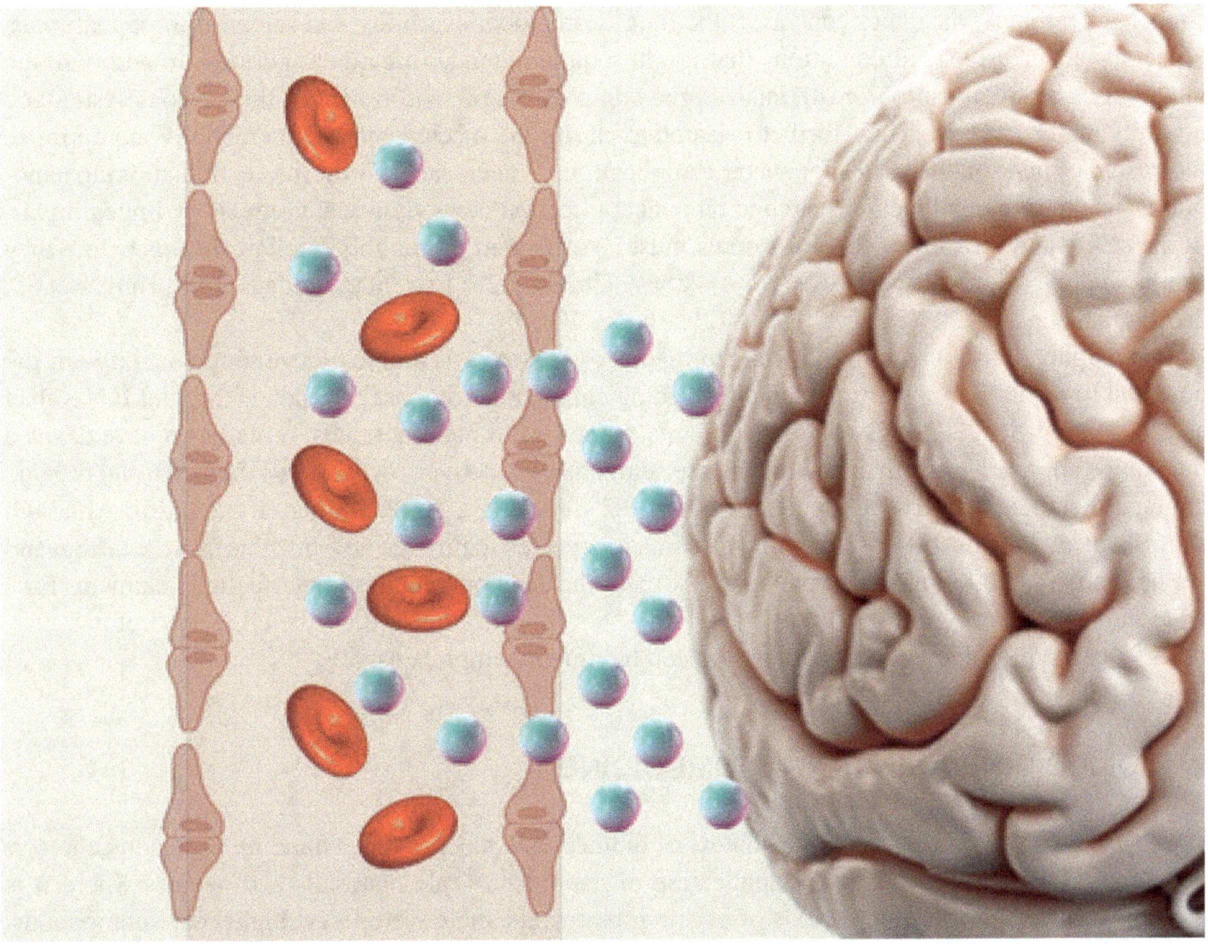

One of the considerable merits of nano-biomaterials in the context of the nervous system is their potential to interact with neural cells and tissues at the nanoscale, which is similar to the scale of biological processes like cell signaling and protein interactions (C.-Y. Yang et al., 2021). This nanoscale interaction allows for precise control over cellular responses, leading to the development of more effective therapies for neurological disorders (C.-Y. Yang et al., 2021). For instance, gold nanoparticles are used to enhance drug delivery to a highly aggressive form of brain cancer named glioblastoma cells, resulting in better therapeutic outcomes (Wang et al., 2021). Similarly, carbon nanotubes have been investigated for their potential in neural tissue engineering by providing a conductive scaffold that supports neural cell differentiation and growth (Sang et al., 2022). Moreover, the ability of nano-biomaterials to interact with various biological molecules, such as peptides, nucleic acids, and proteins, enables the creation of multifunctional platforms that can target multiple pathways involved in disease progression simultaneously (Shahryar Shakeri et al., 2020; Song et al., 2023). This multifunctionality is particularly crucial in treating complex neurological disorders, where a single therapeutic approach may not be sufficient to address the underlying pathology. For instance, liposomal nanoparticles have been designed to co-deliver gene therapy vectors and chemotherapeutic agents to brain tumors, boosting the efficacy of treatment while minimizing side effects (Lo et al., 2020).

Despite the promising potential of nano-biomaterials in the treatment of neurological disorders, several challenges remain. One major concern is the long-term biocompatibility and safety of these materials, especially for neurological conditions that might require extended treatment periods. In addition, the interactions between nano-biomaterials and the complex microenvironment of the nervous system are not fully understood yet. Thus, further research to clarify the mechanisms of interactions and optimize the design of nano-biomaterials for specific applications is necessary. Furthermore, transitioning nano-biomaterials from preclinical research to clinical application faces significant regulatory impediments, as the safety and efficacy of these materials must be evaluated thoroughly in clinical trials before they can be approved for real-world use (Boyes & van Thriel, 2020; Erisen & Uludag, 2024; Kumar et al., 2020; Qian et al., 2021; Vinod & Jena, 2021; Zhang, 2023).

The application of nano-biomaterials in the nervous system is a quickly evolving context with the potential to revolutionize the treatment of neurological disorders (Arora & Baldi, 2024; Dur E Nayab et al., 2023; Patil-Sen, 2021). By using the unique properties of nanomaterials, researchers have devised innovative approaches to repair and regenerate neural tissues, delivery drug across the BBB, and provide targeted therapeutic interventions (Azmi et al., 2023; De et al., 2022; Kumar et al., 2020). Although significant challenges still exist, ongoing research is expected to provide new insights into the design and application of nano-biomaterials, paving the way for the development of more effective treatments for a wide range of neurological conditions. Therefore, this chapter aims to describe all aspects of nanoparticles in nervous system disorders diagnosis and treatment comprehensively.

2. BIOMEDICINE AND NANO-BIOMEDICINE

As we are going to delve into the context of biomedicine, deeper understanding of nanotechnology is essential. Nanotechnology is the application of nanometer-scale materials and devices where it is possible to enhance the effectiveness of existing treatments and develop new highly-safe compounds. Nanoscale materials can display peculiar chemical, physical, and biological features that distinguish them from bulk materials, individual atoms, or molecules (Arcos, 2023). By combining, synthesizing, and manipulating various types of nanomaterials, numerous varieties of materials and structures with unique chemical and physical characteristics can be created (H. Arzaghi et al., 2020). Nanoparticles' physicochemical properties, including their high surface area and nanoscale size, have made them suitable for their application in many fields, including pharmaceuticals (Suradip Das et al., 2016). Nanoparticles, nanosheets, nanotubes, and nanowires, due to their diversity in structure and special properties have been able to gain a special place in various biomedical fields, including drug delivery, gene delivery, tissue engineering, medical implants, biological imaging, etc (Rami et al., 2022). Nanomaterials must show specific characteristics for biomedical applications, such as potential cytotoxicity, biocompatibility, and colloidal stability. These criteria indicate the safety and biocompatibility level and must be maintained under chemical and physical conditions, ideally across a wide pH range (Diez-Pascual & Rahdar, 2022; Eleftheriadou et al., 2020).

According to the different properties and applications of nanomaterials, various types of them are utilized in nanomedicine, including minerals, oxides, metals, salts, and polymers. The size, shape, and chemical structure of nanomaterials can vary, as can the presence or absence of a surface coating; these traits influence these materials' interaction with cells or tissues (Gerardo Nava et al., 2021). Nano-scale materials have the ability to penetrate through cell membranes and proliferate via lymphatic, blood, and nerve cell connections. Therefore, nanomaterials that are being used in close touch with tissues and cells should be very biocompatible for biomedicine applications (D. E. Nayab et al., 2023). To meet these requirements, the surface of the nanomaterial must be modified and functionalized through specific interactions with the target biomolecules. Moreover, functionalization renders nanoparticles compatible with one another in later stages and prevents agglomeration (Ryabchikova, 2021). As a result, nanoparticles have improved pharmacokinetic properties in vivo and can transfer following systemic administration more effectively. By passing through epithelial covering and tiny capillaries, nanoparticles can be extensively injected into tissues, improving the transport of therapeutic agents to the desired site (Shabani et al., 2023). Furthermore, nanoparticles' size amplifies their remarkable physicochemical characteristics, including solubility, immunogenicity, diffusivity, and the capability to target the desired area with minimal diffusion to its surroundings (Salmani-Zarchi et al., 2024; Silva Adaya et al., 2017).

According to the design of the human body, the innate defense system which contains macrophages, and dendritic cells, provides antigens and phagocytic cells that can recognize and respond to external stimuli. So since, nanomaterials are prepared for in vivo applications, it is vital to consider nano-safety interactions as well as their biocompatibility properties. Physical and chemical properties of nanoparticles, such as surface properties, surface charge, structure, shape, composition, size, and crystalline nature, influence their interaction with certain cells and their toxicity. The physiological activity of cells is affected by nanomaterial size directly, which adjusts cellular absorption and the effectiveness of endocytosis (Solangi et al., 2023). Also, the morphology and structure of nanoparticles have an impact on how toxic they are; different structures, such as tubes, fibers, spheres, planes, and polyhydra, may have different effects on toxicity. The surface of nanomaterials can influence toxicity by absorbing ions and biomolecules and regulating cellular or organismal reactions. An organism's reaction through agglomeration or aggregation is influenced by surface charge, which is a crucial factor in determining colloidal behavior. Additionally, the concentration of nanoparticles affects their toxicity (Abbasi et al., 2023; Yang et al., 2020). Modifying the surface of nanomaterials is another crucial aspect of biomedical applications. Surface modulation can influence biodistribution, biological reactions, and particle absorption. By using a specific ligand, surface functionalization can be utilized to enhance tissue and cell targeting, lengthen blood circulation, and decrease nonspecific dispersion. Coating with a biocompatible polymer, conjugating nanomaterials to target particular drugs, fluorescence probes, and cell membrane-penetrating ligands are examples of functionalization techniques (Yoon, 2023).

2.1. Neural Biomedicine

Due to the CNS's extensive protection, disorders of this system are infamously difficult to diagnose and treat (Henna & Pramod, 2020). Larger macromolecules cannot cross the BBB to reach the brain. To create innovative nanostructures with molecular, cellular, or atomic functionalities for both directing and repairing damaged brain circuits, working with materials that are almost at the atomic level required (Liu et al., 2021). Achieving adequate drug concentrations at the disease site besides minimizing side effects and toxicity is a challenging task during targeting and delivering therapeutic macromolecules to the brain. Drug delivery to the brain using nanotechnology has become a novel and exciting strategy for the treatment of neurological disorders and has changed our approach towards the use of CNS-targeted therapies. The wide range of available nanotechnologies allows for the selection of nanoscale materials with appropriate properties for specific CNS diseases (Soni et al., 2016).

Currently, most injuries to the CNS are incurable and lead to disability. This is because, at the injury site, non-permissive ECM components and neuroinflammatory signals are generated to create a neuroinhibitory environment. In order to overcome this challenge, a viscous interfacial self-assembly method has been developed. This method creates a 3D porous bioinspired hybrid nano scaffold structure that can be used to deliver anti-inflammatory molecules and creates a supportive 3D-ECM environment that is able to suppress the neuroinhibitory microenvironment effectively. Improved axonal development from the dual-targeting therapeutic approach in a human induced pluripotent stem cell (hiPSC)-based in vitro model of neuroinflammation is proven by customizing the structural and biochemical characteristics of the 3D porous nano scaffold (Liu et al., 2021). Furthermore, in a spinal cord injury model, nano scaffold-based methods significantly enhanced axonal development and functional recovery in vivo through a unique technique of anti-inflammation-based fibrotic scar reduction. Therefore, therapeutic intervention based on the discovered substances and due to the key role that ECM microenvironments play in neuroinhibitory signaling, a new way to treat CNS injuries is opened (Teleanu et al., 2019). Because of the advancements in neural biomedicine, it is important to understand the underlying causes and histological changes associated with brain and nervous system disorders, paving the way for targeted interventions.

3. ETIOLOGY AND HISTOLOGY OF BRAIN AND NERVOUS SYSTEM DISORDERS

3.1. Pathophysiology of Neurodegenerative Disorders

Neurodegenerative disorders such as Huntington's disease, Parkinson's disease, and Alzheimer's disease, are characterized by the gradual loss of the structure and function of neurons, leading to cognitive and motor impairments. A central feature in the pathophysiology of these disorders is the abnormal accumulation and accumulation of misfolded proteins in nerve tissues. For instance, Alzheimer's disease is characterized by the presence of extracellular amyloid-beta plaques and intracellular neurofibrillary tangles made up of hyperphosphorylated tau protein, which disrupt neuronal communication and induce neurotoxicity (Hardy & Selkoe, 2002; Koszła & Sołek, 2024). Similarly, in Parkinson's disease, there is an accumulation of alpha-synuclein proteins that form Lewy bodies, which result in the degeneration of dopaminergic neurons in the substantia nigra, leading to motor dysfunctions (Spillantini & Goedert, 2000).

Another critical aspect contributing to mitochondrial dysfunction and oxidative stress are key contributors to the development of neurodegenerative diseases. Impaired mitochondrial activity leads to energy deficits and overproduction of reactive oxygen species (ROS), which cause oxidative damage to lipids, proteins, and DNA within neurons (Kathiresan et al., 2024; Lin & Beal, 2006). This oxidative stress exacerbates neuronal injury and promotes apoptotic pathways, further contributing to neuronal loss. Additionally, neuroinflammation mediated by activated microglia and astrocytes plays a crucial role in disease progression by releasing pro-inflammatory cytokines and chemokines, which exacerbate neuronal damage (Kwon & Koh, 2020). Understanding these complex and interrelated pathogenic mechanisms is crucial for expanding effective therapeutic strategies, including the application of nano-biomaterials designed to target and modulate these pathological processes at the molecular level.

3.2. Pathophysiology of Neurodegenerative Disorders

Neurodegenerative disorders are defined by the gradual loss of neuron structure and function, ultimately leading to neuronal death. A central feature in the pathophysiology of these disorders is the abnormal aggregation of misfolded proteins. As we mentioned, in Alzheimer's disease, the accumulation of amyloid-beta plaques and tau tangles disrupt neuronal communication and induces cellular toxicity (Heneka et al., 2015; Sidoryk-Węgrzynowicz et al., 2024). These proteinopathies not only disrupt normal cellular processes but also trigger a cascade of neuroinflammatory responses, exacerbating neuronal damage and accelerating disease progression (Heneka et al., 2015). Beyond protein aggregation, oxidative stress, and mitochondrial dysfunction play pivotal roles in the degeneration of neuronal cells. Mitochondria are essential for energy production and apoptosis regulation, and their impairment leads to decreased adenosine three phosphate (ATP) levels and increased production of ROS (Grünewald et al., 2019). Elevated ROS levels result in oxidative damage to cellular components, including DNA, lipids, and proteins, further compromising neuronal integrity (Butterfield & Halliwell, 2019). Additionally, impaired mitochondrial dynamics and bioenergetics contribute to synaptic failure and neuronal death, hallmark features of neurodegenerative diseases (D. Yang et al., 2021).

3.3. Histological Changes in Brain Disorders

Histological examination of brain tissues in neurodegenerative disorders reveals distinct pathological features that contribute to understanding and the diagnosis of these conditions. In Alzheimer, one of the hallmark histological features is the attendance of amyloid plaques, which are extracellular deposits composed of amyloid-beta peptides. These plaques disrupt neuronal communication and trigger inflammatory responses that lead to neuronal death (Selkoe, 2001). Additionally, neurofibrillary tangles consist of hyperphosphorylated tau proteins, that accumulate within neurons leading to microtubule destabilization and impaired axonal transport (Braak & Braak, 1991). In Parkinson's disease, histological analysis typically reveals the attendance of Lewy bodies, which are intracellular inclusions composed predominantly of aggregated alpha-synuclein proteins. These inclusions are located within the cytoplasm of dopaminergic neurons in the substantia nigra, a critical brain region for motor control (Alfaidi et al., 2024). The attendance of Lewy bodies correlates with the degeneration of these neurons, contributing to the specification of motor symptoms of Parkinson's disease, such as bradykinesia, rigidity, and tremors. Moreover, Lewy bodies are also observed in other related disorders, such as dementia with Lewy bodies, indicating a shared pathogenic mechanism (Koeglsperger et al., 2023) (Table 1).

Table 1. Pathophysiology of neurodegenerative disorders

Disorder	Pathophysiology
Alzheimer	Amyloid-beta plaques and tau tangles
Parkinson	Lewy bodies and alpha-synuclein proteins

3.4. Genetic and Environmental Factors

Besides histological changes that provide us with the structural devastation caused by neurodegenerative disorders, it's crucial to delve into the genetic and environmental factors that underpin these conditions. Neurodegenerative disorders often result from a complex interaction between genetic predispositions and environmental factors. Genetic mutations play a pivotal role in the development of various nervous system disorders. For instance, mutations in the *APP, PSEN1*, and *PSEN2* genes are closely associated with familial Alzheimer's disease, leading to abnormal amyloid-beta production and plaque formation (Lanoiselée et al., 2017; Selkoe, 2001). Similarly, mutations in the *SNCA* gene, which encodes alpha-synuclein, are linked to familial forms of Parkinson's disease, driving the formation of Lewy bodies (Mughal et al., 2024). Additionally, expansions in the *HTT* gene cause Huntington's disease, which is marked by the accumulation of toxic huntingtin protein aggregates (Burgunder, 2024). Environmental factors such as lifestyle, toxins, and traumatic brain injuries can also significantly contribute to the development and progression of nervous system disorders. Exposure to environmental toxins like exposure to pesticides has been associated with an increased risk of Parkinson's disease, by inducing oxidative stress and mitochondrial dysfunction (Santos-Lobato & Schuh, 2024). Furthermore, Traumatic brain injuries have been shown to elevate the risk of developing chronic traumatic encephalopathy (CTE) and other neurodegenerative diseases by triggering inflammatory responses and protein aggregation in the brain (Boulton & Al-Rubaie, 2024). After these topics, understanding how inflammatory responses within the nervous system contribute to these complex conditions is beneficial.

3.5. Inflammatory Responses in the Nervous System

Neuroinflammation plays a pivotal role in the development and progression of neurodegenerative disorders. The activation of microglia, the brain's resident immune cells, is a hallmark of neuroinflammatory responses in conditions like Parkinson's disease and Alzheimer's disease (Heneka et al., 2015). In response to amyloid plaques and Lewy bodies, microglia become activated and release pro-inflammatory cytokines such as tumor necrosis factor-alpha (TNF-α) and interleukin-1beta (IL-1β), which contribute to neuronal damage and worsen disease progression (Cómitre-Mariano et al., 2024). Astrocytes, another type of glial cell, also play a significant role in neuroinflammation. In neurodegenerative disorders, astrocytes can become reactive, characterized by the overexpression of inflammatory mediators and the disruption of their supportive functions for neurons (Ahadiat & Hosseinian, 2023). This reactive gliosis further contributes to the degeneration of neural tissues by promoting oxidative stress and impairing the BBB, allowing peripheral immune cells to infiltrate the brain and amplify the inflammatory response (Ransohoff & Brown, 2012). With being informed about inflammatory mechanisms, it is also essential to get familiar with the impact of neurovascular changes on the progression of neurodegenerative diseases.

3.6. Impact of Neurovascular Changes

Neurovascular changes play a crucial role in the etiology of various brain disorders, particularly in neurodegenerative diseases such as vascular dementia and Alzheimer's. Disruptions in the BBB are a significant factor, as they allow toxic substances, immune cells, and inflammatory molecules to infiltrate the brain, leading to neuronal damage and exacerbating disease progression (Kempuraj et al., 2024). In Alzheimer's disease, vascular abnormalities such as reduced cerebral blood flow and microvascular pathology contribute to the buildup of amyloid-beta and tau proteins, accelerating neurodegeneration (Korte et al., 2020). Cerebral amyloid angiopathy is a condition characterized by the deposition of amyloid-beta in the walls of cerebral blood vessels, further compromising vascular integrity and increasing the risk of hemorrhages, thus contributing to cognitive decline (Sohrabi et al., 2024).

Furthermore, chronic hypoperfusion, or reduced blood supply to the brain, is implicated in the pathogenesis of vascular dementia. This condition leads to white matter lesions and ischemic damage, which impair cognitive function (de La Torre, 2012). The interaction between neurovascular changes and neuroinflammation also plays a critical role in the progression of neurodegenerative diseases. Vascular dysfunction triggers the activation of glial cells, which release pro-inflammatory cytokines, exacerbating neuronal injury and creating a vicious cycle of neurodegeneration (Duncombe et al., 2017). As our understanding of neurovascular impacts has been deepened, the role of molecular biomarkers has become pivotal in diagnosing and tracking the progression of nervous system disorders.

3.7. Molecular Biomarkers in Nervous System Disorders

Molecular biomarkers are crucial in diagnosing, monitoring, and understanding the pathophysiology of various nervous system disturbances. In Alzheimer's disease, biomarkers such as amyloid beta and phosphorylated tau in cerebrospinal fluid (CSF) are widely used to reflect the attendance of neurofibrillary tangles and amyloid plaques, respectively, which are hallmark features of the disease (Duncombe et al., 2017). Furthermore, Decreased levels of amyloid-beta in CSF correlate with amyloid deposition in the brain, making it a reliable marker for early diagnosis and monitoring disease progression (Lehmann et al., 2020). Imaging biomarkers, such as amyloid positron emission tomography (PET) scans, also provide insights into amyloid pathology and are used alongside fluid biomarkers to enhance diagnostic accuracy.

In Parkinson's disease, alpha-synuclein, a protein that aggregates to form Lewy bodies, serves as a key biomarker. Elevated levels of alpha-synuclein in the CSF and blood are associated with disease severity and progression (Mollenhauer et al., 2012). Moreover, inflammatory biomarkers, such as increased levels of cytokines (e.g., IL-6, TNF-α), are observed in various neurodegenerative diseases, reflecting the role of neuroinflammation in these disorders (Roveta et al., 2024). The identification and validation of these biomarkers not only aid in early diagnosis and differentiation of neurodegenerative diseases but also provide targets for therapeutic interventions, enabling personalized treatment strategies. By identifying these molecular biomarkers enhancing our diagnostic capabilities, the advancements in neuroimaging techniques offer a complementary approach to further unravel the complexities of nervous system disorders.

3.8. Advances in Neuroimaging Techniques

Recent advances in neuroimaging techniques have significantly enhanced our ability to visualize and understand histological changes in brain tissues, particularly in the background of neurodegenerative disorders. PET has been instrumental in tau tangles and detecting amyloid-beta plaques in Alzheimer's disease, allowing for primary diagnosis and monitoring of disease progression (Ossenkoppele et al., 2022). PET imaging, combined with novel tracers, has also improved the visualization of neuroinflammation and synaptic activity, providing insights into the underlying mechanisms of neurodegeneration (Nakamoto et al., 2024). Furthermore, advances in magnetic resonance imaging (MRI) techniques, such as diffusion tensor imaging (DTI) and functional MRI (fMRI), allow for detailed mapping of white matter integrity and changes in functional connectivity in the brain, which are crucial for understanding disorders like multiple sclerosis and Parkinson's disease (Agosta et al., 2017; Carnevale & Lembo, 2019).

Another significant development is The use of high-resolution imaging techniques like optical coherence tomography (OCT) and two-photon microscopy offers significant insights, which offer detailed insights into the microstructural and cellular changes within brain tissues (C. Zhang et al., 2023). These techniques are particularly valuable for studying the progression of neurodegenerative diseases at the cellular level and for evaluating the effects of therapeutic interventions in real-time. As neuroimaging technologies continue to advance, they have the potential to not only enhance the diagnosis and management of neurological disorders but also to deepen our understanding of their underlying pathophysiological processes. Although these neuroimaging advancements provide significant insights into brain disorders, addressing the challenges in treating progressive brain disorders is still a critical area (Table 2).

Table 2. Currents' imaging techniques' details

Technique	Resolution	Advantages	Disadvantage
MRI	3 mm	Non-invasive	---
fMRI	3 mm	Non-invasive	Expensive
PET	5 mm	Sensitive, metabolic receptor mapping	Invasive
OCT	5-10 micro m	High-resolution	Limited aspect ratio
MRS	low	Non-invasive	Expensive

4. CHALLENGES IN TREATING PROGRESSIVE BRAIN DISORDERS

As we mentioned, BBB is a critical protective barrier that regulates the movement of substances from the bloodstream into the brain, thereby preserving brain homeostasis and normal function (Viscusi & Viscusi, 2020). This barrier is composed of various brain cells, including brain endothelial cells, astrocytes, neurons, and basal membranes, all of which collaborate to form a tightly sealed capillary network within the brain. The lack of fenestrations in the endothelial cells of these capillaries restricts the passage of small molecules and proteins (Greene & Campbell, 2016). Moreover, these endothelial cells are interconnected by inter-endothelial junctions, creating a continuous barrier that limits the transport of water-soluble substances (Hawkins & Davis, 2005). Surrounding these endothelial cells are the astrocytes, basal lamina, and pericytes, which add an extra layer of protection "BBB", further limiting the entry of therapeutic agents and molecules from the blood into the brain (Lu & Wen, 2024). The efficacy of this barrier is further enhanced by efflux transporters in the brain capillaries that actively return any substances that penetrate the brain back into the bloodstream. BBB's permeability is regulated by inter-endothelial junctions, including gap junctions, tight junctions, and adherens junctions; protein complexes that play a crucial role in maintaining the barrier's integrity (Viscusi & Viscusi, 2020).

Molecules that cross the BBB can do so via either paracellular or transcellular pathways. The ability of compounds to cross the BBB depends on their physicochemical properties, such as size, molecular weight, surface activity, lipid solubility, and charge (Sierri et al., 2025). Some small molecules, such as carbon dioxide (CO_2), ethanol (C_2H_5OH), and barbiturates, can pass through the BBB via passive diffusion.

Additionally, receptor-mediated pass mechanisms involving the transferrin receptor, insulin transporter, and glucose transporter-1 (GLUT-1) facilitate the transport of hydrophilic molecules such as peptides and proteins (S. Arora et al., 2020). Certain pathological conditions can compromise the integrity of the BBB, allowing substances to leak into the brain. Furthermore, the use of specialized drug carriers, such as nanoparticles, has been explored to enhance the delivery of therapeutic agents across the BBB (Bai & Ge, 2024). Although the importance of understanding these mechanisms is undeniable, we must also consider the pharmacokinetic principles that impact how effectively drugs can be delivered to the brain.

4.1. Pharmacokinetic Principles and Their Impact on Brain Drug Delivery

The effectiveness of systemically administered drugs is largely influenced by their pharmacokinetic properties. The journey of a drug from the site of administration to its target location—in this case, the brain—is challenging and often reduces the therapeutic efficacy of the drug (Dong, 2018). One major obstacle is the binding of molecules to plasma proteins in the bloodstream, which limits the amount of free drug available for transport to the brain. Additionally, some drugs are quickly cleared from the bloodstream by major organs, leaving only a small portion available for brain delivery (Wanat, 2020).

The interaction between a drug and its target cells can also impact drug absorption. For example, drug molecules can interact with cells in ways that block channels, alter membrane potential, or change cell conformation, all of which can temporarily affect the cell's response to the drug and its absorption (Zhang et al., 2024). Generally, small lipophilic drug molecules are more favorable and optimal for brain delivery due to their ability to cross the BBB more easily (Hanes et al., 2020). While these pharmacokinetic principles present some challenges for delivering drugs into the brain, nanoparticles offer innovative solutions that have transformed imaging and diagnostic approaches for the brain.

5. NANOPARTICLES AND THEIR ROLE IN IMAGING OF BRAIN

As we mentioned, use of structural imaging techniques, such as MRI, fMRI, computed axial tomography (CAT), PET, magnetoencephalography (MEG), and optical imaging, to assess brain function and diagnose CNS disorders and diseases has been facilitated by the utility of nanoparticles. Superparamagnetic iron oxide (SPIO) nanoparticles (Loo et al., 2004; Lv et al., 2018) are examples of nanosized imaging agents. The other major category of nanoparticles used in brain imaging consists of nanocarriers, which are functionalized with imaging agents like gadolinium and can also carry other functional entities if needed. In theoretical terms, possible to apply drug-delivery nanoparticles to the delivery of diagnostic compounds and imaging agents (Meyers et al., 2013).

The potential applications of nanoparticles for imaging, diagnosis, and treatment are receiving more and more attention. The most researched nanoparticles for brain imaging and diagnosis are those with great structural flexibility and huge carrier capacities, such as magnetic iron nanoparticles, quantum dots (QDs), dendrimers, and gold nanoparticles (Mir et al., 2017) (Figure 2). Materials of many different kinds can be used to create nanoparticles. Nanoparticles can be made of carbon, ceramics, and other metals in addition to polymers. They may be fashioned into a variety of forms, including complicated strands, hollow or solid spheres, tubes, and rods. Certain forms and materials work well in certain situations (Dur E Nayab et al., 2023). As a result, there is no one ideal kind of nanoparticle; rather, the suitability of the material and form chosen depends on the purpose of the particle. A core structure that contains an imaging agent or medicine is a common feature of nanoparticles used for imaging (Habeeb et al., 2024).

Figure 2. Types of nanoparticles used in the nervous system; for imaging and treating purposes.

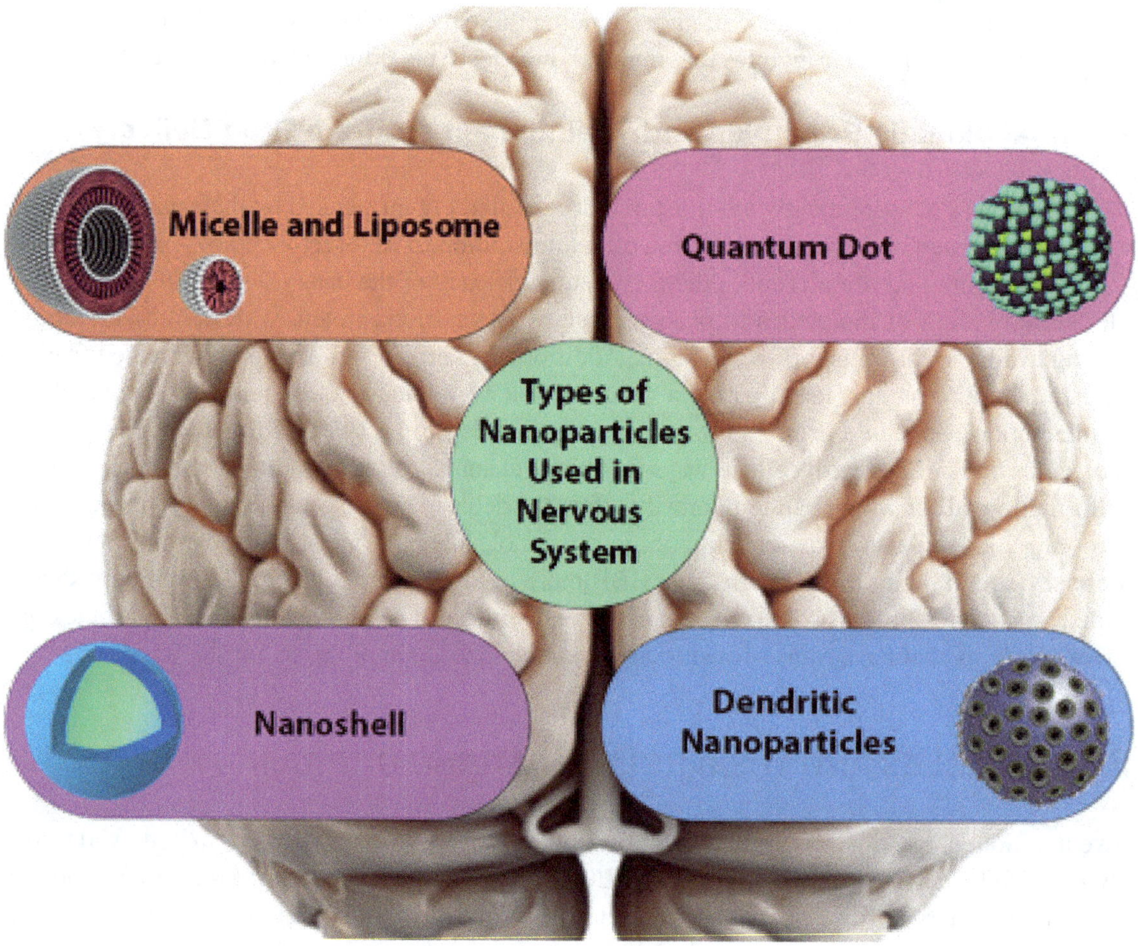

5.1. Nanoshells

Nanoshells are made up of a metallic shell, usually made of gold, covering a dielectric core. Spectral tunability refers to the ability to alter the relative sizes of the core and shell so that the particles emit light with a certain wavelength. These optical characteristics enable the use of nanoshells as optical imaging agents. Furthermore, light in the near-infrared region of the spectrum may be absorbed by gold nanoshells. Because of this, when the right kind of light is applied to the nanoparticles, they may get heated. This heat can then be utilized to thermally ablate tumors (Padmanabhan et al., 2016).

5.2. Iron Oxide Particles

Iron oxide nanoparticles (IONPs) are characterized by their superparamagnetic properties, which have applications in drug delivery and MRI. These nanoparticles composed of ultraminiature paramagnetic iron oxide particles, can enhance signal intensity on T1-weighted images, similar to marketed MRI imaging contrast agents such as gadolinium. They also reduce signal intensity on T2-weighted imaging (El-Sherbiny et al., 2020). MRI imaging agents have been used to image human brain tumors preoperatively and intraoperatively, improving brain tumor contrast for days after administration. Nanoparticles concentrate in reactive astrocytes and macrophages, rather than building up inside tumor cells. Superparamagnetic nanoparticles have been used for brain tumor-targeted treatment, with studies showing that tumors of neuroectodermal origin with membrane-bound matrix metalloproteinase-2 can be targeted by chlorotoxin-conjugated nanoparticles. The quantity of targeted nanoparticles is significantly higher than non-targeted nanoparticles within the tumor (Forouzandehdel et al., 2020; Posadas et al., 2016).

5.3. QDs

QDs are a revolutionary imaging technology with a metalloid crystalline core and a shell that encloses it. They have great brightness, size-tunable narrow emission spectra, and extended photostability, making them ideal for cellular imaging. QDs have a heavy metal core, usually made of cadmium telluride or selenium, and an unreactive metallic shell, usually zinc sulfide, surrounding it. Bioactive molecules can chemically functionalize the particle's exterior coating, allowing for targeted targeting of specific molecules, such as cell membrane receptors (Pathak et al.). QDs have size-dependent light emission, extremely bright signal intensity, and small changes in the semiconductor core that change the wavelength of light. This property allows for longer imaging time than other optical imaging agents, such as organic fluorophores (Qin et al., 2024).

5.4. Liposomes and Micelles

Nanoscale vesicles called "liposomes" have an aqueous core and a bilayer of phospholipid membrane. Therapeutic medications can be sequestered in the aqueous core and transported to target locations, shielding them from degradation and activities at unintended targets (Re et al., 2012). For targeted delivery, liposomes can include ligands that bind biological targets attached to their surface, just like many other forms of nanoparticles. It is possible to make these particles emit their contents over an extended period of time. The Food and Drug Administration (FDA) has authorized liposomal versions of doxorubicin for clinical usage in cancers such multiple myeloma and ovarian cancer; additional formulations are now undergoing feasibility testing for human use.

Although liposomes have just been discovered, their use in treating brain tumors is yet in the experimental phase (Teleanu et al., 2018). Spherical aggregates of molecules known as micelles have the hydrophilic parts of the molecules facing outward and the hydrophobic parts facing inside. Micelles provide a way to sequester a molecule in the hydrophobic core of the micelle, making it soluble even if it would ordinarily be insoluble in that particular solvent. Compared to intravenous delivery of free medicine, the micelle's ability to function as a nanocarrier thanks to its exterior hydrophilic coating enables the delivery of larger doses of medication to the intended region (Bhalodi et al., 2024; Wei et al., 2014).

5.5. Dendritic Nanoparticles

Apart from the copious research conducted on dendrimers in drug administration, there has been significant interest in the dendritic nanoparticle-mediated brain disorders diagnostics development (Zajdel et al., 2023). The advancement of brain imaging and diagnostics mediated by dendritic nanoparticles has garnered a lot of interest. Gadolinium and rhodamine B have been functionalized on polyamidoamine (PAMAM) dendrimer generations 1–8 to enable fluorescence and magnetic resonance imaging of malignant gliomas (Yu et al., 2014).

MRI is a very sensitive diagnostic technique that heavily relies on contrast chemicals for the clinical diagnosis of brain malignancies. Presently, low molecular weight (MW) contrast agents that are often employed, such as gadolinium-diethylene triamine pentaacetic acid (DTPA), have low contrast efficiency, nonspecificity, and quick renal elimination. (Faiz et al., 2022). Using DTPA as the payload and chlorotoxin as a tumor-specific ligand, dendrigraft poly-l-lysines serves as the basis for a macromolecular MRI contrast agent. The MRI pictures demonstrated that the signal increase persisted for a significantly longer period than when employing conventional contrast agents, which might lead to a more precise identification of cancers in animal models with xenografted brains (Dur E Nayab et al., 2023).

6. NANO-BIOMEDICINE IN TREATING BRAIN DISORDERS

The second greatest cause of death worldwide is neurological illness, which includes epilepsy, Parkinson's disease, Alzheimer's disease, stroke, and Huntington's disease (H. Arzaghi et al., 2020). Nerve regeneration is a complex biological phenomenon with limited potential in the CNS due to the complex features of neurons, including morphology and electrical activity (Vashist et al., 2023). Neural regeneration strategies, like direct cell injection and growth factor application, struggle with differentiation efficiency. Nanomaterials can be used to design engineered nerve conduction channels for peripheral nerve damage. Many applications of nanostructures exist, such as suspensions, immobilization on substrates, functionalization with proteins such as asparagine-glycine-aspartate (RGD) peptides, ECM proteins, and monoclonal antibodies, as well as nanorods, nanospheres, nano-scaffolds, and nanoparticles in nanocomposites.

Supplying an appropriate milieu with the ability to control stem cell destiny. These tactics can be used in conjunction with growth factor administration and cell therapy to promote neuronal repair and cure neurological diseases. Nanoparticles have shown good affinity to target tissue, leading to the evaluation of various nano-systems for drug delivery to the brain. One current method is receptor-mediated delivery systems, which use a ligand-receptor mechanism to deliver drugs to specific targets (Hashemi et al., 2023). Ligands on nanoparticles, such as transferrin, apolipoprotein E, B, and A, and antibodies, can pass the blood-brain barrier, enabling targeted carrier delivery. Receptor-mediated endocytosis serves as the primary transport route for these nanoparticles (S. Shakeri et al., 2020).

6.1. Nanomaterials in Tissue Engineering

Nerve regeneration and repair of nerve defects are possible using neural tissue engineering. Nanomaterials can provide an excellent microenvironmental substrate for cell-matrix and cell-cell interactions and direct and manipulate the signaling mechanism of stem cells toward the neural lineage. Proteins and glycoproteins are found in the ECM which a complex biological scaffold that is released by cells. Through interaction with the cellular matrix, it is essential in determining the fate of individual cells. Collagen, laminin, fibronectin, vitronectin, and other ECM proteins and glycoproteins interact directly with integrins and transmembrane adhesive proteins. Under different physiological conditions, their heterodimeric structure provides different functions. Research on the composition and interactions of the ECM can contribute to neural tissue engineering and nerve repair (He et al., 2024; Provenzale & Silva, 2009). For example, aligned nanofiber structures through mimicking the natural ECM play an important role in tissue engineering for brain disorders. This is because aligned nanofibers provide physical guidance for neuronal growth and alignment. By integrating microparticles within these nanofibers, we lead to the controlled release of neuroprotective agents and reduced oxidative stress and inflammation. These two characteristics which provide support and active neuron protection, are promising in neural tissue engineering (Yu et al., 2022; X. Zhang et al., 2023). Zhang et al. in their study fabricated uniaxially aligned nanofiber yarns by using electrospinning and electrospray techniques. They successfully integrated Polycaprolactone as a semi-crystalline, hydrophobic, and biodegradable polymer with microparticles and created preserved structural integrity and optimized conditions for neural regeneration (Jeon et al., 2023; X. Zhang et al., 2023).

Beyond mimicking the ECM, molecular components or products required for CNS repair can be provided by nanoparticles, which can augment the body's natural healing processes. Self-assembling nanodevices, including nanofiber scaffolds that self-assemble in the CNS to encourage neurite formation, can act as platforms for regeneration processes. In a physiological setting, peptide amphiphile molecules self-assemble, enabling active peptide sequences to function as ligands for cell surface receptors. In one experiment, water molecules are trapped by nanofiber scaffolds, creating a gel-like substance that encases brain progenitor cells. The combination of 3D nanostructure and bioactive peptides for signaling on the nanofibers stimulated rapid differentiation of neural progenitor cells into adult neuronal cells (Hamidreza Arzaghi et al., 2020; Vashist et al., 2023).

6.2. Nanomaterials in Drug Delivery

As previously noted, the blood-cerebrospinal fluid barrier (BCSFB) and the BBB limit the bioavailability and efficient transport of medications and other therapeutic chemicals in the neurological system (Sulheim et al., 2016; Vashist et al., 2023). In order to prevent the entry of blood-borne pathogens such as bacteria, viruses, parasites, and toxins, the BBB is essential (Roney et al., 2005). This barrier helps protect the brain, but it also makes many neurological conditions more difficult to treat. Therefore, it is crucial to create a safe and efficient medication delivery method that can pass through the BBB and reach the intended cells without having any negative side effects (Rami et al., 2022; Siddiqi et al., 2018; Vashist et al., 2023). The development of a nanocarrier labeled with specific ligands that can attach to target tissues or cells, boost an implanted drug's effectiveness, and lessen its toxicity is desperately needed (Orive et al., 2009).

The nanoparticles are able to pass through solid barriers and be functionalized to produce regulated medication release. The treatment and management choices for neurological diseases are severely restricted due to BBB which hinders the passage of therapeutic molecules and desirable medications when taken orally. Drug delivery systems based on nanotechnology are an effective way to get drugs into the brain (Alqudah et al., 2024). The nanomaterials can be functionalized to achieve sustained and controlled drug release and modified to cross solid physiological barriers (Cheng et al., 2023; Yang et al.,

2013). Many studies are focusing on drug delivery through nanomaterials. Arora et al. used a liposome altered with a mannose ligand and cell-penetrating to carry and distribute Apolipoprotein E to the brain affected by Alzheimer's disease. In this research high concentration of the gene was transferred to the target tissue safely and efficiently (Sanjay Arora et al., 2020). In addition, it has been shown that gold nanoparticles count as one of the best inorganic nanocomposites in the treatment of high-grade-gliomas. This is because the advantage of having a high surface-to-volume ratio makes them size adjustable and able to cross the BBB (Souri et al., 2022). But, due to the difficulties in targeting the tumor specifically their usage is limited. However recent advancements, confirm that both in vitro and in vivo, the usage of gold-sulfur interactions to stabilize gold nanoparticles has boosted the effectiveness of a DNA aptamer that targets EGFRvIII in glioblastomas (Peng et al., 2020). Also, Karimzadeh et al. used Mesoporous silica nanoparticles that were functionalized by succinic anhydride and 3-aminopropyltriethoxysilane. Then in a separate study, they loaded it with rivastigmine. According to Karimzadeh's research, the formulation could successfully boost stability and bioavailability. However other modification was recommended to decrease the viability loss caused by these nanoparticles (Karimzadeh et al., 2017). Overall, these advancements highlight the potential of nanotechnology in enhancing drug delivery systems for neurological diseases.

6.3. Types of Nanomaterials in Drug Delivery and Tissue Engineering of Neurons

6.3.1. Polymeric Nanomaterials

In order to treat neurological diseases, polymeric nanoparticles are utilized to bind, absorb, dissolve, and encapsulate medications or therapeutic chemicals (Begines et al., 2020). These particles, which offer protection, simple distribution, and great effectiveness, are found in nanospheres and nanocapsules (S. Das et al., 2016). Through their main delivery mechanisms, such as endocytosis, transcytosis, and functionalization with peptides or ligands, they can pass the BBB and enter the CNS (Kashyap & Shukla, 2019).

6.3.2. Dendrimers

Dendrimers are branching polymers that have a core symmetry like a tree and are frequently spheroidal in water. For drug smuggling, their perimeter is densely populated. Polyamidonamine, polypropylenimine, and polyaryl ethers are examples of common dendrimer compounds. PAMAM dendrimers are utilized to treat diseases of the CNS because they are BBB-passing. They can use ligand coupling and endocytosis to get past biological barriers and cell membranes (Kim et al., 2024; Srinageshwar et al., 2017; Xu et al., 2014).

6.3.3. Polymeric Micelles

Polymeric micelles are spherical structures with a hydrophilic shell and a hydrophobic interior that are created when amphiphilic copolymers combine. Their ability to self-assemble is essential. Micelles might possibly improve pharmaceutical bioavailability by providing stability, continuous release, and control over release, as well as by delivering lipophilic and weakly water-soluble chemicals (Wang et al., 2016). across endocytosis or transcytosis, they are able to pass across the BBB, but they are susceptible to destruction by external heat, mechanical stress, ligand conjugation, or enhanced penetrating capacity (Gwak et al., 2016).

6.3.4. Solid Lipid Nanoparticles (SLNs)

Because of their special qualities, capacity to release and stabilize medications without causing cytotoxic effects, and appeal for brain targeting, SLNs are being employed more and more as effective drug carriers (Fong Yen et al., 2015). Their distinctive size and repeatability make them appealing for brain targeting. They lower the danger of contamination and are appropriate for formulations without organic solvents (Gastaldi et al., 2014). SLNs improve medication stability in the blood and entrance via the blood-brain barrier, making them potential for the treatment of cancer and neurological illnesses (Cacciatore et al., 2016).

6.3.5. Liposomes

Liposomes are spherical vesicles composed of cholesterol, phospholipids, and an impenetrable lipid bilayer. Because they are biocompatible and non-toxic, they are regarded as a significant medication delivery medium. They are not immunogenic, carcinogenic, or thrombogenic; nonetheless, the biological system recognizes them as foreign particles without producing any adverse reaction upon their introduction into the system (Senjab et al., 2024). Because liposomes may encapsulate numerous components, they are being employed as bigger transport nanocarriers. In addition, the reticuloendothelial system shields them from enzymatic breakdown and evacuation. The primary attributes are the capacity to merge with biological membranes, traverse cell membranes, and pierce the BBB. Applying polyethylene glycol (PEG) surface treatment to liposomes is an easy way to increase their half-life (Spuch & Navarro, 2011).

6.3.6. Inorganic Nanoparticles

Owing to their surface plasmon behavior and tiny size, inorganic nanoparticles such as magnetic, platinum, chromium, titanium, manganese, zinc, molybdenum, gold, and selenium are commonly used in neurological illnesses treatment (Toader et al., 2024).

6.3.7. Magnetic Nanoparticles

Magnetic nanoparticles offer biomedical applications due to their superparamagnetism, saturation field, high field irreversibility, and additional anisotropy, possibly facilitating the passage of medication through the BBB and potentially curing brain cancer (Konnova & Rozhina, 2024).

6.3.8. Gold Nanoparticles

The favorable characteristics of gold nanoparticles are stability, resistance to oxidation, and biocompatibility (Verma et al., 2024). By decreasing and altering essential proteins, they can be utilized to treat addiction and for drug delivery networks. Gold nanoparticles can effectively transport drugs to treat a variety of disorders, including Alzheimer's disease, by weakly exposing amyloid-beta fibrils and plaques to microwave fields, which breaks them down. Gold nanoparticles have been demonstrated in vitro to slow down the course of Alzheimer's disease, and they can be coupled with substances that obstruct amyloid beta fibrils (Gao et al., 2015; Karimi et al., 2016).

6.3.9. Carbon Nanotubes (CNTs) and Flluerenes

Since CNTs have special mechanical, electrical, and physicochemical characteristics, they are a potential nanomaterial. After brain injuries, they may help restore neurons, which is particularly promising for the treatment of neurological conditions including ischemic stroke, Parkinson's disease, and Alzheimer's disease. Because of their strength, flexibility, affordability, and lightweight, CNTs are perfect for in vivo medication delivery (Kakkar & Dahiya, 2015; Prajapati et al., 2022).

7. ETHICAL CONSIDERATIONS AND FUTURE PERSPECTIVES

Neuronal death occurs in neurodegenerative disorders like Alzheimer's disease and Parkinson's disease. As a result, promoting neurogenesis has become a key therapeutic approach for these conditions. However, effectively delivering drugs to the brain remains a significant challenge due to several critical factors, including the BBB, the lipophilicity of the drug, and its molecular weight. These factors limit the therapeutic potential of drugs, making neurodegenerative diseases particularly difficult to treat. In recent years, nanoparticle-mediated targeted drug delivery has been introduced as a promising strategy for enhancing neurogenesis and improving treatment approaches for these disorders. Despite these potential advantages, nanocarrier-mediated drug delivery also presents challenges related to production and regulatory approval.

The toxicity of nanoparticles is influenced by various factors, including size, surface charge, ionic dissolution, and shape. These characteristics must be carefully considered when developing nanoparticle-based drug delivery systems, adhering to established nanotoxicity guidelines (Zielińska et al., 2020). Furthermore, the approval process for these nanocarriers must thoroughly evaluate their health and environmental impacts (Hofmann-Amtenbrink et al., 2014). The literature suggests several strategies to minimize toxicity associated with nanoparticle size and charge, such as surface modifications using biodegradable or naturally derived biomolecules. Ensuring consistency in size and content from batch to batch is also crucial in the production of nanocarriers. Various production methods, such as high-pressure homogenization, microemulsion, and extrusion, have been documented (Suk et al., 2016).

Moreover, the pharmacokinetic properties of nanotherapeutics play a critical role in determining their toxicity, and efficacy. Therefore, it is essential to study these parameters in relevant animal models. Advanced pharmacokinetic models, such as physiologically based pharmacokinetic (PBPK) models, can be valuable tools for predicting the in vivo behavior of nanotherapeutics. Additionally, the regulatory requirements for the clinical acceptance of nanotherapeutics must be carefully considered (Ji et al., 2017). Nanomedicine offers a promising avenue for overcoming the limitations of current and traditional treatment approaches for nervous diseases (Zhu et al., 2021). Various nanoparticle-based studies targeting different neurodegenerative diseases have been highlighted, which could pave the way for future nanomedicine applications. While developing curative treatments is a long-term goal, early-stage research in this field may provide the foundation needed to combat these disorders effectively. To validate the efficacy of nanoparticle-based therapies against nervous diseases, more comprehensive in vitro and in vivo studies are necessary. Establishing correlations between these investigations will help researchers identify and optimize effective nanoparticles for therapeutic and diagnostic purposes (Jagaran & Singh, 2021).

8. CONCLUSION

Nanomedicine has revolutionized healthcare by enabling targeted drug delivery, crossing the formidable blood-brain barrier, and enhancing the precision of diagnostic imaging. These advancements hold the promise of not only managing but potentially reversing neurodegenerative conditions such as Parkinson's and Alzheimer's disease. Biomedicine and nanomedicine, when combined, offer unprecedented opportunities for the treatment of nervous system disorders. Nanotechnology facilitates the creation of highly specialized materials that interact with biological systems at the molecular level, enabling precise targeting and delivery of treatments. For example, nanoparticles can be engineered to cross the BBB, a major challenge in treating neurological conditions is the direct delivery of therapeutic agents to the affected areas. This approach not only increases the efficacy of treatments but also minimizes side effects typically associated with conventional therapies (Arcos, 2023; Contera et al., 2020).

The etiology and histology of brain and nervous system disorders underline the complexity of these conditions. The intricate architecture of the brain and the diverse pathologies associated with neurological disorders make treatment particularly challenging. Progressive brain disorders such as Parkinson's and Alzheimer's disease are marked by degeneration that is often irreversible, highlighting the need for early and accurate diagnosis. Nanoparticles have a main role in imaging technologies, offering higher resolution and the ability to detect pathological changes at earlier stages (Minchin & Martin, 2010). This capability is vital in developing preventive strategies and in the timely intervention of these debilitating diseases. However, the usage of nano-biomaterials in the nervous system is not without its challenges. The treatment of progressive brain disorders requires addressing the limitations of current drug delivery systems, including issues related to bioavailability, targeting specificity, and potential toxicity. Nanotechnology with its ability to engineer materials at the atomic scale provides solutions to these challenges, such as the expansion of drug delivery vehicles that can navigate the complex environment of the brain and deliver drugs to specific sites of pathology (Arcos, 2023; Contera et al., 2020).

Looking to the future, the integration of nanotechnology in the treatment of nervous system disorders holds immense promise. Continued research and development are likely to lead to even more sophisticated nano-biomaterials that can interact with the nervous system in ways previously unimaginable. These developments could pave the way for new treatments that not only manage symptoms but potentially reverse the damage caused by neurodegenerative diseases. However, realizing this potential will require a multi-disciplinary approach, combining views from biology, materials science, engineering, and ethics (Arcos, 2023; Contera et al., 2020; Patil-Sen, 2021). As a result, bionanomaterials are an excellent approach to address the advanced challenges of nervous system disorders. While significant progress has been made, research and ethical considerations continue to be explored to fully exploit the potential of these technologies. The future of treating brain disorders may lie in the convergence of nanotechnology and biomedicine, which could promise more effective and targeted therapies.

REFERENCES

Abbasi, R., Shineh, G., Mobaraki, M., Doughty, S., & Tayebi, L. (2023). Structural parameters of nanoparticles affecting their toxicity for biomedical applications: A review. *Journal of Nanoparticle Research*, *25*(3), 43. PMID: 36875184

Agosta, F., Galantucci, S., & Filippi, M. (2017). Advanced magnetic resonance imaging of neurodegenerative diseases. *Neurological Sciences*, *38*(1), 41–51. PMID: 27848119

Ahadiat, S.-A., & Hosseinian, Z. (2023). Astrocytes' innate role in neurodegenerative disorders. *Bulletin of the National Research Center*, *47*(1), 105.

Alfaidi, M., Barker, R. A., & Kuan, W. L. (2024). An update on immune-based alpha-synuclein trials in Parkinson's disease. *Journal of Neurology*, *272*(1), 21. DOI: 10.1007/s00415-024-12770-x PMID: 39666171

Alqudah, A., Aljabali, A. A., Gammoh, O., & Tambuwala, M. M. (2024). Advancements in neurotherapeutics: Nanoparticles overcoming the blood–brain barrier for precise CNS targeting. *Journal of Nanoparticle Research*, *26*(6), 1–19.

Arcos, D. (2023). Nanomaterials in Biomedicine 2022. *International Journal of Molecular Sciences*, *24*(10). Advance online publication. DOI: 10.3390/ijms24109026 PMID: 37240372

Arora, R., & Baldi, A. (2024). Revolutionizing Neurological Disorder Treatment: Integrating Innovations in Pharmaceutical Interventions and Advanced Therapeutic Technologies. *Current Pharmaceutical Design*, *30*(19), 1459–1471. PMID: 38616755

Arora, S., Layek, B., & Singh, J. (2020). Design and validation of liposomal ApoE2 gene delivery system to evade blood–brain barrier for effective treatment of Alzheimer's disease. *Molecular Pharmaceutics*, *18*(2), 714–725. PMID: 32787268

Arora, S., Sharma, D., & Singh, J. (2020). GLUT-1: An Effective Target To Deliver Brain-Derived Neurotrophic Factor Gene Across the Blood Brain Barrier. *ACS Chemical Neuroscience*, *11*(11), 1620–1633. DOI: 10.1021/acschemneuro.0c00076 PMID: 32352752

Arzaghi, H., Adel, B., Jafari, H., Askarian-Amiri, S., Dezfuli, A. S., Akbarzadeh, A., & Pazoki-Toroudi, H. (2020). Nanomaterial integration into the scaffolding materials for nerve tissue engineering: A review. *Reviews in the Neurosciences*, *31*(8), 843–872. https://doi.org/doi:10.1515/revneuro-2020-0008 PMID: 32776904

Arzaghi, H., Adel, B., Jafari, H., Askarian-Amiri, S., Shiralizadeh Dezfuli, A., Akbarzadeh, A., & Pazoki-Toroudi, H. (2020). Nanomaterial integration into the scaffolding materials for nerve tissue engineering: A review. *Reviews in the Neurosciences*. Advance online publication. DOI: 10.1515/revneuro-2020-0008 PMID: 32776904

Azmi, N., Tiwari, A., Kaur, N., & Chaskar, A. (2023). Nanotechnology: A Promising Approach in Nerve Regeneration. *Current Nanoscience*, *19*(3), 305–321.

Bai, R., & Ge, X. (2024). Blood-brain barrier disruption following brain injury: Implications for clinical practice.

Begines, B., Ortiz, T., Pérez-Aranda, M., Martínez, G., Merinero, M., Argüelles-Arias, F., & Alcudia, A. (2020). Polymeric Nanoparticles for Drug Delivery: Recent Developments and Future Prospects. *Nanomaterials (Basel, Switzerland)*, *10*(7). Advance online publication. DOI: 10.3390/nano10071403 PMID: 32707641

Bhalodi, K., Kothari, C., & Butani, S. (2024). Next-generation cancer nanotherapeutics: Pluronic® F127 based mixed micelles for enhanced drug delivery. *Naunyn-Schmiedeberg's Archives of Pharmacology*, •••, 1–30. PMID: 39527309

Boulton, M., & Al-Rubaie, A. (2024). Neuroinflammation and neurodegeneration following traumatic brain injuries. *Anatomical Science International*, •••, 1–12. PMID: 38739360

Boyes, W. K., & van Thriel, C. (2020). Neurotoxicology of nanomaterials. *Chemical Research in Toxicology*, *33*(5), 1121–1144. PMID: 32233399

Braak, H., & Braak, E. (1991). Neuropathological stageing of Alzheimer-related changes. *Acta Neuropathologica*, *82*(4), 239–259. PMID: 1759558

Burgunder, J.-M. (2024). Genetics of Huntington's disease and related disorders: beyond triplet repeats.

Butterfield, D. A., & Halliwell, B. (2019). Oxidative stress, dysfunctional glucose metabolism and Alzheimer disease. *Nature Reviews. Neuroscience*, *20*(3), 148–160. PMID: 30737462

Cacciatore, I., Ciulla, M., Fornasari, E., Marinelli, L., & Di Stefano, A. (2016). Solid lipid nanoparticles as a drug delivery system for the treatment of neurodegenerative diseases. *Expert Opinion on Drug Delivery*, *13*(8), 1121–1131. DOI: 10.1080/17425247.2016.1178237 PMID: 27073977

Cao, D., & Ding, J. (2022). Recent advances in regenerative biomaterials. *Regenerative Biomaterials*, *9*, rbac098. PMID: 36518879

Carnevale, L., & Lembo, G. (2019). Innovative MRI techniques in neuroimaging approaches for cerebrovascular diseases and vascular cognitive impairment. *International Journal of Molecular Sciences*, *20*(11), 2656. PMID: 31151154

Chen, B., Xiang, H., Pan, S., Yu, L., Xu, T., & Chen, Y. (2020). Advanced theragenerative biomaterials with therapeutic and regeneration multifunctionality. *Advanced Functional Materials*, *30*(34), 2002621.

Cheng, X., Xie, Q., & Sun, Y. (2023). Advances in nanomaterial-based targeted drug delivery systems. *Frontiers in Bioengineering and Biotechnology*, *11*, 1177151. PMID: 37122851

Cómitre-Mariano, B., Vellila-Alonso, G., Segura-Collar, B., Mondéjar-Ruescas, L., Sepulveda, J. M., & Gargini, R. (2024). Sentinels of neuroinflammation: The crucial role of myeloid cells in the pathogenesis of gliomas and neurodegenerative diseases. *Journal of Neuroinflammation*, *21*, 304. PMID: 39578808

Contera, S., Bernardino de la Serna, J., & Tetley, T. D. (2020). Biotechnology, nanotechnology and medicine. *Emerging Topics in Life Sciences*, *4*(6), 551–554. DOI: 10.1042/etls20200350 PMID: 33295610

Das, S., Carnicer-Lombarte, A., Fawcett, J. W., & Bora, U. (2016). Bio-inspired nano tools for neuroscience. *Progress in Neurobiology, 142*, 1–22. DOI: 10.1016/j.pneurobio.2016.04.008 PMID: 27107796

Das, S., Carnicer-Lombarte, A., Fawcett, J. W., & Bora, U. (2016). Bio-inspired nano tools for neuroscience. *Progress in Neurobiology, 142*, 1–22. PMID: 27107796

De, I., Sharma, P., & Singh, M. (2022). Emerging approaches of neural regeneration using physical stimulations solely or coupled with smart piezoelectric nano-biomaterials. *European Journal of Pharmaceutics and Biopharmaceutics, 173*, 73–91. PMID: 35227856

de La Torre, J. C. (2012). Cardiovascular risk factors promote brain hypoperfusion leading to cognitive decline and dementia. *Cardiovascular Psychiatry and Neurology, 2012*(1), 367516. PMID: 23243502

Diez-Pascual, A. M., & Rahdar, A. (2022). Functional Nanomaterials in Biomedicine: Current Uses and Potential Applications. *ChemMedChem, 17*(16), e202200142. PMID: 35729066

Dong, X. (2018). Current Strategies for Brain Drug Delivery. *Theranostics, 8*(6), 1481–1493. DOI: 10.7150/thno.21254 PMID: 29556336

Duncombe, J., Kitamura, A., Hase, Y., Ihara, M., Kalaria, R. N., & Horsburgh, K. (2017). Chronic cerebral hypoperfusion: A key mechanism leading to vascular cognitive impairment and dementia. Closing the translational gap between rodent models and human vascular cognitive impairment and dementia. *Clinical Science, 131*(19), 2451–2468. PMID: 28963120

El-Sherbiny, I. M., El-Sayed, M., & Reda, A. (2020). Superparamagnetic iron oxide nanoparticles (SPIONs) as multifunctional cancer theranostics. *Magnetic nanoheterostructures: diagnostic, imaging and treatment*, 223-241.

Eleftheriadou, D., Kesidou, D., Moura, F., Felli, E., & Song, W. (2020). Redox-Responsive Nanobiomaterials-Based Therapeutics for Neurodegenerative Diseases. *Small, 16*(43), e1907308. DOI: 10.1002/smll.201907308 PMID: 32940007

Erisen, D. E., & Uludag, K. (2024). Estimating Biosafety of Biodegradable Biomedical Materials from In Vitro Ion Tolerance Parameters and Toxicity of Nanomaterials in Brain. In *Transformative Approaches to Patient Literacy and Healthcare Innovation* (pp. 201–221). IGI Global.

Faiz, K., Lam, F. C., Chen, J., Kasper, E. M., & Salehi, F. (2022). The Emerging Applications of Nanotechnology in Neuroimaging: A Comprehensive Review. *Frontiers in Bioengineering and Biotechnology, 10*, 855195. PMID: 35875504

Fong Yen, W., Basri, M., Ahmad, M., & Ismail, M. (2015). Formulation and evaluation of galantamine gel as drug reservoir in transdermal patch delivery system. *TheScientificWorldJournal, 495271*. Advance online publication. DOI: 10.1155/2015/495271 PMID: 25853145

Forouzandehdel, S., Meskini, M., & Rami, M. R. (2020). Design and application of (Fe3O4)-GOTfOH based AgNPs doped starch/PEG-poly (acrylic acid) nanocomposite as the magnetic nanocatalyst and the wound dress. *Journal of Molecular Structure, 1214*, 128142. https://doi.org/https://doi.org/10.1016/j.molstruc.2020.128142

Furtado, D., Björnmalm, M., Ayton, S., Bush, A. I., Kempe, K., & Caruso, F. (2018). Overcoming the blood–brain barrier: The role of nanomaterials in treating neurological diseases. *Advanced Materials, 30*(46), 1801362. PMID: 30066406

Gao, N., Sun, H., Dong, K., Ren, J., & Qu, X. (2015). Gold-nanoparticle-based multifunctional amyloid-β inhibitor against Alzheimer's disease. *Chemistry (Weinheim an der Bergstrasse, Germany), 21*(2), 829–835. DOI: 10.1002/chem.201404562 PMID: 25376633

García-González, N., Gonçalves-Sánchez, J., Gómez-Nieto, R., Gonçalves-Estella, J. M., & López, D. E. (2024). Advances and Challenges in Gene Therapy for Neurodegenerative Diseases: A Systematic Review. *International Journal of Molecular Sciences, 25*(23), 12485. PMID: 39684197

Gastaldi, L., Battaglia, L., Peira, E., Chirio, D., Muntoni, E., Solazzi, I., Gallarate, M., & Dosio, F. (2014). Solid lipid nanoparticles as vehicles of drugs to the brain: Current state of the art. *European Journal of Pharmaceutics and Biopharmaceutics, 87*(3), 433–444. DOI: 10.1016/j.ejpb.2014.05.004 PMID: 24833004

Gerardo Nava, J. L., Rose, J. C., Altinova, H., Dalton, P. D., De Laporte, L., & Brook, G. A. (2021). Nanofibers and Nanostructured Scaffolds for Nervous System Lesions. *Nanomedicines for Brain Drug Delivery*, 61-101.

Greene, C., & Campbell, M. (2016). Tight junction modulation of the blood brain barrier: CNS delivery of small molecules. *Tissue Barriers, 4*(1), e1138017. DOI: 10.1080/21688370.2015.1138017 PMID: 27141420

Grünewald, A., Kumar, K. R., & Sue, C. M. (2019). New insights into the complex role of mitochondria in Parkinson's disease. *Progress in Neurobiology, 177*, 73–93. PMID: 30219247

Gwak, S. J., Nice, J., Zhang, J., Green, B., Macks, C., Bae, S., Webb, K., & Lee, J. S. (2016). Cationic, amphiphilic copolymer micelles as nucleic acid carriers for enhanced transfection in rat spinal cord. *Acta Biomaterialia, 35*, 98–108. DOI: 10.1016/j.actbio.2016.02.013 PMID: 26873365

Habeeb, M., Vengateswaran, H. T., Tripathi, A. K., Kumbhar, S. T., & You, H. W. (2024). Enhancing biomedical imaging: The role of nanoparticle-based contrast agents. *Biomedical Microdevices, 26*(4), 1–18. PMID: 39441423

Haikal, C., & Weissert, R. (2019). Editorial for Research Topic: Aging, Peripheral Inflammation, and Neurodegeneration. *Frontiers in Aging Neuroscience, 16*, 1529026. PMID: 39720419

Haleem, A., Javaid, M., Singh, R. P., Rab, S., & Suman, R. (2023). Applications of nanotechnology in medical field: A brief review. *Global Health Journal (Amsterdam, Netherlands), 7*(2), 70–77.

Hanes, J., Dobakova, E., & Majerova, P. (2020). Brain drug delivery: Overcoming the blood-brain barrier to treat tauopathies. *Current Pharmaceutical Design, 26*(13), 1448–1465. PMID: 32178609

Hardy, J., & Selkoe, D. J. (2002). The amyloid hypothesis of Alzheimer's disease: Progress and problems on the road to therapeutics. *Science, 297*(5580), 353–356. PMID: 12130773

Hashemi, S. A., Arjmand, M., Lankarani, K. B., Chiang, W.-H., & Ramakrishna, S. (2023). Editorial: Advanced nanotechnological detection and drug delivery configurations [Editorial]. *Frontiers in Pharmacology, 14*. Advance online publication. DOI: 10.3389/fphar.2023.1198100 PMID: 37124201

Hawkins, B. T., & Davis, T. P. (2005). The blood-brain barrier/neurovascular unit in health and disease. *Pharmacological Reviews, 57*(2), 173–185. DOI: 10.1124/pr.57.2.4 PMID: 15914466

He, J., Qiao, L., Li, J., Lu, J., Fu, Z., Chen, J., Zhang, X., & Hu, X. (2024). Advanced strategies for 3D-printed neural scaffolds: Materials, structure, and nerve remodeling. *Bio-Design and Manufacturing, 7*(5), 747–770.

Heneka, M. T., Golenbock, D. T., & Latz, E. (2015). Innate immunity in Alzheimer's disease. *Nature Immunology, 16*(3), 229–236. PMID: 25689443

Henna, T. K., & Pramod, K. (2020). Graphene quantum dots redefine nanobiomedicine. *Materials Science and Engineering C, 110*, 110651. DOI: 10.1016/j.msec.2020.110651 PMID: 32204078

Hersh, A. M., Alomari, S., & Tyler, B. M. (2022). Crossing the blood-brain barrier: Advances in nanoparticle technology for drug delivery in neuro-oncology. *International Journal of Molecular Sciences, 23*(8), 4153. PMID: 35456971

Hofmann-Amtenbrink, M., Hofmann, H., Hool, A., & Roubert, F. (2014). Nanotechnology in medicine: European research and its implications. *Swiss Medical Weekly, 144*, w14044. DOI: 10.4414/smw.2014.14044 PMID: 25473809

Hubbers, J. (2022). *A health economic evaluation of neurological interventions targeting epilepsy, migraine, dementia and Parkinson's disease in Ethiopia, Malawi, and the United Republic of Tanzania* The University of Bergen].

Jagaran, K., & Singh, M. (2021). Nanomedicine for Neurodegenerative Disorders: Focus on Alzheimer's and Parkinson's Diseases. *International Journal of Molecular Sciences, 22*(16). Advance online publication. DOI: 10.3390/ijms22169082 PMID: 34445784

Jeon, H., Kim, M.-S., Park, S. B., Kim, S., Lee, M., Park, S.-A., Hwang, S. Y., Koo, J. M., Oh, D. X., & Park, J. (2023). Improved mechanical properties of biodegradable polycaprolactone nanocomposites prepared using cellulose nanocrystals. *Cellulose (London, England), 30*(18), 11561–11574.

Ji, X., Lu, W., Wu, K., & Cho, W. C. (2017). Influencing Factors of the Pharmacokinetic Characters on Nanopharmaceutics. *Pharmaceutical Nanotechnology, 5*(1), 24–31. DOI: 10.2174/2211738505666161214142755 PMID: 28948908

Kakkar, A. K., & Dahiya, N. (2015). Management of Parkinson s disease: Current and future pharmacotherapy. *European Journal of Pharmacology, 750*, 74–81. DOI: 10.1016/j.ejphar.2015.01.030 PMID: 25637088

Karimi, M., Bahrami, S., Ravari, S. B., Zangabad, P. S., Mirshekari, H., Bozorgomid, M., Shahreza, S., Sori, M., & Hamblin, M. R. (2016). Albumin nanostructures as advanced drug delivery systems. *Expert Opinion on Drug Delivery, 13*(11), 1609–1623. DOI: 10.1080/17425247.2016.1193149 PMID: 27216915

Karimzadeh, M., Rashidi, L., & Ganji, F. (2017). Mesoporous silica nanoparticles for efficient rivastigmine hydrogen tartrate delivery into SY5Y cells. *Drug Development and Industrial Pharmacy*, *43*(4), 628–636. PMID: 28043167

Kashyap, K., & Shukla, R. (2019). Drug Delivery and Targeting to the Brain Through Nasal Route: Mechanisms, Applications and Challenges. *Current Drug Delivery*, *16*(10), 887–901. DOI: 10.2174/1567201816666191029122740 PMID: 31660815

Kathiresan, D. S., Balasubramani, R., Marudhachalam, K., Jaiswal, P., Ramesh, N., Sureshbabu, S. G., Puthamohan, V. M., & Vijayan, M. (2024). Role of Mitochondrial Dysfunctions in Neurodegenerative Disorders: Advances in Mitochondrial Biology. *Molecular Neurobiology*, •••, 1–29. PMID: 39269547

Kempuraj, D., Dourvetakis, K. D., Cohen, J., Valladares, D. S., Joshi, R. S., Kothuru, S. P., Anderson, T., Chinnappan, B., Cheema, A. K., & Klimas, N. G. (2024). Neurovascular unit, neuroinflammation and neurodegeneration markers in brain disorders. *Frontiers in Cellular Neuroscience*, *18*, 1491952. PMID: 39526043

Kim, D., Kim, S., & Na, D. H. (2024). Dendrimer nanoplatforms for oral drug delivery applications. *Journal of Pharmaceutical Investigation*, •••, 1–21.

Koeglsperger, T., Rumpf, S.-L., Schließer, P., Struebing, F. L., Brendel, M., Levin, J., Trenkwalder, C., Höglinger, G. U., & Herms, J. (2023). Neuropathology of incidental Lewy body & prodromal Parkinson's disease. *Molecular Neurodegeneration*, *18*(1), 32. PMID: 37173733

Konnova, S., & Rozhina, E. (2024). *Magnetic Nanoparticles for Biomedical and Imaging Applications* (Vol. 25). MDPI.

Korte, N., Nortley, R., & Attwell, D. (2020). Cerebral blood flow decrease as an early pathological mechanism in Alzheimer's disease. *Acta Neuropathologica*, *140*(6), 793–810. PMID: 32865691

Koszła, O., & Sołek, P. (2024). Misfolding and aggregation in neurodegenerative diseases: Protein quality control machinery as potential therapeutic clearance pathways. *Cell Communication and Signaling : CCS*, *22*(1), 421. PMID: 39215343

Kumar, R., Aadil, K. R., Ranjan, S., & Kumar, V. B. (2020). Advances in nanotechnology and nanomaterials based strategies for neural tissue engineering. *Journal of Drug Delivery Science and Technology*, *57*, 101617.

Kwon, H. S., & Koh, S. H. (2020). Neuroinflammation in neurodegenerative disorders: The roles of microglia and astrocytes. *Translational Neurodegeneration*, *9*(1), 42. DOI: 10.1186/s40035-020-00221-2 PMID: 33239064

Lanoiselée, H. M., Nicolas, G., Wallon, D., Rovelet-Lecrux, A., Lacour, M., Rousseau, S., Richard, A. C., Pasquier, F., Rollin-Sillaire, A., Martinaud, O., Quillard-Muraine, M., de la Sayette, V., Boutoleau-Bretonniere, C., Etcharry-Bouyx, F., Chauviré, V., Sarazin, M., le Ber, I., Epelbaum, S., Jonveaux, T., & Campion, D. (2017). APP, PSEN1, and PSEN2 mutations in early-onset Alzheimer disease: A genetic screening study of familial and sporadic cases. *PLoS Medicine*, *14*(3), e1002270. DOI: 10.1371/journal.pmed.1002270 PMID: 28350801

Lee, A. C. (2009). *Global Burden of Neurological Disorders*. Global Health Education Consortium.

Lehmann, S., Dumurgier, J., Ayrignac, X., Marelli, C., Alcolea, D., Ormaechea, J. F., Thouvenot, E., Delaby, C., Hirtz, C., & Vialaret, J. (2020). Cerebrospinal fluid A beta 1–40 peptides increase in Alzheimer's disease and are highly correlated with phospho-tau in control individuals. *Alzheimer's Research & Therapy*, *12*, 1–12. PMID: 33008460

Lin, M. T., & Beal, M. F. (2006). Mitochondrial dysfunction and oxidative stress in neurodegenerative diseases. *Nature*, *443*(7113), 787–795. PMID: 17051205

Liu, X., Hao, M., Chen, Z., Zhang, T., Huang, J., Dai, J., & Zhang, Z. (2021). 3D bioprinted neural tissue constructs for spinal cord injury repair. *Biomaterials*, *272*, 120771. DOI: 10.1016/j.biomaterials.2021.120771 PMID: 33798962

Lo, Y.-L., Chang, C.-H., Wang, C.-S., Yang, M.-H., Lin, A. M.-Y., Hong, C.-J., & Tseng, W.-H. (2020). PEG-coated nanoparticles detachable in acidic microenvironments for the tumor-directed delivery of chemo-and gene therapies for head and neck cancer. *Theranostics*, *10*(15), 6695. PMID: 32550898

Logothetidis, S. (2011). Nanotechnology: Principles and applications. In *Nanostructured materials and their applications* (pp. 1-22). Springer.

Loo, C., Lin, A., Hirsch, L., Lee, M.-H., Barton, J., Halas, N., West, J., & Drezek, R. (2004). Nanoshell-enabled photonics-based imaging and therapy of cancer. *Technology in Cancer Research & Treatment*, *3*(1), 33–40. PMID: 14750891

Lu, W., & Wen, J. (2024). Crosstalk Among Glial Cells in the Blood–Brain Barrier Injury After Ischemic Stroke. *Molecular Neurobiology*, •••, 1–14. PMID: 38279077

Lv, W., Xu, J., Wang, X., Li, X., Xu, Q., & Xin, H. (2018). Bioengineered boronic ester modified dextran polymer nanoparticles as reactive oxygen species responsive nanocarrier for ischemic stroke treatment. *ACS Nano*, *12*(6), 5417–5426. PMID: 29869497

Meyers, J. D., Doane, T., Burda, C., & Basilion, J. P. (2013). Nanoparticles for imaging and treating brain cancer. *Nanomedicine (London)*, *8*(1), 123–143. DOI: 10.2217/nnm.12.185 PMID: 23256496

Minchin, R. F., & Martin, D. J. (2010). Minireview: Nanoparticles for molecular imaging—an overview. *Endocrinology*, *151*(2), 474–481. PMID: 20016027

Mir, M., Ishtiaq, S., Rabia, S., Khatoon, M., Zeb, A., & Khan, G. M., ur Rehman, A., & ud Din, F. (2017). Nanotechnology: From in vivo imaging system to controlled drug delivery. *Nanoscale Research Letters*, *12*, 1–16. PMID: 28819800

Mollenhauer, B., Trautmann, E., Otte, B., Ng, J., Spreer, A., Lange, P., Sixel-Döring, F., Hakimi, M., VonSattel, J.-P., & Nussbaum, R. (2012). α-Synuclein in human cerebrospinal fluid is principally derived from neurons of the central nervous system. *Journal of Neural Transmission (Vienna, Austria)*, *119*, 739–746. PMID: 22426833

Mughal, K. A., Khan, A., Sarwar, A., Mahmood, R., & Abbasi, A. A. (2024). Comparative protein structural analyses of α-synuclein linked pathogenic variants reveal the role of N-terminally located critical region in Parkinson's disease pathogenesis. *Biologia*, •••, 1–14.

Nakamoto, Y., Inui, Y., Hotta, M., Wakabayashi, H., & Hanaoka, H. (2024). Recent advancements in new tracers from first-in-human studies. *Annals of Nuclear Medicine*, •••, 1–7. PMID: 39325320

Nayab, D. E., & Din, F. U., Ali, H., Kausar, W. A., Urooj, S., Zafar, M., Khan, I., Shabbir, K., & Khan, G. M. (2023). Nano biomaterials based strategies for enhanced brain targeting in the treatment of neurodegenerative diseases: An up-to-date perspective. *Journal of Nanobiotechnology*, *21*(1), 477. PMID: 38087359

Ng, S. Y., & Lee, A. Y. W. (2019). Traumatic brain injuries: Pathophysiology and potential therapeutic targets. *Frontiers in Cellular Neuroscience*, *13*, 528. PMID: 31827423

Orive, G., Anitua, E., Pedraz, J. L., & Emerich, D. F. (2009). Biomaterials for promoting brain protection, repair and regeneration. *Nature Reviews. Neuroscience*, *10*(9), 682–692. DOI: 10.1038/nrn2685 PMID: 19654582

Ossenkoppele, R., van der Kant, R., & Hansson, O. (2022). Tau biomarkers in Alzheimer's disease: Towards implementation in clinical practice and trials. *Lancet Neurology*, *21*(8), 726–734. PMID: 35643092

Padmanabhan, P., Kumar, A., Kumar, S., Chaudhary, R. K., & Gulyás, B. (2016). Nanoparticles in practice for molecular-imaging applications: An overview. *Acta Biomaterialia*, *41*, 1–16. PMID: 27265153

Palanisamy, C. P., Pei, J., Alugoju, P., Anthikapalli, N. V. A., Jayaraman, S., Veeraraghavan, V. P., Gopathy, S., Roy, J. R., Janaki, C. S., & Thalamati, D. (2023). New strategies of neurodegenerative disease treatment with extracellular vesicles (EVs) derived from mesenchymal stem cells (MSCs). *Theranostics*, *13*(12), 4138. PMID: 37554286

Pathak, S., Cao, E., Davidson, M. C., Jin, S., & Silva, G. A. Quantum dot applications to neuroscience: New tools for probing neurons and glia-Supplementary Information. *Journal of Neuroscience*.

Patil-Sen, Y. (2021). Advances in nano-biomaterials and their applications in biomedicine. *Emerging Topics in Life Sciences*, *5*(1), 169–176. PMID: 33825835

Peng, L., Liang, Y., Zhong, X., Liang, Z., Tian, Y., Li, S., Liang, J., Wang, R., Zhong, Y., & Shi, Y. (2020). Aptamer-conjugated gold nanoparticles targeting epidermal growth factor receptor variant III for the treatment of glioblastoma. *International Journal of Nanomedicine*, •••, 1363–1372. PMID: 32184591

Posadas, I., Monteagudo, S., & Ceña, V. (2016). Nanoparticles for brain-specific drug and genetic material delivery, imaging and diagnosis. *Nanomedicine (London)*, *11*(7), 833–849. PMID: 26980585

Prajapati, S. K., Malaiya, A., Kesharwani, P., Soni, D., & Jain, A. (2022). Biomedical applications and toxicities of carbon nanotubes. *Drug and Chemical Toxicology*, *45*(1), 435–450. DOI: 10.1080/01480545.2019.1709492 PMID: 31908176

Provenzale, J. M., & Silva, G. A. (2009). Uses of Nanoparticles for Central Nervous System Imaging and Therapy. *AJNR. American Journal of Neuroradiology*. Advance online publication. DOI: 10.3174/ajnr.A1590 PMID: 19617446

Qian, Y., Lin, H., Yan, Z., Shi, J., & Fan, C. (2021). Functional nanomaterials in peripheral nerve regeneration: Scaffold design, chemical principles and microenvironmental remodeling. *Materials Today*, *51*, 165–187.

Qin, N., Han, H., Guan, G., & Han, M.-Y. (2024). Structurally altered size, composition, shape and interface-dependent optical properties of quantized nanomaterials. *Nano Research*, •••, 1–27.

Rami, M. R., Meskini, M., Qarebaghi, L. M., Salami, M., Forouzandehdel, S., & Cheraghali, M. (2022). Synthesis of magnetic bio-nanocomposites for drug release and adsorption applications. *South African Journal of Chemical Engineering*, *42*, 115–126. https://doi.org/https://doi.org/10.1016/j.sajce.2022.08.003

Ransohoff, R. M., & Brown, M. A. (2012). Innate immunity in the central nervous system. *The Journal of Clinical Investigation*, *122*(4), 1164–1171. PMID: 22466658

Re, F., Moresco, R., & Masserini, M. (2012). Nanoparticles for neuroimaging. *Journal of Physics. D, Applied Physics*, *45*(7), 073001.

Rezghi Rami, M., Meskini, M., & Ebadi Sharafabad, B. (2024). Fungal-mediated nanoparticles for industrial applications: Synthesis and mechanism of action. *Journal of Infection and Public Health*, *17*(10), 102536. https://doi.org/https://doi.org/10.1016/j.jiph.2024.102536 PMID: 39276432

Roney, C., Kulkarni, P., Arora, V., Antich, P., Bonte, F., Wu, A., Mallikarjuana, N. N., Manohar, S., Liang, H. F., Kulkarni, A. R., Sung, H. W., Sairam, M., & Aminabhavi, T. M. (2005). Targeted nanoparticles for drug delivery through the blood-brain barrier for Alzheimer's disease. *Journal of Controlled Release*, *108*(2-3), 193–214. DOI: 10.1016/j.jconrel.2005.07.024 PMID: 16246446

Roveta, F., Bonino, L., Piella, E. M., Rainero, I., & Rubino, E. (2024). Neuroinflammatory Biomarkers in Alzheimer's Disease: From Pathophysiology to Clinical Implications. *International Journal of Molecular Sciences*, *25*(22), 11941. PMID: 39596011

Ryabchikova, E. (2021). *Advances in nanomaterials in biomedicine* (Vol. 11). MDPI.

Salmani-Zarchi, H., Mousavi-Sagharchi, S. M. A., Sepahdoost, N., Ranjbar-Jamalabadi, M., Gross, J. D., Jooya, H., & Samadi, A. (2024). Antimicrobial Feature of Nanoparticles in the Antibiotic Resistance Era: From Mechanism to Application. *Advanced Biomedical Research*, *13*(1). https://journals.lww.com/adbm/fulltext/2024/11300/antimicrobial_feature_of_nanoparticles_in_the.113.aspx PMID: 39717242

Sang, S., Cheng, R., Cao, Y., Yan, Y., Shen, Z., Zhao, Y., & Han, Y. (2022). Biocompatible chitosan/polyethylene glycol/multi-walled carbon nanotube composite scaffolds for neural tissue engineering. *Journal of Zhejiang University. Science. B.*, *23*(1), 58–73. PMID: 35029088

Santos-Lobato, B. L., & Schuh, A. F. S. (2024). Exposure to household pesticides and Parkinson's disease in the Parkinson's Progression Markers Initiative cohort. *Frontiers in Neurology*, *15*, 1411468. PMID: 39188706

Selkoe, D. J. (2001). Alzheimer's disease: Genes, proteins, and therapy. *Physiological Reviews*. PMID: 11274343

Senjab, R. M., AlSawaftah, N., Abuwatfa, W. H., & Husseini, G. A. (2024). Advances in liposomal nanotechnology: From concept to clinics. *RSC Pharmaceutics*.

Shabani, L., Abbasi, M., Azarnew, Z., Amani, A. M., & Vaez, A. (2023). Neuro-nanotechnology: Diagnostic and therapeutic nano-based strategies in applied neuroscience. *Biomedical Engineering Online*, *22*(1), 1. DOI: 10.1186/s12938-022-01062-y PMID: 36593487

Shakeri, S., Ashrafizadeh, M., Zarrabi, A., Roghanian, R., Afshar, E. G., Pardakhty, A., Mohammadinejad, R., Kumar, A., & Thakur, V. K. (2020). Multifunctional Polymeric Nanoplatforms for Brain Diseases Diagnosis, Therapy and Theranostics. *Biomedicines, 8*(1). Advance online publication. DOI: 10.3390/biomedicines8010013 PMID: 31941057

Shakeri, S., Ashrafizadeh, M., Zarrabi, A., Roghanian, R., Afshar, E. G., Pardakhty, A., Mohammadinejad, R., Kumar, A., & Thakur, V. K. (2020). Multifunctional polymeric nanoplatforms for brain diseases diagnosis, therapy and theranostics. *Biomedicines, 8*(1), 13. PMID: 31941057

Sheykhhasan, M., Amini, R., Soleimani Asl, S., Saidijam, M., Hashemi, S. M., & Najafi, R. (2022). Neuroprotective effects of coenzyme Q10-loaded exosomes obtained from adipose-derived stem cells in a rat model of Alzheimer's disease. *Biomedicine and Pharmacotherapy, 152*, 113224. https://doi.org/https://doi.org/10.1016/j.biopha.2022.113224 PMID: 35679720

Shi, J., Zhou, Z., Du, X., Cavagnaro, M. J., & Cai, J. (2024). New insights and perspectives on traumatic brain injury: integration, translation and multidisciplinary approaches. In (Vol. 15, pp. 1427320): Frontiers Media SA.

Siddiqi, K. S., Husen, A., Sohrab, S. S., & Yassin, M. O. (2018). Recent Status of Nanomaterial Fabrication and Their Potential Applications in Neurological Disease Management. *Nanoscale Research Letters, 13*(1), 231. DOI: 10.1186/s11671-018-2638-7 PMID: 30097809

Sidoryk-Węgrzynowicz, M., Adamiak, K., & Strużyńska, L. (2024). Targeting Protein Misfolding and Aggregation as a Therapeutic Perspective in Neurodegenerative Disorders. *International Journal of Molecular Sciences, 25*(22), 12448. PMID: 39596513

Sierri, G., Saenz-de-Santa-Maria, I., Renda, A., Koch, M., Sommi, P., Anselmi-Tamburini, U., Mauri, M., D'Aloia, A., Ceriani, M., & Salerno, D. (2025). Nanoparticle shape is the game-changer for blood–brain barrier crossing and delivery through tunneling nanotubes among glioblastoma cells. *Nanoscale*. PMID: 39588728

Silva Adaya, D., Aguirre-Cruz, L., Guevara, J., & Ortiz-Islas, E. (2017). Nanobiomaterials' applications in neurodegenerative diseases. *Journal of Biomaterials Applications, 31*(7), 953–984. DOI: 10.1177/0885328216659032 PMID: 28178902

Sohrabi, H. R., Greenberg, S. M., & Whiley, L. (2024). Cerebral amyloid angiopathy: From bench to bedside. *Frontiers in Neuroscience, 18*, 1370352. PMID: 38379758

Solangi, N. H., Karri, R. R., Mubarak, N. M., & Mazari, S. A. (2023). *Mechanism of polymer composite-based nanomaterial for biomedical applications*. Advanced Industrial and Engineering Polymer Research.

Song, X., Ding, Q., Wei, W., Zhang, J., Sun, R., Yin, L., Liu, S., & Pu, Y. (2023). Peptide-Functionalized Prussian Blue Nanomaterial for Antioxidant Stress and NIR Photothermal Therapy against Alzheimer's Disease. *Small, 19*(41), 2206959. PMID: 37322406

Soni, S., Ruhela, R. K., & Medhi, B. (2016). Nanomedicine in Central Nervous System (CNS) Disorders: A Present and Future Prospective. *Advanced Pharmaceutical Bulletin, 6*(3), 319–335. DOI: 10.15171/apb.2016.044 PMID: 27766216

Souri, M., Soltani, M., Kashkooli, F. M., & Shahvandi, M. K. (2022). Engineered strategies to enhance tumor penetration of drug-loaded nanoparticles. *Journal of Controlled Release*, *341*, 227–246. PMID: 34822909

Spillantini, M. G., & Goedert, M. (2000). The α-synucleinopathies: Parkinson's disease, dementia with Lewy bodies, and multiple system atrophy. *Annals of the New York Academy of Sciences*, *920*(1), 16–27. PMID: 11193145

Spuch, C., & Navarro, C. (2011). Liposomes for Targeted Delivery of Active Agents against Neurodegenerative Diseases (Alzheimer's Disease and Parkinson's Disease). *Journal of Drug Delivery*, *469679*. Advance online publication. DOI: 10.1155/2011/469679 PMID: 22203906

Srinageshwar, B., Peruzzaro, S., Andrews, M., Johnson, K., Hietpas, A., Clark, B., McGuire, C., Petersen, E., Kippe, J., Stewart, A., Lossia, O., Al-Gharaibeh, A., Antcliff, A., Culver, R., Swanson, D., Dunbar, G., Sharma, A., & Rossignol, J. (2017). PAMAM Dendrimers Cross the Blood-Brain Barrier When Administered through the Carotid Artery in C57BL/6J Mice. *International Journal of Molecular Sciences*, *18*(3). Advance online publication. DOI: 10.3390/ijms18030628 PMID: 28335421

Steinmetz, J. D., Seeher, K. M., Schiess, N., Nichols, E., Cao, B., Servili, C., Cavallera, V., Cousin, E., Hagins, H., & Moberg, M. E. (2024). Global, regional, and national burden of disorders affecting the nervous system, 1990–2021: A systematic analysis for the Global Burden of Disease Study 2021. *Lancet Neurology*, *23*(4), 344–381. PMID: 38493795

Suk, J. S., Xu, Q., Kim, N., Hanes, J., & Ensign, L. M. (2016). PEGylation as a strategy for improving nanoparticle-based drug and gene delivery. *Adv Drug Deliv Rev, 99*(Pt A), 28-51. https://doi.org/DOI: 10.1016/j.addr.2015.09.012

Sulheim, E., Baghirov, H., von Haartman, E., Bøe, A., Åslund, A. K., Mørch, Y., & Davies Cde, L. (2016). Cellular uptake and intracellular degradation of poly(alkyl cyanoacrylate) nanoparticles. *Journal of Nanobiotechnology*, *14*, 1. DOI: 10.1186/s12951-015-0156-7 PMID: 26743777

Teleanu, D. M., Chircov, C., Grumezescu, A. M., Volceanov, A., & Teleanu, R. I. (2018). Impact of nanoparticles on brain health: An up to date overview. *Journal of Clinical Medicine*, *7*(12), 490. PMID: 30486404

Teleanu, D. M., Negut, I., Grumezescu, V., Grumezescu, A. M., & Teleanu, R. I. (2019). Nanomaterials for Drug Delivery to the Central Nervous System. *Nanomaterials (Basel, Switzerland)*, *9*(3). Advance online publication. DOI: 10.3390/nano9030371 PMID: 30841578

Toader, C., Dumitru, A. V., Eva, L., Serban, M., Covache-Busuioc, R.-A., & Ciurea, A. V. (2024). Nanoparticle Strategies for Treating CNS Disorders: A Comprehensive Review of Drug Delivery and Theranostic Applications. *International Journal of Molecular Sciences*, *25*(24), 13302. PMID: 39769066

Vashist, A., Manickam, P., Raymond, A. D., Arias, A. Y., Kolishetti, N., Vashist, A., Arias, E., & Nair, M. (2023). Recent Advances in Nanotherapeutics for Neurological Disorders. *ACS Applied Bio Materials*, *6*(7), 2614–2621. DOI: 10.1021/acsabm.3c00254 PMID: 37368486

Verma, J., Kumar, C., Sharma, M., & Saxena, S. (2024). Biotechnological advances in microbial synthesis of gold nanoparticles: Optimizations and applications. *3 Biotech, 14*(11), 263.

Vinod, C., & Jena, S. (2021). Nano-neurotheranostics: Impact of nanoparticles on neural dysfunctions and strategies to reduce toxicity for improved efficacy. *Frontiers in Pharmacology*, *12*, 612692. PMID: 33841144

Viscusi, E. R., & Viscusi, A. R. (2020). Blood–brain barrier: Mechanisms governing permeability and interaction with peripherally acting μ-opioid receptor antagonists. *Regional Anesthesia and Pain Medicine*, *45*(9), 688–695. PMID: 32723840

Wanat, K. (2020). Biological barriers, and the influence of protein binding on the passage of drugs across them. *Molecular Biology Reports*, *47*(4), 3221–3231. PMID: 32140957

Wang, L., Tang, S., Yu, Y., Lv, Y., Wang, A., Yan, X., Li, N., Sha, C., Sun, K., & Li, Y. (2021). Intranasal delivery of temozolomide-conjugated gold nanoparticles functionalized with anti-EphA3 for glioblastoma targeting. *Molecular Pharmaceutics*, *18*(3), 915–927. PMID: 33417456

Wang, X. J., Gao, Y. P., Lu, N. N., Li, W. S., Xu, J. F., Ying, X. Y., Wu, G., Liao, M. H., Tan, C., Shao, L. X., Lu, Y. M., Zhang, C., Fukunaga, K., Han, F., & Du, Y. Z. (2016). Endogenous Polysialic Acid Based Micelles for Calmodulin Antagonist Delivery against Vascular Dementia. *ACS Applied Materials & Interfaces*, *8*(51), 35045–35058. DOI: 10.1021/acsami.6b13052 PMID: 27750011

Wei, X., Chen, X., Ying, M., & Lu, W. (2014). Brain tumor-targeted drug delivery strategies. *Acta Pharmaceutica Sinica. B*, *4*(3), 193–201. PMID: 26579383

Xu, L., Zhang, H., & Wu, Y. (2014). Dendrimer advances for the central nervous system delivery of therapeutics. *ACS Chemical Neuroscience*, *5*(1), 2–13. DOI: 10.1021/cn400182z PMID: 24274162

Yang, C.-Y., Huang, W.-Y., Chen, L.-H., Liang, N.-W., Wang, H.-C., Lu, J., Wang, X., & Wang, T.-W. (2021). Neural tissue engineering: The influence of scaffold surface topography and extracellular matrix microenvironment. *Journal of Materials Chemistry. B*, *9*(3), 567–584. PMID: 33289776

Yang, D., Ying, J., Wang, X., Zhao, T., Yoon, S., Fang, Y., Zheng, Q., Liu, X., Yu, W., & Hua, F. (2021). Mitochondrial dynamics: A key role in neurodegeneration and a potential target for neurodegenerative disease. *Frontiers in Neuroscience*, *15*, 654785. PMID: 33912006

Yang, L., Conley, B. M., Cerqueira, S. R., Pongkulapa, T., Wang, S., Lee, J. K., & Lee, K. B. (2020). Effective Modulation of CNS Inhibitory Microenvironment using Bioinspired Hybrid-Nanoscaffold-Based Therapeutic Interventions. *Advanced Materials*, *32*(43), e2002578. DOI: 10.1002/adma.202002578 PMID: 32893402

Yang, Z. Z., Zhang, Y. Q., Wang, Z. Z., Wu, K., Lou, J. N., & Qi, X. R. (2013). Enhanced brain distribution and pharmacodynamics of rivastigmine by liposomes following intranasal administration. *International Journal of Pharmaceutics*, *452*(1-2), 344–354. DOI: 10.1016/j.ijpharm.2013.05.009 PMID: 23680731

Yoon, H. (2023). *Polymer Nanomaterials in Biomedicine* (Vol. 24). MDPI.

Yu, C., Morshed, R., Auffinger, B., Tobias, A., & Lesniak, M. (2014). Multifunctional nanoparticles for brain tumor diagnosis and therapy. *Advanced Drug Delivery Reviews*, *66*(1), 42–57. PMID: 24060923

Yu, C., Wang, T., Diao, H., Liu, N., Zhang, Y., Jiang, H., Zhao, P., Shan, Z., Sun, Z., & Wu, T. (2022). Photothermal-triggered structural change of nanofiber scaffold integrating with graded mineralization to promote tendon–bone healing. *Advanced Fiber Materials*, *4*(4), 908–922.

Zajdel, K., Janowska, J., Frontczak-Baniewicz, M., Sypecka, J., & Sikora, B. (2023). Upconverting nanoparticles as a new bio-imaging strategy—Investigating intracellular trafficking of endogenous processes in neural tissue. *International Journal of Molecular Sciences*, *24*(2), 1122. PMID: 36674638

Zhang, C., Yu, B., Lin, F., Samanta, S., Yu, H., Zhang, W., Jing, Y., Shang, C., Lin, D., & Si, K. (2023). Deep tissue super-resolution imaging with adaptive optical two-photon multifocal structured illumination microscopy. *PhotoniX*, *4*(1), 38.

Zhang, N. (2023). Promoting the bench-to-bedside translation of nanomedicines. *Medical Review (Berlin, Germany)*, *3*(1), 1–3. PMID: 37724109

Zhang, X., Guo, M., Guo, Q., Liu, N., Wang, Y., & Wu, T. (2023). Modulating axonal growth and neural stem cell migration with the use of uniaxially aligned nanofiber yarns welded with NGF-loaded microparticles. *Materials Today. Advances*, *17*, 100343.

Zhang, Y., Wang, Y., Wu, C., Zhan, L., Wang, A., Cheng, C., Zhao, J., Zhang, W., Chen, J., & Li, P. (2024). Drug–target interaction prediction by integrating heterogeneous information with mutual attention network. *BMC Bioinformatics*, *25*(1), 361. PMID: 39563226

Zhu, F. D., Hu, Y. J., Yu, L., Zhou, X. G., Wu, J. M., Tang, Y., Qin, D. L., Fan, Q. Z., & Wu, A. G. (2021). Nanoparticles: A Hope for the Treatment of Inflammation in CNS. *Frontiers in Pharmacology*, *12*, 683935. DOI: 10.3389/fphar.2021.683935 PMID: 34122112

Zielińska, A., Costa, B., Ferreira, M. V., Miguéis, D., Louros, J. M. S., Durazzo, A., Lucarini, M., Eder, P., Chaud, M. V., Morsink, M., Willemen, N., Severino, P., Santini, A., & Souto, E. B. (2020). Nanotoxicology and Nanosafety: Safety-By-Design and Testing at a Glance. *International Journal of Environmental Research and Public Health*, *17*(13). Advance online publication. DOI: 10.3390/ijerph17134657 PMID: 32605255

Chapter 5
Application of Responsive Intelligent Biomaterials in Regenerative Medicine

Wenqiang Xie
https://orcid.org/0000-0002-3888-8733
Southern Medical University, China

Yuan Su
Southern Medical University, China

Ruiqi Zhang
Southern Medical University, China

Sijia Li
Foshan University, China

Jia Ni
Southern Medical University, China

Longquan Shao
Southern Medical University, China

ABSTRACT

Responsive intelligent biomaterials, characterized by their ability to sense and respond to external or internal stimuli, have emerged as promising candidates in regenerative medicine (RM). RM aims to replace or regenerate damaged human cells, tissues, or organs to restore normal function. However, their mechanisms of action are not fully understood, especially at the molecular level. Studies on long-term stability, in vivo degradation, and tissue impact are lacking. Translation from research to clinical application faces challenges due to physiological and genetic differences between animal models and humans, and inadequate clinical data. This chapter summarizes the current applications of responsive intelligent biomaterials in RM, encompassing bone regeneration, cartilage regeneration, skin regeneration, nerve regeneration, and myocardial regeneration. This chapter aims to help overcome these hurdles and successfully apply responsive intelligent biomaterials in regenerative medicine.

1.INTRODUCTION

Trauma, tumors, and infectious diseases can lead to damage in various tissues, including bone, cartilage, skin, nerves, and myocardium (Li et al., 2022; Tariq et al., 2022; Kharaziha et al., 2021; Qi et al., 2021; Poongodi et al., 2021), significantly impacting the quality of human life. Restoring the morphological structure and function of these organs has become one of the major challenges faced by modern medicine.

DOI: 10.4018/979-8-3373-0055-9.ch005

Traditionally, reconstructive approaches such as surgical intervention or organ transplantation have been employed to repair damaged tissues; however, these methods are associated with high medical costs and limitations such as secondary injuries, chronic infections, and a shortage of donor organs, resulting in unpredictable treatment outcomes (Zhou et al., 2022; Schraffordt Koops & Hoekstra, 1994). In recent years, the steady development and optimization of biomaterials have shown promising prospects for regenerative medicine (RM) in addressing issues of tissue damage and functional impairment (Gao et al., 2015). RM encompasses cell therapy and tissue engineering, aiming to replace or regenerate human cells, tissues, or organs to restore or establish normal function (Mason & Dunnill, 2008). The utilization of biomaterials that closely mimic the hierarchical structure of natural tissues is a critical component in achieving regeneration, with responsive intelligent biomaterials being regarded as the most promising materials in the field of regenerative medicine (Narkar et al., 2022).

Responsive intelligent biomaterials encompass active, responsive, and autonomous biomaterials (Montoya et al., 2021). Active biomaterials are designed to provide a planned unidirectional effect on biological processes or the surrounding environment, without reacting to external changes (Santin & Phillips, 2012). In contrast, autonomous biomaterials not only deliver targeted and precise therapies upon appropriate stimulation but also possess the ability to actively sense, respond to, and adapt to specific signals, interacting with their environment (Montoya et al., 2021). Currently, autonomous smart biomaterials are still quite rare. In comparison, Responsive intelligent biomaterials have been extensively developed and researched due to their capability to initiate or execute specific biological functions by sensing external or internal stimuli (Morris et al., 2016). These smart materials can dynamically alter their properties in response to various stimuli such as electrical, magnetic, optical, reactive oxygen species (ROS), and pH, thereby optimizing cellular behavior and tissue regeneration processes. This adaptability opens up possibilities for more effective and personalized treatment strategies. However, these advancements have not been comprehensively reviewed.

This article summarizes the current applications of responsive intelligent biomaterials in regenerative medicine, including bone regeneration, cartilage regeneration,skin regeneration, nerve regeneration, and myocardial regeneration. It particularly emphasizes their significant roles in tissue engineering, drug delivery, and personalized medicine, while also looking ahead to future development directions.

2. APPLICATION

2.1 Bone Regeneration

The skeletal system not only provides mechanical support for the body but also plays a crucial role in hematopoiesis, the development of the immune system, mineral storage, and glucose metabolism, among other complex physiological functions (Emmanuelle et al., 2021). Clinically, various diseases such as tumors, severe infections, osteoporosis, osteonecrosis, and some congenital malformations can cause large bone defects (Wei et al., 2022). Bone regeneration typically involves three phases: inflammation, repair, and remodeling. Currently, there are numerous strategies to promote the process of bone regeneration, including the "gold standard" of autologous bone grafting, allografting, and the use of growth factors, bone-conductive scaffolds, and distraction osteogenesis (Figure 1) (Wei et al., 2022; Godoy-Gallardo et al., 2020; Chen & Lv, 2018; Amini et al., 2012). However, these methods often fall short of achieving ideal treatment outcomes due to factors such as high costs, the need for a second surgical site, and uncertain prognoses (Henkel et al., 2013). Consequently, bone regeneration continues to face substantial challenges in both clinical treatment and regenerative medicine research. The rapid advancement of responsive intelligent biomaterials presents an opportunity to address the bottleneck issues in bone regeneration therapies, as they can support and enhance endogenous bone regeneration at various stages of bone reconstruction (Duda et al., 2023).

Figure 1. Strategies for designing and manufacturing responsive intelligent biomaterials with therapeutic and regenerative functions

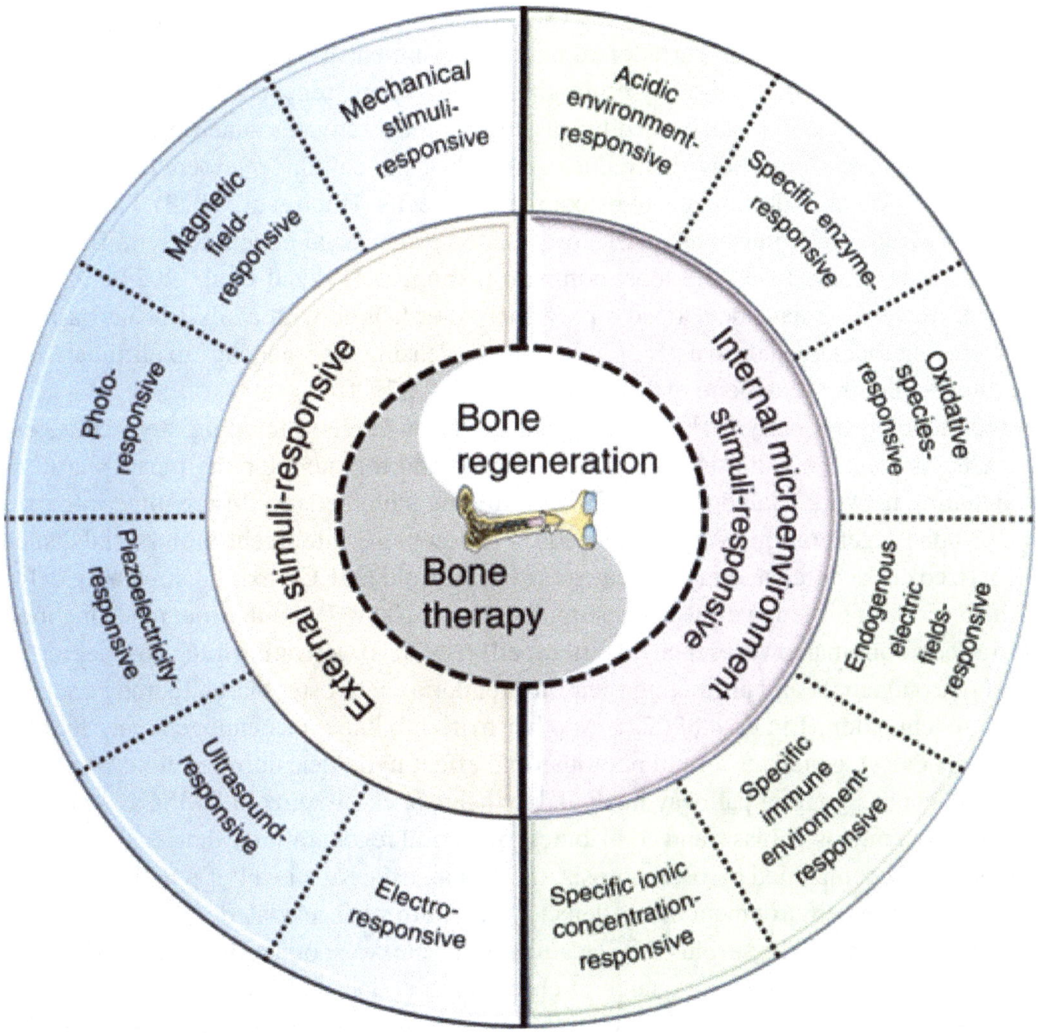

Reprinted in part with permission from Wei et al., 2022

In terms of system development, the process of bone regeneration begins with an inflammatory response. In the early stages of bone injury, cytokines with inflammatory and immune effects, such as IL-6 and TNF-α, recruit the cells necessary for tissue regeneration, thereby initiating the bone repair process (Gentile et al., 2015). However, if pro-inflammatory signals persist abnormally or are accompanied by bacterial infection, the regeneration process may be delayed or even halted (Yang et al., 2022). Currently, researchers have developed various responsive intelligent biomaterials for anti-inflammatory and antibacterial treatment at the site of bone injuries to promote bone regeneration. For instance, a composite hydrogel loaded with the anti-inflammatory drug resveratrol (RES) can respond to low-intensity focused ultrasound (LIFU), triggering the localized release of RES within 24 to 48 hours post-injury, suppressing the peak of the immune response, and blocking the progression of early inflammation, thereby facilitating cranial bone regeneration (Han et al., 2023). In the oral cavity, severe chronic periodontitis leads to alveolar bone resorption and tooth mobility. Additionally, due to the metabolic abnormalities of microbial communities and tissue cells, the periodontal area presents an acidic microenvironment with a pH of 4.5 to 6.5 (Wang et al., 2022). For example, a pH-responsive composite hydrogel (ASA@ tHALLG) releases aspirin in acidic periodontal microenvironments (pH 5.5) to suppress inflammatory cytokines (e.g., TNF-α and IL-6), thereby promoting alveolar bone regeneration (Chen et al., 2022). In cases of long bone injuries, hypoxia caused by vascular rupture activates anaerobic glycolysis, leading to a subsequent lactic acid overflow that reduces the pH of the cellular microenvironment to 4.5–6.5 (Wang et al., 2022), which is detrimental to osteoprogenitor cells (Duda et al., 2023). Furthermore, open wounds resulting from such injuries are often complicated by bacterial infections, with Escherichia coli and Staphylococcus aureus being the most common pathogens (Pakzad et al., 2024). In response to this condition, researchers have developed a polyether ether ketone with a silk fibroin-based CuO/Ag micro-nano coating that responds to a pH of 5.0, rapidly releasing Cu^{2+} and Ag^{+} to eliminate E. coli and S. aureus, thus facilitating long bone regeneration (Yan et al., 2020).

After experiencing the early inflammatory phase, various cells, including stem cells, osteoblast precursor cells, osteoblasts, and osteoclasts, are activated and initiate the bone repair phase through a complex signaling network (Salhotra et al., 2020). Guiding stem cells to differentiate into osteoblasts is crucial for bone regeneration (Wang et al., 2013). Responsive intelligent biomaterials, such as the novel magnetized collagen-coated magnetic responsive scaffold (MCC), exhibit stretching deformation of collagen fibers within the material when stimulated by a 100 mT annular magnet. This mechanical stimulation affects bone marrow mesenchymal stem cells (BMSCs), activating integrin receptor-mediated cytoskeletal rearrangement and promoting their differentiation into osteoblasts, thereby contributing to cranial bone regeneration (Lin et al., 2022). Another material, known as endogenously modified gold nanorods (eP@GNRs), generates a mild photothermal effect upon near-infrared light irradiation, activating the osteogenic signaling pathway mediated by heat shock proteins in BMSCs, which facilitates their differentiation into osteoblasts and contributes to femoral regeneration (Shan et al., 2022). In cases of femoral defects accompanied by osteoporosis, there is an excessive level of reactive oxygen species (ROS) in the local microenvironment of the defect. A titanium surface coated with gelatin/chitosan has been designed to respond to the surplus ROS, leading to the cleavage of amide bonds within the coating and the release of calcium carbonate-quercetin-chromium nanoparticles. Following their endocytosis by bone marrow-derived stem cells (BMSCs), these nanoparticles accelerate glucose metabolism to activate osteogenic signaling pathways, promoting differentiation into osteoblasts (Chen et al., 2022). In addition to stimulating the differentiation of stem cells, responsive intelligent biomaterials can also play a significant role during the differentiation phase of osteoblast precursor cells into osteoblasts. Research has reported a piezoelectric active scaffold (CaP@P@PUF), which can convert mechanical signals from body movements or exercise into electrical signals, thereby activating osteogenic signaling pathways mediated by calcium ion channels on the surface of osteoblast precursor cells, facilitating their differentiation into osteoblasts and aiding in long bone regeneration (Ma et al., 2023). Bone remodeling, as the final stage of bone regeneration, is regulated through a dynamic balance between osteoblasts and osteoclasts (Shakoor et al., 2022). A hydrogel composed of HA-BP can respond to local pH levels ranging from 2.0 to 5.0 in cranial defects, releasing a substantial amount of anti-osteoclast drug bisphosphonates to inhibit osteoclast activity and promote bone remodeling (Li et al., 2022).

It is well established that angiogenesis is integral to the entire process of new bone formation. The recruitment, migration, and proliferation of vascular endothelial cells can enhance the formation of new blood vessels, thereby supplying oxygen and nutrients essential for the repair and regeneration of damaged bone tissue (Li et al., 2020). A pH-responsive PCL/Col/ZIF-8 composite membrane can respond to the localized low pH in cranial defects by releasing Zn^{2+}, which upregulates vascular endothelial growth factor-A (VEGF-A) in osteoblasts (Xue et al., 2021). The latter can recruit endothelial cell precursors or stem cells from surrounding tissues and promote their differentiation into mature endothelial cells (Hu & Olsen, 2016), thereby accelerating neovascularization. Additionally, a thermoresponsive hydrogel incorporated with MgFe-LDH nanosheets can respond to body temperature at 37°C, releasing platelet-derived growth factor-BB (PDGF-BB) to promote endothelial cell migration and proliferation, thereby accelerating in situ angiogenesis and facilitating cranial regeneration (Lv et al., 2023).

As previously mentioned, the majority of current research focuses on how responsive intelligent biomaterials can facilitate the differentiation of stem cells into osteoblasts to promote bone regeneration. However, there is limited reporting on the recruitment of endogenous stem cells to the site of injury. It is understood that there is a deficiency of local stem cell content following bone tissue damage (Hauheur & Gangji, 2010), necessitating recruitment to the injury site through specific signaling molecules such as chemokines and cytokines (Alexeev et al., 2013). Ensuring an adequate supply of stem cells is crucial for achieving efficient bone regeneration (Augustine et al., 2024).Therefore, future research could explore the incorporation of signaling molecules with recruitment capabilities into responsive intelligent biomaterials for controlled release, aiming to recruit stem cells from sources such as bone marrow, peripheral blood, or adipose tissue to the site of injury, thereby enhancing bone regeneration.

2.2 Cartilage Regeneration

Articular cartilage covers and protects the ends of long bones at the joints, serving both load-bearing and lubricating functions (Piluso et al., 2019). Clinically, once articular cartilage damage occurs, conservative and surgical treatments typically provide only temporary improvements in joint function and pain relief, failing to effectively achieve structural and functional reconstruction of the cartilage (Shang et al., 2020). However, responsive intelligent biomaterials can modulate the proliferation and directed differentiation potential of stem cells, offering new and effective avenues for the repair and regeneration of articular cartilage.

For instance, polylactic acid nanofiber scaffolds can generate electrical signals in response to the mechanical stress of joint movement, inducing the migration, proliferation, and chondrogenic differentiation of adipose-derived stem cells (ADSCs) from surrounding adipose tissue, thereby promoting the regeneration of articular cartilage (Liu et al., 2022). This approach serves as a beneficial alternative to address the limited regenerative capacity due to insufficient endogenous chondrocytes. Certain conditions, such as cartilage damage caused by osteoarthritis and rheumatoid arthritis, are often accompanied by complex inflammatory responses, leading to a local microenvironment with a pH typically ranging from 5.0 to 6.5, along with excessive reactive oxygen species (ROS) (Peng et al., 2023). This inflammatory microenvironment is detrimental to the regeneration of chondrocytes or stem cells. Responsive intelligent biomaterials can modulate the inflammatory microenvironment, thereby creating favorable conditions for cartilage regeneration. For instance, nanoprobes loaded with anti-inflammatory drugs and MRC-PPL@PSO nanomicelles can respond to this pH range; the former achieves this by continuously releasing the anti-inflammatory drug hydroxychloroquine, which downregulates IL-1β expression (Chen et al., 2019), while the latter releases the anti-inflammatory herbal compound psoralen, leading to the downregulation of TNF-α, MMP-13, and IL-1β expression (Lan et al., 2020). This process improves the inflammatory microenvironment and promotes cartilage regeneration. Another type of poly(lactic-co-glycolic acid) scaffold responds to excessive ROS in the injury site by degrading via thioester bonds within the material, thereby consuming ROS and downregulating the expression of inflammatory factors such as TNF-α and IL-1β, thus creating space for chondrocyte migration to facilitate articular cartilage regeneration (Wu et al., 2021).

It is noteworthy that existing responsive intelligent biomaterials primarily focus on regulating the inflammatory microenvironment in the context of cartilage regeneration, with limited reports on how to mimic the extracellular matrix (ECM) to enhance cartilage regeneration. However, changes in the composition of the ECM at the injury site are critical for cartilage regeneration (Han et al., 2019). The ECM of natural tissues can self-regulate in response to environmental changes, influencing gene expression through alterations in its nanoscale topological structure and variations in components such as collagen, thereby modulating the proliferation and differentiation of chondrocytes to promote cartilage regeneration (Shao et al., 2017; Ramzan et al., 2022; Kim et al., 2013). Therefore, there remains significant research potential in developing responsive intelligent biomaterials that can simulate the dynamic changes of the ECM to precisely regulate the proliferation of chondrocytes or the directed differentiation potential of stem cells.

2.3 Skin Regeneration

The skin plays a crucial role not only in protecting the body, sensing the environment, regulating temperature, and providing immune defense, but it also has significant functions in endocrine and excretory processes (Peña & Martin, 2024). Clinically, extensive skin defects caused by burns, infections, and cancers cannot heal through normal repair mechanisms, leading to severe disabilities or even death in patients (Wu et al., 2022). Responsive intelligent biomaterials, recognized as one of the most promising biomaterials for skin tissue regeneration, can contribute to the inflammatory and proliferative phases of wound healing, offering a novel therapeutic strategy for skin wound recovery.

The inflammatory response represents an early stage in wound healing; however, the persistence of inflammation or the onset of bacterial infection can cause the healing process to stagnate at the inflammatory phase (Shakya et al., 2019). This phase is characterized by an excess of pro-inflammatory cytokines, insufficient growth factors, accumulation of reactive oxygen species (ROS) (Zhang et al., 2020), and a reduction in pH to 4.5-6.5 following bacterial infection (Albright et al., 2017). It is currently believed that this stagnation is primarily due to macrophages remaining in the M1 phenotype, continuously secreting pro-inflammatory factors (Akinrinmade et al., 2017). Researchers have explored various strategies to reverse this situation, discovering that smart responsive biomaterials can act as carriers that release anti-inflammatory drugs in response to the chronic inflammatory microenvironment, thereby exerting anti-inflammatory effects to promote wound healing. For instance, injectable glycopeptide hydrogels can respond to excessive ROS released from skin wounds by delivering the anti-inflammatory drug diclofenac sodium (Wu et al., 2022). Additionally, a nanocomposite hydrogel composed of catechol-functionalized chitosan and the anti-inflammatory drug curcumin responds to the local pH of 5.5-6.5 in damaged skin, releasing curcumin (Sun et al., 2023). Both approaches have been shown to downregulate the secretion of IL-6 and TNF-α from macrophages, alleviating local inflammation and facilitating skin wound healing. Furthermore, persistent inflammatory responses are often accompanied by bacterial infections, some researchers have found that responsive intelligent biomaterials can directly target damaged skin to inhibit bacterial activity, improve the infection microenvironment, and promote the healing of skin wounds. For instance, copper polyphenol nanoparticles respond to excessive ROS in the microenvironment by releasing antibacterial Cu^{2+}, thereby inhibiting the activity of E. coli and S. aureus, which promotes the healing of infected wounds (Li et al., 2023). Photonic nanosheets generate high-intensity electrical stimulation on their surface in response to external light stimuli, effectively eradicating S. aureus and facilitating the healing of infected wounds (Ren et al., 2022). Furthermore, responsive intelligent biomaterials can serve as carriers for the release of antibacterial agents, suppressing bacterial activity and enhancing skin wound healing. Polylactic acid-glycolic acid and chitosan-modified halloysite nanotube composites can respond to localized pH levels of 5.5-9 in chronic burn wounds, releasing the antibacterial drug minocycline to inhibit the activity of common pathogens such as S. aureus and Pseudomonas aeruginosa, thereby controlling wound infection and accelerating healing (Mohebali & Abdouss, 2020). Amphoteric ionic hydrogels undergo deformation in response to mechanical stress from body movements, releasing the antibacterial drug rifampicin, which inhibits the activity of S. aureus, Staphylococcus epidermidis, and E. coli, thus controlling wound infection and promoting skin wound healing (Fang et al., 2020).

The proliferative phase of skin wound healing encompasses granulation tissue formation and re-epithelialization (Peña & Martin, 2024). Granulation tissue begins to form four days post-injury, during which endothelial cells generate new blood vessels to supply the oxygen and nutrients necessary for cellular metabolism, while fibroblasts produce the ECM to support cell growth (Singer & Clark, 1999). Numerous studies have confirmed that responsive intelligent biomaterials can accelerate granulation tissue formation in response to near-infrared light stimulation, glucose stimulation, and electrical stimulation, thereby promoting skin wound healing. For instance, iron-doped carbon dot nanoenzymes increase in surface temperature to 41-45°C upon near-infrared irradiation, stimulating fibroblast migration, proliferation, and VEGF expression, which in turn accelerates granulation tissue formation (Liu et al., 2022). A glucose-responsive enzymatic biofuel cell patch generates electrical signals to stimulate endothelial cell migration, enhancing granulation tissue formation (Kim et al., 2021). This approach mimics endogenous bioelectrical signals that naturally guide wound healing. Once granulation tissue fills the wound gap, epithelial cells proliferate to form epithelial islands, which then migrate outward to complete the re-epithelialization of the wound. Researchers have developed a 3D biomimetic short fiber scaffold that responds to the localized endogenous electric field of skin wounds, transmitting bioelectrical signals that stimulate epithelial cell migration and proliferation, thereby accelerating the re-epithelialization process (Wang et al., 2022).

We found that responsive intelligent biomaterials appear to play a significant role in regulating the inflammatory and proliferative phases of skin wound healing. However, for the skin to restore its functionality and aesthetic appearance, it is essential to complete the remodeling phase, which involves the rearrangement of newly formed collagen fibers to create uniform and functional new tissue (Tottoli et al., 2020). Factors such as mechanical stress, infection, and genetic predispositions (Lin & Lai, 2024; Ogawa, 2017) often lead to excessive proliferation of fibroblasts and abnormal deposition of collagen, resulting in scar tissue formation that neither restores tissue function nor maintains aesthetic integrity, and may even pose a risk of carcinogenesis (Zhang et al., 2021). Therefore, a promising avenue for future exploration lies in leveraging the responsive characteristics of smart biomaterials to precisely inhibit excessive fibroblast proliferation during the remodeling phase, without adversely affecting fibroblast proliferation during the earlier stages, in order to prevent scar formation (Figure 2).

Figure 2. Strategies for applying responsive intelligent biomaterials to wound healing

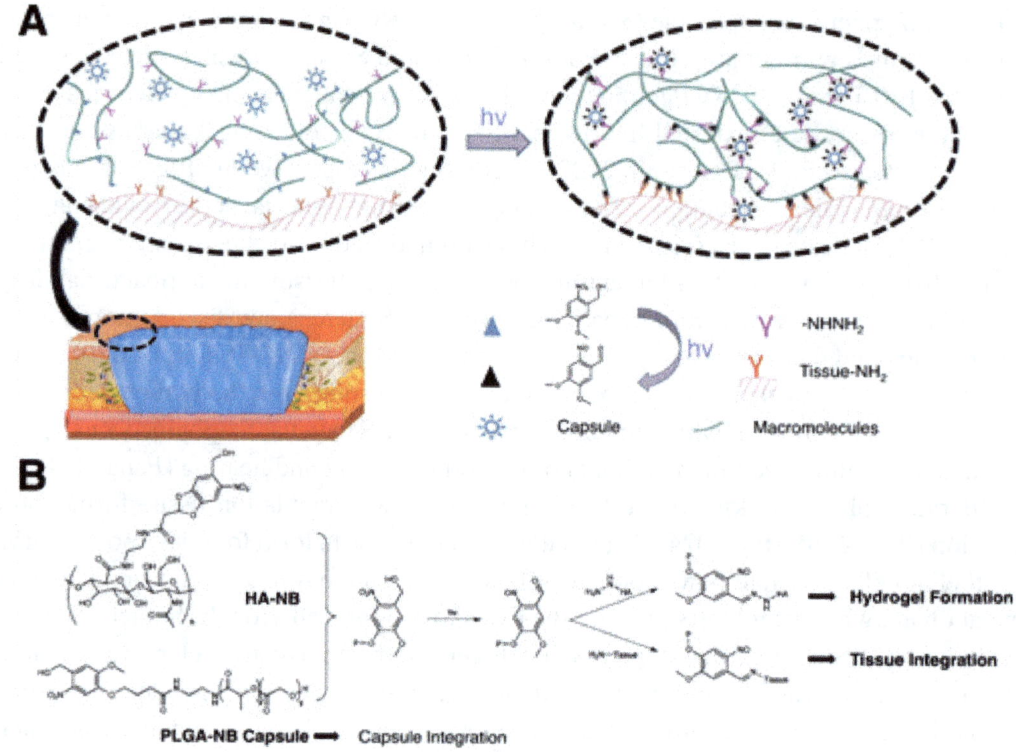

Reprinted in part with permission from Zhang et al., 2021

2.4 Nerve Regeneration

The central nervous system (CNS) and peripheral nervous system (PNS) work in tandem to regulate cognitive processes, behaviors, movements, and sensory functions (Greenshields, 2019). However, in clinical settings, neurological damage caused by tumors, trauma, and infections often results in irreversible injuries, frequently leading to paralysis and disability (Jacobi & Bareyre, 2015). Responsive intelligent biomaterials, due to their ability to sense changes in the injury microenvironment and respond accordingly, exhibit significant potential in inducing the proliferation, differentiation, and functional axon formation of neural stem cells. For instance, spinal cord injury (SCI), a prevalent form of CNS damage, triggers the activation of fibroblasts, immune cells, and astrocytes, which release a substantial amount of matrix metalloproteinases (MMPs) (Alizadeh et al., 2019). Conductive hydrogels containing biomacromolecules (GST-TIMP-bFGF) can create an endogenous electrical microenvironment while responding to the excess MMPs present in the spinal cord injury microenvironment. This interaction leads to the degradation of substrate peptide TIMP and the release of basic fibroblast growth factor (bFGF), thereby promoting the proliferation of neural stem cells (NSCs) and their differentiation into neuronal cells, ultimately facilitating neural regeneration in the injured spinal cord (Fan et al., 2022). Moreover, spinal cord injury (SCI) leads to reduced blood flow and insufficient oxygen supply in the tissues, resulting in the accumulation of metabolic byproducts that acidify the local environment (Foster et al., 2021). The dynamic Schiff base chemical bonds and covalent photopolymerized conductive hydrogels can respond to the local microenvironment of spinal cord injury with a pH of 6.0, causing the cleavage of Schiff base bonds and the release of the anti-inflammatory factor Wnt5a. This process alleviates the inflammatory response, promotes the migration and differentiation of neural stem cells (NSCs), accelerates the formation of functionally myelinated axons in neurons, and enhances the recovery of neural function (Figure 3) (Liu et al., 2024). Composite hydrogels loaded with magnetoelectric nanoparticles convert external magnetic field stimuli into electrical signals, thereby stimulating the elongation of neuronal dendrites and synapses, as well as the formation of functionally myelinated axons, which facilitates neural regeneration in the injured spinal cord and restores the structure and function of the central nervous system (CNS) (Zhang et al., 2021).

Figure 3. Strategies for applying responsive intelligent biomaterials to promote nerve regeneration by reducing the inflammatory response

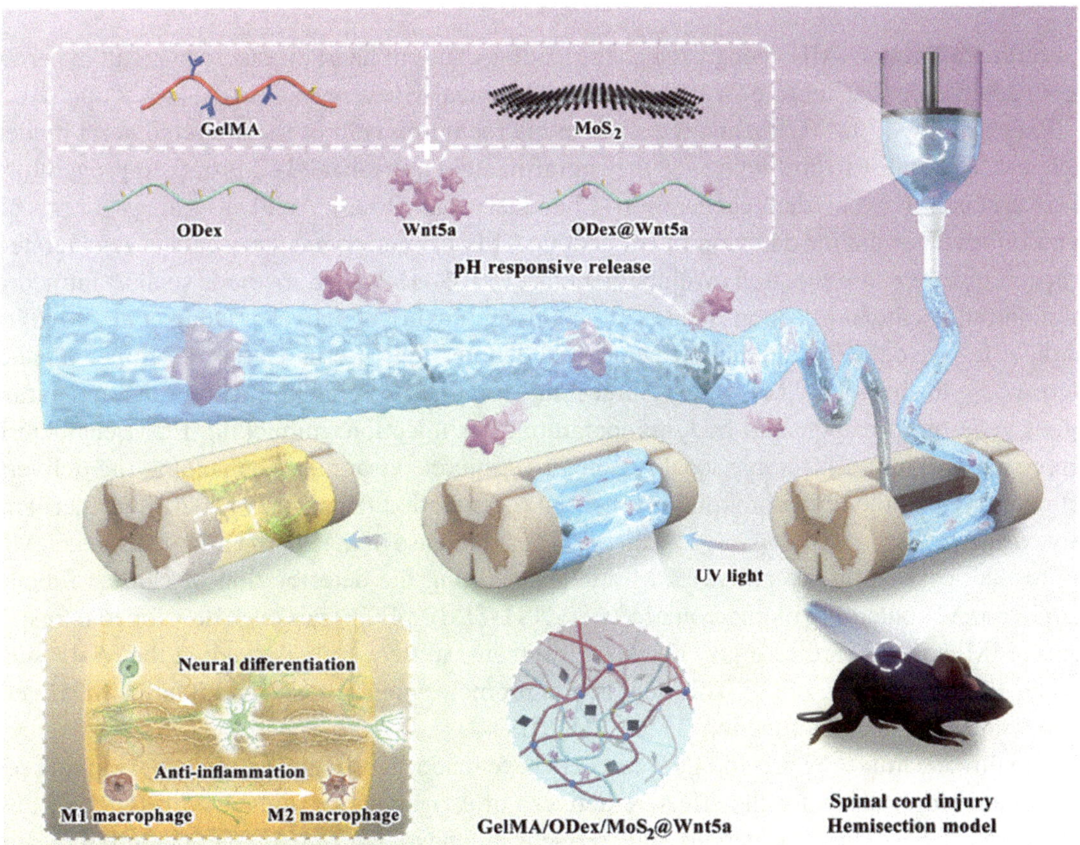

Reprinted in part with permission from Liu et al., 2024

Additionally, there are responsive intelligent biomaterials capable of restoring the structure and function of damaged peripheral nervous system (PNS) tissues. For instance, carbon nanotube/silk fibroin nerve conduits (Li et al., 2020) and conductive polypyrrole/silk fibroin composite scaffolds (Zhao et al., 2020) can induce Schwann cell proliferation at the site of sciatic nerve defects in response to external electrical stimulation, thereby promoting the regeneration of functional myelinated axons and facilitating sciatic nerve repair.

Given that both the CNS and PNS utilize highly conductive nerve fibers to transmit information in the form of electrical impulses (Citi et al., 2008), current responsive intelligent biomaterials are often designed with high electrical conductivity to maintain the endogenous electrical microenvironment. This promotes the differentiation of neural stem cells (NSCs) into neuronal cells and the formation of axons. Although the aforementioned materials have achieved high conductivity through the incorporation of conductive polymers such as polypyrrole or the use of carbon-based nanomaterials like carbon nanotubes, and have demonstrated potential for nerve regeneration in animal studies, the long-term maintenance of such high conductivity in vivo remains to be validated. This aspect is crucial for ensuring successful nerve regeneration and warrants further attention in future material development and in vivo research observations.

2.5 Myocardial Regeneration

Myocardial infarction (MI), recognized as the leading cause of heart disease, can result in permanent damage or death of cardiac muscle (Wei et al., 2022). A typical MI event leads to the loss of approximately one billion cardiomyocytes (CM), and the adverse microenvironment of the infarcted heart hinders the endogenous CM's ability to proliferate and the engraftment of transplanted CMs or their progenitor cells, which are crucial for myocardial regeneration (Chaudhari et al., 2023). An ideal strategy for myocardial regeneration involves the introduction of exogenous CMs or pluripotent stem cellsto regenerate myocardium and restore pump function (Weinberger et al., 2016). However, methods such as intracoronary injection, intramyocardial injection (Hou et al., 2005) and hydrogel delivery (Dong et al., 2016) do not ensure the effective delivery and engraftment of exogenous CMs. Research has developed a nanofiber gel microsphere loaded with CMs, which is injectable as a free-flowing liquid at room temperature and undergoes gelation in response to body temperature after injection, mimicking ECM characteristics. This approach provides a favorable local regenerative microenvironment, enhancing the delivery and engraftment of CMs, thereby promoting myocardial regeneration (Zhao et al., 2020). Furthermore, the extensive death of CMs is accompanied by the significant release of matrix metalloproteinases (MMPs), leading to the degradation of cardiac ECM and accelerating the deterioration of cardiac function. A collagen hydrogel containing biomacromolecules (GST-TIMP-bFGF) has been developed to respond to the excess MMPs in the cardiac injury microenvironment, subsequently degrading the MMP substrate peptide TIMP and releasing bFGF. This dual action promotes angiogenesis and inhibits MMP activity, ultimately improving cardiac function (Fan et al., 2019).

In the realm of cardiac regeneration, researchers have made significant strides in developing responsive intelligent biomaterials for the effective delivery of exogenous cardiomyocytes (CMs). However, the immunogenicity of these exogenous cells remains an unavoidable challenge, potentially leading to their recognition and elimination by the host immune system (Csöbönyeiová et al., 2022). Studies have demonstrated that bone marrow-derived mesenchymal stem cells (MSCs) and adipose tissue-derived stem cells (ADSCs) can differentiate into cardiomyocytes under the influence of transforming growth factor-beta 1(TGF-β1) (Barbuti, 2010; Makino et al., 1999), thereby enhancing cardiac function following MI (Huang et al., 2019). We may be able to leverage responsive intelligent biomaterials as carriers for signaling molecules, enabling localized controlled release at the site of injury to recruit endogenous MSCs and ADSCs, while also regulating their proliferation and differentiation into CMs to repair damaged myocardial tissue. This approach warrants further investigation in future studies.

3 CONCLUSION

In our in-depth exploration of the applications of responsive intelligent biomaterials in the field of regenerative medicine, we have observed that, despite the remarkable outcomes these materials have achieved in tissue engineering and drug delivery due to their unique responsive characteristics, the investigation into their mechanisms of action remains relatively insufficient. Specifically, there is a limited understanding of the molecular mechanisms by which they interact with the cellular microenvironment of damaged tissues or with cell surface receptors. This gap in knowledge may hinder our ability to fully harness the potential advantages of these materials, thereby restricting their further development and enhancement in clinical applications. A comprehensive study of the regulatory mechanisms of responsive intelligent biomaterials at the cellular and molecular levels, aimed at achieving more precise biomedical applications, is one of the key focuses for future research.

From the perspective of material improvement, current research predominantly focuses on enhancing the mechanical properties, biocompatibility, and responsiveness of smart biomaterials. However, studies addressing the long-term stability, in vivo degradation behavior, and the impact on surrounding tissues remain insufficient. Therefore, it is imperative to delve deeper into methods for improving material performance to ensure their safety and efficacy within the human body.

Moreover, despite significant advancements in responsive intelligent biomaterials within laboratory settings, the translation from fundamental research to clinical application continues to face numerous challenges. The pronounced physiological and genetic differences between animal models and humans contribute to a high failure rate in the translation from animal studies to human applications. Additionally, the majority of current research is still confined to animal model stages, lacking adequate clinical data to support further development. These issues collectively hinder the transition of responsive intelligent biomaterials from foundational research outcomes to clinical applications. Moving forward, sustained technological innovation and interdisciplinary collaboration are essential to establish effective translational platforms that will facilitate the successful clinical application of responsive intelligent biomaterials.

REFERENCES

Akinrinmade, O A, CHETTY S, DARAMOLA A K. *et al.* (2017). CD64: An Attractive Immunotherapeutic Target for M1-type Macrophage Mediated Chronic Inflammatory Diseases . *Biomedicines*, *5*(3), 56. PMID: 28895912

Albright, V., Zhuk, I., & Wang, Y.. (2017). Self-defensive antibiotic-loaded layer-by-layer coatings: Imaging of localized bacterial acidification and pH-triggering of antibiotic release . *Acta Biomaterialia*, *61*, 66–74. PMID: 28803214

Alexeev, V., Donahue, A., & Uitto, J.. (2013). Analysis of chemotactic molecules in bone marrow-derived mesenchymal stem cells and the skin: Ccl27-Ccr10 axis as a basis for targeting to cutaneous tissues . *Cytotherapy*, *15*(2), 171–184.e1. PMID: 23321329

Amini, A. R., Laurencin, C. T., & Nukavarapu, S. P. (2012). Bone tissue engineering: Recent advances and challenges . *Critical Reviews in Biomedical Engineering*, *40*(5), 363–408. PMID: 23339648

Augustine, R., Gezek, M., & Nikolopoulos, V. K.. (2024). Stem Cells in Bone Tissue Engineering: Progress, Promises and Challenges . *Stem Cell Reviews and Reports*, *20*(7), 1692–1731. PMID: 39028416

Barbuti, A. (2010). The "hearty" fat: Adipocytes as a source of functional cardiomyocytes . *Cardiovascular Research*, *85*(1), 1–2. PMID: 19887381

Chaudhari, U., Pohjolainen, L., & Ruskoaho, H.. (2023). Genome-wide profiling of miRNA-gene regulatory networks in mouse postnatal heart development-implications for cardiac regeneration . *Frontiers in Cardiovascular Medicine*, *10*, 1148618. PMID: 37283582

Chen, G., & Lv, Y. (2018). Decellularized Bone Matrix Scaffold for Bone Regeneration . *Methods in Molecular Biology (Clifton, N.J.)*, *1577*, 239–254. PMID: 28770492

Chen, H., Qin, Z., & Zhao, J.. (2019). Cartilage-targeting and dual MMP-13/pH responsive theranostic nanoprobes for osteoarthritis imaging and precision therapy . *Biomaterials*, *225*, 119520. PMID: 31586865

Chen, M., Li, M., & Wei, Y.. (2022). ROS-activatable biomimetic interface mediates in-situ bioenergetic remodeling of osteogenic cells for osteoporotic bone repair . *Biomaterials*, *291*, 121878. PMID: 36335716

Chen, P., Zhang, C., & He, P.. (2022). A Biomimetic Smart Nanoplatform as "Inflammation Scavenger" for Regenerative Therapy of Periodontal Tissue . *International Journal of Nanomedicine*, *17*, 5165–5186. PMID: 36388874

Citi, L., Carpaneto, J., & Yoshida, K.. (2008). On the use of wavelet denoising and spike sorting techniques to process electroneurographic signals recorded using intraneural electrodes . *Journal of Neuroscience Methods*, *172*(2), 294–302. PMID: 18534683

Csöbönyeiová, M., Beerová, N., & Klein, M.. (2022). Cell-Based and Selected Cell-Free Therapies for Myocardial Infarction: How Do They Compare to the Current Treatment Options? . *International Journal of Molecular Sciences*, *23*(18), 10314. PMID: 36142245

Dong, R., Zhao, X., & Guo, B.. (2016). Self-Healing Conductive Injectable Hydrogels with Antibacterial Activity as Cell Delivery Carrier for Cardiac Cell Therapy . *ACS Applied Materials & Interfaces*, *8*(27), 17138–17150. PMID: 27311127

Duda, G N, GEISSLER S, CHECA S. *et al.* (2023). The decisive early phase of bone regeneration . *Nature Reviews. Rheumatology*, *19*(2), 78–95. PMID: 36624263

Emmanuelle, N E, MARIE-CÉCILE V, FLORENCE T. *et al.* (2021). Critical Role of Estrogens on Bone Homeostasis in Both Male and Female: From Physiology to Medical Implications . *International Journal of Molecular Sciences*, *22*(4), 1568. PMID: 33557249

Fan, C., Shi, J., & Zhuang, Y.. (2019). Myocardial-Infarction-Responsive Smart Hydrogels Targeting Matrix Metalloproteinase for On-Demand Growth Factor Delivery . *Advanced Materials (Deerfield Beach, Fla.)*, *31*(40), e1902900. PMID: 31408234

Fan, C., Yang, W., & Zhang, L.. (2022). Restoration of spinal cord biophysical microenvironment for enhancing tissue repair by injury-responsive smart hydrogel . *Biomaterials*, *288*, 121689. PMID: 35931574

Fang, K., Wang, R., & Zhang, H.. (2020). Mechano-Responsive, Tough, and Antibacterial Zwitterionic Hydrogels with Controllable Drug Release for Wound Healing Applications . *ACS Applied Materials & Interfaces*, *12*(47), 52307–52318. PMID: 33183010

Foster, V. S., Rash, L. D., & King, G. F.. (2021). Acid-Sensing Ion Channels: Expression and Function in Resident and Infiltrating Immune Cells in the Central Nervous System . *Frontiers in Cellular Neuroscience*, *15*, 738043. PMID: 34602982

Gao, Y., Lim, J., & Teoh, S. H.. (2015). Emerging translational research on magnetic nanoparticles for regenerative medicine . *Chemical Society Reviews*, *44*(17), 6306–6329. PMID: 26505058

Gentile, P, & Nandagiri, V K, PABARI R. *et al.* (2015). Influence of Parathyroid Hormone-Loaded PLGA Nanoparticles in Porous Scaffolds for Bone Regeneration . *International Journal of Molecular Sciences*, *16*(9), 20492–20510. PMID: 26343649

Godoy-Gallardo, M., Portolés-Gil, N., & López-Periago, A. M.. (2020). Immobilization of BMP-2 and VEGF within Multilayered Polydopamine-Coated Scaffolds and the Resulting Osteogenic and Angiogenic Synergy of Co-Cultured Human Mesenchymal Stem Cells and Human Endothelial Progenitor Cells . *International Journal of Molecular Sciences*, *21*(17), 6418. PMID: 32899269

Greenshields, S. (2019). Neurological assessment in children and young people . *British Journal of Nursing (Mark Allen Publishing)*, *28*(16), 1056–1059. PMID: 31518531

Han, P., Frith, J. E., & Gomez, G. A.. (2019). Five Piconewtons: The Difference between Osteogenic and Adipogenic Fate Choice in Human Mesenchymal Stem Cells . *ACS Nano*, *13*(10), 11129–11143. PMID: 31580055

Han, X., Shen, J., & Chen, S.. (2023). Ultrasonic-controlled "explosive" hydrogels to precisely regulate spatiotemporal osteoimmune disturbance . *Biomaterials*, *295*, 122057. PMID: 36805244

Henkel, J., Woodruff, M. A., & Epari, D. R.. (2013). Bone Regeneration Based on Tissue Engineering Conceptions - A 21st Century Perspective . *Bone Research*, *1*(3), 216–248. PMID: 26273505

Hou, D., Youssef, E A S., & Brinton, T. J.. (2005). Radiolabeled cell distribution after intramyocardial, intracoronary, and interstitial retrograde coronary venous delivery: Implications for current clinical trials . *Circulation*, *112*(9, Suppl), I150–I156. PMID: 16159808

Hu, K., & Olsen, B. R. (2016). The roles of vascular endothelial growth factor in bone repair and regeneration . *Bone*, *91*, 30–38. PMID: 27353702

Huang, K., Hu, S., & Cheng, K. (2019). A New Era of Cardiac Cell Therapy: Opportunities and Challenges . *Advanced Healthcare Materials*, *8*(2), e1801011. PMID: 30548836

Jacobi, A., & Bareyre, F. M. (2015). Regulation of axonal remodeling following spinal cord injury . *Neural Regeneration Research*, *10*(10), 1555–1557. PMID: 26692837

Kharaziha, M., Baidya, A., & Annabi, N. (2021). Rational Design of Immunomodulatory Hydrogels for Chronic Wound Healing . *Advanced Materials (Deerfield Beach, Fla.)*, *33*(39), e2100176. PMID: 34251690

Kim, H N, Jiao A, Hwang N S. *et al.* (2013). Nanotopography-guided tissue engineering and regenerative medicine . *Advanced Drug Delivery Reviews*, *65*(4), 536–558. PMID: 22921841

Kim, T H, & Jeon, W Y, JI Y. *et al.* (2021). Electricity auto-generating skin patch promotes wound healing process by activation of mechanosensitive ion channels . *Biomaterials*, *275*, 120948. PMID: 34157562

Lan, Q., Lu, R., & Chen, H.. (2020). MMP-13 enzyme and pH responsive theranostic nanoplatform for osteoarthritis . *Journal of Nanobiotechnology*, *18*(1), 117. PMID: 32854712

Li, D., Li, J., & Wang, S.. (2023). Dually Crosslinked Copper-Poly (tannic acid) Nanoparticles with Microenvironment-Responsiveness for Infected Wound Treatment . *Advanced Healthcare Materials*, *12*(17), e2203063. PMID: 36842067

Li, M., Yin, H., & Yan, Z.. (2022). The immune microenvironment in cartilage injury and repair . *Acta Biomaterialia*, *140*, 23–42. PMID: 34896634

Li, X., Yang, W., & Xie, H.. (2020). CNT/Sericin Conductive Nerve Guidance Conduit Promotes Functional Recovery of Transected Peripheral Nerve Injury in a Rat Model . *ACS Applied Materials & Interfaces*, *12*(33), 36860–36872. PMID: 32649170

Li, Z., Wang, H., & Zhang, K.. (2022). Bisphosphonate-based hydrogel mediates biomimetic negative feedback regulation of osteoclastic activity to promote bone regeneration . *Bioactive Materials*, *13*, 9–22. PMID: 35224288

Li, Z., Zheng, J., & Wan, D.. (2020). Uniaxial Static Strain Promotes Osteoblast Proliferation and Bone Matrix Formation in Distraction Osteogenesis In Vitro . *BioMed Research International*, *2020*, 3906426. PMID: 32855965

Lin, S., Li, J., & Shao, J.. (2022). Anisotropic magneto-mechanical stimulation on collagen coatings to accelerate osteogenesis . *Colloids and Surfaces. B, Biointerfaces*, *210*, 112227. PMID: 34838419

Lin, X., & Lai, Y. (2024). Scarring Skin: Mechanisms and Therapies . *International Journal of Molecular Sciences*, *25*(3), 1458. PMID: 38338767

Liu, Q., Wang, W., & Yang, H.. (2024). Microenvironment-Responsive Injectable Conductive Hydrogel for Spinal Cord Injury Repair . *Advanced Functional Materials, 34*(46), 2406376.

Liu, Y., Dzidotor, G., & Le, T. T.. (2022). Exercise-induced piezoelectric stimulation for cartilage regeneration in rabbits . *Science Translational Medicine, 14*(627), eabi7282. PMID: 35020409

Liu, Y., Xu, B., & Lu, M.. (2022). Ultrasmall Fe-doped carbon dots nanozymes for photoenhanced antibacterial therapy and wound healing . *Bioactive Materials, 12*, 246–256. PMID: 35310377

Lv, Z., Hu, T., & Bian, Y.. (2023). A MgFe-LDH Nanosheet-Incorporated Smart Thermo-Responsive Hydrogel with Controllable Growth Factor Releasing Capability for Bone Regeneration . *Advanced Materials (Deerfield Beach, Fla.), 35*(5), e2206545. PMID: 36426823

Ma, Z., Hu, X., & Zhang, Y.. (2023). Biomineralized Piezoelectrically Active Scaffolds for Inducing Osteogenic Differentiation . *Chemistry (Weinheim an der Bergstrasse, Germany), 29*(15), e202203166. PMID: 36478479

Makino, S., Fukuda, K., & Miyoshi, S.. (1999). Cardiomyocytes can be generated from marrow stromal cells in vitro . *The Journal of Clinical Investigation, 103*(5), 697–705. PMID: 10074487

Mason, C., & Dunnill, P. (2008). A brief definition of regenerative medicine . *Regenerative Medicine, 3*(1), 1–5. PMID: 18154457

Mohebali, A., & Abdouss, M. (2020). Layered biocompatible pH-responsive antibacterial composite film based on HNT/PLGA/chitosan for controlled release of minocycline as burn wound dressing . *International Journal of Biological Macromolecules, 164*, 4193–4204. PMID: 32891643

Montoya, C., Du, Y., & Gianforcaro, A. L.. (2021). On the road to smart biomaterials for bone research: Definitions, concepts, advances, and outlook . *Bone Research, 9*(1), 12. PMID: 33574225

Morris, E., Chavez, M., & Tan, C. (2016). Dynamic biomaterials: Toward engineering autonomous feedback . *Current Opinion in Biotechnology, 39*, 97–104. PMID: 26974245

Narkar, A R, Tong Z, Soman P. *et al.* (2022). Smart biomaterial platforms: Controlling and being controlled by cells . *Biomaterials, 283*, 121450. PMID: 35247636

Ogawa, R. (2017). Keloid and Hypertrophic Scars Are the Result of Chronic Inflammation in the Reticular Dermis . *International Journal of Molecular Sciences, 18*(3), 606. PMID: 28287424

Pakzad, S., Taghavi, R., & Hasanzadeh, A.. (2024). A biocompatible cellulose gum based CMC/PVA/SBA-15 film as a colloidal antibacterial agent against MRSA . *RSC Advances, 14*(49), 36246–36252. PMID: 39539531

Peña, O. A. & Martin, P. (2024). Cellular and molecular mechanisms of skin wound healing . *Nature Reviews. Molecular Cell Biology, 25*(8), 599–616. PMID: 38528155

Peng, Y., Jiang, H., & Zuo, H. D. (2023). Factors affecting osteogenesis and chondrogenic differentiation of mesenchymal stem cells in osteoarthritis . *World Journal of Stem Cells, 15*(6), 548–560. PMID: 37424946

Piluso, S., Li, Y., & Abinzano, F.. (2019). Mimicking the Articular Joint with In Vitro Models . *Trends in Biotechnology*, *37*(10), 1063–1077. PMID: 31000204

Poongodi, R., Chen, Y. L., & Yang, T. H.. (2021). Bio-Scaffolds as Cell or Exosome Carriers for Nerve Injury Repair . *International Journal of Molecular Sciences*, *22*(24), 13347. PMID: 34948144

Qi, J., Yu, T., & Hu, B.. (2021). Current Biomaterial-Based Bone Tissue Engineering and Translational Medicine . *International Journal of Molecular Sciences*, *22*(19), 10233. PMID: 34638571

Ramzan, F., Ekram, S., & Frazier, T.. (2022). Decellularized Human Umbilical Tissue-Derived Hydrogels Promote Proliferation and Chondrogenic Differentiation of Mesenchymal Stem Cells . *Bioengineering (Basel, Switzerland)*, *9*(6), 239. PMID: 35735483

Ren, W., Lin, Z., & Fan, Y.. (2022). Programmable biological state-switching photoelectric nanosheets for the treatment of infected wounds . *Materials Today. Bio*, *15*, 100292. PMID: 35634172

Salhotra, A, & Shah, H N, Levi B. *et al.* (2020). Mechanisms of bone development and repair . *Nature Reviews. Molecular Cell Biology*, *21*(11), 696–711. PMID: 32901139

Santin, M., & Phillips, G. J. (2012). *Biomimetic, bioresponsive, and bioactive materials: an introduction to integrating materials with tissues* [M]. John Wiley & Sons.

Shakoor, S., Kibble, E., & El-Jawhari, J. J. (2022). Bioengineering Approaches for Delivering Growth Factors: A Focus on Bone and Cartilage Regeneration . *Bioengineering (Basel, Switzerland)*, *9*(5), 223. PMID: 35621501

Shan, H., Zhou, X., & Tian, B.. (2022). Gold nanorods modified by endogenous protein with light-irradiation enhance bone repair via multiple osteogenic signal pathways . *Biomaterials*, *284*, 121482. PMID: 35358870

Shang, L., Ma, B., & Wang, F.. (2020). Nanotextured silk fibroin/hydroxyapatite biomimetic bilayer tough structure regulated osteogenic/chondrogenic differentiation of mesenchymal stem cells for osteochondral repair . *Cell Proliferation*, *53*(11), e12917. PMID: 33001510

Shao, X., Lin, S., Peng, Q., ... Tetrahedral DNA Nanostructure: A Potential Promoter for Cartilage Tissue Regeneration via Regulating Chondrocyte Phenotype and Proliferation. Small (Weinheim an Der Bergstrasse, Germany), 2017, 13(12).

Singer, A. J., & Clark, R. A. (1999). Cutaneous wound healing . *The New England Journal of Medicine*, *341*(10), 738–746. PMID: 10471461

Sun, P., Jiao, J., & Wang, X.. (2023). Nanomedicine hybrid and catechol functionalized chitosan as pH-responsive multi-function hydrogel to efficiently promote infection wound healing . *International Journal of Biological Macromolecules*, *238*, 124106. PMID: 36948329

Tariq, U., Gupta, M., & Pathak, S.. (2022). Role of Biomaterials in Cardiac Repair and Regeneration: Therapeutic Intervention for Myocardial Infarction . *ACS Biomaterials Science & Engineering*, *8*(8), 3271–3298. PMID: 35867701

Tottoli, E M, Dorati R, Genta I. *et al.* (2020). Skin Wound Healing Process and New Emerging Technologies for Skin Wound Care and Regeneration . *Pharmaceutics*, *12*(8), 735. PMID: 32764269

Wang, J., Lin, J., & Chen, L.. (2022). Endogenous Electric-Field-Coupled Electrospun Short Fiber via Collecting Wound Exudation . *Advanced Materials (Deerfield Beach, Fla.)*, *34*(9), e2108325. PMID: 34902192

Wang, L., Li, Y., & Ren, M.. (2022). pH and lipase-responsive nanocarrier-mediated dual drug delivery system to treat periodontitis in diabetic rats . *Bioactive Materials*, *18*, 254–266. PMID: 35387157

Wang, X., Shan, M., & Zhang, S.. (2022). Stimuli-Responsive Antibacterial Materials: Molecular Structures, Design Principles, and Biomedical Applications . *Advanced Science (Weinheim, Baden-Wurttemberg, Germany)*, *9*(13), e2104843. PMID: 35224893

Wang, X, & Yu, Y Y, LIEU S. *et al.* (2013). MMP9 regulates the cellular response to inflammation after skeletal injury . *Bone*, *52*(1), 111–119. PMID: 23010105

Wei, H., Cui, J., & Lin, K.. (2022). Recent advances in smart stimuli-responsive biomaterials for bone therapeutics and regeneration . *Bone Research*, *10*(1), 17. PMID: 35197462

Wei, X., Chen, S., & Xie, T.. (2022). An MMP-degradable and conductive hydrogel to stabilize HIF-1α for recovering cardiac functions . *Theranostics*, *12*(1), 127–142. PMID: 34987638

Weinberger, F., Breckwoldt, K., & Pecha, S.. (2016). Cardiac repair in guinea pigs with human engineered heart tissue from induced pluripotent stem cells . *Science Translational Medicine*, *8*(363), 363ra148. PMID: 27807283

Wu, X., Ding, J., & Xu, P.. (2021). A cell-free ROS-responsive hydrogel/oriented poly(lactide-co-glycolide) hybrid scaffold for reducing inflammation and restoring full-thickness cartilage defectsin vivo . *Biomedical Materials (Bristol, England)*, *16*(6). PMID: 34450597

Wu, Y., Wang, Y., & Long, L.. (2022). A spatiotemporal release platform based on pH/ROS stimuli-responsive hydrogel in wound repairing . *Journal of Controlled Release: Official Journal of the Controlled Release Society*, *341*, 147–165. PMID: 34813880

Xue, Y., Zhu, Z., & Zhang, X.. (2021). Accelerated Bone Regeneration by MOF Modified Multifunctional Membranes through Enhancement of Osteogenic and Angiogenic Performance . *Advanced Healthcare Materials*, *10*(6), e2001369. PMID: 33448103

Yan, J., Xia, D., & Zhou, W.. (2020). pH-responsive silk fibroin-based CuO/Ag micro/nano coating endows polyetheretherketone with synergistic antibacterial ability, osteogenesis, and angiogenesis . *Acta Biomaterialia*, *115*, 220–234. PMID: 32777292

Yang, Z., Wu, C., & Shi, H.. (2022). Advances in Barrier Membranes for Guided Bone Regeneration Techniques . *Frontiers in Bioengineering and Biotechnology*, *10*, 921576. PMID: 35814003

Zhang, J., Zheng, Y., & Lee, J.. (2021). A pulsatile release platform based on photo-induced imine-crosslinking hydrogel promotes scarless wound healing . *Nature Communications*, *12*(1), 1670. PMID: 33723267

Zhang, J., Zhou, R., & Xiang, C.. (2020). Huangbai Liniment Accelerated Wound Healing by Activating Nrf2 Signaling in Diabetes . *Oxidative Medicine and Cellular Longevity, 2020*, 4951820. PMID: 32566084

Zhang, Y., Chen, S., & Xiao, Z.. (2021). Magnetoelectric Nanoparticles Incorporated Biomimetic Matrix for Wireless Electrical Stimulation and Nerve Regeneration . *Advanced Healthcare Materials, 10*(16), e2100695. PMID: 34176235

Zhao, C., Tian, S., & Liu, Q.. (2020). Biodegradable nanofibrous temperature-responsive gelling microspheres for heart regeneration . *Advanced Functional Materials, 30*(21), 2000776. PMID: 33071711

Zhao, Y., Liang, Y., & Ding, S.. (2020). Application of conductive PPy/SF composite scaffold and electrical stimulation for neural tissue engineering . *Biomaterials, 255*, 120164. PMID: 32554132

Zhou, Q., Li, T., & Wang, K.. (2022). Current status of xenotransplantation research and the strategies for preventing xenograft rejection . *Frontiers in Immunology, 13*, 928173. PMID: 35967435

Chapter 6
Electroactive Biomaterials for Cell Behavior Regulation and Tissue Regeneration

Zhengao Wang
https://orcid.org/0000-0002-6430-6303
South China Agricultural University, China

Huiyu Zhao
The Third Affiliated Hospital of Southern Medical University, China

Yao Han
South China University of Technology, China

Wuyi Zhou
South China Agricultural University, China

Chengyun Ning
South China University of Technology, China

ABSTRACT

Endogenous electric fields widely occur in bone, skin, cardiac, and nerve tissues, significantly influencing organ development, wound healing, and tissue regeneration. Generated primarily by transmembrane potentials resulting from ion concentration gradients (e.g. Na+, K+, Ca2+), these fields inspire electroactive biomaterials designed with built-in electric fields. Unlike external electrical stimulation, such biomaterials precisely regulate cellular behavior through direct electron/charge interactions and protein-mediated signal transduction. By tailoring physical and chemical properties, these biomimetic materials replicate natural bioelectric cues. This review summarizes endogenous electrical signals, details fabrication methods and regulatory mechanisms of electroactive biomaterials, highlights applications in bone, nerve, and skin regeneration, and discusses clinical challenges and future perspectives.

DOI: 10.4018/979-8-3373-0055-9.ch006

1. INTRODUCTION

Tissue regeneration plays a crucial role in the repair and restoration of damaged or lost tissue functions in the body. It involves a complex interplay of cellular activities, including cell migration, differentiation, proliferation, and extracellular matrix formation, which together contribute to the reconstruction of functional tissue.(Xia et al., 2018) However, current therapeutic strategies for enhancing tissue regeneration often fall short in promoting efficient, natural, and precise healing, particularly in cases of chronic wounds or severe tissue damage. In recent years, the application of electroactive biomaterials has emerged as a promising solution to improve tissue regeneration by leveraging endogenous electrical signals that govern cellular behavior and tissue repair.(Liu et al., 2021; Ning et al., 2018)

Endogenous electric fields are an intrinsic part of biological systems, present across various tissues such as bone, skin, nerve, and cardiac tissues.(Thrivikraman et al., 2018) These electric fields are primarily generated by the transmembrane potential, resulting from the difference in ion concentrations, such as sodium (Na^+), potassium (K^+), and calcium (Ca^{2+}), across the cell membrane.(Nowicka-Bauer & Szymczak-Cendlak, 2021) These fields not only contribute to cellular homeostasis but also regulate key physiological processes, such as wound healing, nerve regeneration, and bone remodeling. For example, during tissue injury, electric fields can guide the migration of cells to the wound site, promote cellular differentiation, and stimulate tissue repair by influencing the behavior of various cell types, including fibroblasts, stem cells, and endothelial cells.(Preetam et al., 2024) These endogenous electrical signals have been recognized for their ability to control cellular activities at a precise spatial and temporal level, thereby playing a pivotal role in tissue regeneration.(Das et al., 2024)

To mimic the beneficial effects of these natural electric cues and enhance tissue repair, electroactive biomaterials have been developed. These materials are designed to generate or respond to electric fields and provide electrical stimuli that directly influence cellular behavior.(Liu et al., 2021) Unlike traditional external electrical stimulation methods, which rely on bulky devices and invasive electrodes, electroactive biomaterials offer several advantages. They can be fabricated with built-in electric fields that operate at the material interface, enabling a more localized, direct, and sustained modulation of cellular functions. Furthermore, electroactive biomaterials are designed to interact with cells at the molecular level through charge-based interactions, triggering signal transduction pathways that regulate cellular processes such as migration, proliferation, and differentiation.(Chen et al., 2024a) The fabrication of electroactive biomaterials requires a careful design process to ensure that their physical, chemical, and electrical properties closely resemble those found in natural tissues.(Chen et al., 2024a; Esmaeili et al., 2022) These materials can be engineered to possess specific characteristics such as conductivity, piezoelectricity, and electrochemical activity, which are essential for eliciting the desired cellular responses.(Casella et al., 2021) By tailoring the composition and structure of electroactive biomaterials, researchers can fine-tune their ability to create electric fields that influence cellular behavior, providing a versatile tool for enhancing tissue regeneration across a wide range of applications.(Deng et al., 2022)

This chapter aims to provide a comprehensive overview of the role of electroactive biomaterials in regulating cell behavior and promoting tissue regeneration. First, we explore the physiological endogenous electric fields that exist in living organisms and tissues, highlighting their importance in regulating cellular processes. Next, we discuss the various fabrication methods used to develop electroactive biomaterials, focusing on how the design of these materials can generate or respond to electric fields. We then delve into the mechanisms by which electroactive biomaterials modulate cellular behavior, discussing how electric cues interact with cell surface receptors, ion channels, and functional proteins to influence cell migration, differentiation, and other regenerative processes. Subsequently, we review the current applications of electroactive biomaterials in tissue regeneration, with a particular focus on bone, nerve, and skin regeneration, and explore the mechanisms by which these materials promote repair and healing in these tissues. Finally, we discuss the challenges and limitations of electroactive biomaterials, as well as the future directions for their development, particularly in clinical settings. The growing understanding of the complex interactions between electric fields and cellular behavior has paved the way for the development of more effective electroactive biomaterials.

2. ENDOGENOUS ELECTRIC FIELDS *IN VIVO*

Endogenous electric fields are naturally occurring electric potentials that are found within living organisms and tissues (Figure 1). At the molecular level, charged biomolecules such as DNA and proteins play a crucial role in cellular functions through electrostatic interactions and biochemical signaling. DNA's negatively charged backbone affects transcription factor binding and chromatin structure, regulating gene expression. Similarly, charged amino acids in proteins influence folding, stability, and enzymatic activity. Additionally, electron transfer processes drive redox reactions, modulating protein function and cellular signaling pathways essential for metabolism and homeostasis. Endogenous electric fields are essential for maintaining cellular function and regulating various biological processes, including development, wound healing, tissue remodeling, and regeneration. The generation of these electric fields is largely driven by the movement of charged particles across biological membranes, which is a fundamental aspect of cellular physiology. The two primary sources of endogenous electric fields are the transmembrane potential and the electric fields that arise during cellular interactions with the extracellular matrix (ECM).(Liu et al., 2025)

One of the most significant sources of endogenous electric fields is the transmembrane potential, which results from the unequal distribution of ions inside and outside the cell. The differential concentrations of ions such as sodium (Na^+), potassium (K^+), calcium (Ca^{2+}), and chloride (Cl^-) across the cell membrane create an electrical potential difference, typically ranging from -30 to -100 mV in most cells. (McCormick, 2014) This electrical gradient is critical for maintaining cellular integrity and is involved in regulating various cellular processes, including signal transduction, nutrient uptake, and waste removal. Moreover, the transmembrane potential plays a vital role in mediating electrical communication between cells, particularly in excitable tissues like nerve and muscle cells.(Yeste et al., 2018)

The role of endogenous electric fields extends beyond the individual cell to the tissue and organ level. For example, during the process of wound healing, electric fields are generated at the site of injury. This phenomenon, known as the “wound electric field,” is characterized by a spatially defined gradient of electrical potential that forms between the wound and the surrounding healthy tissue. (Zhao, 2009) The wound electric field is believed to arise due to the loss of the skin barrier and the difference in ion concentrations at the wound site compared to normal tissue. This electric field plays a crucial role in directing the migration of cells, such as keratinocytes, fibroblasts, and endothelial cells, towards the wound area, thereby promoting tissue repair and regeneration. (Stojanovic et al., 2024) Endogenous electric fields also play a critical role in other regenerative processes, such as bone and nerve repair. In bone tissue, the mechanical loading and deformation of the bone matrix generate piezoelectric signals, which create electric fields that can influence osteoblast activity and promote bone formation. (Khare et al., 2020) Similarly, in nerve regeneration, the endogenous electric fields generated at the injury site have been shown to guide axonal growth and facilitate nerve repair by modulating cellular behavior through electrochemical signaling. (Haan & Song, 2014) These electric fields provide a directional cue for migrating cells and stimulate the repair processes necessary for tissue recovery.

In addition to the wound electric field, endogenous electric fields are also involved in the development of tissues and organs. During embryogenesis, electric fields play a key role in cellular differentiation and tissue morphogenesis. For instance, studies have demonstrated that the electric fields generated during the development of the heart, brain, and other organs contribute to the patterning of cells and the formation of complex tissue structures.(McLaughlin & Levin, 2018) The electrical cues that arise during development influence gene expression and cellular behaviors, thus guiding the formation of functional tissues and organs.(Levin, 2021)

The presence of these endogenous electric fields within living organisms highlights their essential role in regulating physiological processes. They provide cells and tissues with a form of spatial and temporal guidance that is crucial for the proper functioning of biological systems. Understanding the nature of these electric fields, as well as their mechanisms of action, is critical for developing therapeutic strategies that can mimic or modulate these natural signals to promote tissue repair and regeneration.

Figure 1. Schematic illustration of the role of electrical phenomena across multiple biological scales

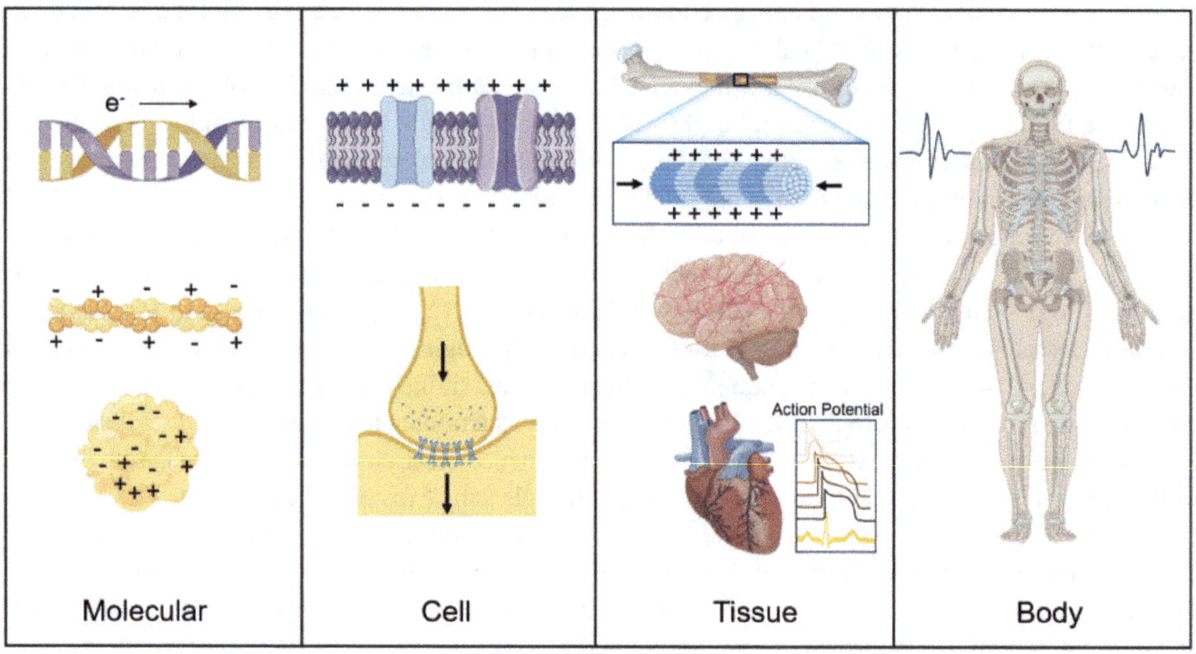

At the molecular level, charged biomolecules (e.g., DNA, proteins) and electron transfer processes influence gene expression and protein function. At the cellular level, transmembrane potentials and synaptic events regulate ion flux, enabling signal transduction and cell-to-cell communication. At the tissue scale, piezoelectric effects in bone, electrical signaling in neural networks, and action potentials in cardiac tissue are critical for tissue development and function. Finally, at the level of the entire organism, these electrical interactions integrate to maintain overall physiological homeostasis and coordinated bodily functions.

3. THE FABRICATION OF ELECTROACTIVE BIOMATERIALS

The design and synthesis of electroactive biomaterials rely on the careful selection of materials, the precise control of structure, and the application of various processing techniques to achieve the desired electroactive properties. These properties include electrical conductivity, piezoelectricity, and electrochemical reactivity, all of which can be tailored to suit specific therapeutic applications.

3.1 Material Selection for Electroactive Biomaterials

Electroactive biomaterials must possess specific electrical, mechanical, and biological properties that allow them to interact effectively with cells and tissues. The choice of materials typically includes conductive polymers, piezoelectric materials, carbon-based materials, and nanomaterials, each offering unique advantages in different tissue regeneration applications.

Conductive polymers are among the most widely used electroactive materials due to their excellent processability, flexibility, and tunable electrical conductivity. Polymers such as polypyrrole (PPy), polyaniline (PANI), and poly(3,4-ethylenedioxythiophene) (PEDOT) have been extensively studied for their ability to support electron and ion conduction. For instance, polypyrrole is commonly used in applications requiring controlled drug release or tissue engineering scaffolds due to its biocompatibility and conductivity. (Liang & Goh, 2020) PEDOT, known for its stability in aqueous environments, has been incorporated into scaffolds for neural tissue regeneration, as it can facilitate electrical signaling between cells, enhancing neuronal growth and differentiation.(Bianchi et al., 2023)

Piezoelectric materials, such as polyvinylidene fluoride (PVDF), its copolymer PVDF-TrFE, and barium titanate ($BaTiO_3$), exhibit the ability to generate electric charges when subjected to mechanical stress, a phenomenon known as the piezoelectric effect. These materials are particularly useful in applications where mechanical forces, such as compression or tension, need to be converted into electrical signals. For example, piezoelectric PVDF films have been integrated into scaffolds for bone regeneration, where mechanical loading from body movements generates piezoelectric signals, promoting osteogenesis and improving bone healing.(Yang et al., 2022)

Carbon-based nanomaterials, including carbon nanotubes (CNTs), graphene, and graphene oxide, offer outstanding electrical conductivity, mechanical strength, and biocompatibility, making them ideal candidates for electroactive biomaterials. CNTs, for example, have been incorporated into hydrogels to form composite materials for neural tissue engineering.(Vashist et al., 2018) Their high aspect ratio and electrical properties facilitate charge transfer, which is essential for neural growth and synaptic communication. (Burnstine-Townley et al., 2020) Graphene and graphene oxide, on the other hand, are gaining attention for their ability to support cell adhesion and proliferation while providing conductive pathways to enhance tissue regeneration. (Zare et al., 2024)

3.2 Structural Design of Electroactive Biomaterials

The structural design of electroactive biomaterials is critical to their performance in regulating cell behavior and promoting tissue regeneration. These materials can be designed in one-dimensional (1D), two-dimensional (2D), and three-dimensional (3D) architectures, each offering distinct advantages and addressing different requirements for tissue engineering applications.

One-dimensional (1D) electroactive materials, such as electrospun nanofibers and nanowires, are widely utilized in tissue engineering due to their high surface area-to-volume ratio. This design enhances cell attachment and facilitates efficient electrical conductivity, making them particularly effective in promoting cellular responses. Electrospun nanofibers, typically made from conductive polymers like polypyrrole (PPy), poly(3,4-ethylenedioxythiophene) (PEDOT), or polyvinylidene fluoride (PVDF), have been used extensively for nerve, bone, and skin regeneration.(Acosta et al., 2022) These nanofibers provide an electrically conductive scaffold that encourages the alignment and growth of cells in specific directions, which is crucial for applications such as neural regeneration where the directional growth of neurons is essential. Furthermore, conductive nanowires made from materials like gold, silver, or carbon-based compounds offer a versatile platform for creating high-resolution electrical interfaces that promote cellular interactions. (Bach-Toledo et al., 2020)

Two-dimensional (2D) electroactive materials, including thin films and graphene-based sheets, are typically used as surface coatings, substrates for cell cultures, or thin scaffolds. These 2D materials are particularly advantageous when uniform electrical stimulation is needed across a surface. Conductive thin films, such as those made from PEDOT:PSS or polypyrrole, are applied to substrates like metals, ceramics, or polymers to impart electrical properties while maintaining the biocompatibility and mechanical stability of the base material. For example, PEDOT:PSS thin films have been used to coat nerve conduits, where their conductive properties facilitate neural growth and guidance. (Pisciotta et al., 2022) Additionally, graphene and its derivatives, such as graphene oxide or reduced graphene oxide, have gained attention for their high conductivity, mechanical strength, and biocompatibility. (Banerjee, 2018) These 2D materials are ideal for applications like neural tissue engineering, where they not only provide electrical stimulation but also support cell adhesion and promote cellular differentiation. Conductive 2D hydrogels, often composed of materials like carbon nanotubes (CNTs) or graphene oxide, have been developed for applications in cardiac tissue engineering. These hydrogels provide a soft, flexible scaffold that delivers electrical cues to improve cell alignment and tissue functionality.

Three-dimensional (3D) electroactive biomaterials represent the most advanced form of structural design, providing the most versatile and complex architectures for tissue regeneration. (Nikolova & Chavali, 2019) 3D structures mimic the intricate organization of native tissues, allowing for better cellular interaction, nutrient diffusion, and the application of electrical signals throughout the material. These scaffolds can be fabricated using techniques such as 3D printing, electrospinning, or stereolithography, offering precise control over material composition, pore size, and architecture. For example, 3D-printed scaffolds composed of conductive polymers like PCL/PEDOT composites have shown promising results in bone regeneration by enhancing osteoblast differentiation through electrical stimulation.(Dixon & Gomillion, 2023) Similarly, 3D electroactive hydrogels made from poly(ethylene glycol)-based polymers and graphene oxide have been used for cardiac tissue engineering, where they support cardiomyocyte growth and synchronize tissue contraction. (Ashtari et al., 2019) The combination of high conductivity and porosity in these 3D structures not only facilitates cell migration and proliferation but also enables the delivery of electrical signals that guide tissue regeneration processes.

The design of electroactive biomaterials in 1D, 2D, and 3D forms offers diverse and tailored solutions for a range of tissue engineering applications. By selecting the appropriate structural architecture, researchers can optimize the electrical properties of the material to enhance cell behavior, support tissue growth, and improve the overall regeneration process. The versatility of these structures in terms of mechanical properties, electrical conductivity, and biocompatibility makes them essential for advancing regenerative medicine, providing effective and targeted therapeutic solutions for damaged tissues.

3.3 Processing Techniques for Electroactive Biomaterials

The fabrication of electroactive biomaterials involves the integration of conductive materials into scaffolds or matrices while maintaining their biocompatibility, mechanical strength, and biological activity. Various processing techniques are used to produce these materials, each with its own advantages in terms of material composition, structure, and electrical performance. These techniques include electrospinning, 3D printing, solvent casting, chemical vapor deposition (CVD), freeze-drying, electrochemical deposition, and solvothermal methods, among others. The choice of processing technique depends on factors such as material properties, the desired structure, and the specific application requirements for tissue regeneration.

Electrospinning is a well-established technique for producing one-dimensional (1D) nanofibers that closely resemble the natural extracellular matrix (ECM). (Su et al., 2021) In this process, a high-voltage electric field is applied to a polymer melt or solution, resulting in the formation of fine fibers that are collected on a substrate. Conductive fillers, such as carbon nanotubes (CNTs), polypyrrole (PPy), or polythiophene, are often incorporated into the polymer solution to impart electroactive properties to the fibers. Electrospun nanofibers offer a high surface area and porosity, making them ideal for cell attachment, migration, and proliferation. The electrical properties of the nanofibers can be fine-tuned by adjusting the concentration of conductive materials or modifying the electrospinning parameters. (Wang et al., 2021) These nanofibers are particularly useful in nerve regeneration, bone repair, and wound healing, where the electrical stimulation of cells is beneficial for promoting tissue regeneration.

3D printing, or additive manufacturing, is increasingly employed for fabricating electroactive biomaterials with complex three-dimensional structures.(Vyas et al., 2023) The technique involves the layer-by-layer deposition of materials, allowing for precise control over scaffold geometry, porosity, and material composition. Conductive polymers (e.g., PEDOT, PPy) or composites containing CNTs or graphene are commonly used as bioinks for 3D printing. These materials allow the scaffolds to transmit electrical signals to stimulate tissue regeneration. 3D printing provides flexibility in combining multiple materials into a single scaffold, resulting in hybrid structures that have both enhanced mechanical and electrical properties.(Li et al., 2024) This technique is particularly useful for creating patient-specific scaffolds for applications in bone, skin, and nerve regeneration.

Solvent casting is a traditional and cost-effective method for producing electroactive thin films or scaffolds. The process involves dissolving a polymer in a solvent, followed by casting the solution onto a substrate and allowing the solvent to evaporate, leaving behind a solid film. Conductive fillers such as graphene oxide, CNTs, or conductive polymers are added to the polymer solution to fabricate electroactive materials. Solvent casting is suitable for producing thin films, coatings, or porous structures. While it may not offer the precision of electrospinning or 3D printing, it is often used for creating electroactive surfaces or coatings on scaffolds. For example, PEDOT:PSS films have been applied as coatings for nerve conduits to guide neural growth through electrical stimulation.(Babaie et al., 2020)

Chemical vapor deposition (CVD) is a method for depositing thin, uniform layers of conductive materials onto substrates. In this process, precursor gases are introduced into a reaction chamber, where they react to form a solid material that is deposited onto the substrate. CVD allows precise control over the thickness and electrical properties of the deposited layers, making it ideal for fabricating electroactive films and coatings. Electrochemical deposition is a versatile technique for fabricating conductive coatings and scaffolds on various substrates. In this process, an electric current is passed through an electrolyte solution containing metal ions or conductive polymer precursors, causing them to be reduced and deposited onto a substrate. Electrochemical deposition allows for precise control over the thickness, morphology, and electrical conductivity of the deposited material. Conductive polymers like PPy and polyaniline (PANI) are commonly electrochemically deposited onto scaffolds to improve their electrical properties and cell-regulating capabilities.(Burnstine-Townley et al., 2020; Li et al., 2016; Stojanovic et al., 2024)

The solvothermal method is a highly versatile technique used for the synthesis of nanostructured materials under high temperature and pressure conditions. This method is particularly effective for producing piezoelectric materials such as barium titanate ($BaTiO_3$), zinc oxide (ZnO), and other metal oxides.(Sood et al., 2023) In a solvothermal process, the precursor material is dissolved in a solvent, which can be water or an organic solvent, and then the solution is heated in a sealed vessel (autoclave) under controlled pressure and temperature. The heat and pressure facilitate the chemical reactions that lead to the formation of highly uniform, crystalline nanoparticles or thin films. The solvothermal method allows precise control over particle size, morphology, and crystallinity, which are crucial for the piezoelectric properties of the materials.

4. THE MECHANISM OF ELECTROACTIVE BIOMATERIALS REGULATING CELL BEHAVIOR

Electroactive biomaterials, due to their ability to generate and respond to electrical signals, play a significant role in regulating cell behavior and promoting tissue regeneration. These materials interact with cells at both the molecular and cellular levels, inducing various responses such as migration, differentiation, proliferation, and alignment.(Li, Yu, Wang, Long, Xiao, Xing, Dong, Zhai, Zhou, Zhou, et al., 2023) The underlying mechanisms by which electroactive biomaterials regulate these cellular behaviors are complex and involve multiple biological pathways. Below, we discuss the main mechanisms through which electroactive biomaterials influence cell behavior, including direct electrical stimulation, charge transfer, mechanical cues, and the role of biomaterial surface properties.

4.1 Electrical Stimulation and Ion Channel Activation

Electrical stimulation of electroactive biomaterials can directly influence the behavior of cells through the modulation of ion channels, which are integral to the regulation of cellular activities. Ion channels are transmembrane proteins that allow the selective passage of ions (e.g., Na^+, K^+, Ca^{2+}, Cl^-) across the cell membrane, and their activity is fundamental to cell signaling and function.(Huang et al.,

2024) In response to an electric field, electroactive biomaterials can create localized changes in the ionic environment surrounding the cells, triggering ion flux through specific ion channels.(Liu et al., 2021)

For example, calcium ion (Ca^{2+}) signaling plays a pivotal role in cell migration, proliferation, and differentiation. (Wei et al., 2012) The application of an electric field can induce localized calcium ion influx by activating voltage-gated calcium channels or other mechanosensitive ion channels. (Li et al., 2014) In mesenchymal stem cells (MSCs), such an influx has been shown to enhance osteogenesis, the process by which stem cells differentiate into bone-forming osteoblasts. Additionally, calcium ions are essential for activating downstream kinases such as Ca^{2+}/calmodulin-dependent protein kinase II and protein kinase C, which further activate transcription factors like Runx2 and Osx that promote osteogenic differentiation. (Zhu et al., 2024) Similarly, the sodium-potassium ATPase pump (Na^+/K^+-ATPase), which regulates the concentration gradient of Na^+ and K^+ ions, can also be modulated by the electrical stimulation of electroactive biomaterials.(Pivovarov et al., 2018) This pump's activity is linked to cellular processes such as membrane potential maintenance, cell migration, and adhesion. Electrical cues can enhance the polarization of this pump, affecting intracellular signaling pathways such as PI3K/Akt and MAPK, which are involved in cell survival, growth, and migration.(Sun et al., 2024)

4.2 Electroactive Materials and Intracellular Signal Transduction Pathways

Beyond ion channel activation, electroactive biomaterials can influence a range of intracellular signal transduction pathways through direct interaction with cellular proteins and receptors. (Jacob et al., 2018) These pathways regulate gene expression, protein synthesis, and cell differentiation. For instance, electrical fields generated by electroactive biomaterials can influence key signaling cascades such as the MAPK/ERK pathway, Wnt/β-catenin pathway, Notch signaling, and TGF-β signaling.(Luo et al., 2024)

The mitogen-activated protein kinase (MAPK) pathway is critical in regulating cellular responses to mechanical and electrical cues. (Mocanu-Dobranici et al., 2023) Electrical stimulation from piezoelectric materials such as barium titanate ($BaTiO_3$) or zinc oxide (ZnO) can activate the MAPK pathway through ion channel activation or direct mechanical strain on the cell membrane.(Yang et al., 2023) Activation of extracellular signal-regulated kinase (ERK) within this pathway can lead to cell proliferation, differentiation, and migration, particularly in tissues such as bone and skin.(Deschênes-Simard et al., 2023)

The Wnt/β-catenin signaling pathway is involved in the regulation of cell fate determination and differentiation.(Liu et al., 2022) Electrical stimulation from electroactive biomaterials can influence this pathway by modulating the phosphorylation state of β-catenin, a central player in this pathway.(Chen et al., 2024b) In bone tissue engineering, electroactive biomaterials that generate electric fields can promote the osteogenic differentiation of stem cells through the activation of β-catenin and subsequent upregulation of osteogenic markers like osteopontin (OPN) and collagen type I.(Luo et al., 2024)

The Notch signaling pathway governs cellular differentiation, proliferation, and apoptosis. (Zhou et al., 2022) It plays an essential role in regulating tissue regeneration, especially in neural and vascular tissues. Electroactive materials can modulate Notch signaling by altering cell-cell interactions via the mechanotransduction of electric fields. The activation of Notch receptors on cell surfaces can lead to the release of the intracellular domain, which translocates to the nucleus to regulate the expression of genes responsible for cellular differentiation. For example, in nerve regeneration, electroactive biomaterials may influence Notch signaling to support the differentiation of neural stem cells (NSCs) into mature neurons.(Mutepfa et al., 2022)

The transforming growth factor-beta (TGF-β) signaling pathway is essential in regulating wound healing, fibrosis, and tissue repair. Electrical cues from electroactive biomaterials can stimulate the TGF-β receptor, which initiates a cascade of intracellular signaling events that result in the activation of SMAD proteins.(Zheng et al., 2020) These proteins translocate to the nucleus and initiate the expression of genes that promote extracellular matrix remodeling and fibroblast activation, which is critical for wound healing and tissue regeneration.

4.3 Piezoelectric Effect and Mechanotransduction Pathways

Another key mechanism by which electroactive biomaterials influence cellular behavior is through the piezoelectric effect, which is the generation of electrical charges in response to mechanical deformation. This mechanical-to-electrical transduction is crucial for regulating mechanotransduction pathways, which are involved in cellular responses to mechanical forces.(Li, Yu, Wang, Long, Xiao, Xing, Dong, Zhai, Zhou, & Zhou, 2023)

When electroactive materials undergo mechanical deformation, they can induce strain on cell-matrix adhesions, leading to the activation of integrins, which are transmembrane receptors that mediate cell adhesion to the ECM. The activation of integrins leads to the recruitment of signaling proteins such as focal adhesion kinase (FAK), which is involved in regulating cell adhesion, migration, and survival. In bone and cartilage engineering, electroactive biomaterials that harness the piezoelectric effect can promote cell adhesion and differentiation via integrin-mediated signaling.

The Yes-associated protein (YAP) and transcriptional co-activator with PDZ-binding motif (TAZ) are key regulators of mechanotransduction, controlling cell proliferation and differentiation in response to physical cues.(Hughes, 2016) The mechanical strain induced by electroactive biomaterials can activate the YAP/TAZ pathway, particularly in soft tissues like skin or muscle. This activation can lead to changes in gene expression that support tissue regeneration and wound healing.(Brusatin et al., 2018)

4.4 Reactive Oxygen Species (ROS) and Redox Signaling

Electroactive biomaterials can also regulate the balance of reactive oxygen species (ROS), which play a critical role in cellular signaling. In response to electrical stimulation, electroactive materials can generate localized ROS, which can serve as secondary messengers in various signaling pathways. ROS are involved in regulating cell migration, proliferation, and differentiation by activating kinases such as MAPKs, PI3K/Akt, and NF-κB, which are key mediators of inflammation and tissue repair.(Aggarwal et al., 2019)

For example, electrospun piezoelectric nanofibers made from materials like ZnO or $BaTiO_3$ can induce the production of ROS, promoting the differentiation of mesenchymal stem cells into osteoblasts in bone regeneration applications.(Liu & Lu, 2024) However, it is crucial to maintain a balance of ROS levels, as excessive ROS production can lead to oxidative stress and cell death. Electroactive biomaterials that modulate ROS levels in a controlled manner are particularly useful in promoting tissue regeneration without inducing inflammation or fibrosis.

In summary, the mechanisms by which electroactive biomaterials regulate cell behavior are multifaceted and involve the modulation of intracellular signaling pathways through electrical, mechanical, and electrochemical cues. The activation of ion channels, the modulation of mechanotransduction pathways, and the control of redox signaling are just a few of the ways in which electroactive materials influence cellular responses. These pathways are essential for promoting processes such as cell migration, proliferation, differentiation, and ECM remodeling, which are all critical for effective tissue regeneration.

5. ELECTROACTIVE BIOMATERIALS ENHANCING TISSUE REGENERATION

Electroactive biomaterials have emerged as an exciting approach in the field of tissue regeneration due to their ability to mimic and interact with the endogenous electrical signals present in the body. These materials can provide spatiotemporal control over cellular behaviors such as migration, proliferation, differentiation, and matrix synthesis, all of which are crucial for tissue repair and regeneration. The electrical properties of these materials offer a unique strategy to enhance tissue healing, particularly in areas such as bone, nerve, skin, and other tissues, where electrical signals play vital roles in their development and function. Below, we discuss how electroactive biomaterials contribute to tissue regeneration, highlighting recent advances and specific examples.

5.1 Bone Regeneration

In the realm of tissue regeneration, bone regeneration is a critical area that has gained significant attention, particularly with the advent of electroactive biomaterials. Bone injuries, particularly those related to trauma, disease, or age-related degeneration, often require advanced strategies to stimulate healing and restoration. Electroactive biomaterials have shown great promise in accelerating bone healing by mimicking the endogenous electric fields that naturally exist within the body. These fields have long been known to play a vital role in guiding cellular activities and tissue regeneration. Bone regeneration, specifically, involves a series of complex biological processes, including the recruitment of stem cells, osteoblast differentiation, and bone matrix synthesis. The introduction of electroactive biomaterials offers a unique approach to influence these processes by providing an external yet bio-compatible stimulus. These materials typically possess built-in electric fields that are designed to interact with cellular signaling pathways, thereby promoting the regeneration of bone tissue.

The role of endogenous electric fields in bone regeneration has been extensively studied. During normal bone development and healing, electric fields generated by ion flux and transmembrane potentials are integral to osteogenesis. These electric fields guide osteoblast differentiation, increase calcium deposition, and accelerate matrix formation. Moreover, the piezoelectricity of bone tissue—where mechanical stress induces electric charge accumulation—is another form of electric field that influences bone remodeling and repair. In bone fracture healing, endogenous electric fields are believed to promote cell migration, enhance cell-cell communication, and stimulate the secretion of growth factors like bone morphogenetic proteins (BMPs) and vascular endothelial growth factor (VEGF). These factors are key to the differentiation of mesenchymal stem cells (MSCs) into osteoblasts and the formation of new bone tissue.

Electroactive biomaterials are designed to replicate the electric fields found naturally within the body. These materials are typically composed of conductive polymers, hydrogels, or composite materials that can produce localized electric fields when stimulated. The built-in electric fields in these biomaterials interact directly with cells, promoting processes such as osteogenesis, angiogenesis, and inflammation modulation—all essential for successful bone regeneration. The design of electroactive biomaterials for bone regeneration requires careful consideration of their mechanical, electrical, and chemical properties. The material must not only be able to produce an electric field but also possess the necessary physical properties to support cellular attachment, proliferation, and differentiation. Conductive polymers like polypyrrole, polyaniline, and poly(3,4-ethylenedioxythiophene) (PEDOT) are commonly used in bone regeneration applications due to their high electrical conductivity and biocompatibility. Furthermore, electroactive hydrogels, which can respond to electrical stimuli, are also gaining popularity for bone tissue engineering. These hydrogels can be engineered to release growth factors or drugs in response to electric fields, providing a dual mechanism of action: electrical stimulation and controlled drug delivery.

The underlying mechanisms by which electroactive biomaterials enhance bone regeneration are diverse and multifactorial. First, the electric fields generated by these materials directly affect the behavior of bone cells. For instance, osteoblasts, osteocytes, and mesenchymal stem cells (MSCs) have been shown to respond to electrical cues by increasing cell migration, enhancing cellular differentiation, and promoting mineralization. These effects are largely mediated through ion channels, integrins, and growth factor signaling pathways that are sensitive to changes in the electrical environment.

The electric fields produced by electroactive materials can alter the membrane potential of cells, influencing their behavior.(Kolosnjaj-Tabi et al., 2019) Studies have demonstrated that electric fields enhance the migration and differentiation of MSCs into osteoblasts, thereby accelerating bone formation.(Wang et al., 2016) The electric field-induced polarization of cells can also influence the alignment and orientation of the cytoskeleton, which is crucial for cellular function and matrix deposition.

One of the most important mechanisms through which electroactive biomaterials promote bone regeneration is the induction of osteogenic differentiation. Electrical stimulation has been shown to enhance the expression of key osteogenic markers, such as alkaline phosphatase (ALP), osteocalcin (OCN), and collagen type I. These markers are critical for bone matrix production and mineralization. In addition to promoting osteogenesis, electric fields can also stimulate angiogenesis, which is crucial for bone regeneration. The electric fields generated by electroactive biomaterials can induce the release of VEGF and other pro-angiogenic factors, thereby enhancing vascularization in the healing bone tissue. Furthermore, electrical stimulation has been shown to modulate inflammation, a critical aspect of the healing process. By controlling the inflammatory response, electroactive biomaterials can ensure that the bone healing process proceeds without excessive scarring or delayed healing.

Electroactive biomaterials are being explored in a variety of experimental settings for their potential to enhance bone healing. One-dimensional $BaTiO_3$ ferroelectric nanorod array coating (NBTO) on the surface of titanium implants were constructed and can generate static electrical stimulations to exhibit osteogenesis properties (Figurer 2A).(Xiao et al., 2022) A 1.5 mm-diameter defect was created in the femoral bone, and four types of implants—titanium (Ti), PBTO, NBTO, and PNBTO—were inserted into the medullary cavity (Figure 2B). After four weeks, microphotography and micro-computed tomography (micro-CT) revealed marked differences in new bone formation among the groups (Firue 2C). While the Ti implant showed relatively modest bone growth, NBTO coatings with one-dimensional nanorod structures demonstrated a greater volume of newly formed bone surrounding the implant. This effect was attributed primarily to mechanoconduction, which promoted osteogenesis around the implant surface. Moreover, the PNBTO-coated group, possessing both the nanorod topography and static electric stimulation, exhibited the highest level of new bone growth and osseointegration. Quantitative micro-CT analysis supported these observations, as the bone volume/tissue volume (BV/TV) ratio in the PNBTO group was 2.3-fold and 1.7-fold higher than in the Ti and PBTO groups, respectively. Similar increases were noted in other parameters, including bone mineral density (BMD), trabecular number (Tb.N), and trabecular thickness (Tb.Th), while trabecular separation (Tb.Sp) decreased. These findings collectively indicate that the synergistic effect of a 1-D nanorod surface topography and a built-in electric field provided an optimal microenvironment for accelerating osteogenic differentiation and new bone deposition. Further histological assessments, such as methylene blue/acid fuchsin staining, corroborated these results by showing abundant newly formed trabecular bone and mature marrow cavities in direct contact with the PNBTO implant. By comparison, the bone-implant contact area in PNBTO was significantly larger than in Ti, PBTO, or NBTO alone, reflecting superior osseointegration. Scanning electron microscopy (SEM) analyses likewise confirmed substantial new bone on the PNBTO implant surface, consistent with in vitro data indicating enhanced BMSC differentiation. These combined findings suggest that, PNBTO develops a static built-in electric field conducive to bone regeneration, while its nanorod architecture promotes the activity of osteoblasts in direct contact with the surface. The integration of these features makes PNBTO a highly effective coating for accelerating bone formation and improving implant–bone interface outcomes *in vivo*.

Although heterojunctions have seen extensive use in fields such as energy conversion, environmental remediation, and photodetection, their potential in regenerative medicine remains largely untapped. In particular, piezoelectric heterojunctions had not been explored for tissue regeneration until a recent study demonstrated that coupling piezoelectric nanocrystals (e.g., Bi_2WO_6) with TiO_2 nanowires could simultaneously enhance bone repair and eradicate pathogenic bacteria through light–cellular force–electric interactions (Figure 2D-F). (Fan et al., 2025) This TiO_2/Bi_2WO_6 heterojunction array forms a biocompatible implant surface that, when exposed to near-infrared light, produces reactive oxygen species (ROS) and local heating. These effects, driven by photodynamic and photothermal mechanisms, serve to kill bacteria and effectively prevent postoperative infections. Concurrently, mechanical forces generated by stem cells on the implant surface trigger localized electric fields in the heterojunctions, promoting osteogenic differentiation and encouraging robust osteointegration (Figure 2G and H). Thus, the dual functionality of antibacterial action and regenerative stimulation addresses common concerns of implant-associated infections while fostering bone regeneration. Taken together, these findings indicate that piezoelectric heterojunctions have tremendous promise in biomedical engineering, particularly as antibacterial, tissue-regenerative implants. By translating established heterojunction principles from energy and photonics research into the biomedical sphere, future developments may yield advanced, multifunctional scaffolds capable of responding to diverse stimuli-light, mechanical forces, and local biological cues-to optimize both tissue repair and infection control.

Figure 2. (A) Schematic illustration of the mechanism by which PNBTO implants promote bone regeneration in vivo. (B) Photograph of the site where the implant was inserted in the rat femur. (C) Micro-CT images and 3D Micro-CT reconstructed images showing bone regeneration around the implants. (Xiao et al., 2022) Copyright 2022, American Chemical Society. (D) Schematic illustration of the in situ construction of TiO2/Bi2WO6 on the titanium implant surface. (E) FE-SEM of TiO_2/Bi_2WO_6. (F) COMSOL simulation of electrical signals generated by TiO_2/Bi_2WO_6 heterojunctions model when strained by cellular force. (G) Scheme of osteogenic differentiation mechanism of mBMSCs co-cultured with piezoelectric heterojunctions. (H) FE-SEM morphology of cells on different substrates including Ti substrates, TiO_2 nanowires on Ti substrates, and TiO_2/Bi_2WO_6 heterojunctions

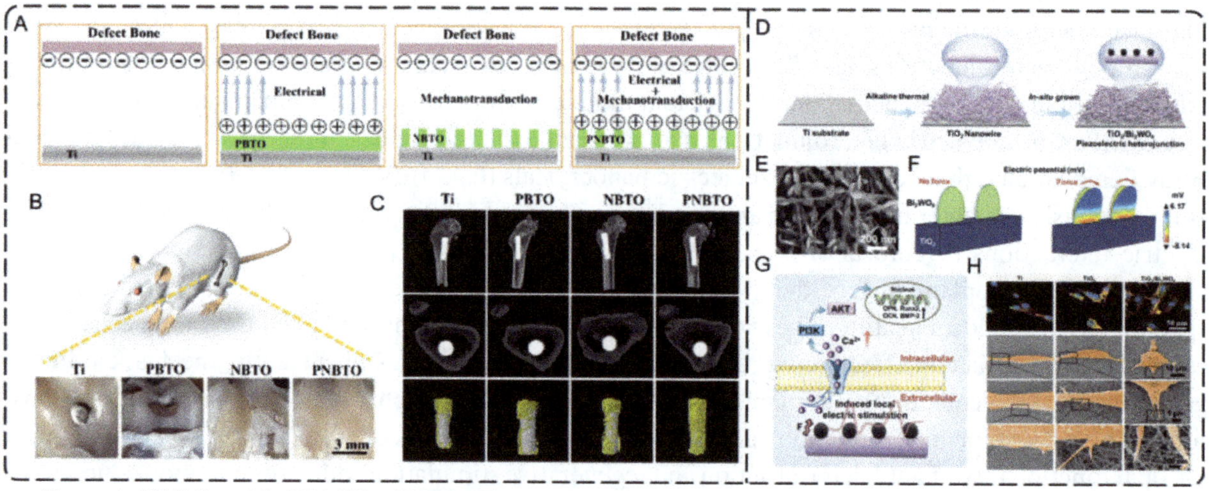

(Fan et al., 2025) Copyright 2024, Wiley-VCH GmbH.

Despite the promising potential of electroactive biomaterials in bone regeneration, several challenges remain. One of the major challenges is the long-term stability of these materials in vivo. Conductive polymers, for example, can degrade over time, which may limit their effectiveness for prolonged tissue regeneration. Additionally, the optimization of material properties, such as conductivity, mechanical strength, and biocompatibility, is still an ongoing area of research. Future directions for electroactive biomaterials in bone regeneration include the development of materials that can provide dynamic electrical signals, mimicking the natural electrical activity of the bone. Another promising avenue is the incorporation of bioactive molecules, such as growth factors or cytokines, into electroactive materials to enhance their regenerative potential. Additionally, advanced fabrication techniques, such as 3D printing, could allow for the creation of highly tailored electroactive scaffolds with precise control over their geometry and electrical properties.

5.2 Nerve Regeneration

The nervous system relies heavily on electrical signals for proper functioning, and nerve regeneration is highly dependent on the ability to guide neuronal growth and support cell survival after injury.(Chu et al., 2022) Electroactive biomaterials have shown great potential in nerve regeneration by providing electrical cues that enhance neural cell differentiation, axonal growth, and synaptic formation. This has led to the development of electroactive scaffolds that mimic the electrical properties of nerve tissue and provide a conducive environment for neuronal repair.

Nerve regeneration is a crucial aspect of tissue repair that presents significant challenges in both clinical and research settings. Nerve injuries, especially those involving the central nervous system (CNS) and peripheral nervous system (PNS), can lead to debilitating functional impairments. The regenerative potential of the nervous system is limited, particularly in the CNS, where axon regeneration is often unsuccessful. This has led to intense interest in developing strategies to enhance nerve regeneration, with electroactive biomaterials emerging as a promising solution. Electroactive biomaterials offer the ability to regulate cell behavior through the application of electrical fields, which can mimic the natural bioelectric signals present in nerve tissues. These signals play an essential role in nerve development, injury response, and repair. Electroactive materials, such as conductive polymers, hydrogels, and composites, can be engineered to provide localized electrical cues that influence various aspects of nerve regeneration, from neuronal growth and differentiation to axon guidance and synaptic formation.

Endogenous electric fields play an essential role in nerve regeneration, especially during development and following injury. In the peripheral nervous system, Schwann cells generate local electric fields that guide the migration of neurons and facilitate axon growth. In response to an injury, such as a nerve cut, these fields can influence the regeneration process by guiding axonal growth toward the target tissue. Moreover, endogenous electric fields have been shown to enhance nerve cell migration, promote the expression of neurotrophic factors (such as nerve growth factor [NGF] and brain-derived neurotrophic factor [BDNF]), and accelerate myelination in the case of PNS injuries. These bioelectric signals are pivotal in regulating neuronal plasticity, synaptic repair, and even the reestablishment of functional connections in the injured nervous system. The challenge lies in effectively replicating these endogenous electric fields with exogenous stimuli provided by electroactive biomaterials, which can aid in stimulating the healing process and promoting more robust nerve regeneration.

Electroactive biomaterials are designed to replicate the natural electrical cues found within the body, providing an effective way to influence cellular behavior and promote nerve regeneration. These materials can deliver electrical signals at various intensities, frequencies, and durations, which can be precisely controlled to stimulate neurogenesis and axon outgrowth. The main classes of electroactive biomaterials used in nerve regeneration include conductive polymers, conductive hydrogels, carbon-based nanomaterials, and piezoelectric materials. These materials can be used as scaffolds to support nerve regeneration, enhance the growth of neurons, or even deliver electrical stimulation in real-time.

Conductive polymers, such as polypyrrole (PPy), polyaniline (PANI), and poly(3,4-ethylenedioxythiophene) (PEDOT), have gained significant attention due to their biocompatibility and electrical conductivity. These polymers can be doped or functionalized with bioactive molecules to enhance their interaction with nerve cells. Studies have shown that conductive polymers can promote neuronal adhesion, differentiation, and axon growth, which are critical for nerve regeneration. These materials are often used in conjunction with scaffolds to create a conductive matrix that mimics the natural extracellular environment of nerve tissues.

Conductive hydrogels are another class of electroactive materials that are particularly attractive for nerve regeneration. Hydrogels are highly hydrated, flexible, and biocompatible, which makes them ideal for soft tissue applications like nerve repair. The incorporation of conductive nanoparticles, such as graphene or gold nanoparticles, into hydrogels allows for the creation of a conductive network capable of transmitting electrical signals. These hydrogels can be tailored to respond to electrical stimuli, which can then promote the alignment of nerve cells, axon growth, and overall tissue regeneration.

Designing hydrogels scaffolds that simultaneously exhibit soft tissue-like mechanical properties, high electrical conductivity, and excellent biocompatibility remains a major challenge in nerve tissue engineering, especially for spinal cord repair. A recent study addressed this limitation by developing a conducting polymer hydrogel (CPH) that combines plant-derived polyphenol (tannic acid, TA) with PPy chains (Figure 3A). (Zhou et al., 2018) The hydrogel's conductivity (ranging from 0.05 to 0.18 S/cm) and mechanical properties (0.3-2.2 kPa) can be precisely tuned via TA concentration, allowing the scaffold to match both the softness and the electrical requirements of native spinal cord tissue. In *in vitro* experiments, CPHs with higher conductivity promoted neuronal differentiation of neural stem cells (NSCs) while minimizing astrocyte formation, suggesting an electric field–mediated bias toward neuronal lineage commitment. *In vivo* tests further showed that, upon implantation in spinal cord injury (SCI) lesions, CPHs with relatively high conductivity could stimulate endogenous NSC neurogenesis, leading to appreciable improvements in locomotor recovery. Remarkably, this tissue repair occurred without the need for additional therapeutic agents or biologics, underscoring the inherent regenerative potential of the hydrogel's physical and electrical properties (Figure 3B-D). These findings highlight the important design principles—namely, biocompatibility, mechanical softness, and elevated electrical conductivity—that can drive functional neural regeneration in SCI. By tailoring the synergy between polyphenol-based crosslinking and conductive polymer doping, such hydrogels open new avenues for SCI therapy and could serve as a platform for further combination treatments targeting enhanced neural repair.

Electroconductive hydrogels are emerging as an attractive platform for spinal cord injury repair due to their ability to closely resemble the electrical conductivity and mechanical softness of native neural tissue. However, the implantation of these hydrogels can potentially exacerbate inflammation, thereby limiting their therapeutic efficacy. To address this challenge, researchers recently developed electroconductive hydrogels loaded with bone marrow stem cell–derived exosomes (BMSC-exosomes), capitalizing on the immunomodulatory and tissue-regenerative properties of exosomes to achieve synergistic SCI treatment (Figure 2E). (Fan et al., 2022) Mechanistically, exosome supplementation within the conductive hydrogel fosters microglial M2 polarization through modulation of the NF-κB pathway, which contributes to an anti-inflammatory environment in the injured spinal cord. In addition, these hydrogel constructs simultaneously stimulate neuronal and oligodendrocyte differentiation of neural stem cells while restraining astrocyte lineage commitment, effectively guiding more beneficial cellular outcomes. The PTEN/PI3K/AKT/mTOR pathway is implicated in enhancing axon outgrowth, further supporting nerve regeneration. *In vivo* studies in an SCI mouse model demonstrated that exosome-loaded electroconductive hydrogels significantly reduce CD68-positive microglia (an indicator of inflammation), promote the recruitment of endogenous NSCs, and facilitate robust neuronal and axonal regeneration at the lesion site (Figure 2F-H). As a consequence, early-stage functional recovery is markedly improved, underscoring the therapeutic advantage of combining electroconductive biomaterials with BMSC-exosomes for neural repair. Together, these findings highlight a powerful immunomodulatory and neuroregenerative strategy, offering substantial promise for future clinical translation in spinal cord regenerative medicine.

*Figure 3. (A) Schematic illustration of soft conducting polymer hydrogels cross-Linked and doped by tannic acid for spinal cord injury repair. (B) Graphical representation of a "C"-shape, semitubular CPH that was implanted as a bridge to cover the spinal cord hemisection gap. (C) Representative images and quantification graphs of transverse spinal cord sections stained with Tuj1. Scale bars = 150 μm. *p < 0.05 and **p < 0.01, ***p < 0.001. (D) Immunohistofluorescence images of transverse spinal cord sections obtained from animals in the sham, SCI, and hydrogel groups at 6 weeks. (Zhou et al., 2018) Copyright 2018, American Chemical Society. (E) Illustration of how the GMPE hydrogel can reduce early inflammation, enhance NSCs recruitment and promote myelin-associated axonal regrowth to synergistically promote locomotor recovery after spinal cord hemisection. (F) HE staining showing the morphology of transverse spinal cord sections of sham and hydrogel implantation after SCI. Scale bars: 500 μm. (G) Representative reconstructions of the transverse spinal cord sections. (H) CMAPs results from normal mice and those in different hydrogel implantation groups 6 weeks post-treatment*

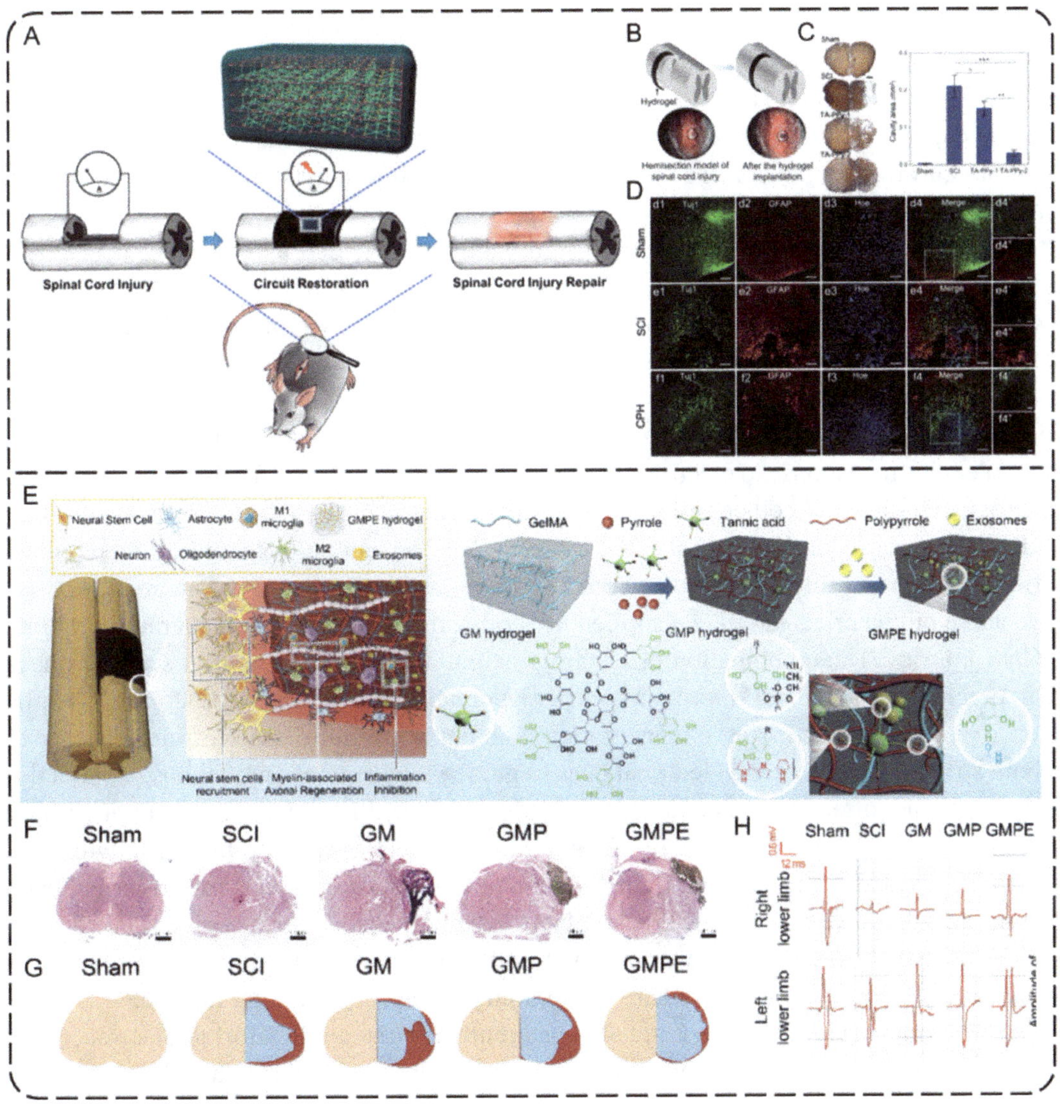

(Fan et al., 2022) Copyright 2022, Wiley-VCH GmbH

Carbon-based nanomaterials, including graphene and carbon nanotubes (CNTs), are increasingly used in nerve regeneration applications due to their excellent electrical conductivity, mechanical properties, and biocompatibility.(Ku et al., 2013) These nanomaterials can be incorporated into scaffolds or hydrogels to create a conductive environment that enhances neuronal differentiation and promotes axon outgrowth. Carbon nanotubes, in particular, have been shown to guide axons along their length, providing a physical and electrical scaffold for nerve cells to grow and repair damaged nerves.

Piezoelectric materials, which generate electrical charges in response to mechanical stress, are particularly interesting for nerve regeneration because they can mimic the mechanical properties of native tissues.(Das et al., 2024) The application of mechanical stress to piezoelectric materials can generate localized electric fields that influence the behavior of neurons and Schwann cells. These materials can be used in applications where mechanical deformation is anticipated, such as in nerve conduits or scaffolds that need to conform to the natural movement of nerve tissues during regeneration.

One of the most significant ways electroactive biomaterials enhance nerve regeneration is by applying electrical fields that stimulate neuronal behavior. Electrical stimulation has been shown to increase neuronal migration, promote axonal growth, and enhance synaptic formation.(Yao & Li, 2016) Electrical cues are known to modulate the activity of ion channels and other signaling pathways that regulate cell growth and differentiation. For example, studies have demonstrated that electric fields increase the expression of neurotrophic factors like NGF and BDNF, which play crucial roles in neuronal survival and axon guidance.

Electroactive biomaterials can mimic the directional cues provided by endogenous electric fields, helping to guide the growth of regenerating axons. This is particularly important in cases of nerve injury, where the guidance of axons to the correct target tissue is essential for restoring function. The electric fields generated by these biomaterials can interact with the cytoskeleton of axons, promoting their elongation and directional growth.

In addition to providing electrical stimulation, electroactive biomaterials can be designed to release bioactive molecules in response to electrical signals. This approach allows for the controlled delivery of growth factors, cytokines, and other molecules that support nerve regeneration. For instance, the release of NGF, BDNF, or vascular endothelial growth factor (VEGF) can be triggered by the application of electric fields, which can further promote axon outgrowth, cell survival, and tissue repair.

Electroactive biomaterials have been applied in various models of nerve regeneration, ranging from PNS to CNS injuries. The combination of electrical stimulation and biomaterial scaffolds has shown promise in preclinical models for improving nerve regeneration. One of the most common applications of electroactive biomaterials in nerve regeneration is in nerve conduits and scaffolds. These scaffolds provide both structural support and electrical cues to guide the regeneration of axons. Conductive polymers or hydrogels are often used to create nerve conduits that facilitate axonal growth and improve the functional recovery of damaged nerves. In addition to using electroactive biomaterials within the scaffolds, external electrical stimulation has been combined with these materials to further enhance nerve regeneration. The use of wearable devices or implants that apply low-frequency electrical currents to the injured nerve site has shown promise in promoting faster and more effective regeneration. This approach can be particularly beneficial in cases where nerve injury extends over a long distance.

Despite the promising results, several challenges remain in the application of electroactive biomaterials for nerve regeneration. The complexity of nerve injuries, particularly those involving the CNS, requires advanced strategies that can overcome barriers such as the limited regenerative capacity of CNS neurons and the formation of scar tissue. Future research should focus on optimizing the electrical properties of electroactive biomaterials to better mimic natural bioelectric cues. The development of more advanced, multifunctional materials that can provide both electrical stimulation and controlled release of growth factors is also a promising avenue for enhancing nerve regeneration. Additionally, exploring the synergistic effects of electroactive biomaterials combined with gene therapy or stem cell therapy could provide a more comprehensive approach to nerve repair.

Electroactive biomaterials represent an exciting frontier in nerve regeneration, offering the potential to enhance neuronal growth, axon regeneration, and functional recovery after nerve injuries. By mimicking the endogenous electrical cues present in nerve tissues, these materials can influence cellular behavior and promote more efficient nerve regeneration. While there are still challenges to overcome, the development of advanced electroactive biomaterials holds great promise for treating nerve injuries and disorders.

5.3 Skin Regeneration and Others

Skin regeneration is a fundamental process in wound healing and tissue repair. The skin serves as the body's first line of defense against environmental damage and pathogens, and its ability to regenerate efficiently is vital for overall health. However, when skin is severely damaged, such as in deep burns or chronic wounds, the regenerative capacity is often compromised. This has driven research into enhancing skin regeneration using electroactive biomaterials, which can provide crucial electrical cues to regulate cell behavior, stimulate wound healing, and promote tissue repair. Apart from skin regeneration, electroactive biomaterials have shown promise in other tissue regeneration applications, including cartilage, muscle, and vascular regeneration. These materials can be designed to provide electrical signals that mimic the body's endogenous bioelectric fields, thus creating a favorable environment for tissue regeneration.

The skin is a highly dynamic organ with a remarkable ability to repair itself after injury. Endogenous electric fields, generated by the electrical potential across the skin membrane, are involved in various stages of skin wound healing, including inflammation, proliferation, and tissue remodeling.(Reid & Zhao, 2014) These bioelectric signals are crucial for orchestrating cellular activities such as migration, proliferation, differentiation, and extracellular matrix (ECM) production, all of which are essential for skin regeneration. In the context of wound healing, electrotaxis refers to the movement of cells in response to an electric field. Cells such as keratinocytes, fibroblasts, and endothelial cells, which are involved in wound healing, exhibit electrotactic behavior when exposed to electric fields. The application of external electric fields has been shown to accelerate the migration of these cells to the wound site, promoting faster healing and tissue closure.

Electric fields also influence cellular proliferation and differentiation during skin regeneration. In response to electrical cues, skin cells can proliferate more rapidly, contributing to the formation of new tissue. For example, electrical stimulation has been shown to enhance the differentiation of stem cells into keratinocytes and fibroblasts, the primary cell types involved in skin repair. Additionally, electrical signals can stimulate collagen production by fibroblasts, helping to restore the ECM and support tissue regeneration.(Das et al., 2024)

Angiogenesis, the formation of new blood vessels, is a critical step in skin regeneration and wound healing. Endogenous electric fields play a role in promoting angiogenesis by stimulating the migration of endothelial cells and the formation of new blood vessels. Electroactive biomaterials can further enhance this process by providing electrical cues that promote endothelial cell migration and capillary network formation, thus improving blood supply to the regenerating tissue.

The application of electroactive biomaterials in skin regeneration has gained significant attention due to their ability to provide controlled electrical stimulation to injured skin tissues.(Zhang et al., 2025) These materials can be used in various forms, such as conductive dressings, scaffolds, and hydrogels, to provide localized electrical signals that mimic natural bioelectric fields. The key benefit of using electroactive biomaterials for skin regeneration is their ability to enhance the body's own repair mechanisms by promoting cellular behaviors that accelerate tissue healing. Conductive dressings, made from materials such as conductive polymers, hydrogels, or carbon-based nanomaterials, have been explored as a means to accelerate wound healing. These dressings can deliver electrical stimulation directly to the wound site, which can enhance cellular migration, proliferation, and ECM production. For example, conductive hydrogels can be incorporated into wound dressings to create a soft, flexible, and biocompatible environment that promotes skin regeneration while delivering controlled electrical signals to the tissue.

Electroactive scaffolds are designed to support the growth of new tissue by providing both structural and electrical cues. These scaffolds can be made from conductive polymers, carbon nanotubes, or graphene composites, and they are capable of transmitting electrical signals to the surrounding cells. When implanted at the site of skin injury, these scaffolds provide a conductive matrix that guides the migration and differentiation of skin cells. Additionally, they can be engineered to release bioactive molecules, such as growth factors or cytokines, that further enhance the healing process. Electrospun nanofibers have emerged as an innovative approach for creating electroactive scaffolds. These nanofibers can be fabricated from conductive polymers or carbon-based nanomaterials, creating highly porous structures that mimic the ECM of the skin. The high surface area and flexibility of electrospun nanofibers allow for efficient cell adhesion, proliferation, and differentiation, while the electrical conductivity of the material promotes the regeneration of skin tissue. Furthermore, the use of electrospun nanofibers can enhance the mechanical properties of the scaffold, providing the necessary strength and elasticity to support the regeneration of skin.

Wound reepithelialization is strongly influenced by endogenous electric fields, which arise from transepithelial potential differences. While negative-pressure wound therapy (NPWT) is widely recognized as an effective clinical intervention for extensive or chronic wounds—primarily by reducing necrotic tissue and infected debris to foster a pro-proliferative microenvironment—it inadvertently promotes electrolyte depletion, thereby weakening the endogenous EF and impeding the reepithelialization process. To overcome this limitation, researchers have developed an electrogenerative dressing (EGD) that combines triboelectric nanogenerators with NPWT (Figure 4A).(Luo et al., 2023) The triboelectric component harnesses the mechanical deformation imposed by negative pressure to generate electricity, thus reestablishing a stable and safe EF at the wound interface. In *in vitro* studies, the EGD-induced EF significantly accelerated epithelial electrotaxis and guided macrophages toward a reparative M2 phenotype, thereby supporting a more efficient and orderly progression of wound healing (Figure 4B). Translational research further confirmed that this approach counteracts the diminished EF caused by continuous negative-pressure drainage *in vivo*, enabling faster wound closure by fostering an earlier transition from the inflammatory to the proliferative phase, and by directing epithelial cell migration and proliferation (Figure 4C). Over extended application, EGD therapy has been shown to improve tissue remodeling, resulting in a mature epithelial layer, well-organized extracellular matrix, and reduced scarring. Thus, compared with NPWT alone, EGD noninvasively integrates and enhances all essential stages of wound repair, demonstrating strong potential for clinical wound management where large or chronic wounds challenge conventional interventions.

Although electrical stimulation is commonly applied in various medical and physiological contexts, its efficacy for inflammatory skin diseases (ISDs) has been hampered by the skin's pathological cuticle barrier and insufficient synergy with traditional chemotherapeutics. To address these challenges, a recent study introduced a wearable, battery-free, multi-component drug-loaded electronic microneedle (mD-eMN) system, composed of flexible triboelectric nanogenerators (TENGs) integrated with remodeled metal microneedles carrying multiple chemical agents (Figure 4D). (Qian et al., 2023) Upon application, this system rapidly delivers these drugs deep into ISD lesions; meanwhile, pulsed electrons generated by TENGs facilitate efficient penetration into cells, driving targeted immunomodulation. Additionally, these electronically delivered pulses help restore skin tissue homeostasis, mitigating inflammation more effectively than either standalone ES or chemotherapy. In a psoriasis model, mD-eMN treatment significantly suppressed skin inflammation, outperforming conventional ES or topical drug therapy (Figure 4F). Importantly, this flexible electronic–chemical platform shows potential for broader applications in ISDs, including diabetic ulcers and cutaneous tumors, highlighting a promising strategy for advanced skin disease management.

Figure 4. (A) Mechanisms of promoting wound healing by the typical EGD system. The EGD accelerates the polarization of macrophages to M2 macrophages, strengthens cell proliferation and migration, augments angiogenesis and reepithelialization, enhances the orientation of collagen, and decreases the scars. (B) Illustration of macrophage polarization from M0 to M2 and photograph of macrophage shape. (C) Multiple effects of the EGD on wounds, including negative pressure and electrical field and Photograph of the EGD–NCD system in a pig. (Luo et al., 2023) Copyright 2023, Wiley-VCH GmbH. (D) Structure diagram of the mD-eMN system: The mD-eMN system is composed of a TENG and a mD-MN, which can be easily worn on the human body to generate physiological synergistic ES to promote drug absorption and tissue homeostasis reconstruction. (F) Photographs of normal and psoriasis model rats and H&E staining of psoriasis rat skin induced by IMQ and normal skin as the control group

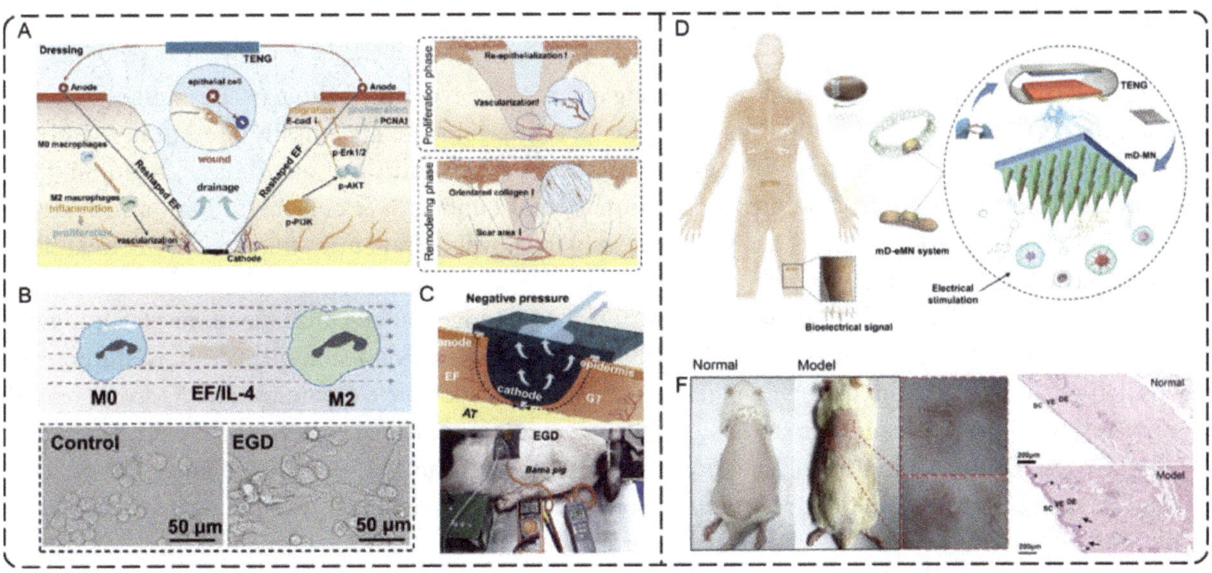

(Qian et al., 2023) Copyright 2023, Wiley-VCH GmbH.

In addition to skin regeneration, electroactive biomaterials are also being explored for use in other tissue regeneration applications, including cartilage, muscle, and vascular regeneration. These materials can provide electrical cues that mimic the natural bioelectric fields present in these tissues, helping to promote regeneration and functional recovery.

Cartilage injuries, particularly those in weight-bearing joints, often fail to heal naturally due to the limited regenerative capacity of cartilage tissue. Electroactive biomaterials, such as conductive hydrogels or scaffolds, can be used to promote chondrogenesis (cartilage formation) by applying electrical stimulation to the affected area.(Zhou et al., 2023) These materials can help guide the differentiation of stem cells into chondrocytes and promote the synthesis of the extracellular matrix that supports cartilage tissue.

Muscle injuries, such as those caused by trauma or degenerative diseases, can lead to long-term functional deficits. Electroactive biomaterials are being investigated for their potential to enhance muscle regeneration by promoting myogenesis (muscle cell formation) and tissue repair.(Park et al., 2025) Electrical stimulation has been shown to improve the alignment and fusion of muscle fibers, accelerating muscle regeneration and improving muscle strength.

Vascular regeneration is essential for restoring blood flow to damaged tissues and organs. Electroactive biomaterials can promote endothelial cell migration and capillary formation, facilitating the creation of new blood vessels in the injured tissue.(Li et al., 2022) This can enhance tissue repair and restore proper function, particularly in the context of ischemic injuries or vascular diseases.

Electroactive biomaterials hold significant promise for advancing tissue regeneration across a wide range of applications, including skin, cartilage, muscle, and vascular regeneration. By mimicking the body's natural bioelectric fields, these materials can provide electrical cues that regulate cell behavior, promote tissue repair, and accelerate healing. While challenges remain in optimizing the design and functionality of these materials, ongoing research and development in this field are paving the way for novel therapeutic strategies to treat a variety of tissue injuries and disorders.

6. CONCLUSION AND OUTLOOK

Electroactive biomaterials represent a promising and rapidly evolving field in tissue engineering and regenerative medicine. By mimicking the endogenous electrical signals naturally present in the body, these materials offer a unique strategy for enhancing cellular behaviors critical to tissue repair and regeneration. Their ability to provide precise, spatiotemporal control over cell migration, proliferation, differentiation, and matrix synthesis has paved the way for novel therapeutic strategies in the treatment of a wide array of diseases and injuries.

In this review, we have highlighted the importance of endogenous electric fields in physiological processes and their role in regulating cellular behavior. The fabrication techniques employed to create electroactive biomaterials, including conductive polymers, piezoelectric ceramics, and hydrogel-based composites, have allowed researchers to design and engineer materials that can effectively simulate these electrical cues. The development of such materials has been complemented by advanced fabrication strategies, such as electrospinning, 3D printing, and solvothermal methods, which enable the creation of biomimetic structures with tailored electrical properties.

The regulation of cell behavior through electroactive biomaterials has demonstrated remarkable potential in several tissue regeneration applications, including bone, nerve, and skin regeneration. In bone regeneration, piezoelectric materials and conductive polymers have shown great promise in promoting osteoblast differentiation and enhancing bone matrix formation. Similarly, electroactive biomaterials have been successfully employed in nerve regeneration by providing electrical cues that guide axonal growth and support neuronal differentiation. In skin regeneration, electroactive hydrogels and nanofiber scaffolds have accelerated wound healing by stimulating fibroblast migration and angiogenesis.

Despite the significant progress made in this field, several challenges remain in the clinical translation of electroactive biomaterials. One of the key issues is the biocompatibility and long-term stability of these materials in vivo. While many electroactive materials demonstrate excellent short-term regenerative effects, ensuring their safety and functionality over prolonged periods is crucial for their widespread clinical adoption. Moreover, the integration of electroactive materials with advanced therapies, such as gene therapy, stem cell therapy, and growth factor delivery systems, offers an exciting avenue for enhancing the therapeutic outcomes of tissue regeneration.

Furthermore, the complexity of tissue architecture and the need for precise control over cellular microenvironments present additional hurdles. The development of 3D scaffolds with highly organized architectures that mimic the native tissue structure and provide dynamic electrical cues is a significant research direction. Additionally, the modulation of electrical properties, such as conductivity, piezoelectricity, and electrochemical activity, needs to be optimized for different tissue types to ensure effective tissue repair.

Looking to the future, several exciting opportunities lie ahead for the advancement of electroactive biomaterials in tissue regeneration. The integration of multifunctional electroactive materials with smart polymers, which respond to multiple external stimuli (e.g., light, pH, temperature, and electrical fields), holds great promise for creating more responsive and adaptable systems. Furthermore, personalized medicine will play a crucial role in optimizing electroactive biomaterials for individual patients. By utilizing patient-specific data, such as genetic and molecular profiles, custom-designed biomaterials could be developed to meet the unique needs of each patient, improving the outcomes of tissue repair and regeneration.

The combination of electroactive biomaterials with other cutting-edge technologies, such as artificial intelligence (AI) and machine learning (ML), offers a novel approach to enhance the design and prediction of optimal biomaterial properties. These tools can assist in the analysis of large datasets, allowing for better material selection, design optimization, and prediction of in vivo behavior, thus accelerating the clinical translation of electroactive biomaterials.

In conclusion, electroactive biomaterials hold tremendous potential in the field of tissue regeneration, with their ability to regulate cellular behavior through electrical signals. The advances in fabrication techniques, material design, and understanding of the underlying mechanisms of electroactivity have opened new frontiers in regenerative medicine. As research continues, addressing the challenges of biocompatibility, long-term stability, and material optimization will be crucial for the clinical translation of these materials. With continued innovation and interdisciplinary collaboration, electroactive biomaterials may soon become integral components of next-generation therapies for tissue regeneration, ultimately improving the quality of life for patients with a variety of injuries and degenerative diseases.

ACKNOWLEDGEMENT

This work was financially supported by Guangdong S&T Programme (2024B0101080001), the National Natural Science Foundation of China (52101285, U21A2055, U22A20160), the National Key Research and Development Program of China (2022YFC2406000).

REFERENCES

Acosta, M., Santiago, M. D., & Irvin, J. A. (2022). Electrospun conducting polymers: Approaches and applications. *Materials (Basel)*, *15*(24), 8820. DOI: 10.3390/ma15248820 PMID: 36556626

Aggarwal, V., Tuli, H. S., Varol, A., Thakral, F., Yerer, M. B., Sak, K., Varol, M., Jain, A., Khan, M. A., & Sethi, G. (2019). Role of reactive oxygen species in cancer progression: Molecular mechanisms and recent advancements. *Biomolecules*, *9*(11), 735. DOI: 10.3390/biom9110735 PMID: 31766246

Ashtari, K., Nazari, H., Ko, H., Tebon, P., Akhshik, M., Akbari, M., Alhosseini, S. N., Mozafari, M., Mehravi, B., Soleimani, M., Ardehali, R., Ebrahimi Warkiani, M., Ahadian, S., & Khademhosseini, A. (2019). Electrically conductive nanomaterials for cardiac tissue engineering. *Advanced Drug Delivery Reviews*, *144*, 162–179. DOI: 10.1016/j.addr.2019.06.001 PMID: 31176755

Babaie, A., Bakhshandeh, B., Abedi, A., Mohammadnejad, J., Shabani, I., Ardeshirylajimi, A., Reza Moosavi, S., Amini, J., & Tayebi, L. (2020). Synergistic effects of conductive PVA/PEDOT electrospun scaffolds and electrical stimulation for more effective neural tissue engineering. *European Polymer Journal*, *140*, 110051. DOI: 10.1016/j.eurpolymj.2020.110051

Bach-Toledo, L., Hryniewicz, B. M., Marchesi, L. F., Dall'Antonia, L. H., Vidotti, M., & Wolfart, F. (2020). Conducting polymers and composites nanowires for energy devices: A brief review. *Materials Science for Energy Technologies*, *3*, 78–90. DOI: 10.1016/j.mset.2019.09.006

Banerjee, A. N. (2018). Graphene and its derivatives as biomedical materials: Future prospects and challenges. *Interface Focus*, *8*(3), 20170056. DOI: 10.1098/rsfs.2017.0056 PMID: 29696088

Bianchi, M., Guzzo, S., Lunghi, A., Greco, P., Pisciotta, A., Murgia, M., Carnevale, G., Fadiga, L., & Biscarini, F. (2023). Synergy of nanotopography and electrical conductivity of PEDOT/PSS for enhanced neuronal development. *ACS Applied Materials & Interfaces*, *15*(51), 59224–59235. DOI: 10.1021/acsami.3c15278 PMID: 38091494

Brusatin, G., Panciera, T., Gandin, A., Citron, A., & Piccolo, S. (2018). Biomaterials and engineered microenvironments to control YAP/TAZ-dependent cell behaviour. *Nature Materials*, *17*(12), 1063–1075. DOI: 10.1038/s41563-018-0180-8 PMID: 30374202

Burnstine-Townley, A., Eshel, Y., & Amdursky, N. (2020). Conductive scaffolds for cardiac and neuronal tissue engineering: Governing factors and mechanisms. *Advanced Functional Materials*, *30*(18), 1901369. DOI: 10.1002/adfm.201901369

Casella, A., Panitch, A., & Leach, J. K. (2021). Endogenous electric signaling as a blueprint for conductive materials in tissue engineering. *Bioelectricity*, *3*(1), 27–41. DOI: 10.1089/bioe.2020.0027 PMID: 34476376

Chen, L., Yang, J., Cai, Z., Huang, Y., Xiao, P., Wang, J., Wang, F., Huang, W., Cui, W., & Hu, N. (2024). Electroactive biomaterials regulate the electrophysiological microenvironment to promote bone and cartilage tissue regeneration. *Advanced Functional Materials, 2314079*. Advance online publication. DOI: 10.1002/adfm.202314079

Chen, L., Yang, J., Cai, Z., Huang, Y., Xiao, P., Wang, J., Wang, F., Huang, W., Cui, W., & Hu, N. (2024). Electroactive biomaterials regulate the electrophysiological microenvironment to promote bone and cartilage tissue regeneration. *Advanced Functional Materials*, *34*(23), 2314079. DOI: 10.1002/adfm.202314079

Chu, X.-L., Song, X.-Z., Li, Q., Li, Y.-R., He, F., Gu, X.-S., & Ming, D. (2022). Basic mechanisms of peripheral nerve injury and treatment via electrical stimulation. *Neural Regeneration Research*, *17*(10), 2185–2193. DOI: 10.4103/1673-5374.335823 PMID: 35259827

Das, K. K., Basu, B., Maiti, P., & Dubey, A. K. (2024). Interplay of piezoelectricity and electrical stimulation in tissue engineering and regenerative medicine. *Applied Materials Today*, *39*, 102332. DOI: 10.1016/j.apmt.2024.102332

Deng, W., Zhou, Y., Libanori, A., Chen, G., Yang, W., & Chen, J. (2022). Piezoelectric nanogenerators for personalized healthcare. *Chemical Society Reviews*, *51*(9), 3380–3435. DOI: 10.1039/D1CS00858G PMID: 35352069

Deschênes-Simard, X., Malleshaiah, M., & Ferbeyre, G. (2023). Extracellular Signal-Regulated Kinases: One Pathway, Multiple Fates. *Cancers (Basel)*, *16*(1). Advance online publication. DOI: 10.3390/cancers16010095 PMID: 38201521

Dixon, D. T., & Gomillion, C. T. (2023). 3D-Printed conductive polymeric scaffolds with direct current electrical stimulation for enhanced bone regeneration. *Journal of Biomedical Materials Research. Part B, Applied Biomaterials*, *111*(7), 1351–1364. DOI: 10.1002/jbm.b.35239 PMID: 36825765

Esmaeili, H., Patino-Guerrero, A., Hasany, M., Ansari, M. O., Memic, A., Dolatshahi-Pirouz, A., & Nikkhah, M. (2022). Electroconductive biomaterials for cardiac tissue engineering. *Acta Biomaterialia*, *139*, 118–140. DOI: 10.1016/j.actbio.2021.08.031 PMID: 34455109

Fan, L., Liu, C., Chen, X., Zheng, L., Zou, Y., Wen, H., Guan, P., Lu, F., Luo, Y., Tan, G., Yu, P., Chen, D., Deng, C., Sun, Y., Zhou, L., & Ning, C. (2022). Exosomes-loaded electroconductive hydrogel synergistically promotes tissue repair after spinal cord injury via immunoregulation and enhancement of myelinated axon growth. *Advancement of Science*, *9*(13), e2105586. DOI: 10.1002/advs.202105586 PMID: 35253394

Fan, Y., Zhai, J., Wang, Z., Yin, Z., Chen, H., Ran, M., Zhu, Z., Ma, Y., Ning, C., Yu, P., & Mao, C. (2025). Piezoelectric heterojunctions as bacteria-killing bone-regenerative implants. *Advanced Materials*, *37*(1), e2413171. DOI: 10.1002/adma.202413171 PMID: 39460412

Haan, N., & Song, B. (2014). Therapeutic application of electric fields in the injured nervous system. *Advances in Wound Care (New Rochelle, N.Y.)*, *3*(2), 156–165. DOI: 10.1089/wound.2013.0450 PMID: 24761356

Huang, J., Pan, X., & Yan, N. (2024). Structural biology and molecular pharmacology of voltage-gated ion channels. *Nature Reviews. Molecular Cell Biology*, *25*(11), 904–925. DOI: 10.1038/s41580-024-00763-7 PMID: 39103479

Jacob, J., More, N., Kalia, K., & Kapusetti, G. (2018). Piezoelectric smart biomaterials for bone and cartilage tissue engineering. *Inflammation and Regeneration*, *38*(1), 2. DOI: 10.1186/s41232-018-0059-8 PMID: 29497465

Khare, D., Basu, B., & Dubey, A. K. (2020). Electrical stimulation and piezoelectric biomaterials for bone tissue engineering applications. *Biomaterials*, *258*, 120280. DOI: 10.1016/j.biomaterials.2020.120280 PMID: 32810650

Kolosnjaj-Tabi, J., Gibot, L., Fourquaux, I., Golzio, M., & Rols, M.-P. (2019). Electric field-responsive nanoparticles and electric fields: Physical, chemical, biological mechanisms and therapeutic prospects. *Advanced Drug Delivery Reviews*, *138*, 56–67. DOI: 10.1016/j.addr.2018.10.017 PMID: 30414494

Ku, S. H., Lee, M., & Park, C. B. (2013). Carbon-based nanomaterials for tissue engineering. *Advanced Healthcare Materials*, *2*(2), 244–260. DOI: 10.1002/adhm.201200307 PMID: 23184559

Levin, M. (2021). Bioelectric signaling: Reprogrammable circuits underlying embryogenesis, regeneration, and cancer. *Cell*, *184*(8), 1971–1989. DOI: 10.1016/j.cell.2021.02.034 PMID: 33826908

Li, C., Xiao, C., Zhan, L., Zhang, Z., Xing, J., Zhai, J., Zhou, Z., Tan, G., Piao, J., & Zhou, Y. (2022). Wireless electrical stimulation at the nanoscale interface induces tumor vascular normalization. *Bioactive Materials*, *18*, 399–408. DOI: 10.1016/j.bioactmat.2022.03.027 PMID: 35415302

Li, C., Yu, P., Wang, Z., Long, C., Xiao, C., Xing, J., Dong, B., Zhai, J., Zhou, L., & Zhou, Z. (2023). Electro-mechanical coupling directs endothelial activities through intracellular calcium ion deployment. *Materials Horizons*, *10*(11), 4903–4913. DOI: 10.1039/D3MH01049J PMID: 37750251

Li, Y., Chen, J., Liu, S., Wang, Z., Zhang, S., Mao, C., & Wang, J. (2024). Biodegradable piezoelectric polymer for cartilage remodeling. *Matter*, *7*(4), 1631–1643. DOI: 10.1016/j.matt.2024.01.034

Li, Y., Wei, Y., Liao, J., Hao, Y., Ning, C., Jiang, L., & Wang, S. (2016). Surface wettability switched cell adhesion and detachment on conducting polymer nanoarray. *Advanced Materials Interfaces*, *3*(19). Advance online publication. DOI: 10.1002/admi.201600598

Li, Y., Yan, X., Liu, J., Li, L., Hu, X., Sun, H., & Tian, J. (2014). Pulsed electromagnetic field enhances brain-derived neurotrophic factor expression through L-type voltage-gated calcium channel-and Erk-dependent signaling pathways in neonatal rat dorsal root ganglion neurons. *Neurochemistry International*, *75*, 96–104. DOI: 10.1016/j.neuint.2014.06.004 PMID: 24937769

Liang, Y., & Goh, J. C. (2020). Polypyrrole-incorporated conducting constructs for tissue engineering applications: A review. *Bioelectricity*, *2*(2), 101–119. DOI: 10.1089/bioe.2020.0010 PMID: 34471842

Liu, J., Xiao, Q., Xiao, J., Niu, C., Li, Y., Zhang, X., Zhou, Z., Shu, G., & Yin, G. (2022). Wnt/β-catenin signalling: Function, biological mechanisms, and therapeutic opportunities. *Signal Transduction and Targeted Therapy*, *7*(1), 3. DOI: 10.1038/s41392-021-00762-6 PMID: 34980884

Liu, S., Manshaii, F., Chen, J., Wang, X., Wang, S., Yin, J., Yang, M., Chen, X., Yin, X., & Zhou, Y. (2025). Unleashing the potential of electroactive hybrid biomaterials and self-powered systems for bone therapeutics. *Nano-Micro Letters*, *17*(1), 44. DOI: 10.1007/s40820-024-01536-9 PMID: 39417933

Liu, Y., & Lu, J. (2024). Research trends of piezoelectric biomaterials in osteochondral tissue engineering. *Materials Today. Communications*, *110264*. Advance online publication. DOI: 10.1016/j.mtcomm.2024.110264

Liu, Z., Wan, X., Wang, Z. L., & Li, L. (2021). Electroactive biomaterials and systems for cell fate determination and tissue regeneration: Design and applications. *Advanced Materials*, *33*(32), 2007429. DOI: 10.1002/adma.202007429 PMID: 34117803

Luo, R., Liang, Y., Yang, J., Feng, H., Chen, Y., Jiang, X., Zhang, Z., Liu, J., Bai, Y., Xue, J., Chao, S., Xi, Y., Liu, X., Wang, E., Luo, D., Li, Z., & Zhang, J. (2023). Reshaping the endogenous electric field to boost wound repair via electrogenerative dressing. *Advanced Materials*, *35*(16). Advance online publication. DOI: 10.1002/adma.202208395 PMID: 36681867

Luo, S., Zhang, C., Xiong, W., Song, Y., Wang, Q., Zhang, H., Guo, S., Yang, S., & Liu, H. (2024). Advances in electroactive biomaterials: Through the lens of electrical stimulation promoting bone regeneration strategy. *Journal of Orthopaedic Translation*, *47*, 191–206. DOI: 10.1016/j.jot.2024.06.009 PMID: 39040489

McCormick, D. A. (2014). Membrane potential and action potential. In *From Molecules to Networks* (pp. 351–376). Elsevier.

McLaughlin, K. A., & Levin, M. (2018). Bioelectric signaling in regeneration: Mechanisms of ionic controls of growth and form. *Developmental Biology*, *433*(2), 177–189. DOI: 10.1016/j.ydbio.2017.08.032 PMID: 29291972

Mocanu-Dobranici, A.-E., Costache, M., & Dinescu, S. (2023). Insights into the molecular mechanisms regulating cell behavior in response to magnetic materials and magnetic stimulation in stem cell (neurogenic) differentiation. *International Journal of Molecular Sciences*, *24*(3), 2028. DOI: 10.3390/ijms24032028 PMID: 36768351

Mutepfa, A. R., Hardy, J. G., & Adams, C. F. (2022). Electroactive scaffolds to improve neural stem cell therapy for spinal cord injury. *Frontiers in Medical Technology*, *4*, 693438. DOI: 10.3389/fmedt.2022.693438 PMID: 35274106

Nikolova, M. P., & Chavali, M. S. (2019). Recent advances in biomaterials for 3D scaffolds: A review. *Bioactive Materials*, *4*, 271–292. DOI: 10.1016/j.bioactmat.2019.10.005 PMID: 31709311

Ning, C., Zhou, Z., Tan, G., Zhu, Y., & Mao, C. (2018). Electroactive polymers for tissue regeneration: Developments and perspectives. *Progress in Polymer Science*, *81*, 144–162. DOI: 10.1016/j.progpolymsci.2018.01.001 PMID: 29983457

Nowicka-Bauer, K., & Szymczak-Cendlak, M. (2021). Structure and function of ion channels regulating sperm motility—An overview. *International Journal of Molecular Sciences*, *22*(6), 3259. DOI: 10.3390/ijms22063259 PMID: 33806823

Park, J., Akbaba, G. E., Sharma, N., Das, R., Vinikoor, T., Liu, Y., Le, D. Q., Angadi, K., & Nguyen, T. D. (2025). Electrically active biomaterials for stimulation and regeneration in tissue engineering. *Journal of Biomedical Materials Research. Part A*, *113*(1), e37871. DOI: 10.1002/jbm.a.37871 PMID: 39806919

Pisciotta, A., Lunghi, A., Bertani, G., Di Tinco, R., Bertoni, L., Orlandi, G., Biscarini, F., Bianchi, M., & Carnevale, G. (2022). PEDOT: PSS promotes neurogenic commitment of neural crest-derived stem cells. *Frontiers in Physiology*, *13*, 930804. DOI: 10.3389/fphys.2022.930804 PMID: 36060701

Pivovarov, A. S., Calahorro, F., & Walker, R. J. (2018). Na(+)/K(+)-pump and neurotransmitter membrane receptors. *Invertebrate Neuroscience*, *19*(1), 1. DOI: 10.1007/s10158-018-0221-7 PMID: 30488358

Preetam, S., Ghosh, A., Mishra, R., Pandey, A., Roy, D. S., Rustagi, S., & Malik, S. (2024). Electrical stimulation: A novel therapeutic strategy to heal biological wounds. *Royal Society of Chemistry Advances*, *14*(44), 32142–32173. DOI: 10.1039/d4ra04258a PMID: 39399261

Qian, L., Jin, F., Wei, Z., Li, T., Sun, Z., Lai, C., Ma, J., Xiong, R., Ma, X., Wang, F., Sun, F., Zheng, W., Dong, W., Sun, K., Wang, T., & Feng, Z. Q. (2023). Wearable, self-powered, drug-loaded electronic microneedles for accelerated tissue repair of inflammatory skin disorders. *Advanced Functional Materials*, *33*(27). Advance online publication. DOI: 10.1002/adfm.202209407

Reid, B., & Zhao, M. (2014). The electrical response to injury: Molecular mechanisms and wound healing. *Advances in Wound Care (New Rochelle, N.Y.)*, *3*(2), 184–201. DOI: 10.1089/wound.2013.0442 PMID: 24761358

Sood, A., Desseigne, M., Dev, A., Maurizi, L., Kumar, A., Millot, N., & Han, S. S. (2023). A comprehensive review on barium titanate nanoparticles as a persuasive piezoelectric material for biomedical applications: Prospects and challenges. *Small*, *19*(12), 2206401. DOI: 10.1002/smll.202206401 PMID: 36585372

Stojanovic, M., Rai, V., & Agrawal, D. K. (2024). Effect of electromagnetic field on proliferation and migration of fibroblasts and keratinocytes: Implications in wound healing and regeneration. *Journal of Biotechnology and Biomedicine*, *7*(3), 387. DOI: 10.26502/jbb.2642-91280162 PMID: 39364330

Su, Y., Toftdal, M. S., Le Friec, A., Dong, M., Han, X., & Chen, M. (2021). 3D electrospun synthetic extracellular matrix for tissue regeneration. *Small Science*, *1*(7), 2100003. DOI: 10.1002/smsc.202100003 PMID: 40213056

Sun, Y., Xiao, Z., Chen, B., Zhao, Y., & Dai, J. (2024). Advances in material-assisted electromagnetic neural stimulation. *Advanced Materials*, *36*, 2400346. DOI: 10.1002/adma.202400346 PMID: 38594598

Thrivikraman, G., Boda, S. K., & Basu, B. (2018). Unraveling the mechanistic effects of electric field stimulation towards directing stem cell fate and function: A tissue engineering perspective. *Biomaterials*, *150*, 60–86. DOI: 10.1016/j.biomaterials.2017.10.003 PMID: 29032331

Vashist, A., Kaushik, A., Vashist, A., Sagar, V., Ghosal, A., Gupta, Y. K., Ahmad, S., & Nair, M. (2018). Advances in carbon nanotubes-hydrogel hybrids in nanomedicine for therapeutics. *Advanced Healthcare Materials*, *7*(9), e1701213. DOI: 10.1002/adhm.201701213 PMID: 29388356

Vyas, J., Shah, I., Singh, S., & Prajapati, B. G. (2023). Biomaterials-based additive manufacturing for customized bioengineering in management of otolaryngology: A comprehensive review. *Frontiers in Bioengineering and Biotechnology*, *11*, 1234340. DOI: 10.3389/fbioe.2023.1234340 PMID: 37744247

Wang, X., Gao, Y., Shi, H., Liu, N., Zhang, W., & Li, H. (2016). Influence of the intensity and loading time of direct current electric field on the directional migration of rat bone marrow mesenchymal stem cells. *Frontiers of Medicine*, *10*, 286–296. DOI: 10.1007/s11684-016-0456-9 PMID: 27324024

Wang, Y., Yokota, T., & Someya, T. (2021). Electrospun nanofiber-based soft electronics. *NPG Asia Materials*, *13*(1), 22. DOI: 10.1038/s41427-020-00267-8

Wei, C., Wang, X., Zheng, M., & Cheng, H. (2012). Calcium gradients underlying cell migration. *Current Opinion in Cell Biology*, *24*(2), 254–261. https://doi.org/https://doi.org/10.1016/j.ceb.2011.12.002 PMID: 22196933

Xia, H., Li, X., Gao, W., Fu, X., Fang, R. H., Zhang, L., & Zhang, K. (2018). Tissue repair and regeneration with endogenous stem cells. *Nature Reviews. Materials*, *3*(7), 174–193. DOI: 10.1038/s41578-018-0027-6

Xiao, C., Fan, L., Zhou, S., Kang, X., Guan, P., Fu, R., Li, C., Ren, J., Wang, Z., Yu, P., Wang, Y., Deng, C., Zhou, L., & Ning, C. (2022). One-dimensional ferroelectric nanoarrays with wireless switchable static and dynamic electrical stimulation for selective regulating osteogenesis and antiosteosarcoma. *ACS Nano*, *16*(12), 20770–20785. DOI: 10.1021/acsnano.2c07900 PMID: 36412574

Yang, C., Ji, J., Lv, Y., Li, Z., & Luo, D. (2022). Application of piezoelectric material and devices in bone regeneration. *Nanomaterials (Basel, Switzerland)*, *12*(24), 4386. DOI: 10.3390/nano12244386 PMID: 36558239

Yang, S., Wang, Y., & Liang, X. (2023). Piezoelectric nanomaterials activated by ultrasound in disease treatment. *Pharmaceutics*, *15*(5), 1338. DOI: 10.3390/pharmaceutics15051338 PMID: 37242580

Yao, L., & Li, Y. (2016). The role of direct current electric field-guided stem cell migration in neural regeneration. *Stem Cell Reviews and Reports*, *12*, 365–375. DOI: 10.1007/s12015-016-9654-8 PMID: 27108005

Yeste, J., Illa, X., Alvarez, M., & Villa, R. (2018). Engineering and monitoring cellular barrier models. *Journal of Biological Engineering*, *12*(1), 18. DOI: 10.1186/s13036-018-0108-5 PMID: 30214484

Zare, I., Mirshafiei, M., Kheilnezhad, B., Far, B. F., Hassanpour, M., Pishbin, E., Eftekhar Vaghefi, S. S., Yazdian, F., Rashedi, H., Hasan, A., Wang, X., Adeli, M., & Makvandi, P. (2024). Hydrogel-integrated graphene superstructures for tissue engineering: From periodontal to neural regeneration. *Carbon*, *223*, 118970. DOI: 10.1016/j.carbon.2024.118970

Zhang, Y., Zheng, Z., Zhu, S., Xu, L., Zhang, Q., Gao, J., Ye, M., Shen, S., Xing, J., & Wu, M. (2025). Electroactive electrospun nanofibrous scaffolds: Innovative approaches for improved skin wound healing. *Advancement of Science, 2416267.* Advance online publication. DOI: 10.1002/advs.202416267 PMID: 40190057

Zhao, C., Zeng, C., Ye, S., Dai, X., He, Q., Yang, B., & Zhu, H. (2020). Yes-associated protein (YAP) and transcriptional coactivator with a PDZ-binding motif (TAZ): A nexus between hypoxia and cancer. *Acta Pharmaceutica Sinica. B, 10*(6), 947–960. DOI: 10.1016/j.apsb.2019.12.010 PMID: 32642404

Zhao, M. (2009). Electrical fields in wound healing—An overriding signal that directs cell migration. *Seminars in Cell & Developmental Biology, 20*(6), 674–682. DOI: 10.1016/j.semcdb.2008.12.009 PMID: 19146969

Zheng, T., Huang, Y., Zhang, X., Cai, Q., Deng, X., & Yang, X. (2020). Mimicking the electrophysiological microenvironment of bone tissue using electroactive materials to promote its regeneration. *Journal of Materials Chemistry. B, 8*(45), 10221–10256. DOI: 10.1039/d0tb01601b PMID: 33084727

Zhou, B., Lin, W., Long, Y., Yang, Y., Zhang, H., Wu, K., & Chu, Q. (2022). Notch signaling pathway: Architecture, disease, and therapeutics. *Signal Transduction and Targeted Therapy, 7*(1), 95. DOI: 10.1038/s41392-022-00934-y PMID: 35332121

Zhou, L., Fan, L., Yi, X., Zhou, Z., Liu, C., Fu, R., Dai, C., Wang, Z., Chen, X., Yu, P., Chen, D., Tan, G., Wang, Q., & Ning, C. (2018). Soft conducting polymer hydrogels cross-linked and doped by tannic acid for spinal cord injury repair. *ACS Nano, 12*(11), 10957–10967. DOI: 10.1021/acsnano.8b04609 PMID: 30285411

Zhou, Z., Zheng, J., Meng, X., & Wang, F. (2023). Effects of electrical stimulation on articular cartilage regeneration with a focus on piezoelectric biomaterials for articular cartilage tissue repair and engineering. *International Journal of Molecular Sciences, 24*(3), 1836. DOI: 10.3390/ijms24031836 PMID: 36768157

Zhu, S., Chen, W., Masson, A., & Li, Y.-P. (2024). Cell signaling and transcriptional regulation of osteoblast lineage commitment, differentiation, bone formation, and homeostasis. *Cell Discovery, 10*(1), 71. DOI: 10.1038/s41421-024-00689-6 PMID: 38956429

Chapter 7 Transformative Approaches in Breast Cancer Treatment: The Role of Polymers, Biomaterials, Nanomaterials, and Metal Ions

Navya Aggarwal
https://orcid.org/0009-0003-7597-0802
Amity University, New Delhi, India

Shinjini Sen
Amity University, New Delhi, India

Tanmay J. Urs
Amity University, New Delhi, India

Shreya Gupta
Amity University, New Delhi, India

Banashree Bondhopadhyay
https://orcid.org/0000-0002-6679-7791
Amity University, New Delhi, India

ABSTRACT

The transformative role of polymers, biomaterials, nanomaterials, and metal ions in breast cancer treatment have highlighted their potential to enhance diagnosis and therapeutic outcomes. Traditional cancer therapies have often fallen short due to their "one-size-fits-all" approach, leading to increased patient non-responsiveness. The integration of interdisciplinary research—from biomedical science to nanotechnology—promises more precise and effective treatments. Polymers enable targeted drug delivery with reduced side effects, while biomaterials enhance regeneration and drug delivery systems. Nanomaterials improve drug accumulation in tumors and combat resistance, and metal ions disrupt cancer cell metabolism and DNA. By exploring case studies and applications of these advanced materials,

DOI: 10.4018/979-8-3373-0055-9.ch007

this paper underscores their significant contributions to personalized cancer therapies and improved patient quality of life. Looking forward, innovations in multifunctional nanomaterials and biosensors are anticipated to further revolutionize breast cancer treatment strategies.

1. INTRODUCTION

In the recent years, polymers, biomaterials, nanomaterials and metal ions have emerged as crucial elements in the domain of science and technology. The unique properties of each material makes them valuable in treatment of breast cancer. In spite of the remarkable achievement in diagnosis and treatment of breast cancer, it remains a major concern. WHO and Globocan have projected that by mid-century 2050, there will be 3.2 million new breast cancer cases and 1.1 million related deaths each year (Bray et al., 2024; Sung et al., 2021a, 2021b; UN, 2024). This amounts to a huge rise in both the diagnostic and therapeutic burden. The traditional therapeutic approaches concentrating on the "one size solution" have, since, become outdated, indicated by the increased in-patient number failing to respond to treatment.

Chemotherapy, radiation, and targeted therapies are examples of conventional breast cancer treatments that are ridden with challenges which restrict their effectiveness and patient outcomes. For instance, chemotherapeutic drugs like doxorubicin and taxanes non-selectively damage both malignant and healthy cells (Mozafari, 2024). They create systemic toxicity, which is still a significant problem. This can result in dose-limiting side effects such as cardiotoxicity, bone marrow suppression, and gastrointestinal upset, some of which may be life threatening (Gupta, 2022; Yu et al., 2022).

Additionally, both acquired and innate drug resistance impair the effectiveness of treatment. Through processes such as accelerated DNA repair, epigenetic changes, and upregulation of drug efflux pumps (e.g., ABC transporters), up to 50% of cases of triple-negative breast cancer (TNBC) develop resistance to anthracyclines and taxanes. Tumor heterogeneity makes resistance worse by allowing certain sub-populations of cancer stem cells to elude treatment, which can result in metastasis and relapse (Kinnel et al., 2023; Prihantono & Faruk, 2021).

In immunosuppressive microenvironments (TME) or thick tumor stroma, where myeloid-derived suppressor cells (MDSCs) and cancer-associated fibroblasts (CAFs) impede medication delivery and immune cell function, conventional therapies frequently fall short. For example, the acidic and hypoxic conditions of the TME reduce the effectiveness of drugs and encourage pro-tumorigenic signalling (J. J. Li et al., 2021; Zhang et al., 2023). Furthermore, real-time monitoring of therapy response is limited by the older techniques' limited theragnostic capabilities (Caballero et al., 2022).

The growing demand of ecofriendly, biodegradable and safe technology has encouraged the development of materials that advances medical treatment and minimize the impact on environment and reduces toxicity. For instance, polymers have emerged as an important component in the future of cancer therapy due to their efficient drug delivery system and targeting of cancer cells, while minimizing side effects. Biomaterials, particularly, are being utilized for regeneration purposes and as nanocarriers for drug delivery, which enhances precision of the breast cancer treatment. Moreover, nanomaterials, offer unique properties that helps in overcoming drug resistance and effectively targeting breast cancer cells. Further, metal ions are exploited for their ability in disturbing metabolism and DNA of cancer cells, with high efficacy and limited drug resistance.

Polymers have emerged as a vital component in the future of cancer treatment due to their versatility and efficiency in drug delivery systems. Polymeric carriers have the ability to increase concentration of therapeutic drugs through penetrating the targeted tissue, increasing the efficiency of treatment. Through these features of polymers, it is possible to reduce the side effects of conventional drugs of chemotherapy by targeting drug delivery to cancer cells. Natural polymers and their derivatives show promise in enhancing drug delivery due to their biocompatibility, ability to increase the bioavailability of therapeutic agents, especially in breast cancer treatment (Adeli et al., 2023).

Development of biomaterials including implants, stents and drug delivery systems, have revolutionized health care and medicine. Among various biomaterials, naturally made biomaterials are the most beneficial and therefore are most widely used. The redefined definition of the biomaterial as a substance that is engineered to take a form in complex system or alone directs the therapeutic and diagnostic procedure through controlled interactions between living system. Biomaterials, owing to their biocompatibility and biodegradability, are being exploited for biomaterial based nanocarriers and drug delivery systems (Shettiwar et al., 2023) .

Nanomaterials offer unique advantages in the treatment of breast cancer by enhancing the drug delivery system, minimizing drug resistance and targeting cancer cells effectively. Their ability to cross biological barriers and accumulate in tumor makes them valuable tools for both, the diagnosis and treatment of breast cancer. They offer a diverse platform, tailoring to different molecular subtypes of the disease, which include HER-2 positive, triple-negative, and hormone-receptor positive breast cancers. Their minute size, biocompatibility, and the ability to deliver theragnostic agents precisely to specific targets, makes them a suitable candidate for developing personalized medicine in breast cancer management. Combination of nanoparticles with multi-omics strategies like genomics and proteomics, enhances the identification of biomarkers and enables personalized therapies (Torres Quintas et al., 2023).

Metal ions are vital for maintaining hemostasis in human body, playing significant role in varied biochemical and physiological processes. Additionally, their unnatural distribution and accumulation in cells can alter these biological processes, damaging or activating biochemical reactions, which ultimately causes irreversible damage to cell or induces cell death. As a result of their action of damaging DNA, interfering signal transduction, affecting metabolism, disturbing osmotic balance and stimulating biocatalyst, they can be used against wide range of cancer with no drug resistance and high efficiency (S. Wang et al., 2022). Moreover, biomedical cancer diagnostic techniques such as MRI and CT include contrasting agents based on metal ions as core component.

This chapter will explore the case studies and real-life applications of polymers, biomaterials, nanomaterials and metal ions, focusing on their role in enhancing diagnosis and therapeutic outcomes in breast cancer treatment. By highlighting key examples, we aim to offer insights into how these technologies are transforming the future of cancer treatment.

2. POLYMERS FOR THERAPY AND DIAGNOSTICS

Polymers have rapidly emerged as transformative tools in the realm of cancer therapy, offering versatility and precision that have reshaped drug delivery systems. These materials are particularly crucial in targeting breast cancer, as they can enhance drug concentration at the site of tumors, significantly improving treatment efficacy while reducing the side effects associated with conventional chemotherapy. The unique properties of polymers—ranging from controlled drug release and targeted delivery to stimuli-responsive behaviors—makes them indispensable in advancing therapeutic strategies.

Natural polymers such as starch, gelatin, chitosan, and gliadin, have long been explored for their potential in breast cancer treatment due to their biocompatibility, biodegradability, and bioavailability. These natural biopolymers, often derived from polysaccharides or proteins, have demonstrated their efficacy in targeting breast cancer cells while minimizing adverse immune responses. Notably, hybrid biopolymer-nanovehicles, which incorporate phytochemicals, are gaining traction as safer, more natural alternatives to traditional drug delivery systems (Chavda et al., 2023; Gulfam et al., 2012).

In addition to their role in direct drug delivery, biopolymeric systems such as polyglutamic acid, hyaluronic acid, and exosomes have shown promise as adjuvants in gene therapies, delivering siRNAs to cancer cells, thus enhancing their therapeutic potential (Subhan & Torchilin, 2023a). Moreover, phloroglucinol-encapsulated starch biopolymers have demonstrated anticancer activity, with research indicating their potential to target breast cancer cell lines effectively (Kumar et al., 2014).

The innovative use of synthetic polymers such as GelMA hydrogels, which combine gelatin and methacrylic anhydride, represents a new frontier in localized chemotherapy delivery. These systems are particularly valuable in procedures like lumpectomies, where targeted delivery is essential to prevent cancer recurrence. Similarly, the use of silver nanoparticle-coated polymers has enhanced the delivery of drugs like tamoxifen and doxorubicin to HER2-positive breast cancer cells, improving therapeutic efficacy and minimizing systemic side effects (Varadharajaperumal et al., 2017).

Moreover, the advent of self-assembling polymers combined with neoantigens holds promise for reducing tumor progression by boosting T-cell infiltration and the immunogenicity of cancer cells, offering novel immunotherapies for breast cancer treatment (Pounraj et al., 2024). Furthermore, polymeric gold nanoparticles have advanced the fields of radiotherapy and phototherapy, highlighting the multifunctional potential of polymers in cancer treatment (Ghaffarlou et al., 2023).

Polymers also serve as invaluable tools in diagnostic imaging, improving real-time monitoring of tumor progression and drug distribution. Polymeric nanoparticles, including those made from hyaluronic acid, are crucial in advancing targeted drug delivery and diagnostic imaging, offering greater precision in chemotherapy and breast cancer diagnostics.

As foundational elements of biological systems, polymers—whether synthetic or natural—play a pivotal role in the advancement of therapeutic and diagnostic technologies. Their ability to be tailored for specific needs and functionalities makes them indispensable in both cancer therapy and diagnostics. This chapter delves into the significant real-world applications of polymers in breast cancer theranostics, showcasing their critical role in overcoming the limitations of conventional treatments and paving the way for more personalized, effective cancer care.

2.1. Polymers in Breast Cancer

Polymers have become crucial in advancing breast cancer treatment and diagnosis, offering innovative solutions for targeted drug delivery, sustained drug release, and improved imaging techniques. Broadly, these materials fall into two main categories: synthetic and natural Polymers, each contributing uniquely to therapeutic advancements. Synthetic polymers, such as Elastin-like Polypeptides (ELP), Polyethylene Glycol (PEG), Poly(lactic-co-glycolic acid) (PLGA), and Polycaprolactone (PCL), are engineered to enhance drug solubility, prolong circulation time, and ensure controlled drug release (Figure 1). These materials have been extensively used in nanoparticle formulations, enabling precise drug targeting and minimizing off-target effects. For instance, ELPs have shown promise in delivering doxorubicin with heat-responsive accumulation at tumor sites, while PEG is widely used to improve the pharmacokinetics of chemotherapy agents (Bhattacharya et al., 2023; Doerksen et al., 2022). Similarly, PLGA-based nanoparticles provide controlled drug release, enhancing the efficacy of breast cancer treatments (Subhan & Torchilin, 2023b).

On the other hand, natural polymers, such as albumin, collagen, gelatin, hyaluronic acid (HA), and chitosan, have gained attention due to their biodegradability, biocompatibility, and ability to interact with biological systems (Figure 2). For example, albumin-based formulations like Abraxane® (albumin-bound paclitaxel) are already in clinical use for treating metastatic breast cancer, demonstrating the potential of protein-based carriers (Talebian et al., n.d.). Collagen-based hydrogels and gelatin-based drug carriers are also being explored for local drug delivery and photothermal therapies, while hyaluronic acid is being leveraged for targeted drug delivery due to its affinity for CD44 receptors, which are overexpressed in breast cancer cells (Gulati et al., 2024; Wu et al., 2024). Additionally, chitosan derivatives have been investigated for delivering small interfering RNA (siRNA) and other therapeutic molecules, enhancing cellular uptake and precision in breast cancer treatment (Alharbi, 2024). The ongoing research in both synthetic and natural polymers continues to push the boundaries of personalized medicine, offering safer, more effective, and targeted approaches for breast cancer therapy.

2.2. Synthetic Polymers and their Applications in Breast Cancer

Poly(lactic-co-glycolic acid) (PLGA) is a widely used biodegradable copolymer that has demonstrated significant potential in breast cancer research, particularly for fabricating three-dimensional (3D) scaffolds that mimic the tumor microenvironment. These scaffolds support breast cancer cell growth and provide a platform for testing anticancer drug efficacy. PLGA-based models better replicate the characteristics of tumors, such as chemoresistance and cell survival pathways, making them valuable for evaluating cancer therapeutics (Rijal et al., 2017).

Polycaprolactone (PCL), a biodegradable polyester, is employed to fabricate 3D scaffolds for studying breast cancer. These scaffolds influence cell morphology, proliferation, and drug response, which makes PCL ideal for mimicking the in vivo tumor environment. PCL-based platforms enhance understanding of tumor biology and improve drug testing by closely replicating the physiological conditions of breast cancer (Rijal et al., 2017).

Polyethylene Glycol (PEG) is extensively used for modifying nanoparticles to enhance their solubility, circulation time, and bioavailability. The PEGylation of drug delivery systems has shown promising results in improving the pharmacokinetics of chemotherapeutic agents, leading to increased therapeutic efficacy against breast cancer while reducing toxicity. This modification has emerged as a key strategy for improving the clinical application of anticancer drugs (Kumar et al., 2023; Sartaj et al., 2021).

Poly(N-isopropylacrylamide) (PNIPAM) is a thermoresponsive polymer that has been explored for controlled drug release in breast cancer treatment. The polymer's ability to release encapsulated drugs at specific temperatures enhances localized drug delivery, offering a more precise and effective treatment option. PNIPAM-based systems hold promise for reducing systemic side effects while maintaining targeted therapeutic effects (Kumar et al., 2023; Sartaj et al., 2021).

Polystyrene (PS) nanoparticles are frequently utilized as carriers for anticancer drugs due to their stability and ease of functionalization. Research has shown that PS nanoparticles can encapsulate a variety of chemotherapeutics, improving the targeted delivery to tumor sites while minimizing off-target effects. This property makes PS an attractive option for enhancing drug delivery systems in breast cancer treatment (M. Zhu et al., 2023).

Polyethylene oxide-block-polypropylene oxide (PEO-PPO) block copolymers are used to create micelles for drug delivery in breast cancer therapies. These micelles can encapsulate hydrophobic drugs, enhancing their solubility and bioavailability. Research demonstrates that PEO-PPO micelles are effective in delivering chemotherapeutic agents to tumors, thus improving their therapeutic outcomes (Ding et al., 2024; Sartaj et al., 2021).

Chitosan Derivatives is a natural polymer, but its synthetic derivatives have been developed for drug delivery applications in breast cancer treatment. These derivatives are particularly useful for delivering small interfering RNA (siRNA) and other therapeutic agents directly to breast cancer cells. The mucoadhesive properties of chitosan derivatives enhance the targeted delivery of drugs, thus improving therapeutic efficacy (Jia et al., 2022a; Madamsetty et al., 2022; Subhan & Torchilin, 2023c, 2023d).

Polyvinyl Alcohol (PVA) is commonly used to create hydrogels for drug delivery systems, owing to its hydrophilicity. PVA-based hydrogels have been shown to encapsulate chemotherapeutic drugs and provide controlled release profiles, enhancing drug delivery to tumor sites. This system improves the solubility and bioavailability of poorly water-soluble drugs, which is particularly beneficial for breast cancer treatments (Asadi et al., 2024).

Polyacrylamide (PAM) is used in breast cancer research to form hydrogels and nanoparticles for drug delivery. Research indicates that PAM-based systems can enhance the delivery of anticancer drugs, providing controlled release mechanisms that improve therapeutic efficacy. The polymer's potential for sustained drug release makes it a promising candidate for treating breast cancer (Ding et al., 2024).

Polylactic Acid (PLA) a biodegradable polymer, is commonly employed in drug delivery systems for breast cancer therapies. PLA-based nanoparticles can encapsulate chemotherapeutic agents, allowing for controlled release and enhanced targeting of tumor sites. PLA's biocompatibility and biodegradability make it an attractive option for developing safer and more effective therapeutic systems (Zhai et al., 2024).

Poly(2-hydroxyethyl methacrylate) (pHEMA) is a synthetic polymer used to create hydrogels for localized drug delivery applications. pHEMA-based hydrogels have shown promise in delivering chemotherapeutic agents directly to tumor sites, enhancing local drug concentrations while minimizing systemic exposure. This localized approach improves therapeutic efficacy and reduces side effects in breast cancer treatment (Senapati et al., 2018; Zhai et al., 2024).

Poly(glycerol-malic acid) (PGMA) is a promising polymer for formulating nanoparticles for drug delivery. Curcumin-loaded PGMA nanoparticles have been found to significantly enhance anticancer activity against breast cancer cell lines like MCF-7 and MDA-MB-231. These nanoparticles show improved solubility and bioavailability, which is crucial for effective treatment, and exhibit high drug entrapment efficiency, indicating their potential in breast cancer therapy (Kumari et al., 2021).

Poly(ethylene imine) (PEI) is a cationic polymer widely studied for gene delivery in breast cancer therapies. Studies show that PEI-based nanoparticles can deliver siRNA to target oncogenes in breast cancer cells, resulting in reduced cell proliferation and increased apoptosis. PEI's ability to form complexes with nucleic acids makes it a promising tool for RNA interference therapies in cancer treatment (Fahira et al., 2022).

Polyvinylpyrrolidone (PVP) is used to enhance the solubility of poorly water-soluble anticancer drugs. PVP-based systems improve the dissolution rate of drugs such as tamoxifen, leading to better absorption and higher therapeutic efficacy. These characteristics make PVP an important polymer for improving drug delivery and efficacy in breast cancer treatment (Odeh et al., 2023; Rusdin et al., 2024) (Mehta et al., 2020).

Polyaniline (PANI) is a conductive polymer being investigated for use in photothermal therapy in breast cancer. Upon exposure to near-infrared light, PANI nanoparticles generate heat, selectively targeting and destroying breast cancer cells while sparing healthy tissue. This property presents PANI as a promising candidate for non-invasive cancer treatments (Alamdari et al., 2022; Vines et al., 2018).

Poly(N-vinylcaprolactam) PVCL is a thermoresponsive polymer that responds to temperature changes, making it suitable for controlled drug release applications. Research suggests that PVCL-based systems can release drugs in response to local temperature shifts, enhancing targeted treatment strategies. This approach has the potential to improve the precision and effectiveness of breast cancer therapies (Romero et al., 2021).

Poly(2-ethyl-2-oxazoline) (PEOz) is a synthetic polymer used in drug delivery applications for breast cancer treatment. Research indicates that PEOz-based formulations effectively deliver chemotherapeutic agents to breast cancer cells, enhancing therapeutic efficacy while minimizing systemic toxicity. This polymer shows promise in improving the clinical outcomes of cancer therapies (Wan et al., 2019).

Poly(ethylene glycol)-poly(lactic acid) (PEG-PLA) combines the properties of PEG and PLA to enhance drug solubility and provide controlled release characteristics. PEG-PLA nanoparticles can encapsulate anticancer drugs, improving targeting to tumor sites and enhancing the overall therapeutic efficacy of breast cancer treatments (Jiaying et al., 2023; Wadhawan et al., 2022).

Figure 1. Monomers of synthetic polymers, commonly used in breast cancer

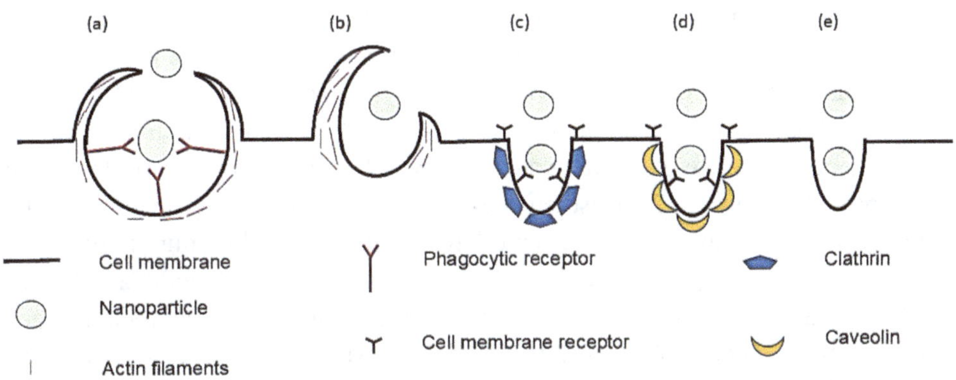

(This figure was created using Marvin JS on 1st February 2025)

2.3. Natural Polymers Used in Breast Cancer Therapeutics and Diagnostics

Natural polymers have emerged as a promising tool in breast cancer therapeutics and diagnostics due to their biocompatibility, biodegradability, and ease of functionalization. They have been explored for use in drug delivery systems, providing enhanced solubility, targeted delivery, and controlled release. Below are some key natural polymers and their applications in breast cancer treatment.

Chitosan, a biopolymer derived from chitin, is commonly utilized in drug delivery systems due to its mucoadhesive properties and biocompatibility. It has been studied for its ability to encapsulate anticancer drugs, such as doxorubicin, leading to enhanced bioavailability and targeted delivery to breast cancer cells. Studies suggest that chitosan nanoparticles can effectively reduce systemic toxicity, providing a more efficient and safer delivery platform for cancer therapeutics (X. Gao et al., 2024; Herdiana et al., 2023; Madamsetty et al., 2022).

Hyaluronic acid (HA) is a polysaccharide that plays a critical role in cell signaling and tissue hydration. It is widely used in drug delivery systems due to its ability to target CD44 receptors, which are overexpressed on many breast cancer cells. HA-based nanoparticles have shown improved cellular uptake and therapeutic efficacy, making them a promising candidate for targeted cancer therapies. These systems can significantly enhance the drug's anticancer potential while reducing off-target effects (M. Fatima et al., 2022; Jia et al., 2022b).

Alginate is a gel-forming polysaccharide derived from brown seaweed, widely utilized in drug delivery systems. It has been shown to encapsulate anticancer agents effectively, providing controlled release and improving the stability and bioavailability of drugs in breast cancer treatment. Research indicates that alginate-based formulations can enhance drug targeting, increasing the therapeutic efficacy of chemotherapeutic agents while minimizing systemic side effects (Ghaffarlou et al., 2023; Mondal et al., 2023).

Gelatin, a protein obtained from collagen, is known for its ability to form hydrogels and its high biocompatibility. Gelatin-based formulations have been explored for their potential to encapsulate chemotherapeutic agents, such as doxorubicin, offering controlled release and targeting capabilities. These formulations have demonstrated improved drug efficacy in targeting breast cancer cells while reducing adverse side effects (Jiang et al., 2023; Nascimento et al., 2024)

Dextran, a polysaccharide produced by certain bacteria, has been explored for its ability to form nanoparticles suitable for drug delivery applications. Studies suggest that dextran-based nanoparticles can effectively deliver anticancer drugs to breast cancer cells, improving therapeutic outcomes and reducing systemic toxicity. These nanoparticles offer a promising platform for enhancing the targeting and bioavailability of chemotherapeutics (Curcio et al., 2019; Khan et al., 2023; M. Li et al., 2014).

Cellulose, a natural polymer found in plant cell walls, has shown promise in drug delivery systems for breast cancer therapeutics. Research indicates that cellulose-based nanoparticles can encapsulate anticancer drugs, providing sustained release profiles and improved targeting capabilities. These formulations can enhance the therapeutic effect of chemotherapeutic agents while minimizing the side effects associated with conventional drug delivery systems (Mondal et al., 2023; Nief et al., 2021a)

Pectin, a polysaccharide found in fruits, is commonly used as a gelling agent and has been explored for its potential in drug delivery applications. Pectin-based formulations have demonstrated the ability to improve the solubility and bioavailability of anticancer agents. These systems enhance the targeting of breast cancer cells and offer a promising approach to overcoming the limitations of traditional drug delivery methods (Attallah et al., 2020; Mondal et al., 2023).

Cyclodextrins are cyclic oligosaccharides that can form inclusion complexes with various drugs, improving their solubility and stability. These formulations have been utilized in drug delivery systems for breast cancer, facilitating enhanced targeting and controlled release of therapeutic agents. Cyclodextrin-based drug delivery systems are particularly beneficial for improving the bioavailability of poorly soluble anticancer drugs (Nief et al., 2021a; Velhal et al., 2022; P. Zhou et al., 2019).

Sodium alginate, a derivative of alginate, is known for its gel-forming capabilities and biocompatibility. It has been widely studied in drug delivery applications for encapsulating chemotherapeutic agents, providing controlled release, and enhancing the stability and bioavailability of drugs. Sodium alginate-based hydrogels have shown promise in improving therapeutic efficacy and reducing side effects in breast cancer treatment (Iravani & Varma, 2022; Nief et al., 2021b).

Fibrin, a protein involved in blood clotting, has been explored for its potential in localized drug delivery systems for breast cancer. Fibrin-based hydrogels can support cell growth and tissue regeneration while delivering anticancer drugs directly to tumor sites. Research suggests that fibrin hydrogels can provide controlled release of chemotherapeutic agents, enhancing their effectiveness and promoting localized treatment (Chen et al., 2022; Heilala et al., 2023; Kirtane et al., 2017).

Collagen, a major structural protein in the extracellular matrix, is widely studied for its potential in drug delivery systems. Collagen-based nanoparticles have demonstrated the ability to encapsulate various chemotherapeutic agents, improving drug targeting to breast cancer cells. Studies suggest that collagen's biocompatibility and ability to mimic the tumor microenvironment make it an effective carrier for targeted therapies, enhancing therapeutic outcomes (Kumar et al., 2023; Passos et al., 2023).

Chondroitin sulfate, a glycosaminoglycan found in cartilage, has shown potential in drug delivery systems for breast cancer. Research indicates that chondroitin sulfate can enhance the targeting of anticancer drugs to breast cancer cells by exploiting specific receptors overexpressed on these cells. This targeting capability improves the therapeutic outcomes and minimizes off-target effects, making chondroitin sulfate a promising candidate for breast cancer treatment (Nadanaka et al., 2022; Yu et al., 2023).

Figure 2. Monomers of natural polymers, commonly used in breast cancer

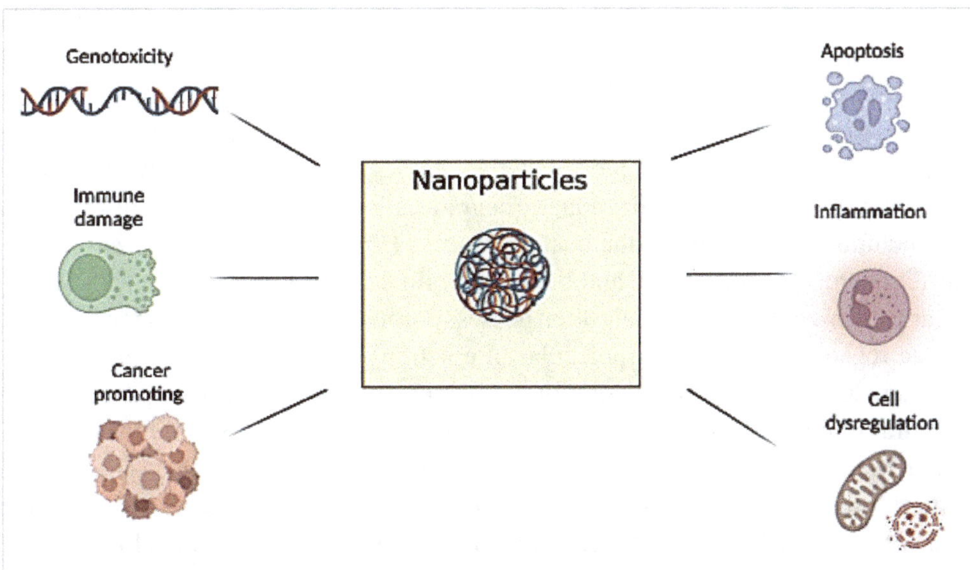

(This figure was created using Marvin JS on 1st February 2025. The structures of hyaluronic acid, alginate, dextran, cellulose, galacturonic acid and chondroitin were sourced from https://commons.wikimedia.org/wiki/, which states they are in the public domain, and image of chitosan is released under the CC0 1.0 Universal Public Domain Dedication indicating it is free to use.)

Polymers, both synthetic and natural, are integral to advancing breast cancer treatment, particularly in drug delivery systems. Synthetic polymers like PLA and PLGA are valued for their tunable properties, allowing precise control over drug release and degradation rates. These features are ideal for creating targeted treatments with reduced side effects. However, concerns about their biocompatibility, potential toxicity, and environmental impact arise, especially due to the solvents used in their production. In contrast, natural polymers such as chitosan and alginate offer excellent biocompatibility and biodegradability, reducing the risk of adverse immune responses. They also possess beneficial biological properties like anti-inflammatory effects, enhancing treatment outcomes. However, their lack of precise control over structure and degradation, along with cost concerns, poses challenges for consistent drug release and large-scale application. Combining synthetic and natural polymers in hybrid systems can overcome these limitations. These hybrid formulations leverage the stability of synthetic polymers and the biocompatibility of natural polymers, promising more effective, safer treatments with controlled drug delivery. In summary, while both synthetic and natural polymers offer distinct advantages, their integration into hybrid systems could lead to more personalized, efficient, and sustainable breast cancer therapies, minimizing side effects while improving patient outcomes.

3. BIOMATERIALS FOR BREAST CANCER

Biomaterials are substances engineered to interact with biological systems for medical purposes, such as therapeutic or diagnostic applications. They can be natural, synthetic, or hybrid materials (Cao & Ding, 2022). These are designed to replace, regenerate, or augment bodily functions. Common examples include prosthetics, implants, and drug delivery systems. (Fogg et al., 2022).

Biomaterials can be briefly divided into four categories (Figure 3). The first generation consists of bioinert materials, such as certain metals (like stainless steel and titanium), ceramics (like alumina), and polymers (like silicone rubber) (Farag, 2023). These materials do not elicit any significant biological responses when implanted. They function by forming a fibrous capsule around them to minimize immune reactions. The second generation consists of bioactive and absorbable materials, which interact positively with biological environments. These include bioactive glasses and resorbable polymers like polylactide (PLA). They promote tissue bonding and gradually degrade, allowing for natural tissue regeneration while maintaining structural integrity during the healing process (*Biomaterials Generations*, n.d.). The third generation further combines bioactivity and bioresorbability, featuring materials designed to stimulate specific cellular responses. These materials often possess porous structures that support cellular activities, enhancing tissue regeneration through molecular interactions (*Biomaterials Generations*, n.d.; Demir et al., 2023; Farag, 2023). The fourth generation is characterized by biomimetic properties, engineered to replicate natural biological processes. These advanced materials manipulate bioelectric signals and monitor cellular behaviour, significantly improving integration with host tissues (Demir et al., 2023). Hence, biomaterials have undergone significant evolution through several generations, each marked by distinct properties and functionalities that enhance their application in biomedical fields.

Figure 3. There are four generations of biomaterials. The first generation focused on the development of bioinert materials including alumina and silicon-derivative products. The second generation gave way to bioactive and absorbable materials, while the third focused on bioresorbability. The fourth generation introduced biomimetic materials, that could effectively mimic the biological conditions, when supplied to the cells

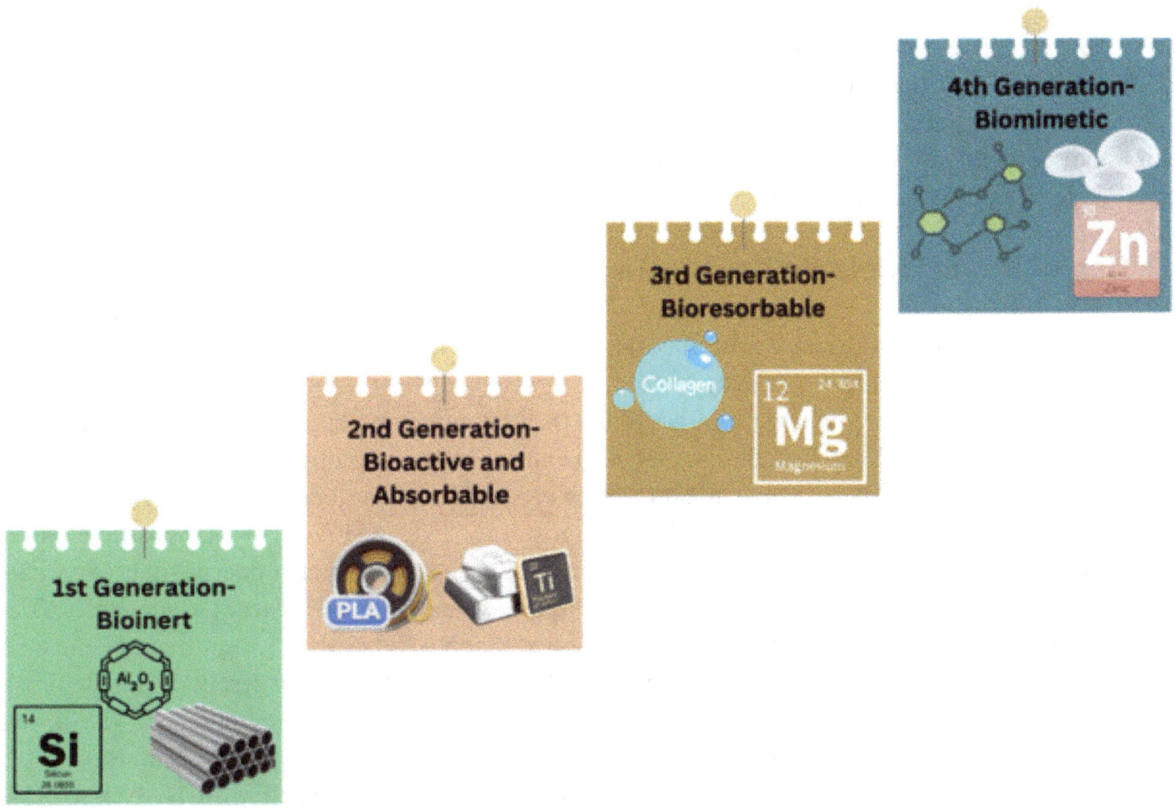

(This figure was created on Canva.com on 26th January 2025)

3.1. Use of Biomaterials for Breast Reconstruction

In the domain of breast cancer, the use of biomaterials has especially improved the quality of healthcare being provided to patients. In the past few years, the demand for breast reconstruction has gone higher. Often times, a bilateral mastectomy is followed by an immediate procedure for reconstruction (Teixeira & Martins, 2023). This is mostly seen as a result of the psychological satisfaction of the patient, the need heightened by societal norms and pressures. Thus, with the growing demand, various types of biomaterials are entering the market, and are being used actively for breast reconstruction (Park et al., 2021).

Some of the natural biomaterials used for breast reconstruction include alginate, chitosan, decellularized extracellular matrix, collagen and hyaluronic acid (Mayer et al., 2024). The decellularized extracellular matrix provides an appropriate, biomimetic microenvironment. This is particularly due to the presence of polysaccharides, protein and glycosaminoglycans in it, that make it more favourable for the adhesion, proliferation and differentiation of the adipocytes (Vindigni et al., 2024). Similarly, chitosan and alginate also provide a suitable microenvironment for the viability of adipocytes, as well as adipose-derived stem cells (Park et al., 2021). A three-dimensional acellular dermal matrix made out of collagen and commercially referred to as BRAXON® *Fast* was evaluated comprehensively, for prepectoral breast reconstruction after mastectomy. This innovative material enhances the surgical process by providing a pre-shaped, curved interface that simplifies the wrapping of breast implants. This helps to reduce exposure time and for maintaining sterility. The study involved 23 patients who underwent nipple-sparing mastectomy and BRAXON® *Fast*-assisted prepectoral breast reconstruction at Ca' Foncello Hospital in Treviso, Italy. The overall complication rate was reported to be at 11.1%, with minor complications such as seroma and necrosis. The only major complication involved implant infection, leading to its removal (Berna et al., 2022).

3.2. Use of Biomaterials for Drug Delivery in Breast Cancer

Drug delivery is one of the major processes that involves the use of biomaterials, such as alginate, in the management of breast cancer. For instance, for the targeted delivery of Paclitaxel, alginate nanoparticles were formulated. Upon comparison, it was found that these alginate-based nanoparticles had higher solubility than paclitaxel alone. In addition, they also enhanced the apoptotic effect on Paclitaxel on the malignant cells, and induced cell cycle arrest (Shaikh et al., 2022). Another study developed chitosan/alginate polymeric nanoparticles loaded with α-mangostin (NANO-AMCAL). α-Mangostin, a natural compound derived from the mangosteen fruit, has demonstrated anticancer properties but faces challenges related to solubility and bioavailability when administered in its pure form (Alam et al., 2023). The in-vitro evaluations on MCF-7 breast cancer revealed that NANO-AMCAL significantly enhanced cell death. This is attributed to increased cellular uptake facilitated by the positively charged chitosan nanoparticles, which electrostatically interact with negatively charged cell membranes. The in-vivo model of a DMBA-induced breast cancer rat showed a marked reduction in tumor volume compared to controls. Additionally, body weight monitoring indicated that rats treated with NANO-AMCAL experienced less weight loss compared to those receiving pure α-mangostin, thus suggesting reduced systemic toxicity (Muchtaridi et al., 2023). An in-vitro investigation was carried out to encapsulate of turmeric oil (TO) in chitosan/alginate nanocapsules (CS/Alg-NCs). Turmeric oil is known for its various biological activities but faces limitations due to instability, volatility, and poor water solubility (J. Zhu et al., 2024). The resulting TO-loaded CS/Alg-NCs exhibited favorable characteristics, including an average size of approximately 200 nm, homogeneous distribution, and spherical shapes with smooth surfaces. The encapsulation efficiency was notably high at 70%, indicating effective loading of turmeric oil within the nanocapsules. Release studies demonstrated a sustained release profile, with about 50% of TO released after 12 hours at physiological pH levels (7.4) and slightly lower at pH 5.5, which mimics the acidic environment of tumor tissues. Additionally, the cytotoxicity assays revealed that TO-CS/Alg-NCs displayed significantly enhanced cytotoxic effects against the breast cancer cell lines MDA-MB-231 and MCF-7, in comparison to unencapsulated turmeric oil. This suggests that the chitosan/alginate nano-capsules not only improve the stability and solubility of turmeric oil, but also enhance its anticancer activity (San et al., 2022).

Hyaluronic acid (HA), a natural polysaccharide abundant in the extracellular matrix, exhibits excellent biocompatibility, biodegradability, and non-immunogenic properties (Abatangelo et al., 2020; Iaconisi et al., 2023). Its interaction with the CD44 receptor, which is overexpressed in breast cancer cells, allows for targeted delivery of therapeutic agents directly to tumor sites (Jia et al., 2022c). Quercetin-loaded hyaluronic acid nanogels combined with everolimus, an immunosuppressant, demonstrated improved biosafety and treatment efficacy against hormone-responsive breast cancer cells. Furthermore, the use of reactive oxygen species (ROS) inducers like shikonin within HA-encapsulated liposomes showed promise in increasing oxidative stress specifically in cancer cells, thereby promoting apoptosis (Jia et al., 2022c). Another study aimed to develop hyaluronic acid-loaded Harmine polymeric micelles with properties targeting CD 44 and to evaluate their anti-breast cancer effects in vitro. The developed polymeric micelles self-assembled to encapsulate Harmine within a hydrophobic core. Results indicated that the polymer micelles exhibited a uniform size of approximately 200 nm, with good dispersion. They achieved an encapsulation rate of 87%. The micelles demonstrated effective enzymatic degradation by hyaluronidase, with a hemolysis rate below 1%. In cytotoxicity assays, the micelles displayed dose-dependent toxicity towards both CD 44-positive MDA-MB-231 cells and CD 44-negative MCF-7 cells. Notably, there was an increased uptake of the polymer micelles by the CD 44-positive MDA-MB-231 cells, when compared to the CD 44-negative MCF-7 cells, hence indicating enhanced targeting (Tang et al., 2023).

Silk fibroin is a biomaterial obtained from the cocoon of the *Bombyx morii*. It is used after the processing and removal of the sericin component present in it. Silk fibroin is being increasingly evaluated for its use as a vehicle for drug delivery mechanisms. It promised high specificity, increased cytotoxicity and better therapeutic effects (Tulay et al., 2018). For example, a study developed silk fibroin-based nanoparticle delivery system for paclitaxel. In-vitro analysis showed that docetaxel trapped in the developed nanoparticles had better cellular uptake of docetaxel and enhanced cytotoxicity, as compared to docetaxel alone. It achieved this by inducing a G2/M phase cell cycle arrest, followed by the subsequent induction of apoptosis, hence increasing the antitumor effects of the delivered drug (Saqr et al., 2021).

The use of graphene oxide nanosheets as biomaterials in breast cancer is also an emerging technique. A study evaluated their potential with the breast cancer cell line, MDA-MB-231. The results revealed that the graphene oxide nanosheets did exhibit a high amount of cytotoxicity on the cell line. It was also found to be concentration dependent, as per effect of different dosages on the cells. Thus, graphene oxide can be used for carbon-based anti-cancerous therapeutic developments, and is being actively researched for its potential use (Mohanta et al., 2021).

Despite these promising advancements, several challenges remain in the application of biomaterials for breast cancer therapeutics. One major hurdle is the scalability of manufacturing these advanced materials while ensuring consistent quality and performance. As these technologies transition from laboratory research to clinical applications, it is important to develop standardized protocols for production that meet regulatory requirements without compromising efficacy and safety. Besides, extensive clinical testings need to be conducted to demonstrate their safety and effectiveness, and to demonstrate their value in real-world applications. There is a significant dearth of clinical trials in this context. Hence, it is imperative that the pre-clinical research translates into clinical evaluations. The integration of biomaterials with other therapeutic modalities such as gene therapy or combination treatments can also be explored.

4. NANOPARTICLES FOR BREAST CANCER

Nanoparticles typically range from 1 to 100 nanometres in size. These materials are adaptable and are engineered to exhibit required physical and chemical properties. Their nano size, shape and morphology is potentially suitable for biomedical applications in breast cancer treatment (Ganesan et al., 2021).

Nanoparticles are entirely made up of organic and inorganic biomaterials. Some of the raw materials include lipids, polymers, carbon and inorganic metals such as gold and silver. Different types of nanoparticles are utilized in different areas of cancer treatment, such as tumor imaging, biomarker detection, therapeutic delivery and diagnostic fields (Harish et al., 2022).

4.1. Biomedical Applications of Nanoparticles in Breast Cancer Treatment

4.1.1. Quantum Dots for Tumor Imaging and Biomarker Detection

Quantum Dots (QDs) are semiconducting nanoparticles engineered to emit fluorescence with sharp, adjustable peaks when exposed to high-intensity light. Unlike the conventional contrast agents, QDs have better optical properties. The property includes fluorescence in the near-infrared (NIR) range, which enables deeper tissue penetration (L. W. Wang et al., 2015). It produces high resolution imaging in vivo, in modalities like Magnetic Resonance Imaging (MRI) and computed tomography (CT). QDs efficiently map axillary lymphatic systems, metabolic pathways and lymph node involvement, which provide insights for breast tumor staging (Robe et al., 2008).

When combined with Human Epidermal Growth Factor Receptor (HER-2) specific antibodies, QDs selectively bind to cells overexpressing HER-2. These conjugates generate strong fluorescent signals for tumor visualization. They yield high-resolution images that can trace and delineate tumor boundaries, allowing small and invasive tumor detection. QDs are capable of three-dimensional (3D) tumor imaging for precise mapping of TME. QD-Trastuzumab conjugates provide clear images of $HER2^{+}$ tumors. The resulting images are invaluable for surgical planning, as they ensure accurate excision of tumor (I. Fatima et al., 2021).

The QDs are also employed in biosensors. They identify cancer-associated non-coding microRNAs (miRNA-155 and miRNA-182), which regulate tumor progression and resist therapy in TNBC. The sensors detect the presence of miRNA using fluorescence resonance energy transfer (FRET) mechanisms, with high sensitivity (Díaz-García et al., 2024) Nitrogen-doped QDs conjugated with gold nanoparticles, enhance sensitivity to detect low concentrations of miRNAs (H. Wang, 2024). Additionally, QDs identify multiplexed biomarkers such as HER2 and Ki-67 within a single assay (Omidian et al., 2024).

For therapeutic delivery, QDs encapsulate and release therapeutic agents like doxorubicin (DOX) effectively. In TNBC models, QD-based drug delivery systems (DDS) exhibit tumor cell death, with real-time tracking of drug distribution (Zeng et al., 2023). Furthermore, QDs convert NIR light into heat, producing reactive oxygen species (ROS) for selective cytotoxicity. These properties are being explored in photothermal and photodynamic therapy (Dar et al., 2023; Z. Zhou et al., 2016).

The translation of QDs into clinical settings is duly limited by toxicity concerns, despite their superior optical properties, deep-tissue imaging and biomarker analysis. The worry arises primarily from heavy metals like Cadmium and lead, which pose fatal risks such as poor systemic clearance and foetal damage. The biocompatibility of QDs remain inadequate, despite of many coating options. However, newer low- or non-toxic alternatives like carbon, graphene, silicon, and silver QDs show promise

4.1.2. Metallic Nanoparticles for Molecular Targeting and Imaging

Metallic Nanoparticles (MNPs) including gold and silver (AuNPs and AgNPs), have distinct advantages like size, high surface area and functionalized morphology. The surface of MNPs can be modified with biomolecules such as antibodies, aptamers, ligands and drugs to achieve specific targeting (Khursheed et al., 2022).

For instance, AuNPs functionalized with DNA-aptamers are used to detect and molecularly target human matrix metalloproteinase-9 (hMMP-9) in MDA-MB-231 cells. The hMMP-9 enzyme plays key role in ECM remodelling, strongly associated with tumour invasion and metastasis. The conjugation stabilises the AuNPs in biological environments, simultaneously enabling precise hMMP-9 detection through spectral signatures for spectral photoacoustic tomography (SPAT). The surface modified AuNPs with MMP-9 specific aptamers, allow aggregation in the TME through DNA displacement and hybridization. The accumulation leads to increased optical absorption using imaging techniques. (J. Kim et al., 2021).

AgNPs on the other hand, selectively induce proteotoxic stress in TNBC by disrupting protein homeostasis, with minimal oxidative stress pathway activation. The disturbance activates stress response pathways such as, Heat Shock Response (HSR) and Integrated Stress Response (ISR), AgNPs suppress HSP70 and HSP27 and promote phosphorylation of eIF2α. The mechanism promotes the accumulation of misfolded proteins and activation of pro-apoptotic factors. Additionally, the combination of AgNPs with Salubrinal, an ISR modulator, increases AgNP cytotoxicity in TNBC cells, causing no harm to healthy mammary tissues (Snyder et al., 2024; J. Kim et al., 2021).

4.1.3. Magnetic Nanoparticles for Magnetic Resonance Imaging

Magnetic Nanoparticles (MgNPs), particularly Iron Oxide Nanoparticles (IONPs) have significant strengths to improve MRI for breast cancer diagnosis. Cross-linked nanocluster formulations, referred as NFGs show high stability and functionality for transverse relaxivity (T2)-weighted MRI, improving contrast for tumor and lymph node imaging. The positively charged MgNPs interact better with water and enhance T2 values (Seo et al., 2021).

In one study, biocompatible and biodegradable Fe_3O_4 was conjugated with Herceptin for the dual-imaging and therapeutic function. Under an external magnetic field, the conjugate selectively targeted $HER2^+$ SK-BR-3 cells. It improved the MRI imaging and displayed significant anti-cancer effects, when combined with paclitaxel (PTX) (Guo et al., 2021). Fe_3O_4 NPs were utilised as stable drug delivery system for their beneficial magnetic properties. The magnetic NP enabled both passive targeting via the EPR effect and active magnetic targeting. The surface of Fe_3O_4 was modified with Herceptin for specific binding to $HER2^+$ cells. The mechanism increased selective cytotoxicity and reduced off-target toxicity. Moreover, the system superiorly induced apoptosis, through increased cleaved-PARP and Caspase-3 expression. The precise, magnetically guided delivery with enhanced T2-weighted MRI contrast was simultaneously enabled (Guo et al., 2021).

Another example includes the functionalization of folic acid to IONPs. IONPs were synthesized using polyol method and was coated with silica (SiO_2) and hydroxypropyl-grafted (HPG) polymers. Folic acid was additionally conjugated to enhance the targeting efficiency of the nanoparticle. Partial cytotoxicity was observed in higher concentrations, however, the functionalization improved selectivity for MCF-7 cells. It enhanced the receptor-mediated endocytosis and imaging precision (Hossein et al., 2020).

4.1.4. Liposomes for Localised and Targeted Drug Delivery

Liposomes, with diameters under 100nm are designed for controlled drug release, triggered by external stimuli. Focused-ultrasound (FUS)-responsive liposomes loaded with DOX showed substantial tumor suppression in MDA-MB-231 bearing mice. *in vitro* studies suggested that the drug release ratio was twice to that of conventional liposomal formulations (Y. S. Kim et al., 2022).

To overcome the drug resistance occurring in breast cancer, paclitaxel-loaded functional ginsenoside Rg3 liposome have been developed in another study. The formulation targets glucose transporter GLUT-1 on MCF-7/T cells, to facilitate selective uptake. Upon internalisation, the system inhibits angiogenesis and induces TME remodelling, blocking IL-6/STAT3/p-STAT3 signalling pathway. The system down-regulates drug resistance proteins such as, P-glycoprotein and PD-L1, responsible for reversing MDR and increasing intracellular drug retention. Additionally, it reprogrammed tumour-associated macrophages (TAMs) towards an anti-tumour phenotype, and simultaneously promoted apoptosis. (Y. Zhu et al., 2022).

4.1.5. Polymeric Nanoparticles for Stimulus-Responsive Drug Delivery

Hypoxia-responsive polymersomes have been engineered to encapsulate DOX for targeted drug delivery in TNBC cell lines. A tumor penetrating peptide iRGD, was incorporated to these polymersomes. The conjugation guided the polymersomes to interact with neuropilin-1 (NRP-1) receptors, overexpressed in TNBC under hypoxic conditions. *in vitro* studies highlight, targeted polymersomes increased cytotoxicity against TNBC in hypoxia. It reduced the cell's viability in both monolayer and 3-D spheroid cultures. The integration of a hypoxia-responsive modality enabled selective drug release within the TME, maximizing therapeutic efficacy and minimizing off-target toxicity (Mamnoon et al., 2020).

Similarly, chitosan soluplus Nanoparticles (CS-SP NPs) provide dual stimuli drug delivery, triggered by changes in pH and temperature, to MCF-7 and MDA-MB-231 cells. Further, PEGylated nano systems enable pH-sensitive DOX release. The stealth coating enhanced therapeutic efficacy and reduced systemic toxicity (Twal et al., 2024).

Polymeric nanoparticles like PLGA have pH-sensitive release profiles, delivering chemotherapeutic DOX to acidic tumor sites (Fernandes et al., 2022). Glutathione-sensitive nanoparticles enhanced drug release, responding to the elevated intracellular glutathione levels in tumor. Nano systems co-deliver DOX and siRNA in synergy. They enhance DOX accumulation via P-gp inhibition and target oncogenic pathways like STAT3 and PI3K/Akt with siRNA, to suppress EMT and reduce metastasis (Ashrafizadeh et al., 2020).

4.1.6. Lipid Nanoparticles for Gene Delivery

Nanoparticles have overcome low transfection efficiency and enzymatic degradation issues by delivering gene and small interference RNA into breast tumor. LNPs such as PL-1 encapsulate siRNA and mRNA, protecting their nucleic acids for cellular uptake (H. L. Wang et al., 2022). The development of PLGA-b-PEG polymeric NPs conjugated with the sTN145 aptamer, is an example of lipid-based delivery system for gene therapy. The system selectively targets TNBC and releases encapsulated siRNA against PD-L1. It is achieved by exploiting a pH-responsive surface charge reversal to escape endo-lysosomal compartments. This mechanism effectively silences PD-L1 expression, reducing immune evasion and chemoresistance (Camorani et al., 2022). Similarly, phospholipid PL-1 nanoparticles loaded with mRNA encoding tumor-associated antigens, trigger immune responses and inhibit tumor growth. Polyethylenimine (PEI)-coated nanoparticles carrying CRISPR-Cas9 gene-editing complexes show precise genome editing, silencing oncogenes or repairing tumor-suppressor genes in preclinical settings (H. L. Wang et al., 2022; Camorani et al., 2022).

4.1.7. Dendrimers and Mesoporous Silica NPs for Multi-Drug Resistance Therapy

The dendritic polymeric framework of the organic dendrimers, allows them to encapsulate or attach drugs to their surfaces. For instance, dendrimers loaded with phthalocyanine and interacting with poly(ethylene glycol)-b-poly(L-lysine) block copolymers (PEG-PLL) formed complexes with micelles that effectively deliver DOX into the nuclei of multidrug-resistant (MDR) MCF-7 cells and xenograft models (Martinelli & Biglietti, 2020).

In comparison, mesoporous silica nanomaterials (MSNs) are characterized by high pore volume and tuneable structures. They offer significant drug loading and targeted delivery. Doxorubicin-loaded MSNs reverse MDR by downregulating P-glycoprotein (P-gp). Rod-shaped MSNs co-deliver PTX and tetrandrine to overcome MDR through pH-sensitive release mechanisms (X. Wang et al., 2014). Furthermore, MSNs which are designed for targeted nucleus delivery of DOX disrupt p53-dependent tumor suppressor pathways. They altered the expression of MDR-related genes. Functionalized MSNs carrying DOX alongside siRNAs targeting P-gp have synergistic effects in tumor growth inhibition. It resulted in significant P-gp reduction in MDR xenograft models (Meng et al., 2013).

4.1.8. Multifunctional Nanoparticles

Multifunctional Nanoparticle are the advanced solutions for a holistic breast cancer therapy approach. This was shown in a recent study, with the development of multifunctional IDNPs for chemo-photothermal therapy. It was mediated to target primary TNBC and induce systemic anti-tumor immunity. The IDNPs integrated double therapeutic agents with descent tumoral accumulation. Upon laser activation, it induced chemo-photothermal effect, triggering immunogenic cell death (ICD). The ICD combined with anti-PD-1 therapy to activate systemic immune response to inhibit distant tumor growth. The imaging modalities confirmed targeted accumulation and therapy precision (H. Zhu et al., 2023). Nanoparticles offer variety of biomedical applications in each stage of breast cancer treatment. Several nanosystems are advantageous in tumor imaging, biomarker detection, drug delivery and real-time monitoring of theranostics . Recent research has characterized nanoparticles to perform multiple functions in a single assay, simplifying treatment options.

Figure 4. MOF/MOC-based drug delivery to breast cancer cells, highlighting its role in stimulating macrophages polarization and T lymphocyte activation by the cytokines, released during polarisation of macrophages, for an enhanced immune response

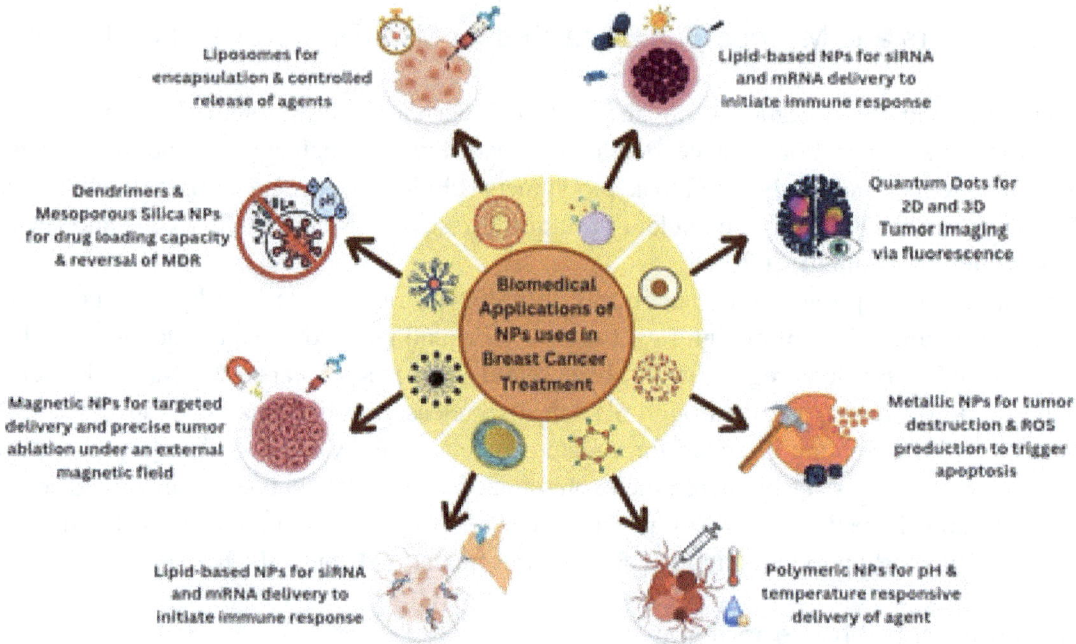

(This figure was created on Canva.com on 27th January 2025)

5. METAL IONS IN BREAST CANCER

Metal-organic complexes (MOCs) are 2D and 3D molecular structures that are formed by the self-assembly of meta centers with abundant ligands binding sites. These metallic structures provide adjustable dimensions and geometries, which makes them valuable in medical applications. 3D structures like metal prism, cages and capsules, serve as drug carriers. Commonly used metals are platinum, ruthenium, rhodium and iridium. For example, metal cages containing platinum, formed with TPPE not only have anti-cancer properties but also enable tissue imaging (Kuntamung et al., 2021). Additionally, MOCs can be integrated with chemotherapy and photodynamic therapy, which enhances cancer treatment efficacy and precision imaging with techniques like MRI and PET. Moreover, MOCs improve the stability of drugs, lower toxicity and reduce resistance issues. Polymeric nanoparticle embedding is an innovative technique for drug loading of MOCs and therapeutic applications. Metal capsules, metal barrels and metal prisms have been modified for dual drug delivery and release systems (Pimalai et al., 2021; Zhao et al., 2021).

Crowley and Casini have worked on Pd_2L_4 MOCs cages, which have shown effective encapsulating drugs such as cisplatin for targeted drug delivery. MOC 75 is a functionalized MOC, designed by Casini that when loaded with cisplatin has significantly increased the cytotoxicity. This helps overcome drug resistance and improve drug delivery by interacting with cancer cells (Moreno-Alc Antar et al., 2023). MOCs 88-91 is a platinum-based MOCs that is created with pH responsive system for controlled drug release, hence enabling targeted drug delivery. MOCs also facilitate synergistic therapies, for instance, MOCs 97 engineered by Therrien's, delivers the photosensitizer, which achieves the effective phototoxicity in breast cancer cells with no extracellular damage, ensuring targeted drug delivery and improved outcomes (Sun et al., 2024). Moreover, MOCs 101, loaded with chemotherapeutic agents and photosensitizers such as $a\gamma_{\beta_3}$ integrin, which are overexpressed in cancer cells. Superior anti tumor effect was demonstrated by the combination therapy in resistant breast cancer models. Tumor-specific accumulation, solubility and stability was enhanced by encapsulation in nanoparticles. Moreover, the synthesis of various supramolecular metallacycles, such as MOCs 42,43,44 and 45 have changed the direction of research. They were designed by Das and their subordinates, demonstrating significant antitumor activity. For instance, MOCs 42 and 43 have shown increased apoptosis in A549 cell lines compared to the drug alone. Additionally, MOCs 44 and 45 have shown growth inhibitory effect in MCF-7 and MDA-MB-231 cell lines. MOCs 44 and 45 owing to their ability to reduce cell growth and proliferation and induce apoptosis might be the future of breast cancer treatment (Z. Gao et al., 2023; Sharma et al., 2024).

Microfluidic rapid mixing is an innovative technique, which can counteract the limitations of traditional therapy. It involves adjusting the flow rate to maintain the composition of nanoparticles. Microfluidic assisted core shell ZIF-8 nanoparticle, modified with silk fibroin and polydopamine for controlled drug release in breast cancer treatment. Silk fibrosis improves the capacity of ZIF-8 as the reservoir of zinc. At the same time, the reaction(self-etching) between ZIF-8 and polydopamine enables the pH stimulated release of the zinc ions and curcumin. After testing this mechanism on the MDA-MB-231 and MCF-7 cell lines, the results showed increased cytotoxicity, cellular uptake a cell cycle arrest (Zhong et al., 2023). Moreover, zinc based metallic organic framework (MOFs) is an innovative platform for the drug delivery of β-estradiol. These MOFs offer high porosity and total pore volume, which enables better encapsulation of β-estradiol. Techniques like scanning electron microscopy and transmission electron microscopy were used to confirm the strong bond formation between β-estradiol and organic ligands. MOFs achieved controlled and sustained drug release of β-estradiol. 8.2 percent of the drug was released at a pH of 8.9 over 8 hours. Hence, the ZN MOFs solve the problem of delivering low solubility drugs such as β-estradiol. The combination of high surface area, adjustable porosity and controlled drug release properties have made them address challenges of drug solubility and bioavailability (Kazimir et al., 2023). Similarly, the development of MIL 101 an iron-based nanocarrier are designed to target tumor associated macrophages. Tumor associated macrophages are important for the activation of the tumor microenvironment. However, the activation is limited by insufficient exposure to iron and other activating agents (Mitchell et al., 2021; Ranjbar et al., 2018). The HM/Ef/LNT-MOFMIL-101(Fe) system, integrates multiple components to overcome these challenges. This framework comprises of nano-sized MIL-101 core with $NaHCO_3$ as a pH-sensitive airbag, and a membrane shell developed from 4T1 cells and RBCs for reduction in non-specific arrest in the blood vessels. Once at the tumor site, the acidic tumor microenvironment stimulates the release of CO_2 bubbles, leading to the MOFMIL-101 disassembly and secretion of Fe^{2+}/Fe^{3+} and lentinan. Lentinan activates interferon-β, which boosts ferroptosis and a Fenton-like reaction, leading to polarization of tumor associated macrophages for the M1 phenotype through the IFM-β-ferroptosis-ROS-Caspase-3 pathway. This polarization improves anti-tumor immune responses by enhancing T lymphocyte infiltration and lowering regulatory T cells, thereby amplifying TME activation. The MIL-101 framework has strong anti-cancer activity by combining TAM polarization, ferroptosis, and immune modulation. The innovative airbag strategy assures controlled delivery of Fe^{2+}/Fe^{3+} and lentinan (Han et al., 2024).

Cu $(DDC)_2$, metal-organic nanoparticles synthesized using copper ions and diethyldithiocarbamate $(DDC)_2$ with bovine serum albumin (BSA) as a stabilizing agent. Moreover, BSA not only stabilizes the Cu $(DDC)_2$ MONs but also serves as a targeting ligand for albumin-binding proteins such as SPARC, which are overexpressed in tumors. The successful formation of Cu $(DDC)_2$ was confirmed by specific spectroscopic peaks, with the nanoparticles expressing a unique dark brown color and a size of approximately 63nm. The anti-cancer potential of BSA Cu $(DDC)_2$ MONs was determined using MDA-MB-231 and 4T1 breast cancer cell lines (Bashir et al., 2022; Yao et al., 2021). The nanoparticles were significant, with cytotoxicity with IC_{50} values of 200nM and 170nM, respectively, compared to nominal effects from therapies with $CuCl_2$ or DDC-Na. In 3-dimensional tumor spheroid models, SA Cu $(DDC)_2$ MONs was accumulated efficiently at tumor sites within 30 minutes post-injection, for up to 24 hours. These nanoparticles significantly inhibited tumor growth at a dose of 3mg/kg, outperforming lower dose or monotherapies. Tumors in the treated group were notably smaller, with no observed toxicity or weight loss in mice, indicating good tolerance (Chang et al., 2020).

Metal ions and their networks such as MOCs and MOFs offer a futuristic approach to traditional therapies. Conventional treatments such as chemotherapy and surgery have various limitations. Chemotherapy has limitations like high toxicity, poor specificity and multidrug resistance. Surgical method lacks in addressing the metastatic area. Similarly, other conventional methods also have several limitations such as bioavailability, weakens immune system, kills healthy tissues. However, metal ions prove to be better than conventional therapy, due to its features like controlled drug release, opportunity for synergistic effect and enhanced tumor penetration. Several metal ions like copper, gold and zinc have selective toxicity for tumor through its various mechanism such as by generating ROS and redox reaction. To enhance the stability and functionality of metal ions they are incorporated into MOCs and MOFs.

In spite of offering great advantages over tradition treatment methods, metal ions have several challenges. Several metal ions such as platinum and gadolinium are cytotoxic(Z. Zhu et al., 2014). They can damage healthy tissues by disrupting their function and metabolism. Moreover, some metal ions have been found to trigger immune cells resulting in inflammation or allergic reaction. Metal ions are designed for controlled and stimulus responsive drug release. However, they can be triggered by unexpected change in pH or enzymes, leading to nonspecific drug release (C. Wang et al., 2020). Designing and manufacturing metal ion complexes requires highly skilled labors, significant investment and time. Reproducing precise replication of the original is extremely difficult. Even a small difference in size and coordination has potential to induce major changes in the pharmacological mechanism (Liu et al., 2021).

6. FUTURE PERSPECTIVES

While biomaterials, nanoparticles, polymers and metal ions are being rapidly utilized for drug delivery purposes, there are still various challenges that need to be addressed. One of the primary concerns is systemic toxicity, associated with polymeric and nanoparticle-based systems. These delivery methods can inadvertently affect healthy tissues, leading to unwanted side effects that compromise patient safety. To mitigate this issue, there is a pressing need for the development of effective fabrication methods that prioritize biocompatibility and reduce systemic exposure to toxic agents. Sustainable, biocompatible and biodegradable materials can be utilized, along with the formulation of eco-friendly methods. Another critical challenge in breast cancer treatment is the prevalence of multiple drug resistance (MDR), which significantly hampers the effectiveness of conventional therapies. Many breast cancer patients develop resistance to standard chemotherapeutic agents, necessitating the exploration of novel drug delivery systems that can circumvent these mechanisms. Furthermore, combining multiple therapeutic agents within a single nanoparticle platform could provide a synergistic effect that overcomes resistance and improves treatment outcomes. The integration of diagnostic and therapeutic functions into single nanoparticle platforms, termed as theragnostic systems are being developed. These multifunctional platforms allow for real-time monitoring of treatment effectiveness while simultaneously delivering therapeutic agents. The ability to visualize drug distribution and tumor response enables the use of personalized medicine by enabling clinicians to adjust treatment protocols based on individual patient needs. Also, despite the promising advancements in these drug delivery systems, there remains a significant dearth of clinical trials investigating biomaterial, polymeric, and nanoparticle-based approaches specifically for breast cancer treatment. It is imperative that pre-clinical research translate to clinical outcomes, so that their real-world applications can be successfully demonstrated.

Scalability of these advanced material solely depends on development of affordable and reproducible synthesise methods. Development of safe chemical procedure is necessary for large scale synthesis of metal ions and their complexes. On the other hand, synthesis of polymers is already well developed. However, issues like reproducibly and variations are to solved. Methods like green chemistry can reduce production cost and environmental damage. Advancement of bioconjugation and microreactor design can product high quality biopolymers. Nanomaterial synthesis is more complex due their shape, size and surface functionalisation. Top-down technique can increase scalability and automation. Scalability of biomaterial depends on bio-fabrication techniques like 3D printing. Real-time monitoring and automation of process will be key in increasing quality and quantity of manufactured product, minimizing human error. Standardization of procedure ensures regulatory compliance. Using advanced techniques like microfluids and flow chemistry helps in reducing cost and enhance efficiency of production.

7. CONCLUSION

The fight against breast cancer is continuously evolving, and innovative materials like polymers, biomaterials, nanoparticles, and metal ions are playing a transformative role in both treatment and diagnostics. Synthetic and natural polymers have proven to be powerful tools in drug delivery, with hybrid systems offering the best of both worlds—combining stability with biocompatibility for safer, more controlled treatments. However, translating these materials from research to real-world clinical use remains a challenge, requiring careful attention to scalability, safety, and regulatory approval. Biomaterials hold immense promise, but their full potential can only be realized through rigorous clinical validation and seamless integration with other therapies, such as gene therapy and combination treatments. Nanoparticles, with their ability to aid in tumor imaging, biomarker detection, and real-time monitoring, are revolutionizing how breast cancer is diagnosed and managed, making treatments more precise and efficient.Metal-based therapies, including metal-organic cages (MOCs) and metal-organic frameworks (MOFs), are opening new frontiers in targeted drug delivery.

By leveraging their adaptable structures, these materials enhance drug solubility, minimize toxicity, and enable precise control over drug release, making treatment more patient-friendly and effective. Despite these promising advances, the journey from laboratory research to real-world application is ongoing. Continued research, clinical trials, and collaboration across disciplines will be key to ensuring these breakthroughs translate into accessible, life-changing therapies. By harnessing the power of these innovative technologies, we move closer to a future where breast cancer treatments are not only more effective but also safer, more personalized, and ultimately, more hopeful for patients worldwide.

REFERENCES

Abatangelo, G., Vindigni, V., Avruscio, G., Pandis, L., & Brun, P. (2020). Hyaluronic Acid: Redefining Its Role. Cells 2020, Vol. 9, Page 1743, 9(7), 1743. https://doi.org/DOI: 10.3390/CELLS9071743

Adeli, F., Abbasi, F., Ghandforoushan, P., Külahlı, H. E., Meran, M., Abedi, F., Ghamkhari, A., & Afif, S. (2023). Recent advances in formulation and application of molecular polymer brushes in biomedicine: Therapeutic, diagnostic, and theranostics capabilities. In Nano Today (Vol. 53). https://doi.org/DOI: 10.1016/j.nantod.2023.102010

Alam, M., Rashid, S., Fatima, K., Adnan, M., Shafie, A., Akhtar, M. S., Ganie, A. H., Eldin, S. M., Islam, A., Khan, I., & Hassan, M. I. (2023). Biochemical features and therapeutic potential of α-Mangostin: Mechanism of action, medicinal values, and health benefits. *Biomedicine and Pharmacotherapy*, *163*, 114710. DOI: 10.1016/J.BIOPHA.2023.114710 PMID: 37141737

Alamdari, S. G., Amini, M., Jalilzadeh, N., Baradaran, B., Mohammadzadeh, R., Mokhtarzadeh, A., & Oroojalian, F. (2022). Recent advances in nanoparticle-based photothermal therapy for breast cancer. *Journal of Controlled Release : Official Journal of the Controlled Release Society*, *349*, 269–303. DOI: 10.1016/J.JCONREL.2022.06.050 PMID: 35787915

Alharbi, H. M. (2024). Exploring the Frontier of Biopolymer-Assisted Drug Delivery: Advancements, Clinical Applications, and Future Perspectives in Cancer Nanomedicine. *Drug Design, Development and Therapy*, *18*, 2063–2087. DOI: 10.2147/DDDT.S441325 PMID: 38882042

Asadi, K., Samiraninezhad, N., Akbarizadeh, A. R., Amini, A., & Gholami, A. (2024). Stimuli-responsive hydrogel based on natural polymers for breast cancer. *Frontiers in Chemistry*, *12*, 1325204. DOI: 10.3389/FCHEM.2024.1325204/BIBTEX PMID: 38304867

Ashrafizadeh, M., Zarrabi, A., Hushmandi, K., Hashemi, F., Rahmani Moghadam, E., Raei, M., Kalantari, M., Tavakol, S., Mohammadinejad, R., Najafi, M., Tay, F. R., & Makvandi, P. (2020). Progress in Natural Compounds/siRNA Co-delivery Employing Nanovehicles for Cancer Therapy. *ACS Combinatorial Science*, *22*(12), 669. DOI: 10.1021/ACSCOMBSCI.0C00099 PMID: 33095554

Attallah, O. A., Shetta, A., Elshishiny, F., & Mamdouh, W. (2020). Essential oil loaded pectin/chitosan nanoparticles preparation and optimization via Box–Behnken design against MCF-7 breast cancer cell lines. *RSC Advances*, *10*(15), 8703–8708. DOI: 10.1039/C9RA10204C PMID: 35496538

Bashir, M., Yousuf, I., & Prakash Prasad, C. (2022). Mixed Ni(II) and Co(II) complexes of nalidixic acid drug: Synthesis, characterization, DNA/BSA binding profile and in vitro cytotoxic evaluation against MDA-MB-231 and HepG2 cancer cell lines. *Spectrochimica Acta. Part A: Molecular and Biomolecular Spectroscopy*, *271*, 120910. DOI: 10.1016/J.SAA.2022.120910 PMID: 35077983

Berna, G., De Grazia, A., Antoniazzi, E., Romeo, M., Dell'Antonia, F., Lovero, S., Marchica, P., Rizzetto, C., & Burelli, P. (2022). Novel three-dimensional acellular dermal matrix for prepectoral breast reconstruction: First year in review with BRAXON®Fast. *Frontiers in Surgery*, *9*, 970053. DOI: 10.3389/FSURG.2022.970053/BIBTEX PMID: 36132205

Bhattacharya, T., Preetam, S., Ghosh, B., Chakrabarti, T., Chakrabarti, P., Samal, S. K., & Thorat, N. (2023). Advancement in Biopolymer Assisted Cancer Theranostics. *ACS Applied Bio Materials*, *6*(10), 3959–3983. DOI: 10.1021/ACSABM.3C00458/ASSET/IMAGES/LARGE/MT3C00458_0010.JPEG PMID: 37699558

Biomaterials generations. (n.d.).

Bray, F., Laversanne, M., Sung, H., Ferlay, J., Siegel, R. L., Soerjomataram, I., & Jemal, A. (2024). Global cancer statistics 2022: GLOBOCAN estimates of incidence and mortality worldwide for 36 cancers in 185 countries. *CA: a Cancer Journal for Clinicians*. Advance online publication. DOI: 10.3322/CAAC.21834 PMID: 38572751

Caballero, D., Abreu, C. M., Lima, A. C., Neves, N. N., Reis, R. L., & Kundu, S. C. (2022). Precision biomaterials in cancer theranostics and modelling. In Biomaterials (Vol. 280). https://doi.org/DOI: 10.1016/j.biomaterials.2021.121299

Camorani, S., Tortorella, S., Agnello, L., Spanu, C., d'Argenio, A., Nilo, R., Zannetti, A., Locatelli, E., Fedele, M., Comes Franchini, M., & Cerchia, L. (2022). Aptamer-Functionalized Nanoparticles Mediate PD-L1 siRNA Delivery for Effective Gene Silencing in Triple-Negative Breast Cancer Cells. *Pharmaceutics*, *14*(10), 2225. DOI: 10.3390/PHARMACEUTICS14102225/S1 PMID: 36297659

Cao, D., & Ding, J. (2022). Recent advances in regenerative biomaterials. *Regenerative Biomaterials*, *9*. Advance online publication. DOI: 10.1093/RB/RBAC098 PMID: 36518879

Chang, Y., Jiang, J., Chen, W., Yang, W., Chen, L., Chen, P., Shen, J., Qian, S., Zhou, T., Wu, L., Hong, L., Huang, Y., & Li, F. (2020). Biomimetic metal-organic nanoparticles prepared with a 3D-printed microfluidic device as a novel formulation for disulfiram-based therapy against breast cancer. *Applied Materials Today*, *18*. Advance online publication. DOI: 10.1016/J.APMT.2019.100492 PMID: 34746366

Chavda, V. P., Vuppu, S., Bezbaruah, R., Nalla, L. V., Gajula, S. N. R., Balar, P. C., Mishra, T., Sharma, N., Kamaraj, S., Suresh, T., Sairam, A., & Bhattacharjee, B. (2023). Phytochemical Loaded Nanovehicles of Biopolymer for Breast Cancer: A Systemic Review. *Clinical Complementary Medicine and Pharmacology*, *3*(4), 100114. DOI: 10.1016/J.CCMP.2023.100114

Chen, J., Zhang, Z., Li, Y., Zeng, H., Li, Z., Wang, C., Xu, C., Deng, Q., Wang, Q., Yang, X., & Li, Z. (2022). Precise fibrin decomposition and tumor mechanics modulation with hydroxyethyl starch-based smart nanomedicine for enhanced antitumor efficacy.

Curcio, M., Cirillo, G., Tucci, P., Farfalla, A., Bevacqua, E., Vittorio, O., Iemma, F., & Nicoletta, F. P. (2019). Dextran-Curcumin Nanoparticles as a Methotrexate Delivery Vehicle: A Step Forward in Breast Cancer Combination Therapy. *Pharmaceuticals*, *13*(1), 2. DOI: 10.3390/PH13010002 PMID: 31881645

Dar, M. S., Tabish, T. A., Thorat, N. D., Swati, G., & Sahu, N. K. (2023). Photothermal therapy using graphene quantum dots. *APL Bioengineering*, *7*(3), 031502. DOI: 10.1063/5.0160324 PMID: 37614868

Demir, D., Bolgen, N., & Vaseashta, A. (2023). Electrospun Nanofibers for Biomedical, Sensing, and Energy Harvesting Functions. Polymers 2023, Vol. 15, Page 4253, 15(21), 4253. https://doi.org/DOI: 10.3390/POLYM15214253

Díaz-García, D., Díaz-Sánchez, M., Álvarez-Conde, J., & Gómez-Ruiz, S. (2024). Emergence of Quantum Dots as Innovative Tools for Early Diagnosis and Advanced Treatment of Breast Cancer. *ChemMedChem*, *19*(16), e202400172. DOI: 10.1002/CMDC.202400172 PMID: 38724442

Ding, L., Agrawal, P., Singh, S. K., Chhonker, Y. S., Sun, J., & Murry, D. J. (2024). Polymer-Based Drug Delivery Systems for Cancer Therapeutics. Polymers 2024, Vol. 16, Page 843, 16(6), 843. https://doi.org/DOI: 10.3390/POLYM16060843

Do Nascimento, S. da P., de Souza, R. R. M., Sobral, M. V., Xavier-Junior, F. H., da Silva, M. V. S., Viana, M. M., da Silva, F. F., Serpe, M. J., & de Souza, A. L. (2024). Gelatin-Oxidized Alginate and Chitosan-Coated Zein Nanoparticle Hydrogel Composite to Enhance Breast Cancer Cytotoxicity in Dual-Drug Delivery. *ACS Omega*. Advance online publication. DOI: 10.1021/ACSOMEGA.4C06404/ASSET/IMAGES/LARGE/AO4C06404_0011.JPEG PMID: 39554428

Doerksen, R. J., Osborn, H., Socorro, S., Dragojevic, S., Su Ryu, J., Hall, M. E., & Raucher, D. (2022). Targeted Drug Delivery Biopolymers Effectively Inhibit Breast Tumor Growth and Prevent Doxorubicin-Induced Cardiotoxicity. *Molecules (Basel, Switzerland)*, *27*(11), 3371. DOI: 10.3390/MOLECULES27113371 PMID: 35684309

Fahira, A. I., Abdulah, R., Barliana, M. I., Gatera, V. A., & Amalia, R. (2022). Polyethyleneimine (PEI) as a Polymer-Based Co-Delivery System for Breast Cancer Therapy. *Breast Cancer (Dove Medical Press)*, *14*, 71. DOI: 10.2147/BCTT.S350403 PMID: 35422657

Farag, M. M. (2023). Recent trends on biomaterials for tissue regeneration applications: review. Journal of Materials Science 2022 58:2, 58(2), 527–558. https://doi.org/DOI: 10.1007/S10853-022-08102-X

Fatima, I., Rahdar, A., Sargazi, S., Barani, M., Hassanisaadi, M., & Thakur, V. K. (2021). Quantum Dots: Synthesis, Antibody Conjugation, and HER2-Receptor Targeting for Breast Cancer Therapy. Journal of Functional Biomaterials 2021, Vol. 12, Page 75, 12(4), 75. https://doi.org/DOI: 10.3390/JFB12040075

Fatima, M., Sheikh, A., Abourehab, M. A. S., & Kesharwani, P. (2022). Advancements in Polymeric Nanocarriers to Mediate Targeted Therapy against Triple-Negative Breast Cancer. *Pharmaceutics*, *14*(11), 2432. DOI: 10.3390/PHARMACEUTICS14112432 PMID: 36365249

Fernandes, R. S., Arribada, R. G., Silva, J. O., Silva-Cunha, A., Townsend, D. M., Ferreira, L. A. M., & Barros, A. L. B. (2022). In Vitro and In Vivo Effect of pH-Sensitive PLGA-TPGS-Based Hybrid Nanoparticles Loaded with Doxorubicin for Breast Cancer Therapy. Pharmaceutics 2022, Vol. 14, Page 2394, 14(11), 2394. https://doi.org/DOI: 10.3390/PHARMACEUTICS14112394

Fogg, K., Tseng, N. H., Peyton, S. R., Holeman, P., Mc Loughlin, S., Fisher, J. P., Sutton, A., Shikanov, A., Gnecco, J. S., Knight, K. M., Slaby, E. M., Weaver, J. D., Hashemi, N. N., Zhang, Y., House, M. D., Vogt, B. J., Aguado, B. A., Bradford, J. C., Robinson, J. L., & Oyen, M. L. (2022). Roadmap on biomaterials for women's health. *JPhys Materials*, *6*(1), 012501. DOI: 10.1088/2515-7639/AC90EE

Ganesan, K., Wang, Y., Gao, F., Liu, Q., Zhang, C., Li, P., Zhang, J., & Chen, J. (2021). Targeting Engineered Nanoparticles for Breast Cancer Therapy. *Pharmaceutics*, *13*(11), 1829. DOI: 10.3390/PHARMACEUTICS13111829 PMID: 34834243

Gao, X., Caruso, B. R., & Li, W. (2024). Advanced Hydrogels in Breast Cancer Therapy. Gels 2024, Vol. 10, Page 479, 10(7), 479. https://doi.org/DOI: 10.3390/GELS10070479

Gao, Z., Mansor, M. H., Winder, N., Demiral, S., Maclnnes, J., Zhao, X., & Muthana, M. (2023). Microfluidic-Assisted ZIF-Silk-Polydopamine Nanoparticles as Promising Drug Carriers for Breast Cancer Therapy. *Pharmaceutics*, *15*(7). Advance online publication. DOI: 10.3390/PHARMACEUTICS15071811 PMID: 37513998

Ghaffarlou, M., Rashidzadeh, H., Mohammadi, A., Mousazadeh, N., Barsbay, M., Shara, A., Gharbavi, M., Danafar, H., & Golestan, S. J. (2023). Catechol-Conjugated Alginate Biopolymer for Eco-Friendly Synthesis of Gold Nanoparticles: A Promising Approach for Photothermal and Radiotherapy in Breast Cancer Treatment. https://doi.org/DOI: 10.21203/RS.3.RS-3496758/V1

Gulati, S., Ansari, N., Moriya, Y., Joshi, K., Prasad, D., Sajwan, G., Shukla, S., Kumar, S., & Varma, R. S. (2024). Nanobiopolymers in cancer therapeutics: Advancing targeted drug delivery through sustainable and controlled release mechanisms. *Journal of Materials Chemistry. B*, *12*(46), 11887–11915. DOI: 10.1039/D4TB00599F PMID: 39502076

Gulfam, M., Kim, J. E., Lee, J. M., Ku, B., Chung, B. H., & Chung, B. G. (2012). Anticancer drug-loaded gliadin nanoparticles induce apoptosis in breast cancer cells. *Langmuir*, *28*(21). Advance online publication. DOI: 10.1021/la300691n PMID: 22568862

Guo, L., Zhang, H., Liu, P., Mi, T., Ha, D., Su, L., Huang, L., Shi, Y., & Zhang, J. (2021). Preclinical Assessment of Paclitaxel- and Trastuzumab-Delivering Magnetic Nanoparticles Fe3O4 for Treatment and Imaging of HER2-Positive Breast Cancer. *Frontiers in Medicine*, *8*, 738775. DOI: 10.3389/FMED.2021.738775/FULL PMID: 34778301

Gupta, S. (2022, February 4). Side effects of breast cancer treatment | LBBC. https://www.lbbc.org/about-breast-cancer/side-effects

Han, T., Sun, Y., Jiang, X., Gong, C., Kong, F., Luo, Y., Ge, C., Liu, C., Liu, Y., Mou, Y., Zhang, H., Ju, J., Chen, Y., & Qu, D. (2024). Air bag-embedded MIL-101(Fe) metal-organic frameworks for an amplified tumor microenvironment activation loop through strategic delivery of iron ions and lentinan. *Theranostics*, *14*(15), 5883–5902. DOI: 10.7150/THNO.99303 PMID: 39346539

Harish, V., Tewari, D., Gaur, M., Yadav, A. B., Swaroop, S., Bechelany, M., & Barhoum, A. (2022). Review on Nanoparticles and Nanostructured Materials: Bioimaging, Biosensing, Drug Delivery, Tissue Engineering, Antimicrobial, and Agro-Food Applications. *Nanomaterials (Basel, Switzerland)*, *12*(3), 457. DOI: 10.3390/NANO12030457 PMID: 35159802

Heilala, M., Lehtonen, A., Arasalo, O., Peura, A., Pokki, J., & Ikkala, O., Nonappa, Klefström, J., & Munne, P. M. (2023). Fibrin Stiffness Regulates Phenotypic Plasticity of Metastatic Breast Cancer Cells. *Advanced Healthcare Materials*, *12*(31), 2301137. DOI: 10.1002/ADHM.202301137 PMID: 37671812

Herdiana, Y., Husni, P., Nurhasanah, S., Shamsuddin, S., & Wathoni, N. (2023). Chitosan-Based Nano Systems for Natural Antioxidants in Breast Cancer Therapy. In Polymers (Vol. 15, Issue 13). https://doi.org/DOI: 10.3390/polym15132953

Hossein, H. H. S., Jabbari, I., Zarepour, A., Zarrabi, A., Ashrafizadeh, M., Taherian, A., & Makvandi, P. (2020). Functionalization of Magnetic Nanoparticles by Folate as Potential MRI Contrast Agent for Breast Cancer Diagnostics. *Molecules (Basel, Switzerland)*, *25*(18), 4053. DOI: 10.3390/MOLECULES25184053 PMID: 32899812

Iaconisi, G. N., Lunetti, P., Gallo, N., Cappello, A. R., Fiermonte, G., Dolce, V., & Capobianco, L. (2023). Hyaluronic Acid: A Powerful Biomolecule with Wide-Ranging Applications—A Comprehensive Review. International Journal of Molecular Sciences 2023, Vol. 24, Page 10296, 24(12), 10296. https://doi.org/DOI: 10.3390/IJMS241210296

Iravani, S., & Varma, R. S. (2022). Alginate-Based Micro- and Nanosystems for Targeted Cancer Therapy. Marine Drugs 2022, Vol. 20, Page 598, 20(10), 598. https://doi.org/DOI: 10.3390/MD20100598

Jia, Y., Chen, S., Wang, C., Sun, T., & Yang, L. (2022). Hyaluronic acid-based nano drug delivery systems for breast cancer treatment: Recent advances. *Frontiers in Bioengineering and Biotechnology*, *10*, 990145. DOI: 10.3389/FBIOE.2022.990145/BIBTEX PMID: 36091467

Jiang, X., Du, Z., Zhang, X., Zaman, F., Song, Z., Guan, Y., Yu, T., & Huang, Y. (2023). Gelatin-based anticancer drug delivery nanosystems: A mini review. *Frontiers in Bioengineering and Biotechnology*, *11*, 1158749. DOI: 10.3389/FBIOE.2023.1158749/BIBTEX PMID: 37025360

Jiaying, Y., Bo, S., Xiaolu, W., Yanyan, Z., Hongjie, W., Nan, S., Bo, G., Linna, W., Yan, Z., Wenya, G., Keke, L., Shan, J., Chuan, L., Yu, Z., Qinghe, Z., & Haiyu, Z. (2023). Arenobufagin-loaded PEG-PLA nanoparticles for reducing toxicity and enhancing cancer therapy. *Drug Delivery*, *30*(1), 2177362. DOI: 10.1080/10717544.2023.2177362 PMID: 36772846

Kazimir, A., Schwarze, B., Lönnecke, P., Jelača, S., Mijatović, S., Maksimović-Ivanić, D., & Hey-Hawkins, E. (2023). Metallodrugs against Breast Cancer: Combining the Tamoxifen Vector with Platinum(II) and Palladium(II) Complexes. Pharmaceutics 2023, Vol. 15, Page 682, 15(2), 682. https://doi.org/DOI: 10.3390/PHARMACEUTICS15020682

Khan, M. S., Gowda, B. H. J., Nasir, N., Wahab, S., Pichika, M. R., Sahebkar, A., & Kesharwani, P. (2023). Advancements in dextran-based nanocarriers for treatment and imaging of breast cancer. *International Journal of Pharmaceutics*, *643*, 123276. DOI: 10.1016/J.IJPHARM.2023.123276 PMID: 37516217

Khursheed, R., Dua, K., Vishwas, S., Gulati, M., Jha, N. K., Aldhafeeri, G. M., Alanazi, F. G., Goh, B. H., Gupta, G., Paudel, K. R., Hansbro, P. M., Chellappan, D. K., & Singh, S. K. (2022). Biomedical applications of metallic nanoparticles in cancer: Current status and future perspectives. *Biomedicine and Pharmacotherapy*, *150*, 112951. DOI: 10.1016/J.BIOPHA.2022.112951 PMID: 35447546

Kim, J., Yu, A. M., Kubelick, K. P., & Emelianov, S. Y. (2021). Gold nanoparticles conjugated with DNA aptamer for photoacoustic detection of human matrix metalloproteinase-9. *Photoacoustics*, *25*, 100307. DOI: 10.1016/J.PACS.2021.100307 PMID: 34703762

Kim, Y. S., Ko, M. J., Moon, H., Sim, W., Cho, A. S., Gil, G., & Kim, H. R. (2022). Ultrasound-Responsive Liposomes for Targeted Drug Delivery Combined with Focused Ultrasound. *Pharmaceutics*, *14*(7), 1314. DOI: 10.3390/PHARMACEUTICS14071314/S1 PMID: 35890210

Kinnel, B., Singh, S. K., Oprea-Ilies, G., & Singh, R. (2023). Targeted Therapy and Mechanisms of Drug Resistance in Breast Cancer. *Cancers (Basel)*, *15*(4), 1320. DOI: 10.3390/CANCERS15041320 PMID: 36831661

Kirtane, A. R., Sadhukha, T., Kim, H., Khanna, V., Koniar, B., & Panyam, J. (2017). Fibrinolytic enzyme co-therapy improves tumor perfusion and therapeutic efficacy of anticancer nanomedicine. *Cancer Research*, *77*(6), 1465. DOI: 10.1158/0008-5472.CAN-16-1646 PMID: 28108516

Kumar, P., Mangla, B., Javed, S., Ahsan, W., Musyuni, P., Sivadasan, D., Alqahtani, S. S., & Aggarwal, G. (2023). A review of nanomaterials from synthetic and natural molecules for prospective breast cancer nanotherapy. *Frontiers in Pharmacology*, *14*, 1149554. DOI: 10.3389/FPHAR.2023.1149554 PMID: 37274111

Kumar, P., Senthamilselvi, S., Govindaraju, M., & Sankar, R. (2014). Unraveling the caspase-mediated mechanism for phloroglucinol-encapsulated starch biopolymer against the breast cancer cell line MDA-MB-231. *RSC Advances*, *4*(86), 46157–46163. DOI: 10.1039/C4RA06664B

Kumari, M., Sharma, N., Manchanda, R., Gupta, N., Syed, A., Bahkali, A. H., & Nimesh, S. (2021). PGMD/curcumin nanoparticles for the treatment of breast cancer. *Scientific Reports*, *11*(1), 3824. DOI: 10.1038/S41598-021-81701-X PMID: 33589661

Kuntamung, K., Jakmunee, J., & Ounnunkad, K. (2021). A label-free multiplex electrochemical biosensor for the detection of three breast cancer biomarker proteins employing dye/metal ion-loaded and antibody-conjugated polyethyleneimine-gold nanoparticles. *Journal of Materials Chemistry. B*, *9*(33), 6576–6585. DOI: 10.1039/D1TB00940K PMID: 34279016

Li, J. J., Tsang, J. Y., & Tse, G. M. (2021). Tumor Microenvironment in Breast Cancer—Updates on Therapeutic Implications and Pathologic Assessment. *Cancers (Basel)*, *13*(16), 4233. DOI: 10.3390/CANCERS13164233 PMID: 34439387

Li, M., Tang, Z., Zhang, Y., Lv, S., Yu, H., Zhang, D., Hong, H., & Chen, X. (2014). LHRH-peptide conjugated dextran nanoparticles for targeted delivery of cisplatin to breast cancer. *Journal of Materials Chemistry. B*, *2*(22), 3490–3499. DOI: 10.1039/C4TB00077C PMID: 32261469

Madamsetty, V. S., Tavakol, S., Moghassemi, S., Dadashzadeh, A., Schneible, J. D., Fatemi, I., Shirvani, A., Zarrabi, A., Azedi, F., Dehshahri, A., Aghaei Afshar, A., Aghaabbasi, K., Pardakhty, A., Mohammadinejad, R., & Kesharwani, P. (2022). Chitosan: A versatile bio-platform for breast cancer theranostics. *Journal of Controlled Release : Official Journal of the Controlled Release Society*, *341*, 733–752. DOI: 10.1016/J.JCONREL.2021.12.012 PMID: 34906606

Mamnoon, B., Loganathan, J., Confeld, M. I., De Fonseka, N., Feng, L., Froberg, J., Choi, Y., Tuvin, D. M., Sathish, V., & Mallik, S. (2020). Targeted polymeric nanoparticles for drug delivery to hypoxic, triple-negative breast tumors. *ACS Applied Bio Materials*, *4*(2), 1450. DOI: 10.1021/ACSABM.0C01336 PMID: 33954285

Martinelli, C., & Biglietti, M. (2020). Nanotechnological approaches for counteracting multidrug resistance in cancer. *Cancer Drug Resistance (Alhambra, Calif.)*, *3*(4), 1003. DOI: 10.20517/CDR.2020.47 PMID: 35582219

Mayer, H. F., Coloccini, A., & Viñas, J. F. (2024). Three-Dimensional Printing in Breast Reconstruction: Current and Promising Applications. Journal of Clinical Medicine 2024, Vol. 13, Page 3278, 13(11), 3278. https://doi.org/DOI: 10.3390/JCM13113278

Meng, H., Mai, W. X., Zhang, H., Xue, M., Xia, T., Lin, S., Wang, X., Zhao, Y., Ji, Z., Zink, J. I., & Nel, A. E. (2013). Co-delivery of an Optimal Drug/siRNA Combination Using Mesoporous Silica Nanoparticle to Overcome Drug Resistance in Breast Cancer In Vitro and In Vivo. *ACS Nano*, *7*(2), 994. DOI: 10.1021/NN3044066 PMID: 23289892

Mitchell, M. J., Billingsley, M. M., Haley, R. M., Wechsler, M. E., Peppas, N. A., & Langer, R. (2021). Engineering precision nanoparticles for drug delivery. *Nature Reviews. Drug Discovery*, *20*(2), 101–124. DOI: 10.1038/S41573-020-0090-8 PMID: 33277608

Mohanta, Y. K., Biswas, K., Rauta, P. R., Mishra, A. K., De, D., Hashem, A., Al-Arjani, A. B. F., Alqarawi, A. A., Abd-Allah, E. F., Mahanta, S., & Mohanta, T. K. (2021). Development of Graphene Oxide Nanosheets as Potential Biomaterials in Cancer Therapeutics: An In-Vitro Study Against Breast Cancer Cell Line. *Journal of Inorganic and Organometallic Polymers and Materials*, *31*(11), 4236–4249. DOI: 10.1007/S10904-021-02046-6/METRICS

Mondal, A., Nayak, A. K., Chakraborty, P., Banerjee, S., & Nandy, B. C. (2023). Natural Polymeric Nanobiocomposites for Anti-Cancer Drug Delivery Therapeutics: A Recent Update. *Pharmaceutics*, *15*(8), 2064. DOI: 10.3390/PHARMACEUTICS15082064 PMID: 37631276

Moreno-Alc Antar, G., Casini, A., Moreno-Alc, G., & Casini, A. (2023). Bioinorganic supramolecular coordination complexes and their biomedical applications. *FEBS Letters*, *597*(1), 191–202. DOI: 10.1002/1873-3468.14535 PMID: 36345593

Mozafari, M. (2024). Editorial Note: Biomaterials in Cancer - From Research Breakthroughs to Clinical Implementation. *Translational Oncology*, *48*, 102061. DOI: 10.1016/J.TRANON.2024.102061 PMID: 39134454

Muchtaridi, M., Suryani, A. I., Wathoni, N., Herdiana, Y., Mohammed, A. F. A., Gazzali, A. M., Lesmana, R., & Joni, I. M. (2023). Chitosan/Alginate Polymeric Nanoparticle-Loaded α-Mangostin: Characterization, Cytotoxicity, and In Vivo Evaluation against Breast Cancer Cells. Polymers 2023, Vol. 15, Page 3658, 15(18), 3658. https://doi.org/DOI: 10.3390/POLYM15183658

Nadanaka, S., Tamura, J. I., & Kitagawa, H. (2022). Chondroitin Sulfates Control Invasiveness of the Basal-Like Breast Cancer Cell Line MDA-MB-231 Through ROR1. *Frontiers in Oncology*, *12*, 914838. DOI: 10.3389/FONC.2022.914838/BIBTEX PMID: 35712490

Nief, C., Morhard, R., Chelales, E., Alvarez, D. A., Ioanna Bourla, B. S., Lam, C. T., Sag, A. A., Crouch, B. T., Mueller, J. L., Katz, D., Dewhirst, M. W., Everitt, J. I., & Ramanujam, N. (2021). Polymer-assisted intratumoral delivery of ethanol: Preclinical investigation of safety and efficacy in a murine breast cancer model. *PLoS One*, *16*(1), e0234535. DOI: 10.1371/JOURNAL.PONE.0234535 PMID: 33507942

Odeh, A. B., El-Sayed, B., Knopp, M. M., Rades, T., & Blaabjerg, L. I. (2023). Influence of Polyvinylpyrrolidone Molecular Weight and Concentration on the Precipitation Inhibition of Supersaturated Solutions of Poorly Soluble Drugs. Pharmaceutics 2023, Vol. 15, Page 1601, 15(6), 1601. https://doi.org/DOI: 10.3390/PHARMACEUTICS15061601

Omidian, H., Wilson, R. L., & Cubeddu, L. X. (2024). Quantum Dot Research in Breast Cancer: Challenges and Prospects. Materials 2024, Vol. 17, Page 2152, 17(9), 2152. https://doi.org/DOI: 10.3390/MA17092152

Park, J. H., Park, J. U., & Chang, H. (2021). Advances in Biomaterials for Breast Reconstruction. Applied Sciences 2021, Vol. 11, Page 7493, 11(16), 7493. https://doi.org/DOI: 10.3390/APP11167493

Passos, J. S., Lopes, L. B., & Panitch, A. (2023). Collagen-Binding Nanoparticles for Paclitaxel Encapsulation and Breast Cancer Treatment. *ACS Biomaterials Science & Engineering*, *9*(12), 6805–6820. DOI: 10.1021/ACSBIOMATERIALS.3C01332/SUPPL_FILE/AB3C01332_SI_001.PDF PMID: 37982792

Pimalai, D., Putnin, T., Waiwinya, W., Chotsuwan, C., Aroonyadet, N., & Japrung, D. (2021). Development of electrochemical biosensors for simultaneous multiplex detection of microRNA for breast cancer screening. *Microchimica Acta*, *188*(10), 1–10. DOI: 10.1007/S00604-021-04995-8/METRICS PMID: 34495394

Pounraj, S., Chen, S., Triscott, H., Lam, A. K., Ma, L., & Rehm, B. H. A. (2024). Engineering neoantigens to form immunogenic biopolymer particles targeting metastatic breast cancer. *Applied Materials Today*, *38*, 102238. DOI: 10.1016/J.APMT.2024.102238

Prihantono, & Faruk, M. (2021). Breast cancer resistance to chemotherapy: When should we suspect it and how can we prevent it? Annals of Medicine and Surgery, 70, 102793. https://doi.org/DOI: 10.1016/J.AMSU.2021.102793

Ranjbar, M., Pardakhty, A., Amanatfard, A., & Asadipour, A. (2018). Efficient drug delivery of β-estradiol encapsulated in Zn-metal-organic framework nanostructures by microwave-assisted coprecipitation method. *Drug Design, Development and Therapy*, *12*, 2635–2643. DOI: 10.2147/DDDT.S173324 PMID: 30214152

Rijal, G., Bathula, C., & Li, W. (2017). Application of Synthetic Polymeric Scaffolds in Breast Cancer 3D Tissue Cultures and Animal Tumor Models. *International Journal of Biomaterials*, *8074890*. Advance online publication. DOI: 10.1155/2017/8074890 PMID: 29599800

Robe, A., Pic, E., Lassalle, H. P., Bezdetnaya, L., Guillemin, F., & Marchal, F. (2008). Quantum dots in axillary lymph node mapping: Biodistribution study in healthy mice. *BMC Cancer*, *8*, 111. DOI: 10.1186/1471-2407-8-111 PMID: 18430208

Romero, J. F., Diaz-Barrios, A., & Gonzalez, G. (2021). Biocompatible thermo-responsive N-vinylcaprolactam based hydrogels for controlled drug delivery systems. *Bionatura*, *6*(2), 1712–1719. DOI: 10.21931/RB/2021.06.02.8

Rusdin, A., Mohd Gazzali, A., Ain Thomas, N., Megantara, S., Aulifa, D. L., Budiman, A., & Muchtaridi, M. (2024). Advancing Drug Delivery Paradigms: Polyvinyl Pyrolidone (PVP)-Based Amorphous Solid Dispersion for Enhanced Physicochemical Properties and Therapeutic Efficacy. *Polymers, 16*(2), 286. DOI: 10.3390/POLYM16020286 PMID: 38276694

San, H. H. M., Alcantara, K. P., Bulatao, B. P. I., Chaichompoo, W., Nalinratana, N., Suksamrarn, A., Vajragupta, O., Rojsitthisak, P., & Rojsitthisak, P. (2022). Development of Turmeric Oil—Loaded Chitosan/Alginate Nanocapsules for Cytotoxicity Enhancement against Breast Cancer. Polymers 2022, Vol. 14, Page 1835, 14(9), 1835. https://doi.org/DOI: 10.3390/POLYM14091835

Saqr, A. Al, Wani, S. U. D., Gangadharappa, H. V., Aldawsari, M. F., Khafagy, E. S., & Abu Lila, A. S. (2021). Enhanced Cytotoxic Activity of Docetaxel-Loaded Silk Fibroin Nanoparticles against Breast Cancer Cells. Polymers 2021, Vol. 13, Page 1416, 13(9), 1416. https://doi.org/DOI: 10.3390/POLYM13091416

Sartaj, A., Qamar, Z., Qizilbash, F. F., Annu, Md, S., Alhakamy, N. A., Baboota, S., & Ali, J. (2021). Polymeric Nanoparticles: Exploring the Current Drug Development and Therapeutic Insight of Breast Cancer Treatment and Recommendations. Polymers 2021, Vol. 13, Page 4400, 13(24), 4400. https://doi.org/DOI: 10.3390/POLYM13244400

Senapati, S., Mahanta, A. K., Kumar, S., & Maiti, P. (2018). Controlled drug delivery vehicles for cancer treatment and their performance. Signal Transduction and Targeted Therapy 2018 3:1, 3(1), 1–19. https://doi.org/DOI: 10.1038/s41392-017-0004-3

Seo, Y., Ghazanfari, L., Master, A., Vishwasrao, H. M., Wan, X., Sokolsky-Papkov, M., & Kabanov, A. V. (2021). Poly(2-oxazoline)-Magnetite NanoFerrogels: Magnetic Field Responsive Theranostic Platform for Cancer Drug Delivery and Imaging. *Nanomedicine; Nanotechnology, Biology, and Medicine, 39*, 102459. DOI: 10.1016/J.NANO.2021.102459 PMID: 34530163

Shaikh, M. A. J., Alharbi, K. S., Almalki, W. H., Imam, S. S., Albratty, M., Meraya, A. M., Alzarea, S. I., Kazmi, I., Al-Abbasi, F. A., Afzal, O., Altamimi, A. S. A., Singh, Y., Singh, S. K., Dua, K., & Gupta, G. (2022). Sodium alginate based drug delivery in management of breast cancer. *Carbohydrate Polymers, 292*, 119689. DOI: 10.1016/J.CARBPOL.2022.119689 PMID: 35725179

Sharma, A., Debik, J., Naume, B., Ohnstad, H. O., Sahlber, K. K., Borgen, E., Børresen-Dale, A. L., Engebråten, O., Fritzman, B., Garred, Ø., Geisler, J., Geitvik, G. A., Hofvind, S., Kristensen, V. N., Kåresen, R., Langerød, A., Lingjærde, O. C., Mælandsmo, G. M., Russnes, H. G., … Giskeødegård, G. F. (2024). Comprehensive multi-omics analysis of breast cancer reveals distinct long-term prognostic subtypes. Oncogenesis 2024 13:1, 13(1), 1–13. https://doi.org/DOI: 10.1038/s41389-024-00521-6

Shettiwar, A., Gupta, U., Paul, P., Nair, R., Aalhate, M., Mahajan, S., Maji, I., Guru, S. K., Mehra, N. K., & Singh, P. K. (2023). A comprehensive review of the biomaterial-based multifunctional nanocarriers for therapeutic applications in breast cancer. In Journal of Drug Delivery Science and Technology (Vol. 89). https://doi.org/DOI: 10.1016/j.jddst.2023.104990

Snyder, C. M., Mateo, B., Patel, K., Fahrenholtz, C. D., Rohde, M. M., Carpenter, R., & Singh, R. N. (2024). Enhancement of Triple-Negative Breast Cancer-Specific Induction of Cell Death by Silver Nanoparticles by Combined Treatment with Proteotoxic Stress Response Inhibitors. *Nanomaterials (Basel, Switzerland), 14*(19), 1564. DOI: 10.3390/NANO14191564 PMID: 39404291

Subhan, M. A., & Torchilin, V. P. (2023a). Biopolymer-Based Nanosystems for siRNA Drug Delivery to Solid Tumors including Breast Cancer. Pharmaceutics 2023, Vol. 15, Page 153, 15(1), 153. https://doi.org/DOI: 10.3390/PHARMACEUTICS15010153

Sun, D., Feng, X., Zhu, X., Wang, Y., & Yang, J. (2024). Anticancer agents based on metal organic cages. *Coordination Chemistry Reviews*, *500*, 215546. DOI: 10.1016/J.CCR.2023.215546

Sung, H., Ferlay, J., Siegel, R. L., Laversanne, M., Soerjomataram, I., Jemal, A., & Bray, F. (2021a). Global Cancer Statistics 2020: GLOBOCAN Estimates of Incidence and Mortality Worldwide for 36 Cancers in 185 Countries. *CA: a Cancer Journal for Clinicians*, *71*(3), 209–249. DOI: 10.3322/CAAC.21660 PMID: 33538338

Sung, H., Ferlay, J., Siegel, R. L., Laversanne, M., Soerjomataram, I., Jemal, A., & Bray, F. (2021b). Global Cancer Statistics 2020: GLOBOCAN Estimates of Incidence and Mortality Worldwide for 36 Cancers in 185 Countries. *CA: a Cancer Journal for Clinicians*, *71*(3), 209–249. DOI: 10.3322/CAAC.21660 PMID: 33538338

Talebian, S., Foroughi, J., Wade, S. J., Vine, K. L., Dolatshahi-Pirouz, A., Mehrali, M., Conde, J., Wallace, G. G., Talebian, S., & Foroughi, J. (n.d.). Article Type: Review Biopolymers for Anti-Tumor Implantable Drug Delivery Systems: Recent Advances and Future Outlook. https://doi.org/DOI: 10.1002/adma.201706665

Tang, X., Kurban, M., Hafiz, I., Shen, Q., & Wang, M. (2023). Preparation of hyaluronic acid-loaded Harmine polymeric micelles and in vitro effect anti-breast cancer. *European Journal of Pharmaceutical Sciences*, *183*, 106388. DOI: 10.1016/J.EJPS.2023.106388 PMID: 36758771

Teixeira, A. M., & Martins, P. (2023). A review of bioengineering techniques applied to breast tissue: Mechanical properties, tissue engineering and finite element analysis. *Frontiers in Bioengineering and Biotechnology*, *11*. Advance online publication. DOI: 10.3389/FBIOE.2023.1161815 PMID: 37077233

Torres Quintas, S., Canha-Borges, A., Oliveira, M. J., Sarmento, B., & Castro, F. (2023). Special Issue: Nanotherapeutics in Women's Health Emerging Nanotechnologies for Triple-Negative Breast Cancer Treatment. In Small. https://doi.org/DOI: 10.1002/smll.202300666

Tulay, P., Galam, N., & Adali, T. (2018). The Wonders of Silk Fibroin Biomaterials in the Treatment of Breast Cancer. Critical Reviews™ in Eukaryotic Gene Expression, 28(2), 129–134. https://doi.org/DOI: 10.1615/CRITREVEUKARYOTGENEEXPR.2018021331

Twal, S., Jaber, N., Al-Remawi, M., Hamad, I., Al-Akayleh, F., & Alshaer, W. (2024). Dual stimuli-responsive polymeric nanoparticles combining soluplus and chitosan for enhanced breast cancer targeting. *RSC Advances*, *14*(5), 3070. DOI: 10.1039/D3RA08074A PMID: 38239437

UN. G. perspective H. stories. (2024, February). Breast cancer cases projected to rise by nearly 40 per cent by 2050, WHO warns | UN News. Webpage. https://news.un.org/en/story/2025/02/1160391

Varadharajaperumal, P., Subramanian, B., & Santhanam, A. (2017). Biopolymer mediated nanoparticles synthesized from Adenia hondala for enhanced tamoxifen drug delivery in breast cancer cell line. *Advances in Natural Sciences: Nanoscience and Nanotechnology*, *8*(3), 035011. DOI: 10.1088/2043-6254/AA7253

Velhal, K., Barage, S., Roy, A., Lakkakula, J., Yamgar, R., Alqahtani, M. S., Yadav, K. K., Ahn, Y., & Jeon, B. H. (2022). A Promising Review on Cyclodextrin Conjugated Paclitaxel Nanoparticles for Cancer Treatment. Polymers 2022, Vol. 14, Page 3162, 14(15), 3162. https://doi.org/DOI: 10.3390/POLYM14153162

Vindigni, V., Marena, F., Zanettin, C., & Bassetto, F. (2024). Breast Reconstruction: The Oncoplastic Approach. Journal of Clinical Medicine 2024, Vol. 13, Page 4718, 13(16), 4718. https://doi.org/DOI: 10.3390/JCM13164718

Vines, J. B., Lim, D. J., & Park, H. (2018). Contemporary Polymer-Based Nanoparticle Systems for Photothermal Therapy. Polymers 2018, Vol. 10, Page 1357, 10(12), 1357. https://doi.org/DOI: 10.3390/POLYM10121357

Wadhawan, A., Singh, J., Sharma, H., Handa, S., Singh, G., Kumar, R., Barnwal, R. P., Pal Kaur, I., & Chatterjee, M. (2022). Anticancer Biosurfactant-Loaded PLA–PEG Nanoparticles Induce Apoptosis in Human MDA-MB-231 Breast Cancer Cells. *ACS Omega*, *7*(6), 5231. DOI: 10.1021/ACSOMEGA.1C06338 PMID: 35187338

Wan, X., Beaudoin, J. J., Vinod, N., Min, Y., Makita, N., Bludau, H., Jordan, R., Wang, A., Sokolsky, M., & Kabanov, A. V. (2019). Co-delivery of paclitaxel and cisplatin in poly(2-oxazoline) polymeric micelles: Implications for drug loading, release, pharmacokinetics and outcome of ovarian and breast cancer treatments. *Biomaterials*, *192*, 1–14. DOI: 10.1016/J.BIOMATERIALS.2018.10.032 PMID: 30415101

Wang, C., Zhang, R., Wei, X., Lv, M., & Jiang, Z. (2020). Metalloimmunology: The metal ion-controlled immunity. *Advances in Immunology*, *145*, 187–241. DOI: 10.1016/BS.AI.2019.11.007 PMID: 32081198

Wang, H. (2024). A Review of Nanotechnology in microRNA Detection and Drug Delivery. *Cells*, *13*(15), 1277. DOI: 10.3390/CELLS13151277 PMID: 39120308

Wang, H. L., Wang, Z. G., & Liu, S. L. (2022). Lipid Nanoparticles for mRNA Delivery to Enhance Cancer Immunotherapy. *Molecules (Basel, Switzerland)*, *27*(17), 5607. DOI: 10.3390/MOLECULES27175607 PMID: 36080373

Wang, L. W., Peng, C. W., Chen, C., & Li, Y. (2015). Quantum dots-based tissue and in vivo imaging in breast cancer researches: Current status and future perspectives. *Breast Cancer Research and Treatment*, *151*(1), 7. DOI: 10.1007/S10549-015-3363-X PMID: 25833213

Wang, S., Yin, N., Li, Y., Xiang, T., Jiang, W., Zhao, X., Liu, W., Zhang, Z., Shi, J., Zhang, K., Guo, X., Si, P., & Liu, J. (2022). Copper-based metal–organic framework impedes triple-negative breast cancer metastasis via local estrogen deprivation and platelets blockade. *Journal of Nanobiotechnology*, *20*(1). Advance online publication. DOI: 10.1186/s12951-022-01520-8 PMID: 35794596

Wang, X., Teng, Z., Wang, H., Wang, C., Liu, Y., Tang, Y., Wu, J., Sun, J., Wang, H., Wang, J., & Lu, G. (2014). Increasing the cytotoxicity of doxorubicin in breast cancer MCF-7 cells with multidrug resistance using a mesoporous silica nanoparticle drug delivery system. *International Journal of Clinical and Experimental Pathology*, *7*(4), 1337. https://pmc.ncbi.nlm.nih.gov/articles/PMC4014214/ PMID: 24817930

Wu, X., Xin, Y., Zhang, H., Quan, L., & Ao, Q. (2024). Biopolymer-Based Nanomedicine for Cancer Therapy: Opportunities and Challenges. *International Journal of Nanomedicine*, *19*, 7415. DOI: 10.2147/IJN.S460047 PMID: 39071502

Yao, Y., Sangani, C. B., Duan, Y. T., Bhadja, P., & Ameta, R. K. (2021). Molecular modelling, thermal, adsorption and biological studies of conjugate Cu2+-BSA nanoparticles. *Journal of Molecular Liquids*, *331*, 115732. DOI: 10.1016/J.MOLLIQ.2021.115732

Yu, J., Mu, Q., Fung, M., Xu, X., Zhu, L., & Ho, R. J. Y. (2022). Challenges and Opportunities in Metastatic Breast Cancer Treatments: Nano-Drug Combinations Delivered Preferentially to Metastatic Cells May Enhance Therapeutic Response. *Pharmacology & Therapeutics*, *236*, 108108. DOI: 10.1016/J.PHARMTHERA.2022.108108 PMID: 34999182

Yu, J., Xie, X., Wang, L., Liu, W., Xu, H., Lu, X., Li, X., Ren, J., & Li, W. (2023). Smart Chondroitin Sulfate Micelles for Effective Targeted Delivery of Doxorubicin Against Breast Cancer Metastasis. *International Journal of Nanomedicine*, *18*, 663–677. DOI: 10.2147/IJN.S398802 PMID: 36798532

Zeng, W., Luo, Y., Gan, D., Zhang, Y., Deng, H., & Liu, G. (2023). Advances in Doxorubicin-based nano-drug delivery system in triple negative breast cancer. *Frontiers in Bioengineering and Biotechnology*, *11*, 1271420. DOI: 10.3389/FBIOE.2023.1271420 PMID: 38047286

Zhai, M., Wu, P., Liao, Y., Wu, L., & Zhao, Y. (2024). Polymer Microspheres and Their Application in Cancer Diagnosis and Treatment. International Journal of Molecular Sciences 2024, Vol. 25, Page 6556, 25(12), 6556. https://doi.org/DOI: 10.3390/IJMS25126556

Zhang, L., Ren, Z., Lü, J., Mo, X., Lin, J., Li, Y., Ma, W., Liu, P., Shen, Y., Zhao, Q., Qian, L., Cheng, X., Yu, Z., & Zhang, B. (2023). Nanoparticles carrying paclitaxel and anti-miR-221 for breast cancer therapy triggered by ultrasound. *Cell Death Discovery*, *9*(1), 1–11. DOI: 10.1038/S41420-023-01594-9;SUBJMETA=1347,152,54,61,631,67;KWRD=BREAST+CANCER,DRUG+DELIVERY PMID: 37582832

Zhao, X., Campbell, S., El-Khoury, P. Z., Jia, Y., Wallace, G. Q., Claing, A., Bazuin, C. G., & Masson, J. F. (2021). Surface-Enhanced Raman Scattering Optophysiology Nanofibers for the Detection of Heavy Metals in Single Breast Cancer Cells. *ACS Sensors*, *6*(4), 1649–1662. DOI: 10.1021/ACSSENSORS.1C00332/SUPPL_FILE/SE1C00332_SI_001.PDF PMID: 33847111

Zhong, Y., Peng, Z., Peng, Y., Li, B., Pan, Y., Ouyang, Q., Sakiyama, H., Muddassir, M., & Liu, J. (2023). Construction of Fe-doped ZIF-8/DOX nanocomposites for ferroptosis strategy in the treatment of breast cancer. *Journal of Materials Chemistry. B*, *11*(27), 6335–6345. DOI: 10.1039/D3TB00749A PMID: 37350051

Zhou, P., Liang, X., Zhou, C., Qin, J., Hou, C., Zhu, Z., Zhang, W., Wang, S., & Zhong, D. (2019). Glutamine-β-cyclodextrin for targeted doxorubicin delivery to triple-negative breast cancer tumors via the transporter ASCT2. *Journal of Materials Chemistry. B*, *7*(35), 5363–5375. DOI: 10.1039/C9TB01225G PMID: 31403158

Zhou, Z., Song, J., Nie, L., & Chen, X. (2016). Reactive Oxygen Species Generating Systems Meeting Challenges of Photodynamic Cancer Therapy. *Chemical Society Reviews*, *45*(23), 6597. DOI: 10.1039/C6CS00271D PMID: 27722328

Zhu, H., Yang, K., Yao, H., Chen, X., Yan, S., He, Y., Cao, Y., Luo, J., & Wang, D. (2023). Multifunctional Nanoplatform-Mediated Chemo-Photothermal Therapy Combines Immunogenic Cell Death with Checkpoint Blockade to Combat Triple-Negative Breast Cancer and Distant Metastasis. *International Journal of Nanomedicine*, *18*, 3109. DOI: 10.2147/IJN.S408855 PMID: 37323948

Zhu, J., Li, Q., Wu, Z., Xu, Y., & Jiang, R. (2024). Curcumin for Treating Breast Cancer: A Review of Molecular Mechanisms, Combinations with Anticancer Drugs, and Nanosystems. Pharmaceutics 2024, Vol. 16, Page 79, 16(1), 79. https://doi.org/DOI: 10.3390/PHARMACEUTICS16010079

Zhu, M., Wang, X., Zhou, Y., Li, M., Wang, Y., Wang, Y., Li, T., Song, S. Y., Zhang, B., Hezam, K., & Zhang, J. (2023). Breast tumor-targeted drug delivery via polymer nanocarriers: Endogenous and exogenous strategies. *Journal of Applied Polymer Science*, *140*(31), e54227. DOI: 10.1002/APP.54227

Zhu, Y., Wang, A., Zhang, S., Kim, J., Xia, J., Zhang, F., Wang, D., Wang, Q., & Wang, J. (2022). Paclitaxel-loaded ginsenoside Rg3 liposomes for drug-resistant cancer therapy by dual targeting of the tumor microenvironment and cancer cells. *Journal of Advanced Research*, *49*, 159. DOI: 10.1016/J.JARE.2022.09.007 PMID: 36167294

Zhu, Z., Wang, X., Li, T., Aime, S., Sadler, P. J., & Guo, Z. (2014). Platinum(II)–Gadolinium(III) Complexes as Potential Single-Molecular Theranostic Agents for Cancer Treatment. *Angewandte Chemie*, *126*(48), 13441–13444. DOI: 10.1002/ANGE.201407406 PMID: 25255766

Chapter 8

Therapeutic Applications of Palladium Nanoparticles (PdNPs)

Aditya Shrivastav
https://orcid.org/0009-0000-8461-8171
Ajeenkya D.Y. Patil University, India

Sunil Sankathala
https://orcid.org/0009-0004-7809-7647
Ajeenkya D.Y. Patil University, India

Kashvi Chaturvedi
https://orcid.org/0009-0005-3192-3400
Ajeenkya D.Y. Patil University, India

Susanta Das
https://orcid.org/0000-0002-9314-3988
Ajeenkya D.Y. Patil University, India

ABSTRACT

PdNPs have received much attention in nanomedicine due to their unique physicochemical values such as high thermal and chemical stability, high surface area, & tuneable optical properties. These characteristics make them suitable for many biomedical applications, including cancer therapy, antibacterial treatments, drug delivery, & biosensing. In this chapter, synthesis methods for PdNPs are reviewed from the environmental friendliness view point, & the properties are described. Enhanced PdNPs' stability, biocompatibility, and efficacy in biological systems are explored through the development of functionalization strategies. PdNPs possess great promise for photothermal therapy in targeted cancer treatment, antibacterial effectiveness against multidrug-resistant pathogens, & drug delivery vehicles. They also offer support for early disease detection in biosensing technologies. Finally, safety and toxicity issues are also examined, emphasizing the need for careful evaluation of safe and effective clinical applications of PdNPs.

DOI: 10.4018/979-8-3373-0055-9.ch008

1. INTRODUCTION

The emergence of nanotechnology has been, and continues to be a transformative field of modern science in the development of novel therapeutic agents. Metal nanostructures such as palladium nanoparticles (PdNPs) have received great attention for their peculiar physicochemical properties and their possible biomedical applications (Phan et al., 2019; Yaqoob et al., 2020; Leso & Iavicoli, 2018). PdNPs display excellent chemical, and thermal stability, and profound optical and photocatalytic properties, with relative cost effectiveness (Phan et al., 2019). Nanotechnology received formal recognition in the late 20th century but researchers from the 4th century used the Lycurgus Cup to demonstrate nanoscale optical effects. The scientific investigation of nanomaterials began when Michael Faraday examined colloidal gold in 1857 while the field experienced significant development after scientists invented the Scanning Tunnelling Microscope in 1981 (Bayda et al., 2019). This instrument enabled atomic-level manipulation. Advancements in nanomaterials research during this period resulted in discovering such as fullerenes in 1985 and carbon nanotubes in 1991 as well as graphene in 2004. These materials demonstrated exceptional mechanical properties and outstanding electrical functionality (Bayda et al., 2019). The US National Nanotechnology Initiative (2000) used its accelerated research funding to expand nanotechnology into electronics and energy systems as well as environmental science and advanced materials which revolutionised miniaturisation and energy efficiency and pollution cleanup and smart materials development. Nanotechnology has enabled better results in medicine through its use in targeted drug delivery systems as well as diagnostic biosensors and nanomedicine-based treatment methods which lead to more effective personalised medicine (Bayda et al., 2019). Figure 1 shows a clear representation of evolution of nanotechnology and its expansion in medicine.

Figure 1. Evolution of Nanotechnology

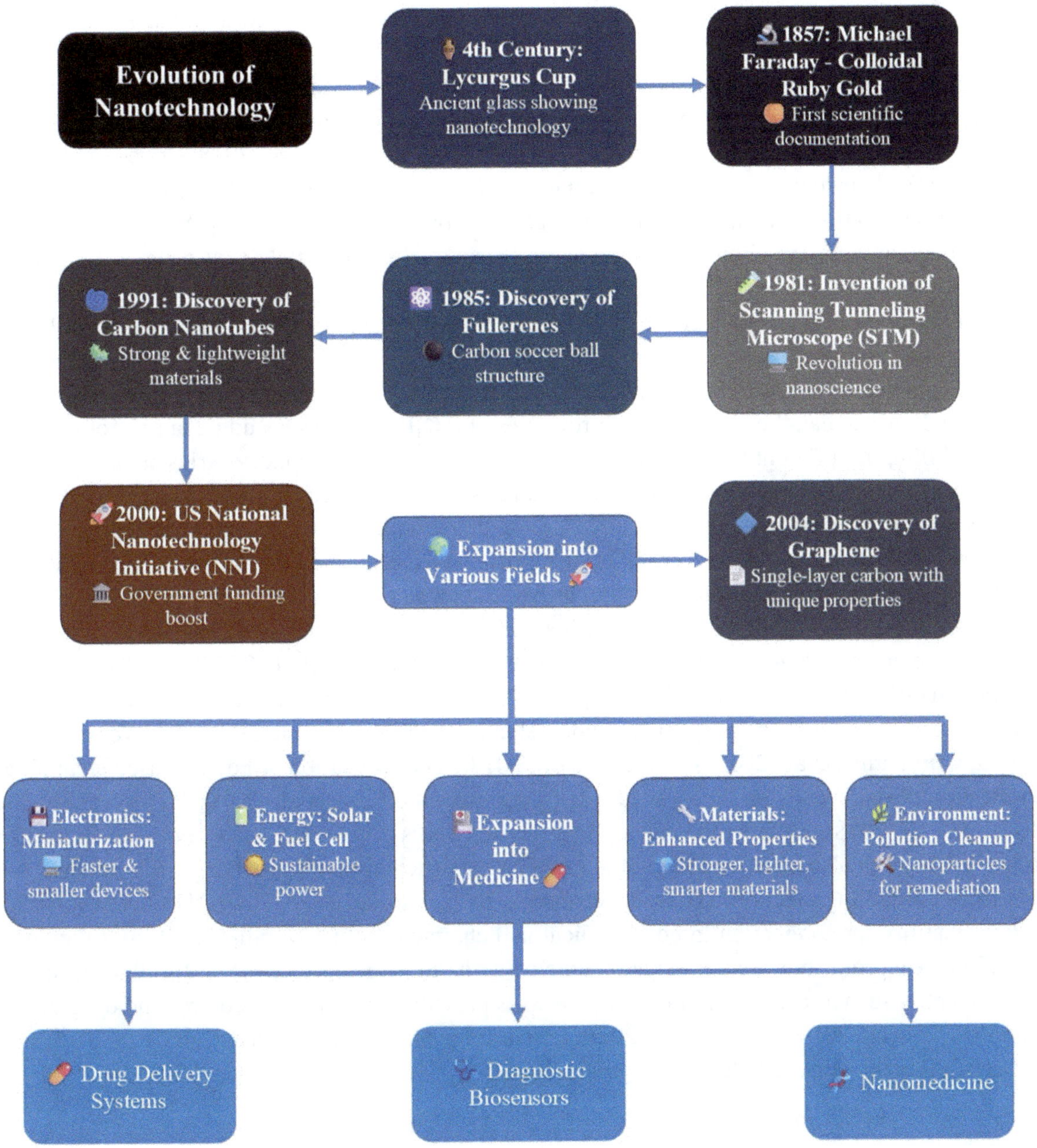

Their compatibility for application can also be improved with biocompatible polymers or molecules coated on these nanoparticles, which can be synthesised in various sizes and shape (Phan et al., 2019; Leso & Iavicoli, 2018; Gurunathan et al., 2015; Vinodhini et al., 2022). PdNPs can be synthesised via chemical reduction, metal vaporisation, sol-gel processing, laser ablation, sonochemical, and biological methods (Gurunathan et al., 2015; Vinodhini et al., 2022; Adams et al., 2014; Fernandez-Arias et al., 2022). Green synthesis techniques using plant extract have become a sustainable for the environment and also size and shape of particles can be controlled by this techniques (Gurunathan et al., 2015; Vinodhini et al., 2022).

The potential of PdNPs in many parts of therapy, including photothermal and photoacoustic therapy, drug and gene delivery, pro-drug activation, anticancer therapy, and antimicrobial therapy, has been towed by recent research (Phan et al. 2019; Dumas & Couvreur 2015; Thapa et al., 2018; Rokade et al., 2018; Yin et al., 2023). Due to their strong absorption in the Near infrared range (NIR), these materials enable deeper light penetration in tissues for photothermal therapy (Thapa et al., 2018; Yin et al., 2023) and improve imaging quality in photoacoustic imaging (Phan et al., 2019). PdNPs as drug carriers in drug delivery have advantages such as high surface area, stability, and low toxicity, thus are efficient carriers for therapeutic agents (Phan et al., 2019; Yetisgin et al., 2020). For example, PdNPs have been shown to activate prodrugs, including *5-fluorouracil* and *gemcitabine hydrochloride*, selectively in cancer cells, and possess potential as targeted therapeutic agents (Phan et al., 2019; Dumas & Couvreur, 2015; Rokade et al., 2018). In particular, the antimicrobial properties of PdNPs have been studied in previous research, capable of killing in different bacterial strains (Adams et al., 2014; Fernandez-Arias et al., 2022). This becomes very relevant as attention to antimicrobial resistance increases, and a different treatment strategy is necessary (Fernandez-Arias et al., 2022). Dual therapeutic mechanism combined approaches of PdNPs have been proven in cancer therapy as such PdNPs have been shown to have enhanced selectively as it both thermalizes photothermal treatment as well as act as a chemotherapy agent (Yin et al., 2023; Gurunathan et al., 2015). Furthermore, other innovative applications, e.g. PdNPs loaded hydrogels in post-surgical cancer therapy and wound healing, have been built (Yin et al., 2023).

Although PdNPs have many appealing properties for biomedical applications, studies on their properties, toxicity, distribution in the body, pharmacokinetics, and pharmacodynamics are yet to come (Phan et al., 2019; Gurunathan et al., 2015; Leso & Iavicoli, 2018). The realisation of PdNPs based technologies in clinical practise depends on the knowledge of these aspects (Phan et al., 2019; Yaqoob et al., 2020; Leso & Iavicoli, 2018). The therapeutic applications of PdNPs do not only involve classic applications in medicine but can also be extended to other applications such as catalysis of organic synthesis, notably in carbon-coupling reactions (Deraedt & Astruc, 2013; Sarmah et al., 2019). However, this shows PdNPs versatility for application in both medical and chemical field. Overall, PdNPs as a therapeutic application are a promising frontier in nanomedicine with unique nanoscale properties that can develop clinical benefits such as cancer treatment or processes providing by infection control and drug delivery systems (Phan et al., 2019; Yaqoob et al., 2020; Leso & Iavicoli, 2018; Dumas & Couvreur, 2015; Thapa et al., 2018; Rokade et al., 2019; Yetisgin et al., 2020).

2. SYNTHESIS TECHNIQUES FOR PALLADIUM NANO PARTICLES

PdNPs are becoming more and more important in several fields in particular for therapeutic applications. Because PdNPs possess considerable performance and suitability for medical uses, the synthesis methods, and the characteristics of PdNPs, including size, shape, structure, and surface modifications, have significant roles in determining their performance. PdNPs can be synthesised by physical, chemical, and green (biogenic) methods. Each of these provide some benefits and allow for control at the particle level to optimise applicability for these particles (Chen & Ostrom, 2015).

Metal precursors are used in physical synthesis techniques that do not form new substances through molecular rearrangements. Pd nanostructures can be prepared using commonly employed physical methods, e.g. ion or electron beam deposition, sputtering, and laser ablation. Another method is hydrothermal synthesis, a cost-effective and simple approach that permits changes in temperature, pressure, and precursor concentration (Chen & Ostrom, 2015). Chemical synthesis techniques are widely used to achieve specific nanoparticle properties by controlling conditions such as precursor concentration, additives, surfactants, temperature, and pH. The chemical approach is effective however, the use of toxic reducing agents and stabilisers limits its utility in biomedicine. Because this is not a viable method, greener alternatives, such as atmospheric pressure alcohol plasma and dielectric barrier discharge plasmas, have been developed. These methods give rise to small, surfactant-free PdNPs with high crystallinity that allow for eco-friendly PdNPs production (Phan et al., 2019).

Green synthesis methods is an environment friendly PdNPs synthesis route, which involves extract from plants, algae, fungi, or bacteria as reducing and capping agents. For example, mixing a plant extract in methanol with palladium acetate [$Pd(OAc)_2$] about 60°C resulted in spherical PdNPs with a UV-visible absorption maximum in the range of 360-400 nm. PdNPs in the 2.5-14 nm in size range with spherical shape were produced from plant leaves with high content of polyphenols such as those from *Euphorbia condylocarpa* using advanced techniques such as transmission electron microscopy (TEM), scanning electron microscopy (SEM), and X-ray diffraction (XRD) (Siddiqi & Husen, 2016; Alaqarbeh et al., 2023). Stability, effectiveness, and reduced toxicity of PdNPs synthesised by green methods are important for various therapeutic applications. For example, PdNPs synthesised using white tea extract are spherical and contain desirable compounds (phenols, flavonoids, etc.). *Diospyros kaki* leaves based PdNPs are larger and possess strong antibacterial activity against *Escherichia coli* and *Staphylococcus aureus*. Additionally, extracts from chaga mushrooms can yield porous PdNPs with anticancer properties that could make them good candidates for applications in near infrared (NIR) cancer therapy (Phan et al., 2019). For example, marine algae such as *Sargassum bovinum* are also effective in reducing Pd^{2+} ions to Pd (visible colour change from yellow to dark brown) and UV-vis spectra from 300-800 nm. Green synthesis of PdNPs is influenced by the biochemical composition of the plant extract, temperature, pH, and chemical composition of the reducing and capping agents. These green synthesized PdNPs present eco-friendly production methods and stability, therefore they are highly suitable for catalytic applications (Siddiqi & Husen, 2016). Table 1 gives us the comparative analysis of all the synthesis of PdNPs.

Synthesis methods for PdNPs evaluate their costs through the analysis of energy utilisation alongside reagent expenses and system scalability requirements and waste disposal measures. Physical Vapour Deposition (PVD) and laser ablation synthesis methods under the physical synthesis category produce the highest costs because of their needs for intense temperature and pressure as well as sophisticated equipment and considerable power usage. Laboratory techniques that synthesise PdNPs cause various environmental regulatory expenses through the creation of harmful waste products (Joudeh et al., 2022). The utilisation of sodium borohydride as a reducing chemical for chemical synthesis leads to a medium-level cost structure. The use of expensive chemicals and strict safety standards associated with toxic substances make chemical synthesis more expensive in terms of operational costs (Osonga et al., 2020; Bhardwaj et al., 2020).

Green synthesis stands as the most economical and green approach when compared to other alternatives. Naturally derived plant extract reducing agents replace chemical reagents of high cost in this synthetic method which works under standard temperature conditions and pressure to lower energy needs. The production of PdNPs through green synthesis creates minimal toxic waste products which helps organizations save money on waste disposal practices while fulfilling environmental and economic standards for large-scale manufacturing (Singh et al., 2023).

Table 1. Comparative Analysis of PdNPs Synthesis Methods

Synthesis Method	Advantage	Disadvantage	Limitations
Green Synthesis	Environmentally friendly, Cost-effective, Biocompatible	Size variation, Agglomeration risk	Stability and reproducibility require optimization
Physical Synthesis	Precise control, Industrial scalability	High energy consumption, Expensive equipment	Needs vacuum/inert gas environments
Chemical Synthesis	High-purity nanoparticles, Controlled characteristics	Toxic precursors, Cost-intensive	Requires meticulous reaction control

(Joudeh et al., 2022)

3. FUNCTIONALIZATION, STABILITY, AND EMERGING BIOMEDICAL APPLICATIONS OF PALLADIUM NANO PARTICLES

The modification of PdNPs stands essential for their improved biocompatibility and precise targeting functions in medical applications. The attachment of polyethylene glycol (PEG) to PdNPs through PEGylation serves two purposes by making them more soluble and reducing their detectability by immune systems which improves circulation longevity and benefits biocompatibility. The attachment of specific ligands to PdNPs allows cells to target receptors on cancer tumours or disease sites through receptor-specific drug delivery systems. Bioconjugation of PdNPs through antibody or peptide bonding results in targeted medical interventions alongside imaging possibilities. By modifying PdNPs scientists gain improved biocompatibility along with the ability to develop specific drug delivery platforms (Phan et al., 2019). The factors which affect PdNPs stability include pH strength as well as ionic conditions and environmental temperature. Stable pH levels and controlled ionic strength combination serve to prevent PdNPs aggregation and create better dispersion quality. Stable PdNPs require operating at proper temperatures to avoid deterioration and incorporation of stabilising agents such as PVP and chitosan prevents agglomeration and improves stability (Ayogu et al., 2025; Phan et al., 2019). The stability of these systems can be improved by enclosing them in porous substrates or by using solid platforms for support and these methods enable the recovery and reuse of the materials (Yang et al., 2018). Biological extraction from fungi bacteria or plant extracts serves as a superior alternative to conventional chemical production methods for synthesis purposes. The methods lead to decreased usage of toxic solvents and reagents making the technology more environmentally friendly while preserving biological compatibility. The production of nanoparticles through biogenic processes results in stable and well-dispersed structures compared to chemical synthesis methods (Phan et al., 2019).

PdNPs show promise for biomedical use in different medical applications. Through immunomodulatory agent and genetic material delivery PdNPs serve as carriers for improved therapeutic benefits in immunotherapy and gene therapy systems. Additionally, their integration into combination therapies for cancer and infectious diseases leverages their photothermal and catalytic properties (Yaqoob et al., 2020). The study by Phan et al. (2019) demonstrates PdNPs apply to photoacoustic imaging and photothermal therapy as well as drug delivery applications. An investigation of biodistribution and efficacy and toxicity using animal models presents essential information for in vivo studies. Strategic dosing studies need development to achieve the best treatment outcomes and reduce harmful side effects (Gurunathan et al., 2021). Measurements against standard medical procedures need to evaluate the uses and bounds of PdNPs (Yaqoob et al., 2020). The assessment of PdNPs alongside AuNPs, AgNPs, and PtNPs shows how PdNPs offer specific advantages as well as unique set of limitations for use. The pharmaceutical and sensor detection fields employ PdNPs because of their advanced catalytic behaviour which enables activation of drugs and assessments (Ayogu et al., 2025; Yaqoob et al., 2020). The toxicity level of PdNPs falls beneath that of AgNPs while requiring additional precaution when handling like AuNPs (Yaqoob et al., 2020). The synthesis process for PdNPs produces bulk quantities that reduces their manufacturing cost compared to PtNPs (Ayogu et al., 2025). The balance between scalable effectiveness and affordable production should be considered alongside biological system environmental implications and extended operational stability. The biomedical field widely recognises PdNPs because their diverse uses and catalytic capabilities provide significant promise for targeted therapeutic applications. Additional studies must be conducted to determine PdNPs clinical performance along with their stability characteristics and toxicity properties. The ongoing development of sustainable synthesis techniques combined with enhanced functionalization approaches positions PdNPs as pivotal components in forthcoming biomedical breakthroughs (Ayogu et al., 2025; Yaqoob et al., 2020).

4. THERAPEUTIC APPLICATIONS OF PALLADIUM NANO PARTICLES

PdNPs hold great promise as an enabling targeted therapeutic platform due to their tunable physicochemical properties. PdNPs are often designed to release therapeutic agents at targeted sites within the body in biomedical settings. Once inside blood circulatory system, PdNPs may be opsonized, i.e., bound by circulating proteins, thereby circulating the body for immune clearance via filtration in the liver (also known as phagocytosis), spleen, and kidney. Thus, PdNPs surfaces can be modified with polymers, such as polyethylene glycol (PEG), acetyl groups, carbohydrates, or specific proteins, so as to improve circulation time. PdNPs can be modified such that they bypass the immune system and eventually accumulate in targeted tissues (Yetisgin et al., 2020).

PdNPs with different Size, Shape, and Charge are critical for their therapeutic efficacy. For example, the kidneys can quickly remove smaller particles from the bloodstream, while immune cells would be more successful at removing larger particles. Retention times in the blood stream tend to be better for nanoparticles less than 100 nm in size, which benefits accumulation at target sites such as tumours. The shape of PdNPs also affect how they interact with cells, rod-like PdNPs are more likely to be recognised by immune cells. In addition, surface charge modulates immune response, PdNPs with a positive charge induce stronger immune responses than neutral or slightly negative PdNPs, which evade immune detection and prolong circulation (Yetisgin et al., 2020). Since PdNPs are synthesised through Green Methods, PdNPs thus are often preferred for therapeutic applications because of their ecofriendly synthesis, stability, and various chemical reactions utility. PdNPs are nowadays frequently used as catalysts for organic reactions, dye degradation, and reduction reactions and then are green synthesised. The bio-synthesised PdNPs, for example, ones obtained from *Euphorbia condylocarpa*, also allow for reusability in catalytic cycles, especially when these PdNPs are incorporated into magnetic composites, such as the $PdFe_3O_4$ composites which facilitate easy separation and recycling. The stability, biocompatibility, and efficiency of these PdNPs makes them appropriate for use in catalytic processes in drug delivery where catalytic properties of PdNPs can facilitate drug release in targeted tissues (Siddiqi & Husen, 2016; Alaqarbeh et al., 2023).

4.1 Cancer

Photothermal therapy (PTT) is a promising therapeutic approach to treat cancers and infections owing to the use of photo-absorbing agents to convert light energy to heat. This heated light can selectively and non-invasively kill primary and metastatic tumours across a range of cancer cell types. Metal based nanoparticles are advantageously used among the materials in PTT due to their localised surface plasmon resonance (LSPR), which provides efficient photon absorption and heat conversion (Phan et al., 2019). Although PdNPs are not oxidizable, they display strong photothermal conversion efficiency, good photostability, and can be made of various shapes and sizes to show flexibility. Phan et al. (2019), in his research found that ultrathin hexagonal Pd nanosheets are an excellent photothermal converter with high biocompatibility and stability under irradiation. To increase their blood circulation times and tumour accumulation, modifications have been made including use of small Pd nanosheets (SPNS) coated with reduced glutathione (GSH) (Phan et al., 2019). Gold nanoparticles have been popular as PT agents due to biocompatibility and high extinction in NIR region. This however is often accompanied with marginal thermal stability, and thus deactivation on exposure to the heat generated in the process of irradiation (Dumas & Couvreur, 2015). Previously unconsidered for PT applications, Pd nanostructures were found to perform very well, using ultrathin (1.8 nm) hexagonal Pd nanosheets. Strong, tunable NIR absorption with high size dependent PT conversion efficiency were demonstrated by these nanosheets. For example, 41 nm Pd nanosheets fully eliminated liver cancer cells following 5 minutes of 808 nm laser irradiation with high biocompatibility in the unirradiated state. Continued improvements including surface coating have improved cellular uptake and PT efficiency (Dumas & Couvrouw, 2015).

Innovative coatings of PdNPs have been used to enhance the biocompatibility and targeting capabilities of PdNPs. For instance, RGD peptide functionalized PdNPs coated with chitosan oligosaccharide showed high biocompatibility and very good targeted accumulation in breast cancer cells, which indicate their great application in targeted cancer therapy. Beyond cancer treatment, PdNPs are attractive for overcoming infections, which have risen with antibiotic resistance. Specifically, photothermal responsive membranes have been fabricated for treatment of infected wounds. The biocompatibility, high porosity, moisture retention, and photothermal performances of PdNPs embedded in a chitosan/polyvinyl alcohol membrane were also demonstrated. These membranes exhibited good efficiency for killing *Escherichia coli* bacteria under the irradiation of NIR laser, which might be used for wound care (Phan et al., 2019). PdNPs can be synthesised in eco-friendly methods with controllable shapes and sizes, such as Pulsed Laser Ablation in Liquid (PLAL). Using high power pulses on a material's liquid surface is an efficient way to produce high purity materials. PLAL produced PdNPs have demonstrated a high antibacterial activity especially against Gram-negative bacteria (e.g. *Escherichia coli* and *Staphylococcus aureus*) and were shown to be more effective than Gram-negative strains (Alaqarbeh et al., 2023). Moreover, PdNPs can serve as anticancer agents using photothermal method on the grounds of their large NIR absorption, photothermal stability, high photothermal conversion efficiency and biocompatibility. In the photothermal approach, PdNPs interact with the cancer cells by binding to DNA and proteins and disrupting cell functions using physicochemical bonding. Also, they produce reactive oxygen species (ROS), reactive nitrogen species (RNS), which damage the cancer cell lipid membranes, causing the cell death. Hypoxia is a major obstacle in cancer therapy, promoting growth of cancer cells and making them therapy resistant. We have engineered PdNPs supported on graphdiyne surfaces (GDY) as oxygen generators by hydrolysing H_2O_2 within tumours to elevate oxygen levels and thwart cancer cell proliferation. Antitumour effects of PdNPs/GDY in combination with chemotherapeutic agent DOX were enhanced (Alaqarbeh et al., 2023). PdNPs coated with polyvinylpyrrolidone (PVP) were thus synthesised in breast cancer research, and were stable owing to the interaction of Pd ions with PVPs nitrogen and oxygen groups. The PVP-PdNPs were shown to be effective against human breast cancer MCF7 cells without being toxic to normal cells, indicating the potential for them as targeted treatment options (Alaqarbeh et al., 2023).

The use of PdNPs also provides multi modal therapies which provide enhanced therapeutic effect. As a specific example, doxorubicin loaded mesoporous silica nanoparticles are used in combination with Pd nanosheets for dual photothermal and chemo therapy whereby heat from NIR laser irradiation promotes the release of doxorubicin upon cellular uptake. The ability to target cells as in this approach makes it more effective for killing cancer cells. Pd nanostructures also can serve as a stand-alone or a combined photothermal and photodynamic therapy with a single laser, which are useful for simple clinical applications. Researches get reactive oxygen species (ROS) from the Pd based systems with the photosensitizers, which actually destroy the cancer cells. Clinical advantage of using a single laser for PT and PD therapies lies in the fact that it obviates the necessity of putting two separate laser beams onto the same target area as required in the case of conventional approaches (Dumas & Couvreur, 2015). Compared with PdNPs, PdO NPs hold great potential for the applications of photothermal therapy in cancer and infection therapy. Due to the efficient NIR absorption, biocompatibility, high photothermal conversion efficiency, stability under light exposure and mini-invasiveness, they have the potentials as effective, minimally invasive agents for clinical use (Phan et al., 2019; Dumas & Couvreur, 2015; Alaqarbeh et al., 2023).

Experimental studies demonstrate that PdNPs could potentially treat cancer cells including breast cancer but researchers have performed limited research on their effects within live subjects. Plant extract-based PdNPs fall within the nanometre scale from 10-20 nm as demonstrated by Boutou et al. (2019). The in vitro experiments which use PdNPs require dosages measured in micrograms per millilitre (μg/mL) but there is insufficient documentation about specific doses required to treat breast cancer (Ramirez-Rios et al., 2020). The standard delivery method for cancer treatment uses intravenous administration of nanoparticles which provides effective tumour-targeting mechanisms. PdNPs stimulate both cell death processes called apoptosis and autophagy through cancer cells and demonstrate potential activity as prodrug activators along with phototherapeutic properties. High PdNPs dosages cause oxidative stress that results in detrimental effects although they show generalised compatibility with biological systems. PdNPs deliver treatments more specifically and reduce toxic side effects throughout the body (Ramirez-Rios et al., 2020; Boutou et al., 2019).

The application of PdNPs extends to progressive research areas like cancer immunotherapy and gene editing where they assist in delivering CRISPR-Cas9 systems. These nanoparticles demonstrate high photothermal conversion efficiency which makes them potential candidates in theragnostic applications. PdNPs stand out from other nanoparticles because they possess both high stability and excellent biocompatibility features. Complete exploitation of PdNPs opportunities requires additional research to determine their performance match with alternative nanoparticles (Liu et al., 2020).

4.2 Antibacterial Therapy

PdNPs gained attention recently due to their promising antibacterial properties, making them strong candidates for antimicrobial applications. These are properties resulting from both physical and chemical antibacterial mechanisms. By breaking themselves into nanocrystal facets, PdNPs disrupt the bacterial membranes and produce ROS that ultimately disrupt the bacterial membranes, DNA, and proteins as well as interferes with the electron transport systems that result in bacterial death. More importantly, PdNPs showed better antibacterial activity than Pd^{2+} ions (Phan et al., 2019). Particle size and shape have a strong influence on the antibacterial efficiency of PdNPs. However, ultra small PdNPs (where the size difference is only 1 nm) are shown in studies to be more than 10s of time more toxic to *Escherichia coli* than larger PdNPs, even at very low concentrations (10^{-9} M), so there are more than robust antibacterial effects at very low concentrations. In addition, PdNPs possess crystal shape dependent antibacterial activity, wherein Pd nanocrystals like cubes and octahedrons do, for instance, have different effects on Gram-positive and Gram-negative bacteria. Pd cubes are potent against Gram-positive bacteria, whereas Pd octahedrons penetrate Gram-negative bacterial membranes more efficiently leading to increased antibacterial activity. Pd nanocrystals generate ROS, with this shape dependent oxidase and peroxidase like activity contributing to their antibacterial activity (Phan et al., 2019; Alaqarbeh et al., 2023). There has also been strong antibacterial capability of PdNPs biogenically synthesised by biogenic methods. For example, PdNPs prepared using naturally occurring reducing and capping reagents derived from biomass waste of *Moringa oleifera* petals displayed bactericidal activity against *Enterococcus faecalis*. PdNPs synthesised with white tea extract (Pd@W.tea NPs), demonstrated heightened antibacterial activity to *Staphylococcus epidermidis* and *Escherichia coli* in comparison to antibacterial performance of white tea extract alone, which may be a result of strong interaction between Pd@W.tea NPs and bacterial cells (Phan et al., 2019). In addition, the combination of PdNPs with other metals increases their antibacterial effects. For instance, hydroxyapatite/chitosan/Ag@Pd scaffolds show greater bactericidal activity, which will be useful for dental surgeries for prevention of infection (Phan et al., 2019). Defining the nanotoxicological profiles of NPs requires understanding how NPs interact with microorganisms. Assessment of the PdNPs impact on both simple biological entities and more complex organisms in different environments is provided by such studies. The antibacterial properties of PdNPs against both Gram-negative bacteria, and Gram-positive bacteria, have been demonstrated in the research of Leso & Iavicoli, (2018). Interestingly, PdNPs appear to exert more inhibitory effect on Gram-positive strains where cell wall energy contributes to resistance. Furthermore, PdNPs display an antibacterial efficacy that varies according to particle size and the exposure duration (Leso & Iavicoli, 2018). In addition, PdNPs have antibacterial effects against multiple drug-resistant strains. In one instance, PdNPs were effective in killing *Cronobacter sakazakii*, a multidrug resistant strain of this opportunistic pathogen where PdNPs eradicated biofilms responsible for its resistance to various drugs. Leso & Iavicoli, (2018) in his research, further corroborate that PdNPs are versatile and effective in antibacterial therapy by showing bimetallic Pd-iron nanoparticles exert dose dependent inhibition to specific bacterial strains. PdNPs have been incorporated into materials such as polyaniline and shown to enhance antibacterial properties against Gram-positive and Gram-negative bacteria. Interestingly, Pt-Pd metal colloids by themselves did not show antibacterial activity, however, when using them in the polyaniline matrix, their antibacterial activity notably improved. This enhanced effect is probably due to the interaction between the metal NPs and the polyaniline matrix and to the high reactivity of smaller NPs in the composite (Leso & Iavicoli, 2018).

Environmental impact of PdNPs was also investigated in marine environments and soil ecosystem. The results obtained in studies of marine bacteria and bio-remediating microbial communities demonstrated that PdNPs do not exert cytotoxic effects on these ecosystems and do not disturb the biodiversity or respiratory abilities of microbial communities. PdNPs also do not significantly alter soil microbial community structure, despite varying PdNPs concentration, perhaps because PdNPs become immobilised on organic soil compounds, such as humic acids (Leso & Iavicoli, 2018).

Another strategy that enhances the antibacterial activities of PdNPs is through addition of PTT. In PTT, near infrared (NIR) light absorbed by PdNPs is converted to heat, that effectively destroys bacterial infections. Alaqarbeh et al. (2023) in their study used 20 mg/L PdNPs which completely killed 99.99% of *Staphylococcus aureus* and *E. coli* within only 10 minutes under NIR irradiation. The PdNPs rapid sterilisation capability, in turn, makes them potential agents for fast sterilisation applications (Alaqarbeh et al., 2023). Additionally, methods for synthesising PdNPs in an ecofriendly manner which are also advancing, with pulsed laser ablation in liquid (PLAL) being shown to be efficient and low-cost methods to produce high purity PdNPs in various shapes and sizes. PLAL produced PdNPs have been studied in which the nanoparticles show a greater antibacterial effect against Gram-negative bacteria than Gram-positive, supporting the broad potential of PdNPs as antimicrobial agents (Alaqarbeh et al., 2023). With minimal ecotoxic effects on environmental microbiomes, PdNPs exhibited remarkable antibacterial properties, which make them a promising candidate for antibacterial therapies and environmental applications with little risk to surrounding ecological microbial ecosystems. By carefully optimizing their size, shape, and even with other metals or materials, its effectiveness is further enhanced in antibacterial treatment and infection control (Phan et al., 2019; Leso & Iavicoli, 2018; Alaqarbeh et al., 2023).

4.3 Drug Delivery

The high porosity of PdNPs enhances their advantages as drug delivery agents, including improved bioavailability, targeted delivery, and decreased systemic toxicity (Phan et al., 2019; Yetisgin et al., 2020). Applications of these antibodies include infectious diseases, autoimmune conditions, cardiovascular diseases, neurodegenerative disorders, ophthalmic treatments, and pulmonary diseases. In infectious diseases, PdNPs are employed as carriers for antimicrobial agents, enhancing drug efficacy against resistant bacterial strains, while minimising the side effects. They can also provide controlled release of drugs that helps prevent formation of biofilm on medical devices and surface. PdNPs enable targeted drug delivery in the case of autoimmune diseases with reduced required dosing and systemic exposure. But in AIDS treatment, PdNPs improve the delivery of antiretroviral drugs from encapsulating the drugs, retaining them longer and providing targeted drug delivery to HIV infected cells. PdNPs based formulations enhance drug stability and bioavailability with controlled release of drugs and improved therapeutic outcomes in cardiovascular disease management. PdNPs have promise for crossing the blood-brain barrier in neurodegenerative diseases such as Alzheimer's and Parkinson's, to deliver neuroprotective drugs across the blood brain barrier and to remove amyloid plaque. PdNPs also provide possible means to overcome drug delivery issues in the eye for glaucoma, and to enable local drug delivery to the lungs for pulmonary diseases (Yetisgin et al., 2020).

4.4 Biosensing and Diagnostics

Recently, biosensing technologies have seen widespread use, including with the inclusion of nanoparticles. PdNPs are one of these, as they offer not only high catalytic activity, but also offer stability, and biocompatibility. Due to their remarkable catalytic, electronic, and optical properties, palladium nanoparticles are characterised. They have high surface area to volume ratio, which allows for reactivity and target biomolecule interaction. The ability to synthesise PdNPs in a broad range of shapes and sizes provides the opportunity to tailor the applications of these particles in biosensing and diagnostic (Alaqarbeh et al., 2023). PdNPs also possess excellent thermal stability as well as chemical resistance, making them suitable for use in many environmental conditions found in biomedical applications (Alaqarbeh et al., 2023). PdNPs incorporated into the biosensors have significantly increased sensitivity and selectivity. As an example, a Pd nanoparticle decorated nanoporous gold (Pd/NPG) wire biosensor for dopamine detection was investigated. Yi et al. (2017) reported that this biosensor had a high sensitivity of 1.19 μA μM^{-1}, a wide detection range of 1-220 μM, and a low detection limit of 1 μM. By virtue of its unique structure based on interconnected nanoporous channels giving an enlarged surface area and fast electron transfer ability, it can offer a high sensor performance (Yi et al., 2017). PdNPs have also been examined for pathogen identification as well as biomolecule detection. As antimicrobial properties can be used against numerous bacterial species, such as *Staphylococcus aureus* and *Escherichia coli*, their use as components of biosensors for clinical diagnostics is possible. PdNPs not only increase the specificity of these sensors, but also add an additional layer of protection against contamination (Tang et al., 2019).

The sensing mechanisms involving PdNPs usually depend on electrochemical reactions associated with their catalytic properties. For example, PdNPs can perform redox reactions when interacting with target analytes giving rise to measurable signals. The effect of this principle is employed in sensing hydrogen through catalysis of hydrogen to palladium hydride by the PdNPs (Tang et al., 2019; Alaqarbeh et al., 2023). PdNPs have much greater potential in early diagnosis through rapid and sensitive diagnostic tests. This is particularly useful as it provides quick signal generation, which is essential in biosensors for timely disease diagnosis, by virtue of their ability to catalyse reactions efficiently. Hybrid sensors that combine optical and electrochemical techniques for improved detection capabilities are recent advancements (Tang et al., 2019). Additionally, research indicates that PdNPs can be customised to improve interaction with specific biomolecules, increasing diagnostic accuracy. In other words, the versatility in synthesising PdNPs with well-controlled size and shape gives the ability to tailor properly for specific diagnostic needs (Shkotova et al., 2019; Alaqarbeh et al., 2023). While PdNPs have shown promising applications in biosensing, some issues need to be addressed. The stability and reproducibility of PdNPs are critical issues that must be addressed prior to the sensor performance being reliable over time (Alaqarbeh et al., 2023). More importantly, the long-term biocompatibility as well as the toxicity of PdNPs should be considered before applications in clinical settings. Further research should aim to optimise synthesis methods to get PdNPs that have consistent properties, and explore other combinations of materials that could further increase the sensor performance. Integration of the advanced nanomaterials will facilitate the development of multifunctional biosensors capable of detecting multiple biomarkers simultaneously (Shkotova et al. 2019; Tang et al. 2019).

4.5 PdNPs and the Gut Microbiome: Interactions and Systemic Effects

The gut microbiome shows interaction with PdNPs which may modify its structural composition patterns. The correct understanding of gut microbiota composition stands essential since these changes often produce dysbiosis associated with many medical problems. Silver nanoparticles and other nanomaterials demonstrate their ability to change the Firmicutes/Bacteroidetes ratio in intestinal microbiota which reflects gut health condition. The limited research on PdNPs demonstrates parallel evidence that nanoparticles may disrupt microbial equilibrium leading to disruption of microbial harmony (Qiao et al., 2024; Joudeh et al., 2022).

The immune system receives its regulation through the gut microbiome which gets affected when microbial composition changes. PdNPs function like other nanoparticles by connecting with the intestinal mucosal layer which subsequently changes the immune system functions (Qiao et al., 2024; Joudeh et al., 2022). Intestinal ROS levels function as an element which nanoparticles regulate through their influence on gut microbiota composition thus affecting immune responses (Qiao et al., 2024). An evaluation of PdNPs immune system modifications requires a clear comprehension of their behaviours toward the gut microbial community. Results concerning PdNPs impacts on gut microbiome require further investigation through specific in vivo or in vitro experiments because current research focuses on other nanoparticles to reveal potential mechanisms. Research on silver nanoparticles has established the modification of both in vitro and in vivo microbial communities according to studies. Plant-derived nanomaterials have proven to affect gut bacteria as they influence metabolic pathways while possibly modifying the structure of gut microbes (Teng et al., 2025).

4.6 Economic Feasibility of PdNPs-Based Therapies vs. Conventional Treatments

PdNPs display significant promise in cancer treatment together with antibacterial applications because of their unique stability characteristics along with optical features that enable photothermal procedures and antimicrobial functions. The effectiveness of high-end treatments in terms of costs must be tested against normal procedures to achieve widespread usage. Researchers are currently investigating PdNPs-based photothermal cancer therapy because it shows great potential for precise tumor targeting which could lead to fewer adverse effects than standard chemotherapy treatments. Scientific studies indicate that PdNPs kill different strains of bacteria effectively thereby providing a substitute for traditional antibiotics. The economic viability of PdNPs faces limitations because of palladium costs together with synthesis complexity (Joudeh et al., 2022). The process to make PdNPs involves chemical and physical synthesis as well as biogenic production methods (Joudeh et al., 2022; Phan et al., 2019). Plant extracts and microorganisms enable biogenic synthesis of nanoparticles through environmentally conscious techniques which potentially lower production costs more than standard chemical methods. Production at an industrial scale meets challenges when using these methods because yield and consistency remain inconsistent (Law et al., 2022; Phan et al., 2019). The extensive increase in Pd prices throughout the past decade creates economic barriers for industrial PdNPs manufacturing at scale. The market orientation and production expenses of commercial products depend heavily on scalability since it determines both unit affordability and demand satisfaction (Joudeh et al., 2022; Phan et al., 2019).

Current nanomedicine regulations including PdNPs-based therapeutic medicines face ongoing development issues that create significant obstacles. Complete therapeutic implementation of PdNPs demands strict regulatory guidelines for demonstrating their safety as well as efficacy and consistency characteristics. The price related factors affecting PdNPs manufacturing expenses consist of synthetic method selection and material expenses along with modifications needed to enhance biocompatibility and targeting properties. These production expenses together with regulatory requirements function as barriers to the clinical adoption of PdNPs-based therapeutic solutions (Law et al., 2022; Joudeh et al., 2022).

4.7 Clinical and Preclinical Studies on PdNPs

The biomedical applications of PdNPs include cancer treatment as well as antibacterial therapy along with drug delivery functions. Studies have recently demonstrated the clinical usage potential of PdNPs as well as preclinical therapeutic benefits. The medical community investigates PdNPs for their potential to treat cancer because of their anticancer effects. The cancer-fighting properties of PVP-PdNPs make them highly efficient as anticancer agents when applied to human breast cancer MCF7 cells. Through their use as quercetin carriers PdNPs demonstrate effective inhibitory mechanisms toward colorectal cancer cells (Pinon-Castillo et al., 2021). Experimental approaches for such works depend on in vitro cell cultures for toxicity assessments and viability measurements and in vivo xenograft experiments to evaluate tumor suppression. Research is currently focused on creating better toxicological profiles and determining optimal dosages of PdNPs as promising anticancer agents despite their encouraging cancer-fighting effects. The drug delivery system offers performance benefits comparable to standard chemotherapy by providing targeted delivery and minimising side effects (Ramalingam et al., 2020).

The antimicrobial properties of PdNPs successfully reduce the growth of *Staphylococcus aureus* and *Escherichia coli* bacteria according to Alaqarbeh et al., 2023. The effectiveness of these nanoparticles extends to breaking through resistance methods used by microbes. The experimental protocol for antibacterial tests consists of subjecting multiple bacterial cultures grown in the lab to different PdNPs concentrations for determining the minimum inhibitory concentration (MIC). Extensive in vivo toxicity evaluations must be performed on PdNPs before their utilisation in medical applications due to their demonstrated antibacterial strength. Direct use of PdNPs for drug and genetic material delivery became possible because of their stable structure with high surface area availability. Nanoparticles improve therapeutic payload delivery which may lead to better treatment effects. Studies using experimental methods inside test tubes and living animals analyse how drugs attach to PdNPs and how these drug-PdNPs move throughout the body. Research on PdNPs must continue to determine safe long-term use parameters and appropriate dosage methods (Alaqarbeh et al., 2023). The application of PdNPs over traditional treatments offers targeted therapeutic methods that lead to decreased side effects and improved treatment outcomes. Their diverse properties enable PdNPs to perform well in imaging procedures as well as therapy services. Despite the benefits of PdNPs more work needs to be done because they face limitations in scalability as well as instability problems and toxicity concerns. Substantial clinical trials need to take place before medical professionals can safely and effectively use PdNPs with human patients (Alaqarbeh et al., 2023; Ramalingam et al., 2020).

5. SAFETY AND TOXICITY CONSIDERATIONS OF PALLADIUM NANOPARTICLES

PdNPs are now widely exploited in numerous applications and biomedical areas, including diagnostics and therapeutics. But there are a constant stream of safety and toxicity issues with their widespread use. Studies of the toxicology of PdNPs are essential to elucidate risks. Research shows that PdNPs have cytotoxic effects across a number of biological systems. This was illustrated by a review of Aarzoo et al. (2021), in which he found that PdNPs were cytotoxic and caused proinflammatory responses in human bronchial epithelial cells and lung carcinoma cells. On the other hand, exposure to PdNPs specifically affected cell viability as well as inflammatory cytokines secretion (IL8 and PGE2), in a dose dependent way (Leso & Iavicoli, 2018; Aarzoo et al., 2021). In another study by Leso & Iavicoli (2018) and Aarzoo et al. (2021), radiation was used to evaluate the toxic effects of PdNPs on murine macrophage cell lines. Results showed that at lower concentrations (25-200 μg/mL), cell viability was well above 95%, down to 50% at 400-500 μg/mL following 24 hours. Compared to lower concentrations, Aarzoo et al. (2021) also observed moderate apoptosis induction and increased reactive oxygen species (ROS) production at elevated concentrations of TPx. In addition, a systematic review suggested that PdNPs could also influence ecosystem processes and trigger early changes in many landmark target organs in vivo models. However, according to Leso & Iavicoli (2018) and Aarzoo et al. (2021), further study is necessary to understand more precisely the role that physicochemical properties of nanoparticles play in their toxicological behaviour. This is also a point of interest when regarding the use of palladium nanoparticles in biomedical applications because biocompatibility is a must. Biocompatibility assessments are of the interaction of PdNPs with biological systems for, cellular responses and immune reactions. PdNPs have shown to provoke immune responses. Phan et al. (2019) in his study found that excess exposure to PdNPs has been related to enhanced cytokine secretion, indicative of chronic inflammation with continuous exposure. Additionally, the biocompatibility of PdNPs was found to be enhanced with modifications to the size and surface chemistry of PdNPs. For instance, biocompatible polymers were explored for coating PdNPs to frustrate toxicity without forfeiting its functional properties (Feng et al., 2018). Specifically, a high concentration of PdNPs was found to have a high cytotoxic effect on human primary bronchial epithelial cells, whereas lower concentrations do not significantly affect cell viability. This gives an additional reason why the dosage deserves careful consideration when the biocompatibility is being assessed (Feng et al., 2018).

The therapeutic and industrial use of PdNPs requires an investigation of their final destinations after completion of service. PdNPs remain in the environment while awaiting biodegradation because their persistence becomes a serious challenge due to their entry into ecosystems through industrial operations and vehicle exhaust. The basic PdNPs show moderate safety toward fungi and plants but their released ions become environmental threats (Maryska et al., 2023). The main mechanisms through which PdNPs leave biological systems include renal excretion and phagocytosis by immune cells because these metal nanoparticles demonstrate limited breakdown capacity (Leso & Iavicoli, 2018). PdNPs have gained significant attention due to their widespread applications in catalysis, electronics, and biomedicine. However, their increasing presence in the environment and potential exposure to humans have raised concerns about their toxicity and elimination pathways in biological systems (Petrarca et al., 2014; Aarzoo et al., 2021). Understanding their biodegradability, excretion mechanisms, bioaccumulation, and interactions with biological components is essential to assess their safety profile. PdNPs are generally considered non-biodegradable, meaning they do not naturally break down in the environment or biological systems, leading to their persistence and accumulation (Leso & Iavicoli, 2018). Renal excretion plays a crucial role in nanoparticle elimination, but its efficiency largely depends on size and surface properties. Smaller nanoparticles are more likely to be cleared via the kidneys, whereas larger PdNPs tend to accumulate in organs such as the liver and lungs. Additionally, PdNPs can induce renal toxicity, potentially impairing kidney function and affecting their excretion (Leso & Iavicoli, 2018). Another key elimination mechanism is phagocytosis, where macrophages engulf PdNPs, which can lead to oxidative stress, inflammation, and cellular apoptosis (Aarzoo et al., 2021). This accumulation within immune cells can further complicate the clearance of PdNPs from biological systems. Nanoparticle persistence depends heavily on size together with surface modifications and smaller nanoparticles tend to undergo more changes in environmental matrices (Leso & Iavicoli, 2018). Green synthesis techniques represent emerging approaches to lower the environmental hazards that occur during PdNPs production according to Law et al. (2022).

Comparatively, PdNPs offer unique advantages over other nanomaterials. The catalytic properties and environmental remediation capabilities of PdNPs surpass the optical traits along with biocompatibility that gold nanoparticles possess. Organic pollutant degradation performance of PdNPs matches that of other transition metal nanoparticles but this group shows better efficiency in reaction-specific applications (Kalathil & Chaudhuri, 2016; Alaqarbeh et al., 2023). The antibacterial as well as anticancer potential from PdNPs matches silver along with platinum nanoparticles yet provides extra benefits because of their photothermal qualities (Chen et al., 2022; Alaqarbeh et al., 2023).

The effects of PdNPs occur through reactive oxygen species (ROS) generation under near-infrared (NIR) irradiation which causes bacterial cell death (Chen et al., 2022). The substance promotes apoptosis and hinders cell growth through alterations of signaling pathways between cells (Alaqarbeh et al., 2023). The chemical properties of their surface play a crucial role in how PdNPs interact with biological components since citrate-based stabilisation enhances their compatibility with cells and uptake mechanism (Maryska et al., 2023). Precision medical applications benefit from PdNPs which developers customise using functionalization and size along with shape control methods to boost targeting performance. The functionalization of PdNPs allows targeted tissue accumulation including tumours which makes photothermal therapy or drug delivery treatment possible without causing substantial side effects (Alaqarbeh et al., 2023). Nanotechnology advances permit scientists to manufacture multi-purpose PdNPs which unite diagnostic tools with therapeutic capabilities on one unified platform (Karnwal et al., 2024).

The safe application of PdNPs depends on a detailed comprehension of their dose-dependent cytotoxicity together with their biodistribution behaviour and clearance pathways (Phan et al., 2019). Research findings demonstrate that PdNPs have the ability to concentrate inside body tissues where they could trigger immune system responses that become a concern for extended periods of use. The evaluation of PdNPs should include toxicity assessments compared to reference metallic nanoparticles including gold (AuNPs) or silver (AgNPs) to establish proper toxicity rankings. AuNPs demonstrate good biocompatibility but studies show AgNPs possess elevated toxic effects (Yaqoob et al., 2020).

Table 2 briefly demonstrates Comparative Toxicity of PdNPs and Other Metallic Nanoparticles.

Table 2. Comparative Toxicity of PdNPs and Other Metallic Nanoparticles

Nanoparticle Type	Test Organism/Cell Line	Size & Concentration	Exposure Duration	Observed Toxic Effects	Key Findings	Reference
PdNPs	Female Wistar Rats	20 nm, 80 nm; 1, 10 µg/kg	28 days	Increased oxidative stress, inflammation, histological changes in lungs. Higher cytotoxicity for 20 nm PdNPs.	PdNPs show significant toxicity, especially at smaller sizes. Naringin mitigates toxicity.	Aarzoo et al. (2024)
PdNPs	J774 cell lines	4 nm; 25-500 µg/mL	24 hours	Dose-dependent cytotoxicity, cell membrane injury, apoptosis induction, ROS generation.	PdNPs exhibit adverse effects in vitro.	Leso & Iavicoli (2018)
PdNPs	Various cell lines	-	-	Cytotoxic and pro-inflammatory effects. Pd ions more toxic than PdNPs.	PdNPs pose health risks; further studies needed on physicochemical properties.	Talarska et al. (2021)
AuNPs	In vitro studies	-	-	Generally, less toxic than AgNPs. Potential for diagnostic and therapeutic applications.	AuNPs are promising for biomedical use due to lower toxicity.	Aarzoo et al. (2021)
AuNPs	Mice	4.5 nm; 15-25 µg/kg	8 days	Anti-inflammatory effects in macrophages; accumulation primarily in kidneys.	AuNPs can induce anti-inflammatory effects and accumulate mainly in kidneys.	Yaqoob et al. (2020)
AgNPs	Various cell lines	5 nm, 15 nm, 30 nm, 55 nm; up to 75 µg/mL	-	Higher toxicity compared to AuNPs. Size-dependent effects on cell viability and ROS production.	AgNPs are more toxic than AuNPs, with size influencing toxicity.	Yaqoob et al. (2020)

Bioaccumulation is another critical concern, as PdNPs have been detected in various tissues, including the liver, lungs, and kidneys. Their persistent nature raises the possibility of long-term toxicity, which can manifest as chronic inflammation or organ damage (Leso & Iavicoli, 2018). Cytotoxicity studies indicate that PdNPs generate oxidative stress by disrupting the balance between reactive oxygen species (ROS) production and antioxidant defenses, ultimately leading to cellular damage and apoptosis (Petrarca et al., 2014). In vitro studies confirm that PdNP exposure reduces cell viability, particularly at higher concentrations, further emphasizing their dose-dependent toxicity (Aarzoo et al., 2021). At the molecular level, PdNPs interact with cellular components, including DNA and proteins, potentially causing inflammation, DNA damage, and epigenetic alterations (Petrarca et al., 2014; Leso & Iavicoli, 2018). Their toxicity is strongly influenced by surface properties, size, and functionalization, which determine their biodistribution and cellular uptake. Studies show that PdNPs accumulate in different organs based on these physicochemical properties, making it imperative to assess their impact across various biological systems. Comparatively, PdNPs exhibit distinct toxicity profiles when assessed alongside other metal nanoparticles like gold (AuNPs) and silver (AgNPs). AuNPs are generally regarded as biocompatible, whereas AgNPs possess strong antimicrobial properties but also exhibit high toxicity due to ion release. PdNPs, like AgNPs, have been shown to induce oxidative stress and inflammation, yet their exact toxicity mechanisms remain underexplored (Aarzoo et al., 2021; Leso & Iavicoli, 2018). Recent research highlights the potential risks of PdNPs in both in vitro and in vivo models. In vivo studies on rats have demonstrated that PdNPs induce lung inflammation and oxidative stress, while in vitro experiments on macrophage cell lines confirm dose-dependent cytotoxicity and apoptosis (Aarzoo et al., 2021). These findings emphasize the importance of assessing PdNPs toxicity in biomedical applications, particularly for their use in biosensing and drug delivery. While PdNPs hold promise for breast cancer therapy due to their catalytic properties, their potential to induce oxidative stress and inflammation necessitates careful evaluation. For biosensing applications, ensuring biocompatibility and stability is crucial, as PdNPs must interact with biological systems without causing harm (Aarzoo et al., 2021).

Rapid advances in applications for PdNPs are leading to evolution of their regulatory landscape. Guidelines regarding the safety and efficacy of the nanomaterials in medical applications are recognised by the regulatory agencies. Perhaps, the European Medicines Agency (EMA) and the U.S. Food and Drug Administration (FDA) have started their frameworks of assessment of nanomaterials based on overall risk-based approach. First, regulatory considerations, including evaluation of physicochemical properties of PdNPs, PdNPs toxic potential, and impact on the environment, are considered. Need for standardised methods of testing safety of nanoparticle is essential for providing public health protection (Phan et al., 2019; Feng et al., 2018). Moreover, continuous discussions between researchers, industry stakeholders and regulatory bodies are needed to build robust guidelines for travelling PdNPs from laboratory to clinical applications (Feng et al., 2018). Overall, PdNPs appear to be good candidates for use in diagnostics and therapeutics, but safety and toxicity considerations are essential. Potential cytotoxicity and pro-inflammatory properties were revealed by toxicological studies of PdNPs while biocompatibility evaluations reveal potential for biomodification to increase safety profile. These materials are reassessed under regulatory perspectives that are evolving so that they are evaluated rigorously before use in clinical settings. The mechanisms of toxicity need to be further elucidated, and strategies to increase biocompatibility developed (Phan et al., 2019).

The biomedical uses of PdNPs have increased substantially while environmental impact considerations remain unresolvable. The research on PdNPs toxic effects in aquatic systems compares less to toxicological studies conducted on gold (AuNPs) and silver (AgNPs) and platinum (PtNPs). Research has established that PdNPs demonstrate matching toxicity properties with other metal nanoparticles according to their size and shape and modification status. PdNPs along with AgNPs face greater toxicity risk than AuNPs since they display both elevated reactivity properties and capacity to accumulate in living organisms. The production methods for nanomaterials require sustainability assessment so biogenic synthesis presents a superior approach since it reduces exposure to toxic chemicals (Bi et al., 2024). Results from various real-life studies show that nanoparticles stay inside biological environments which leads to build-up in organisms along with harmful effects on their health. Research examining the environmental behavior of nanoparticles demands more information about their continued survival and tendency to accumulate inside biological systems and environments. Studies of PdNPs environmental impact lack specific data due to their minimal biodegradability aspect comparable to other noble metal nanoparticles (Chen & Ostrom, 2015). PdNPs have many benefits along with specific drawbacks compared to AuNPs, AgNPs, and PtNPs due to their applications in medicine, their toxicity characteristics and their monetary costs and manufacturing capabilities. PdNPs hold potential in biomedical applications because their catalytic behaviour combined with ROS generation makes them suitable for cancer therapy and drug delivery systems (Ayogu et al., 2025). Photothermal therapy relies heavily on AuNPs but AgNPs serve as top antimicrobial agents and PtNPs show potential use in biosensing systems. The toxicity levels of PdNPs and AgNPs tend to exceed those of AuNPs though data shows PtNPs present minimal toxicity. AuNPs stand as relatively nontoxic metals whereas PdNPs and AgNPs demonstrate potential risks but studies indicate PtNPs display low toxic properties (Chen & Ostrom, 2015). The price of PdNPs lies below PtNPs but above AgNPs thereby AuNPs maintain a moderate price scale as commercial applications frequently use them (Ayogu et al., 2025). Biogenic synthesis methods for PdNPs enable sustainable and cost-effective operations which outperform traditional chemical synthesis techniques when it comes to industrial scalability (Bi et al., 2024).

The therapeutic impacts of PdNPs result from ROS generation besides enzyme inhibition drug carrier potential and apoptosis induction processes. The catalytic activity of PdNPs produces ROS that triggers cellular apoptosis through oxidative stress in cancer cells. PdNPs demonstrate excellent antitumor capabilities because they produce reactive oxygen species. The therapeutic effect of PdNPs occurs through enzyme inhibition mechanisms which improve their overall therapeutic performance. PdNPs serve as efficient platforms for drug delivery performed in a targeted manner because their functionalization capabilities enable improved treatment effectiveness. Despite their biomedical potential, PdNPs face significant regulatory and economic challenges. In order to receive approval for PdNPs-based nanomedicines manufacturers must meet the precise requirements defined by regulatory organisations such as FDA and EMA and present detailed safety and efficacy data. The nanomedicine approval process faces ongoing challenges because regulatory frameworks continue evolving (Phan et al., 2019). The production scale-up of PdNPs faces barriers that consist of maintaining stability alongside uniformity because these properties determine their market cost-efficiency (Bi et al., 2024). The economic feasibility of PdNPs requires additional cost-analysis despite their beneficial therapeutic properties when compared to standard medical approaches (Ayogu et al., 2025). The synthesis of pharmaceutical products demands optimised strategies because operational scalability and decreased expenses remain vital for leading-edge industrial applications. Plant extract-based approaches for green synthesis have shown promise in reducing both production expenses and environmental strain of PdNPs (Bi et al., 2024). Research also shows that using biogenic synthesis techniques helps reduce the potential hazards associated with PdNPs (Bi et al., 2024). The clinical application of PdNPs encounters two major obstacles because regulatory hurdles block their medical translation and their mass production presents technical difficulties (Phan et al., 2019). Additional research and development efforts will address prevailing issues which might make PdNPs viable for future biomedical applications.

6. CONCLUSION AND FUTURE DIRECTIONS

PdNPs demonstrate strong applications in nanomedicine through their potential uses for cancer therapy and antibacterial treatment and drug delivery and biosensing. Palladium nanoparticles offer attractive potential as therapeutic platforms because of their high surface area together with thermal stability and biocompatible properties. Research has shown that PTT utilising PdNPs successfully eliminates cancer cells while protecting unafflicted tissues. The antibacterial properties of these materials create an essential treatment opportunity against the escalating global issue of drug-resistant pathogens. PdNPs function as effective drug delivery systems by optimising the availability of drugs while managing their release schedule for better therapeutic processes. Biosensing technologies enable the use of PdNPs as high-speed diagnostic tools because they provide sensitive detection capabilities for early disease discovery.

PdNPs hold great promise but the main issues in their safety profile and toxicity profile and biological interaction duration need further attention. Further clinical investigations should be conducted to evaluate their distribution patterns in the body as well as their clearance systems and immune response patterns. The successful clinical implementation of PdNPs requires solution of regulatory issues combined with large-scale production optimization through affordable synthesis methods that guarantee reproducibility. Future prospects reveal promising clinical applications of PdNPs designed for cancer immune therapy and genetic adjustment techniques and theragnostic medical fields. The therapeutic value of platinum-nanoparticles could be improved through research that develops drug activating PdNPs and biodegradable nanocarrier technologies. PdNPs demonstrate great promise to transform modern medical practises through precision therapy applications and diagnostics by continuing research and developing advanced technologies.

REFERENCES

Aarzoo, N., Naqvi, S., Agarwal, N. B., Singh, M. P., & Samim, M. (2021). Bio-engineered palladium nanoparticles: Model for risk assessment study of automotive particulate pollution on macrophage cell lines. *RSC Advances*, *11*(3), 1850–1861. DOI: 10.1039/d0ra09336j PMID: 35424083

Aarzoo, N., Siddiqui, M. A., Hasan, M., Nidhi, N., Khan, H. A., Rastogi, S., Arora, I., & Samim, M. (2024). Palladium Nanoparticles and Lung Health: Assessing Morphology-Dependent subacute toxicity in rats and toxicity modulation by Naringin – paving the way for cleaner vehicular emissions. *ACS Omega*. Advance online publication. DOI: 10.1021/acsomega.4c02269 PMID: 39100302

Adams, C. P., Walker, K. A., Obare, S. O., & Docherty, K. M. (2014). Size-Dependent Antimicrobial Effects of Novel Palladium Nanoparticles. *PLoS One*, *9*(1), e85981. DOI: 10.1371/journal.pone.0085981 PMID: 24465824

Alaqarbeh, M., Adil, S. F., Ghrear, T., Khan, M., Bouachrine, M., & Al-Warthan, A. (2023). Recent Progress in the Application of Palladium Nanoparticles: A Review. *Catalysts*, *13*(10), 1343. DOI: 10.3390/catal13101343

Ayogu, J. I., Elahi, N., & Zeinalipour-Yazdi, C. D. (2025). Emerging Trends in palladium Nanoparticles: Sustainable Approaches for Enhanced Cross-Coupling Catalysis. *Catalysts*, *15*(2), 181. DOI: 10.3390/catal15020181

Bayda, S., Adeel, M., Tuccinardi, T., Cordani, M., & Rizzolio, F. (2019). The history of Nanoscience and Nanotechnology: From Chemical–Physical applications to Nanomedicine. *Molecules (Basel, Switzerland)*, *25*(1), 112. DOI: 10.3390/molecules25010112 PMID: 31892180

Bhardwaj, B., Singh, P., Kumar, A., Kumar, S., & Budhwar, V. (2020). Eco-Friendly greener synthesis of nanoparticles. *Advanced Pharmaceutical Bulletin*, *10*(4), 566–576. PMID: 33072534

Bi, S., Srivastava, R., & Ahmad, T. (2024). The potential antifungal activity of the developed palladium nanoparticles. *Biomedical & Pharmacology Journal*, *17*(4), 2617–2628. DOI: 10.13005/bpj/3053

Boutou, A. K., Pitsiou, G. G., Stanopoulos, I., Kontakiotis, T., Papakosta, D., & Kioumis, I. (2019). Regular Treatment and Disease Outcomes in Patients with Chronic Obstructive Pulmonary Disease Group A. *Respiration*, *99*(5), 458. DOI: 10.1159/000504142 PMID: 31694036

Chen, A., & Ostrom, C. (2015). Palladium-Based Nanomaterials: Synthesis and Electrochemical Applications. *Chemical Reviews*, *115*(21), 11999–12044. DOI: 10.1021/acs.chemrev.5b00324 PMID: 26402587

Chen, Y., Chen, Z., Yang, D., Zhu, L., Liang, Z., Pang, Y., & Zhou, L. (2022). Novel Microbial Palladium Nanoparticles with a High Photothermal Effect for Antibacterial Applications. *ACS Omega*, *8*(1), 1534–1541. DOI: 10.1021/acsomega.2c07037 PMID: 36643470

Deraedt, C., & Astruc, D. (2013). "Homeopathic" Palladium Nanoparticle Catalysis of Cross Carbon–Carbon Coupling Reactions. *Accounts of Chemical Research*, *47*(2), 494–503. DOI: 10.1021/ar400168s PMID: 24215156

Dumas, A., & Couvreur, P. (2015). Palladium: A future key player in the nanomedical field? *Chemical Science (Cambridge)*, *6*(4), 2153–2157. DOI: 10.1039/c5sc00070j PMID: 28694948

Feng, C., Wang, Y., & Yang, J. (2018). Effects of Reorientation of Graphene Platelets (GPLs) on Young's Modulus of Polymer Composites under Bi-Axial Stretching. *Nanomaterials (Basel, Switzerland)*, *8*(1), 27. DOI: 10.3390/nano8010027 PMID: 29316669

Fernandez-Arias, M., Vilas, A. M., Boutinguiza, M., Rodriguez, D., Arias-Gonzalez, F., Pou-Alvarez, P., Riveiro, A., Gil, J., & Pou, J. (2022). Palladium Nanoparticles Synthesized by Laser Ablation in Liquids for Antimicrobial Applications. *Nanomaterials (Basel, Switzerland)*, *12*(15), 2621. DOI: 10.3390/nano12152621 PMID: 35957051

Gurunathan, S., Kang, M., Jeyaraj, M., & Kim, J. (2021). Palladium Nanoparticle-Induced oxidative stress, endoplasmic reticulum stress, apoptosis, and immunomodulation enhance the biogenesis and release of exosome in human leukemia monocytic cells (THP-1). *International Journal of Nanomedicine*, *16*, 2849–2877. DOI: 10.2147/ijn.s305269 PMID: 33883895

Gurunathan, S., Kim, E., Han, J., Park, J., & Kim, J. (2015). Green Chemistry Approach for Synthesis of Effective Anticancer Palladium Nanoparticles. *Molecules (Basel, Switzerland)*, *20*(12), 22476–22498. DOI: 10.3390/molecules201219860 PMID: 26694334

Joudeh, N., Saragliadis, A., Koster, G., Mikheenko, P., & Linke, D. (2022). Synthesis methods and applications of palladium nanoparticles: A review. *Frontiers in Nanotechnology*, *4*. Advance online publication. DOI: 10.3389/fnano.2022.1062608

Kalathil, S., & Chaudhuri, R. (2016). Hollow palladium nanoparticles facilitated biodegradation of an azo dye by electrically active biofilms. *Materials (Basel)*, *9*(8), 653. DOI: 10.3390/ma9080653 PMID: 28773775

Karnwal, A., Jassim, A. Y., Mohammed, A. A., Sharma, V., Al-Tawaha, A. R. M. S., & Sivanesan, I. (2024). Nanotechnology for Healthcare: Plant-Derived nanoparticles in disease treatment and regenerative medicine. *Pharmaceuticals*, *17*(12), 1711. DOI: 10.3390/ph17121711 PMID: 39770553

Law, C. K. Y., Bonin, L., De Gusseme, B., Boon, N., & Kundu, K. (2022). Biogenic synthesis of palladium nanoparticles: New production methods and applications. *Nanotechnology Reviews*, *11*(1), 3104–3124. DOI: 10.1515/ntrev-2022-0482

Leso, V., & Iavicoli, I. (2018). Palladium Nanoparticles: Toxicological Effects and Potential Implications for Occupational Risk Assessment. *International Journal of Molecular Sciences*, *19*(2), 503. DOI: 10.3390/ijms19020503 PMID: 29414923

Liu, Y., Li, J., Chen, M., Chen, X., & Zheng, N. (2020). Palladium-based nanomaterials for cancer imaging and therapy. *Theranostics*, *10*(22), 10057–10074. DOI: 10.7150/thno.45990 PMID: 32929334

Maryska, L., Jindrichova, B., Siegel, J., Zaruba, K., & Burketova, L. (2023). Impact of palladium nanoparticles on plant and its fungal pathogen. A case study: Brassica napus–Plenodomus lingam. *AoB Plants*, *15*(2). Advance online publication. DOI: 10.1093/aobpla/plad004 PMID: 36970187

Osonga, F. J., Kalra, S., Miller, R. M., Isika, D., & Sadik, O. A. (2020). Synthesis, characterization and antifungal activities of eco-friendly palladium nanoparticles. *RSC Advances*, *10*(10), 5894–5904. DOI: 10.1039/c9ra07800b PMID: 35497427

Petrarca, C., Clemente, E., Di Giampaolo, L., Mariani-Costantini, R., Leopold, K., Schindl, R., Lotti, L. V., Mangifesta, R., Sabbioni, E., Niu, Q., Bernardini, G., & Di Gioacchino, M. (2014). Palladium nanoparticles induce disturbances in cell cycle entry and progression of peripheral blood mononuclear cells: Paramount role of ions. *Journal of Immunology Research*, *2014*, 1–8. DOI: 10.1155/2014/295092 PMID: 25105151

Phan, T. T. V., Huynh, T., Manivasagan, P., Mondal, S., & Oh, J. (2019). An Up-To-Date Review on Biomedical Applications of Palladium Nanoparticles. *Nanomaterials (Basel, Switzerland)*, *10*(1), 66. DOI: 10.3390/nano10010066 PMID: 31892149

Pinon-Castillo, H. A., Martinez-Chamarro, R., Reyes-Martinez, R., Salinas-Vera, Y. M., Manjarrez-Nevarez, L. A., Munoz-Castellanos, L. N., Lopez-Camarillo, C., & Orrantia-Borunda, E. (2021). Palladium Nanoparticles Functionalized with PVP-Quercetin Inhibits Cell Proliferation and Activates Apoptosis in Colorectal Cancer Cells. *Applied Sciences (Basel, Switzerland)*, *11*(5), 1988. DOI: 10.3390/app11051988

Qiao, X., Bao, L., Liu, G., & Cui, X. (2024). Nanomaterial journey in the gut: From intestinal mucosal interaction to systemic transport. *Nanoscale*. Advance online publication. DOI: 10.1039/d4nr02480j PMID: 39347780

Ramalingam, V., Raja, S., & Harshavardhan, M. (2020). In situone-step synthesis of polymer-functionalized palladium nanoparticles: An efficient anticancer agent against breast cancer. *Dalton Transactions (Cambridge, England)*, *49*(11), 3510–3518. DOI: 10.1039/c9dt04576g PMID: 32107506

Ramirez-Rios, S., Michallet, S., Peris, L., Barette, C., Rabat, C., Feng, Y., Fauvarque, M., Andrieux, A., Sadoul, K., & Lafanechère, L. (2020). A new quantitative Cell-Based assay reveals unexpected microtubule stabilizing activity of certain kinase inhibitors, clinically approved or in the process of approval. *Frontiers in Pharmacology*, *11*. Advance online publication. DOI: 10.3389/fphar.2020.00543 PMID: 32425788

Rokade, S. S., Joshi, K. A., Mahajan, K., Patil, S., Tomar, G., Dubal, D. S., Parihar, V. S., Kitture, R., Bellare, J. R., & Ghosh, S. (2018). Gloriosa superba Mediated Synthesis of Platinum and Palladium Nanoparticles for Induction of Apoptosis in Breast Cancer. *Bioinorganic Chemistry and Applications*, *2018*, 1–9. DOI: 10.1155/2018/4924186 PMID: 30057593

Sarmah, M., Neog, A. B., Boruah, P. K., Das, M. R., Bharali, P., & Bora, U. (2019). Effect of Substrates on Catalytic Activity of Biogenic Palladium Nanoparticles in C–C Cross-Coupling Reactions. *ACS Omega*, *4*(2), 3329–3340. DOI: 10.1021/acsomega.8b02697 PMID: 31459548

Shkotova, L., Bohush, A., Voloshina, I., Smutok, O., & Dzyadevych, S. (2019). Amperometric biosensor modified with platinum and palladium nanoparticles for detection of lactate concentrations in wine. *SN Applied Sciences*, *1*(4). Advance online publication. DOI: 10.1007/s42452-019-0315-9

Siddiqi, K. S., & Husen, A. (2016). Green Synthesis, Characterization and Uses of Palladium/Platinum Nanoparticles. *Nanoscale Research Letters*, *11*(1). Advance online publication. DOI: 10.1186/s11671-016-1695-z PMID: 27807824

Singh, H., Desimone, M. F., Pandya, S., Jasani, S., George, N., Adnan, M., Aldarhami, A., Bazaid, A. S., & Alderhami, S. A. (2023). Revisiting the green synthesis of nanoparticles: Uncovering influences of plant extracts as reducing agents for enhanced synthesis efficiency and its biomedical applications. *International Journal of Nanomedicine*, *18*, 4727–4750. DOI: 10.2147/ijn.s419369 PMID: 37621852

Talarska, P., Boruczkowski, M., & Zurawski, J. (2021). Current knowledge of Silver and Gold Nanoparticles in Laboratory Research Application, toxicity, Cellular Uptake. *Nanomaterials (Basel, Switzerland)*, *11*(9), 2454. DOI: 10.3390/nano11092454 PMID: 34578770

Tang, X., Haddad, P., Mager, N., Geng, X., Reckinger, N., Hermans, S., Debliquy, M., & Raskin, J. (2019). Chemically deposited palladium nanoparticles on graphene for hydrogen sensor applications. *Scientific Reports*, *9*(1). Advance online publication. DOI: 10.1038/s41598-019-40257-7 PMID: 30842583

Teng, Y., Luo, C., Qiu, X., Mu, J., Sriwastva, M. K., Xu, Q., Liu, M., Hu, X., Xu, F., Zhang, L., Park, J. W., Hwang, J. Y., Kong, M., Liu, Z., Zhang, X., Xu, R., Yan, J., Merchant, M. L., McClain, C. J., & Zhang, H. (2025). Plant-nanoparticles enhance anti-PD-L1 efficacy by shaping human commensal microbiota metabolites. *Nature Communications*, *16*(1). Advance online publication. DOI: 10.1038/s41467-025-56498-2 PMID: 39900923

Thapa, R. K., Soe, Z. C., Ou, W., Poudel, K., Jeong, J., Jin, S. G., Ku, S. K., Choi, H., Lee, Y. M., Yong, C. S., & Kim, J. O. (2018). Palladium nanoparticle-decorated 2-D graphene oxide for effective photodynamic and photothermal therapy of prostate solid tumors. *Colloids and Surfaces. B, Biointerfaces*, *169*, 429–437. DOI: 10.1016/j.colsurfb.2018.05.051 PMID: 29843117

Vinodhini, S., Vithiya, B. S. M., & Prasad, T. A. (2022). Green synthesis of palladium nanoparticles using aqueous plant extracts and its biomedical applications. *Journal of King Saud University. Science*, *34*(4), 102017. DOI: 10.1016/j.jksus.2022.102017

Yang, J., Yuan, M., Xu, D., Zhao, H., Zhu, Y., Fan, M., Zhang, F., & Dong, Z. (2018). Highly dispersed ultrafine palladium nanoparticles encapsulated in a triazinyl functionalized porous organic polymer as a highly efficient catalyst for transfer hydrogenation of aldehydes. *Journal of Materials Chemistry. A, Materials for Energy and Sustainability*, *6*(37), 18242–18251. DOI: 10.1039/c8ta07502f

Yaqoob, S. B., Adnan, R., Khan, R. M. R., & Rashid, M. (2020). Gold, Silver, and Palladium Nanoparticles: A Chemical Tool for Biomedical Applications. *Frontiers in Chemistry*, *8*. Advance online publication. DOI: 10.3389/fchem.2020.00376 PMID: 32582621

Yetisgin, A. A., Cetinel, S., Zuvin, M., Kosar, A., & Kutlu, O. (2020). Therapeutic Nanoparticles and Their Targeted Delivery Applications. *Molecules (Basel, Switzerland)*, *25*(9), 2193. DOI: 10.3390/molecules25092193 PMID: 32397080

Yi, X., Wu, Y., Tan, G., Yu, P., Zhou, L., Zhou, Z., Chen, J., Wang, Z., Pang, J., & Ning, C. (2017). Palladium nanoparticles entrapped in a self-supporting nanoporous gold wire as sensitive dopamine biosensor. *Scientific Reports*, *7*(1). Advance online publication. DOI: 10.1038/s41598-017-07909-y PMID: 28801614

Yin, X., Fan, T., Zheng, N., Yang, J., Yan, L., He, S., Ai, F., & Hu, J. (2023). Palladium nanoparticle based smart hydrogels for NIR light-triggered photothermal/photodynamic therapy and drug release with wound healing capability. *Nanoscale Advances*, *5*(6), 1729–1739. DOI: 10.1039/d2na00897a PMID: 36926581

ADDITIONAL READING

Kora, A. J., & Rastogi, L. (2015). Green synthesis of palladium nanoparticles using gum ghatti (Anogeissus latifolia) and its application as an antioxidant and catalyst. *Arabian Journal of Chemistry*, *11*(7), 1097–1106. DOI: 10.1016/j.arabjc.2015.06.024

Lakshmipathy, R., Reddy, B. P., Sarada, N. C., Chidambaram, K., & Pasha, S. K. (2014). Watermelon rind-mediated green synthesis of noble palladium nanoparticles: Catalytic application. *Applied Nanoscience*, *5*(2), 223–228. DOI: 10.1007/s13204-014-0309-2

Shaik, M., Ali, Z., Khan, M., Kuniyil, M., Assal, M., Alkhathlan, H., Al-Warthan, A., Siddiqui, M., Khan, M., & Adil, S. (2017). Green Synthesis and Characterization of Palladium Nanoparticles Using Origanum vulgare L. Extract and Their Catalytic Activity. *Molecules (Basel, Switzerland)*, *22*(1), 165. DOI: 10.3390/molecules22010165 PMID: 28106856

Un, S. S., Unlu, A., Un, I., & Ok, S. (2020). Green synthesis, characterization and catalytic activity evaluation of palladium nanoparticles facilitated by Punica granatum peel extract. *Inorganic and Nano-Metal Chemistry*, 1–9. https://doi.org/DOI: 10.1080/24701556.2020.1832118

Yang, X., Li, Q., Wang, H., Huang, J., Lin, L., Wang, W., Sun, D., Su, Y., Opiyo, J. B., Hong, L., Wang, Y., He, N., & Jia, L. (2009). Green synthesis of palladium nanoparticles using broth of Cinnamomum camphora leaf. *Journal of Nanoparticle Research*, *12*(5), 1589–1598.

KEY TERMS AND DEFINITIONS

Antibacterial Properties: The ability of PdNPs to inhibit or kill bacteria, making them useful in treating bacterial infections and preventing biofilm formation.

Biocompatibility: The ability of PdNPs to interact with biological systems without causing adverse reactions, a critical factor for safe medical use.

Biosensing: The use of PdNPs in devices that detect biological molecules, enabling precise diagnostics through electrochemical or optical signals.

Cancer Therapy: Medical treatments designed to combat cancer cells, where PdNPs are explored for their potential in photothermal therapy, drug delivery, and as imaging agents.

Drug Delivery: The technique of transporting therapeutic agents to specific parts of the body, where PdNPs can act as carriers to release drugs in a controlled manner.

Functionalization: Modifying the surface of PdNPs with chemical groups or molecules to enhance their compatibility, stability, or targeting ability in biological systems.

Nanomedicine: A branch of medicine that uses nanotechnology to diagnose, treat, and prevent diseases, often through targeted drug delivery and innovative therapies at the nanoscale level.

Palladium Nanoparticles (PdNPs): Tiny particles of palladium metal, typically less than 100 nanometres in size, valued for their unique chemical properties and used in medical and industrial applications.

Synthesis: The process of creating palladium nanoparticles through various chemical, physical, or biological methods to control size, shape, and surface characteristics.

Toxicity: The potential harmful effects of PdNPs on living organisms, which are carefully studied to ensure safe therapeutic applications.

Chapter 9
Therapeutic Implications of Chitosan and Its Derivatives:
Focus on Cancer Treatment

Youssef Ait Hamdan
https://orcid.org/0009-0005-8616-8852
Cadi Ayyad University, Morocco & University of Rennes, France

Imade Armadi
https://orcid.org/0009-0003-3063-5058
Cadi Ayyad University, Morocco

khawla Rachmoune
Cadi Ayyad University, Morocco

Fatima Ezzahra Elamrani
https://orcid.org/0009-0009-0248-1964
Cadi Ayyad University, Morocco

Maryem Mekouar
Faculté de Médecine et de Pharmacie de Rabat, Morocco

Adil El Housseini
https://orcid.org/0009-0007-9294-2685
Ibn Zohr University, Morocco

Aicha Moufakkir
Chouaib Doukkali University, Morocco

Samia Elouali
https://orcid.org/0009-0008-8014-9249
Cadi Ayyad University, Morocco

Souad Loqman
Cadi Ayyad University, Morocco

Anass Belbachir
https://orcid.org/0009-0002-4828-0251
Laboratory of Innovations in Biotechnology, Pharmaceutical Bioengineering, and Artificial Intelligence, Morocco

Mohammed Rhazi
https://orcid.org/0000-0003-2736-1802
Cadi Ayyad University, Morocco

ABSTRACT

Chitosan and its derivatives have garnered increasing interest in oncology due to their unique properties, such as biocompatibility, bioactivity and their ability to interact with molecules. These characteristics make them promising candidates for the development of novel anticancer therapies. Chitosan can be employed as drug delivery systems, enhancing the availability and targeted release of therapeutic agents into cancer cells. Additionally, they have demonstrated the ability to inhibit cell proliferation, induce apoptosis, and modulate immune responses, positioning them as multifunctional tools in the fight

DOI: 10.4018/979-8-3373-0055-9.ch009

against cancer. Recent research has focused on the development of chitosan-based nanoparticles for targeted drug delivery in chemotherapy, reducing side effects on healthy cells. Furthermore, Chitosan systems have been used to deliver siRNA and other therapeutic molecules, inhibiting tumor growth at the genetic level. These advancements highlight chitosan's potential in developing more effective and less toxic cancer treatments.

1. INTRODUCTION

Chitosan, a natural polysaccharide derived from chitin, has remarkable physicochemical and biological properties, making it a material of choice for oncology applications (Mawazi et al., 2024). Its biocompatibility, essential for any material in contact with living tissue, means it can be integrated into a variety of medical devices without provoking an undesirable immune response. What's more, its biodegradability ensures that chitosan and its derivatives break down naturally in the body into non-toxic products, reducing the risk of accumulation and long-term toxicity, a crucial aspect for anti-cancer treatments. Discovery of chitosan and its many applications, explored as early as the mid-twentieth century, have enabled scientists to envisage innovative uses for this multifunctional substance. Although chitin has long been recognized as a versatile material, recent advances have enabled chitosan to distinguish itself in the biomedical field (Ahmad et al., 2020).

Oncology research shows that chitosan and its derivatives are particularly promising for targeted drug delivery. As therapeutic delivery systems, chitosan derivatives offer the ability to transport therapeutic agents in a targeted and controlled manner to cancer cells, maximizing their bioavailability and reducing effects on healthy cells (Herdiana et al., 2023). This targeted approach is essential in chemotherapy, as it reduces the side effects of traditional treatments, which are often associated with toxicity to healthy cells. In addition to their role as vectors, these derivatives have direct biological effects, such as inhibiting cell proliferation by disrupting growth signals and inducing apoptosis, or programmed cell death, which is essential for specifically targeting cancer cells without harming healthy tissue.

The bioactivity of chitosan has been investigated using a variety of dosage forms, including chitosan nanoparticles (El-Baz et al., 2023), chitosan hydrogels (Kong et al., 2023) and chitosan films (Xu et al., 2005). Its anti-inflammatory and antimicrobial properties, as well as its ability to stimulate cell growth and blood vessel formation, give it a notable advantage in the field of wound healing. Chitosan has also been used as a support material in tissue engineering to promote cell adhesion and proliferation (Machałowski et al., 2021). A major innovation in nanomedicine is the use of chitosan-based nanoparticles. These nanoparticles are designed to reach tumor cells and progressively release chemotherapeutic agents, thereby increasing treatment efficacy and reducing the dose required to minimize adverse effects. In addition, chitosan-based systems are used for the delivery of small molecule interference (siRNA) and other gene therapies (Al-Absi et al., 2023). By targeting specific genes associated with cancer cell growth, these siRNAs can inhibit tumor proliferation at the genetic level, offering a personalized, precision approach in the fight against cancer.

These advances underscore chitosan's potential as a multifunctional tool in the design of innovative, more effective and less toxic oncology treatments. With its versatility and capacity for chemical modification, chitosan paves the way for targeted and personalized cancer therapies, optimizing clinical outcomes while reducing adverse effects for patients. This chapter explores the use of chitosan and its derivatives in the development of targeted anticancer therapies, improving drug delivery, inhibiting cell proliferation, and reducing side effects.

2. PROPERTIES AND MECHANISMS OF ACTION OF CHITOSAN AND ITS DERIVATIVES IN ONCOLOGY

2.1. Physicochemical and Biological Properties of Chitosan in Oncology

According to numerous studies, it is widely accepted that chitosan is a biocompatible, non-toxic biomaterial that can be used safely in a variety of applications, including oncology. In addition, other unique properties, particularly in biomedical applications, are also important, such as antimicrobial, antitumor and antioxidant activities... promoting the development of more effective and less toxic therapies for cancer patients.

2.1.1. Biocompatibility and Biodegradability

Biocompatibility and biodegradability are essential properties for any biomedical application, particularly in oncology, where treatment safety is a priority to preserve patient health. Chitosan, a naturally-occurring biomaterial derived from chitin, is characterized by these two features. Its biocompatibility means it can be safely interchanged with human tissue, reducing side effects. This property is therefore extremely important for anticancer therapies. On the other hand, chitosan's biodegradability represents a very strong plus point. Indeed, its importance lies in the fact that this natural polymer is degraded by the body itself, which reduces the risk of post-therapeutic complications, unlike other synthetic materials. (Crini, 2019). In oncology, the combination of these two properties - biocompatibility and biodegradability - provides the basis for innovative applications, with remarkable potential for the development of drug delivery systems and other oncology applications. Moreover, these qualities enable it to be used in a variety of forms (films, hydrogels, sponges, etc.) or as drug carriers: thiochitosan, trimethyl chitosan, carboxymethyl chitosan, ethylene glycol chitosan, etc.(Grosso et al., 2021; Hu et al., 2020; Li et al., 2021).

2.1.2. Solubility and Viscosity

Generally speaking, the biomedical properties of chitosan depend mainly on its molecular weight and its degree of deacetylation (DD), two key parameters that directly influence its efficacy in various therapeutic applications (Park et al., 2011). The stability of chitosan-based complexes depends on their molecular weight, particularly in the field of targeted drug delivery. Indeed, a high molecular weight increases the stability of chitosan-based complexes and the circulation time of nanoparticles in the blood, but delays their dissociation and subsequent effect in cells, thus assuring high tumor selectivity, whereas a lower molecular weight has the opposite effects (Ding et al., 2022). Obtaining molecular weight by viscometry remains a classic, simple and rapid method for calculating the average molecular weight of chitosan (Wang et al., 1994). However, alternative techniques such as steric exclusion chromatography (SEC) or MALDI-TOF mass spectrometry are regularly used for more precise determination, especially for advanced clinical applications. Furthermore, the work of (Rodrigues et al., 2021), clearly shows that the degree of deacetylation influences biomedical properties. Chitosan is divided into four main groups according to its DD: low DD of chitosan, medium DD of chitosan, high DD of chitosan and ultra-high DD of chitosan. However, a high DD makes it possible to produce nanoparticles with a high surface charge density, which favors cellular absorption and antitumor effect. However, when developing chitosan-based biomaterials for clinical applications, it is essential to bear in mind not only the molecular weight and degree of deacetylation, but also the source and extraction method used to obtain the chitosan. Depending on specific therapeutic objectives, the choice of preparation and characterization methods can significantly influence clinical results and expected efficacy.

2.1.3. Antioxidant Activity

Chitosan's antioxidant activity, although not considered its main property, plays an important role in the development of its applications, particularly in oncology. Several studies have demonstrated that this activity is closely linked to its structural characteristics, such as its molecular weight and degree of N-deacetylation. (Sun et al., 2007) found that the antioxidant activity of chitosan increases as its molecular weight decreases, suggesting that low-molecular-weight chitosans are more effective at neutralizing free radicals, a crucial property in the fight against oxidative stress associated with degenerative diseases and cancer. (Yen et al., 2008) quantified the antioxidant activity of chitosan, observing values ranging from 58.3% to 70.2% for moderate activities, and up to 85.2% for higher activities, underlining its potential for cellular protection against oxidative damage. Furthermore, the study by (Lingnert et al., 1979) revealed that the intensity of antioxidant activity is proportional to chitosan concentration, confirming that a higher concentration enhances chitosan's effectiveness in neutralizing free radicals. Another determining factor is N-deacetylation time, a process that alters the structure of chitosan by removing acetyl groups. This change can either improve or reduce chitosan's ability to fight oxidation, depending on the degree of modification of the polysaccharide chains.

In all this context, the use of chitosan nanoparticles (CSNPs) for the targeted delivery of antioxidants offers great potential. Nanoparticles, thanks to their biodegradability, biocompatibility and non-toxicity properties, enhance the bioactivity of encapsulated phytochemicals by optimizing their pharmacokinetics, targeted delivery and reducing toxicity. Particle size, polydispersity index (PDI) and zeta potential (ZP) are key parameters influencing the efficacy of CSNPs. Optimum particle size is necessary for effective drug release, while low PDI ensures better colloidal stability, essential for efficient phytochemical loading (Herdiana et al., 2022). Furthermore, a high molecular weight chitosan generates larger, more viscous particles, while a lower molecular weight chitosan enables the formation of smaller nanoparticles, thus promoting faster release and better loading efficiency (Iacob et al., 2021). A higher concentration of chitosan increases loading efficiency up to the maximum loading capacity (Mikušová et al., 2021a). Properly tuned, it is possible to design CSNPs with ideal characteristics (size, PDI, ZP) to maximize their therapeutic performance in antioxidant delivery.

Although antioxidant activity is not the main function of chitosan, it represents a considerable advantage for its biomedical applications, particularly in the prevention and treatment of cancer, where management of oxidative stress is essential. Research aimed at optimizing these properties, notably by adjusting molecular weight and degree of N-deacetylation, is paving the way for more effective chitosan formulations.

2.1.4. Antitumor Activity

In oncology, anti-tumoral activity occupies a very important place, despite its limitations. High-molecular-weight chitosan has low solubility in acid-free aqueous media, which limits its absorption into the human body (Lu et al., 2000). To overcome this, it is preferable to work with a low-molecular-weight chitosan, which is more easily assimilated. Evaluation of the effect of chitosan on dendritic cell (DC) and natural killer (NK) cell activation, as well as its antitumor potential, has shown that chitosan stimulates DCs to produce pro-inflammatory cytokines. This stimulation activates NK cells via the STAT4 and NF-kB signaling pathways, enhancing IFN-γ production and NK cell cytotoxicity against tumor cells. The study suggests that chitosan may have clinical applications in cancer prevention or treatment (Fathy et al., 2018).

Chitosan has also been used as a vaccine adjuvant to modulate the tumor microenvironment (TME), which plays a crucial role in the control and elimination of cancer (Katsuta et al., 2020). Recent studies have shown that co-encapsulation of chitosan with IL-12 strongly inhibits tumor growth (Eslahi et al., 2021; Yang et al., 2013). In addition, chitosan nanoparticles have been studied for their ability to deliver DNA vaccines and enhance the antitumor immune response via macrophage stimulation and interleukin-2 production (Yang et al., 2019b). The use of chitosan solutions or nanoparticles as adjuvants has also shown immunostimulatory activities, inhibiting tumor growth by inducing macrophage polarization, thereby enhancing antitumor activity (Nandgude et al., 2021). Another study demonstrated that chitosan intervention in lung cancer patients improves their immune response during radiotherapy (Ma et al., 2015).

Chitosan could also accumulate at the tumor site and initiate polarization of M1 macrophages, thus transforming an immunosuppressive tumor microenvironment into a state favorable to antitumor immune response. This approach has shown a promising antitumor effect and may enhance the efficacy of cancer immunotherapy (Ding et al., 2020; Zeng et al., 2021). By activating innate immune responses, chitosan exerts an anticancer effect by directly targeting tumor cells and modifying the local microenvironment (Chen et al., 2020; Li et al., 2018; Liang et al., 2021; Lima et al., 2021; Van Hees et al., 2020; Wu et al., 2019; Yang et al., 2019a).

In addition to its effects as a monotherapy, research is exploring the combination of chitosan with conventional treatments such as chemotherapy and radiotherapy. For example, chitosan could improve the sensitivity of tumor cells to these treatments by inhibiting their ability to repair DNA damage after radiation exposure, thereby enhancing the efficacy of radiotherapeutic treatments (Deng et al., 2015). Furthermore, the combination of chitosan with immunomodulatory agents or monoclonal antibodies could modulate the immune response and enhance the antitumor efficacy of chitosan by stimulating the activation of immune system cells such as macrophages and dendritic cells (Chen et al., 2014).

These advances open up new prospects for the use of chitosan and its derivatives in cancer treatment, not only as stand-alone antitumor compounds, but also as therapeutic adjuvants to enhance the efficacy of existing treatments. However, more clinical research is needed to validate these approaches and to better understand the underlying molecular mechanisms governing the interaction between chitosan, tumor cells and the immune system.

2.1.5. Anti-Inflammatory Activity

Inflammation is a protective biological response aimed at repairing damaged tissue, eliminating pathogens and limiting collateral damage (Gallo et al., 2017). However, defective regulation of this response can lead to chronic diseases, such as autoimmune or inflammatory disorders. In this context, the anti-inflammatory activity of chitosan, a polysaccharide derived from chitin, is attracting growing interest due to its potential beneficial effects in the management of inflammation.

Chitosan's properties make it an attractive polymer for biomedical, gene delivery and tissue engineering applications, where it was previously considered an inactive ingredient. Several studies have been published on the anti-inflammatory properties of chitosan or its oligosaccharides in chronic inflammatory bowel disease (IBD), showing that chitosan or its oligosaccharides demonstrate therapeutic efficacy in experimental models of colitis (Azuma et al., 2015; Rabišková et al., 2012; Yousef et al., 2012). A study by (Davydova et al., 2016) showed that chitosan induced an anti-inflammatory response by increasing the production of the anti-inflammatory cytokine IL-10 and inhibiting the progression of colitis in animals. The researchers compared two chitosan samples, one with a high molecular weight (115 kDa) and the other with a low molecular weight (5.5 kDa). The results suggested that the anti-inflammatory impact of chitosan was more related to its structural characteristics than to its molar mass, indicating that changes in its structure, independently of its molecular weight, could influence its therapeutic efficacy. In addition, work by (Friedman et al., 2013) revealed that chitosan effectively inhibited the production of inflammatory cytokines and chemokines in human monocyte and keratinocyte models following stimulation by Propionibacterium acnes. This activity showed a dose-dependent response, underlining the importance of dose in optimizing the anti-inflammatory effect of chitosan.

Furthermore, (Li et al., 2014) demonstrated that chitosan reduced the production of pro-inflammatory cytokines and attenuated knee swelling in a mouse model of osteoarthritis. These results confirm its potential in the treatment of joint inflammation. Another study by (Li et al., 2018) showed that low-molecular-weight chitosan significantly inhibited the production of nitric oxide, a key mediator of inflammation, without inducing toxicity. This property is essential for its application as a therapeutic biomaterial, as non-toxicity is crucial for the development of long-term treatments.

However, work by (Jhundoo et al., 2020) suggests that neither molecular weight (MW) nor degree of deacetylation (DD) has a major direct impact on chitosan's anti-inflammatory activity. Although chitosan with a low molecular weight (20-100 kDa) and a high degree of deacetylation (95%) showed a trend towards greater anti-inflammatory efficacy, this influence remains modest and requires further research to better understand the underlying mechanisms. These results suggest that, while the chemical structure of chitosan is important, other factors, such as formulation, concentration and administration methods, may also play a key role in optimizing its therapeutic effects.

The **table** below summarizes the main properties of chitosan relevant to oncology, and these key features are described to highlight their role in improving anticancer therapies (Ait Hamdan et al., 2023):

Table 1. Summary of Chitosan Properties and Functional Characteristics

Properties	Description
Biocompatibility	Biocompatible in nature and does not possess any toxicity profiles. It also did not trigger any inflammatory or rejection reaction
Structure	Strong hydrogen bonding groups (i.e., -OH, -COOH, and -NH2) are present in the chitosan structure. Presence of flexibility thanks to its macromolecular chains.
Mucoadhesive Property	Surface energy properties favor spreading into mucus
Stability	Due to the surface charge

continued on following page

Table 1. Continued

Properties	Description
Nature	Chitosan can carry a large number of pharmacological activities
Biodegradability	Easily eliminated from/by the body
Antioxidant And Antitumoral	Reducing oxidative stress
Porosity	Highly permeable. It can interact with mucosal cells of the respiratory tract, opening the tight junction and increasing the drug's permeation
Mechanical Properties	High absorbing property, flexibility and elasticity.

2.2. Mechanisms of Activity of Chitosan and its Derivatives

2.2.1. Induction of apoptosis

Chitosan exerts its anticancer effects primarily by inducing apoptosis, which is triggered through the activation of pro-aspartic proteases, thereby promoting cell death signals (Salehi et al., 2017). The cytotoxicity of chitosan is influenced by its molecular weight, with low molecular weight chitosan (LMWC) exhibiting stronger anticancer potential due to its positive charge. This positive charge enhances its interaction with the negatively charged cell membranes of cancer cells. LMWC has been shown to induce cell cycle arrest at the G1 and S phases, thereby inhibiting DNA synthesis and inducing cell death (Adhikari et al., 2018). Additionally, chitosan nanoparticles demonstrate anticancer activity, particularly in liver and colon cancer cells, by disrupting cell membranes and inducing lipid peroxidation (Zhao et al., 2019). Furthermore, chitooligosaccharides (COS) have been reported to reduce tumor growth by inhibiting critical enzymes involved in metastasis and angiogenesis, along with reducing inflammatory responses. These findings underscore that the anticancer effects of chitosan are closely tied to its molecular structure, and further research is necessary to better understand the mechanisms at play, especially concerning high molecular weight chitosan (Li et al., 2020).

2.2.2. Modulation of Immune Response

Chitosan and its derivatives offer significant potential in both cancer and antiviral treatments. In cancer therapy, chitosan activates immune responses, induces apoptosis, and inhibits tumor growth by enhancing immune cell activity, such as natural killer (NK) cells and T lymphocytes. Low molecular weight chitosan improves targeting of tumor cells, reduces drug resistance, and disrupts pro-tumor enzymes (Adhikari et al., 2018). For antiviral applications, chitosan serves as an effective adjuvant in vaccines, boosting immune responses through enhanced cytokine production and T-cell activation (Zhang et al., 2023). When conjugated with mannose, it improves vaccine stability and targeting (Yao et al., 2013). Its mucoadhesive properties and electrostatic interactions with viral membranes also enhance antiviral effects (Jaber et al., 2022). Chitosan's physicochemical properties further improve drug delivery and vaccine efficacy (Cheung et al., 2015).

2.2.3. Inhibition of Cell Proliferation

Chitosan, a biopolymer derived from chitin, has demonstrated considerable promise in both promoting cellular adhesion and suppressing the proliferation of cancer cells. Due to its inherent hydrophilic nature, chitosan's native structure often limits its ability to support efficient cell attachment. However, various chemical modifications, such as the incorporation of polyethylene glycol (PEG) or hexanoyl chloride, have been shown to enhance its physicochemical properties, thereby improving its biocompatibility for use in tissue engineering and regenerative medicine applications (Liu et al., 2016). In addition to its role in promoting cell adhesion, chitosan exhibits potent anticancer activity. It inhibits the proliferation of malignant cells by disrupting mitochondrial membrane potential, elevating the levels of reactive oxygen species (ROS), and inducing DNA damage (Salehi et al., 2017). These alterations in cellular homeostasis trigger cell cycle arrest at the S phase, followed by apoptotic cell death through caspase-3 activation. Moreover, chitosan has a direct interaction with DNA, potentially interfering with DNA replication, which further halts cell division (Li et al., 2018). Importantly, chitosan shows selectivity for cancerous cells, exhibiting no significant toxicity to normal, healthy cells, highlighting its potential as an effective therapeutic agent for cancer treatment (Zhang et al., 2017). The dual capability of chitosan to enhance tissue biocompatibility while simultaneously inhibiting tumor cell growth exemplifies its versatility and potential for broader biomedical applications.

CS induces apoptosis by causing DNA damage, which activates apoptotic signaling pathways, leading to the death of cancer cells. Additionally, it inhibits cell proliferation by interfering with the cell cycle, blocking the growth progression of cancer cells through key stages of division and thereby preventing tumor growth (Salehi et al., 2017). Moreover, CS enhances immune responses by stimulating immune cells, such as macrophages and lymphocytes, improving the body's ability to detect and destroy cancer cells, and supporting the immune system's role in combating tumors (**Figure 1**).

Figure 1. Chitosan-Based Mechanisms in Cancer Therapy: Cell Proliferation Inhibition, Apoptosis Induction, and Immune Modulation

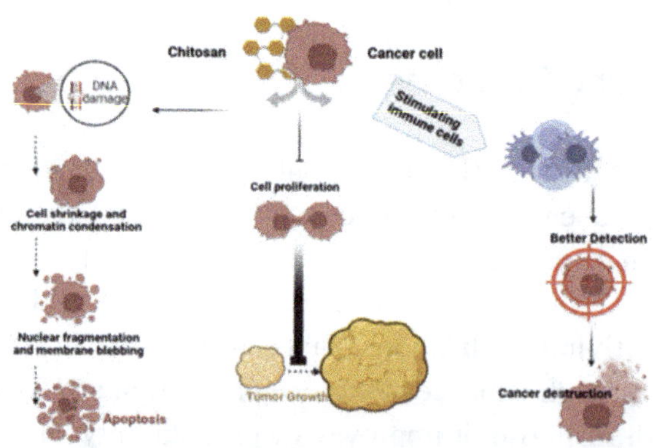

3. APPLICATION OF CHITOSAN AS A TARGETED DRUG DELIVERY SYSTEM

3.1. Importance of Targeted Delivery Systems in Oncology

Cancer, the disease characterized by an uncontrolled proliferation of transformed cells subject to evolution by natural selection (Brown, 2023), represents to date a global burden due to its complexity which stems from the molecular heterogeneity between cancers of the same type, between primary tumors and their metastatic foci and even between the cells that make up individual tumors (Webster, 2006), this complexity causes cells to escape treatment which generally leads to the administration of therapeutic combinations often associated with several disadvantages (**Figure 2**).

Figure 2. Essential Disadvantages of conventional cancer treatments

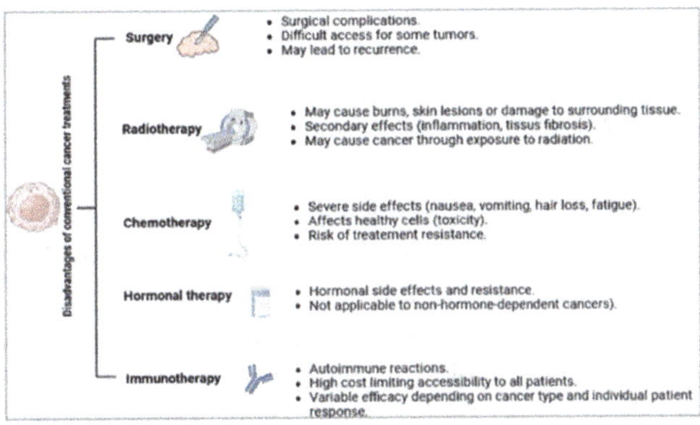

To overcome these problems and ensure a better safety profile to antitumoral drug delivery, scientists turned to targeted delivery of anti-cancer treatments thanks to nanomedicine, which has emerged as the gold standard in health science (Zarrintaj, 2017), applying the principle of German researcher Paul Ehrlich (Nobel Prize in 1908) (Strebhardt, 2008) ; in this sense, Chitosan nanoparticles CSNPs are developed to play the role of nanocarriers offering advantageous characteristics that facilitate drug delivery to cellular compartments, even those difficult to access (nucleus, blood-brain barrier and heterogeneous tumor cell microenvironments) (Moro, 2023). They are biocompatible, biodegradable, with a better tolerance rate and can reduce the side effects of chemotherapy molecules. They have helped overcome problems of salting-out, and above all guarantee specific absorption associated with an excellent anticancer effect compared to traditional formulations, with improved biodistribution (Wei, 2015) and pharmacokinetics (Sharma, 2019). Also, they are weakly immunogenic, which preserves the integrity of the drug and optimizes the targeting delivery to the desired body site thanks to their higher surface/volume ratio (Kashkooli, 2020) and they can be synthesized in large quantities at relatively low cost. To date, a number of CSNPs based nanomedicine products have been approved by the FDA and placed on the market (Rodríguez, 2022) and many patents are deposed for a large spectrum of cancers especially from China, United states of America, Turkey and India (Silva, 2024).

Figure 3. Nanoparticulate Drug Delivery Systems (NDDS) for Cancer Therapy

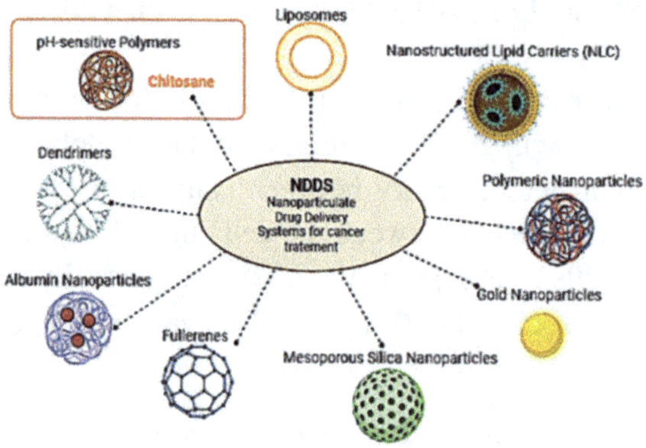

3.2. Chitosan as a Chemotherapy Carrier

Chitosan (CS) is gaining in popularity in precision medicine; In its native state, it presents targeting problems, unwrapping capabilities, solubility and low buffering capacity. It is the main polymer widely used in targeted drug delivery employed to fabricate various delivery systems, taking advantage of its immunosuppressives characteristics, hemocompatibility (hemolytic character, low level of platelet aggregation without effects on the complement and coagulation systems), antitumoral effect helping to enhance the activity of conventional chemotherapeutic drugs, higher blood circulation and significant accumulation in tumor tissues (AKHDAR), effect on humoral and cellular immune responses (Shakil, 2021).

Although metal and carbon NPs are promising structures in tumor elimination, their biocompatibility is low, and this can minimize their application at the clinical level. Modified CS by nanoarchitecture has shown versatility in many applications such as wound treatment, tissue engineering, real-time cancer detection and monitoring (Moramkar, 2021). Also, CS can be used for the coating of heart valves, the manufacture of artificial skins for the formation of hydrogels and for the delivery of DNA, RNA or proteins into cells (Declèves, 2009), (Clark, 2015). Additionally, it is used in oral, parenteral and pain therapy (Dristant, 2023) and for the targeted delivery of chemotherapeutic drugs to a broad spectrum of cancers (**Table 2**), where it leads to increase therapeutic efficacy and decrease side effects (El-Leithy, 2019), and when incorporate with a reactive amino group on its backbones improve its physiochemical and biological properties, and it can overcome the hurdle of poor aqueous solubility.

Chitosan nanoparticles (CSNPs) have been reported as drug delivery systems that can control the release and load hydrophobic and hydrophilic drugs, improve bioavailability, enhance their cytotoxic activity (Yun, 2019), target a single receptor if used with a monoclonal antibody on the surface, improve solubility, facilitates specific passive and active targeting of nanoparticles to cancer cells (Cavalli, 2014). The passive targeting relies on the ability of chitosan nanocomplexes (CSNCs) to transport antineoplastic drugs into tumor cells via angiogenic vessels and accumulate at its location due to pathophysiological variations between tumor and normal tissues, while the active targeting involves molecular recognition (Ag-Ac, ligand-receptor) by attaching targeting motifs to the nanoparticle surface to deliver them specifically to the cells, tissues and organ of interest. CS NPs have been shown to have an improved tumor inhibition rate (88%) (Yan, 2020). Studies have shown the ability of CS NPs to induce 90% silencing of the MDR1 gene, responsible for chemoresistance, and the possibility of realizing this silencing phenotypically opens the horizon to in-depth study of the role of nanoparticles in chemosensitizing cancer cells in vitro and in vivo (Benhammadi, 2014). Recently, researchers suggested that (CSNS) are safe and robust smart nanocarriers in delivering drug selectively to tumor cells based on dual stimuli-triggered response with high cellular uptake and cytotoxicity (Twal, 2024).

Table 2. Advantages of NDDS based on CS validated for the DDT

Type of cancer	Benefits of validated CS-based for cancer treatment	Reference
Cervix	● Extended-release profile. ● Improve cellular absorption of anti-cancer drugs. ● Effective targeting verified by obvious tumor accumulation. ● Minimal toxicity. ● Good immunomodulatory capacity. ● Genome-editing capabilities for gene therapy. ● Effective tumor regulation. ● Protects siRNA from degradation and production of chemosensitive phenotype in multi-drug-resistant tumor cells. ● Demonstrates ability to deliver to tumor cells.	(Naghibi Beidokhti, 2017) (Zhang, 2020) (Heidari, 2021) (Nemati, 2021)
Brain	● Precise targeting of cells to be treated. ● Promising for this kind of hard-to-reach cancer.	(Lara-Velazquez, 2020)
Lung	● Enhancing the immunosuppressive microenvironment to promote antitumor activity. ● Immunomodulatory effect. ● Inhibition of tumor development. ● Reduces cytotoxicity and improves therapeutic efficacy, with extended circulation time.	(Ding, 2018) (ŞENEL, 2019) (Cheng, 2019) (Zhu, 2021)
Hepatocellular carcinoma	● Accumulation in the tumor site and efficacy in suppressing tumor cell development. ● Prolonged retention in the bloodstream, and reduced biodistribution in the heart and kidneys. ● Activity induces tumor cell death. ● Enhances intracellular delivery and absorption to increase anti-tumor efficacy and safety.	(Siahmansouri, 2016) (Chi, 2019) (Hefnawy, 2020)
Breast	● Great localization of siRNA in the tumor. ● Biodistribution of NPs on brain, heart, spleen, liver, lung, muscle, kidney and localized to the tumor after i.v. ● Tumor volume reduction. ● Minimizes side effects and increases therapeutic efficacy. ● Selective, long-lasting release. ● Controlled drug release. ● Inhibitory effects on proliferation and migration *in vitro*, and on angiogenesis and tumor development *in vivo*.	(Han, 2011) (Şalva E. K., 2012) (Şalva E. T., 2015) (Zavareh, 2020) (Nikkhoo, 2020) (Karuppaiah, 2021) (Helmi, 2021)
Ovaries	● Maximizes anti-cancer and anti-metastasis effects and reduces toxicity. ● Increased apoptosis and reduced tumor proliferation and angiogenesis. ● Controlled release and improved therapeutic effect. ● Inhibits tumor progression.	(Zhou, 2010) (Lee, 2011) (Aftab, 2020) (Qu, 2020)
Prostate	● Inhibits proliferation and migration of PC-3 cells *in vivo.*	(Afkham, 2018)
Pancreas	● Superior antiproliferative activity and prolonged intracellular retention.	(Arya, 2011)
Melanoma	● Inhibits tumor angiogenesis and induces apoptosis of melanoma cells via the mitochondrial apoptotic pathway.	(Shan, 2020)

Recent study approves that CS based nanogels have a potential in carrying drugs offering promising prospects for advancing intelligent drug delivery, by augmenting the drug absorption and mitigate drug-related toxicity (Damiri, 2023). Although CS and its derivatives have performed, very well in vitro and preclinical studies, clinical translation is hindered as CS has not received a GRAS status from US FDA for interventional biomedical applications (Kumar, 2024), and the CS delivery of natural products has been largely ignored.

3.3. Chitosan for Delivery of Interfering RNA (siRNA) and Other Genetic Molecules

Initially developed as vectors for RNA-based drugs, NPs are now being used more for non-viral delivery by encapsulating siRNAs (Kozielski, 2019) (Mousazadeh, 2021). The positive charge of CS can bind and condense negatively charged nucleic acids to form NPs for targeted delivery of siRNA which are small non-coding interfering RNAs, double-stranded and form part of therapeutic oligonucleotides that can specifically silencing genes as they are recognized and bound to protein complexes that drive them to a single mRNA (Kruspe, 2017). Due to their polyanionic characteristics, they have biological obstacles in systematic administration case which prevent them from crossing the lipid bilayer such as the short half-life, the degradation by serum enzymes, rapid renal clearance or immune-mediated elimination. To stabilize them they are chemically synthesized or generated by enzymatic digestion (Ichim, 2004) which are both costly and time-consuming (Amarzguioui, 2003). Researchers approved that siRNA hybrid delivery system could stop tumor growth and decrease side effects (Kordbacheh, 2024).

NPs are promoting for the targeted siRNA delivery, they can escape nuclease degradation with improved transfection efficiency thanks to their binding capacity (Santos J. D.-F., 2021). This delivery power and safety make CS a non-toxic transfection agent for cells competing with lipid delivery systems that encounter obstacles in crossing biological barriers (Benhammadi, 2014), and it can avoid the rapid renal clearance as well (Gao, 2009). (CSNPs) have promising stimuli-responsive properties and therefore can be used to enhance siRNA delivery platforms to undruggable metastatic cancer cells (Subhan, 2023). mRNA has difficulty crossing the endosomal barrier but has promising potential in gene therapy; it is being applied in cancer vaccination, defective protein replacement, genome editing and stem cell reprogramming's. Researchers proved that (CSNPs) can escape endosomal sequestration if incorporated into an endosmolytic agent (Jeandupeux, 2018).

Researchers in some studies have used chitosan for in vivo DNA delivery orally (Bivas-Benita, 2003), and in others (Khatri, 2008) nasally, and the treatment strategy has been verified. The cytotoxicity and transfection efficiency of CS was verified by a cell viability greater than 90% compared with other NPs (Lu, 2014). Two patents on the use of CS biopolymers as non-viral vectors for gene delivery were filed by Li et al. (Li, 2010) and Oskuee et al., both of which provide methods for preparing non-viral gene vectors using CS nanoconjugates. And a patent filed by (Chen, 2015) describes a CS gene vector with dual response in the microenvironment and its preparation method, this polymer exhibits good biocompatibility and gene transfection efficiency (Santos J. D.-F., 2021).

The main challenges for the use of chitosan in targeted drug delivery are the problems of passive targeting (limiting nanoparticle accumulation), active targeting (problems of immune detection and large-scale production), stimulus reactivity to the efficacy of treatment responses (Kashkooli, 2020), its limited exploration as a gene carrier with superior biocompatibility and efficacy, and its availability at high purity. Gene drugs are still at a preliminary stage and are fertile ground for further exploration to efficiently deliver nucleic acid molecules as FDA-approved formulations in the future (Gholap, 2024). Moreover, despite huge progress in CS based nanoparticles, the combination of CS with other polymers biodegradable to make novel drug carriers is unexplored (Virmani, 2023), and safety assessments for formulating chitosan-based drug carriers and evaluation of their nanotheranostic platforms are critical which can encourage new researches in the future (Bhattacharjee, 2023).

4. ADVANTAGES AND LIMITATIONS OF CHITOSAN APPLICATIONS IN CANCER THERAPY

4.1. Benefits of Chitosan over Conventional Systems

4.1.1. Versatility of Chitosan in Different Therapeutic Strategies

Chitosan is a natural, biocompatible polymer widely exploited in a variety of therapeutic strategies, notably in oncology (Virmani et al., 2023). Thanks to its great versatility, it can be used for drug delivery, gene therapy and immunotherapy. Its unique capacity for chemical modification means it can be tailored to specific drugs or biological targets, enhancing its therapeutic efficacy. In particular, chitosan nanoparticles (ChNPs) have shown promising results in anticancer drug formulations, enabling targeted delivery and reducing systemic toxicity. This polymer can also be functionalized with ligands or antibodies, increasing specificity and uptake by tumor cells, representing a significant advantage over conventional delivery systems (Babu et al., 2017a; Shanmuganathan et al., 2019a).

In addition to anti-cancer treatments, chitosan has wide-ranging applications in tissue engineering (Ge, 2004; Lee et al., 2000). As a non-viral vector, it protects biological molecules from degradation, thus prolonging their activity in the biological system (Agnihotri et al., 2004; Prabaharan et al., 2004). Its intrinsic antimicrobial properties reinforce its therapeutic interest, particularly in wound healing, by stimulating the activity of macrophages and neutrophils, thereby accelerating the healing process (Ueno et al., 1999). Chitosan's ability to form various structures, such as films, gels and nanoparticles, makes it particularly well suited to biomedical supports for tissue regeneration.

In the fields of bone and cartilage engineering, for example, composites of chitosan with calcium phosphate and its formulations enriched with growth factors promote cell regeneration, facilitating improved tissue healing and repair. Its adaptability to specific therapeutic needs and ability to limit adverse reactions make chitosan ideal for a variety of regenerative strategies. This versatility underlines its promising role in addressing a variety of medical challenges, supporting both targeted therapeutic delivery and regenerative medicine initiatives.

4.1.2. Enhanced Safety Profile and Customization Potential

Chitosan's biocompatibility and low toxicity make it a particularly promising solution for long-term therapeutic applications. Its unique characteristics, such as biodegradability and good tolerance by the body, make it an attractive option for the development of innovative and safe treatments. In addition, its natural degradation involves renal elimination, which limits the risk of cumulative toxicity. As a result, this degradation capacity reduces the undesirable side effects associated with other synthetic polymers and ensures long-term use without compromising patient safety. Chitosan's gradual, controlled elimination makes it an ideal candidate for long-term therapies suitable for widespread medical applications (Babu et al., 2017b) (Ghaz-Jahanian et al., 2015). Chitosan can be modified to improve its pharmacokinetic properties, for example by coupling it with polymers such as polyethylene glycol (PEG), thereby increasing its solubility and circulation time in vivo. This modification circumvents certain limitations linked to its low solubility at physiological pH, making chitosan more effective as a vector for drugs or genes. In addition to improving solubility, the addition of PEG helps to reduce recognition by the reticuloendothelial system (RES), prolonging nanoparticle half-life and enabling more efficient delivery of targeted therapeutic agents. Such modifications are not without consequences, however, as they can also influence nanoparticle stability and their ability to target specific cells, particularly cancer cells, which can adversely affect therapeutic efficacy if they are poorly targeted (Shanmuganathan et al., 2019b). Consequently, optimizing these modifications is crucial to ensure the successful clinical application of modified chitosan (Mikušová et al., 2021b).

Although no clinical trials have yet been conducted specifically for the use of siRNA-loaded chitosan nanoparticles in cancer therapy, research shows that nanocarriers can be effective and well tolerated in cancer patients. For example, a Phase II clinical trial demonstrated that BIND-014, a docetaxel-loaded nanocarrier, effectively targeted prostate cancer cells through antigen modification of the prostate-specific membrane, improving treatment specificity (Autio et al., 2018). In addition, nanoparticles show an excellent tolerance and safety profile in cancer patients, highlighting their potential for clinical applications (Futamura et al., 2017), (Ota et al., 2019). Chitosan, as a nanocarrier base, exhibits high biocompatibility and has already been successfully used in the treatment of diseases, with a high safety profile (Aksungur et al., 2004; Pavis et al., 2002). In anti-cancer therapy, chitosan can also protect siRNA from degradation, facilitating its uptake by tumor cells, and enable co-delivery of siRNA and anti-tumor agents for enhanced therapeutic effect (Ashrafizadeh et al., 2021). The next step could be to load siRNA onto chitosan-based nanoarchitectures for clinical use in cancer treatment, opening up new perspectives in targeted gene therapy.

4.2. Limitations and Challenges Associated With the Use of Chitosan in Oncology

Chitosan, although a promising polymer for anticancer therapeutic applications, presents several major challenges in its clinical use. One of the main problems is its low solubility, particularly at physiological pH (pH 7.4), which limits its effectiveness as a carrier for drugs or genes. Its high hydrophilicity, low degree of ductility, high level of swelling and lower thermal stability are also additional constraints on its use (Safdar et al., 2019).

To enhance pharmacokinetic properties and overcome this low solubility, chemical modifications of chitosan can be performed, such as its conjugation with polymers like polyethylene glycol (PEG). Such modifications increase its solubility and prolong its circulation time in vivo. However, these modifications can sometimes affect nanoparticle stability and disrupt the distribution of therapeutic agents (Shanmuganathan et al., 2019b; G. Wang et al., 2021). This can impair the systems' ability to specifically target cancer cells, with an increased risk of elimination through the reticuloendothelial system (RES) or accumulation in non-target organs, thereby reducing therapeutic efficacy (Cong et al., 2024).

In economic terms, although chitosan is relatively abundant as a natural polymer, its use in oncology remains expensive. Producing chitosan nanoparticles for clinical applications involves complex processes, including self-assembly and chemical modifications, which significantly increase costs (Wang et al., 2021).

Moreover, the quality and regulatory compliance requirements for clinical approval of these nanoparticles necessitate substantial financial investment, which hampers their widespread adoption for cancer treatments (Hobson, 2016). Chitosan has a number of important facets, as already mentioned. **Table 3**, illustrates the most significant challenges and advantages, highlighting the strengths and limitations of chitosan in cancer therapy. These elements provide a better understanding of the current prospects and obstacles concerning the use of chitosan in cancer therapy.

Recent research offers solutions to the challenges of stability, biodistribution and cost-effectiveness of chitosan-based therapies. By combining chitosan with smart, biodegradable materials such as pH-sensitive polymers, researchers are looking to improve the targeted release of therapeutic agents and enhance the stability of chitosan formulations in the body (Singh et al., 2023). Such advances not only enable more precise drug delivery to cancer cells, but also minimize off-target effects, making chitosan a more viable option for large-scale clinical use. In parallel, new synthesis methods are being explored to simplify the production of chitosan nanoparticles, thereby reducing costs and making these treatments more accessible. In turn, these advances could encourage the wider adoption of chitosan-based treatments in oncology.

Table 3. Advantages and limitations of Chitosan in Cancer Therapy

Aspect	Advantages of CS	Disadvantages of CS	References
Therapeutic versatility	Wide range of applications: drug delivery, gene therapy, immunotherapy. Possibility of chemical modification for specific customization.	Precise targeting challenges due to non-selective biodistribution	(Babu et al., 2017b; Shanmuganathan et al., 2019b)
Safety profile	Biodegradable and biocompatible. Low toxicity. Renal elimination of nanoparticles.	The stability of nanoparticles under varying biological conditions can be problematic.	(Babu et al., 2017b; Ghaz-Jahanian et al., 2015)
Solubility and modification	Improved solubility with functional groups such as PEG. Improved circulation in vivo.	Limited solubility in water, requiring chemical modifications to improve it.	(Ghaz-Jahanian et al., 2015; Shanmuganathan et al., 2019b)
Production costs	Chitosan is cheap and natural. Easily available.	Complex, costly manufacturing process. High cost of specific formulations.	(Quiñones et al., 2018; Wang et al., 2021)
Biodistribution challenges	Can be modified with ligands for enhanced tumor cell targeting.	Non-targeted biodistribution, accumulation in non-target organs.	(Shanmuganathan et al., 2019b; Wang et al., 2021)

5. CONCLUSION

Chitosan and its derivatives are attracting considerable interest in oncology due to their unique physicochemical and biological properties, such as biocompatibility, biodegradability, and antitumor and immunomodulatory activities. These natural polymers, derived from chitin, show particular promise for the development of innovative anti-cancer therapies, offering potentially safer and more effective solutions for the treatment of tumors.

One of the main applications of chitosan in oncology lies in its ability to be used in a variety of forms, such as nanoparticles, hydrogels, films and microspheres, enabling targeted and controlled drug delivery. Chitosan-based nanoparticles, for example, show significant advantages in terms of tumor targeting. Their small size and ability to encapsulate therapeutic agents mean that these nanoparticles can effectively penetrate tumor tissue while reducing systemic side effects, a crucial feature for minimizing undesirable impacts on healthy tissue. By acting as vectors, they not only improve the bioavailability of anti-cancer drugs, but also facilitate the prolonged and controlled release of therapeutic agents, thereby increasing their efficacy and reducing the risk of drug relapse. In addition, the structural versatility of chitosan allows for a variety of chemical modifications to enhance its therapeutic impact. For enhanced immunostimulant and antitumor properties, for example, sulfation and phosphorylation can be applied to chitosan, improving the immune response against cancer cells. Conjugation with specific ligands can also increase targeting precision, enabling nanoparticles to bind preferentially to specific receptors present on the surface of tumor cells, thus limiting drug dispersion into other tissues. However, despite this significant progress, challenges remain in optimizing chitosan formulations for clinical applications. Improved solubility under physiological conditions, long-term stability of formulations, and reproducibility of their large-scale production remain major objectives. The low solubility of chitosan in neutral media, for example, limits its use in certain applications, prompting researchers to explore chemical modifications, such as quaternization, to increase its solubility and facilitate its use in a variety of body environments. Improving targeting precision through the development of more sophisticated chitosan nanocarriers would also maximize therapeutic effects while reducing dose requirements. The exploration of chitosan as a gene carrier also opens up new prospects for cancer therapy. Indeed, chitosan's ability to transport and protect DNA or RNA from enzymatic degradation makes it an ideal candidate for gene therapy, enabling the transfer of therapeutic genes targeting cancer cells and promoting the expression of specific proteins to induce an anti-tumor response. In addition, combining chitosan with biodegradable polymers, such as poly(lactide-co-glycolide) (PLGA) or polyethylene glycol (PEG), could lead to even more effective drug delivery platforms, by enhancing their stability, load-bearing capacity and penetration into tumors. Additionally, future research should focus on refining chitosan nanocarriers to maximize therapeutic benefits. Hybrid nanostructures, for example, could enable chitosan to be combined with other materials to combine the benefits of each component. Research efforts such as these are essential not only to improve the efficacy of cancer therapies, but also to minimize treatment side-effects, thus offering milder treatment options for cancer patients.

Chitosan and its derivatives offer remarkable potential for the development of more targeted, less invasive cancer treatments. However, to fully realize this potential, in-depth research is still needed to optimize formulations and manufacturing processes, while validating the long-term safety and efficacy of chitosan in clinical settings.

REFERENCES

Adhikari, H. S., & Yadav, P. N. (2018). Anticancer Activity of Chitosan, Chitosan Derivatives, and Their Mechanism of Action. *International Journal of Biomaterials, 2018*, 1–29. DOI: 10.1155/2018/2952085 PMID: 30693034

Afkham, A. A.-M. (2018). Chitosan (CMD)-mediated co-delivery of SN38 and Snail-specific siRNA as a useful anticancer approach against prostate cancer. *Pharmacological Reports*, *70*(3), 418–425. PMID: 29626645

Aftab, A. B. (2020). Microfluidic platform for encapsulation of plant extract in chitosan microcarriers embedding silver nanoparticles for breast cancer cells. *Applied Nanoscience*, *10*, 2281–2293.

Agnihotri, S. A., Mallikarjuna, N. N., & Aminabhavi, T. M. (2004). Recent advances on chitosan-based micro- and nanoparticles in drug delivery. *Journal of Controlled Release*, *100*(1), 5–28. DOI: 10.1016/j.jconrel.2004.08.010 PMID: 15491807

Ahmad, S. I., & Ahmad, R., Khan, Mohd. S., Kant, R., Shahid, S., Gautam, L., Hasan, G. M., & Hassan, Md. I. (2020). Chitin and its derivatives: Structural properties and biomedical applications. *International Journal of Biological Macromolecules*, *164*, 526–539. DOI: 10.1016/j.ijbiomac.2020.07.098 PMID: 32682975

Ait Hamdan, Y., El Amerany, F., Desbrières, J., Aghrinane, A., Oudadesse, H., & Rhazi, M. (2023). The evolution of the global COVID-19 epidemic in Morocco and understanding the different therapeutic approaches of chitosan in the control of the pandemic. *Polymer Bulletin*, *80*(10), 10633–10659. DOI: 10.1007/s00289-022-04579-3 PMID: 36466080

AKHDAR, R. &. (n.d.). Potentiel biologique des polysaccharides issus de Bifurcaria bifurcata R. Ross: Activité anti complément .

Aksungur, P., Sungur, A., Ünal, S., Iskit, A. B., Squier, C. A., & Şenel, S. (2004). Chitosan delivery systems for the treatment of oral mucositis: In vitro and in vivo studies. *Journal of Controlled Release*, *98*(2), 269–279. DOI: 10.1016/j.jconrel.2004.05.002 PMID: 15262418

Al-Absi, M. Y., Caprifico, A. E., & Calabrese, G. (2023). Chitosan and Its Structural Modifications for siRNA Delivery. *Advanced Pharmaceutical Bulletin*, *13*(2), 275–282. DOI: 10.34172/apb.2023.030 PMID: 37342385

Amarzguioui, M. H. (2003). Tolerance for mutations and chemical modifications in a siRNA. *Nucleic Acids Research*, *31*(2), 589–595. PMID: 12527766

Arya, G. V. (2011). Enhanced antiproliferative activity of Herceptin (HER2)-conjugated gemcitabine-loaded chitosan nanoparticle in pancreatic cancer therapy. *Nanomedicine; Nanotechnology, Biology, and Medicine*, *7*(6), 859–870. PMID: 21550422

Ashrafizadeh, M., Delfi, M., Hashemi, F., Zabolian, A., Saleki, H., Bagherian, M., Azami, N., Farahani, M. V., Sharifzadeh, S. O., Hamzehlou, S., Hushmandi, K., Makvandi, P., Zarrabi, A., Hamblin, M. R., & Varma, R. S. (2021). Biomedical application of chitosan-based nanoscale delivery systems: Potential usefulness in siRNA delivery for cancer therapy. *Carbohydrate Polymers*, *260*(February), 117809. DOI: 10.1016/j.carbpol.2021.117809 PMID: 33712155

Autio, K. A., Dreicer, R., Anderson, J., Garcia, J. A., Alva, A., Hart, L. L., Milowsky, M. I., Posadas, E. M., Ryan, C. J., Graf, R. P., Dittamore, R., Schreiber, N. A., Summa, J. M., Youssoufian, H., Morris, M. J., & Scher, H. I. (2018). Safety and Efficacy of BIND-014, a Docetaxel Nanoparticle Targeting Prostate-Specific Membrane Antigen for Patients with Metastatic Castration-Resistant Prostate Cancer: A Phase 2 Clinical Trial. *JAMA Oncology*, *4*(10), 1344–1351. DOI: 10.1001/jamaoncol.2018.2168 PMID: 29978216

Azuma, K., Osaki, T., Kurozumi, S., Kiyose, M., Tsuka, T., Murahata, Y., Imagawa, T., Itoh, N., Minami, S., Sato, K., & Okamoto, Y. (2015). Anti-inflammatory effects of orally administered glucosamine oligomer in an experimental model of inflammatory bowel disease. *Carbohydrate Polymers*, *115*, 448–456. DOI: 10.1016/j.carbpol.2014.09.012 PMID: 25439918

Babu, A., & Ramesh, R. (2017). Multifaceted Applications of Chitosan in Cancer Drug Delivery and Therapy. *Marine Drugs*, *15*(4), 96. DOI: 10.3390/md15040096 PMID: 28346381

Benhammadi, M., . (2014). Évaluation de l'efficacité des nanoparticules chitosane-siARN à inhiber, in vitro, l'expression du gène MDR1 responsable de la résistance à la chimiothérapie.

Bhattacharjee, R. N. (2023). Nanotheranostics to target antibiotic-resistant bacteria: Strategies and applications. *OpenNano*, *11*, 100138.

Bivas-Benita, M. L. (2003). Generation of Toxoplasma gondii GRA1 protein and DNA vaccine loaded chitosan particles: Preparation, characterization, and preliminary in vivo studies. *International Journal of Pharmaceutics*, *266*(1-2), 17–27. PMID: 14559390

Brown, J. S. (2023). Updating the definition of cancer. *Molecular Cancer Research*, *21*(11), 1142–1147. PMID: 37409952

Cavalli, R. (2014). New chitosan nanospheres for the delivery of 5-fluorouracil. *Current Drug Delivery*, *11*, 270–278. PMID: 24499357

Chen, C., Wu, W., Xu, X., Zhang, L., Liu, Y., & Wang, K. (2014). Chain conformation and anti-tumor activity of derivatives of polysaccharide from Rhizoma Panacis Japonici. *Carbohydrate Polymers*, *105*, 308–316. DOI: 10.1016/j.carbpol.2014.01.089 PMID: 24708985

Chen, H. C. (2015). Grafting chitosan with polyethylenimine in an ionic liquid for efficient gene delivery. *PLoS One*, *10*(4), e0121817. PMID: 25875475

Chen, J., Dou, Y., Tang, Y., & Zhang, X. (2020). Folate receptor-targeted RNAi nanoparticles for silencing STAT3 in tumor-associated macrophages and tumor cells. *Nanomedicine; Nanotechnology, Biology, and Medicine*, *25*, 102173. DOI: 10.1016/j.nano.2020.102173 PMID: 32084593

Cheng, B. G. (2019). Repurposing suramin for the treatment of breast cancer lung metastasis with glycol chitosan-based nanoparticles. *Acta Biomaterialia*, *84*, 378–390. PMID: 30528604

Cheung, R. C. F., Ng, T. B., & Wong, J. H. (2015). Chitosan: An overview of its applications in antiviral treatment. *Applied Biochemistry and Biotechnology*, *177*(2), 357–367.

Chi, J. J. (2019). Antitumor evaluation of carboxymethyl chitosan based norcantharidin conjugates against gastric cancer as novel polymer therapeutics. *International Journal of Biological Macromolecules*, *136*, 1–12. PMID: 31158420

Cong, X., Zhang, Z., Li, H., Yang, Y. G., Zhang, Y., & Sun, T. (2024). Nanocarriers for targeted drug delivery in the vascular system: Focus on endothelium. *Journal of Nanobiotechnology*, *22*(1), 620. DOI: 10.1186/s12951-024-02892-9 PMID: 39396002

Crini, G. (2019). Historical review on chitin and chitosan biopolymers. *Environmental Chemistry Letters*, *17*(4), 1623–1643. DOI: 10.1007/s10311-019-00901-0

Damiri, F. F. (2023). Smart stimuli-responsive polysaccharide nanohydrogels for drug delivery: A review. *Journal of Materials Chemistry. B*. PMID: 37909361

Davydova, V. N., Kalitnik, A. A., Markov, P. A., Volod'ko, A. V., Popov, S. V., & Ermak, I. M. (2016). Cytokine-inducing and anti-inflammatory activity of chitosan and its low-molecular derivative. *Applied Biochemistry and Microbiology*, *52*(5), 476–482. DOI: 10.1134/S0003683816050070 PMID: 29513410

Declèves, X. L. (2009). Transport des anticancéreux par les transporteurs ABC et implications en hématologie. La Lettre du cancérologue (Boulogne),, 18(2), 115-125.

Deng, C., Fu, H., Xu, J., Shang, J., & Cheng, Y. (2015). Physiochemical and biological properties of phosphorylated polysaccharides from Dictyophora indusiata. *International Journal of Biological Macromolecules*, *72*, 894–899. DOI: 10.1016/j.ijbiomac.2014.09.053 PMID: 25316421

Ding, F., Fu, J., Tao, C., Yu, Y., He, X., Gao, Y., & Zhang, Y. (2020). Recent Advances of Chitosan and its Derivatives in Biomedical Applications. *Current Medicinal Chemistry*, *27*(18), 3023–3045. DOI: 10 .2174/0929867326666190405151538 PMID: 30961477

Ding, J., & Guo, Y. (2022). Recent Advances in Chitosan and its Derivatives in Cancer Treatment. *Frontiers in Pharmacology*, *13*. Advance online publication. DOI: 10.3389/fphar.2022.888740 PMID: 35694245

Ding, Y. F. (2018). Highly biocompatible chlorin e6-loaded chitosan nanoparticles for improved photodynamic cancer therapy. *ACS Applied Materials & Interfaces*, *10*(12), 9980–9987. PMID: 29498260

Dristant, U. M. (2023). An overview of polymeric nanoparticles-based drug delivery system in cancer treatment. *Technology in Cancer Research & Treatment*. PMID: 36718541

El-Baz, F. K., Salama, A., Ali, S. I., & El-Hashemy, H. A. (2023). Dunaliella salina chitosan nanoparticles as a promising wound healing vehicles: In-vitro and in-vivo study. *OpenNano*, *12*, 100165. DOI: 10.1016/j.onano.2023.100165

El-Leithy, E. S.-R. (2019). Tamoxifen citrate/coenzyme Q10 as smart nanocarriers bitherapy for breast cancer: Cytotoxicity, genotoxicity, and antioxidant activity. *Journal of Drug Delivery Science and Technology*, *51*, 36–44.

Eslahi, M., Dana, P. M., Asemi, Z., Hallajzadeh, J., Mansournia, M. A., & Yousefi, B. (2021). The effects of chitosan-based materials on glioma: Recent advances in its applications for diagnosis and treatment. *International Journal of Biological Macromolecules*, *168*, 124–129. DOI: 10.1016/j.ijbiomac.2020.11.180 PMID: 33275978

Fathy, M. M., Mohamed, F. S., Elbialy, N., & Elshemey, W. M. (2018). Multifunctional Chitosan-Capped Gold Nanoparticles for enhanced cancer chemo-radiotherapy: An invitro study. *Physica Medica*, *48*, 76–83. DOI: 10.1016/j.ejmp.2018.04.002 PMID: 29728233

Friedman, A. J., Phan, J., Schairer, D. O., Champer, J., Qin, M., Pirouz, A., Blecher-Paz, K., Oren, A., Liu, P. T., Modlin, R. L., & Kim, J. (2013). Antimicrobial and Anti-Inflammatory Activity of Chitosan–Alginate Nanoparticles: A Targeted Therapy for Cutaneous Pathogens. *The Journal of Investigative Dermatology*, *133*(5), 1231–1239. DOI: 10.1038/jid.2012.399 PMID: 23190896

Futamura, M., Nagao, Y., Ishihara, K., Takeuchi, M., Nakada, T., Kawaguchi, Y., Asano, M., Kumazawa, I., Shiroko, T., Morimitsu, K., Mori, R., Nawa, M., Shimokawa, T., & Yoshida, K. (2017). Preoperative neoadjuvant chemotherapy using nanoparticle albumin-bound paclitaxel followed by epirubicin and cyclophosphamide for operable breast cancer: A multicenter phase II trial. *Breast Cancer (Tokyo, Japan)*, *24*(4), 615–623. DOI: 10.1007/s12282-016-0748-6 PMID: 28050738

Gallo, J., Raska, M., Kriegova, E., & Goodman, S. B. (2017). Inflammation and its resolution and the musculoskeletal system. *Journal of Orthopaedic Translation*, *10*, 52–67. DOI: 10.1016/j.jot.2017.05.007 PMID: 28781962

Gao, S. D.-H. (2009). The effect of chemical modification and nanoparticle formulation on stability and biodistribution of siRNA in mice. *Molecular Therapy : the Journal of the American Society of Gene Therapy*, *17*(7), 1225–1233. PMID: 19401674

Ge, Z. (2004). Hydroxyapatite–chitin materials as potential tissue engineered bone substitutes. *Biomaterials*, *25*(6), 1049–1058. DOI: 10.1016/S0142-9612(03)00612-4 PMID: 14615170

Ghaz-Jahanian, M. A., Abbaspour-Aghdam, F., Anarjan, N., Berenjian, A., & Jafarizadeh-Malmiri, H. (2015). Application of Chitosan-Based Nanocarriers in Tumor-Targeted Drug Delivery. *Molecular Biotechnology*, *57*(3), 201–218. DOI: 10.1007/s12033-014-9816-3 PMID: 25385004

Gholap, A. D. (2024). Exploring modified chitosan-based gene delivery technologies for therapeutic advancements. *International Journal of Biological Macromolecules*, *260*, 129581. PMID: 38266848

Grosso, R., & de-Paz, M.-V. (2021). Thiolated-Polymer-Based Nanoparticles as an Avant-Garde Approach for Anticancer Therapies—Reviewing Thiomers from Chitosan and Hyaluronic Acid. *Pharmaceutics*, *13*(6), 854. DOI: 10.3390/pharmaceutics13060854 PMID: 34201403

Han, H. D. (2011). Chitosan hydrogel for localized gene silencing. *Cancer Biology & Therapy*, *11*(9), 839–845. PMID: 21358280

Hefnawy, A. K.-S. (2020). Dual-ligand functionalized core-shell chitosan-based nanocarrier for hepatocellular carcinoma-targeted drug delivery. *International Journal of Nanomedicine*, •••, 821–837. PMID: 32103939

Heidari, R. A.-R. (2021). A systematic review of long non-coding RNAs with a potential role in Breast Cancer. *Mutation Research/Reviews in Mutation Research, 787*, 108375. PMID: 34083033

Helmi, O. E. (2021). Targeted doxorubicin delivery and release within breast cancer environment using PEGylated chitosan nanoparticles labeled with monoclonal antibodies. *International Journal of Biological Macromolecules, 184*, 325–338. PMID: 34119547

Herdiana, Y., Husni, P., Nurhasanah, S., Shamsuddin, S., & Wathoni, N. (2023). Chitosan-Based Nano Systems for Natural Antioxidants in Breast Cancer Therapy. *Polymers, 15*(13), 2953. DOI: 10.3390/polym15132953 PMID: 37447598

Hobson, D. W. (2016). *The Commercialization of Medical Nanotechnology for Medical Applications.*, DOI: 10.1007/978-3-319-43525-1_17

Hu, H., Qi, Q., Dong, Z., Yu, X., Mo, Y., Luo, J., Wang, Y., Du, S., & Lu, Y. (2020). Albumin coated trimethyl chitosan-based targeting delivery platform for photothermal/chemo-synergistic cancer therapy. *Carbohydrate Polymers, 241*, 116335. DOI: 10.1016/j.carbpol.2020.116335 PMID: 32507205

Iacob, A. T., Lupascu, F. G., Apotrosoaei, M., Vasincu, I. M., Tauser, R. G., Lupascu, D., Giusca, S. E., Caruntu, I.-D., & Profire, L. (2021). Recent Biomedical Approaches for Chitosan Based Materials as Drug Delivery Nanocarriers. *Pharmaceutics, 13*(4), 587. DOI: 10.3390/pharmaceutics13040587 PMID: 33924046

Ichim, T. E. (2004). RNA interference: A potent tool for gene-specific therapeutics. *American Journal of Transplantation : Official Journal of the American Society of Transplantation and the American Society of Transplant Surgeons, 4*(8), 1227–1236. PMID: 15268723

Jaber, H. A., Ibrahim, S. A., & Elmansi, M. (2022). The role of chitosan as an antiviral agent. *Antiviral Chemistry & Chemotherapy, 30*, 204020662210991.

Jeandupeux, E. (2018). *Utilisation De l'Acide Poly (Propyl Acrylique) Pour Augmenter l'Efficacité in Vitro Des Nanoparticules à Base De Chitosane Et ARN Messager*. Ecole Polytechnique.

Jhundoo, H. D., Siefen, T., Liang, A., Schmidt, C., Lokhnauth, J., Béduneau, A., Pellequer, Y., Larsen, C. C., & Lamprecht, A. (2020). Anti-Inflammatory Activity of Chitosan and 5-Amino Salicylic Acid Combinations in Experimental Colitis. *Pharmaceutics, 12*(11), 1038. DOI: 10.3390/pharmaceutics12111038 PMID: 33138176

Karuppaiah, A. B. (2021). Building and behavior of a pH-stimuli responsive chitosan nanoparticles loaded with folic acid conjugated gemcitabine silver colloids in MDA-MB-453 metastatic breast cancer cell line and pharmacokinetics in rats. *European Journal of Pharmaceutical Sciences, 165*, 105938. PMID: 34256103

Kashkooli, F. M. (2020). Controlled anti-cancer drug release through advanced nano-drug delivery systems: Static and dynamic targeting strategies. *Biotechnology Reports (Amsterdam, Netherlands).*

Katsuta, E., Rashid, O. M., & Takabe, K. (2020). Clinical relevance of tumor microenvironment: Immune cells, vessels, and mouse models. *Human Cell, 33*(4), 930–937. DOI: 10.1007/s13577-020-00380-4 PMID: 32507979

Khatri, K. G. (2008). Plasmid DNA loaded chitosan nanoparticles for nasal mucosal immunization against hepatitis B. *International Journal of Pharmaceutics*, *354*(1-2), 235–241. PMID: 18182259

Kong, X., Chen, H., Li, F., Zhang, F., Jiang, Y., Song, J., Sun, Y., Zhao, B., & Shi, J. (2023). Three-dimension chitosan hydrogel loading melanin composite nanoparticles for wound healing by anti-bacteria, immune activation and macrophage autophagy promotion. *International Journal of Biological Macromolecules*, *237*, 124176. DOI: 10.1016/j.ijbiomac.2023.124176 PMID: 37023589

Kordbacheh, H. B. (2024). Co-delivery of Bcl-2 siRNA and doxorubicin using liposome-incorporated poly (ε-caprolactone)/chitosan nanofibers for the treatment of lung cancer. *Journal of Drug Delivery Science and Technology*, *99*, 105994.

Kozielski, K. L.-V.-C. (2019). Cancer-selective nanoparticles for combinatorial siRNA delivery to primary human GBM in vitro and in vivo. *Biomaterials*, *209*, 79–87. PMID: 31026613

Kruspe, S. (2017). Aptamer-siRNA Chimeras: Discovery, Progress, and Future Prospects. *Biomedicines*, *209*, 79–87. PMID: 28792479

Kumar, A. (2024). Chitosan-based drug conjugated nanocomposites: Advances and innovation in cancer therapy. *Regenerative Engineering and Translational Medicine*, *10*(1), 1–8.

Lara-Velazquez, M. A.-L.-C.-E. (2020). Chitosan-based non-viral gene and drug delivery systems for brain cancer. *Frontiers in Neurology*. PMID: 32849207

Lee, S. J. (2011). Comparative study of photosensitizer loaded and conjugated glycol chitosan nanoparticles for cancer therapy. *Journal of Controlled Release*, *152*(1), 21–29. PMID: 21457740

Lee, Y., Park, Y., Lee, S., Ku, Y., Han, S., Klokkevold, P. R., & Chung, C. (2000). The Bone Regenerative Effect of Platelet-Derived Growth Factor-BB Delivered With a Chitosan/Tricalcium Phosphate Sponge Carrier. *Journal of Periodontology*, *71*(3), 418–424. DOI: 10.1902/jop.2000.71.3.418 PMID: 10776929

Li, M., Zhao, Y., Zhang, W., Zhang, S., & Zhang, S. (2021). Multiple-therapy strategies via polysaccharides-based nano-systems in fighting cancer. *Carbohydrate Polymers*, *269*, 118323. DOI: 10.1016/j.carbpol.2021.118323 PMID: 34294335

Li, W., Zhang, M., & Zhang, L. (2020). Chitosan and its derivatives in cancer therapy: Mechanisms of action and potential applications. *Current Drug Targets*, *21*(3), 253–263.

Li, X., Dong, W., Nalin, A. P., Wang, Y., Pan, P., Xu, B., Zhang, Y., Tun, S., Zhang, J., Wang, L.-S., He, X., Caligiuri, M. A., & Yu, J. (2018). The natural product chitosan enhances the anti-tumor activity of natural killer cells by activating dendritic cells. *OncoImmunology*, *7*(6). Advance online publication. DOI: 10.1080/2162402X.2018.1431085 PMID: 29872557

Li, Yali, Chen, L., Liu, Y., Zhang, Y., Liang, Y., & Mei, Y. (2018). Anti-inflammatory effects in a mouse osteoarthritis model of a mixture of glucosamine and chitooligosaccharides produced by bi-enzyme single-step hydrolysis. *Scientific Reports*, *8*(1), 5624. DOI: 10.1038/s41598-018-24050-6 PMID: 29618773

Li, Yu., Liu, H., Xu, Q.-S., Du, Y.-G., & Xu, J. (2014). Chitosan oligosaccharides block LPS-induced O-GlcNAcylation of NF-κB and endothelial inflammatory response. *Carbohydrate Polymers*, *99*, 568–578. DOI: 10.1016/j.carbpol.2013.08.082 PMID: 24274545

Li, Z. T. (2010). Chitosan-graft-polyethylenimine with improved properties as a potential gene vector. *Carbohydrate Polymers*, *80*(1), 254–259.

Liang, X., Li, L., Li, X., He, T., Gong, S., Zhu, S., Zhang, M., Wu, Q., & Gong, C. (2021). A spontaneous multifunctional hydrogel vaccine amplifies the innate immune response to launch a powerful antitumor adaptive immune response. *Theranostics*, *11*(14), 6936–6949. DOI: 10.7150/thno.58173 PMID: 34093863

Lima, B. V., Oliveira, M. J., Barbosa, M. A., Gonçalves, R. M., & Castro, F. (2021). Immunomodulatory potential of chitosan-based materials for cancer therapy: A systematic review of *in vitro, in vivo* and clinical studies. *Biomaterials Science*, *9*(9), 3209–3227. DOI: 10.1039/D0BM01984D PMID: 33949372

Lingnert, H., Vallentin, K., & Eriksson, C. E. (1979). Measurement of antioxidative effect in model system. *Journal of Food Processing and Preservation*.

Liu, J., Yang, C., & Zhou, S. (2016). Polyethylene glycol-modified chitosan nanoparticles for tissue engineering and cancer therapy applications. *Bioengineering & Translational Medicine*, *1*(2), 153–162.

Lu, H. D. (2014). Chitosan-graft-polyethylenimine/DNA nanoparticles as novel non-viral gene delivery vectors targeting osteoarthritis. *PLoS One*. PMID: 24392152

Lu, R., Yoshida, T., Nakashima, H., Premanathan, M., Aragaki, R., Mimura, T., Kaneko, Y., Yamamoto, N., Miyakoshi, T., & Uryu, T. (2000). Specific biological activities of Chinese lacquer polysaccharides. *Carbohydrate Polymers*, *43*(1), 47–54. DOI: 10.1016/S0144-8617(99)00198-8

Ma, J.-X., Qian, L., & Zhou, Y. (2015). Stimulation effect of chitosan on the immunity of radiotherapy patients suffered from lung cancer. *International Journal of Biological Macromolecules*, *72*, 195–198. DOI: 10.1016/j.ijbiomac.2014.08.027 PMID: 25175732

Machałowski, T., Rusak, A., Wiatrak, B., Haczkiewicz-Leśniak, K., Popiel, A., Jaroszewicz, J., Żak, A., Podhorska-Okołów, M., & Jesionowski, T. (2021). Naturally Formed Chitinous Skeleton Isolated from the Marine Demosponge Aplysina fistularis as a 3D Scaffold for Tissue Engineering. *Materials (Basel)*, *14*(11), 2992. DOI: 10.3390/ma14112992 PMID: 34205950

Mawazi, S. M., Kumar, M., Ahmad, N., Ge, Y., & Mahmood, S. (2024). Recent Applications of Chitosan and Its Derivatives in Antibacterial, Anticancer, Wound Healing, and Tissue Engineering Fields. *Polymers*, *16*(10), 1351. DOI: 10.3390/polym16101351 PMID: 38794545

Mikušová, V., & Mikuš, P. (2021). Advances in Chitosan-Based Nanoparticles for Drug Delivery. *International Journal of Molecular Sciences*, *22*(17), 9652. DOI: 10.3390/ijms22179652 PMID: 34502560

Moramkar, N. (2021). Insight into chitosan derived nanotherapeutics for anticancer drug delivery and imaging. *European Polymer Journal*, *154*, 110540.

Moro, S. (2023). *Développement d'une nouvelle approche théranostique ciblant la réponse télomérique dans les cellules cancéreuses radio-résistantes du poumon.*

Mousazadeh, H. P.-S. (2021). Cyclodextrin based natural nanostructured carbohydrate polymers as effective non-viral siRNA delivery systems for cancer gene therapy. *Journal of Controlled Release*, *330*, 1046–1070. PMID: 33188829

Naghibi Beidokhti, H. R. (2017). Preparation, characterization, and optimization of folic acid-chitosan-methotrexate core-shell nanoparticles by box-behnken design for tumor-targeted drug delivery. *AAPS PharmSciTech, 18*, 115–129. PMID: 26896317

Nandgude, T., & Pagar, R. (2021). Plausible role of chitosan in drug and gene delivery against resistant breast cancer cells. *Carbohydrate Research, 506*, 108357. DOI: 10.1016/j.carres.2021.108357 PMID: 34146935

Nemati, M. B. (2021). Unraveling the effect of breast cancer patients' plasma on the targeting ability of folic acid-modified chitosan nanoparticles. *Molecular Pharmaceutics, 18*(12), 4341–4353. PMID: 34779630

Nikkhoo, A. R.-N. (2020). Codelivery of STAT3 siRNA and BV6 by carboxymethyl dextran trimethyl chitosan nanoparticles suppresses cancer cell progression. *International Journal of Pharmaceutics, 581*, 119236. PMID: 32240809

Ota, D., Akatsuka, S., Nishi, T., Kato, T., Takeuchi, M., Tsuji, M., & Fukuuchi, A. (2019). Phase I study of combination therapy with weekly nanoparticle albumin-bound paclitaxel and cyclophosphamide in metastatic breast cancer patients. *Anticancer Research, 39*(12), 6903–6907. DOI: 10.21873/anticanres.13910 PMID: 31810960

Park, J. K., Chung, M. J., Choi, H. N., & Park, Y. Il. (2011). Effects of the Molecular Weight and the Degree of Deacetylation of Chitosan Oligosaccharides on Antitumor Activity. *International Journal of Molecular Sciences, 12*(1), 266–277. DOI: 10.3390/ijms12010266 PMID: 21339986

Pavis, H., Wilcock, A., Edgecombe, J., Carr, D., Manderson, C., Church, A., & Fisher, A. (2002). Pilot study of nasal morphine-chitosan for the relief of breakthrough pain in patients with cancer. *Journal of Pain and Symptom Management, 24*(6), 598–602. DOI: 10.1016/S0885-3924(02)00522-5 PMID: 12551810

Prabaharan, M., & Mano, J. F. (2004). Chitosan-Based Particles as Controlled Drug Delivery Systems. *Drug Delivery, 12*(1), 41–57. DOI: 10.1080/10717540590889781 PMID: 15801720

Qu, D. J. (2020). Anisamide-functionalized pH-responsive amphiphilic chitosan-based paclitaxel micelles for sigma-1 receptor targeted prostate cancer treatment. *Carbohydrate Polymers, 229*, 115498. PMID: 31826492

Quiñones, J. P., Peniche, H., & Peniche, C. (2018). Chitosan based self-assembled nanoparticles in drug delivery. *Polymers, 10*(3), 1–32. DOI: 10.3390/polym10030235 PMID: 30966270

Rabišková, M., Bautzová, T., Gajdziok, J., Dvořáčková, K., Lamprecht, A., Pellequer, Y., & Spilková, J. (2012). Coated chitosan pellets containing rutin intended for the treatment of inflammatory bowel disease: In vitro characteristics and in vivo evaluation. *International Journal of Pharmaceutics, 422*(1–2), 151–159. DOI: 10.1016/j.ijpharm.2011.10.045 PMID: 22079717

Rodrigues, M. Á. V., Marangon, C. A., & Martins, V. da C. A., & Plepis, A. M. de G. (2021). Chitosan/gelatin films with jatobá resin: Control of properties by vegetal resin inclusion and degree of acetylation modification. *International Journal of Biological Macromolecules, 182*, 1737–1745. DOI: 10.1016/j.ijbiomac.2021.05.160 PMID: 34051262

Rodríguez, F. C. (2022). Nano-based approved pharmaceuticals for cancer treatment: Present and future challenges. *Biomolecules*, *12*(6), 784. PMID: 35740909

Rondon-Cavanzo, E. P. (2022). Biocompatibilité des nanoparticules de chitosane-siRNA in vitro pour la thérapie génique.

Safdar, R., Omar, A. A., Arunagiri, A., Regupathi, I., & Thanabalan, M. (2019). Potential of Chitosan and its derivatives for controlled drug release applications – A review. *Journal of Drug Delivery Science and Technology, 49*(July 2018), 642–659. doi: DOI: 10.1016/j.jddst.2018.10.020

Salehi, F., Behboudi, H., Kavoosi, G., & Ardestani, S. K. (2017). Chitosan promotes ROS-mediated apoptosis and S phase cell cycle arrest in triple-negative breast cancer cells: Evidence for intercalative interaction with genomic DNA. *RSC Advances*, *7*(68), 43141–43150. DOI: 10.1039/C7RA06793C

Salehi, M., Khodadadi, A., & Sadeghi, F. (2017). The potential of chitosan nanoparticles in anticancer therapy: A review of biological activities. *Cancer Nanotechnology*, *22*(4), 369–382.

Şalva, E. K. (2012). Local delivery of chitosan/VEGF siRNA nanoplexes reduces angiogenesis and growth of breast cancer in vivo. *Nucleic Acid Therapeutics*, *22*(1), 40–48. PMID: 22217324

Şalva, E. T. (2015). The enhancement of gene silencing efficiency with chitosan-coated liposome formulations of siRNAs targeting HIF-1α and VEGF. *International Journal of Pharmaceutics*, *478*(1), 147–154. PMID: 25445537

Santos, J. D.-F. (2021). Biopolymeric materials used as nonviral vectors: A review. *Polysaccharides*, *2*(1), 100–109.

ŞENEL, B. (2019). In vitro preliminary studies of chitooligosaccharide coated nanostructured lipidic nanoparticles for efficient gene delivery. *Journal of Research in Pharmacy*, *23*(4).

Shakil, M. S. (2021). Using chitosan or chitosan derivatives in cancer therapy. *Polysaccharides*, *2*(4), 795–816.

Shan, X. Y. (2020). Effect of chitosan magnetic nanoparticles loaded with Ang2-siRNA plasmids on the growth of melanoma xenografts in nude mice. *Cancer Management and Research*, •••, 7475–7485. PMID: 32904466

Shanmuganathan, R., & Edison, T. N. J. I., LewisOscar, F., Kumar, P., Shanmugam, S., & Pugazhendhi, A. (2019a). Chitosan nanopolymers: An overview of drug delivery against cancer. *International Journal of Biological Macromolecules*, *130*, 727–736. DOI: 10.1016/j.ijbiomac.2019.02.060 PMID: 30771392

Shanmuganathan, R., & Edison, T. N. J. I., LewisOscar, F., Kumar, P., Shanmugam, S., & Pugazhendhi, A. (2019b). Chitosan nanopolymers: An overview of drug delivery against cancer. *International Journal of Biological Macromolecules*, *130*, 727–736. DOI: 10.1016/j.ijbiomac.2019.02.060 PMID: 30771392

Sharma, A. T. (2019). Commercial application of cellulose nano-composites–A review. *Biotechnology Reports (Amsterdam, Netherlands)*, *21*, e00316. PMID: 30847286

Siahmansouri, H. S.-N. (2016). Effects of HMGA2 siRNA and doxorubicin dual delivery by chitosan nanoparticles on cytotoxicity and gene expression of HT-29 colorectal cancer cell line. *The Journal of Pharmacy and Pharmacology*, *68*(9), 1119–1130. PMID: 27350211

Silva, A. C. (2024). The Role of Inhaled Chitosan-Based Nanoparticles in Lung Cancer Therapy. *Pharmaceutics*, *16*(8), 969. PMID: 39204314

Singh, J., & Nayak, P. (2023). pH-responsive polymers for drug delivery: Trends and opportunities. *Journal of Polymer Science*, *61*(22), 2828–2850. DOI: 10.1002/pol.20230403

Strebhardt, K. (2008). Paul Ehrlich's magic bullet concept: 100 years of progress. *Nature Reviews. Cancer*, *8*(6), 473–480. PMID: 18469827

Subhan, M. A. (2023). Biopolymer-based nanosystems for siRNA drug delivery to solid tumors including breast cancer. *Pharmaceutics*, *15*(1), 153. PMID: 36678782

Sun, T., Zhou, D., Mao, F., & Zhu, Y. (2007). Preparation of low-molecular-weight carboxymethyl chitosan and their superoxide anion scavenging activity. *European Polymer Journal*, *43*(2), 652–656. DOI: 10.1016/j.eurpolymj.2006.11.014

Twal, S. J.-R.-A. (2024). Dual stimuli-responsive polymeric nanoparticles combining soluplus and chitosan for enhanced breast cancer targeting. *RSC Advances*, *14*(5), 3070–3084. PMID: 38239437

Ueno, H., Yamada, H., Tanaka, I., Kaba, N., Matsuura, M., Okumura, M., Kadosawa, T., & Fujinaga, T. (1999). Accelerating effects of chitosan for healing at early phase of experimental open wound in dogs. *Biomaterials*, *20*(15), 1407–1414. DOI: 10.1016/S0142-9612(99)00046-0 PMID: 10454012

Van Hees, S., Elbrink, K., De Schryver, M., Delputte, P. L., & Kiekens, F. (2020). Improving Cellular Uptake and Cytotoxicity of Chitosan-Coated Poly(Lactic- *Co* -Glycolic Acid) Nanoparticles in Macrophages. *Nanomedicine (London)*, *15*(27), 2671–2688. DOI: 10.2217/nnm-2020-0317 PMID: 33112210

Virmani, T. K. (2023). Amelioration of cancer employing chitosan, its derivatives, and chitosan-based nanoparticles: Recent updates. *Polymers*, *15*(13), 2928. PMID: 37447573

Virmani, T., Kumar, G., Sharma, A., Pathak, K., Akhtar, M. S., Afzal, O., & Altamimi, A. S. A. (2023). Amelioration of Cancer Employing Chitosan, Its Derivatives, and Chitosan-Based Nanoparticles: Recent Updates. *Polymers*, *15*(13), 2928. DOI: 10.3390/polym15132928 PMID: 37447573

Wang, G., Li, R., Parseh, B., & Du, G. (2021). Prospects and challenges of anticancer agents' delivery via chitosan-based drug carriers to combat breast cancer: A review. *Carbohydrate Polymers*, *268*(May). Advance online publication. DOI: 10.1016/j.carbpol.2021.118192 PMID: 34127212

Wang, W., & Xu, D. (1994). Viscosity and flow properties of concentrated solutions of chitosan with different degrees of deacetylation. *International Journal of Biological Macromolecules*, *16*(3), 149–152. DOI: 10.1016/0141-8130(94)90042-6 PMID: 7981162

Webster, T. J. (2006). Nanomedicine: What's in a definition? *International Journal of Nanomedicine*, *1*(2), 115–116. PMID: 17722527

Wei, T. (2015). *Systèmes innovants de délivrance de médicaments basés sur des nanomicelles pour le traitement du cancer.*

Wu, C., Dai, Y., Yuan, G., Su, J., & Liu, X. (2019). Immunomodulatory Effects and Induction of Apoptosis by Different Molecular Weight Chitosan Oligosaccharides in Head Kidney Macrophages From Blunt Snout Bream (Megalobrama amblycephala). *Frontiers in Immunology, 10*. Advance online publication. DOI: 10.3389/fimmu.2019.00869 PMID: 31156612

Xu, Y. X., Kim, K. M., Hanna, M. A., & Nag, D. (2005). Chitosan–starch composite film: Preparation and characterization. *Industrial Crops and Products*, *21*(2), 185–192. DOI: 10.1016/j.indcrop.2004.03.002

Yan, T. Z. (2020). Chitosan based pH-responsive polymeric prodrug vector for enhanced tumor targeted co-delivery of doxorubicin and siRNA. *Carbohydrate Polymers*, *250*, 116781. PMID: 33049806

Yang, J. M. (2004). Evaluation of chitosan/PVA blended hydrogel membranes. journal of Membrane Science,, 236(1-2), 39-51.

Yang, L., & Zaharoff, D. A. (2013). Role of chitosan co-formulation in enhancing interleukin-12 delivery and antitumor activity. *Biomaterials*, *34*(15), 3828–3836. DOI: 10.1016/j.biomaterials.2013.02.031 PMID: 23453060

Yang, Y., Xing, R., Liu, S., Qin, Y., Li, K., Yu, H., & Li, P. (2019a). Hydroxypropyltrimethyl ammonium chloride chitosan activates RAW 264.7 macrophages through the MAPK and JAK-STAT signaling pathways. *Carbohydrate Polymers*, *205*, 401–409. DOI: 10.1016/j.carbpol.2018.10.101 PMID: 30446121

Yang, Y., Xing, R., Liu, S., Qin, Y., Li, K., Yu, H., & Li, P. (2019b). Immunostimulatory Effects of Chitooligosaccharides on RAW 264.7 Mouse Macrophages via Regulation of the MAPK and PI3K/Akt Signaling Pathways. *Marine Drugs*, *17*(1), 36. DOI: 10.3390/md17010036 PMID: 30626153

Yao, Z., Li, H., & Zhang, L. (2013). Mannosylated chitosan nanoparticles enhance DNA vaccine efficacy in fish vaccination. *Fish & Shellfish Immunology*, *34*(4), 981–988. PMID: 22960101

Yen, M.-T., Yang, J.-H., & Mau, J.-L. (2008). Antioxidant properties of chitosan from crab shells. *Carbohydrate Polymers*, *74*(4), 840–844. DOI: 10.1016/j.carbpol.2008.05.003

Yousef, M., Pichyangkura, R., Soodvilai, S., Chatsudthipong, V., & Muanprasat, C. (2012). Chitosan oligosaccharide as potential therapy of inflammatory bowel disease: Therapeutic efficacy and possible mechanisms of action. *Pharmacological Research*, *66*(1), 66–79. DOI: 10.1016/j.phrs.2012.03.013 PMID: 22475725

Yun, U. J. (2019). Anti-cancer effect of doxorubicin is mediated by downregulation of HMG-Co A reductase via inhibition of EGFR/Src pathway. *Laboratory Investigation*, *99*(8), 1157–1172. PMID: 30700846

Zarrintaj, P. M. (2017). Can regenerative medicine and nanotechnology combine to heal wounds? The search for the ideal wound dressing. *Nanomedicine (London)*, *12*(19), 2403–2422. PMID: 28868968

Zavareh, H. S. (2020). Chitosan/carbon quantum dot/aptamer complex as a potential anticancer drug delivery system towards the release of 5-fluorouracil. *International Journal of Biological Macromolecules*, *165*, 1422–1430. PMID: 32987067

Zeng, Z., Liu, Y., Wen, Q., Li, Y., Yu, J., Xu, Q., Wan, W., He, Y., Ma, C., Huang, Y., Yang, H., Jiang, O., & Li, F. (2021). Experimental study on preparation and anti-tumor efficiency of nanoparticles targeting M2 macrophages. *Drug Delivery*, *28*(1), 943–956. DOI: 10.1080/10717544.2021.1921076 PMID: 33988472

Zhang, B. C. (2020). Efficient CRISPR/Cas9 gene-chemo synergistic cancer therapy via a stimuli-responsive chitosan-based nanocomplex elicits anti-tumorigenic pathway effect. *Chemical Engineering Journal*, *393*, 124688.

Zhang, L., Luo, J., & Zhang, Z. (2017). Chitosan as a promising agent for selective cancer cell targeting and inhibition of tumor growth. *Scientific Reports*, *7*(1), 10262. PMID: 28860532

Zhang, W., Yu, X., & Cheng, S. (2023). Mannosylated chitosan as an effective adjuvant for DNA vaccine development. *Vaccine*, *41*(6), 862–871.

Zhao, Y., Zheng, J., & Wang, X. (2019). Chitosan nanoparticles for targeted anticancer drug delivery: Design and development. *Journal of Nanobiotechnology*, *17*(1), 10. PMID: 30670021

Zhou, L. D. (2010). The antitumor and antimetastatic effects of N-trimethyl chitosan-encapsulated camptothecin on ovarian cancer with minimal side effects. *Oncology Reports*, *24*(4), 941–948. PMID: 20811674

Zhu, X. Y. (2021). Chitosan-based nanoparticle co-delivery of docetaxel and curcumin ameliorates anti-tumor chemoimmunotherapy in lung cancer. *Carbohydrate Polymers*, *268*, 118237. PMID: 34127219

Chapter 10
Radioprotective Effects of Chitosan Polysaccharides and Biomedical Applications

Khawla Rachmoune
https://orcid.org/0009-0001-6483-1060
Université Claude Bernard Lyon, France

Saloua Mabsor-Zgandaoui
https://orcid.org/0009-0006-4601-9059
Biology and Health Laboratory, Department of Biology, Faculty of Sciences, Ibn Tofail University, Morocco

Adil El Housseini
https://orcid.org/0009-0007-9294-2685
Laboratory Marine and Continental Environments (AQUAMAR), Faculty of Sciences, Ibn Zohr University, Morocco

Youssef Ait Hamdan
https://orcid.org/0009-0005-8616-8852
Cadi Ayyad University, Morocco

Maryam Bousaid
Laboratory of Materials Science and Composite Polymers(SMPC), UMONS, Belgium

Adil Aknouch
https://orcid.org/0000-0003-4340-6687
LICPM Laboratory, Department of Physics, Faculty of Sciences and Techniques, Sultan Moulay Slimane University, Morocco

Mohammed Rhazi
https://orcid.org/0000-0003-2736-1802
Higher Normal School, Cadi Ayyad University, Morocco

Adem Gharsallaoui
Université Claude Bernard Lyon, France

Abdelghani Iddar
https://orcid.org/0000-0001-6111-0607
Biotechnology and Biomolecules Engineering Unit, CNESTEN, Morocco

Mohammed El Mzibri
Biotechnology and Biomolecules Engineering Unit, CNESTEN, Morocco

Ahmed Moussaif
Biotechnology and Biomolecules Engineering Unit, CNESTEN, Morocco

El Hassan El Mouden
Laboratory of Water, Biodiversity and Climate Change, Faculty of Sciences Semlalia, Morocco

Adnane Moutaouakkil
https://orcid.org/0000-0002-1762-881X
Biotechnology and Biomolecules Engineering Unit, CNESTEN, Morocco

DOI: 10.4018/979-8-3373-0055-9.ch010

ABSTRACT

Ionizing radiation (IR) plays a central role in the biomedical field, both for diagnosis and for the treatment of pathologies such as cancer. However, exposure to IR can cause significant cellular damage, affecting both tumor cells and neighbouring healthy tissue. In this context, the use of effective radioprotective agents is a necessity. Chitosan (CS), a natural biopolymer derived from chitin, stands out for its antioxidant, anti-inflammatory and reparative properties, as well as its biocompatibility and biodegradability. These characteristics make it a promising candidate for radiation protection. This chapter examines the radio-protective mechanisms of chitosan, in particular its ability to trap reactive oxygen species and enhance the activity of antioxidant enzymes. It also highlights its many biomedical applications, particularly in tissue engineering and controlled drug release. Finally, the limits to its clinical use and recent advances aimed at improving its physico-chemical properties are discussed, paving the way for new therapeutic strategies in radiation protection.

INTRODUCTION

Ionizing radiation (IR) is widely used in the biomedical field, both for diagnostic purposes, such as medical imaging, and for therapeutic purposes, such as radiotherapy for the treatment of cancer. However, such exposure can cause significant cellular damage, affecting not only tumor cells but also surrounding healthy tissue. In this context, radioprotectors play a crucial role in helping to limit the side effects of radiation while preserving therapeutic efficacy.

Among the natural compounds studied for their radioprotective potential, chitosan (CS), a biopolymer derived from the deacetylation of chitin, is attracting growing interest. Its antioxidant, anti-inflammatory and reparative properties make it a promising biocompatible alternative to traditional synthetic agents. Because of its biocompatibility, biodegradability and unique bioactive characteristics, chitosan is now emerging as a prime candidate for protecting cells and tissues against the deleterious effects of IR. Its integration into innovative biomedical strategies opens the way to new approaches to protection against radiation-induced damage, both in patients and exposed healthcare professionals (Omer., 2021; Frolova et al., 2020).

CS's radioprotective effects have been ascribed to its potent antioxidant activity, scavenging ROS and reducing oxidative damage to cellular components such as DNA, proteins and lipids (Frolova et al. 2020). Moreover, its cationic nature allowed it to interact with negatively charged cellular structures, stabilizing membranes and shielding against radiation-induced fragmentation. CS has been shown to enhance the body's natural defense mechanisms by stimulating the activity of key antioxidant enzymes such as superoxide dismutase (SOD) and catalase (CAT). This dual action not only mitigates immediate radiation damage, but also supports long-term cellular repair processes (Wang et al. 2024).

Alongside its radioprotective properties, CS has been extensively used in the biomedical field. In drug delivery field, nanoparticles, hydrogels and films have enabled targeted, controlled release of therapeutic agents. These systems improve drug bioavailability and stability while minimizing side effects, making them ideal for a wide range of treatments, including those requiring preside release into damaged or diseased tissue (Tao et al. 2020). In tissue engineering, CS scaffolds and membranes mimic the extra-cellular matrix, promoting cell adhesion, proliferation and differentiation (Georgopoulou et al. 2018).

Hence, their effectiveness in promoting the regeneration of skin, bone and neural tissue, particularly in applications involving radiation-induced damage that has hampered the healing process (Tian et al. 2022).

However, the clinical use of CS was fraught with difficulties, including the variability of its properties depending on source and processing methods, as well as the limitations of large-scale production. Advances in nanotechnology and chemical modifications have opened up new avenues for improving its solubility, stability and bioactivity. These innovations pave the way for developing CS-based solutions tailored to specific clinical needs, from adjuvants for cancer treatment to protective measures in the event of exposure to occupational or cosmic radiation (Jiménez-Gómez and Cecilia 2020).

This chapter explores the potential of chitosan as a natural radioprotective agent, while highlighting its multiple applications in the medical field. This interest is based on its remarkable properties, notably its biocompatibility, biodegradability and antioxidant activity, which give it the ability to effectively protect tissues and cells against the deleterious effects of IR. To develop this theme, this chapter is structured as follows: it begins with an overview of ionizing radiation and its biological effects. This is followed by a presentation of the general characteristics of chitosan, including its physicochemical and biological properties. The next section discusses in detail the radioprotective mechanisms of chitosan and reviews the experimental evidence supporting its effectiveness. Finally, current biomedical applications of chitosan in radioprotection are examined, along with future perspectives in this field.

1. CHITOSAN - PROPERTIES AND GENERAL APPLICATIONS

1.1 Chitosan

Chitin, recognized as the second most prevalent biopolymer on Earth after cellulose, is mainly present in a range of eukaryotic organisms such as crustaceans, insects, and fungi (Crini, 2019). Structurally, chitin consists of long chains of N-acetyl-β-D-glucosamine monomers, which are linked together by β-(1→4) glycosidic bonds, as illustrated in **Figure 1**. This structure is formally named poly β-(1→4)-2-acetamido-2-deoxy-D-glucopyranose. When chitin undergoes deacetylation, it transforms into its derivative, chitosan, which is composed of β-(1→4)-D-glucosamine units. Chitosan is characterized by three reactive functional groups: an amine group located at the C2 position, along with hydroxyl groups at the C3 and C6 positions (Kean and Thanou, 2011; Younes and Rinaudo, 2015).

Numerous researchers and major enterprises across various sectors including food, pharmaceuticals, medicine, cosmetics, agriculture, textiles (Morin-Crini et al., 2019), and water treatment (Bhatt et al., 2023) have shown great interest in the biopolymer chitosan due to its remarkable and unique properties. Its solubility and other characteristics are influenced by the presence of functional groups such as hydroxyl and amine groups, which can be protonated, making chitosan soluble under dilute acidic conditions (Daraghmeh et al., 2011; Hakima et al., 2017; Bonilla et al., 2014). Studies have shown that chitosan, while insoluble in water and most organic solvents, dissolves effectively in several acidic aqueous media such as acetic, lactic, formic, and citric acids, as well as in certain other specific solvents (Shajahan et al., 2017).

In addition to these considerations, it has been shown that reducing the molecular weight and controlling the degree of deacetylation enhance chitosan's solubility, albeit at the cost of yield. Solubility can also be improved through physical and chemical modifications, such as reacetylation or the use of adjuvants capable of disrupting hydrogen bonds (Hamedi et al., 2018). Furthermore, the generation of low-molecular-weight derivatives or the incorporation of small reactive groups contributes to increased solubility. Beyond solubility, biocompatibility and biodegradability are key features for therapeutic and biomedical applications. Owing to these properties, chitosan represents a strong candidate for medical use, as it can safely interact with human tissues and is progressively degraded and eliminated from the body, thereby minimizing the risk of adverse reactions (Crini, 2019).

Furthermore, in several studies, Chitosan has continuously shown outstanding biocompatibility. When applied to human cells, such as HaCaT keratinocytes(Zhao et al., 2023), chitosan has been shown to be non-toxic, making it a viable option for use in medical devices and wound dressings (Cano-Vicent et al., 2023). Furthermore, chitosan's capacity to engage in constructive interactions with immune cells raises the possibility that it could promote tissue regeneration and healing, thus expanding its therapeutic potential (Jaber et al., 2022). Additionally, according to other research, chitosan breaks down into non-toxic byproducts, which makes it a perfect material for short-term uses in the human body like controlled drug delivery systems and wound dressings (Foster et al., 2015).

Figure 1. Chemical structure of chitin and chitosan

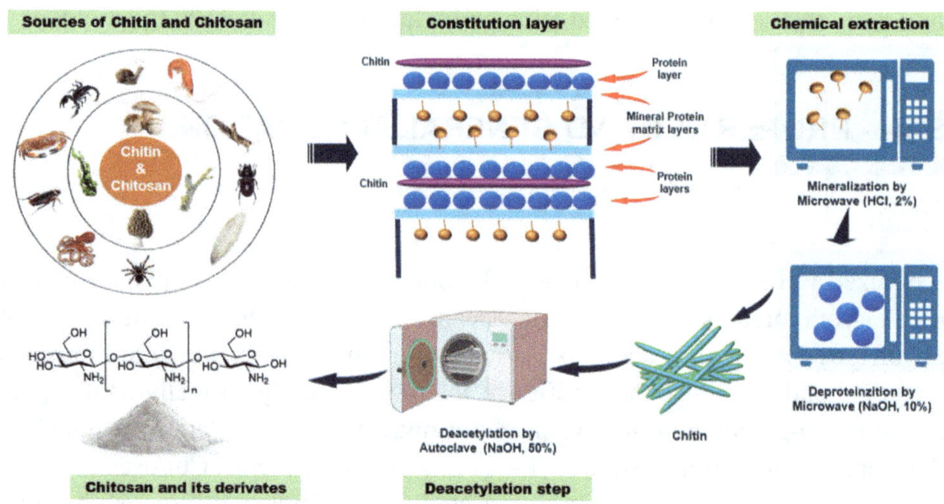

1.2 Antimicrobial Properties

There is a lot of interest in chitosan's antimicrobial properties because of the growing need for natural alternatives to fight off bacteria, fungi, and yeasts. Due to variables like degree of deacetylation (DD) and molecular weight (MW), this biopolymer has shown significant antimicrobial and antifungal properties. Despite its incomplete understanding, chitosan's antibacterial action has been connected to the promotion of Ca^{++} ion migration from bacterial membranes, which causes cellular damage (Husain et al., 2017). It has demonstrated efficacy against oral pathogens such as *Porphyromonas gingivalis*, and *Actinobacillus actinomycetemcomitans*, as well as Gram-positive and Gram-negative bacteria (Jeon,

2001). Furthermore, species like *Candida parapsilosis* have been shown to be susceptible to its antifungal activity (Hussain et al., 2012).

The degree of deacetylation (DD) and molecular weight (MW) of chitosan are important variables that affect its antimicrobial efficacy. Higher levels of deacetylation cause more disruption because they improve electrostatic interactions with bacterial membranes (Brugnerotto et al., 2001). Because it can more easily enter bacterial cells, chitosan with a lower molecular weight has a stronger antibacterial effect. With minimum inhibitory concentrations (MIC) ranging from 10 to 1000 ppm, studies have demonstrated that chitosan inhibits a broad range of microorganisms, including *Salmonella typhimurium*, and *Escherichia coli*. Gram-positive bacteria are typically more sensitive than Gram-negative ones (No et al., 2002). Additionally, chitosan has antifungal qualities; spore development is hindered by high molecular weight chitosan, whereas fungal mycelial growth is inhibited by low molecular weight chitosan. Disrupting RNA and protein synthesis, chelating vital nutrients, destabilizing proteins, and rupturing cell membranes are the main mechanisms behind its antimicrobial action. These versatile properties render chitosan a promising candidate for diverse applications, including agriculture, medicine, and food preservation (Kołodziejska et al., 2021).

1.3 Antioxidant Properties

The potential uses of chitosan are largely driven by its antioxidant activity, especially in the food, cosmetics, and biomedical sectors. Its structural parameters, such as degree of deacetylation (DD) and molecular weight (MW), have a significant impact on this activity. Research has shown that chitosan's antioxidant capacity increases significantly with decreasing molecular weight, emphasizing the critical role of structural optimization in enhancing its functional properties (Schreiber et al., 2013). For instance, Sun et al. (2007) reported that chitosan oligosaccharides (OS-C) exhibit stronger antioxidant activity than higher-molecular-weight chitosan, with activity further improving as chain length decreases. On the other hand, high-MW chitosan shows little protection against reactive species such as hydroxyl radicals and superoxide. Additionally, chitosan derivatives, particularly those grafted with phenolic substances like gallic acid, exhibit significantly enhanced antioxidant properties.

Chitosan can effectively reduce oxidative damage because of its antioxidant activity, which ranges from moderate levels (58.3% to 70.2%) to as high as 85.2% at higher concentrations (Yen et al., 2008). According to the DPPH assay, ultrasonic deacetylation procedures have also been demonstrated to increase its free radical scavenging activity, with rates varying from 18.27% to 44.17% based on concentration (0.25 to 1 mg/mL). This outperforms chitosan made using traditional chemical deacetylation techniques (Yen et al., 2008). Depending on the concentration, chitosan also exhibits a notable inhibition of lipid peroxidation, with rates ranging from 11.7% to 77.39%. This characteristic is essential for shielding cell membranes from oxidative damage and stress. The antioxidant mechanism of chitosan is primarily attributed to the interaction of its amine groups with free radicals, whereby harmful species are neutralized through the formation of stable compounds (Yen et al., 2008; Curcio et al., 2009). These multifunctional antioxidant properties make chitosan a versatile agent with wide-ranging applications. In the biomedical field, its capacity to scavenge free radicals and mitigate oxidative stress renders it suitable for therapeutic strategies targeting conditions associated with oxidative damage. In the food industry, chitosan's antioxidant activity can help extend product shelf life by preventing oxidative degradation. Likewise, in the cosmetics sector, it shows promise for the development of protective skin formulations designed to minimize oxidative damage and promote skin health (Lee et al., 2014).

1.4 Anti-Inflammatory Properties

Having said that, chitosan's anti-inflammatory properties are just as significant in treatment as those of any therapeutic biomaterial (Crini, 2019). Chitosan and its derivatives have been shown in numerous studies to have the capacity to regulate inflammatory responses. Studies on chitosan-alginate nanoparticles showed that they were effective against the acne-causing bacteria Propionibacterium acnes. These nanoparticles prevented human monocytes and keratinocytes from producing pro-inflammatory cytokines. For instance, even at low doses, they virtually suppressed IL-6 in keratinocytes and dose-dependently decreased IL-12p40 in monocytes (Friedman et al., 2013).

Research has shown that chitosan can reduce inflammation through multiple mechanisms. It decreases the expression of membrane metalloproteinases, which degrade the extracellular matrix and exacerbate inflammatory responses, while also inhibiting the activation of the transcription factor NF-κB, a key regulator of inflammatory pathways (Gallo et al., 2017). Chitosan interacts with Toll-like receptors (TLRs) to modulate signaling pathways and reduce the production of inflammatory cytokines. Inhibition of NF-κB further suppresses the expression of genes involved in chronic inflammation. Its previously described antioxidant properties allow chitosan to limit oxidative stress and associated tissue damage by scavenging reactive oxygen species (ROS). Additionally, its antimicrobial activity against skin pathogens such as P. acnes contributes to a less inflammatory environment (Friedman et al., 2013). These combined anti-inflammatory, antioxidant, and antimicrobial properties position chitosan as a promising material for tissue engineering and biomedical applications. While some mechanisms underlying its anti-inflammatory effects have been partially elucidated, further research is needed to optimize formulations and define precise clinical applications. Tailoring chitosan's chemical structure, adjusting its concentration, and refining its mode of administration could further enhance its therapeutic potential (Crini, 2019).

2. MECHANISM OF ACTION OF IR

IR radiation interacts with atoms or molecules in exposed cellular structures, leading to two primary types of reactions: ionization and excitation (Mabsor-Zgandaoui, 2025) (Figure 2). The energy absorbed from IR can damage cellular molecular structures through two main mechanisms: direct action and indirect action.

2.1. Direct Action

IR causes direct ionization of cellular macromolecules, notably DNA, proteins and lipids. This ionization results in a loss of electrons, leading to the breaking of chemical bonds and functional alterations in the molecules concerned. This mechanism is particularly dominant in the case of radiation with a high linear energy transfer (LET), such as alpha (α) particles and neutrons, as well as during exposure to high doses. Direct DNA damage, such as single and double strand breaks, can result in mutations, chromosomal abnormalities and, potentially, cell death (Jia et al., 2021).

2.2. Indirect Action

In this case, IRs interact mainly with water molecules rather than cellular macromolecules. Given that water makes up around 70-80% of the cell mass, the majority of ionization reactions induced by IRs occur at the level of water molecules (Dowlath et al., 2021). This ionization generates highly reactive free radicals, such as hydroxyl radicals (-OH) and hydrogen ions (H^+). Hydroxyl radicals can cause significant damage to macromolecules, particularly DNA, by inducing breaks or mutations. Furthermore, recombination of these radicals can form hydrogen peroxide (H_2O_2), an unstable molecule that can react with cellular organic compounds to form stable organic peroxides (Jomova et al., 2023).

2.3. Bystander Effect

In addition to direct damage to irradiated cells, IR can affect neighbouring non-irradiated cells through the bystander effect. This phenomenon results in the appearance of genomic instabilities and chromosomal rearrangements (Rusin et al., 2018), caused by the release of free radicals, growth factors, and cytokines such as TNF-α (tumor necrosis factor alpha), inosine nucleotides, IL-8 (interleukin-8), TGF-β (transforming growth factor beta), as well as the end products of lipid peroxidation. These signalling molecules participate in the amplification of cellular damage by creating a pro-inflammatory and pro-oxidative environment in surrounding tissues (Burgio et al., 2018).

The effects of IR are not limited to immediate damage. They can also cause long-term alterations, such as genomic instability transmitted to subsequent cell generations, stimulation of oncogenic pathways, and disruptions in DNA repair. This can lead to serious pathologies, including cancers and degenerative diseases, even years after initial exposure. Ionizing radiation thus represents a major risk factor for cellular health, requiring effective approaches to limit its harmful effects, notably through the use of radioprotectants such as chitosan.

Figure 2. Mechanism of action of IR

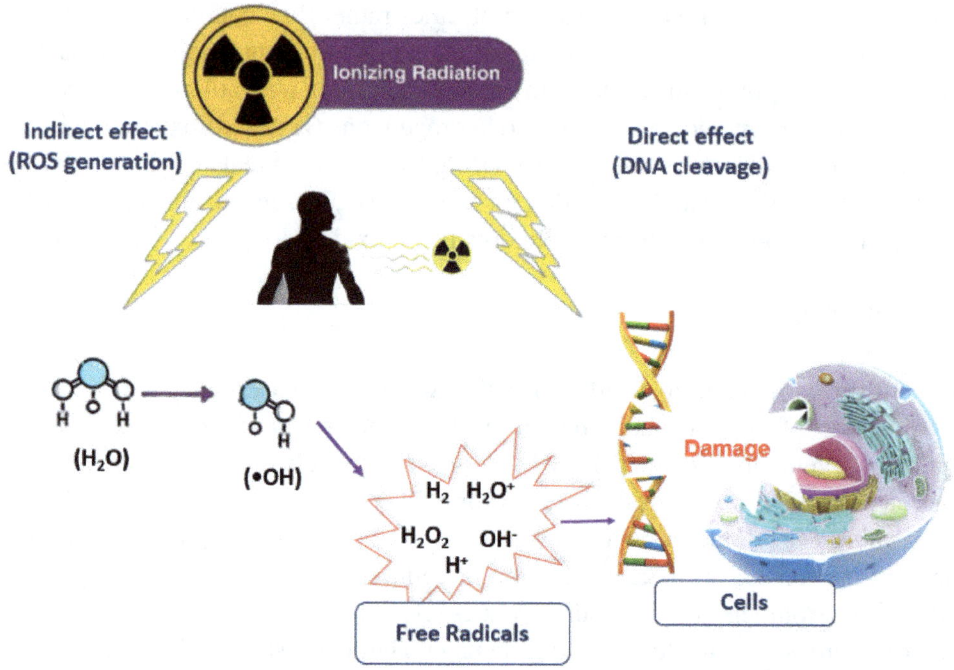

3. RADIOPROTECTIVE EFFECTS OF CHITOSAN

Given the current limitations of radiotherapy in the treatment of malignant tumors limitations of radiotherapy in the treatment of malignant tumors, including the unintended destruction of healthy cells along with cancer cells (Xie et al., 2018), new therapeutic approaches are emerging. Among these, the use of natural molecules, such as biopolymers like chitosan, is attracting growing interest due to their non-toxicity and promising efficacy.

3.1 Mechanisms of Action

Chitosan as a radioprotector can have different modalities of action, in this section, we focus mainly on oxidative stress, the latter which can be defined as an excessive production of reactive oxygen species (ROS) in a way that is uncontrollable by the antioxidant systems of the cell (Sies., 2017). Chitosan exerts a notable protective action by neutralizing reactive oxygen species (ROS) generated by ionizing radiation. These ROS, such as hydroxyl radicals (·OH\cdot OH·OH), peroxide radicals (ROO·ROO \cdotROO·), and superoxide anions (O2−O_2^-O2−), are byproducts of oxidative stress and cause cascading damage to cellular components, including membrane lipids, proteins, and DNA. The functional groups of chitosan, particularly amines (−NH2-NH_2−NH2) and hydroxyls (−OH-OH−OH), have a strong ability to interact with these free radicals through an electron donation mechanism, thus neutralizing their reactivity.

Studies have revealed that the efficacy of chitosan is highly dependent on its degree of deacetylation (DD) and molecular weight. High DD (>85%) and low molecular weight chitosan has demonstrated better free radical scavenging ability due to a higher density of free functional groups; for example, in vitro models, chitosan treatment reduced malondialdehyde (MDA) levels, a key marker of lipid peroxidation, by more than 70%, indicating a significant decrease in oxidative stress (Aranaz et al., 2021).

This scavenger property can be optimized by formulating chitosan into nanoparticles, allowing for increased surface area and improved interaction with ROS (Khan et al., 2021). These nanoparticles can also be functionalized with additional antioxidant agents, further enhancing their efficacy.

In addition, curcumin-loaded chitosan nanoparticles (Cur-CsNPs) have been explored for their radioprotective effect in rats, particularly against liver damage resulting from ionizing radiation, in this sense, researchers have exploited the antioxidant as well as anti-inflammatory activity of curcumin on one side, and the drug delivery power of chitosan on the other side, to mitigate the harmful effects of ionizing radiation on the liver. The characterization of Cur-CsNPs was carried out by different techniques such as: TEM, XRD, DLS and FTIR. The results revealed a remarkable radioprotective effect of Cur-CsNPs (Wu et al., 2020)

These antioxidant properties linked to chitosan, allow it to confer a remarkable radioprotective potential to counterbalance the generation of ROS and free radicals, such as superoxide anion ($O_2^{\bullet -}$), peroxides (H_2O_2) and hydroxyl radicals (•OH), these key substances are markers of oxidative stress in living cells. And consequently, this type of stress leads to several harmful results in the cell, in particular on major macromolecules; lipids, proteins and nucleic acids, which aggravates the pathophysiological situation resulting in multiple genetic mutations and significant damage which gives cellular dysfunctions which will end in neuronal death (apoptosis) (Zhang et al., 2021).

The chemical structure of chitosan gives it this antioxidant power, in particular the presence of glucosamine and N-acetylglucosamine chains, which promotes the trapping of free radicals. This can be explained by the presence of the amine group and the hydroxyl group, which allows this biopolymer to better capture and trap free radicals **(Figure 3)**, thus limiting their propagation and the oxidation of essential macromolecules (Ai et al., 2012). Chitosan also can control the prolonged release of vitamin C, a key element of the non-enzymatic antioxidant response (Hu et al., 2020). This antioxidant power gives chitosan a key role in protecting cells against harmful damage that can be linked to attack by free radicals. For example, the protection of DNA against single and double strand breaks is an essential aspect of the radioprotective properties of chitosan. When cells are exposed to ionizing radiation, chain reactions triggered by ROS lead to structural damage to DNA, causing mutations, chromosomal abnormalities, and sometimes cell apoptosis. Chitosan, through its ability to interact directly with DNA, forms a protective barrier around the double helix. This interaction is based on the cationic properties of chitosan, which allow it to bind to the negative phosphates of the DNA structure, thus stabilizing the strands and limiting breaks. Experiments have shown that administration of chitosan before radiation exposure reduces DNA fragmentation levels by up to 50% in irradiated human fibroblast models (Kim, 2018). In addition, chitosan facilitates the repair of damaged DNA by activating signaling pathways associated with repair proteins, such as Ataxia Telangiectasia Mutated (ATM) and Chk2 kinase. These proteins play a crucial role in detecting and correcting DNA breaks.

Formulations combining chitosan with protective agents such as antioxidant vitamins (e.g. vitamin E) have also shown promise in improving the preservation of genetic integrity in highly radioactive environments. Chitosan also acts indirectly by stimulating the natural antioxidant defenses of cells. In particular, it promotes the activation of key antioxidant enzymes, such as superoxide dismutase (SOD), catalase (CAT) and glutathione peroxidase (GPx) (Herdiana et al., 2023). These enzymes are essential for neutralizing endogenous free radicals and protecting cells from prolonged oxidative damage (**Figure 3).**

Recent research has shown that chitosan influences cellular signaling pathways involved in oxidative stress responses, including the Nrf2/ARE (Nuclear factor erythroid 2–related factor 2/Antioxidant Response Element) pathway. Activation of Nrf2 leads to increased expression of antioxidant enzymes and improved cellular responsiveness to stressful environments (Liu et al., 2019). In animal studies, chitosan treatment before irradiation has been shown to significantly increase levels of glutathione (GSH), a crucial intracellular antioxidant, while reducing levels of oxidative stress markers, such as malondialdehyde (MDA) and hydrogen peroxide (H_2O_2). In addition, chitosan plays an immunomodulatory role, enhancing the activity of macrophages and T lymphocytes, which contribute to tissue repair and the clearance of radiation-damaged cells. These properties make it an ideal candidate for use in integrated radiation protection therapies, particularly in contexts such as oncology radiation therapy or space radiation protection.

Figure 3. Mechanism through which chitosan exerts its radioprotective effect.

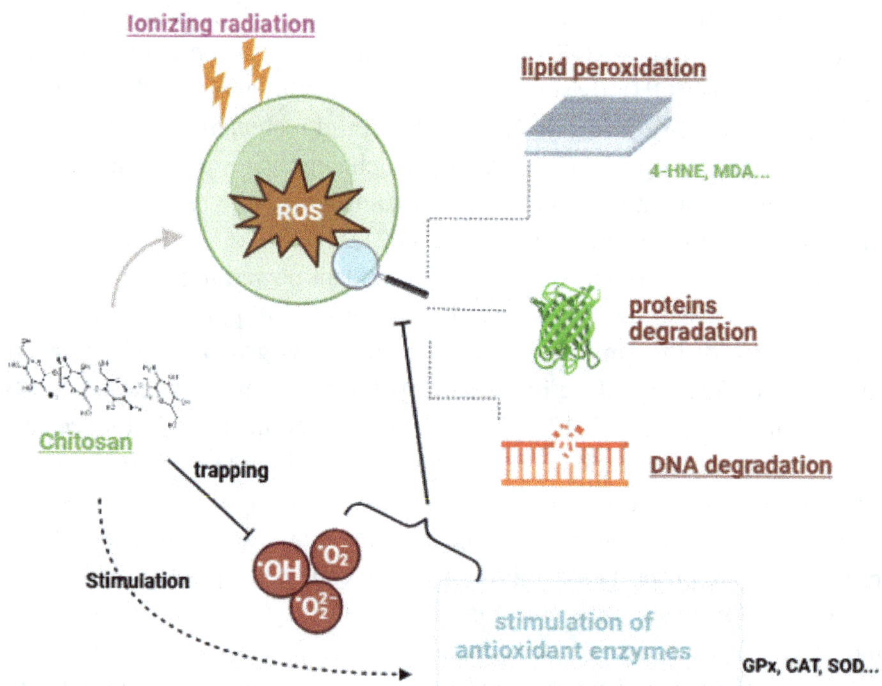

3.2 Radioprotective Applications of Chitosan

Chitosan has demonstrated significant potential as a radioprotective agent, through its ability to interact effectively with cell membranes. Its action manifests itself in antioxidant and anti-inflammatory effects, which play a crucial role in reducing radiation damage. Used in various formulations, such as topical films and gels, chitosan helps limit the impact of radiation on skin tissue, while promoting its regeneration. In fact, because of their small size, chitosan nanoparticles enable targeted delivery of radioprotective agents, increasing their efficacy and minimizing side effects. These properties make them a promising ally for the protection of exposed tissues, whether in medical treatments or situations of environmental exposure (Di et al., 2021).

3.2.1 Topical Films and Gels

In particular, chitosan-based topical films and gels are used to protect the skin during radiotherapy treatments. Radiotherapy, while effective in treating various types of cancer, often entails side effects such as skin burns, inflammation and premature aging of the skin. Topical chitosan-based formulations act as a physical barrier, preventing direct skin contact with ionizing radiation. In addition to their protective role, they are enriched with antioxidants and anti-inflammatory agents, such as vitamin E and plant extracts, which help neutralize the free radicals generated by oxidative stress, a common side effect of radiation exposure (Xia et al., 2022).

Chitosan, through its ability to interact with cells and accelerate cell regeneration processes, promotes the repair of damaged skin tissue and stimulates healing after exposure. The results of numerous clinical studies have shown that these topical formulations significantly reduce the incidence of skin burns, while promoting skin repair, which is particularly important for patients undergoing radiotherapy. In addition, their direct application facilitates rapid, targeted action, reducing the impact of radiation on healthy tissue. Such topical films and gels are used in radiotherapy and astronaut skin care due to the high radiation levels in space (Che et al., 2024).

3.2.2 Nanoparticles for Targeted Delivery

A further promising application for chitosan in the field of radiation protection lies in the targeted delivery of active ingredients via chitosan nanoparticles. Due to their small size (of the order of a few nanometers), these nanoparticles have a large specific surface area, enabling them to transport and deliver various radioprotective agents to radiation-exposed tissues in a targeted manner. The use of these nanoparticles enables the protective effects to be concentrated where they are most needed, while minimizing the systemic side effects often associated with the administration of radioprotective substances by the systemic route (Mikušová and Mikuš, 2021).

Chitosan nanoparticles can be formulated to encapsulate active substances such as antioxidants (vitamin C, vitamin E, polyphenols) or radioprotective agents such as selenium, which plays a crucial role in reducing oxidative damage. These agents can be released in a controlled, targeted manner according to therapeutic needs, depending on the type of radiation and dose administered. The ability to modulate the release of protective agents, as well as to adapt the bioavailability of nanoparticles, is a major feature of this approach (Detsi et al., 2020 ; Dai et al., 2024).

A major advantage of nanoparticles is their ability to increase the efficacy of radioprotective agents, enabling smaller quantities to be used to achieve optimum results. In addition, the surface of chitosan nanoparticles can be modified, for example, by the addition of specific ligands, enabling them to bind to specific cell receptors or penetrate cell membranes more easily. This increases their ability to concentrate in areas of the body that need protection, while reducing the need for over-delivery (Kwatra et al., 2013 ; Kumbhar et al., 2023).

Research into the use of chitosan nanoparticles has intensified in recent years, with promising results in experimental models and preclinical studies. They have shown great potential for preventing and reducing the effects of radiation, both in medical contexts, such as radiotherapy, and in environmental situations, such as exposure to radiation in space (Mikušová and Mikuš, 2021).

3.3 Biomedical Applications

Naturally-occurring biopolymers, such as collagen/gelatin, alginate, chitosan, silk, hyaluronic acid, elastin, glycosaminoglycans (GAGs), and peptides, are widely studied for their biomedical applications (Rahimi et al., 2022 ; Zarrintaj et al., 2023). These include polysaccharides such as chitosan and its precursor chitin, which are particularly prized for their complex structures that can form highly complex supramolecular networks (Mittal et al., 2018). Moreover, these biopolymers are not only non-toxic, biodegradable and biocompatible, but their versatility means that they can be transformed into a variety of forms adapted to the specific needs of biomedical applications (Baranwal et al., 2022). Such forms encompass membranes, sponges, gels, scaffolds, microparticles, nanoparticles and nanofibers, and find applications in fields as diverse as drug delivery, gene therapy, tissue engineering, wound healing, and many other therapeutic treatments (Öztürk et al., 2024).

3.3.1 Drug Delivery and Gene Therapy

The unique ability of chitosan to encapsulate a wide range of therapeutic agents, including hydrophilic and hydrophobic drugs, proteins and nucleic acids, means it is an ideal matrix for drug delivery. Thanks to its cationic properties, studies have shown that it interacts effectively with cell membranes, facilitating the controlled release of active substances into the body (Jafernik et al., 2023). This is particularly useful in gene therapy, where chitosan can protect DNA or messenger RNA (mRNA) from enzymatic degradation while promoting intracellular transfer (Cao et al., 2019). Recent studies have demonstrated that chitosan nanoparticles enable targeted delivery, reducing the side effects associated with conventional treatments, chiefly in oncology. In another area, the chemical modification of chitosan can improve the efficacy and stability of therapeutic formulations, paving the way for new strategies in advanced drug delivery systems (Herdiana et al., 2021).

3.3.2 Tissue Engineering

In tissue engineering, chitosan plays a crucial role due to its ability to replicate the extracellular matrix. It generates an environment propitious to cell growth, cohesion and differentiation. In the form of gels, membranes or three-dimensional scaffolds, it facilitates the recovery of soft and hard tissues such as bone, cartilage and muscle (Samrot et al., 2023). By contrast, chitosan scaffolds, often combined with other biopolymers or nanomaterials, have shown exceptional ability to promote osteogenesis in bone regeneration applications (Sindhusha and Doraiswamy, 2024). In recent years, chitosan gels loaded with growth factors or stem cells have also been used to regenerate damaged tissue, thereby enhancing the results of tissue therapies (Kim et al.,2023). Innovations in the fabrication of chitosan-based nanofibers and nanostructures enable us to better control the porosity and degradation of scaffolds, optimizing their mechanical and their biological properties. This makes them an ideal prospect for applications in regenerative medicine, where biomimetic structures are needed to replace or repair damaged tissues (Islam et al., 2020).

3.4 Healing Wounds

The hemostatic and antimicrobial properties of chitosan make it ideal for wound healing. It can be transformed into membranes, sponges or dressings to control infection while promoting tissue regeneration (Rajkumar et al., 2024). Chitosan accelerates clot formation, reducing the risk of bleeding, and inhibits the growth of pathogenic micro-organisms thanks to its natural antibacterial properties (Gheorghi ă et al., 2023). Chitosan-enriched dressings, often combined with other bioactive agents such as growth promoters or antimicrobials, have proved effective in treating complex wounds, including chronic ulcers and severe burns. Such materials not only protect the wound from infection, they also stimulate cell proliferation and promote extracellular matrix synthesis, which is essential for a rapid and effective healing process (Iqbal et al., 2024).

3.5 Treatment of Radiogenic Injuries

Radiogenic lesions, resulting from exposure to ionizing radiation, cause cellular, tissue and organ damage that can permanently affect the function of irradiated tissues (Azzam et al., 2012). Studies have revealed that chitosan, thanks to its antioxidant and anti-inflammatory properties, has proved effective in treating these lesions. Its wound-healing properties promote tissue repair by stimulating cell regeneration and reducing oxidative stress generated by exposure to radiation (Fadilah et al., 2023). Studies have revealed that chitosan helps reduce scarring and accelerates the regeneration of skin, muscle and even deeper tissues, such as internal organs. During radiotherapy treatments, the application of chitosan has been reported to improve the healing of irradiated lesions, by reducing inflammation and facilitating the repair of damaged cells (Sierra-Sánchez et al., 2021).

3.5.1 Integrated Applications for Internal Organ Protection

Chitosan also has potential systemic applications in protecting the internal organs of patients undergoing radiotherapy. During radiotherapy treatments for cancer, internal organs can be severely damaged by radiation, leading to complications such as infertility, digestive disorders or damage to the central nervous system (Arif et al., 2023 ; Hamdan et al., 2024). Systemic chitosan-based products, such as nanoparticles or chitosan complexes combined with radioprotective agents, can be administered orally or by injection. These systems aim to provide protection for internal organs by reducing inflammation and decreasing oxidative damage at the cellular level (Jafernik et al., 2023). Thanks to its controlled-release properties, chitosan can be used to deliver radioprotectants directly to sensitive areas of the body, offering targeted protection. Recent research has shown that systemic administration of chitosan during radiotherapy could reduce internal organ toxicity, protecting structures such as the liver, kidneys and lungs from the adverse effects of radiation (Zhou et al ., 2015).

3.6 Tissue Engineering and Bone Regeneration

Chitosan also has an important role to play in tissue engineering, particularly in bone regeneration. Its ability to promote healing and cell growth is being exploited to repair bone damaged by radiation, injury or degenerative disease. Thanks to its bioactive properties, chitosan is used in the manufacture of biomaterial matrices for bone regeneration. These matrices support the growth of bone cells and facilitate the reconstruction of bone tissue after radiation therapy or fracture (Kim et al., 2023). In addition, chitosan helps reduce inflammation in surrounding tissue, a key factor in fostering an environment conducive to cell regeneration. Research has shown that the use of chitosan in bone implants or in membranes for bone regeneration increases healing efficiency, reduces the risk of infection and accelerates the formation of new bone tissue (Islam et al., 2020).

The biomedical applications of chitosan depend on the integration of nanotechnologies and multidisciplinary approaches to optimize its properties and extend its uses. The development of functionalized chitosan-based nanoparticles could improve the precision of medical treatments, particularly in oncology and gene therapy. The use of chitosan in intelligent implantable devices, capable of releasing drugs in response to specific stimuli such as pH or temperature, represents a promising avenue. In short, chitosan's versatility and intrinsic properties make it a material of choice for innovative biomedical applications. Its ability to be transformed into various functional formats, combined with its biocompatibility and biodegradability characteristics, make it a key player in modern medicine, with immense potential to meet emerging therapeutic needs (Das et al., 2024 ; Iqbal et al., 2024).

4. FUTURE PERSPECTIVES AND DEVELOPMENT OF CHITOSAN

4.1 Chitosan's Versatility in Therapeutic Strategies

Biocompatible chitosan is a natural polymer with remarkable flexibility, enabling it to be adapted to a variety of therapeutic strategies in medicine. It can be used for drug delivery, gene therapy, immunotherapy, and even in the treatment of neurological pathologies. Its ability to be chemically modified means it can be tailored to specific therapeutic agents or biological targets, enhancing its efficacy (Grewal and Salar, 2024).

For example, in neurological diseases, studies have shown that chitosan is an excellent vector for the delivery of drugs to treat neurodegenerative disorders such as Alzheimer's and Parkinson's disease (Hamdan et al., 2024). Thanks to its ability to cross the blood-brain barrier (BBB), it offers a unique opportunity to deliver active compounds to the central nervous system (CNS). Chitosan nanoparticles (ChNPs) loaded with neuropeptides, growth factors or antioxidants have been shown to improve cognitive function and reduce oxidative stress in preclinical models (Wihadmadyatami., 2023).

In the field of bone regeneration, chitosan, combined with bioactive agents or minerals such as hydroxyapatite, promotes cell proliferation and osteoblast differentiation. Chitosan-based drug delivery systems have been used to accelerate bone healing, showing antimicrobial properties while promoting tissue reconstruction (Khodaverdi et al., 2024).

In the context of radiation protection, chitosan and its derivatives play an important role in preventing damage caused by ionizing radiation. Their antioxidant properties help reduce radiation-induced levels of oxidative stress, thereby minimizing cellular and tissue damage. For example, chitosan-based formulations enriched with natural antioxidants, such as vitamins C and E, have been shown to effectively protect cells against the effects of radiation under in vitro and in vivo conditions (Sen et al., 2021).

4.2 Enhanced Safety Profile and Customization Potential

Chitosan's high biocompatibility and low toxicity make it a particularly promising solution for long-term therapeutic applications. Its unique characteristics, such as biodegradability and good tolerance by the body, make it attractive for the development of innovative, safe treatments. Its naturally occurring degradation, involving renal elimination, limits the risk of cumulative toxicity. This ability to degrade reduces the undesirable side effects associated with other synthetic polymers and guarantees prolonged use without compromising patient safety (Yadav et al., 2024).

In the treatment of neurological diseases, the safety profile of chitosan is particularly beneficial. Chitosan nanoparticles loaded with siRNAs or therapeutic proteins protect these agents from enzymatic degradation and facilitate their uptake by neurons and glial cells, reducing the need for repeated doses and minimizing side effects (Omidian et al., 2024).

Chitosan can be modified to improve its pharmacokinetic properties, for example by coupling it with polymers such as polyethylene glycol (PEG), thereby increasing its solubility and circulation time in vivo. This modification circumvents certain limitations linked to its low solubility at physiological pH, making chitosan more effective as a drug or gene carrier. In addition, the addition of PEG reduces recognition by the reticuloendothelial system (RES), prolonging nanoparticle half-life and enabling more effective delivery of targeted therapeutic agents (Kesharwani et al., 2024).

In the area of regenerative medicine, chitosan-based hydrogels are also used for soft tissue repair and skin regeneration. These hydrogels act as a support for stem cells, enabling them to multiply and differentiate in controlled environments. They have been successfully applied in the treatment of severe skin lesions, showing accelerated healing while limiting infections thanks to their intrinsic antimicrobial properties (Kim et al., 2023). In the context of radiation protection, recent studies have explored the use of chitosan-antioxidant complexes for the protection of sensitive tissues, such as hematopoietic cells and gastrointestinal mucosa. These complexes reduce DNA damage and improve cell survival after exposure to radiation (Zhang et al., 2023).

In emerging clinical applications, chitosan has also been used to enhance the delivery of anticancer drugs and radiopharmaceutical agents, optimizing their accumulation in tumors while minimizing systemic side effects (Ding et al., 2022). With the expansion of long-duration space missions, radiation protection for astronauts is becoming a crucial priority. Cosmic rays, composed mainly of protons and heavy nuclei (HZE rays), represent a major threat to human health due to their ability to cause DNA damage and chronic side effects, such as cancer and neurodegenerative diseases (Chancellor et al., 2018).

In this context, chitosan offers unique potential thanks to its antioxidant and radioprotective properties. Chitosan-based formulations, enriched with natural antioxidants such as polyphenols or vitamin E, could neutralize the free radicals generated by cosmic radiation, thereby limiting cellular damage (Herdiana et al., 2023). In addition, its controlled delivery properties make it possible to develop systems capable of releasing protective agents in a prolonged and targeted manner, thus improving their effectiveness during prolonged exposure to radiation in the space environment.

Experimental studies have shown that chitosan nanoparticles effectively protect cells against DNA damage induced by ionizing radiation. These preliminary results open up prospects for the integration of chitosan into radiation protection countermeasures for space missions. For example, combinations of chitosan nanoparticles and metal compounds such as zinc or selenium, known for their radioprotective properties, could offer enhanced protection against high-energy radiation (Aibani et al., 2021).

These developments demonstrate the versatility and unique potential of chitosan in a wide range of medical fields, from neurology and bone and tissue regeneration to radiation protection and oncology. With ongoing research and improvements in manufacturing techniques, chitosan is well placed to play a key role in next-generation medical technologies.

CONCLUSION

Chitosan stands out as a natural and versatile radioprotective agent, offering innovative solutions for reducing the harmful effects of ionizing radiation on biological systems. Thanks to its antioxidant, anti-inflammatory and cell membrane stabilizing properties, it helps to limit the damage caused to DNA and healthy tissues during radiological exposure. Its exceptional biocompatibility, low toxicity and ease of chemical modification make it a material of choice for the development of advanced biomedical devices, particularly in the fields of targeted drug delivery, tissue engineering and cell regeneration. Recent research has also highlighted its ability to act as a carrier for other radioprotective agents, reinforcing its interest in combined protection and treatment approaches. Nevertheless, despite these promising advantages, further studies are essential to better understand its interactions with complex biological environments, improve its stability and optimize its formulations. Continuing this work will enable us to overcome the current technological challenges and fully exploit the potential of chitosan in radiation protection, regenerative medicine and, more broadly, in the entire biomedical field.

REFERENCES

Ai, H., Wang, F., Xia, Y., Chen, X., & Lei, C. (2012). Antioxidant, antifungal and antiviral activities of chitosan from the larvae of housefly, Musca domestica L. *Food Chemistry*, *132*(1), 493–498. PMID: 26434321

Aibani, N., Rai, R., Patel, P., Cuddihy, G., & Wasan, E. K. (2021). Chitosan nanoparticles at the biological interface: Implications for drug delivery. *Pharmaceutics*, *13*(10), 1686. PMID: 34683979

Aranaz, I., Alcántara, A. R., Civera, M. C., Arias, C., Elorza, B., Heras Caballero, A., & Acosta, N. (2021). Chitosan: An overview of its properties and applications. *Polymers*, *13*(19), 3256. PMID: 34641071

Arif, M., Nawaz, A. F., Mueen, H., Rashid, F., Hemeg, H. A., & Rauf, A. (2023). Nanotechnology-based radiation therapy to cure cancer and the challenges in its clinical applications. *Heliyon*, *9*(6). PMID: 37389057

Azzam, E. I., Jay-Gerin, J. P., & Pain, D. (2012). Ionizing radiation-induced metabolic oxidative stress and prolonged cell injury. *Cancer Letters*, *327*(1-2), 48–60. PMID: 22182453

Baranwal, J., Barse, B., Fais, A., Delogu, G. L., & Kumar, A. (2022). Biopolymer: A sustainable material for food and medical applications. *Polymers*, *14*(5), 983. PMID: 35267803

Bhatt, P., Joshi, S., Urper Bayram, G. M., Khati, P., & Simsek, H. (2023). Developments and application of chitosan-based adsorbents for wastewater treatments. *Environmental Research, 226*(November 2022), 115530. https://doi.org/DOI: 10.1016/j.envres.2023.115530

Bonilla, J., Fortunati, E., Atarés, L., Chiralt, A., & Kenny, J. M. (2014). Physical, structural and antimicrobial properties of poly vinyl alcohol–chitosan biodegradable films. *Food Hydrocolloids*, *35*, 463–470. DOI: 10.1016/j.foodhyd.2013.07.002

Brugnerotto, J., Lizardi, J., Goycoolea, F., Argüelles-Monal, W., Desbrières, J., & Rinaudo, M. (2001). An infrared investigation in relation with chitin and chitosan characterization. *Polymer*, *42*(8), 3569–3580. DOI: 10.1016/S0032-3861(00)00713-8

Cano-Vicent, A., Tuñón-Molina, A., Martí, M., & Serrano-Aroca, Á. (2023). Biocompatible Chitosan Films Containing Acetic Acid Manifested Potent Antiviral Activity against Enveloped and Non-Enveloped Viruses. *International Journal of Molecular Sciences*, *24*(15), 12028. DOI: 10.3390/ijms241512028 PMID: 37569404

Cao, Y., Tan, Y. F., Wong, Y. S., Liew, M. W. J., & Venkatraman, S. (2019). Recent advances in chitosan-based carriers for gene delivery. *Marine Drugs*, *17*(6), 381. PMID: 31242678

Chancellor, J. C., Blue, R. S., Cangel, K. A., Aunon-Chancellor, S. M., Rubins, K. H., Katzgraber, H. G., & Kennedy, A. R. (2018). Limitations in predicting the space radiation health risk for exploration astronauts. npj Microgravity 4.

Crini, G. (2019). Historical review on chitin and chitosan biopolymers. *Environmental Chemistry Letters*, *17*(4), 1623–1643. DOI: 10.1007/s10311-019-00901-0

Curcio, M., Puoci, F., Iemma, F., Parisi, O. I., Cirillo, G., Spizzirri, U. G., & Picci, N. (2009). Covalent insertion of antioxidant molecules on chitosan by a free radical grafting procedure. *Journal of Agricultural and Food Chemistry*, *57*(13), 5933–5938. DOI: 10.1021/jf900778u PMID: 19566085

Dai, Y., Guo, Y., Tang, W., Chen, D., Xue, L., Chen, Y., & Wang, S. (2024). Reactive oxygen species-scavenging nanomaterials for the prevention and treatment of age-related diseases. *Journal of Nanobiotechnology*, *22*(1), 252. PMID: 38750509

Daraghmeh, N. H., Chowdhry, B. Z., Leharne, S. A., Al Omari, M. M., & Badwan, A. A. (2011). Chitin (pp. 35–102). https://doi.org/DOI: 10.1016/B978-0-12-387667-6.00002-6

Das, A., Ghosh, S., & Pramanik, N. (2024). Chitosan biopolymer and its composites: Processing, properties and applications-A comprehensive review. *Hybrid Advances*, 100265.

Detsi, A., Kavetsou, E., Kostopoulou, I., Pitterou, I., Pontillo, A. R. N., Tzani, A., & Zoumpoulakis, P. (2020). Nanosystems for the encapsulation of natural products: The case of chitosan biopolymer as a matrix. *Pharmaceutics*, *12*(7), 669. PMID: 32708823

Ding, J., & Guo, Y. (2022). Recent advances in chitosan and its derivatives in cancer treatment. *Frontiers in Pharmacology*, *13*, 888740. PMID: 35694245

Dowlath, M. J. H., Karuppannan, S. K., Sinha, P., Dowlath, N. S., Arunachalam, K. D., Ravindran, B., & Nguyen, D. D. (2021). Effects of radiation and role of plants in radioprotection: A critical review. *The Science of the Total Environment*, *779*, 146431. PMID: 34030282

Fadilah, N. I. M., Phang, S. J., Kamaruzaman, N., Salleh, A., Zawani, M., Sanyal, A., & Fauzi, M. B. (2023). Antioxidant biomaterials in cutaneous wound healing and tissue regeneration: A critical review. *Antioxidants*, *12*(4), 787. PMID: 37107164

Foster, L. J. R., Ho, S., Hook, J., Basuki, M., & Marçal, H. (2015). Chitosan as a Biomaterial: Influence of Degree of Deacetylation on Its Physiochemical, Material and Biological Properties. *PLoS One*, *10*(8), e0135153. DOI: 10.1371/journal.pone.0135153 PMID: 26305690

Friedman, A. J., Phan, J., Schairer, D. O., Champer, J., Qin, M., Pirouz, A., Blecher-Paz, K., Oren, A., Liu, P. T., Modlin, R. L., & Kim, J. (2013). Antimicrobial and Anti Inflammatory Activity of Chitosan–Alginate Nanoparticles: A Targeted Therapy for Cutaneous Pathogens. *The Journal of Investigative Dermatology*, *133*(5), 1231–1239. DOI: 10.1038/jid.2012.399 PMID: 23190896

Frolova, M. A., Gryn, S. A., Kovaleva, E. I., Albulov, A. I., Melnik, R. N., Eremets, V. I., ... & Matveeva, I. N. (2020). Study of radioprotective properties of chitosan at gamma-irradiation. *International journal of research in pharmaceutical sciences, 11*(2), 1734-1737.. https://doi.org/.DOI: 10.26452/ijrps.v11i2.2076

Gallo, J., Raska, M., Kriegova, E., & Goodman, S. B. (2017). Inflammation and its resolution and the musculoskeletal system. *Journal of Orthopaedic Translation*, *10*, 52–67. DOI: 10.1016/j.jot.2017.05.007 PMID: 28781962

Georgopoulou, A., Papadogiannis, F., Batsali, A., Marakis, J., Alpantaki, K., Eliopoulos, A. G., & Chatzinikolaidou, M. (2018). Chitosan/gelatin scaffolds support bone regeneration. *Journal of Materials Science. Materials in Medicine*, *29*, 1–13. DOI: 10.1007/s10856-018-6064-2 PMID: 29730855

Gheorghi ă, D., Moldovan, H., Robu, A., Bi a, A. I., Grosu, E., Antoniac, A., & Băcilă, C. I. (2023). Chitosan-Based Biomaterials for Hemostatic Applications: A Review of Recent Advances. *International Journal of Molecular Sciences*, *24*(13), 10540. PMID: 37445718

Grewal, A. K., & Salar, R. K. (2024). Chitosan nanoparticle delivery systems: an effective approach to enhancing efficacy and safety of anticancer drugs. *Nano TransMed*, 100040.

Hakima, el K., Raja, B., Redouan, E. K., Ali, L., A., A., & Ahmed, lahsini. (2017). Physicochemical characterization of chitin and chitosan producted from Parapenaeus Longirostris shrimp shell wastes. Journal of Materials and Environmental Sciences .

Hamdan, Y. A., El-Mansoury, B., Elouali, S., Rachmoune, K., Belbachir, A., Oudadesse, H., & Rhazi, M. (2024). A review of chitosan polysaccharides: Neuropharmacological implications and tissue regeneration. *International Journal of Biological Macromolecules*, •••, 135356. PMID: 39244136

Hamedi, H., Moradi, S., Hudson, S. M., & Tonelli, A. E. (2018). Chitosan based hydrogels and their applications for drug delivery in wound dressings: A review. *Carbohydrate Polymers*, *199*, 445–460. DOI: 10.1016/j.carbpol.2018.06.114 PMID: 30143150

Herdiana, Y., Husni, P., Nurhasanah, S., Shamsuddin, S., & Wathoni, N. (2023). Chitosan-based nano systems for natural antioxidants in breast cancer therapy. *Polymers*, *15*(13), 2953. PMID: 37447598

Herdiana, Y., Wathoni, N., Shamsuddin, S., Joni, I. M., & Muchtaridi, M. (2021). Chitosan-based nanoparticles of targeted drug delivery system in breast cancer treatment. *Polymers*, *13*(11), 1717. PMID: 34074020

Hu, X., Wang, Y., Zhang, L., & Xu, M. (2020). Formation of self-assembled polyelectrolyte complex hydrogel derived from salecan and chitosan for sustained release of Vitamin C. *Carbohydrate Polymers*, *234*, 115920. PMID: 32070539

Husain, S., Al-Samadani, K. H., Najeeb, S., Zafar, M. S., Khurshid, Z., Zohaib, S., & Qasim, S. B. (2017). Chitosan Biomaterials for Current and Potential Dental Applications. *Materials (Basel)*, *10*(6), 602. DOI: 10.3390/ma10060602 PMID: 28772963

Hussain, I., Singh, T., & Chittenden, C. (2012). Preparation of chitosan oligomers and characterization: Their antifungal activities and decay resistance. *Holzforschung*, *66*(1), 119–125. DOI: 10.1515/HF.2011.130

Iqbal, Y., Amin, F., Fatima, M., Khalid, M., Kanwal, N., Hafez, A. A. A., & Alam, M. (2024). Chitosan-based functional materials combined with plant extract: A promising strategy in the stimulation of wound healing process. *Journal of Drug Delivery Science and Technology*, •••, 106314.

Islam, M. M., Shahruzzaman, M., Biswas, S., Sakib, M. N., & Rashid, T. U. (2020). Chitosan based bioactive materials in tissue engineering applications-A review. *Bioactive Materials*, *5*(1), 164–183. PMID: 32083230

Jaber, N., Al-Remawi, M., Al-Akayleh, F., Al-Muhtaseb, N., Al-Adham, I. S. I., & Collier, P. J. (2022). A review of the antiviral activity of Chitosan, including patented applications and its potential use against COVID-19. Journal of Applied Microbiology, 132(1), 41 58. https://doi.org/DOI: 10.1111/jam.15202

Jafernik, K., Ładniak, A., Blicharska, E., Czarnek, K., Ekiert, H., Wiącek, A. E., & Szopa, A. (2023). Chitosan-based nanoparticles as effective drug delivery systems—A review. *Molecules (Basel, Switzerland)*, *28*(4), 1963. PMID: 36838951

Jia, C., Wang, Q., Yao, X., & Yang, J. (2021). The role of DNA damage induced by low/high dose ionizing radiation in cell carcinogenesis. *Exploratory Research and Hypothesis in Medicine*, *6*(4), 177–184.

Jiménez-Gómez, C. P., & Cecilia, J. A. (2020). Chitosan: A natural biopolymer with a wide and varied range of applications. *Molecules (Basel, Switzerland)*, *25*(17), 3981. DOI: 10.3390/molecules25173981 PMID: 32882899

Jomova, K., Raptova, R., Alomar, S. Y., Alwasel, S. H., Nepovimova, E., Kuca, K., & Valko, M. (2023). Reactive oxygen species, toxicity, oxidative stress, and antioxidants: Chronic diseases and aging. *Archives of Toxicology*, *97*(10), 2499–2574. PMID: 37597078

Kean, T., & Thanou, M. (2011). Chitin and Chitosan: Sources, Production and Medical Applications. In Renewable Resources for Functional Polymers and Biomaterials (pp. 292–318). The Royal Society of Chemistry. https://doi.org/DOI: 10.1039/9781849733519 00292

Kesharwani, P., Halwai, K., Jha, S. K., Al Mughram, M. H., Almujri, S. S., Almalki, W. H., & Sahebkar, A. (2024). Folate-engineered chitosan nanoparticles: Next-generation anticancer nanocarriers. *Molecular Cancer*, *23*(1), 244. PMID: 39482651

Khodaverdi, K., Naghib, S. M., Mozafari, M. R., & Rahmanian, M. (2024). Chitosan/hydroxyapatite hydrogels for localized drug delivery and tissue engineering: A Review. *Carbohydrate Polymer Technologies and Applications*, •••, 100640.

Kim, S. (2018). Competitive biological activities of chitosan and its derivatives: Antimicrobial, antioxidant, anticancer, and anti-inflammatory activities. *International Journal of Polymer Science*, *2018*(1), 1708172.

Kim, Y., Zharkinbekov, Z., Raziyeva, K., Tabyldiyeva, L., Berikova, K., Zhumagul, D., & Saparov, A. (2023). Chitosan-based biomaterials for tissue regeneration. *Pharmaceutics*, *15*(3), 807. PMID: 36986668

Kołodziejska, M., Jankowska, K., Klak, M., & Wszoła, M. (2021). Chitosan as an underrated polymer in modern tissue engineering. *Nanomaterials (Basel, Switzerland)*, *11*(11), 1–44. DOI: 10.3390/nano11113019 PMID: 34835782

Kumbhar, P. R., Kumar, P., Lasure, A., Velayutham, R., & Mandal, D. (2023). An updated landscape on nanotechnology-based drug delivery, immunotherapy, vaccinations, imaging, and biomarker detections for cancers: Recent trends and future directions with clinical success. *Discover Nano*, *18*(1), 156. PMID: 38112935

Kwatra, D., Venugopal, A., & Anant, S. (2013). Nanoparticles in radiation therapy: A summary of various approaches to enhance radiosensitization in cancer. *Translational Cancer Research*, *2*(4).

Lee, D. S., Woo, J. Y., Ahn, C. B., & Je, J. Y. (2014). Chitosan-hydroxycinnamic acid conjugates: Preparation, antioxidant and antimicrobial activity. *Food Chemistry*, *148*, 97–104. DOI: 10.1016/j.foodchem.2013.10.019 PMID: 24262532

Mabsor-Zgandaoui, S., Rachmoune, K., El Hessni, A., El Mzibri, M., Iddar, A., & Moutaouakkil, A. (2025). Effect of hydrogen peroxide pre-treatment on enhancing the cellular resistance of Tetrahymena pyriformis exposed to ionizing radiation. *Toxicological and Environmental Chemistry*, *107*(2), 294–309.

Mikušová, V., & Mikuš, P. (2021). Advances in chitosan-based nanoparticles for drug delivery. *International Journal of Molecular Sciences*, *22*(17), 9652. PMID: 34502560

Mittal, H., Ray, S. S., Kaith, B. S., Bhatia, J. K., Sharma, J., & Alhassan, S. M. (2018). Recent progress in the structural modification of chitosan for applications in diversified biomedical fields. *European Polymer Journal*, *109*, 402–434.

Morin-Crini, N., Lichtfouse, E., Torri, G., & Crini, G. (2019). Applications of chitosan in food, pharmaceuticals, medicine, cosmetics, agriculture, textiles, pulp and paper, biotechnology, and environmental chemistry. *Environmental Chemistry Letters*, *17*(4), 1667–1692. DOI: 10.1007/s10311-019-00904-x

No, H. K., Young Park, N., Ho Lee, S., & Meyers, S. P. (2002). Antibacterial activity of chitosans and chitosan oligomers with different molecular weights. *International Journal of Food Microbiology*, *74*(1–2), 65–72. DOI: 10.1016/S0168-1605(01)00717-6 PMID: 11929171

Omer, H. (2021). Radiobiological effects and medical applications of non-ionizing radiation. *Saudi Journal of Biological Sciences*, *28*(10), 5585–5592. PMID: 34588869

Omidian, H., Gill, E. J., Dey Chowdhury, S., & Cubeddu, L. X. (2024). Chitosan Nanoparticles for Intranasal Drug Delivery. *Pharmaceutics*, *16*(6), 746. PMID: 38931868

Öztürk, A. B., Alarcin, E., Yaşayan, G., Avci-Adali, M., Khosravi, A., Zarepour, A., & Zarrabi, A. (2024). Innovative approaches in skin therapy: Bionanocomposites for skin tissue repair and regeneration. *Materials Advances*.

Rahimi, M., Mir, S. M., Baghban, R., Charmi, G., Plummer, C. M., Shafiei-Irannejad, V., & Pietrasik, J. (2022). Chitosan-based biomaterials for the treatment of bone disorders. *International Journal of Biological Macromolecules*, *215*, 346–367. PMID: 35718150

Rajkumar, D. S. R., Keerthika, K., & Vijayaragavan, V. (2024). Chitosan-Based Biomaterial in Wound Healing: A Review. *Cureus*, *16*(2). PMID: 38562272

Rusin, A., Seymour, C., & Mothersill, C. (2018). Chronic fatigue and immune deficiency syndrome (CFIDS), cellular metabolism, and ionizing radiation: A review of contemporary scientific literature and suggested directions for future research. *International Journal of Radiation Biology*, *94*(3), 212–228. PMID: 29297728

Sahani, R. M., & Dixit, A. (2022). A comprehensive review on zinc oxide bulk and nano-structured materials for ionizing radiation detection and measurement applications. *Materials Science in Semiconductor Processing*, *151*, 107040.

Samrot, A. V., Sathiyasree, M., Rahim, S. B. A., Renitta, R. E., Kasipandian, K., Krithika Shree, S., & Chinni, S. V. (2023). Scaffold using chitosan, agarose, cellulose, dextran and protein for tissue engineering—A review. *Polymers*, *15*(6), 1525. PMID: 36987305

Schreiber, S. B., Bozell, J. J., Hayes, D. G., & Zivanovic, S. (2013). Introduction of primary antioxidant activity to chitosan for application as a multifunctional food packaging material. *Food Hydrocolloids*, *33*(2), 207–214. DOI: 10.1016/j.foodhyd.2013.03.006

Sen, V. D., Balakina, A. A., Stupina, T. S., Mumyatova, V. A., Kulikov, A. V., Tikhonov, I. V., & Pliss, E. M. (2021). Amphiphilic chitosan–polyaminoxyls loaded with daunorubicin: Synthesis, antioxidant activity, and drug delivery capacity. *International Journal of Biological Macromolecules*, *193*, 965–979. PMID: 34751143

Shajahan, A., Shankar, S., Sathiyaseelan, A., Narayan, K. S., Narayanan, V., Kaviyarasan, V., & Ignacimuthu, S. (2017). Comparative studies of chitosan and its nanoparticles for the adsorption efficiency of various dyes. *International Journal of Biological Macromolecules*, *104*, 1449–1458. DOI: 10.1016/j.ijbiomac.2017.05.128 PMID: 28554794

Sierra-Sánchez, Á., Kim, K. H., Blasco-Morente, G., & Arias-Santiago, S. (2021). Cellular human tissue-engineered skin substitutes investigated for deep and difficult to heal injuries. *NPJ Regenerative Medicine*, *6*(1), 35. PMID: 34140525

Sies, H., Berndt, C., & Jones, D. P. (2017). Oxidative stress. *Annual Review of Biochemistry*, *86*(1), 715–748. PMID: 28441057

Sindhusha, V. B., & Doraiswamy, J. N. (2024). The Role of Chitosan and Gelatin-Based Scaffolds in Bone Regeneration: A Systematic Review. *Cureus*, *16*(9), e69793. PMID: 39435228

Sun, T., Zhou, D., Mao, F., & Zhu, Y. (2007). Preparation of low-molecular-weight carboxymethyl chitosan and their superoxide anion scavenging activity. *European Polymer Journal*, *43*(2), 652–656. DOI: 10.1016/j.eurpolymj.2006.11.014

Tao, J., Zhang, Y., Shen, A., Yang, Y., Diao, L., Wang, L., & Hu, Y. (2020). Injectable chitosan-based thermosensitive hydrogel/nanoparticle-loaded system for local delivery of vancomycin in the treatment of osteomyelitis. *International Journal of Nanomedicine*, •••, 5855–5871. DOI: 10.2147/IJN.S247088 PMID: 32848394

Tian, Y., Wu, D., Wu, D., Cui, Y., Ren, G., Wang, Y., & Peng, C. (2022). Chitosan-based biomaterial scaffolds for the repair of infected bone defects. *Frontiers in Bioengineering and Biotechnology*, *10*, 899760. DOI: 10.3389/fbioe.2022.899760 PMID: 35600891

Wang, Z., Yan, Y., Zhang, Z., Li, C., Mei, L., Hou, R., Liu, X., & Jiang, H. (2024). Effect of Chitosan and Its Water-Soluble Derivatives on Antioxidant Activity. *Polymers*, *16*(7). Advance online publication. DOI: 10.3390/polym16070867 PMID: 38611124

Wihadmadyatami, H., Zulfikar, M. A., Herawati, H., Pratama, D. A. O., Saragih, G. R., Kustiati, U., & Handayani, N. (2023). Chitosan hydrogel nanoparticle enhance therapeutic effect of bovine umbilical mesenchymal stem cell conditioned medium on canine cognitive dysfunction or canine Alzheimer's like mediated by inhibition of neuronal apoptotic. *Open Veterinary Journal*, *13*(12), 1504. PMID: 38292722

Wu, S. Y., Parasuraman, V., Arunagiri, V., Gunaseelan, S., Chou, H. Y., Anbazhagan, R., & Prasad, R. (2020). Radioprotective effect of self-assembled low molecular weight Fucoidan–Chitosan nanoparticles. *International Journal of Pharmaceutics*, *579*, 119161. PMID: 32081800

Xia, Y., Wang, D., Liu, D., Su, J., Jin, Y., Wang, D., & Liu, B. (2022). Applications of chitosan and its derivatives in skin and soft tissue diseases. *Frontiers in Bioengineering and Biotechnology*, *10*, 894667. PMID: 35586556

Xie, J., Wang, C., Zhao, F., Gu, Z., & Zhao, Y. (2018). Application of multifunctional nanomaterials in radioprotection of healthy tissues. *Advanced Healthcare Materials*, *7*(20), 1800421. PMID: 30019546

Yadav, H., Malviya, R., & Kaushik, N. (2024). Chitosan in biomedicine: A comprehensive review of recent developments. *Carbohydrate Polymer Technologies and Applications*, •••, 100551.

Yen, M.-T., Yang, J.-H., & Mau, J.-L. (2008). Antioxidant properties of chitosan from crab shells. *Carbohydrate Polymers*, *74*(4), 840–844. DOI: 10.1016/j.carbpol.2008.05.003

Younes, I., & Rinaudo, M. (2015). Chitin and Chitosan Preparation from Marine Sources. Structure, Properties and Applications. *Marine Drugs*, *13*(3), 1133–1174. DOI: 10.3390/md13031133 PMID: 25738328

Zarrintaj, P., Seidi, F., Azarfam, M. Y., Yazdi, M. K., Erfani, A., Barani, M., & Mozafari, M. (2023). Biopolymer-based composites for tissue engineering applications: A basis for future opportunities. *Composites. Part B, Engineering*, *258*, 110701.

Zhang, X., Liang, S., Gao, X., Huang, H., Lao, F., & Dai, X. (2021). Protective effect of chitosan oligosaccharide against hydrogen peroxide-mediated oxidative damage and cell apoptosis via activating Nrf2/ARE signaling pathway. *Neurotoxicity Research*, *39*, 1708–1720. PMID: 34622385

Zhang, Y., Huang, Y., Li, Z., Wu, H., Zou, B., & Xu, Y. (2023). Exploring Natural products as Radioprotective agents for Cancer Therapy: Mechanisms, challenges, and opportunities. *Cancers (Basel)*, *15*(14), 3585. PMID: 37509245

Zhao, K., Pu, S., Sun, L., & Zhou, D. (2023). Gentiopicroside-Loaded Chitosan Nanoparticles Inhibit TNF-α-Induced Proliferation and Inflammatory Response in HaCaT Keratinocytes and Ameliorate Imiquimod-Induced Dermatitis Lesions in Mice. *International Journal of Nanomedicine*, *18*, 3781–3800. DOI: 10.2147/IJN.S406649 PMID: 37457802

Zhou, Y., Hua, S., Yu, J., Dong, P., Liu, F., & Hua, D. (2015). A strategy for effective radioprotection by chitosan-based long-circulating nanocarriers. *Journal of Materials Chemistry. B*, *3*(15), 2931–2934. PMID: 32262491

Chapter 11

Chitosan Polysaccharides for Osteoporosis Therapy in Postmenopausal Women

Youssef Ait Hamdan
https://orcid.org/0009-0005-8616-8852
Cadi Ayyad University, Morocco

Imade Armadi
https://orcid.org/0009-0003-3063-5058
Cadi Ayyad University, Morocco

Saloua Mabsor-Zgandaoui
https://orcid.org/0009-0006-4601-9059
Ibn Tofail University, Morocco

Fatima Ezzahra Elamrani
https://orcid.org/0009-0009-0248-1964
Cadi Ayyad University, Morocco

khawla Rachmoune
Cadi Ayyad University, Morocco

Adil El Housseini
https://orcid.org/0009-0007-9294-2685
AQUAMAR Laboratory, Faculty of Sciences, Ibn Zohr University, Morocco

Meriem Mekouar
Faculté de Médecine et de Pharmacie de Rabat, Morocco

Amina El Kaoui
Moulay Ismail University, Meknes, Morocco

Maysae Chehber
Cadi Ayyad University, Morocco

Priscilla Barbosa Sales de Albuquerque
https://orcid.org/0000-0002-6310-1365
University of Pernambuco, Brazil

Anass Belbachir
https://orcid.org/0009-0002-4828-0251
Laboratory of Innovations in Biotechnology, Pharmaceutical Bioengineering, and Artificial Intelligence, Morocco

Mohammed Rhazi
https://orcid.org/0000-0003-2736-1802
Cadi Ayyad University, Morocco

ABSTRACT

Osteoporosis is a common degenerative disease in postmenopausal women, characterized by a loss of bone density and skeletal fragility, increasing fracture risk. This chapter examines the use of chitosan, a natural biopolymer derived from chitin, as a therapeutic agent in osteoporosis treatment. Chitosan's unique properties, such as biocompatibility, biodegradability, and the ability to form three-dimensional

DOI: 10.4018/979-8-3373-0055-9.ch011

structures, make it a promising candidate for bone therapies. The chapter also analyzes chemical and functional modifications to chitosan that enhance its bioactive effects and drug delivery capabilities, offering promising perspectives for non-hormonal osteoporosis treatments.

I. INTRODUCTION

Bone disorders encompass any disease affecting human bones, leading to abnormalities in the skeletal system. Osteoporosis, a skeletal disorder, is characterized by an increased risk of fractures, resulting in significant morbidity and mortality (Santhosh et al., 2019). Several factors contribute to this condition, including aging, insufficient calcium intake, reduced androgen levels, and genetic predisposition (Satarug et al., 2003). Alongside conventional therapies (such as parathyroid hormone and its analogs) and/or anti-resorptive treatments (bisphosphonates, hormone replacement therapy, selective estrogen receptor modulators, and calcitonin), a challenge lies in identifying biomaterials that can simultaneously provide detoxification against heavy metals and promote bone regeneration. Currently available bone biomaterials on the market do not fully satisfy these requirements and often involve substantial production costs.

Natural-origin polymers are gaining increasing attention for various biomedical and biological applications, including drug delivery systems, tissue engineering, wound dressings, and food packaging (Juncos Bombin et al., 2020). Among the most extensively studied polymers in the biomedical field are collagen, gelatin, alginate, chitosan, silk, hyaluronic acid, elastin, glycosaminoglycans (GAGs), and peptides. These materials play a vital role in numerous biological processes, such as cell differentiation, proliferation, and adhesion, as well as in inflammatory and immune responses (Gim et al., 2019). Their diversity in molecular weight, size, and charge, combined with their stability and low in vivo toxicity, makes them promising candidates for a range of applications. The presence of functional groups such as hydroxyl, carboxylic acid, and amine groups in the structure of polysaccharides provides several in vivo advantages, including prolonged half-life in the body, the ability to support cell growth, and the formation of hydrogels, while minimizing inflammatory or immune responses (Ferreira et al., 2015).

In this context, the novelty of this research lies in its innovative approach to investigating chitosan not only as a regenerative biomaterial but also as a therapeutic agent for postmenopausal bone disorders, specifically osteoporosis, by targeting both bone repair and heavy metal detoxification simultaneously (Khan & Ahmad, 2013). This dual-function role of chitosan remains largely underexplored in current literature. Thus, the objective of this chapter is to investigate the therapeutic potential of chitosan in treating postmenopausal disorders, with a particular focus on osteoporosis.

II. PATHOPHYSIOLOGY OF OSTEOPOROSIS IN POSTMENOPAUSAL WOMEN

Bone is a living tissue, and therefore, it is constantly renewed. This renewal process is called **bone remodeling**. During this biological action, two types of cells intervene successively, first the osteoclasts to destroy the old bone, and the osteoblasts which will rebuild the new bone. When there is an imbalance in this balance of bone remodeling, where bone resorption mediated by osteoclasts exceeds bone formation mediated by osteoblasts, this case is called osteoporosis (H. Li et al., 2021). It is a pathophysiology that results in the loss of bone mass, which often leads to bone fragility as well as fractures (H. Li et al., 2021), in fact, more than 8.9 million fractures are caused by osteoporosis each year (Noh et al., 2020).

It is a chronic metabolic disease that affects more than 200 million people worldwide, more specifically older men and postmenopausal women, where one in three postmenopausal women is affected by osteoporosis (Noh et al., 2020).

1. Biological Mechanisms of Osteoporosis

1.1. Bone Physiology

The skeleton is composed of cortical bone (present in long bones) and trabecular or cancellous bone, characterized by irregular and loose trabeculae, like those of the vertebrae. Bone tissue fulfills several roles, including supporting and protecting the body, as well as maintaining muscles. It also plays a crucial metabolic role by regulating calcium homeostasis, a set of processes that ensure the balance of calcium flows in the body. During human development, bone mass fluctuates at different stages, reaching its peak level around the age of 20, or even slightly earlier in women (Emmanuel Jeanpierre, 2014). This optimal peak bone mass is essential to prevent bone loss and reduce the risk of fractures, which is increased with aging. Between the ages of 35 and 40, bone loss intensifies. This phenomenon, described as physiological, results in a reduction in bone mass of approximately 0.5% each year (Munhoz et al., 2018).

Despite bone loss, which is a normal physiological process, bone tissue is constantly renewed, and is therefore in perpetual remodeling. This renewal is known as "bone remodeling", which is manifested by a continuous cycle between bone formation carried out by osteoblasts and bone resorption carried out by osteoclasts. This process helps replace aging bone tissue and maintain bone vitality. Osteoclasts and osteoblasts play an essential role at each stage of bone remodeling, thus contributing to the balance between bone formation and resorption (Lepage et al., 2024).

Osteoclasts create cavities called "lacunae" in the bone during the resorption phase. These cavities will then be filled by proteins secreted by osteoblasts, which constitutes the formation phase. This stage leads to mineralization, during which bone tissue, referred to as "osteoid tissue," is formed and allows calcium deposition (Clunie et al., 2009). These different phases are represented in **Figure 1** below:

Figure 1. Bone remodeling cycle

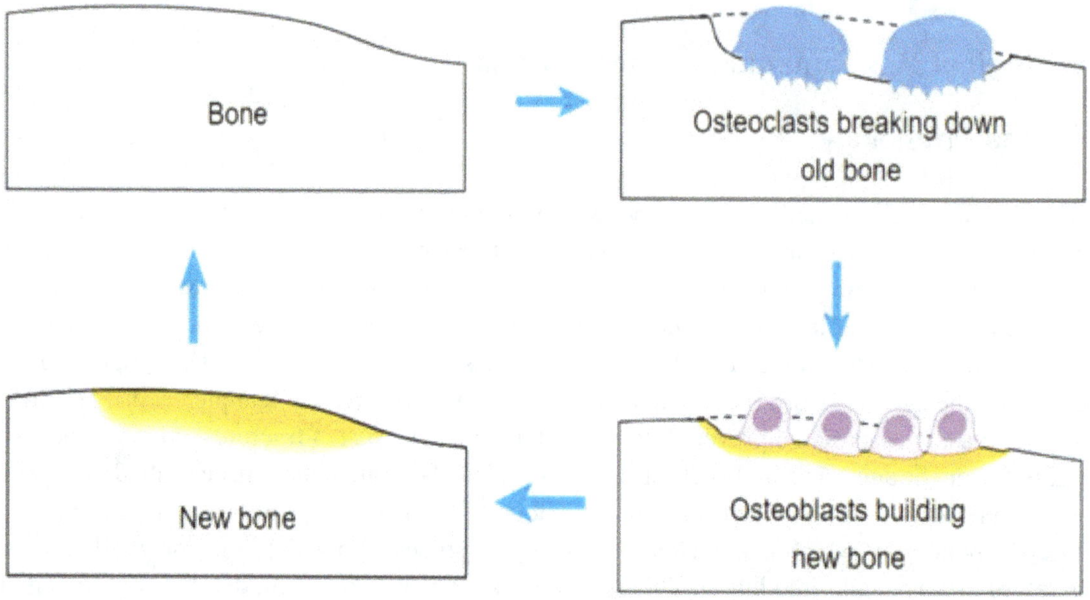

In adulthood, the balance between these cells allows bone formation to be adapted to resorption, thus helping the renewal and repair of bone tissue. However, imbalances in cellular activity can set in, these imbalances can be linked to aging or to hormonal disturbances observed during menopause. In these cases, bone resorption would be more important than bone formation, which leads to weakening and loss of bone mass, thus advancing the onset of this pathology (Lepage et al., 2024). To understand how an imbalance between osteoclasts and osteoblasts can weaken the bone system and promote the development of osteoporosis, it is essential to look at the pathophysiology of the latter.

1.2. Pathophysiology of Osteoporosis

Osteoporosis could be due to different factors, as shown in **Figure 2**, including hormonal imbalance, in fact, several hormones in the body play a crucial role in maintaining bone structure, among these hormones, we note estrogen, testosterone and parathyroid hormone (PTH). These hormones inhibit bone degradation and therefore, they promote the formation of bone tissue. The level of these hormones changes with age, in fact, women between 25 and 30 years old, experience their peak bone mass, then a decrease in estrogen production that induces the loss of their bone mass (Noh et al., 2020).

In addition, osteoporosis is a good illustration of a multifactorial disease, resulting from the complex interaction of genetic, biological, environmental and lifestyle factors. Traditional pathophysiological models have often focused on endocrine mechanisms, such as decreased estrogen, secondary hyper-parathyroidism in the elderly, and vitamin D deficiency and inadequate nutritional intake, which were considered the primary causes of postmenopausal osteoporosis (Clarke & Khosla, 2010).

Figure 2. Risk factors for osteoporosis

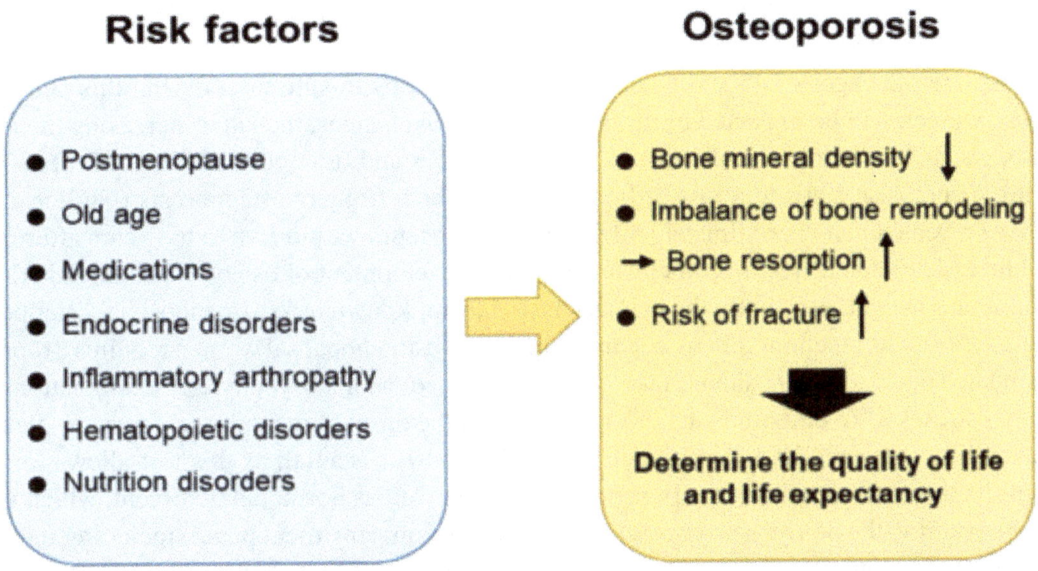

(Noh et al., 2020)

The interaction between osteoclasts and osteoblasts is essential to the bone remodeling process. However, certain imbalances can disrupt this mechanism of bone tissue renewal and repair. On a pathophysiological level, rapid bone loss is explained by an estrogen deficiency, which accelerates bone remodeling and causes an imbalance leading to excessive bone resorption, not compensated by the formation of new bone tissue (Lepage et al., 2024). On a cellular level, estrogens act on the three types of cells in bone tissue. They inhibit osteocyte apoptosis and reduce their production of RANKL, thereby decreasing osteoclast differentiation. In addition, estrogens cause osteoclast apoptosis, which limits resorption. They also preserve osteoblasts by inhibiting their apoptosis, which maintains bone formation. Finally, they stimulate the production of osteoprotegerin (OPG) while reducing that of RANKL by osteoblasts, thus slowing down the differentiation of osteoclasts (Dessay et al., 2022). In case of estrogen deficiency after menopause, a change in the RANKL/OPG ratio is observed, promoting a significant increase in osteoclast differentiation, while reducing bone formation. This deregulation of bone remodeling activation leads to variable bone loss. These effects lead to a decrease in bone mineral density (BMD) and a degradation of cortical and trabecular microarchitecture, characteristics of osteoporosis, making bones increasingly fragile. This secondary weakening is responsible for the fractures observed in this condition, which can vary according to the clinical forms of the disease, whether primary or secondary.

2. Impact of Osteogenic Deficiency on Osteoporosis

The problem of menopause is caused by estrogen deficiency due to the cessation of ovarian function. This process begins with a transition period called perimenopause, this phase is characterized by hormonal disorders, temporary or lasting, can appear in 80% of women, affecting their quality of life in 20 to 25% of cases (Lepage et al., 2024). Symptoms related to the drop in estrogen include vasomotor manifestations such as hot flashes and night sweats, vulvovaginal atrophy (inflammation, vaginal dryness), sleep disturbances, changes in mood, also impacts on sexual activity, as well as remarkable bone loss (Palacios et al., 2015).

Women between the ages of 45 and 55 are usually affected by this disease, during this period, the body of these women will be impacted by aging and hormonal changes, the latter increasing the risk of serious diseases such as cardiovascular diseases, breast cancer and osteoporosis. The lack of estrogen levels during menopause leads to a loss of bone density, which triggers osteoporosis (Guillot, 2015). Statistics have revealed that approximately 50% of this loss occurs within five to ten years after menopause (Goldberg & Fidler, 2015). Factors influencing the development of osteoporosis include the age of menopause and weight, especially the primary type I form, known as postmenopausal osteoporosis (Akdeniz et al., 2009). Protection from osteoporosis in these postmenopausal women requires a thorough risk assessment. This assessment should take into account medical history, physical examinations, and various diagnostic tests. To determine the risk of fracture at the onset of menopause, three key elements must be taken into account. DXA densitometric examination is essential, as this test allows for early diagnosis of the disease and helps identify people at risk. As well as hormonal treatment, which is also effective in preventing the bone tissue transformations that accompany menopause, including increased bone loss and degradation of bone microarchitecture (Pouillès et al., 2021). And finally, other therapeutic avenues, such as raloxifene, bisphosphonates, denosumab, and teriparatide, can also be used to reduce the risk of fractures (Pouillès et al., 2021).

III. CONVENTIONAL TREATMENTS FOR OSTEOPOROSIS

1. Commonly Used Anti-Osteoporotic Drugs

Bisphosphonates (BPs) are well-known for their pivotal role in reducing bone resorption, primarily by targeting osteoclast activity. This mechanism is crucial for maintaining bone density and preventing bone loss. Recent developments have focused on novel bisphosphonate conjugates designed to enhance their specificity and therapeutic efficacy (Mbese & Aderibigbe, 2021). The clinical efficacy of bisphosphonates is attributed to two primary properties: their strong affinity for bone mineral and their ability to inhibit mature osteoclasts. The selective binding of bisphosphonates to bone mineral gives them the unique capability of targeted delivery to their intended site of action. At these bone resorption sites, bisphosphonates enter and exert inhibitory effects on osteoclasts, effectively reducing bone degradation (Baron et al., 2011).

Bisphosphonate-Vitamin D Conjugates

Researchers have combined bisphosphonate structures with vitamin D to create innovative conjugates aimed at improving bone health. Among these, compound 17 demonstrated the most notable activity by mimicking vitamin D's biological functions in specific cell models such as HL-60 and ROS. This highlights its potential as a promising candidate for supporting bone regeneration (Baron et al., 2011; Mbese & Aderibigbe, 2021).

Estrogen-Conjugated Bisphosphonates

Hybrid molecules, such as the estrogen-bisphosphonate conjugate (compound 20), have been synthesized to facilitate targeted estrogen delivery to bone tissue. This compound proved effective in reducing bone loss while specifically localizing its therapeutic action to the skeletal system (Baron et al., 2011; Mbese & Aderibigbe, 2021).

Methotrexate-Bisphosphonate Conjugates

The methotrexate-alendronate conjugate (compound 21) was designed to combat bone diseases by preventing excessive bone resorption and suppressing osteoclastogenesis. This dual-functional compound demonstrated the ability to bind to bone surfaces and improve drug concentration in diseased areas, making it particularly suitable for targeted therapies (Baron et al., 2011; Mbese & Aderibigbe, 2021).

The efficacy of newly developed bisphosphonate (BP) derivatives has been extensively evaluated, yielding varied results depending on the compounds. Initial trials with BP conjugates (compounds 10–14) produced suboptimal outcomes compared to existing anti-osteoporotic drugs, emphasizing the need for further innovation in BP-based therapies. However, compounds 15–19 demonstrated moderate efficacy, with compound 17 standing out for its vitamin D-like activity in promoting bone health, marking it as a promising candidate for future studies. A targeted delivery approach using hybrid conjugates, such as the estrogen-bisphosphonate conjugate (compound 20), showed significant potential by selectively delivering estrogen to skeletal tissues and effectively mitigating bone loss. Similarly, the methotrexate-alendronate conjugate (compound 21) exhibited enhanced efficacy over individual components, reducing osteoclast formation and preventing bone degradation, thereby offering a more comprehensive treatment. Additionally, radiolabeled bisphosphonate derivatives like the zoledronate conjugate (compound 22), tagged with 99mTc, demonstrated superior bone-targeting capabilities, with preferential accumulation in bone tissue and rapid clearance from blood and soft tissues (Baron et al., 2011; Mbese & Aderibigbe, 2021). Despite their effectiveness, traditional bisphosphonates are associated with side effects such as gastrointestinal irritation, atypical fractures, and osteonecrosis of the jaw, particularly with long-term use. These challenges highlight the necessity of developing safer therapeutic options. Advances in safety have been achieved with novel conjugates like compound 21, whose bone-specific binding properties reduce systemic exposure and toxicity. Likewise, the radiolabeled zoledronate conjugate (compound 22) demonstrated efficient bone targeting with minimal retention in soft tissues, suggesting improved safety and reduced off-target effects (Baron et al., 2011; Mbese & Aderibigbe, 2021).

Estrogen receptors (ERα and ERβ) play a pivotal role in regulating various physiological processes, including reproduction, cardiovascular health, and bone maintenance, through direct DNA interactions or partnerships with transcription factors. Dysregulated estrogen signaling is implicated in conditions such as osteoporosis, hormone-responsive cancers, and Alzheimer's disease (Motlani et al., 2023). Selective Estrogen Receptor Modulators (SERMs) leverage tissue-specific estrogenic or anti-estrogenic effects to address issues like osteoporosis and hormone-related cancers, offering protective benefits for bones and lipids while reducing risks to breast and uterine tissues. A notable example, raloxifene, enhances bone density by acting as an estrogen agonist in bones and an antagonist in breast and uterine tissues, presenting a balanced therapeutic approach for postmenopausal women. Despite its benefits, raloxifene carries a risk of venous thromboembolism and warrants further exploration of its cardiovascular and neuroprotective potential (Motlani et al., 2023).

Hormone Replacement Therapy (HRT), previously a popular intervention for menopausal symptoms and osteoporosis prevention, saw a decline after the Women's Health Initiative (WHI) trials highlighted increased cardiovascular and cancer risks. However, subsequent studies suggest that initiating HRT early during menopause may offer benefits with fewer risks, particularly for cardiovascular and arterial health. While combination HRT slightly raises breast cancer risk, estrogen-only therapy might reduce risks in older women. Adjusting hormone components, such as replacing medroxyprogesterone acetate with micronized progesterone, could further influence safety profiles (Bolton, 2016). Inhibitors of bone resorption, like denosumab, offer a modern alternative for managing excessive osteoclast activity in conditions such as rheumatoid arthritis (RA). Denosumab, a monoclonal antibody, inhibits RANKL from activating osteoclast precursors, effectively reducing bone resorption. Unlike bisphosphonates, which have a slow, long-lasting effect, denosumab provides a rapid, reversible action cleared within its six-month dosing cycle. Combined with methotrexate or other csDMARDs, denosumab proves effective and safe, even for RA patients on corticosteroids or anti-TNF therapies, offering a distinct therapeutic advantage in managing systemic and localized bone loss (Chiu & Ritchlin, 2017; Kendler et al., 2022).

2. Limitations of Existing Treatments

Prolonged high-dose glucocorticoid use is linked to significant long-term effects on bone health, notably an increased risk of fractures and reduced bone mineral density (BMD). While short-term use poses lower risks, chronic glucocorticoid therapy is a leading cause of glucocorticoid-induced osteoporosis (GIOP), particularly in patients with conditions such as rheumatoid arthritis or Duchenne muscular dystrophy. Younger patients, especially those with Duchenne muscular dystrophy, are also at heightened fracture risk (Laurent et al., 2022). Managing and preventing GIOP can be expensive, with treatments like bisphosphonates contributing to high costs, though generic alternatives may provide more affordable options for high-risk individuals (Laurent et al., 2022). Despite the availability of treatments, underdiagnosis and undertreatment are common, as fewer than 40% of long-term glucocorticoid users undergo BMD testing or receive osteoporosis medications. Barriers to treatment adherence include costs, delayed diagnoses, medication side effects, and unclear treatment responsibilities (Laurent et al., 2022).

Additional risks associated with long-term glucocorticoid use include vertebral compression fractures, for which procedures like vertebroplasty and kyphoplasty are not recommended due to limited benefits and an increased risk of further fractures. Older adults face a heightened fall risk from glucocorticoid-induced muscle atrophy, and prolonged bisphosphonate use may lead to osteonecrosis of the jaw (Laurent et al., 2022). Glucocorticoid excess also adversely impacts muscle, cartilage, and fat mass, potentially lowering quality of life, although regular exercise and a healthy diet may mitigate some of these effects (Laurent et al., 2022).

In light of these challenges, there is a growing need for safer and more natural therapeutic alternatives. Non-pharmacological approaches have garnered attention for addressing the limitations of conventional treatments. Nutritional support through increased intake of calcium, vitamin D, and magnesium plays a vital role in maintaining bone health and reducing osteoporosis risk. Physical activity, particularly weight-bearing exercises, has proven effective in enhancing bone formation, strengthening muscles, and improving balance to lower fracture risk. Additionally, natural compounds such as green tea catechins, lycopene, and resveratrol show promise in reducing oxidative stress and promoting bone health, offering a safer alternative to synthetic drugs with fewer long-term risks (Hsu et al., 2024). By integrating these holistic strategies, the focus is shifting toward minimizing dependence on pharmacological treatments while reducing side effects and long-term risks (Hsu et al., 2024).

IV. CHITOSAN - PROPERTIES AND GENERAL APPLICATIONS

1. Chitosan

Chitin, the most abundant biopolymer on earth after cellulose, is found in a variety of eukaryotic organisms, mainly crustaceans, insects and fungi. (Crini, 2019). Chemically, it is shown in Figure 1 as a chain of N-acetyl-β-D-glucosamine monomers linked by a (1→4) glycosidic bond, and the nomenclature defines this structure as that of a poly β-(1→4)-2-acetamido-2-deoxy-D-glucopyranose. Following deacetylation, its derivative is prepared as chitosan, illustrated by a chain of ß (1-4) -D-glucosamine monomers. It has three reactive functional groups - an amine in C2 position and two hydroxyl groups in C3 and C6 respectively (Kean & Thanou, 2011; Younes & Rinaudo, 2015).

Figure 3. Chemical structure of chitin and chitosan

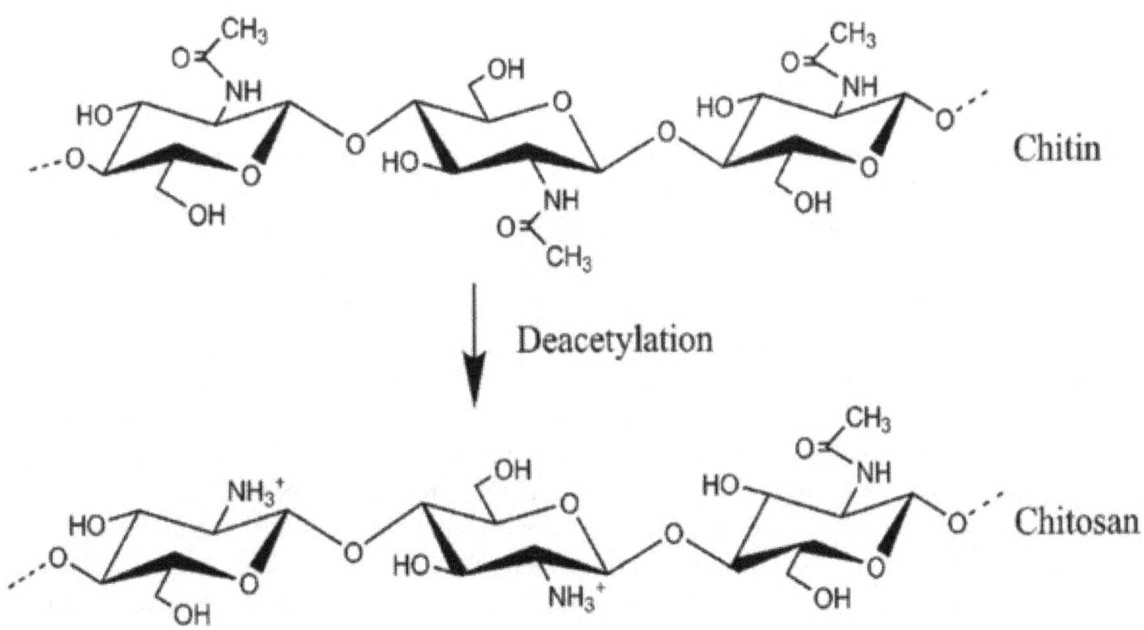

Our biopolymer, chitosan, is endowed with enormous and unique properties, which have aroused the interest of many researchers and major industries in various fields (pharmaceutical/biomedical/agricultural/water treatment...). Solubility, among other properties, is influenced by the presence of functional groups such as hydroxyl and amine groups, which can be protonated, making chitosan soluble in dilute acidic media. (Bonilla et al., 2014; Daraghmeh et al., 2011; Hakima et al., 2017). Although chitosan is insoluble in water and most organic solvents, studies confirm that it dissolves well in certain acidic aqueous media, such as lactic, acetic, formic and citric acids, as well as other specific solvents. (Shajahan et al., 2017).

In addition to these factors, controlling deacetylation and reducing molecular weight are found to increase solubility, although this may reduce yield. By adopting chemical and physical approaches, such as reacetylation or the use of adjuvants that disrupt hydrogen bonds, solubility is further increased. (Hamedi et al., 2018). Furthermore, the addition of certain smaller reactive groups or the synthesis of low-molecular-weight compounds also improves chitosan's solubility. (Jayakumar, Prabaharan, Nair, Tokura, et al., 2010).

In osteoporosis treatment and biomedical applications in general, biocompatibility and biodegradability are essential properties. These qualities make chitosan a prime candidate for medical therapies. Its ability to interact safely with human tissues, while being progressively degraded and eliminated by the body, limits the risk of undesirable reactions. (Crini, 2019). In addition, our biopolymer has demonstrated excellent biocompatibility in several studies. For example, chitosan has been shown to be non-toxic when applied to human cells such as HaCaT keratinocytes, making it suitable for applications such as wound dressings and medical devices (Cano-Vicent et al., 2023). What's more, its ability to interact positively with immune cells suggests that it could promote tissue healing and regeneration. (Jaber et al., 2022). Other studies show that this biopolymer breaks down into non-toxic products, making it suitable for temporary applications in the human body, such as dressings or drug delivery. (Foster et al., 2015).

2. Antimicrobial Properties

Nowadays, the focus on antimicrobial activity is growing significantly, especially with the urgent need to find a natural-source alternative against bacteria, filamentous fungi and yeasts. (Husain et al., 2017). Although the mechanism of action is not precisely described, its antibacterial action remains to be elucidated. Two studies, (Dilamian et al., 2013; Husain et al., 2017), clearly suggest that chitosan, as much as a biopolymer, stimulates the migration of Ca++ ions from the anionic zones of bacterial membranes, thus causing cellular damage. It has also demonstrated efficacy against oral pathogens such as Porphyromonas gingivalis, Prevotella intermedia and Actinobacillus actinomycetemcomitans. More pertinently, while remaining non-toxic to mammals, chitosan applies its bacterial activity against both gram-positive and gram-negative microorganisms (Jeon, 2001). In addition to its efficacy against bacteria, chitosan has also demonstrated antifungal activity against strains such as Candida albicans and Candida parapsilosis (No & Meyers, 1995; Tolaimate et al., 2003). As influencing factors, chitosan deacetylated by conventional chemical or ultrasonic methods showed significantly higher antimicrobial activities than non-deacetylated chitin. Deacetylation therefore increases the antimicrobial efficacy of chitosan (No & Meyers, 1995). Another influential parameter is the degree of deacetylation. The more chitosan is deacetylated, the greater its efficacy against microorganisms, underlining the importance of this parameter in the development of chitosan-based antimicrobial agents (Brugnerotto et al., 2001; GALED et al., 2005).

3. Antioxidant Properties

Beyond its antibacterial effects, the antioxidant properties of chitosan are also crucial for bone health. This activity is closely linked to its structural characteristics, such as molecular weight and degree of deacetylation. Like the anti-bacterial property, the latter is also influenced by chitosan parameters, so that as its molecular weight decreases, antioxidant activity becomes more important (Sun et al., 2007). Moreover, according to (Yen et al., 2008), chitosan displays antioxidant activity quantified between 58.3% and 70.2% for moderate activities, reaching up to 85.2% for higher concentrations, underlining its potential to protect cells from oxidative damage. Ultrasonic deacetylation revealed a free radical scavenging capacity ranging from 18.27% to 44.17%, depending on concentration (0.25 to 1 mg/mL), as measured by the DPPH assay. These properties surpass those of chitosan deacetylated by conventional chemical methods (Lingnert et al., 1979). In terms of inhibition of lipid peroxidation, chitosan demonstrated inhibition rates ranging from 11.7% to 77.39%, depending on concentration, thus protecting cell membranes from oxidative damage. This mechanism of action relies primarily on the interaction of chitosan's amine groups with free radicals, neutralizing their destructive potential and forming stable compounds.

These qualities position chitosan and its derivatives as promising antioxidant agents, both in the biomedical field and in the food and cosmetics industries. Their ability to scavenge free radicals, inhibit lipid peroxidation and protect against oxidative stress opens up new prospects for extending the shelf life of food products, developing protective skin formulations and designing effective osteoporosis therapies.

4. Anti-Inflammatory Properties

Speaking now of an equally important property in osteoporosis treatment, that is the anti-inflammatory activity of chitosan as much as a therapeutic biomaterial (Crini, 2019).

Several studies have demonstrated the ability of chitosan and its derivatives to modulate inflammatory responses. Research on chitosan-alginate nanoparticles revealed their efficacy against Propionibacterium acnes, a bacterium associated with acne. These nanoparticles inhibited the production of pro-inflammatory cytokines in human monocytes and keratinocytes. For example, they dose-dependently reduced IL-12p40 in monocytes and virtually suppressed IL-6 in keratinocytes, even at low doses (Friedman et al., 2013). Other studies have shown that chitosan inhibits activation of the transcription factor NF-κβ, a central player in inflammatory pathways, while reducing expression of membrane metalloproteinases, responsible for degradation of the extracellular matrix and aggravation of inflammatory responses. (Gallo et al., 2017).

Chitosan exerts its anti-inflammatory effects via several mechanisms. It can interact with Toll-like receptors (TLRs), modulating signaling pathways and reducing the production of inflammatory cytokines. Inhibition of NF-κβ activation reduces the expression of genes involved in chronic inflammation. Thanks to its aforementioned antioxidant properties, chitosan scavenges reactive oxygen species (ROS), thereby limiting oxidative stress and associated damage. In addition, its antimicrobial activity against skin pathogens such as P. acnes promotes a skin environment less conducive to inflammation. (Friedman et al., 2013).

Chitosan's anti-inflammatory properties make it a promising candidate for applications in tissue engineering and biomedical therapies. However, although its mechanisms of action are partially elucidated, further research is needed to optimize its formulations and define its clinical applications. Modulation of its chemical structure, concentration and mode of administration could also help maximize its therapeutic effects. (Crini, 2019).

5. Biomedical Applications of Chitosan

Polymers of natural origin such as collagen/gelatin, alginate, chitosan, silk, hyaluronic acid, elastin, glycosaminoglycans (GAGs), peptides and others are the most commonly studied for biomedical applications (K. Y. Lee & Mooney, 2012; Nguyen et al., 2019; Zhao et al., 2018). Among them, polysaccharides are the most common natural polymers, with complex structures that tend to form supramolecular networks (Rahimi et al., 2022a). Chitosan and its precursor, chitin, are natural biopolymers with the enormous properties mentioned above: non-toxic, biodegradable and biocompatible. Their versatility enables them to be easily transformed into a variety of forms, responding to urgent and/or specific needs, notably in the form of membranes, sponges, gels, scaffolds, microparticles, nanoparticles and nanofibers, suitable for a wide range of biomedical applications. These applications include drug delivery, gene therapy, tissue engineering, osteoporosis treatment and wound healing figure 2.B (Anitha et al., 2014).

In tissue engineering, these biopolymers mark an important trace due to their ability to be prepared as gels (Roughley et al., 2006), membranes (Ehrlich et al., 2006), nanofibres (Jayakumar, Prabaharan, Nair, & Tamura, 2010), nanofibrilles, beads, microparticles, nanoparticles, scaffolds and sponge-like shapes figure 2.A. This flexibility enables the creation of three-dimensional structures that mimic the extracellular matrix, providing a favorable environment for cell growth and proliferation (Tığlı & Gümüşderelioğlu, 2009; Venkatesan & Kim, 2010). From tissue engineering to wound healing. Their hemostatic, antimicrobial and biocompatible properties make them the materials of choice for wound dressings. They can be used to create membranes, sponges and bandages that promote tissue regeneration and accelerate the healing process (Kumar et al., 2010; Radhakumary et al., 2011).

Figure 4. Chitosan derivatives in various forms (A) and their biomedical applications (B)

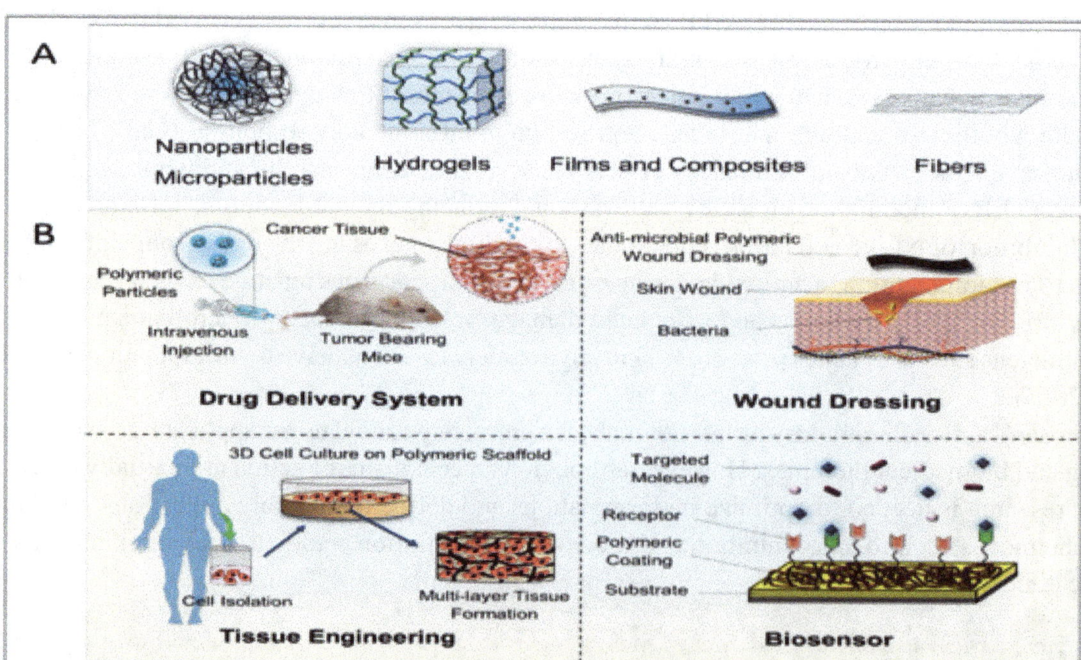

(Rahimi et al., 2022a)

Not forgetting their incredible role in drug delivery. Their mucoadhesive properties and ability to form nanoparticles and microparticles enable drugs to be encapsulated and released in a controlled manner, improving efficacy and reducing side effects (Dev et al., 2010). Finally, in the treatment of post-menopausal osteoporosis, our biopolymer, chitosan, could be considered as an adjunct to traditional treatments such as bisphosphonates or hormone therapies. Although drugs such as denosumab and romosozumab are commonly used for this condition thanks to their proven efficacy in reducing the risk of fractures, chitosan, illustrated in Figure 3 (Rahimi et al., 2022a) with a list of its derivatives applied to the treatment of osteoporosis as well as a comparison between healthy bone and bone with osteoporosis, could represent an alternative or complementary approach due to its previously described regenerative properties.

Figure 5. Chitosan derivatives for osteoporosis treatment and bone health comparison

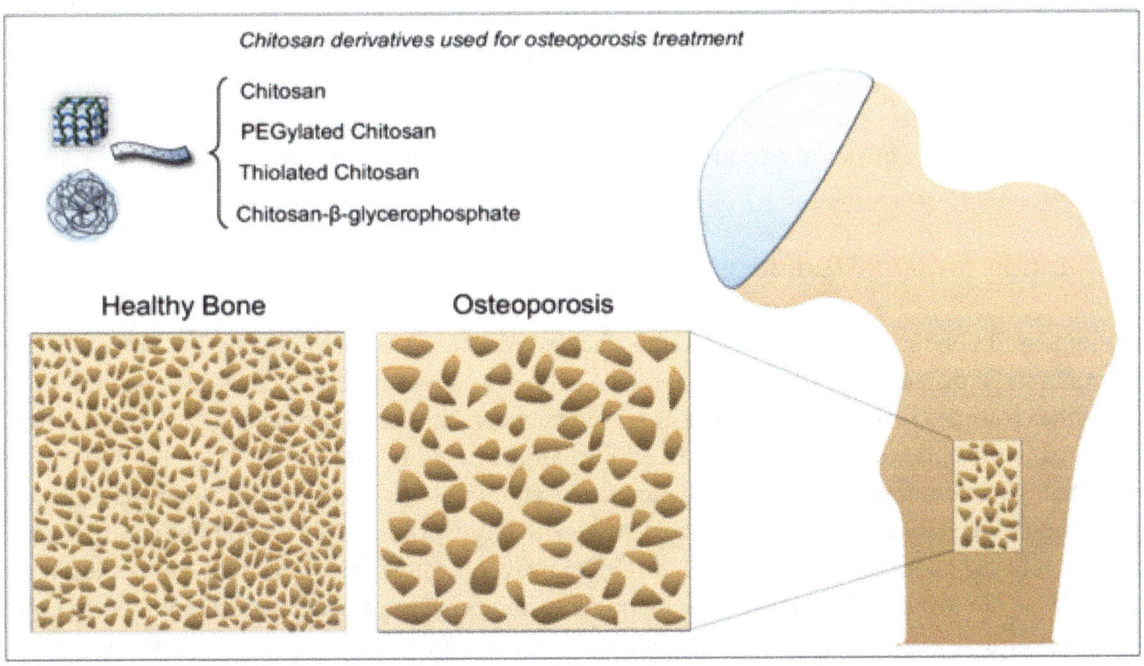

V. ROLE OF CHITOSAN IN BONE REMODELING PROCESS

Bone remodeling is a permanent and dynamic process that is essential for maintaining bone integrity, repairing damage caused by stress or fractures and remodeling bone during growth. This process involves the coordinated action of specialized cells, the osteoclasts (OC), which resorb bone, and the osteoblasts (OB), which form new bone. As described, bone is mainly formed by two types of bone: cancellous bone which contains a greater number of cells like OB and OC, creating a trabecular network that supplies a wide surface area for cellular activity, and compact bone, formed by dense outer shell with a richly mineralized structure and less cells (Šromová et al., 2023).

Chitosan (CS) was broadly recognized for its potential to enhance bone metabolism, due to its exceptional properties. Its biocompatibility, biodegradability, antibacterial activity and osteo-conductivity, make it a valuable material for bone tissue engineering. CS promoted bone remodeling by supporting the differentiation, proliferation and mineralization of OB. The CS's cationic nature enhanced its hemostatic and adherence properties to mucous membranes, contributing to scar healing and facilitating bone repair. Its solubility, which depended on the degree of deacetylation and pH, had a considerable influence on its interaction with cellular mechanisms. CS's amino groups enabled it to form ionic complexes with metal ions and negatively charged compounds, which further enhanced its bioactivity (Y. Zhu et al., 2022). As a scaffolding material, CS provided a supportive matrix for cell attachment and growth, as well as acting as a carrier for osteogenic bioactive. Combined with materials such as hydroxyapatite, CS improves mechanical properties, promoted protein adsorption and facilitated apatite deposition, all of which are essential for effective bone integration. It also stimulated the expression of osteoblast-related genes, favoring cell maturation. CS's role in vascularization and granulation tissue generation ensured an efficient delivery of nutrients to the repair site, contributed to bone healing. These combined features highlighted the usefulness of CS in bone tissue engineering, providing a multifunctional platform that actively supported bone metabolism and regeneration through cellular and molecular interactions (Tian et al., 2022).

1. *In Vivo* and *In Vitro* Studies

CS improved bone remodeling through multifaceted mechanisms supported by in vivo and in vitro evidence. In in vivo studies using animals' models, CS demonstrate an accelerated bone healing of rat's tibias, with significant differences in repair observed between one- and four- weeks after surgery in CS-treated sites compared with control sites (Ezoddini-Ardakani et al., 2012). Munhoz et al. 2018 study, showed that the biocompatibility of the collagen-chitosan mixtures used in the grafting process enabled the formation of new bone at the edges of the defect, with larger volumes observed in the grafted groups compared with the controls (Munhoz et al., 2018). However, sponge mineralization did not effectively stimulate bone neoformation, which lead to incomplete bone repair over the experimental period. Also, the addition of hydroxyapatite did not significantly improve bone repair, indicating that further study of its effects on mineralization is warranted. The study concluded that while collagen-chitosan sponges could stimulate osteogenesis in cranial defects, the process was slow and insufficient for complete bone generation in the short time frame studies (Munhoz et al., 2018). The scaffold was tested both in vitro and in rat model of calvarial bone failure, demonstrating its ability to enhance osteogenic differentiation of human adipose-derived stem cells. This was evidenced by increasing expression of key osteogenic markers, including collagen, sialoprotein, osteocalcin, alkaline phosphate and osteopontin (Huang et al., 2020).

An injectable hydrogel developed by Lee et al. from glycolic chitosan and oxidized hyaluronic acid, have optimized the delivery of graphene oxide to promote osteogenesis in cell cultures and rat models (S. J. Lee et al., 2020). Georgopoulou et al. further proved that CS-gelatin scaffolds enhanced the adhesion and proliferation of pre-osteoblastic cells and bone marrow mesenchymal stem cells promoting osteogenic differentiation and the formation of new extracellular matrix in a mouse femur model (Georgopoulou et al., 2018). Cui et al. generated an injectable hydrogel composed of methacrylated glycolic CS and montmorillonite, which permitted bone tissue regeneration in a mouse calvarial defect model with no requirement for extra bioactive agents (Z. K. Cui et al., 2019). In contrast, Tao et al. engineered a heat-sensitive CS-glycerol phosphate hydrogel that effectively delivered vancomycin in a chronic osteomyelitis model, markedly reducing inflammation and improving bone repair after eight weeks in rabbit (Tao et al., 2020). Finally, Su et al. modified CS membranes to increase their durability, resulting in improved bone healing in rat calvarial model, with new bone formation that closely resembled natural bone after eight weeks (Su et al., 2021). Collectively, these studies highlight the effectiveness of various CS formulations in supporting bone repair and regeneration though their unique properties and mechanisms of action.

In other hand, the in vitro studies of CS-based biomaterials have demonstrated their significant potential to enhance bone regeneration through a range of formulations, including injectable hydrogels, nanofiber membranes and dried scaffolds. Study conducted by Hang et al. developed a composite scaffold consisting of eggshell particles coated with magnesium (Mg) oxide nanoparticles and chitosan was engineered to deliver Mg ions and bone morphogenic protein 2 to promote osteogenesis (Huang et al., 2020). In addition, Zhu et al. pointed out that CS/polycaprolactone/metformin nanofibrous membranes facilitated the controlled release of metformin for more than 3 weeks, also promoted stem cell bioactivity, proliferation and adhesion of bone marrow mesenchymal stem cells (J. Zhu et al., 2020). Ahmadi et al. study reported that CS/gelatin nanofiber membranes permitted sustained cinnamon liberation for more than 2 weeks and exhibited antibacterial and osteogenic properties (Ahmadi et al., 2021). Biomimetic CS and gelatin scaffolds have been shown by Bakopoulou et al. to promote the differentiation and survival of dental pulp stem cells, leading to the development of calcium phosphate-rich tissue. Hydrogel injectables of CS glycerophosphate (CS/GP), reinforced with graphene oxide (GO), showed promising potential for the repair of bone defects. When graphene oxide was added, the porosity, protein absorption, degradation, biomineralization and swelling properties of the hydrogel were improved. These results make CS/GP/GO hydrogels a viable platform for bone tissue regeneration (Saravanan et al., 2018).

Injectable chitosan-based hydrogels and 3D scaffolds combining chitosan and bioactive glass are promising solutions for bone regeneration. Hydrogels, which promoted the attachment of mesenchymal stem cells, can accelerated the repair of vertebral fractures and other bone lesions, by creating a favorable environment for regeneration. At the same time, 3D scaffolds with chitosan and bioactive glass promote bone growth and mineralization, supporting the healing of large bone defects. These technologies offer significant advantages for the treatment of fractures and complex bone defects (Ghavimi et al., 2019; L. Wang et al., 2019).

The role of chitosan-based biomaterials in bone regeneration and the treatment of osteoporosis. It showed the transition from osteoporotic to healthy bone, facilitated by the activation of osteocytes, the inhibition of osteoclast proliferation, and the promotion of osteoblast proliferation. Diverse forms of chitosan biomaterials, including scaffolds, membranes, hydrogels, and nanoparticles, are depicted alongside their in vitro (bird, mouse, fish, rabbit) and in vivo (eggshell, dental enamel, bacteria) applications in bone-related studies (figure 1).

Figure 6. The effect of chitosan-based biomaterials in enhancing bone regeneration and combating Osteoporosis through cellular modulation

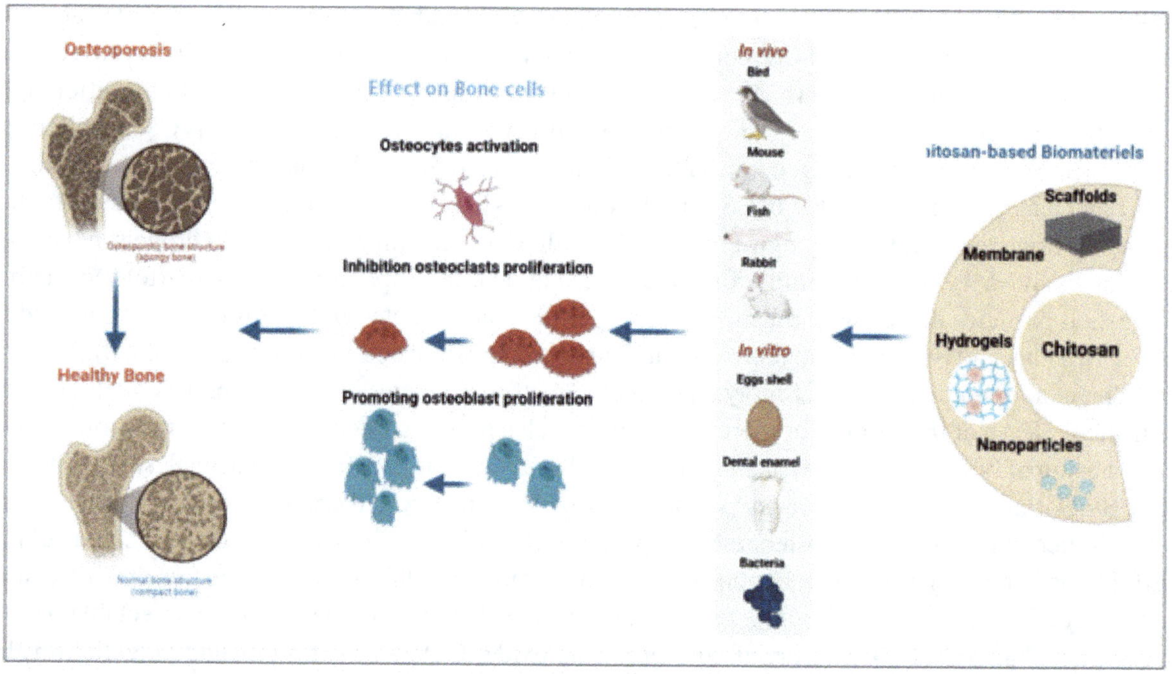

2. Preclinical and Clinical Studies on Chitosan in the Treatment of Osteoporosis

Preclinical and clinical research has demonstrated the potential of CS in treating osteoporosis and promoting bone regeneration. Chitosan had anti-inflammatory and antioxidant properties, and had been shown to reduce bone loss, promote bone formation and inhibit bone resorption. However, chitosan contributed to improving bioavailability and intestinal absorption of calcium, an essential mineral to maintain bone strength, by increasing its uptake into bone tissue bone tissue (Santhosh et al., 2019). These effects confirm its potential as a complementary treatment for disorders linked to excessive bone such as post-menopausal osteoporosis. Its beneficial effects don't stop there. Chitosan also possessed anti-inflammatory and antioxidant properties, which played a crucial role in bone health. It modulated pro-inflammatory cytokines, chemicals that promote inflammation, and limits oxidative stress, a major factor contributing to bone tissue degradation. This anti-inflammatory action is particularly important in preventing diseases such as osteoporosis, where inflammation contributes to loss of bone density (Y. Li et al., 2024).

The study showed that in animal models subjected to a hyperlipidic diet, chitosan demonstrated its efficacy in preserving bone density by reducing systemic inflammation. This result highlighted its protective potential against the deleterious effects of high-fat diets, often associated with impaired bone health (Ghavimi et al., 2019). Chitosan also acted on the RANK/RANKL/OPG pathway, a key signalling pathway involved in bone metabolism. By regulating this pathway, chitosan helped to decrease osteoclast activity, thereby reducing bone resorption and promoting the preservation of bone mass (Peng et al., 2019). These multiple effects make chitosan a promising agent for the prevention and treatment of bone disorders, particularly in inflammatory and calcium deficiency contexts. Preclinical studies on chitosan have revealed promising results in the prevention and treatment of osteoporosis, thanks to its multiple properties beneficial to bone health.

Functionalized chitosan nanoparticles have emerged as an innovative therapeutic solution for the treatment of osteoporosis. These nanoparticles are capable of encapsulating therapeutic agents such as risedronate, a drug used to inhibit bone resorption. This encapsulation improves the bioavailability of the agents, enabling more effective and targeted administration, while reducing the side effects associated with these treatments. Research has shown that these chitosan nanoparticles contribute to an increase in bone mineral density and an improvement in bone microarchitecture, reinforcing their potential in the management of osteoporosis (Santhosh et al., 2019). Research into the encapsulation of chitosan, for example with Shilajit extract, has shown that this approach enhances the absorption and targeted release of active substances, thereby boosting their therapeutic effects, particularly against osteoporosis. The use of chitosan nanoparticles (SWE/NCT) increases calcium, phosphorus and osteocalcin levels, while reducing oxidative stress and inflammation, two major factors contributing to bone degradation. This combination of biopolymers and bioactive substances, such as Shilajit, opens up promising prospects for fracture prevention and osteoporosis treatment (Alshubaily & Jambi, 2022).

Another promising advance is based on the bio-conjugate of chitosan and quercetin, a flavonoid with antioxidant and anti-inflammatory properties. This bio-conjugate has shown an enhanced osteogenic effect in cell culture studies and in vivo models. It promoted bone calcification while inhibiting the activity of osteoclasts, the cells responsible for bone resorption. This combination of chitosan and quercetin could therefore play a key role in bone regeneration and in the treatment of pathologies linked to altered bone metabolism, such as osteoporosis (Y. Li et al., 2024). These applications open up new avenues for more effective, less invasive treatments for bone health.

3. Mechanisms Underlying Therapeutic Action

3.1. Anti-Inflammatory and Antioxidant Effects

Chitosan's anti-inflammatory properties play an essential role in the prevention and treatment of bone pathologies. It acts by regulating pro-inflammatory cytokines, such as tumor necrosis factor alpha (TNF-α) and interleukin-6 (IL-6), which are often overexpressed in chronic inflammatory diseases affecting bone health (Y. Li et al., 2024). In addition, its antioxidant effects help limit the damage caused by oxidative stress, a key factor in cellular aging and bone degradation. By increasing the activity of osteoclasts and inhibiting that of osteoblasts, oxidative stress contributes to the alteration of osseous balance. By reducing free radicals and boosting antioxidant defenses, chitosan effectively protects bone cells against these deleterious effects. Chitosan has been shown to mitigate inflammation by inhibiting pro-inflammatory cytokines such as IL-1 and TNF-α. These cytokines, widely implicated in inflammatory cascades, are significantly reduced after treatment with chitosan-based complexes, particularly when combined with bioactive compounds such as quercetin. LPS-stimulated RAW264.7 macrophages, a model commonly used to assess anti-inflammatory effects, showed a notable reduction in levels of these cytokines after treatment with the chitosan-quercetin complex(S. Chen et al., 2023; MOON et al., 2020). These results suggested that chitosan acts by modulating inflammatory mediators at the cellular level.

In addition to its anti-inflammatory properties, chitosan has antioxidant capacities that help reduce oxidative stress. Oxidative stress is often an aggravating factor in inflammatory conditions and tissue damage. Thanks to its amino group-rich polymeric structure, chitosan is able to scavenge free radicals and reduce hydrogen peroxide levels in LPS-activated inflammatory cells. These properties were enhanced when combined with natural antioxidant compounds such as quercetin, which improves the neutralization of reactive oxygen species (ROS) and inflammation biomarkers such as IL-6 (Watanabe et al., 2019; Zupan et al., 2012).

3.2. Molecular Mechanisms

Chitosan interacts with cellular signalling pathways crucial to bone metabolism, notably the RANK/RANKL/OPG pathway. This pathway regulates the balance between osteoclasts and osteoblasts, playing a central role in bone resorption and formation. Chitosan acts by promoting the expression of osteoprotegerin (OPG), a natural inhibitor of RANKL (osteoclast-activating ligand), while reducing osteoclast activity. Studies on animal models have shown that chitosan can also mitigate the adverse effects of high-fat diets, often associated with systemic inflammation and degradation of bone health. These findings highlight chitosan's role in protecting against metabolic imbalances impacting bone (Ghavimi et al., 2019; Peng et al., 2019).

3.3. Combined Inhibition of Oxidative Stress and Inflammatory Cytokines

Chitosan and its derivatives, such as the chitosan-quercetin complex, offer a synergistic approach to reducing inflammation and oxidative stress. By inhibiting inflammatory mediators (TNF-α, IL-1, IL-6) and regulating the activities of enzymes such as TRAP, this complex helps limit bone degradation processes and promotes tissue regeneration. These combined mechanisms make chitosan particularly effective for biomedical applications related to chronic inflammatory conditions and oxidative disorders (MOON et al., 2020; Tsukamoto et al., 2022).

3.4. Apoptosis and Menopause Effects

Chitosan has also demonstrated significant potential in the management of apoptotic changes associated with menopause. In studies using animal models of ovariectomy (OVX), which mimic menopausal conditions, chitosan combined with calcium and vitamin D nanoparticles showed an ability to reduce the expression of pro-apoptotic genes such as BAX and caspase-3. At the same time, it increased the expression of anti-apoptotic genes such as BCL-2, helping to reduce cell death. These effects, combined with a reduction in oxidative stress, position chitosan as a promising therapeutic complement to counter bone loss associated with menopause (Dehghan-Manshadi et al., 2022). Although results concerning chitosan's efficacy in weight management are contradictory, its potential as a bone-protecting agent is promising. Studies have shown that, even without strict dietary restrictions, chitosan can help preserve bone density and reduce the risks associated with osteoporosis. Its combination with other essential nutrients such as vitamin D and calcium could further enhance its beneficial effects, offering an integrated, non-pharmacological strategy for bone disorders related to age or inflammatory conditions (Trivedi et al., 2015).

VI. FORMULATION AND ADMINISTRATION OF CHITOSAN FOR OSTEOPOROSIS

1. Chitosan Formulations for the Treatment of Osteoporosis

Chitosan is the keystone in bone tissue engineering, playing a significant role in developing next-generation treatments for osteoporosis. Its versatility and properties make it an ideal material for designing advanced pharmaceutical forms adapted to targeted and effective treatments. Combined with therapeutic molecules thanks to their properties to improve osteoporosis treatment formulations, chitosan is administered by various routes (oral, mucosal, rectal, nasal, parenteral, topical, etc.).

Its nanocomposite formulation combined with shilajit water has been highly recommended as a potential innovative treatment for (Alshubaily & Jambi, 2022). Nanoparticles are the ideal platform for the design of combined therapies that slowly release orally administered therapeutic molecules thanks to their ability to open the junctions between epithelial cells and facilitate the paracellular transport of drugs such as biophosphonates with -1% bioavailability and enable a long-term effect of bone regeneration (PATHAK et al., 2021). Researchers have validated the efficacy of a residronate-loaded chitosan nanoparticle in treating osteoporosis compared with current drug therapy. The resulting formulation can block osteoclastogenesis and inhibit bone resorption, generating new bone formation and increasing bone density(Santhosh et al., 2019). A study confirms the efficacy of treatment with chitosan nanoparticles loaded with alendronate (Sandhya et al., 2022). It can also be formulated as an injectable hydrogel or as a topical treatment. This formulation provides mechanical strength similar to that of vertebral bone, which can transport molecules and cells (Rahimi et al., 2022b) Alendronate-loaded chitosan and β-glycerophosphate thermogels enabled rapid gel formation and localization with no signs of inflammation at the injection site (Rahimi et al., 2022a), while a transdermal gel containing residronate-chitosan nanoparticles showed efficacy with significant cellular uptake and penetration of released drug exceeding 70% (Pathak et al., 2024).Nasal administration of raloxifene with a limited bioavailability of 2% has also been validated for treating osteoporosis (Abdel-Salam et al., 2020; Rahimi et al., 2022b).

The scaffold, chitosan helps target a specific region of bone and gradually release the drug, it facilitates tissue growth as a conductive environment support framework while degrading during healthy tissue substitution (Ahmed et al., 2018) providing a practical solution for bone regeneration (Zhang et al., 2024). This formulation accelerates post-operative recovery, offering properties of improved degradation, mechanical strength, controlled release and enhances osteogenic differentiation of adipose tissue-derived stem cells (Y.-M. Wang & Shen, 2024).

Several types of film formulation have been tested; mucoadhesive films improve therapeutic efficacy by escaping metabolism, and the drug is absorbed directly through the venous system of the cheek (Baus et al., 2019). Researchers have used films in the form of a bracelet loaded with quercetin formulated as a transformant, which has shown beneficial effects when applied topically to treat secondary osteoporosis (Pandit et al., 2020). An optimized transdermal delivery system for residronate has demonstrated translational clinical efficacy by offering improved penetration via skin lipids, transport, retention and distribution to the target site (Pathak et al., 2024). Chitosan is combined with all these formulations to remedy the major problem in osteoporosis treatments of low bioavailability when administered orally due to instabilities in the gastrointestinal tract, which reduce the effectiveness of formulations that escape presystematic absorption and adhere to cells.

As an encapsulating agent, chitosan in nanoparticles protects the active ingredients and improves their absorption (Abdel-Ghany & Salem, 2020). It has been shown that the bioavailability of phytochemicals(Rahat et al., 2021), liposomes (Elsayad et al., 2021) and biophosphonate has been improved, while maintaining the desired therapeutic effect combined with prolonged drug release, in addition to mitigating the negative consequences of alendronate sodium in oral administration (Jing et al., 2021).

2. Optimization of Chitosan-Based Treatments

The highly organised crystalline structure of chitosan is the cause of its limited solubility at neutral pH. Its poor stability limits its applications, leading to difficulty in developing formulations with an appropriate shelf life due to poor long-term stability where it undergoes progressive degradation of the chain during storage. The degree of purity, degree of acetylation, moisture content and molecular weight of CS influence its solubility and stability (Grewal & Salar, 2024).

Macromolecular engineering techniques can be used to remedy these problems. Studies suggest that chemical modifications such as thiolation, alkylation, acetylation, incorporation of hydrophilic groups to improve swelling behavior, cross-linking, adhesion, permeability, controllable drug release, enzyme inhibition and antioxidant properties can improve the solubility of chitosan. Another option for improving properties is the use of appropriate solvents as stabilizing agents, the use of chemical or ionic crosslinkers (Rashed et al., 2023) or changing their chemical composition (Grewal & Salar, 2024) by polyelectrolyte complexation and gelation which improves colloidal stability and encapsulation efficiency (Guadarrama-Escobar et al., 2023). As a result, these modified biomaterials exhibit varying degrees of non-toxicity, biocompatibility, stability, target specificity and biodegradability, making them more effective delivery methods for the treatment of osteoporosis (Y.-M. Wang & Shen, 2024). Also, the optimization of chitosan-based treatments can be achieved by minimizing the invasiveness of applications thanks to the potential of materials (injectable hydrogels and composite sponges) as scaffolds (Carla, 2019). In addition, a new paradigm restoring bone homeostasis from multiple targets shows promise for the treatment of osteoporosis (Y. Cui et al., 2024).

Studies have shown the potential of chitosan to transport vitamins, fatty acids and phytochemicals (Abdel-Ghany & Salem, 2020). The biocompatibility of chitosan scaffolds integrated with vitamin D has been demonstrated (Y.-M. Wang & Shen, 2024). Phytochemicals such as flavonoids are phytoestrogens (Pandit et al., 2020) that stimulate osteoblastic activity and suppress osteoclastic activity, and it has been shown that flavonoids can be mixed individually with the chitosan matrix (Bouhadjer & Zabar, 2023). Structural modifications to chitosan improve its phytochemical delivery potential and mucoadhesive characteristics, as well as intestinal permeation and phytochemical bioactivity (Imam et al., 2021).

3. Safety and Side Effects of Chitosan

Although chitosan is widely preferred because of its favorable properties, opinions differ as to its potential toxicity, while some studies validate its safety in vitro and in vivo (Y.-M. Wang & Shen, 2024), others advise low-dose administration(Sandomierski et al., 2022), and others assert toxicity that differs depending on the organ (Zoe et al., 2023). For certain cell types (liver, platelets, blood) it shows greater cytotoxicity influenced by the size of the particles; if they are small, they are associated with greater toxicity. On the other hand, chitosan concentration, stirring duration and pH did not directly affect cytotoxicity (Zoe et al., 2023).

Researchers state that chitosan can cause undesirable effects such as allergic reactions(Guadarrama-Escobar et al., 2023), by triggering an immune response, which may limit its use in clinical applications. Aspects of modulation of innate and adaptive immunity in vivo and in vitro have been revealed (Grewal & Salar, 2024). Clinical follow-up of chitosan applications seems safe to date, since it has revealed safety and significant clinical improvement after 24 months as new-scaffold-based cartilage regeneration technique (Poggi et al., 2023), after 14 weeks as a gel for chronic diabetic foot ulcers or post-surgical wound healing (Slivnik et al., 2024) but further studies are needed to confirm the safety and efficacity in osteoporosis treatments. To date, there are still restrictions on the clinical use of chitosan; fate and clearance after administration are a major challenge; there may be interactions between chitosan particles and immune cells or serum proteins. In addition, chitosan particles may accumulate in tissues or organs that are not intended targets, such as the kidneys, liver, spleen or lungs (Rashed et al., 2023).

4. Limitations and Challenges of Chitosan in Therapeutic Use for Osteoporosis

For effective clinical application, chitosan must face several limitations and challenges in its treatment of osteoporosis, despite its promising potential:

Stability and Long-Term Durability

Chitosan suffers from limited stability over time, due to the progressive degradation of its polymeric chains and the loss of functional groups during storage. This instability compromises its physicochemical properties and reduces its shelf life, impacting the long-term performance of chitosan-based formulations (Rahman & Mondal, 2024).

Hemocompatibility Issues

The hemocompatibility of chitosan remains a variable property. This variability can give rise to unpredictable immune responses and requires thorough evaluation to ensure its safety in systemic therapies, including bone applications.however Although globally biocompatible and biodegradable, chitosan exhibits variable hemocompatibility (Rahimi et al., 2022b).

Variability in Commercial Production

The irregularity of industrial production of chitosan leads to heterogeneity in its quality, molecular weight, degree of deacetylation, and other physicochemical properties. This lack of standardization is a major obstacle to its pharmaceutical use and to the reproducibility of results (Rashed et al., 2023).

Drug Delivery Challenges

With the advances in research, we are now talking about the use of chitosan nanoparticles, which are very satisfactory in results terms. However, challenges persist in this area, particularly in bioavailability and bone targeting (especially for molecules such as risedronate or raloxifene). In addition, potential cytotoxic effects have been observed, warranting long-term safety studies (Y.-M. Wang & Shen, 2024).

Low Bioavailability of Anti-Osteoporotic Drugs

Current osteoporosis treatments often have low oral bioavailability. Chitosan-based formulations (notably nasal or oral mucoadhesive systems) offer a solution to improve absorption. However, maintaining constant and effective systemic levels remains a challenge (Rahman & Mondal, 2024).

Limited Clinical Translation

Despite numerous encouraging preclinical studies, clinical validation of chitosan in the treatment of osteoporosis remains insufficient. Obstacles include the difficulty of scaling up to industrial scale, stringent regulatory requirements, and the demonstration of a clear clinical benefit (Rahman & Mondal, 2024).

VII. FUTURE PERSPECTIVES AND DEVELOPMENT OF CHITOSAN IN OSTEOPOROSIS TREATMENT

1. Innovation in Therapeutic Formulations

The application of nanomaterials in bone regeneration has gained significant attention in recent years, particularly nano-hydroxyapatite (nano-HAp), which has been extensively studied for its positive impact on the biocompatibility and bioactivity of biomaterials. Nano-HAp, typically ranging from 1 to 100 nanometers, is a promising material for enhancing bone repair. However, challenges remain in controlling the size, crystallinity, and surface characteristics of nano-HAp crystals to meet specific requirements. Integrating hydroxyapatite (HAp) with a biopolymer like chitosan offers an effective approach to improve the performance of nano-HAp composites. Research has shown that chitosan/HAp composites promote osteoconductivity in bone defects and have potential as drug delivery systems. Several studies have focused on developing chitosan-HAp nanoparticles for tissue engineering, highlighting the role of chitosan in improving the properties of HAp nanoparticles, making them more suitable for bone regeneration applications.

Bisphosphonates are the most commonly prescribed treatments for osteoporosis, with extensive clinical trials supporting their ability to reduce osteoporosis-related risks. Their strong affinity for hydroxyapatite, relatively low toxicity, and regulatory approval make them ideal for targeting bone tissue (Gendler et al., 2015). Alendronate, recognized as the most potent bisphosphonate, is the preferred drug for treating osteoporosis in clinical settings (Giger et al., 2013). The use of drug nanocarriers has been shown to reduce toxicity, improve solubility and bioavailability, facilitate controlled release, and provide better formulation options for medications (Khadilkar & Mandlik, 2015; Rumian et al., 2021). In a study by Pathak et al., chitosan-based nanocarriers were developed to enhance the therapeutic potential of osteoporosis treatments. Alendronate-loaded chitosan nanoparticles (CS NPs) were prepared to improve the drug's bioavailability and its integration into biological systems. The study demonstrated that Alendronate-loaded chitosan nanoparticles offer an effective treatment for osteoporosis. The results of this research suggest a novel strategy for drug development and delivery, with Alendronate encapsulated in the nanoscale range. Alendronate-loaded chitosan NPs, produced via the ionic gelation technique, exhibited notable drug release profiles and improved therapeutic efficacy for osteoporosis treatment (Sandhya et al., 2022).

Abouelseoud et al. aimed to assess the effectiveness of Lepidium sativum (LS) extract encapsulated in chitosan nanoparticles for the treatment of osteoporosis in rats. After inducing osteoporosis in the rats, the animals were treated with LS, chitosan nanoparticles, or a combination of both. The results revealed improvements in bone biochemical markers and changes in the expression of microRNAs (miR-23a and miR-142-3p) associated with bone remodeling. Histopathological analysis supported these findings, indicating the therapeutic potential of this approach. Furthermore, the treatment was found to be safe and free of side effects (Abouelseoud et al., 2024). Another study explored the pulsatile drug delivery of Teriparatide, a drug used for osteoporosis, through a chitosan hydrogel loaded with near-infrared light-sensitive reduced graphene oxide. In vitro and in vivo experiments showed that this biomimetic delivery system enhanced bone regeneration by regulating drug release, stimulating osteogenesis, and promoting angiogenesis (X. Wang et al., 2021).

In the context of bone tissue engineering, research on BMP-2 (Bone Morphogenetic Protein-2) revealed that sulfated and carboxylated cellulose derivatives, along with sulfated chitosan, could significantly enhance the osteogenic activity of BMP-2. This suggests that these derivatives could be used as potential modulators in osteoporosis treatment (Peschel et al., 2012). Risedronate sodium, a nitrogen-containing bisphosphonate commonly used to treat osteoporosis, has a very low oral bioavailability (less than 1%), leading to the development of alternative formulations. Chitosan nanoparticles, formed by ionic gelation with tripolyphosphate (TPP), have been used to improve the drug's delivery. Molecular modeling and in vivo studies have shown that these risedronate-chitosan nanoparticles selectively target and inhibit human farnesyl diphosphate synthase, an enzyme involved in the mevalonate pathway. This targeted inhibition suppresses osteoclastogenesis, prevents bone resorption, and promotes bone regeneration, ultimately increasing bone density (Cardozo et al., 2021; Kim et al., 2016; Santhosh et al., 2019).

2. Regulation and Commercialization

The regulatory challenges for approving chitosan as a treatment for osteoporosis are multifaceted and demand comprehensive clinical and preclinical studies to ensure its safety and efficacy. Regulatory authorities must guarantee that chitosan meets rigorous safety standards and clinical efficacy, while also adhering to stringent manufacturing and quality control guidelines. If these challenges are successfully addressed, chitosan could offer a promising alternative for osteoporosis treatment, particularly for patients seeking natural or less invasive therapies. These challenges can be classified into several key areas: safety and toxicity assessment, clinical efficacy and trials, manufacturing and quality standards, pharmacokinetic and pharmacodynamic properties, ethical considerations, patient acceptance, and regulations specific to biological products or medical devices.

The commercial potential of chitosan as a treatment for osteoporosis is driven by its appeal in the growing market for safer, less invasive alternatives. As demand for natural and biological solutions increases, chitosan could attract patients seeking alternatives to traditional medications. Additionally, if chitosan proves effective and safe, it could offer a more cost-effective option compared to conventional treatments such as bisphosphonates or hormone injections, benefiting both patients and healthcare systems. Furthermore, the rising awareness of osteoporosis in developing countries could present significant market opportunities for affordable treatments like chitosan, particularly as osteoporosis is increasingly recognized as a public health concern.Bone defects and injuries significantly affect patients' quality of life, necessitating advanced therapeutic solutions. Traditional approaches include bone graft transplants (autologous, allogeneic, and xenogenic), bone transport methods like the Ilizarov technique, and implants made from various materials. Each year, over 2.2 million bone graft procedures are performed globally, generating over $2.5 billion in revenue (Golubovsky et al., 2021; Lewandrowski et al., 2000; Lu et al., 2019). Autografts, the gold standard, are highly effective but pose challenges such as donor-site morbidity, limited tissue availability, significant blood loss, and high costs (Y.-C. Chen et al., 2006; Muramatsu et al., 2003; Zermatten & Wettstein, 2012). To address these limitations, chitosan-based scaffolds have emerged as a promising alternative in bone tissue engineering. Due to its biocompatibility, biodegradability, and mechanical properties, chitosan facilitates osteogenic differentiation of mesenchymal stem cells, promoting bone regeneration. Various production techniques, including freeze-drying, solvent exchange, and electrospinning, have been employed to develop these scaffolds, with both in vitro and in vivo studies demonstrating their potential to enhance bone repair (Costa-Pinto et al., 2011). Furthermore, carboxymethyl chitosan (CM-chitosan) has shown efficacy in knee osteoarthritis treatment, with clinical data from 133 injections in 112 patients indicating symptom relief lasting up to 35 months (Van Overschelde et al., 2023). Chitosan also exhibits potential in treating bone fractures and bone cancer by stimulating osteoblast proliferation, inhibiting osteoclastogenesis, reducing endothelial cell invasion, and inducing pro-apoptotic effects in osteotropic cancer cells, as demonstrated in recent studies (Tan et al., 2014).

3. Recent Chitosan-Based Osteoporosis Therapeutic Approaches

In this context, innovative therapeutic approaches based on chitosan have recently been developed for the treatment of osteoporosis. The table below summarizes some highly promising strategies, their mechanisms of action, and their current development status (Afra et al., 2024; Sandhya et al., 2022; Y.-M. Wang & Shen, 2024; Zhang et al., 2024).

Table 1. Recent therapeutic approaches based on chitosan for the treatment of osteoporosis

Approach	Description & Findings	Status
Risedronate-loaded chitosan NPs	Improved targeting, inhibition of osteoclasts, enhanced bone density in vivo	Preclinical, promising
Alendronate-loaded chitosan NPs	Oral delivery optimization, controlled release, potential effective osteoporosis treatment	Preclinical, formulation
Injectable chitosan hydrogels	Sustained release of anti-osteoporotic agents (e.g., salvianolic acid B), bone formation promotion	Preclinical animal models
3D-printed chitosan scaffolds	Localized drug delivery, mechanical strength, osteoclast inhibition, biocompatibility	Preclinical, advanced stage
Raloxifene chitosan NPs (intranasal)	Enhanced bioavailability via nasal mucosa, sustained release	Preclinical
Transdermal risedronate-chitosan gel	Good skin penetration and cellular uptake, potential non-invasive delivery	Ex vivo/in vitro studies

VIII. CONCLUSION

Chitosan, a natural biopolymer derived from chitin, emerges as a promising therapeutic option for the treatment of osteoporosis, particularly in postmenopausal women. With its distinctive characteristics, such as biocompatibility, biodegradability, and the ability to form three-dimensional structures, this biopolymer shows significant potential for promoting bone regeneration and regulating the biological mechanisms responsible for bone density loss. Current research suggests that chitosan can specifically influence the processes of bone formation and resorption, notably by reducing inflammation and oxidative stress. These properties could offer an effective solution for increasing bone density and reducing the risk of fractures, while minimizing the side effects associated with traditional treatments such as bisphosphonates and hormone therapies.

However, although preclinical and clinical studies have highlighted the potential benefits of chitosan, further research is needed to confirm its long-term efficacy and safety. Improving chitosan formulations, through techniques such as nano-encapsulation or combining it with other nutrients and biomolecules, could optimize its bioavailability and effectiveness. Furthermore, challenges related to the regulation and commercialization of chitosan as an osteoporosis treatment require careful attention to facilitate its clinical adoption.

REFERENCES

Abdel-Ghany, H. M., & Salem, M. E. (2020). Effects of dietary chitosan supplementation on farmed fish; a review. *Reviews in Aquaculture, 12*(1), 438–452. DOI: 10.1111/raq.12326

Abdel-Salam, F. S., Elkheshen, S. A., Mahmoud, A. A., Basalious, E. B., Amer, M. S., Mostafa, A. A., & Elkasabgy, N. A. (2020). In-situ forming chitosan implant-loaded with raloxifene hydrochloride and bioactive glass nanoparticles for treatment of bone injuries: Formulation and biological evaluation in animal model. *International Journal of Pharmaceutics, 580*, 119213. DOI: 10.1016/J.IJPHARM.2020.119213 PMID: 32165229

Abouelseoud, I., Salama, S., Mahmoud, B., Dawood, H., Moustafa, M., & Mahmoud, S. (2024). The Effect of Lepidium Sativum L. Extract Loaded on Chitosan Nanoparticles in Experimental Rat Model of Osteoporosis. *Journal of the Medical Research Institute, 45*(1), 28–39. DOI: 10.21608/jmalexu.2024.265274.1011

Afra, S., Koch, M., Żur-Pińska, J., Dolatshahi, M., Bahrami, A. R., Sayed, J. E., Moradi, A., Matin, M. M., & Włodarczyk-Biegun, M. K. (2024). Chitosan/Nanohydroxyapatite/Hydroxyethyl-cellulose-based printable formulations for local alendronate drug delivery in osteoporosis treatment. *Carbohydrate Polymer Technologies and Applications, 7*, 100418. DOI: 10.1016/j.carpta.2023.100418

Ahmadi, S., Hivechi, A., Bahrami, S. H., Milan, P. B., & Ashraf, S. S. (2021). Cinnamon extract loaded electrospun chitosan/gelatin membrane with antibacterial activity. *International Journal of Biological Macromolecules, 173*, 580–590. DOI: 10.1016/j.ijbiomac.2021.01.156 PMID: 33513421

Ahmed, S., Annu, Ali, A., & Sheikh, J. (2018). A review on chitosan centred scaffolds and their applications in tissue engineering. *International Journal of Biological Macromolecules, 116*, 849–862. DOI: 10.1016/J.IJBIOMAC.2018.04.176 PMID: 29730001

Akdeniz, N., Akpolat, V., Kale, A., Erdemoglu, M., Kuyumcuoglu, U., & Celik, Y. (2009). Risk factors for postmenopausal osteoporosis: Anthropometric measurements, age, age at menopause and the time elapsed after menopause onset. *Gynecological Endocrinology, 25*(2), 125–129. DOI: 10.1080/09513590802549817 PMID: 19253109

Alshubaily, F. A., & Jambi, E. J. (2022). Correlation between Antioxidant and Anti-Osteoporotic Activities of Shilajit Loaded into Chitosan Nanoparticles and Their Effects on Osteoporosis in Rats. *Polymers, 14*(19), 3972. DOI: 10.3390/polym14193972 PMID: 36235920

Anitha, A., Sowmya, S., Kumar, P. T. S., Deepthi, S., Chennazhi, K. P., Ehrlich, H., Tsurkan, M., & Jayakumar, R. (2014). Chitin and chitosan in selected biomedical applications. *Progress in Polymer Science, 39*(9), 1644–1667. DOI: 10.1016/J.PROGPOLYMSCI.2014.02.008

Baron, R., Ferrari, S., & Russell, R. G. G. (2011). Denosumab and bisphosphonates: Different mechanisms of action and effects. *Bone, 48*(4), 677–692. DOI: 10.1016/J.BONE.2010.11.020 PMID: 21145999

Baus, R. A., Zahir-Jouzdani, F., Dünnhaupt, S., Atyabi, F., & Bernkop-Schnürch, A. (2019). Mucoadhesive hydrogels for buccal drug delivery: In vitro-in vivo correlation study. *European Journal of Pharmaceutics and Biopharmaceutics, 142*, 498–505. DOI: 10.1016/J.EJPB.2019.07.019 PMID: 31330258

Bolton, J. L. (2016). Menopausal Hormone Therapy, Age, and Chronic Diseases: Perspectives on Statistical Trends. *Chemical Research in Toxicology*, *29*(10), 1583–1590. DOI: 10.1021/acs.chemrestox.6b00272 PMID: 27636306

Bonilla, J., Fortunati, E., Atarés, L., Chiralt, A., & Kenny, J. M. (2014). Physical, structural and antimicrobial properties of poly vinyl alcohol–chitosan biodegradable films. *Food Hydrocolloids*, *35*, 463–470. DOI: 10.1016/j.foodhyd.2013.07.002

Bouhadjer, R., & Zabar, L. (2023). *Étude bibliographique sur l'activité des flavonoides sur l'organisme humain*.

Brugnerotto, J., Lizardi, J., Goycoolea, F., Argüelles-Monal, W., Desbrières, J., & Rinaudo, M. (2001). An infrared investigation in relation with chitin and chitosan characterization. *Polymer*, *42*(8), 3569–3580. DOI: 10.1016/S0032-3861(00)00713-8

Cano-Vicent, A., Tuñón-Molina, A., Martí, M., & Serrano-Aroca, Á. (2023). Biocompatible Chitosan Films Containing Acetic Acid Manifested Potent Antiviral Activity against Enveloped and Non-Enveloped Viruses. *International Journal of Molecular Sciences*, *24*(15), 12028. DOI: 10.3390/ijms241512028 PMID: 37569404

Cardozo, B., Karatza, E., & Karalis, V. (2021). Osteoporosis treatment with risedronate: A population pharmacokinetic model for the description of its absorption and low plasma levels. *Osteoporosis International*, *32*(11), 2313–2321. DOI: 10.1007/s00198-021-05944-0 PMID: 34002251

Carla, P. D. (2019). *Hydrogels injectables et éponges à base de complexe polyélectrolytes (chitosane/polymère de cyclodextrine) pour une application en ingénierie tissulaire osseuse*.

Chen, S., Saeed, A. F. U. H., Liu, Q., Jiang, Q., Xu, H., Xiao, G. G., Rao, L., & Duo, Y. (2023). Macrophages in immunoregulation and therapeutics. *Signal Transduction and Targeted Therapy*, *8*(1), 207. DOI: 10.1038/s41392-023-01452-1 PMID: 37211559

Chen, Y.-C., Chen, C.-H., Chen, P.-L., Huang, I.-Y., Shen, Y.-S., & Chen, C.-M. (2006). Donor site morbidity after harvesting of proximal tibia bone. *Head & Neck*, *28*(6), 496–500. DOI: 10.1002/hed.20452 PMID: 16673417

Chiu, Y. G., & Ritchlin, C. T. (2017). Denosumab: Targeting the RANKL pathway to treat rheumatoid arthritis. *Expert Opinion on Biological Therapy*, *17*(1), 119–128. DOI: 10.1080/14712598.2017.1263614 PMID: 27871200

Clarke, B. L., & Khosla, S. (2010). Physiology of Bone Loss. *Radiologic Clinics of North America*, *48*(3), 483–495. DOI: 10.1016/j.rcl.2010.02.014 PMID: 20609887

Clunie, G. P., Clark, A., Mortimer, C. J., Stephenson, S., Aitken, J., Smith, C., Sherwin, E., & Archer, T. J. (2009). Evaluating bone health in women with oestrogen receptor positive breast cancer (ERBC) starting aromatase inhibitors. [EJSO]. *European Journal of Surgical Oncology*, *35*(5), 475–480. DOI: 10.1016/j.ejso.2008.08.001 PMID: 18950981

Costa-Pinto, A. R., Reis, R. L., & Neves, N. M. (2011). Scaffolds Based Bone Tissue Engineering: The Role of Chitosan. *Tissue Engineering. Part B, Reviews, 17*(5), 331–347. DOI: 10.1089/ten.teb.2010.0704 PMID: 21810029

Crini, G. (2019). Historical review on chitin and chitosan biopolymers. *Environmental Chemistry Letters, 17*(4), 1623–1643. DOI: 10.1007/s10311-019-00901-0

Cui, Y., Lv, B., Li, Z., Ma, C., Gui, Z., Geng, Y., Liu, G., Sang, L., Xu, C., Min, Q., Kong, L., Zhang, Z., Liu, Y., Qi, X., & Fu, D. (2024). Bone-Targeted Biomimetic Nanogels Re-Establish Osteoblast/Osteoclast Balance to Treat Postmenopausal Osteoporosis. *Small, 20*(6). Advance online publication. DOI: 10.1002/smll.202303494 PMID: 37794621

Cui, Z. K., Kim, S., Baljon, J. J., Wu, B. M., Aghaloo, T., & Lee, M. (2019). Microporous methacrylated glycol chitosan-montmorillonite nanocomposite hydrogel for bone tissue engineering. *Nature Communications, 10*(1). Advance online publication. DOI: 10.1038/s41467-019-11511-3 PMID: 31388014

Daraghmeh, N. H., Chowdhry, B. Z., Leharne, S. A., Al Omari, M. M., & Badwan, A. A. (2011). *Chitin* (pp. 35–102). https://doi.org/DOI: 10.1016/B978-0-12-387667-6.00002-6

Dehghan-Manshadi, M., Azarbayjani, M.-A., Atashak, S., Peeri, M., & Rahmati-Ahmadabad, S. (2022). Effect of resistance training with and without vitamin D calcium chitosan nanoparticle supplements on apoptosis markers in ovariectomized rats: An experimental study. *International Journal of Reproductive Biomedicine, 20*(7), 549–560. DOI: 10.18502/ijrm.v20i7.11557 PMID: 36187741

Dessay, M., Couture, E., Maaroufi, H., Fournier, F., Gagnon, E., Droit, A., Brown, J. P., & Michou, L. (2022). Attenuated clinical and osteoclastic phenotypes of Paget's disease of bone linked to the p.Pro-392Leu/SQSTM1 mutation by a rare variant in the DOCK6 gene. *BMC Medical Genomics, 15*(1), 41. DOI: 10.1186/s12920-022-01198-9 PMID: 35241069

Dev, A., Binulal, N. S., Anitha, A., Nair, S. V., Furuike, T., Tamura, H., & Jayakumar, R. (2010). Preparation of poly(lactic acid)/chitosan nanoparticles for anti-HIV drug delivery applications. *Carbohydrate Polymers, 80*(3), 833–838. DOI: 10.1016/j.carbpol.2009.12.040

Dilamian, M., Montazer, M., & Masoumi, J. (2013). Antimicrobial electrospun membranes of chitosan/poly(ethylene oxide) incorporating poly(hexamethylene biguanide) hydrochloride. *Carbohydrate Polymers, 94*(1), 364–371. DOI: 10.1016/j.carbpol.2013.01.059 PMID: 23544550

Ehrlich, H., Krajewska, B., Hanke, T., Born, R., Heinemann, S., Knieb, C., & Worch, H. (2006). Chitosan membrane as a template for hydroxyapatite crystal growth in a model dual membrane diffusion system. *Journal of Membrane Science, 273*(1–2), 124–128. DOI: 10.1016/J.MEMSCI.2005.11.050

Elsayad, M. K., Mowafy, H. A., Zaky, A. A., & Samy, A. M. (2021). Chitosan caged liposomes for improving oral bioavailability of rivaroxaban: *In vitro* and *in vivo* evaluation. *Pharmaceutical Development and Technology, 26*(3), 316–327. DOI: 10.1080/10837450.2020.1870237 PMID: 33356742

Emmanuel Jeanpierre. (2014). *L'ostéoporose et son traitement à l'épreuve des faits. Etude rétrospective des pratiques cliniques de la prise en charge médicale proposée à 222 patients, recrutés en 4 ans dans la filière "pour la Qualité de l'Os en Lorraine"*. QOL.

Ezoddini-Ardakani, F., Navabazam, A., Fatehi, F., Danesh-Ardekani, M., Khadem, S., & Rouhi, G. (2012). Histologic evaluation of chitosan as an accelerator of bone regeneration in microdrilled rat tibias. *Dental Research Journal (Isfahan)*, *9*(6), 694–699. PMID: 23559943

Ferreira, S. S., Passos, C. P., Madureira, P., Vilanova, M., & Coimbra, M. A. (2015). Structure–function relationships of immunostimulatory polysaccharides: A review. *Carbohydrate Polymers*, *132*, 378–396. DOI: 10.1016/J.CARBPOL.2015.05.079 PMID: 26256362

Foster, L. J. R., Ho, S., Hook, J., Basuki, M., & Marçal, H. (2015). Chitosan as a Biomaterial: Influence of Degree of Deacetylation on Its Physiochemical, Material and Biological Properties. *PLoS One*, *10*(8), e0135153. DOI: 10.1371/journal.pone.0135153 PMID: 26305690

Friedman, A. J., Phan, J., Schairer, D. O., Champer, J., Qin, M., Pirouz, A., Blecher-Paz, K., Oren, A., Liu, P. T., Modlin, R. L., & Kim, J. (2013). Antimicrobial and Anti-Inflammatory Activity of Chitosan–Alginate Nanoparticles: A Targeted Therapy for Cutaneous Pathogens. *The Journal of Investigative Dermatology*, *133*(5), 1231–1239. DOI: 10.1038/jid.2012.399 PMID: 23190896

Galed, G., Miralles, B., Panos, I., Santiago, A., & Heras, A. (2005). -Deacetylation and depolymerization reactions of chitin/chitosan: Influence of the source of chitin. *Carbohydrate Polymers*, *62*(4), 316–320. DOI: 10.1016/j.carbpol.2005.03.019

Gallo, J., Raska, M., Kriegova, E., & Goodman, S. B. (2017). Inflammation and its resolution and the musculoskeletal system. *Journal of Orthopaedic Translation*, *10*, 52–67. DOI: 10.1016/j.jot.2017.05.007 PMID: 28781962

Gendler, P. E., Coviak, C. P., Martin, J. T., Kim, K. K., Dankers, J. K., Barclay, J. M., & Sanchez, T. A. (2015). Revision of the Osteoporosis Knowledge Test. *Western Journal of Nursing Research*, *37*(12), 1623–1643. DOI: 10.1177/0193945914537565 PMID: 24923463

Georgopoulou, A., Papadogiannis, F., Batsali, A., Marakis, J., Alpantaki, K., Eliopoulos, A. G., Pontikoglou, C., & Chatzinikolaidou, M. (2018). Chitosan/gelatin scaffolds support bone regeneration. *Journal of Materials Science. Materials in Medicine*, *29*(5). Advance online publication. DOI: 10.1007/s10856-018-6064-2 PMID: 29730855

Ghavimi, S. A. A., Lungren, E. S., Faulkner, T. J., Josselet, M. A., Wu, Y., Sun, Y., Pfeiffer, F. M., Goldstein, C. L., Wan, C., & Ulery, B. D. (2019). Inductive co-crosslinking of cellulose nanocrystal/chitosan hydrogels for the treatment of vertebral compression fractures. *International Journal of Biological Macromolecules*, *130*, 88–98. DOI: 10.1016/J.IJBIOMAC.2019.02.086 PMID: 30779988

Giger, E. V., Castagner, B., & Leroux, J. C. (2013). Biomedical applications of bisphosphonates. *Journal of Controlled Release*, *167*(2), 175–188. DOI: 10.1016/J.JCONREL.2013.01.032 PMID: 23395668

Gim, S., Zhu, Y., Seeberger, P. H., & Delbianco, M. (2019). Carbohydrate-based nanomaterials for biomedical applications. *Wiley Interdisciplinary Reviews. Nanomedicine and Nanobiotechnology*, *11*(5). Advance online publication. DOI: 10.1002/wnan.1558 PMID: 31063240

Goldberg, T., & Fidler, B. (2015). Conjugated Estrogens/Bazedoxifene (Duavee): A Novel Agent for the Treatment of Moderate-to-Severe Vasomotor Symptoms Associated With Menopause And the Prevention of Postmenopausal Osteoporosis. *P&T : a Peer-Reviewed Journal for Formulary Management*, *40*(3), 178–182. PMID: 25798038

Golubovsky, J. L., Ejikeme, T., Winkelman, R., & Steinmetz, M. P. (2021). Osteobiologics. *Operative Neurosurgery (Hagerstown, Md.)*, *21*(Supplement_1), S2–S9. DOI: 10.1093/ons/opaa383 PMID: 34128067

Grewal, A. K., & Salar, R. K. (2024). Chitosan nanoparticle delivery systems: An effective approach to enhancing efficacy and safety of anticancer drugs. *Nano TransMed*, *3*, 100040. DOI: 10.1016/J.NTM.2024.100040

Guadarrama-Escobar, O. R., Serrano-Castañeda, P., Anguiano-Almazán, E., Vázquez-Durán, A., Peña-Juárez, M. C., Vera-Graziano, R., Morales-Florido, M. I., Rodriguez-Perez, B., Rodriguez-Cruz, I. M., Miranda-Calderón, J. E., & Escobar-Chávez, J. J. (2023). Chitosan Nanoparticles as Oral Drug Carriers. *International Journal of Molecular Sciences*, *24*(5), 4289. DOI: 10.3390/ijms24054289 PMID: 36901719

Guillot, X. (2015). Les différents types d'ostéoporose et leurs stratégies thérapeutiques. *Actualites Pharmaceutiques*, *54*(549), 18–22. DOI: 10.1016/j.actpha.2015.07.004

Hakima, el K., Raja, B., Redouan, E. K., Ali, L., A., A., & Ahmed, lahsini. (2017). Physicochemical characterization of chitin and chitosan producted from Parapenaeus Longirostris shrimp shell wastes. *Journal of Materials and Environmental Sciences* .

Hamedi, H., Moradi, S., Hudson, S. M., & Tonelli, A. E. (2018). Chitosan based hydrogels and their applications for drug delivery in wound dressings: A review. *Carbohydrate Polymers*, *199*, 445–460. DOI: 10.1016/j.carbpol.2018.06.114 PMID: 30143150

Hsu, S.-H., Chen, L.-R., & Chen, K.-H. (2024). Primary Osteoporosis Induced by Androgen and Estrogen Deficiency: The Molecular and Cellular Perspective on Pathophysiological Mechanisms and Treatments. *International Journal of Molecular Sciences*, *25*(22), 12139. DOI: 10.3390/ijms252212139 PMID: 39596206

Huang, Y. zhuo, Ji, Y. rong, Kang, Z. wen, Li, F., Ge, S. fang, Yang, D. P., Ruan, J., & Fan, X. qun. (2020). Integrating eggshell-derived CaCO3/MgO nanocomposites and chitosan into a biomimetic scaffold for bone regeneration. *Chemical Engineering Journal*, *395*. Advance online publication. DOI: 10.1016/j.cej.2020.125098

Husain, S., Al-Samadani, K. H., Najeeb, S., Zafar, M. S., Khurshid, Z., Zohaib, S., & Qasim, S. B. (2017). Chitosan Biomaterials for Current and Potential Dental Applications. *Materials (Basel)*, *10*(6), 602. DOI: 10.3390/ma10060602 PMID: 28772963

Imam, S. S., Alshehri, S., Ghoneim, M. M., Zafar, A., Alsaidan, O. A., Alruwaili, N. K., Gilani, S. J., & Rizwanullah, M. (2021). Recent Advancement in Chitosan-Based Nanoparticles for Improved Oral Bioavailability and Bioactivity of Phytochemicals: Challenges and Perspectives. *Polymers*, *13*(22), 4036. DOI: 10.3390/polym13224036 PMID: 34833334

Jaber, N., Al-Remawi, M., Al-Akayleh, F., Al-Muhtaseb, N., Al-Adham, I. S. I., & Collier, P. J. (2022). A review of the antiviral activity of Chitosan, including patented applications and its potential use against COVID-19. *Journal of Applied Microbiology, 132*(1), 41–58. DOI: 10.1111/jam.15202 PMID: 34218488

Jayakumar, R., Prabaharan, M., Nair, S. V., & Tamura, H. (2010). Novel chitin and chitosan nanofibers in biomedical applications. *Biotechnology Advances, 28*(1), 142–150. DOI: 10.1016/J.BIOTECHADV.2009.11.001 PMID: 19913083

Jayakumar, R., Prabaharan, M., Nair, S. V., Tokura, S., Tamura, H., & Selvamurugan, N. (2010). Novel carboxymethyl derivatives of chitin and chitosan materials and their biomedical applications. *Progress in Materials Science, 55*(7), 675–709. DOI: 10.1016/j.pmatsci.2010.03.001

Jeon, Y. (2001). Antimicrobial effect of chitooligosaccharides produced by bioreactor. *Carbohydrate Polymers, 44*(1), 71–76. DOI: 10.1016/S0144-8617(00)00200-9

Jing, C., Li, B., Tan, H., Zhang, C., Liang, H., Na, H., Chen, S., Liu, C., & Zhao, L. (2021). Alendronate-Decorated Nanoparticles as Bone-Targeted Alendronate Carriers for Potential Osteoporosis Treatment. *ACS Applied Bio Materials, 4*(6), 4907–4916. DOI: 10.1021/acsabm.1c00199 PMID: 35007039

Juncos Bombin, A. D., Dunne, N. J., & McCarthy, H. O. (2020). Electrospinning of natural polymers for the production of nanofibres for wound healing applications. *Materials Science and Engineering C, 114*, 110994. DOI: 10.1016/J.MSEC.2020.110994 PMID: 32993991

Kean, T., & Thanou, M. (2011). Chitin and Chitosan: Sources, Production and Medical Applications. In *Renewable Resources for Functional Polymers and Biomaterials* (pp. 292–318). The Royal Society of Chemistry. https://doi.org/DOI: 10.1039/9781849733519-00292

Kendler, D. L., Cosman, F., Stad, R. K., & Ferrari, S. (2022). Denosumab in the Treatment of Osteoporosis: 10 Years Later: A Narrative Review. *Advances in Therapy, 39*(1), 58–74. DOI: 10.1007/s12325-021-01936-y PMID: 34762286

Khadilkar, A., & Mandlik, R. (2015). Epidemiology and treatment of osteoporosis in women: An Indian perspective. *International Journal of Women's Health, 841*. Advance online publication. DOI: 10.2147/IJWH.S54623 PMID: 26527900

Khan, F., & Ahmad, S. R. (2013). Polysaccharides and Their Derivatives for Versatile Tissue Engineering Application. *Macromolecular Bioscience, 13*(4), 395–421. DOI: 10.1002/mabi.201200409 PMID: 23512290

Kim, J. S., Jang, S. W., Son, M., Kim, B. M., & Kang, M. J. (2016). Enteric-coated tablet of risedronate sodium in combination with phytic acid, a natural chelating agent, for improved oral bioavailability. *European Journal of Pharmaceutical Sciences, 82*, 45–51. DOI: 10.1016/J.EJPS.2015.11.011 PMID: 26594027

Kumar, P. T. S., Abhilash, S., Manzoor, K., Nair, S. V., Tamura, H., & Jayakumar, R. (2010). Preparation and characterization of novel β-chitin/nanosilver composite scaffolds for wound dressing applications. *Carbohydrate Polymers, 80*(3), 761–767. DOI: 10.1016/j.carbpol.2009.12.024

Laurent, M. R., Goemaere, S., Verroken, C., Bergmann, P., Body, J.-J., Bruyère, O., Cavalier, E., Rozenberg, S., Lapauw, B., & Gielen, E. (2022). Prevention and Treatment of Glucocorticoid-Induced Osteoporosis in Adults: Consensus Recommendations From the Belgian Bone Club. *Frontiers in Endocrinology, 13*. Advance online publication. DOI: 10.3389/fendo.2022.908727 PMID: 35757436

Lee, K. Y., & Mooney, D. J. (2012). Alginate: Properties and biomedical applications. *Progress in Polymer Science, 37*(1), 106–126. DOI: 10.1016/j.progpolymsci.2011.06.003 PMID: 22125349

Lee, S. J., Nah, H., Heo, D. N., Kim, K. H., Seok, J. M., Heo, M., Moon, H. J., Lee, D., Lee, J. S., An, S. Y., Hwang, Y. S., Ko, W. K., Kim, S. J., Sohn, S., Park, S. A., Park, S. Y., & Kwon, I. K. (2020). Induction of osteogenic differentiation in a rat calvarial bone defect model using an In situ forming graphene oxide incorporated glycol chitosan/oxidized hyaluronic acid injectable hydrogel. *Carbon, 168*, 264–277. DOI: 10.1016/j.carbon.2020.05.022

Lepage, L., Poivret, D., & Lemétayer, F. (2024). Relationships between prescribed anti-osteoporotic treatment and necessity-concerns assessment: A psychological network analysis. *Current Psychology (New Brunswick, N.J.), 43*(5), 4098–4112. DOI: 10.1007/s12144-023-04611-3

Lewandrowski, K. U. D., Gresser, J., Wise, D. L., & Trantolo, D. J. (2000). Bioresorbable bone graft substitutes of different osteoconductivities: A histologic evaluation of osteointegration of poly(propylene glycol-co-fumaric acid)-based cement implants in rats. *Biomaterials, 21*(8), 757–764. DOI: 10.1016/S0142-9612(99)00179-9 PMID: 10721744

Li, H., Xiao, Z., Quarles, L. D., & Li, W. (2021). Osteoporosis: Mechanism, Molecular Target and Current Status on Drug Development. *Current Medicinal Chemistry, 28*(8), 1489–1507. DOI: 10.2174/0929867327666200330142432 PMID: 32223730

Li, Y., Selvaraj, V., Saravanan, S., Abullais, S. S., & Wankhade, V. (2024). Exploring the osteogenic potential of chitosan-quercetin bio-conjugate: In vitro and in vivo investigations in osteoporosis models. *International Journal of Biological Macromolecules, 274*, 133492. DOI: 10.1016/J.IJBIOMAC.2024.133492 PMID: 38944072

Lingnert, H., Vallentin, K., & Eriksson, C. E. (1979). Measurement of antioxidative effect in model system. *Journal of Food Processing and Preservation.*

Lu, Y., Li, M., & Long, Z., Di Yang, Guo, S., Li, J., Liu, D., Gao, P., Chen, G., Lu, X., Lu, J., & Wang, Z. (2019). Collagen/ β -TCP composite as a bone-graft substitute for posterior spinal fusion in rabbit model: A comparison study. *Biomedical Materials (Bristol, England), 14*(4), 045009. DOI: 10.1088/1748-605X/ab1caf PMID: 31022713

Mbese, Z., & Aderibigbe, B. A. (2021). Bisphosphonate-Based Conjugates and Derivatives as Potential Therapeutic Agents in Osteoporosis, Bone Cancer and Metastatic Bone Cancer. *International Journal of Molecular Sciences, 22*(13), 6869. DOI: 10.3390/ijms22136869 PMID: 34206757

MOON, S.-H., JI, S. H., SON, J.-L., SHIN, S.-J., OH, S., KIM, S. H., & BAE, J.-M. (2020). Antibacterial, anti-inflammatory, and anti-osteoclastogenic activities of Potential applications in preventing and treating periodontal diseases. *Dental Materials Journal, 39*(6), 1096–1102. https://doi.org/DOI: 10.4012/dmj.2020-157

Motlani, G., Motlani, V., Acharya, N., Dave, A., Pamnani, S., Somyani, D., & Agrawal, S. (2023). Novel Advances in the Role of Selective Estrogen Receptor Modulators in Hormonal Replacement Therapy: A Paradigm Shift. *Cureus*. Advance online publication. DOI: 10.7759/cureus.49079 PMID: 38125238

Munhoz, M. A. S., Hirata, H. H., Plepis, A. M. G., Martins, V. C. A., & Cunha, M. R. (2018). Use of collagen/chitosan sponges mineralized with hydroxyapatite for the repair of cranial defects in rats. *Injury*, *49*(12), 2154–2160. DOI: 10.1016/j.injury.2018.09.018 PMID: 30268514

Muramatsu, K., Doi, K., Ihara, K., Shigetomi, M., & Kawai, S. (2003). Recalcitrant posttraumatic nonunion of the humerus: 23 patients reconstructed with vascularized bone graft: 23 patients reconstructed with vascularized bone graft. *Acta Orthopaedica Scandinavica*, *74*(1), 95–97. DOI: 10.1080/00016470310013734 PMID: 12635801

Nguyen, T. P., Nguyen, Q. V., Nguyen, V.-H., Le, T.-H., Huynh, V. Q. N., Vo, D.-V. N., Trinh, Q. T., Kim, S. Y., & Van Le, Q. (2019). Silk Fibroin-Based Biomaterials for Biomedical Applications: A Review. *Polymers*, *11*(12), 1933. DOI: 10.3390/polym11121933 PMID: 31771251

No, H. K., & Meyers, S. P. (1995). Preparation and Characterization of Chitin and Chitosan—A Review. *Journal of Aquatic Food Product Technology*, *4*(2), 27–52. DOI: 10.1300/J030v04n02_03

Noh, J.-Y., Yang, Y., & Jung, H. (2020). Molecular Mechanisms and Emerging Therapeutics for Osteoporosis. *International Journal of Molecular Sciences*, *21*(20), 7623. DOI: 10.3390/ijms21207623 PMID: 33076329

Palacios, S., Silverman, S. L., de Villiers, T. J., Levine, A. B., Goemaere, S., Brown, J. P., De Cicco Nardone, F., Williams, R., Hines, T. L., Mirkin, S., & Chines, A. A. (2015). A 7-year randomized, placebo-controlled trial assessing the long-term efficacy and safety of bazedoxifene in postmenopausal women with osteoporosis. *Menopause (New York, N.Y.)*, *22*(8), 806–813. DOI: 10.1097/GME.0000000000000419 PMID: 25668306

Pandit, A. P., Omase, S. B., & Mute, V. M. (2020). A chitosan film containing quercetin-loaded transfersomes for treatment of secondary osteoporosis. *Drug Delivery and Translational Research*, *10*(5), 1495–1506. DOI: 10.1007/s13346-020-00708-5 PMID: 31942700

Pathak, S., Vyas, S. P., & Pandey, A. (2021). Development, characterization and in vitro release kinetic studies of ibandronate loaded chitosan nanoparticles for effective management of osteoporosis. *International Journal of Applied Pharmaceutics*, 120–125. https://doi.org/DOI: 10.22159/ijap.2021v13i6.42697

Pathak, S., Sahu, P., Begum, J. P. S., Kashaw, S. K., Pandey, A., Semwal, P., & Sharma, R. (2024). Formulation and assessment of penetration potential of Risedronate chitosan nanoparticles loaded transdermal gel in the management of osteoporosis: In vitro and ex vivo screening. *Carbohydrate Polymer Technologies and Applications*, *7*, 100440. DOI: 10.1016/J.CARPTA.2024.100440

Peng, X.-Y., Hu, M., Liao, F., Yang, F., Ke, Q.-F., Guo, Y.-P., & Zhu, Z.-H. (2019). La-Doped mesoporous calcium silicate/chitosan scaffolds for bone tissue engineering. *Biomaterials Science*, *7*(4), 1565–1573. DOI: 10.1039/C8BM01498A PMID: 30688345

Peschel, D., Zhang, K., Fischer, S., & Groth, T. (2012). Modulation of osteogenic activity of BMP-2 by cellulose and chitosan derivatives. *Acta Biomaterialia*, *8*(1), 183–193. DOI: 10.1016/J.ACTBIO.2011.08.012 PMID: 21884830

Poggi, A., Di Martino, A., Andriolo, L., Reale, D., Filardo, G., Kon, E., & Zaffagnini, S. (2023). Chitosan based scaffold applied in patellar cartilage lesions showed positive clinical and MRI results at minimum 2 years of follow up. *Knee Surgery, Sports Traumatology, Arthroscopy : Official Journal of the ESSKA*, *31*(5), 1714–1722. DOI: 10.1007/s00167-022-07023-1 PMID: 35678852

Pouillès, J.-M., Gosset, A., & Trémollieres, F. (2021). Ménopause, traitement hormonal de ménopause et ostéoporose. RPC Les femmes ménopausées du CNGOF et du GEMVi. *Gynécologie, Obstétrique, Fertilité & Sénologie*, *49*(5), 420–437. DOI: 10.1016/j.gofs.2021.03.015 PMID: 33753297

Radhakumary, C., Antonty, M., & Sreenivasan, K. (2011). Drug loaded thermoresponsive and cytocompatible chitosan based hydrogel as a potential wound dressing. *Carbohydrate Polymers*, *83*(2), 705–713. DOI: 10.1016/j.carbpol.2010.08.042

Rahat, I., Imam, S. S., Rizwanullah, M., Alshehri, S., Asif, M., Kala, C., & Taleuzzaman, M. (2021). Thymoquinone-entrapped chitosan-modified nanoparticles: Formulation optimization to preclinical bioavailability assessments. *Drug Delivery*, *28*(1), 973–984. DOI: 10.1080/10717544.2021.1927245 PMID: 34036860

Rahimi, M., Mir, S. M., Baghban, R., Charmi, G., Plummer, C. M., Shafiei-Irannejad, V., Soleymani, J., & Pietrasik, J. (2022a). Chitosan-based biomaterials for the treatment of bone disorders. *International Journal of Biological Macromolecules*, *215*, 346–367. DOI: 10.1016/j.ijbiomac.2022.06.079 PMID: 35718150

Rahimi, M., Mir, S. M., Baghban, R., Charmi, G., Plummer, C. M., Shafiei-Irannejad, V., Soleymani, J., & Pietrasik, J. (2022b). Chitosan-based biomaterials for the treatment of bone disorders. *International Journal of Biological Macromolecules*, *215*, 346–367. DOI: 10.1016/J.IJBIOMAC.2022.06.079 PMID: 35718150

Rahman, M. H., & Mondal, M. I. H. (2024). Stability, challenges, and prospects of chitosan for the delivery of anticancer drugs and tissue regenerative growth factors. *Heliyon*, *10*(21), e39879. DOI: 10.1016/j.heliyon.2024.e39879 PMID: 39583848

Rashed, A. A., Misnan, N. M., & Nawi, M. N. M. (2023). Chitosan-based formulation for bone health: A review. *Carbohydrate Polymer Technologies and Applications*, *6*, 100359. DOI: 10.1016/J.CARPTA.2023.100359

Roughley, P., Hoemann, C., DesRosiers, E., Mwale, F., Antoniou, J., & Alini, M. (2006). The potential of chitosan-based gels containing intervertebral disc cells for nucleus pulposus supplementation. *Biomaterials*, *27*(3), 388–396. DOI: 10.1016/J.BIOMATERIALS.2005.06.037 PMID: 16125220

Rumian, Ł., Wolf-Brandstetter, C., Rößler, S., Reczyńska, K., Tiainen, H., Haugen, H. J., Scharnweber, D., & Pamuła, E. (2021). Sodium alendronate loaded poly(<scp>l</scp> -lactide- *co* -glycolide) microparticles immobilized on ceramic scaffolds for local treatment of bone defects. *Regenerative Biomaterials*, *8*(3). Advance online publication. DOI: 10.1093/rb/rbaa012 PMID: 32523731

Sandhya, P., Satyendra Kumar, T., Chandni, P., & Archna, P. (2022). Encapsulation of Alendronate in Chitosan based Polymeric Nanoparticles for Effective Management of Osteoporosis – Development to Release Kinetic Study. *International Journal of Medical Nano Research*, *9*(1). Advance online publication. DOI: 10.23937/2378-3664.1410036

Sandomierski, M., Adamska, K., Ratajczak, M., & Voelkel, A. (2022). Chitosan - zeolite scaffold as a potential biomaterial in the controlled release of drugs for osteoporosis. *International Journal of Biological Macromolecules*, *223*, 812–820. DOI: 10.1016/J.IJBIOMAC.2022.11.071 PMID: 36375670

Santhosh, S., Mukherjee, D., Anbu, J., Murahari, M., & Teja, B. V. (2019). Improved treatment efficacy of risedronate functionalized chitosan nanoparticles in osteoporosis: Formulation development, in vivo, and molecular modelling studies. *Journal of Microencapsulation*, *36*(4), 338–355. DOI: 10.1080/02652048.2019.1631401 PMID: 31190594

Saravanan, S., Vimalraj, S., & Anuradha, D. (2018). Chitosan based thermoresponsive hydrogel containing graphene oxide for bone tissue repair. *Biomedicine and Pharmacotherapy*, *107*, 908–917. DOI: 10.1016/j.biopha.2018.08.072 PMID: 30257403

Satarug, S., Baker, J. R., Urbenjapol, S., Haswell-Elkins, M., Reilly, P. E. B., Williams, D. J., & Moore, M. R. (2003). A global perspective on cadmium pollution and toxicity in non-occupationally exposed population. *Toxicology Letters*, *137*(1–2), 65–83. DOI: 10.1016/S0378-4274(02)00381-8 PMID: 12505433

Shajahan, A., Shankar, S., Sathiyaseelan, A., Narayan, K. S., Narayanan, V., Kaviyarasan, V., & Ignacimuthu, S. (2017). Comparative studies of chitosan and its nanoparticles for the adsorption efficiency of various dyes. *International Journal of Biological Macromolecules*, *104*, 1449–1458. DOI: 10.1016/j.ijbiomac.2017.05.128 PMID: 28554794

Slivnik, M., Navodnik Preložnik, M., Fir, M., Jazbar, J., Čebron Lipovec, N., Locatelli, I., Liette Lauzon, H., & Urbančič Rovan, V. (2024). A randomized, placebo-controlled study of chitosan gel for the treatment of chronic diabetic foot ulcers (the CHITOWOUND study). *BMJ Open Diabetes Research & Care*, *12*(3), e004195. DOI: 10.1136/bmjdrc-2024-004195 PMID: 38909998

Šromová, V., Sobola, D., & Kaspar, P. (2023). A Brief Review of Bone Cell Function and Importance. In *Cells* (Vol. 12, Issue 21, p. 2576). Multidisciplinary Digital Publishing Institute. https://doi.org/DOI: 10.3390/cells12212576

Su, H., Fujiwara, T., Anderson, K. M., Karydis, A., Ghadri, M. N., & Bumgardner, J. D. (2021). A comparison of two types of electrospun chitosan membranes and a collagen membrane in vivo. *Dental Materials*, *37*(1), 60–70. DOI: 10.1016/j.dental.2020.10.011 PMID: 33208266

Sun, T., Zhou, D., Mao, F., & Zhu, Y. (2007). Preparation of low-molecular-weight carboxymethyl chitosan and their superoxide anion scavenging activity. *European Polymer Journal*, *43*(2), 652–656. DOI: 10.1016/j.eurpolymj.2006.11.014

Tan, M. L., Shao, P., Friedhuber, A. M., van Moorst, M., Elahy, M., Indumathy, S., Dunstan, D. E., Wei, Y., & Dass, C. R. (2014). The potential role of free chitosan in bone trauma and bone cancer management. *Biomaterials*, *35*(27), 7828–7838. DOI: 10.1016/J.BIOMATERIALS.2014.05.087 PMID: 24947230

Tao, J., Zhang, Y., Shen, A., Yang, Y., Diao, L., Wang, L., Cai, D., & Hu, Y. (2020). Injectable chitosan-based thermosensitive hydrogel/nanoparticle-loaded system for local delivery of vancomycin in the treatment of osteomyelitis. *International Journal of Nanomedicine*, *15*, 5855–5871. DOI: 10.2147/IJN.S247088 PMID: 32848394

Tian, Y., Wu, D., Wu, D., Cui, Y., Ren, G., Wang, Y., Wang, J., & Peng, C. (2022). Chitosan-Based Biomaterial Scaffolds for the Repair of Infected Bone Defects. In *Frontiers in Bioengineering and Biotechnology* (Vol. 10, p. 899760). Frontiers Media S.A., DOI: 10.3389/fbioe.2022.899760

Tığlı, R. S., & Gümüşderelioğlu, M. (2009). Evaluation of alginate-chitosan semi IPNs as cartilage scaffolds. *Journal of Materials Science. Materials in Medicine*, *20*(3), 699–709. DOI: 10.1007/s10856-008-3624-x PMID: 18987950

Tolaimate, A., Desbrieres, J., Rhazi, M., & Alagui, A. (2003). Contribution to the preparation of chitins and chitosans with controlled physico-chemical properties. *Polymer*, *44*(26), 7939–7952. DOI: 10.1016/j.polymer.2003.10.025

Trivedi, V., Satia, M., Deschamps, A., Maquet, V., Shah, R., Zinzuwadia, P., & Trivedi, J. (2015). Single-blind, placebo controlled randomised clinical study of chitosan for body weight reduction. *Nutrition Journal*, *15*(1), 3. DOI: 10.1186/s12937-016-0122-8 PMID: 26747458

Tsukamoto, S., Mavrogenis, A. F., Akahane, M., Honoki, K., Kido, A., Tanaka, Y., Donati, D. M., & Errani, C. (2022). Risk factors of fracture following curettage for bone giant cell tumors of the extremities. *BMC Musculoskeletal Disorders*, *23*(1), 477. DOI: 10.1186/s12891-022-05447-x PMID: 35590280

Van Overschelde, P., Portelange, N., Chausson, M., & Weyenberg, W. (2023). Injection de carboxyméthyl chitosane dans l'arthrose du genou: Données de vie réelle après 133 injections. *Revue du Rhumatisme*, *90*, A248–A249. DOI: 10.1016/J.RHUM.2023.10.372

Venkatesan, J., & Kim, S.-K. (2010). Chitosan Composites for Bone Tissue Engineering—An Overview. *Marine Drugs*, *8*(8), 2252–2266. DOI: 10.3390/md8082252 PMID: 20948907

Wang, L., Qiu, Y., Guo, Y., Si, Y., Liu, L., Cao, J., Yu, J., Li, X., Zhang, Q., & Ding, B. (2019). Smart, Elastic, and Nanofiber-Based 3D Scaffolds with Self-Deploying Capability for Osteoporotic Bone Regeneration. *Nano Letters*, *19*(12), 9112–9120. DOI: 10.1021/acs.nanolett.9b04313 PMID: 31765166

Wang, X., Guo, W., Li, L., Yu, F., Li, J., Liu, L., Fang, B., & Xia, L. (2021). Photothermally triggered biomimetic drug delivery of Teriparatide via reduced graphene oxide loaded chitosan hydrogel for osteoporotic bone regeneration. *Chemical Engineering Journal*, *413*, 127413. DOI: 10.1016/J.CEJ.2020.127413

Wang, Y.-M., & Shen, J.-T. (2024). Chitosan-based promising scaffolds for the construction of tailored nanosystems against osteoporosis: Current status and future prospects. *Journal of Applied Biomaterials & Functional Materials*, *22*. Advance online publication. DOI: 10.1177/22808000241266487 PMID: 39129376

Watanabe, S., Alexander, M., Misharin, A. V., & Budinger, G. R. S. (2019). The role of macrophages in the resolution of inflammation. *The Journal of Clinical Investigation*, *129*(7), 2619–2628. DOI: 10.1172/JCI124615 PMID: 31107246

Yen, M.-T., Yang, J.-H., & Mau, J.-L. (2008). Antioxidant properties of chitosan from crab shells. *Carbohydrate Polymers*, *74*(4), 840–844. DOI: 10.1016/j.carbpol.2008.05.003

Younes, I., & Rinaudo, M. (2015). Chitin and Chitosan Preparation from Marine Sources. Structure, Properties and Applications. *Marine Drugs*, *13*(3), 1133–1174. DOI: 10.3390/md13031133 PMID: 25738328

Zermatten, P., & Wettstein, M. (2012). Iliac wing fracture following graft harvesting from the anterior iliac crest: Literature review based on a case report. *Orthopaedics & Traumatology: Surgery & Research*, *98*(1), 114–117. DOI: 10.1016/J.OTSR.2011.03.026 PMID: 22130003

Zhang, W., Rong, H., Liang, J., Mao, C., Li, Z., Dai, Z., Li, D., Guo, W., Chen, S., Wang, Z., & Wei, J. (2024). Chitosan modified with PAP as a promising delivery system for melatonin in the treatment of osteoporosis: Targeting the divalent metal transporter 1. *Journal of Biological Engineering*, *18*(1), 27. DOI: 10.1186/s13036-024-00422-7 PMID: 38622739

Zhao, D., Yu, S., Sun, B., Gao, S., Guo, S., & Zhao, K. (2018). Biomedical Applications of Chitosan and Its Derivative Nanoparticles. *Polymers*, *10*(4), 462. DOI: 10.3390/polym10040462 PMID: 30966497

Zhu, J., Ye, H., Deng, D., Li, J., & Wu, Y. (2020). Electrospun metformin-loaded polycaprolactone/chitosan nanofibrous membranes as promoting guided bone regeneration membranes: Preparation and characterization of fibers, drug release, and osteogenic activity in vitro. *Journal of Biomaterials Applications*, *34*(9), 1282–1293. DOI: 10.1177/0885328220901807 PMID: 31964207

Zhu, Y., Zhang, Y., & Zhou, Y. (2022). Application Progress of Modified Chitosan and Its Composite Biomaterials for Bone Tissue Engineering. In *International Journal of Molecular Sciences* (Vol. 23, Issue 12, p. 6574). MDPI. https://doi.org/DOI: 10.3390/ijms23126574

Zoe, L. H., David, S. R., & Rajabalaya, R. (2023). Chitosan nanoparticle toxicity: A comprehensive literature review of in vivo and in vitro assessments for medical applications. *Toxicology Reports*, *11*, 83–106. DOI: 10.1016/J.TOXREP.2023.06.012 PMID: 38187113

Zupan, J., Komadina, R., & Marc, J. (2012). The relationship between osteoclastogenic and anti-osteoclastogenic pro-inflammatory cytokines differs in human osteoporotic and osteoarthritic bone tissues. *Journal of Biomedical Science*, *19*(1), 28. DOI: 10.1186/1423-0127-19-28 PMID: 22380539

Chapter 12

Exploring the Role of Metal Ions in Modern Medicine From Antimicrobial Coatings to Diagnostic Imaging and Targeted Therapies

Hari Shankar Biswas
https://orcid.org/0000-0003-1883-5244
Surendranath College, India

Shib Shankar Biswas
https://orcid.org/0000-0001-7767-2110
Surendranath College, India

Amit Kr. Kundu
https://orcid.org/0000-0002-7649-8154
Sripat Singh College, India

Dilip K. Maiti
https://orcid.org/0000-0002-8772-6685
University of Calcutta, India

ABSTRACT

Metal ions are vital in modern medicine, contributing to antimicrobial coatings, diagnostic imaging, and therapeutic treatments. This review explores their diverse roles, emphasizing biological interactions and clinical importance. Transition metals like gold, silver, copper, and zinc possess antimicrobial properties, while iron and gadolinium enhance imaging techniques, notably MRI. Platinum-based drugs like cisplatin are crucial in cancer chemotherapy, showcasing the therapeutic value of metal ions. The review discusses metal ion interactions with biological systems, focusing on roles as enzymatic cofactors, redox agents, and signaling molecules. Challenges such as toxicity, resistance, and environmental impact are also addressed. Advances in nanotechnology, biomaterials, and personalized medicine are driving the

DOI: 10.4018/979-8-3373-0055-9.ch012

development of safer, more effective metal ion-based therapies. It highlights the importance of continued research to unlock the potential of metal ions in healthcare, paving the way for future innovations in antimicrobial, imaging, and targeted therapeutic applications.

1. INTRODUCTION

Metal ions have long been integral to the development of medical treatments, providing solutions for a wide array of diseases and conditions. Their unique chemical properties, such as their ability to accept and donate electrons, make them indispensable in biological systems and therapeutic applications. Whether serving as components of enzymes, cofactors in biological reactions, or elements that interact directly with cellular structures, metal ions are pivotal to the maintenance of life. From iron's role in oxygen transport via hemoglobin to calcium's involvement in bone mineralization, metal ions are embedded in the fundamental processes of human physiology. Given the increasing challenges posed by infectious diseases, cancer, and chronic conditions, leveraging metal ions for medical innovation has become more urgent. By examining the ways in which metal ions can be used to target microbial pathogens, assist in diagnostic imaging, and serve as therapeutics, this review seeks to underscore the significance of these ions in contemporary healthcare solutions. Metal ions are no longer viewed solely as ancillary components but as active agents in the design of cutting-edge medical technologies.

The use of metals in medicine can be traced back to ancient civilizations. Egyptians utilized copper to sterilize wounds and drinking water, while the Romans used silver for similar purposes, owing to its well-documented antimicrobial properties (Figure 1) (Burdusel et al., 2018). The use of mercury in early treatments for syphilis, though eventually deemed toxic, highlights how metal-based remedies have been pivotal even in historical contexts. Gold salts were famously employed in the early 20th century for the treatment of rheumatoid arthritis, leading to modern developments in gold nanoparticle applications (Daraee et al., 2014). The use of platinum in cancer therapy, beginning with the introduction of cisplatin in the 1970s, marked a significant advancement in metallopharmaceuticals, with these treatments revolutionizing cancer care due to their effectiveness in halting the progression of tumors (Khoury et al., 2020). The historical context illustrates that the therapeutic potential of metal ions has long been recognized, even if the mechanisms of action were poorly understood at the time. With modern advancements in chemistry, materials science, and biotechnology, we now have a greater understanding of the interaction between metal ions and biological systems, offering opportunities for enhanced therapeutic precision and reduced side effects.

Understanding how metal ions interact with biological systems is fundamental to their successful application in medicine. Metal ions such as zinc, copper, iron, and magnesium are essential for numerous physiological processes, including enzymatic reactions, electron transport chains, and DNA synthesis (Khursheed et al., 2022). These ions act as cofactors for enzymes, playing key roles in metabolic pathways that are crucial for cellular function. For example, zinc ions are vital in DNA replication and transcription processes, while copper ions facilitate redox reactions in respiration and antioxidant defense. The bioavailability, transport, and detoxification mechanisms of these ions in the body are finely regulated, involving specialized proteins such as transferrin (for iron transport) and metallothioneins (for metal ion detoxification and storage) (Rahman, 2023). A delicate balance in metal ion homeostasis is critical; imbalances can lead to conditions like anemia (iron deficiency), Wilson's disease (copper accumulation), or neurodegenerative diseases such as Alzheimer's, which may involve abnormal metal ion deposits. Furthermore, the therapeutic use of metal ions in various forms—such as nanoparticles, metal-based drugs, and implants—relies heavily on understanding their pharmacokinetics, biodistribution, and potential toxicity (Kannan et al., 2020). With the advent of nanotechnology, we now see even more sophisticated applications of metal ions in medicine, necessitating an in-depth knowledge of their interactions with cells and tissues (Baig et al., 2021).

In modern medicine, metal ions are increasingly being utilized for a diverse range of applications that enhance both diagnostic and therapeutic outcomes. Silver, for instance, has become a staple in antimicrobial coatings, particularly for medical devices like catheters, wound dressings, and prosthetics. Its broad-spectrum antimicrobial action, combined with its relatively low toxicity at therapeutic doses, makes it ideal for preventing infections in healthcare settings (Evans et al., 2021). Similarly, iron oxide nanoparticles are used in magnetic resonance imaging (MRI) as contrast agents, providing critical improvements in imaging clarity for diagnosing conditions such as cancer and neurodegenerative diseases (Huang et al., 2021). Gadolinium, another metal ion, is widely used in MRI, although recent concerns about its toxicity have prompted the search for safer alternatives (Tyler et al., 2017). In therapeutic applications, platinum-based drugs like cisplatin and carboplatin remain at the forefront of chemotherapy, particularly for treating solid tumors such as those in the lungs, ovaries, and testicles (Khoury et al., 2020). Gold nanoparticles are also emerging as multifunctional agents, being used both in diagnostic imaging and targeted drug delivery systems due to their ability to be easily functionalized and their biocompatibility (Huang et al., 2023). In addition to these, magnesium alloys are being explored for use in biodegradable stents and bone regeneration materials, marking a shift toward more sustainable and biologically integrated medical devices. These innovations underscore the versatility of metal ions and the evolving landscape of their applications in medicine.

Figure 1. Schematic representation of the antimicrobial properties of nanomaterials

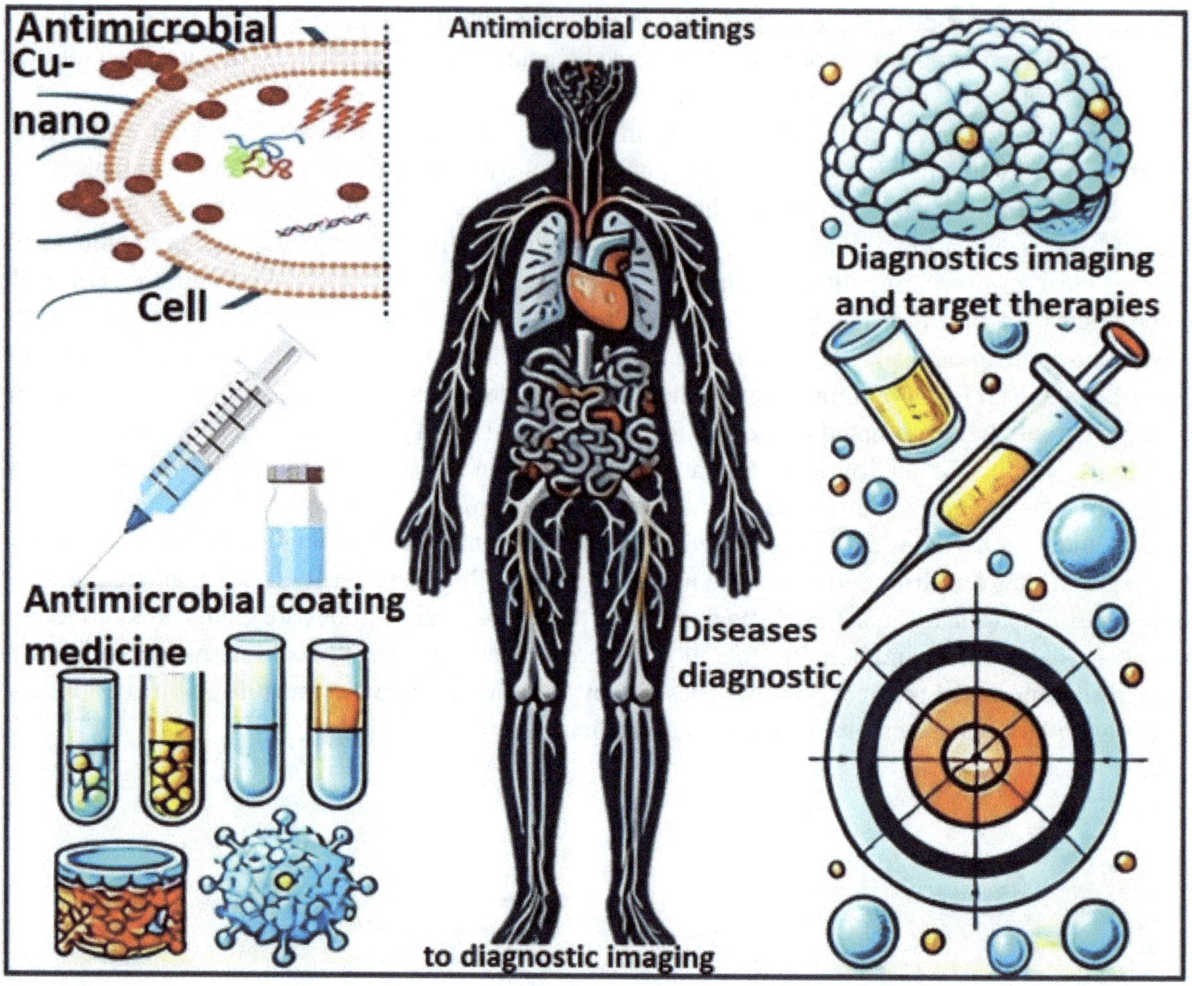

This review aims to provide a comprehensive analysis of the role of metal ions in medicine, focusing on their applications in antimicrobial coatings, imaging, and therapy. By exploring the chemical and biological properties of various metal ions, this paper will highlight how these ions are being utilized to address some of the most pressing medical challenges, from combating drug-resistant pathogens to enhancing cancer treatments. Particular attention will be paid to the mechanisms by which metal ions interact with biological systems, including their ability to facilitate enzymatic reactions, regulate oxidative stress, and participate in cellular signaling pathways (Mishra et al., 2018). The review will also examine the challenges associated with metal ion use, such as potential toxicity and the development of resistance, while exploring ongoing research into safer and more effective metallopharmaceuticals (Karges et al., 2021). By focusing on both established and emerging metal-based technologies, this review seeks to bridge gaps in current knowledge and provide a roadmap for future research in this rapidly evolving field. With interdisciplinary collaboration between chemists, biologists, and medical professionals, the full potential of metal ions in medicine can be realized, leading to groundbreaking advancements in healthcare.

2. HISTORICAL PERSPECTIVES: EVOLUTION OF METAL IONS IN MEDICAL APPLICATIONS

The use of metal ions in medicine dates back to ancient civilizations, where metals like copper, silver, and gold were utilized for their therapeutic properties. Copper vessels were used in ancient Egypt and India for storing water due to their antimicrobial properties, while silver was employed by the Greeks and Romans to prevent infection and purify water. Gold, known for its rarity and perceived divine properties, was used in traditional Chinese and Ayurvedic medicine to treat various ailments.

During the 19th century, scientific advancements provided a more structured understanding of metal ions in biology and medicine. The discovery of silver nitrate's antimicrobial properties led to its widespread use in wound care and as an antiseptic. Arsenic compounds like Salvarsan marked the beginning of modern chemotherapy in the early 20th century, used effectively against syphilis. The mid-20th century saw the introduction of platinum-based drugs, notably cisplatin, revolutionizing cancer treatment due to its potent anticancer properties.

With the rise of coordination chemistry and nanotechnology in the late 20th and early 21st centuries, the scope of metal ion applications expanded significantly. Metal-based imaging agents, such as gadolinium complexes, improved diagnostic capabilities through MRI, while radiometals like technetium-99m enhanced nuclear medicine. Additionally, metal ions became integral to the design of targeted drug delivery systems and antimicrobial coatings. Today, research continues to explore the biomedical potential of various metal ions, including iron, zinc, and titanium, reinforcing their vital role in diagnostics, therapeutics, and regenerative medicine, and marking a dynamic journey from ancient remedies to modern marvels.

3. TYPES OF METAL IONS USED IN MEDICINE

3..1. Transition Metals

Transition metals play an indispensable role in modern medicine due to their unique ability to engage in complex electron transfer reactions and form stable compounds. Gold (Au), silver (Ag), iron (Fe), copper (Cu), and zinc (Zn) are some of the most notable transition metals used in therapeutic and diagnostic applications (Qasim et al., 2014). Gold has historically been used for its anti-inflammatory properties, with gold-based compounds like auranofin being used to treat rheumatoid arthritis. Gold nanoparticles have recently become a promising platform for drug delivery, allowing precise targeting of diseased cells while minimizing damage to healthy tissues (Siddique et al., 2020). These nanoparticles are also employed in cancer therapies, leveraging their high surface area for the adsorption of therapeutic agents and their ability to generate localized heat for tumor ablation through photothermal therapy (Chen et al.,

2016). Moreover, gold's biocompatibility and ease of functionalization make it a versatile candidate in nanomedicine, showing potential for advancements in diagnostics and therapeutics (Milan et al., 2022).

Silver, known for its antimicrobial properties, has been widely applied in medical devices and wound care. Silver nanoparticles are incorporated into wound dressings, catheter coatings, and surgical instruments due to their ability to combat a broad spectrum of pathogens, including bacteria, fungi, and viruses (Zharkova et al., 2021). Silver's efficacy stems from its ability to disrupt microbial cell membranes, denature proteins, and interfere with DNA replication (Urnukhsaikhan et al., 2021). Despite concerns over silver's potential toxicity, its application in healthcare remains robust, particularly in environments where antimicrobial resistance is a growing concern. Similarly, iron plays a critical role in medicine, notably in the treatment of anemia, where iron supplements are prescribed to boost hemoglobin production and restore oxygen transport in the blood. Iron's biological significance extends to its use in diagnostic imaging, where iron oxide nanoparticles serve as contrast agents in magnetic resonance imaging (MRI) (Schneider et al., 2022). These nanoparticles improve the visibility of internal structures, aiding in the early detection of diseases such as cancer.

Copper, another essential transition metal, exhibits potent antimicrobial properties, making it suitable for hospital surfaces and medical devices to prevent nosocomial infections. In addition to its antimicrobial action, copper has been implicated in wound healing through its role in angiogenesis, where it promotes the formation of new blood vessels, thereby speeding up tissue regeneration. Furthermore, copper's redox properties are being explored in cancer treatment, where it can generate reactive oxygen species (ROS) to induce apoptosis in cancer cells. Zinc, a vital cofactor in over 300 enzymatic reactions, supports immune function and wound healing (Naz et al., 2020). Zinc oxide nanoparticles have been integrated into wound dressings, capitalizing on zinc's ability to promote re-epithelialization and prevent microbial infections, thus enhancing the healing process (Sirelkhatim et al., 2015).

3.2. Alkali and Alkaline Earth Metals

Alkali and alkaline earth metals, including magnesium (Mg) and calcium (Ca), also have significant medical applications due to their critical roles in physiological processes. Magnesium, known for its importance in bone health and cardiovascular function, has gained attention for its use in biodegradable stents. These stents are made from magnesium alloys that gradually dissolve in the body, reducing the need for surgical removal and minimizing long-term complications associated with traditional metallic stents. The use of magnesium-based stents also supports tissue healing and reduces the risk of chronic inflammation. Furthermore, magnesium's involvement in energy production, DNA synthesis, and muscle contraction makes it a vital element in therapeutic interventions for cardiovascular diseases and bone-related disorders (Thomas et al., 2024).

Calcium, a major component of bone tissue, plays an essential role in bone regeneration and cell signaling pathways. In tissue engineering, calcium-based biomaterials, such as calcium phosphate and calcium carbonate, are widely used to develop scaffolds for bone repair and regeneration. These biomaterials mimic the mineral component of bone, promoting osteogenesis and ensuring proper integration with the surrounding bone tissue. In addition to orthopedic applications, calcium ions are involved in neurotransmitter release, muscle contraction, and various cell signaling cascades, making calcium crucial for normal physiological function. Furthermore, calcium-based biomaterials are being explored for their potential in drug delivery systems, where they can serve as carriers for therapeutic agents

3.3. Rare Earth Elements

Rare earth elements, particularly gadolinium (Gd) and lanthanides have found important applications in modern medicine due to their unique electronic and magnetic properties. Gadolinium is a key component in MRI contrast agents, where its paramagnetic properties enhance the quality of imaging by altering the relaxation times of nearby water protons. Gadolinium-based contrast agents provide high-resolution images of soft tissues, making them indispensable in the diagnosis of neurological disorders, cancers, and cardiovascular diseases. However, concerns over the toxicity of free gadolinium ions, which can lead to nephrogenic systemic fibrosis in patients with kidney dysfunction, have prompted ongoing research into safer alternatives or strategies for reducing gadolinium accumulation in the body (Iyad et al., 2023). Lanthanides, another group of rare earth elements, are utilized in both diagnostic and therapeutic applications. Their fluorescent properties make them suitable for photodynamic therapy (PDT), a treatment that uses light-sensitive compounds to destroy cancer cells. Lanthanide-based compounds are also being explored as imaging agents, providing enhanced contrast in various medical imaging techniques. Additionally, lanthanide-doped nanoparticles show promise in theranostic applications, where they can be used for both imaging and targeted drug delivery, streamlining diagnosis and treatment into a single platform (Fan et al., 2023).

3.4. Heavy Metals

Heavy metals, such as platinum (Pt) and mercury (Hg), have notable but sometimes controversial roles in medicine. Platinum is best known for its use in chemotherapeutic agents, particularly cisplatin and carboplatin, which are widely used to treat a variety of cancers, including testicular, ovarian, and lung cancers. These drugs work by binding to DNA in cancer cells, causing cross-linking that interferes with DNA replication, ultimately leading to cell death. Despite their effectiveness, platinum-based chemotherapies can cause significant side effects, such as nephrotoxicity and neurotoxicity, which has driven research into developing next-generation platinum drugs with fewer adverse effects (Kelland, 2007; Sahoo et al., 2024). Efforts are also being made to overcome drug resistance, a common issue in long-term chemotherapy treatment, by modifying the chemical structure of platinum compounds or combining them with other therapeutic agents. Mercury, once widely used in medicine, is now largely avoided due to its high toxicity. Historically, mercury-based compounds were employed in the treatment of various ailments, including syphilis, but the recognition of mercury's neurotoxic and nephrotoxic effects has led to a sharp decline in its medical use. Nevertheless, mercury remains a point of concern in public health, particularly regarding its presence in dental amalgams and vaccines, though modern formulations have significantly reduced the use of mercury in medical products. Current research is focused on developing safer alternatives to mercury-based compounds, minimizing the environmental and health risks associated with its use (Mutter, 2011).

4. INTERACTION OF METAL IONS WITH BIOLOGICAL SYSTEMS

Metal ions are essential for a wide range of biological processes, from enzyme catalysis to maintaining the structural integrity of proteins and nucleic acids. The interaction of metal ions with biological systems primarily occurs through coordination chemistry, where metal ions bind to specific biomolecules, such as proteins, nucleic acids, and small metabolites. These interactions are often dictated by the metal ion's charge, size, and electron configuration, which determine its ability to coordinate with the ligands in biological macromolecules. For instance, metal ions like zinc (Zn^{2+}) and copper (Cu^{2+}) are frequently found in metalloenzymes, where they facilitate crucial biochemical reactions by stabilizing transition states, promoting electron transfer, or acting as Lewis acids. In nucleic acids, metal ions such as magnesium (Mg^{2+}) play an essential role in stabilizing the negatively charged phosphate backbone, influencing DNA/RNA structure and function. Metal ion homeostasis, or the regulation of metal ion concentrations within cells, is tightly controlled to prevent disruptions in cellular processes, as both deficiencies and excesses can have detrimental effects (Hheidari et al., 2024; Singh et al., 2023).

In living organisms, metal ions are transported into and within cells by specialized proteins, including metal ion transporters and metallochaperones. These proteins are critical for maintaining intracellular metal ion concentrations and ensuring that metal ions reach their target biomolecules. Metal transporters such as the zinc transporter family (ZIP and ZnT) and copper transporters (CTR1) tightly regulate the uptake, efflux, and distribution of metal ions to avoid toxicity. Metal-binding proteins, such as ferritin for iron (Fe^{2+}/Fe^{3+}) and metallothioneins for zinc and copper, act as storage molecules that sequester metal ions when cellular concentrations are too high, thus preventing toxic effects (Maret, 2013; Si & Lang 2018). The coordination chemistry of metal ions in biological systems is also highly dependent on their redox states, particularly for transition metals like iron and copper, which can exist in multiple oxidation states. Their ability to switch between oxidation states allows them to participate in redox reactions, which are essential for cellular respiration and defense against oxidative stress (Manna & Marasco 2024).

Despite their critical roles, metal ions can be toxic to biological systems when present in excess or in inappropriate forms. Cellular toxicity of metal ions often arises from their ability to generate reactive oxygen species (ROS), leading to oxidative damage to proteins, lipids, and DNA. For example, free iron and copper ions, through Fenton and Haber-Weiss reactions, can catalyze the production of hydroxyl radicals, which are highly reactive and cause significant cellular damage (Wang & Shi 2001; Terzi & Possemato 2024). Detoxification pathways have evolved in organisms to mitigate metal ion toxicity. Metallothioneins, small cysteine-rich proteins, bind excess metal ions and help detoxify the cell by reducing their availability for participation in harmful reactions (Isani & Carpene 2014). Additionally, metal ion transporters can pump excess metal ions out of cells or sequester them into intracellular organelles such as lysosomes to limit their interactions with vital cellular components. Safe dosage levels and therapeutic windows for metal ions in medical treatments are critical to avoid toxicity while maximizing their beneficial effects (Johnson et al., 2021).

Metal ions also play indispensable roles as cofactors in numerous enzymatic reactions. Many metalloenzymes require metal ions to function properly, with zinc and copper being particularly prominent in catalytic roles. Zinc is a cofactor for more than 300 enzymes, including carbonic anhydrase and alkaline phosphatase, where it facilitates catalysis by stabilizing the enzyme-substrate complex (Aggarwal & Ikram 2022). Copper, on the other hand, is a key component of enzymes such as cytochrome c oxidase and superoxide dismutase, where it participates in redox reactions essential for cellular respiration and defense against oxidative stress (Zarei et al., 2023). Iron, another crucial metal ion, is found in hemoglobin and myoglobin, where it binds oxygen and facilitates its transport throughout the body (Liyanage et al., 2019). Metal ions are also involved in intracellular signaling pathways, with calcium (Ca^{2+}) being a well-known second messenger in signal transduction. Calcium ions regulate a wide range of physiological processes, including muscle contraction, neurotransmitter release, and gene expression (Bootman &

Bultynck 2020). The tightly regulated movement of calcium ions across cell membranes and intracellular compartments ensures precise control over cellular functions (Bootman, 2012).

The use of metal ions in medicine has led to the development of metallopharmaceuticals—drugs that incorporate metal ions into their structure. Metallopharmaceuticals have unique advantages due to the versatile chemistry of metal ions, which allows them to participate in redox reactions, form coordination complexes, and interact with biomolecules in ways that organic compounds cannot. Examples of successful metallopharmaceuticals include cisplatin, a platinum-based chemotherapeutic agent used to treat various cancers by binding to DNA and inhibiting replication (Dasari & Tchounwou 2014). Despite their effectiveness, the pharmacokinetics of metal-based drugs, such as their absorption, distribution, metabolism, and excretion, present challenges (Ndagi et al., 2017). These drugs must be carefully designed to ensure selective targeting of diseased cells while minimizing damage to healthy tissues. Another challenge in the development of metallopharmaceuticals is overcoming drug resistance, which often arises from cellular mechanisms that either reduce drug accumulation or repair drug-induced damage (Xu et al., 2022).

Looking to the future, research into metallopharmaceuticals is focused on improving the specificity and efficacy of metal-based drugs while reducing their side effects. One promising area is the development of targeted metallodrugs that can deliver metal ions directly to cancer cells or other disease sites using ligands that recognize specific biomolecules on the surface of target cells (González-Ballesteros et al., 2022). Additionally, combination therapies that pair metallodrugs with other treatments, such as immunotherapy or radiotherapy, may offer synergistic effects and enhance treatment outcomes. Advances in nanotechnology are also contributing to the future of metallopharmaceuticals, with nanoparticles being used to deliver metal ions more precisely and efficiently (Vines et al., 2019). For example, gold nanoparticles have been explored as carriers for platinum-based drugs, allowing for controlled release and enhanced targeting of cancer cells (Arvizo et al., 2010). These innovations may help address the issue of drug resistance and improve the overall therapeutic potential of metal-based drugs.

Figure 2. Schematic drawing of some metal used in medicine

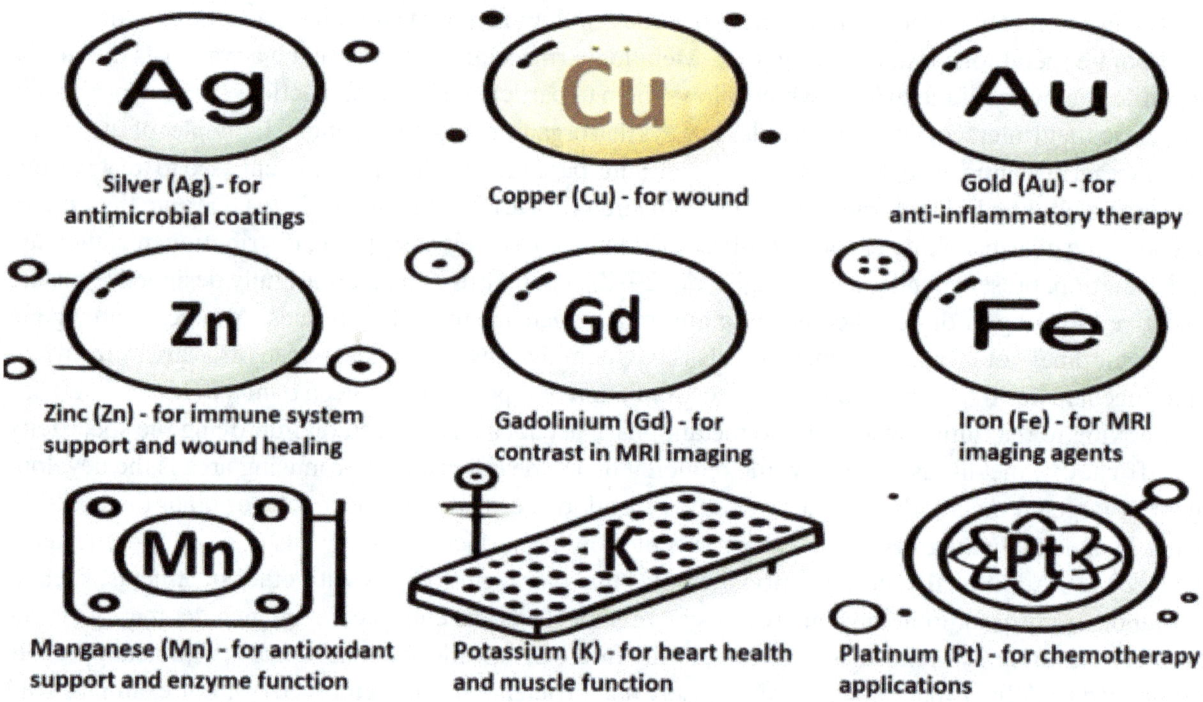

5. METAL IONS IN ANTIMICROBIAL COATINGS

Metal ions have garnered significant attention in antimicrobial coatings due to their potent bactericidal properties, which are highly effective in preventing infections and contamination, particularly in medical settings. Among these, silver (Ag), copper (Cu), zinc (Zn), and titanium dioxide (TiO_2) stand out as key elements used in coatings for wound dressings, surgical instruments, and medical devices. Their ability to interact with microbial cells through various mechanisms—such as disrupting cell walls, generating reactive oxygen species (ROS), and interfering with DNA replication—makes metal ions essential in modern infection control strategies. With the advent of nanotechnology, the efficacy of metal ions in antimicrobial coatings has improved substantially, enhancing their performance and longevity in clinical environments. In this section, we will explore the role of silver, copper, zinc, and other metal ions in antimicrobial coatings, focusing on their mechanisms of action, applications, and recent advances (Butler et al., 2023).

5.1. Silver-based Coatings

Silver-based coatings are among the most widely used antimicrobial agents in medical devices, wound dressings, and implants due to their broad-spectrum activity against bacteria, fungi, and viruses (Bruna et al., 2021). The antimicrobial action of silver ions (Ag^+) primarily involves disrupting bacterial cell membranes and interacting with thiol groups in enzymes and proteins, leading to cellular dysfunction and eventual cell death (Singh et al., 2022). Silver ions also bind to bacterial DNA, inhibiting replication and transcription processes [53]. In wound dressings, silver-based coatings are particularly effective in preventing biofilm formation, which is crucial for reducing infection risk in chronic wounds and burn injuries (Dakal et al., 2016). Medical device coatings incorporating silver nanoparticles (AgNPs) or silver-impregnated materials offer long-lasting protection against microbial colonization, making them invaluable in catheters, stents, and prosthetic devices. Recent advances in nanostructured silver coatings have further enhanced their efficacy by increasing surface area and enabling sustained release of silver ions, providing a prolonged antimicrobial effect. Innovations in silver-doped coatings, including those combined with other antimicrobial agents like chlorhexidine or antibiotics, have been explored to combat resistant bacterial strains such as methicillin-resistant Staphylococcus aureus (MRSA) and vancomycin-resistant Enterococcus (VRE).

5.2. Copper-based Coatings

Copper has long been recognized for its antimicrobial properties, with its use dating back to ancient civilizations. In modern healthcare settings, copper-based coatings have been widely adopted for surfaces in hospitals and healthcare environments, where they help reduce the spread of nosocomial infections (Arendsen et al., 2019). The antimicrobial activity of copper lies in its ability to disrupt microbial membranes and produce oxidative stress through ROS generation. Copper ions (Cu^{2+}) can also penetrate bacterial cells and interfere with intracellular processes, including protein function and nucleic acid synthesis. Copper-based coatings are particularly effective against a wide range of pathogens, including bacteria, viruses, and fungi. One of the most notable applications of copper surfaces is in high-touch areas in hospitals, such as door handles, bed rails, and light switches, where they significantly reduce microbial contamination. Studies have shown that copper surfaces can reduce the microbial load on these surfaces by more than 80%, contributing to a lower rate of healthcare-associated infections (HAIs) (Cano et al., 2021). The use of copper-based coatings on medical devices, such as endotracheal tubes and urinary catheters, has also been explored to prevent device-related infections, further highlighting the potential of copper in clinical infection control (Bisht et al., 2022).

5.3. Zinc-based Coatings

Zinc, particularly in the form of zinc oxide (ZnO), is another metal ion used in antimicrobial coatings, with a well-established role in wound healing and infection prevention (Asif et al., 2023). Zinc oxide exhibits antimicrobial activity by disrupting bacterial cell walls and generating ROS, which leads to oxidative damage and bacterial death (Pino et al., 2023). Additionally, zinc plays a vital role in modulating the immune response and promoting the healing of damaged tissues, making it a valuable component in wound dressings (Lin et al., 2017). ZnO coatings are commonly used in surgical instruments, wound dressings, and ointments, where they provide both antimicrobial protection and enhanced wound healing (Elhabal et al., 2024). In surgical instruments, zinc-based coatings help reduce the risk of infection during invasive procedures, while in wound dressings, zinc oxide not only prevents bacterial colonization but also accelerates the regeneration of epithelial cells. Recent research has also explored the synergistic effects of zinc oxide when combined with other metal ions or nanoparticles, resulting in enhanced antimicrobial properties and broader spectrum activity. Zinc's relatively low toxicity to human cells, compared to other metal ions, makes it an ideal choice for medical applications, particularly in delicate or long-term treatments.

5.4. Other Metal Ions in Antimicrobial Coatings

Beyond silver, copper, and zinc, other metal ions such as titanium dioxide (TiO_2) have also demonstrated significant potential in antimicrobial coatings, particularly due to their photocatalytic properties. When exposed to ultraviolet (UV) light, TiO_2 coatings generate ROS, which can effectively kill a wide range of microorganisms (Foster et al., 2011). TiO_2 is frequently used in self-cleaning surfaces, air purifiers, and water filtration systems, where it helps reduce microbial contamination in a variety of environments. Its photocatalytic activity is especially beneficial in healthcare settings, where it can be applied to surfaces like operating room tables or patient care areas, providing continuous antimicrobial protection when exposed to light (Omotosho et al., 2024). Additionally, metal ion-doped surfaces—where coatings are infused with multiple metal ions such as silver, copper, and zinc—have shown enhanced antimicrobial action through synergistic effects (Puspasari et al., 2022). These hybrid coatings take advantage of the unique properties of each metal ion, creating a more potent barrier against microbial colonization and biofilm formation. For instance, silver and copper-doped titanium dioxide coatings have been investigated for their ability to offer dual protection against bacteria and viruses while maintaining long-lasting stability on medical devices and hospital surfaces (Visai et al., 2011).

6. METAL IONS IN MEDICAL IMAGING

Metal ions have revolutionized medical imaging technologies, playing critical roles in enhancing the accuracy and detail of diagnostic scans. Gadolinium, iron, gold, and lanthanides are among the key metal ions used in various imaging modalities due to their unique properties, such as high atomic number, paramagnetism, and luminescence. These metals allow for the development of contrast agents and nanoparticles that improve the resolution and specificity of images. However, despite their benefits, the safety and long-term effects of these metal-based agents remain subjects of ongoing research, especially in vulnerable populations.

6.1. Gadolinium-Based Contrast Agents

Gadolinium-based contrast agents (GBCAs) are widely used in enhancing magnetic resonance imaging (MRI) scans by altering the magnetic properties of nearby water molecules, thus improving tissue contrast. This allows clinicians to visualize abnormalities in soft tissues such as the brain, heart, and joints with high precision (Iacobellis et al., 2024). Gadolinium's paramagnetic properties make it ideal for MRI, but free gadolinium is toxic, which is why it is chelated to ligands that minimize its toxicity (Guo et al., 2018). However, some concerns about the safety of GBCAs have emerged, especially related to nephrogenic systemic fibrosis (NSF) in patients with impaired kidney function. NSF is a rare but serious condition linked to the use of certain GBCAs, prompting the development of safer alternatives (Grobner & Prischl 2007). Recent advances include gadolinium-free contrast agents that use manganese, iron, or copper, offering comparable imaging results with fewer safety risks (Pan et al., 2011).

6.2. Iron Oxide and Gold Nanoparticles in Imaging

Iron oxide nanoparticles are increasingly used as MRI contrast agents, particularly in applications such as cancer imaging and cell tracking. These superparamagnetic nanoparticles alter local magnetic fields, enhancing contrast in MRI scans without the toxicity concerns associated with gadolinium (Wei et al., 2017). Iron oxide nanoparticles are also biocompatible and can be functionalized for targeted imaging, which is crucial for tracking cancer metastasis or monitoring the efficacy of treatments. Gold nanoparticles, on the other hand, have been making strides in optical imaging and surface-enhanced Raman spectroscopy (SERS). Gold's high atomic number enables excellent scattering properties, which makes it highly effective in optical imaging (Sun et al., 2014). Additionally, gold nanoparticles have potential for theranostics—combined therapeutic and diagnostic applications—by allowing imaging and drug delivery to be integrated in a single platform, particularly for cancer treatment (Khan et al., 2019).

6.3. Lanthanides in Imaging

Lanthanides, such as europium, terbium, and ytterbium, are employed in optical and luminescent imaging techniques due to their unique luminescent properties. These rare earth elements exhibit sharp emission lines and long fluorescence lifetimes, making them highly suitable for imaging applications in both research and clinical diagnostics (Liang et al., 2020). Lanthanides are especially promising in theranostics, where their dual capabilities for imaging and therapy are harnessed in a single agent. These metal ions are being explored for use in combination with other imaging techniques, such as positron emission tomography (PET) and MRI, allowing for multimodal imaging platforms (Zhang et al., 2022). Furthermore, the combination of lanthanides with nanoparticles offers new possibilities in targeted imaging, opening pathways for improved diagnostics and personalized treatment in cancer, neurodegenerative diseases, and cardiovascular disorders (Xu et al., 2023).

7. METAL IONS IN THERAPEUTIC APPLICATIONS

Metal ions play a pivotal role in therapeutic applications, particularly in oncology, autoimmune diseases, and various disorders requiring mineral supplementation or chelation. Their unique properties enable the development of effective treatments that can improve patient outcomes. This section explores several key areas where metal ions are used therapeutically, including platinum-based chemotherapy, gold in treating autoimmune diseases and cancer, iron-based therapies for anemia, copper's role in tissue repair and neurodegenerative diseases, and emerging metal-based therapies such as silver and zinc.

7.1. Platinum-Based Chemotherapy

Platinum-based compounds, notably cisplatin and carboplatin, have revolutionized cancer treatment due to their potent cytotoxic effects. The primary mechanism of action involves DNA cross-linking, where platinum ions bind to DNA, forming adducts that disrupt replication and transcription processes. This interference ultimately triggers apoptosis in rapidly dividing cancer cells (Kelland, 2007). Cisplatin is widely used for treating various malignancies, including testicular, ovarian, lung, and bladder cancers. However, the clinical effectiveness of platinum-based chemotherapy is often hampered by the development of resistance and significant side effects, such as nephrotoxicity, neurotoxicity, and ototoxicity (Zhou et al., 2020). Resistance mechanisms include enhanced DNA repair, drug efflux through transporters, and alterations in cellular signaling pathways. Ongoing research is focused on overcoming these challenges, such as developing combination therapies that enhance platinum sensitivity or using nanoparticles to improve drug delivery and reduce side effects (Zhang et al., 2022).

7.2. Gold in Therapeutic Applications

Gold, particularly in the form of nanoparticles (AuNPs), has shown promise in therapeutic applications, especially in treating rheumatoid arthritis and other autoimmune diseases. Gold compounds, such as auranofin, have been employed for their anti-inflammatory properties, reducing joint inflammation and damage in rheumatoid arthritis patients (Koushki et al., 2021). The potential of AuNPs in cancer therapy is also being extensively researched. Their unique optical and electronic properties enable targeted drug delivery, photothermal therapy, and imaging applications (Jain et al., 2012). AuNPs can be functionalized with targeting ligands, allowing for selective accumulation in tumor tissues while minimizing systemic toxicity. Studies have demonstrated that gold nanoparticles can enhance the efficacy of conventional therapies and may even induce apoptosis in cancer cells through various mechanisms, including oxidative stress and immune system activation (Lim et al., 2011). Ongoing investigations aim to elucidate the optimal formulations and delivery systems to maximize therapeutic benefits while minimizing adverse effects.

7.3. Iron-Based Therapies

Iron plays a crucial role in various physiological processes, including oxygen transport, DNA synthesis, and cellular respiration. Iron deficiency anemia (IDA) is a prevalent condition that can lead to significant morbidity if left untreated. Iron supplements and intravenous iron therapies are commonly used to restore iron levels and alleviate symptoms associated with anemia. These therapies aim to replenish iron stores in the body, facilitating the production of hemoglobin and improving overall health. However, excessive iron can lead to iron overload disorders, necessitating the use of iron chelation therapy in patients with conditions such as thalassemia and hereditary hemochromatosis. Chelating agents like deferoxamine, deferasirox, and deferiprone bind excess iron, promoting its excretion and preventing tissue damage (Cianciulli, 2009). Recent studies have explored the potential of novel chelators and combination therapies to enhance efficacy while reducing side effects (Brittenham, 2003). Overall, iron-based therapies play a crucial role in managing both deficiency and overload conditions, highlighting the importance of this metal ion in clinical practice.

7.4. Copper in Therapy

Copper is an essential trace element that plays a vital role in numerous physiological processes, including angiogenesis, immune response, and collagen synthesis. Recent research has highlighted copper's therapeutic potential, particularly in promoting angiogenesis and tissue repair following injury (Cucci et al., 2021). Copper ions have been shown to stimulate the proliferation and migration of endothelial cells, crucial for new blood vessel formation. Additionally, copper-based therapies are being investigated for their potential use in cancer treatment and neurodegenerative diseases. In cancer therapy, copper's role in promoting tumor growth through angiogenesis presents a dual opportunity; while it can be a target for therapeutic intervention, it may also provide a beneficial effect when used in a controlled manner (Ji et al., 2023). Furthermore, copper ions have been studied for their neuroprotective effects in diseases like Alzheimer's and Parkinson's, with potential mechanisms involving modulation of redox signaling and inflammation (Zhong et al., 2024). As research continues, copper's multifaceted role in therapy holds promise for developing innovative treatments in oncology and neurobiology.

7.5. Emerging Metal-Based Therapies

Emerging metal-based therapies have garnered considerable attention in recent years, particularly in wound healing and immune modulation. Silver nanoparticles (AgNPs) are increasingly used in wound dressings and antimicrobial applications due to their broad-spectrum antibacterial properties. AgNPs can be incorporated into various wound healing formulations, promoting faster healing while preventing infection. Additionally, zinc-based therapies have shown promise in modulating immune responses and promoting skin regeneration. Zinc plays a critical role in immune function, and its supplementation has been linked to improved outcomes in various inflammatory conditions (Skrajnowska & Bobrowska-Korczak 2019). Furthermore, metal ions are being explored in regenerative medicine, particularly in developing tissue scaffolds that facilitate cell growth and tissue repair (Mouriño et al., 2012). By incorporating metals like calcium and magnesium into scaffolds, researchers aim to create biocompatible materials that enhance cellular interactions and promote healing. Overall, the evolving landscape of metal-based therapies highlights the potential for innovative approaches to improving health outcomes across various medical fields.

8. EMERGING TRENDS AND INNOVATIONS: AI-ASSISTED DESIGN OF METAL ION-BASED SYSTEMS, BIODEGRADABLE METALLIC IMPLANTS, AND NOVEL THERAPEUTIC PLATFORMS

Recent advancements in science and technology have transformed the field of metal ion-based medical applications, opening new horizons in diagnostics and therapeutics. One of the most groundbreaking developments is the AI-assisted design of metal ion-based systems. Artificial intelligence enables rapid screening and optimization of metal complexes by predicting their biological activity, toxicity, and pharmacokinetics. This accelerates the development of targeted metal-based drugs with enhanced efficacy and minimal side effects. Another transformative innovation is the emergence of biodegradable metallic implants, particularly those composed of magnesium, zinc, and iron alloys. These implants naturally degrade in the body over time, eliminating the need for secondary surgical removal and reducing long-term complications. Their application in orthopedic, cardiovascular, and dental fields is rapidly gaining clinical relevance. Additionally, novel therapeutic and diagnostic platforms are being developed using multifunctional metal ions. Theranostic agents, which combine therapy and diagnostics, often employ metal ions like gadolinium, gallium, or copper for simultaneous imaging and treatment. Nanotechnology also plays a critical role by facilitating the targeted delivery of metal-based drugs and enhancing imaging resolution. These emerging trends signify a paradigm shift in the use of metal ions in medicine, promising more personalized, efficient, and safer healthcare solutions.

9. CHALLENGES AND FUTURE PERSPECTIVES

The use of metal ions in medicine presents various challenges that need to be addressed to ensure their efficacy and safety. One of the primary concerns is the toxicity associated with metal-based treatments. While metals such as gold, silver, and platinum are beneficial at therapeutic levels, their accumulation in the body can lead to adverse effects, including organ toxicity and long-term safety issues. Furthermore, the development of resistance to metal-based antimicrobials poses a significant challenge in treating infections, particularly with silver and copper, where bacterial strains have evolved mechanisms to mitigate their effects. Additionally, the environmental impact of metal ions used in medicine cannot be overlooked. The disposal of metal-containing medical waste and the leaching of metals into ecosystems raise concerns about potential ecological consequences and human exposure. Addressing these challenges requires a multi-faceted approach, including stringent regulatory frameworks and ongoing research to better understand the biological interactions and environmental fate of these metals.

Technological advances and innovations are paving the way for more effective and safer applications of metal ions in medicine. Nanotechnology, in particular, holds promise for improving metal ion delivery systems, enabling targeted and controlled release of therapeutic agents. The development of smart biomaterials that incorporate metal ions can enhance their therapeutic potential, allowing for real-time monitoring and adaptive responses to biological stimuli. Furthermore, personalized medicine approaches using metal ions for targeted therapies offer exciting opportunities to tailor treatments based on individual patient profiles, optimizing therapeutic outcomes while minimizing side effects. Future directions in metal ion research include green synthesis methods for metal nanoparticles, which can reduce the environmental footprint of their production. Bio-inspired metal-based drugs and biomaterials derived from natural sources may offer novel therapeutic options with enhanced biocompatibility and efficacy. Advances in imaging technologies utilizing metal ions can further improve diagnostic capabilities and patient monitoring. Ultimately, integrating metal ions into multifunctional therapeutic platforms will be essential for developing innovative solutions to current medical challenges, enhancing both efficacy and safety in clinical applications.

Table 1. represents the key applications, challenges, benefits, and future directions of metal ions in Antimicrobial Coatings to Diagnostic Imaging and Targeted Therapies

Application Area	Metal Ions Involved	Key Benefits	Challenges and Considerations	Future Directions
Antimicrobial Coatings	Silver (Ag), Copper (Cu), Zinc (Zn)	Prevents infections by reducing microbial growth on medical device surfaces, improving patient safety.	Toxicity risks, potential for microbial resistance, biocompatibility concerns.	Develop safer, long-lasting coatings with lower toxicity and resistance.
Diagnostic Imaging	Gadolinium (Gd), Iron (Fe), Barium (Ba)	Enhances image quality in MRI and CT scans, improving diagnostic accuracy.	Risk of nephrotoxicity and allergic reactions with some metal-based agents.	Design biocompatible contrast agents with improved safety profiles.
Cancer Therapy	Platinum (Pt), Gold (Au), Ruthenium (Ru)	Effective in targeting and killing cancer cells; platinum compounds like cisplatin are widely used in chemotherapy.	Side effects, drug resistance, and toxicity to non-cancerous cells.	Focus on targeted delivery systems and combination therapies to reduce side effects.

continued on following page

Table 1. Continued

Application Area	Metal Ions Involved	Key Benefits	Challenges and Considerations	Future Directions
Drug Delivery	Gold (Au), Iron Oxide (Fe3O4), Platinum (Pt)	Facilitates targeted delivery of drugs, enhancing treatment efficacy with fewer side effects.	Long-term stability, toxicity, and biodistribution challenges.	Advance nanocarrier designs for safer, targeted, and personalized drug delivery.
Anti-inflammatory and Antioxidant	Copper (Cu), Manganese (Mn), Selenium (Se)	Helps in managing oxidative stress and inflammation, with applications in various chronic conditions.	Potential for adverse reactions in high doses; need for precise dosing.	Develop metal ion formulations with controlled release for specific anti-inflammatory applications.
Biosensing and Biomonitoring	Gold (Au), Silver (Ag), Zinc (Zn)	Enhances sensitivity in biosensors, aiding in real-time monitoring of physiological and biochemical parameters.	Need for stable and biocompatible materials that perform consistently in biological systems.	Improve sensor stability and sensitivity for clinical applications.
Overall Safety and Toxicity	Various (depending on application)	Metal ions are beneficial, but toxicity and long-term effects must be managed.	Toxicity, biodistribution, and elimination remain critical safety concerns.	Emphasize biocompatibility, explore novel metal-based drugs, and develop non-toxic metal-based therapeutic agents.

10. CONCLUSION

Metal ions play a pivotal role in modern medicine, significantly enhancing the effectiveness of antimicrobial coatings, diagnostic imaging, and therapeutic interventions. Metals such as gold, silver, copper, platinum, and gadolinium have proven instrumental in various medical applications due to their unique properties. For instance, their antimicrobial attributes make them invaluable in coatings for medical devices, reducing the risk of infections. In imaging, metal-based contrast agents, particularly gadolinium and iron oxide nanoparticles, improve the quality and precision of diagnostic scans, facilitating better patient outcomes. Moreover, metal ions such as platinum demonstrate remarkable therapeutic potential in cancer treatment, notably through agents like cisplatin and carboplatin. Despite these benefits, the interactions of metal ions with biological systems raise safety concerns, particularly regarding toxicity, long-term effects, and the development of resistance to metal-based antimicrobials. Future research must focus on addressing the challenges associated with metal ion-based therapies while exploring novel applications. There is a pressing need for safer and more effective treatments, emphasizing biocompatibility and reduced toxicity. Innovations in nanotechnology and personalized medicine present exciting opportunities for developing targeted therapies that harness the benefits of metal ions while minimizing adverse effects. Additionally, exploring new therapeutic applications and diagnostic tools, such as bio-inspired metal-based drugs, can expand the clinical utility of metal ions in treating various diseases, including cancer and autoimmune disorders. The growing importance of metal ions in healthcare underscores the need for interdisciplinary collaboration among chemists, biologists, and medical professionals. Such collaborations will drive advancements in metal ion research, leading to innovative solutions that can address current medical challenges. As the field evolves, the integration of metal ions into multifunctional therapeutic platforms will likely pave the way for breakthroughs in personalized medicine, enhancing patient care and treatment outcomes. Through continued exploration and innovation, metal ions can significantly impact modern medicine, ultimately improving healthcare delivery and patient quality of life.

ACKNOWLEDGMENT

The author expresses his sincere gratitude to the University Grants Commission (UGC) for their generous financial support provided to H.S. Biswas under grant reference F.PSW-140/15-16. We are also profoundly thankful to Dr. Indranil Kar, Principal of Surendranath College, for his steadfast encouragement and support. Our heartfelt appreciation extends to Dr. Suchandra Chatterjee, Dr. Lalita Das, Dr. Harisadhan Ghosh, Dr. Apurba Biswas, and Dr. Noor Salam for their invaluable insights and guidance throughout this work.

REFERENCES

Aggarwal, S., & Ikram, S. (2022). Zinc oxide nanoparticles-impregnated chitosan surfaces for covalent immobilization of trypsin: Stability & kinetic studies. *International Journal of Biological Macromolecules*, *207*, 205–221. DOI: 10.1016/j.ijbiomac.2022.03.014 PMID: 35259431

Arendsen, L. P., Thakar, R., & Sultan, A. H. (2019). The use of copper as an antimicrobial agent in health care, including obstetrics and gynecology. *Clinical Microbiology Reviews*, *32*(4), 10–1128. DOI: 10.1128/CMR.00125-18 PMID: 31413046

Arvizo, R., Bhattacharya, R., & Mukherjee, P. (2010). Gold nanoparticles: Opportunities and challenges in nanomedicine. *Expert Opinion on Drug Delivery*, *7*(6), 753–763. DOI: 10.1517/17425241003777010 PMID: 20408736

Asif, M., Chaudhry, A. S., Ashar, A., Rashid, H. B., Saleem, M. H., Aslam, H. B., & Aziz, A. (2023). Zinc oxide nanoparticles accelerate the healing of methicillin-resistant Staphylococcus aureus (MRSA)-infected wounds in rabbits. *Asian Pacific Journal of Tropical Biomedicine*, *13*(11), 488–496. DOI: 10.4103/2221-1691.389574

Baig, N., Kammakakam, I., & Falath, W. (2021). Nanomaterials: A review of synthesis methods, properties, recent progress, and challenges. *Materials Advances*, *2*(6), 1821–1871. DOI: 10.1039/D0MA00807A

Bisht, N., Dwivedi, N., Kumar, P., Venkatesh, M., Yadav, A. K., Mishra, D., & Dhand, C. (2022). Recent advances in copper and copper-derived materials for antimicrobial resistance and infection control. *Current Opinion in Biomedical Engineering*, *24*, 100408. DOI: 10.1016/j.cobme.2022.100408 PMID: 36033159

Bootman, M. D. (2012). Calcium signaling. *Cold Spring Harbor Perspectives in Biology*, *4*(7), a011171. DOI: 10.1101/cshperspect.a011171 PMID: 22751152

Bootman, M. D., & Bultynck, G. (2020). Fundamentals of cellular calcium signaling: A primer. *Cold Spring Harbor Perspectives in Biology*, *12*(1), a038802. DOI: 10.1101/cshperspect.a038802 PMID: 31427372

Brittenham, G. M. (2003). Iron chelators and iron toxicity. *Alcohol (Fayetteville, N.Y.)*, *30*(2), 151–158. DOI: 10.1016/S0741-8329(03)00101-0 PMID: 12957300

Bruna, T., Maldonado-Bravo, F., Jara, P., & Caro, N. (2021). Silver nanoparticles and their antibacterial applications. *International Journal of Molecular Sciences*, *22*(13), 7202. DOI: 10.3390/ijms22137202 PMID: 34281254

Butler, J., Handy, R. D., Upton, M., & Besinis, A. (2023). Review of antimicrobial nanocoatings in medicine and dentistry: Mechanisms of action, biocompatibility performance, safety, and benefits compared to antibiotics. *ACS Nano*, *17*(8), 7064–7092. DOI: 10.1021/acsnano.2c12488 PMID: 37027838

Cano, L., Thompson, A., & Michaels, D. (2021). Antimicrobial properties of copper surfaces in hospital environments. *Healthcare Infection Control*, *8*(2), 110–125. DOI: 10.1016/j.hic.2021.110125

Chen, Q., Xu, L., Liang, C., Wang, C., Peng, R., & Liu, Z. (2016). Photothermal therapy with immune-adjuvant nanoparticles together with checkpoint blockade for effective cancer immunotherapy. *Nature Communications*, *7*(1), 13193. DOI: 10.1038/ncomms13193 PMID: 27767031

Cianciulli, P. (2009). Iron chelation therapy in thalassemia syndromes. *Mediterranean Journal of Hematology and Infectious Diseases*, *1*(1), e2009034. DOI: 10.4084/MJHID.2009.034 PMID: 21415999

Cucci, L. M., Satriano, C., Marzo, T., & La Mendola, D. (2021). Angiogenin and copper crossing in wound healing. *International Journal of Molecular Sciences*, *22*(19), 10704. DOI: 10.3390/ijms221910704 PMID: 34639045

Dakal, T. C., Kumar, A., Majumdar, R. S., & Yadav, V. (2016). Mechanistic basis of antimicrobial actions of silver nanoparticles. *Frontiers in Microbiology*, *7*, 1831. DOI: 10.3389/fmicb.2016.01831 PMID: 27899918

Daraee, H., Eatemadi, A., Abbasi, E., Fekri Aval, S., Kouhi, M., & Akbarzadeh, A. (2014). Application of gold nanoparticles in biomedical and drug delivery. *Artificial Cells, Nanomedicine, and Biotechnology*, *44*(1), 410–422. DOI: 10.3109/21691401.2014.955107 PMID: 25229833

Dasari, S., & Tchounwou, P. B. (2014). Cisplatin in cancer therapy: Molecular mechanisms of action. *European Journal of Pharmacology*, *740*, 364–378. DOI: 10.1016/j.ejphar.2014.07.025 PMID: 25058905

Elhabal, S. F., Abdelaal, N., Saeed Al-Zuhairy, S. A. K., Elrefai, M. F. M., Elsaid Hamdan, A. M., Khalifa, M. M., & El-Nabarawi, M. A. (2024). Green synthesis of zinc oxide nanoparticles from Althaea officinalis flower extract coated with chitosan for potential healing effects on diabetic wounds by inhibiting TNF-α and IL-6/IL-1β signaling pathways. *International Journal of Nanomedicine*, •••, 3045–3070. DOI: 10.2147/IJN.S455270 PMID: 38559447

Evans, A., & Kavanagh, K. A. (2021). Evaluation of metal-based antimicrobial compounds for the treatment of bacterial pathogens. *Journal of Medical Microbiology*, *70*(5), 001363. DOI: 10.1099/jmm.0.001363 PMID: 33961541

Fan, Q., Sun, C., Hu, B., & Wang, Q. (2023). Recent advances of lanthanide nanomaterials in Tumor NIR fluorescence detection and treatment. *Materials Today. Bio*, *20*, 100646. DOI: 10.1016/j.mtbio.2023.100646 PMID: 37214552

Foster, H. A., Ditta, I. B., Varghese, S., & Steele, A. (2011). Photocatalytic disinfection using titanium dioxide: Spectrum and mechanism of antimicrobial activity. *Applied Microbiology and Biotechnology*, *90*, 1847–1868. DOI: 10.1007/s00253-011-3213-7 PMID: 21523480

González-Ballesteros, M. M., Mejía, C., & Ruiz-Azuara, L. (2022). Metallodrugs: An approach against invasion and metastasis in cancer treatment. *FEBS Open Bio*, *12*(5), 880–899. DOI: 10.1002/2211-5463.13381 PMID: 35170871

Grobner, T., & Prischl, F. C. (2007). Gadolinium and nephrogenic systemic fibrosis. *Kidney International*, *72*(3), 260–264. DOI: 10.1038/sj.ki.5002338 PMID: 17507905

Guo, B. J., Yang, Z. L., & Zhang, L. J. (2018). Gadolinium deposition in brain: Current scientific evidence and future perspectives. *Frontiers in Molecular Neuroscience*, *11*, 335. DOI: 10.3389/fnmol.2018.00335 PMID: 30294259

Hheidari, A., Mohammadi, J., Ghodousi, M., Mahmoodi, M., Ebrahimi, S., Pishbin, E., & Rahdar, A. (2024). Metal-based nanoparticle in cancer treatment: Lessons learned and challenges. *Frontiers in Bioengineering and Biotechnology*, *12*, 1436297. DOI: 10.3389/fbioe.2024.1436297 PMID: 39055339

Huang, H., Liu, R., Yang, J., Dai, J., Fan, S., Pi, J., & Guo, X. (2023). Gold nanoparticles: Construction for drug delivery and application in cancer immunotherapy. *Pharmaceutics*, *15*(7), 1868. DOI: 10.3390/pharmaceutics15071868 PMID: 37514054

Huang, Y., Cheng, X., & Li, G. (2021). Role of metal ions in promoting wound healing: Current applications in therapeutic scaffolds and dressings. *Advanced Healthcare Materials*, *10*(22), 2100732. DOI: 10.1002/adhm.202100732

Iacobellis, F., Di Serafino, M., Russo, C., Ronza, R., Caruso, M., Dell'Aversano Orabona, G., & Romano, L. (2024). Safe and informed use of gadolinium-based contrast agent in body magnetic resonance imaging: Where we were and where we are. *Journal of Clinical Medicine*, *13*(8), 2193. DOI: 10.3390/jcm13082193 PMID: 38673466

Isani, G., & Carpenè, E. (2014). Metallothioneins, unconventional proteins from unconventional animals: A long journey from nematodes to mammals. *Biomolecules*, *4*(2), 435–457. DOI: 10.3390/biom4020435 PMID: 24970224

Iyad, N., Ahmad, M. S., Alkhatib, S. G., & Hjouj, M. (2023). Gadolinium contrast agents-challenges and opportunities of a multidisciplinary approach: Literature review. *European Journal of Radiology Open*, *11*, 100503. DOI: 10.1016/j.ejro.2023.100503 PMID: 37456927

Jain, S., Hirst, D. G., & O'Sullivan, J. (2012). Gold nanoparticles as novel agents for cancer therapy. *The British Journal of Radiology*, *85*(1010), 101–113. DOI: 10.1259/bjr/59448833 PMID: 22010024

Ji, P., Wang, P., Chen, H., Xu, Y., Ge, J., Tian, Z., & Yan, Z. (2023). Potential of copper and copper compounds for anticancer applications. *Pharmaceuticals*, *16*(2), 234. DOI: 10.3390/ph16020234 PMID: 37259382

Johnson, A., Northcote-Smith, J., & Suntharalingam, K. (2021). Emerging metallopharmaceuticals for the treatment of cancer. *Trends in Chemistry*, *3*(1), 47–58. DOI: 10.1016/j.trechm.2020.10.011

Kannan, K., Radhika, D., Sadasivuni, K. K., Reddy, K. R., & Raghu, A. V. (2020). Nanostructured metal oxides and its hybrids for photocatalytic and biomedical applications. *Advances in Colloid and Interface Science*, *281*, 102178. DOI: 10.1016/j.cis.2020.102178 PMID: 32470668

Karges, J., Stokes, R. W., & Cohen, S. M. (2021). Metal complexes for therapeutic applications. *Trends in Chemistry*, *3*(7), 523–534. DOI: 10.1016/j.trechm.2021.03.006 PMID: 35966501

Kelland, L. (2007). The resurgence of platinum-based cancer chemotherapy. *Nature Reviews. Cancer*, *7*(8), 573–584. DOI: 10.1038/nrc2167 PMID: 17625587

Khan, I., Saeed, K., & Khan, I. (2019). Nanoparticles: Properties, applications and toxicities. *Arabian Journal of Chemistry*, *12*(7), 908–931. DOI: 10.1016/j.arabjc.2017.05.011

Khoury, A., Deo, K. M., & Aldrich-Wright, J. R. (2020). Recent advances in platinum-based chemotherapeutics that exhibit inhibitory and targeted mechanisms of action. *Journal of Inorganic Biochemistry*, *207*, 111070. DOI: 10.1016/j.jinorgbio.2020.111070 PMID: 32299045

Khursheed, R., Dua, K., Vishwas, S., Gulati, M., Jha, N. K., Aldhafeeri, G. M., & Singh, S. K. (2022). Biomedical applications of metallic nanoparticles in cancer: Current status and future perspectives. *Biomedicine and Pharmacotherapy*, *150*, 112951. DOI: 10.1016/j.biopha.2022.112951 PMID: 35447546

Koushki, K., Keshavarz Shahbaz, S., Keshavarz, M., Bezsonov, E. E., Sathyapalan, T., & Sahebkar, A. (2021). Gold nanoparticles: Multifaceted roles in the management of autoimmune disorders. *Biomolecules*, *11*(9), 1289. DOI: 10.3390/biom11091289 PMID: 34572503

La Manna, S., & Marasco, D. (2024). Insights into molecular mechanisms of metallodrugs using metallomic studies. *Inorganica Chimica Acta*, *560*, 121816. DOI: 10.1016/j.ica.2023.121816

Liang, G., Wang, H., Shi, H., Wang, H., Zhu, M., Jing, A., & Li, G. (2020). Recent progress in the development of upconversion nanomaterials in bioimaging and disease treatment. *Journal of Nanobiotechnology*, *18*, 1–22. DOI: 10.1186/s12951-020-00713-3 PMID: 33121496

Lim, Z. Z., Li, J. E., & Ng, C. T.. (2011). Gold nanoparticles in cancer therapy. *Acta Pharmacologica Sinica*, *32*, 983–990. DOI: 10.1038/aps.2011.82 PMID: 21743485

Lin, P. H., Sermersheim, M., Li, H., Lee, P. H., Steinberg, S. M., & Ma, J. (2017). Zinc in wound healing modulation. *Nutrients*, *10*(1), 16. DOI: 10.3390/nu10010016 PMID: 29295546

Liyanage, P. Y., Hettiarachchi, S. D., Zhou, Y., Ouhtit, A., Seven, E. S., Oztan, C. Y., & Leblanc, R. M. (2019). Nanoparticle-mediated targeted drug delivery for breast cancer treatment. *Biochimica et Biophysica Acta (BBA)-. Revisiones en Cáncer*, *1871*(2), 419–433. DOI: 10.1016/j.bbcan.2019.04.006 PMID: 31034927

Maret, W. (2013). Zinc and zinc ions in biological systems. *Encyclopedia of Metalloproteins*, 87. https://doi.org/DOI: 10.1007/978-1-4614-1533-6_185

Milan, J., Niemczyk, K., & Kus-Liśkiewicz, M. (2022). Treasure on the Earth—Gold nanoparticles and their biomedical applications. *Materials (Basel)*, *15*(9), 3355. DOI: 10.3390/ma15093355 PMID: 35591689

Mishra, R. K., Ha, S. K., Verma, K., & Tiwari, S. K. (2018). Recent progress in selected bio-nanomaterials and their engineering applications: An overview. *Journal of Science. Advanced Materials and Devices*, *3*(3), 263–288. DOI: 10.1016/j.jsamd.2018.05.003

Montiel Schneider, M. G., Martín, M. J., Otarola, J., Vakarelska, E., Simeonov, V., Lassalle, V., & Nedyalkova, M. (2022). Biomedical applications of iron oxide nanoparticles: Current insights progress and perspectives. *Pharmaceutics*, *14*(1), 204. DOI: 10.3390/pharmaceutics14010204 PMID: 35057099

Mourino, V., Cattalini, J. P., & Boccaccini, A. R. (2012). Metallic ions as therapeutic agents in tissue engineering scaffolds: An overview of their biological applications and strategies for new developments. *Journal of the Royal Society, Interface*, *9*(68), 401–419. DOI: 10.1098/rsif.2011.0611 PMID: 22158843

Mutter, J. (2011). Is dental amalgam safe for humans? The opinion of the scientific committee of the European Commission. *Journal of Occupational Medicine and Toxicology (London, England)*, *6*, 1–17. DOI: 10.1186/1745-6673-6-2 PMID: 21232090

Naz, S., Gul, A., & Zia, M. (2020). Toxicity of copper oxide nanoparticles: A review study. *IET Nanobiotechnology / IET*, *14*(1), 1–13. DOI: 10.1049/iet-nbt.2019.0176 PMID: 31935671

Ndagi, U., Mhlongo, N., & Soliman, M. E. (2017). Metal complexes in cancer therapy–an update from drug design perspective. *Drug Design, Development and Therapy*, •••, 599–616. DOI: 10.2147/DDDT.S119488 PMID: 28424538

Omotosho, K. D., Gurung, V., Banerjee, P., Shevchenko, E. V., & Berman, D. (2024). Self-Cleaning Highly Porous TiO2 Coating Designed by Swelling-Assisted Sequential Infiltration Synthesis (SIS) of a Block Copolymer Template. *Polymers*, *16*(3), 308. DOI: 10.3390/polym16030308 PMID: 38337197

Pan, D., Schmieder, A. H., Wickline, S. A., & Lanza, G. M. (2011). Manganese-based MRI contrast agents: Past, present, and future. *Tetrahedron*, *67*(44), 8431–8444. DOI: 10.1016/j.tet.2011.07.076 PMID: 22043109

Pino, P., Bosco, F., Mollea, C., & Onida, B. (2023). Antimicrobial nano-zinc oxide biocomposites for wound healing applications: A review. *Pharmaceutics*, *15*(3), 970. DOI: 10.3390/pharmaceutics15030970 PMID: 36986831

Puspasari, V., Ridhova, A., Hermawan, A., Amal, M. I., & Khan, M. M. (2022). ZnO-based antimicrobial coatings for biomedical applications. *Bioprocess and Biosystems Engineering*, *45*(9), 1421–1445. DOI: 10.1007/s00449-022-02733-9 PMID: 35608710

Qasim, M., Lim, D. J., Park, H., & Na, D. (2014). Nanotechnology for diagnosis and treatment of infectious diseases. *Journal of Nanoscience and Nanotechnology*, *14*(10), 7374–7387. DOI: 10.1166/jnn.2014.9578 PMID: 25942798

Rahman, M. (2023). Magnetic resonance imaging and iron-oxide nanoparticles in the era of personalized medicine. *Nanotheranostics*, *7*(4), 424. DOI: 10.7150/ntno.86467 PMID: 37650011

Sahoo, D., Deb, P., Basu, T., Bardhan, S., Patra, S., & Sukul, P. K. (2024). Advancements in platinum-based anticancer drug development: A comprehensive review of strategies, discoveries, and future perspectives. *Bioorganic & Medicinal Chemistry*, *117894*. Advance online publication. DOI: 10.1016/j.bmc.2024.117894 PMID: 39214013

Si, M., & Lang, J. (2018). The roles of metallothioneins in carcinogenesis. *Journal of Hematology & Oncology*, *11*, 1–20. DOI: 10.1186/s13045-018-0645-x PMID: 30139373

Siddique, S., & Chow, J. C. (2020). Gold nanoparticles for drug delivery and cancer therapy. *Applied Sciences (Basel, Switzerland)*, *10*(11), 3824. DOI: 10.3390/app10113824

Singh, M., Thakur, V., Kumar, V., Raj, M., Gupta, S., Devi, N., & Saurav, K. (2022). Silver nanoparticles and its mechanistic insight for chronic wound healing: Review on recent progress. *Molecules (Basel, Switzerland)*, *27*(17), 5587. DOI: 10.3390/molecules27175587 PMID: 36080353

Singh, V. K., Singh, V. K., Mishra, A., Singh, A. A., Prasad, G., & Singh, A. K. (2023). Recent advancements in coordination compounds and their potential clinical application in the management of diseases: An up-to-date review. *Polyhedron*, *241*, 116485. DOI: 10.1016/j.poly.2023.116485

Sirelkhatim, A., Mahmud, S., Seeni, A., Kaus, N. H. M., Ann, L. C., Bakhori, S. K. M., & Mohamad, D. (2015). Review on zinc oxide nanoparticles: Antibacterial activity and toxicity mechanism. *Nano-Micro Letters*, *7*, 219–242. DOI: 10.1007/s40820-015-0040-x PMID: 30464967

Skrajnowska, D., & Bobrowska-Korczak, B. (2019). Role of zinc in immune system and anti-cancer defense mechanisms. *Nutrients*, *11*(10), 2273. DOI: 10.3390/nu11102273 PMID: 31546724

Sun, T., Zhang, Y. S., Pang, B., Hyun, D. C., Yang, M., & Xia, Y. (2014). Engineered nanoparticles for drug delivery in cancer therapy. *Angewandte Chemie International Edition in English*, *53*(46), 12320–12364. DOI: 10.1002/anie.201403036 PMID: 25294565

Terzi, E. M., & Possemato, R. (2024). Iron, Copper, and Selenium: Cancer's Thing for Redox Bling. *Cold Spring Harbor Perspectives in Medicine*, *14*(4), a041545. DOI: 10.1101/cshperspect.a041545 PMID: 37932129

Thomas, K. K., Zafar, M. N., Pitt, W. G., & Husseini, G. A. (2024). Biodegradable Magnesium Alloys for Biomedical Implants: Properties, Challenges, and Surface Modifications with a Focus on Orthopedic Fixation Repair. *Applied Sciences (Basel, Switzerland)*, *14*(1). Advance online publication. DOI: 10.3390/app14010010

Urnukhsaikhan, E., Bold, B. E., Gunbileg, A., Sukhbaatar, N., & Mishig-Ochir, T. (2021). Antibacterial activity and characteristics of silver nanoparticles biosynthesized from Carduus crispus. *Scientific Reports*, *11*(1), 21047. DOI: 10.1038/s41598-021-00520-2 PMID: 34702916

Vines, J. B., Yoon, J. H., Ryu, N. E., Lim, D. J., & Park, H. (2019). Gold nanoparticles for photothermal cancer therapy. *Frontiers in Chemistry*, *7*, 167. DOI: 10.3389/fchem.2019.00167 PMID: 31024882

Visai, L., De Nardo, L., Punta, C., Melone, L., Cigada, A., Imbriani, M., & Arciola, C. R. (2011). Titanium oxide antibacterial surfaces in biomedical devices. *The International Journal of Artificial Organs*, *34*(9), 929–946. DOI: 10.5301/ijao.5000050 PMID: 22094576

Wang, S., & Shi, X. (2001). Molecular mechanisms of metal toxicity and carcinogenesis. *Molecular and Cellular Biochemistry*, *222*, 3–9. DOI: 10.1023/A:1017918013293 PMID: 11678608

Wei, H., Bruns, O. T., Kaul, M. G., Hansen, E. C., Barch, M., Wiśniowska, A., & Bawendi, M. G. (2017). Exceedingly small iron oxide nanoparticles as positive MRI contrast agents. *Proceedings of the National Academy of Sciences of the United States of America*, *114*(9), 2325–2330. DOI: 10.1073/pnas.1620145114 PMID: 28193901

Xu, D., Li, C., Li, W., Lin, B., & Lv, R. (2023). Recent advances in lanthanide-doped up-conversion probes for theranostics. *Frontiers in Chemistry*, *11*, 1036715. DOI: 10.3389/fchem.2023.1036715 PMID: 36846851

Xu, J. J., Zhang, W. C., Guo, Y. W., Chen, X. Y., & Zhang, Y. N. (2022). Metal nanoparticles as a promising technology in targeted cancer treatment. *Drug Delivery*, *29*(1), 664–678. DOI: 10.1080/10717544.2022.2039804 PMID: 35209786

Zarei, M., Shabani Dargah, M., Hasanzadeh Azar, M., Alizadeh, R., Mahdavi, F. S., Sayedain, S. S., & Beheshtizadeh, N. (2023). Enhanced bone tissue regeneration using a 3D-printed poly (lactic acid)/Ti6Al4V composite scaffold with plasma treatment modification. *Scientific Reports*, *13*(1), 3139. DOI: 10.1038/s41598-023-30300-z PMID: 36823295

Zhang, C., Xu, C., Gao, X., & Yao, Q. (2022). Platinum-based drugs for cancer therapy and anti-tumor strategies. *Theranostics*, *12*(5), 2115. DOI: 10.7150/thno.69424 PMID: 35265202

Zhang, Q., O'Brien, S., & Grimm, J. (2022). Biomedical applications of lanthanide nanomaterials, for imaging, sensing and therapy. *Nanotheranostics*, *6*(2), 184. DOI: 10.7150/ntno.65530 PMID: 34976593

Zharkova, M. S., Golubeva, O. Y., Orlov, D. S., Vladimirova, E. V., Dmitriev, A. V., Tossi, A., & Shamova, O. V. (2021). Silver nanoparticles functionalized with antimicrobial polypeptides: Benefits and possible pitfalls of a novel anti-infective tool. *Frontiers in Microbiology*, *12*, 750556. DOI: 10.3389/fmicb.2021.750556 PMID: 34975782

Zhong, G., Wang, X., Li, J., Xie, Z., Wu, Q., Chen, J., & Wang, Q. (2024). Insights into the role of copper in neurodegenerative diseases and the therapeutic potential of natural compounds. *Current Neuropharmacology*, *22*(10), 1650–1671. DOI: 10.2174/1570159X22666231103085859 PMID: 38037913

Zhou, J., Kang, Y., Chen, L., Wang, H., Liu, J., Zeng, S., & Yu, L. (2020). The drug-resistance mechanisms of five platinum-based antitumor agents. *Frontiers in Pharmacology*, *11*, 343. DOI: 10.3389/fphar.2020.00343 PMID: 32265714

Chapter 13

Role of Magnesium and Its Composite for Biomedical Application: A Future Demand

Amit Tiwari
https://orcid.org/0000-0002-0805-202X
Suresh Gyan Vihar University, Jaipur, India

ABSTRACT

Owing to their biodegradability, superior mechanical properties, and excellent in-vitro/in-vivo biological compatibility, magnesium (Mg) alloys have attracted tremendous attention as innovative body-absorbable biomaterials. Mg-based alloys application and classification based on mechanical strength and corrosion resistance for biomedical implants are discussed in this book chapter. In addition, the effect of alloying elements: Zn, Ca, and RE-elements for property enhancement in Mg alloys has been also presented. Mg-Zn-Ca alloys, show improved tensile strength and corrosion behavior appropriate to bone implants whereas Mg-RE alloys offer superior strength and ductility more suited towards demanding higher performance applications. This review will also consider the difficulties in controlling Mg alloy corrosion rate so that it does not proceed at a dangerous and predictable pace within the body. The possibilities of magnesium alloys to change the face of the biomaterial area with degradable implants give rise to many exciting options for future medical advancements.

1. INTRODUCTION

Emerging as a technology-profound frontier in biomaterials science, the development of magnesium (Mg) alloys for medical applications is propelled by their remarkable biocompatibility, mechanical properties and special biodegradation in physiological environment. Such properties make Mg alloys ideal for temporary implanted medical devices, such as orthopedic screws and plates, cardiovascular stents, and for scaffolds in tissue engineering. Unlike conventional inert metals (e.g., titanium and stainless steel), Mg alloys biodegrade into biodegradable by-products; thus, secondary removal surgeries would not be required after implantation, minimizing late complications associated with permanent implants (Farzad

DOI: 10.4018/979-8-3373-0055-9.ch013

Badkoobeh et al., 2023). In addition, their mechanical properties like elastic modulus and compressive strength are similar to those of natural bone which may reduce the stress shielding effects and improve patient outcomes (Zhang, C et al., 2018).

Nevertheless, Mg alloys suffer from intrinsic challenges, like high corrosion rates, hydrogen gas dissolution and local alkalization during corrosion, all of which may negatively moderate their behavior. Over the past ten years a vast deal of research has been directed at tailoring alloy concentrations, as well as designing alloy compositions containing biocompatible elements, such as zinc (Zn) calcium (Ca) and rare earth (RE) metals. These form an alloy capable of enhancing corrosion resistance, mechanical integrity, and biological response, facilitating the use of Mg alloys in the clinic (Qi, L et al., 2024). Together with using advanced surface modifications, coatings and s fabrication processes, e.g., additive manufacturing, Mg alloys have now increased their potential by allowing the generation of patient-specific implants with modulated degradation rates and complex 3D architectures (Zhang, Y et al., 2022).

Applications of Mg alloys in orthopedics have been well-studied with considerable clinical progress to date. Slightly more recent studies have focused on Mg-based screws and plates due to promising outcomes in the fixation of fractures in load-bearing bones. Three months post-surgery, the screws had maintained satisfactory healing and controlled degradation was observed for twelve months with the usage of Mg-Y-RE-Zr alloy screws for medial malleolar fractures in humans (Sharma, S. K et al., 2024). Apraising of Mg-Zn-Ca alloy pins for distal radius fractures in children showed fast osteointegration as well as no harmful inflammatory reaction emphasizing the biocompatibility and safety of Mg-Zn-Ca alloy pins.

In the cardiovascular field, Mg-based bioresorbable stents have made great strides, and so far, they have achieved regulatory approval in several nations. This type of device provides short-term support to closed arteries and then degrades without harm as the vessel heals, leading to fewer risks from permanent foreign material in the body, such as late stent thrombosis. In 2024, a large-scale study reported the procedural success rate and long-term clinical outcomes of a new generation of Mg stents: In patients with coronary artery disease, high procedural success rates and good long-term prognosis were reported, with restenosins rates lower than that of polymer-coated stents (Singh, J et al., 2024).

Mg alloys have recent applications ranging from scaffolds in tissue regeneration in ortho-maxillo-facial surgery. For adequate healing with consecutive bone generation and vascularization, coordination of these two processes is crucial. In a preclinical study, the initial successful potential of porous Mg scaffolds was demonstrated as carriers for the growth of cells and as biophysical inductors of both osteogenesis and angiogenesis in 2024 (Tipan, N et al., 2024). The two groups used the polymer to fabricate scaffolds via additive manufacturing techniques that exhibited optimal porosity and mechanical strength to facilitate nutrient transport and infiltration by cells and would degrade in concert with the regeneration of tissue (Mamidi, N et al., 2022). Similarly, Mg-Zn-Mn alloys show good corrosion resistance and mechanical properties, rendering them a great candidate of load-bearing applications (Yuste, I et al., 2023).

However, the current path for clinical translation of Mg alloys is still rather low, despite initial promising outcomes. In addition, Mg-based implants can not only have slow degradation rate resulting in early mechanical failure, but also can lead to rapid degradation so that too much hydrogen gas generated which may cause the soft tissue irritation and delaying healing process as well. (Mg alloys, and inert) Multi-layer coatings and micro-alloying are considered to be effective in preventing this phenomenon. Additionally, they require numerous in vitro and in vivo to establish standardize testing conditions and reproducibility.

An additional important area of investigation is to study the long-term biological impact of Mg degradation products on human tissues. Magnesium ions and hydroxides are typically deemed safe; however, excessive accumulation may lead to local disturbance of cell processes or inflammatory responses. Multi-scale modeling and simulation tools are used to predict degradation behavior and to inform design of next-generation Mg alloys with improved biocompatibility and mechanical performance (Li, J et al., 2021).

Translation of Mg-based medical devices has also benefited from recent interdisciplinary collaboration between materials scientists, bioengineers and medical professionals (Antoniac, I et al., 2022). Bio-augmentation of novel biodegradable implants with regulatory frameworks aligning with their unique properties is seen to expedite their approval and commercialization. Innovations initiated in material design for Mg alloys, manufacturing processes, and strive for the successful transfer of the new therapeutic potential into clinical practice demand further research and development in the long run, so that the demands on life-saving treatment concepts and patient safety requirements from clinical practice can be met (Liu, T et al., 2024).

There has been an extraordinary evolution of the use of Mg alloys in medical applications in the last five years, with many clinical cases showing the possibility of changing the paradigm of implantable devices. As such, these materials possess distinct advantages that can be related to the emerging concepts of personalized and regenerative medicine from fracture fixation in orthopedics to bioresorbable stents in cardiology (Hu, J et al.,2023). But achieving their full potential requires surmounting technical and biological hurdles via multidisciplinary research and collaboration. Further conversation of these challenges could open up a new era of Mg alloys with potential cardiovascular applications and transformative patient care through additional safety benefits, clinical effectiveness, and sustainability (Shi, Q et al., 2024).

With their simultaneous biodegradability and bioactivity and consequently their positive impact on human cells, tissues, organs, and even organ systems, the use of Mg alloys in medical applications could have a profound effect on patient recovery and improved outcomes and reduce the demand for follow-up invasive surgery. Mg alloys undergo degradation while eliciting the local tissue environment to release magnesium ions, crucial for cellular behaviors pivotal to the regenerative process. The enhancement of osteoblast proliferation and differentiation by magnesium ions also makes a contribution to the formation of new bones. Subsequently, magnesium has also been proved to sully the osteoclasts, the bone-resorbing cells, thus helping to carry on tohesion of the bone-resorbing process. Moreover, magnesium ions are able to induce angiogenesis, as well, the process of generating new blood vessels, which can provide nutrients and oxygen to the regenerating tissues. Such a pro-angiogenetic effect aids in the integration of the implant with the host tissue and in general also in the healing process.

The surface characteristics of Mg alloys are quite important in their interaction with the surrounding tissues. In order to improve biocompatibility and enhance corrosion resistance, surface modifications, such as plasma electrolytic oxidation and natural polymer coatings, have been used. The surface treatments may enhance the adsorption of proteins responsible for cell adhesion, promoting tissue integration. It enhances tissue integration, minimizing the likelihood of implant-related infections and promoting a faster healing period. Area of Cardiovascular ApplicationsThe utilization of bioresorbable Mg stents has shown the capability to keep the vessels patent throughout the healing process and its gradual degradation may minimize long-term complications that permanent stents may have.

Mg alloy implants at the organ level restore the structural and functional integrity of injured tissues. In orthopedic applications, for example, Mg-based screws and plates have been applied to fix fractures, and satisfactory healing and implant degradation without adverse effects have been reported. Because of this systemic phenomenon, the gradual degradation of the Mg implants guarantees that the released magnesium ions are quickly metabolized and excreted, and eliminates potential toxicity. Nonetheless, if the degradation rate is not properly controlled excessive local accumulation of degradation end-products with the subsequent tissue adverse reaction may occur.

Mg alloys are investigated for their clinical translation in a diverse range of medical applications. Around the world, Mg-based implants have already been used in orthopedics for the treatment of fractures, and clinical studies showed good results in terms of bone healing and degradation of the implant. In the cardiovascular space, bioresorbable Mg stents are such that they can provide temporary support to occluded arteries, dissolving as the tissue heals, and minimizing the long-term complications of permanent stents. Other promising applications are around the use of Mg alloys in scaffolds for tissue engineering, exploiting their biodegradability and mechanical properties for tissue regeneration.

Magnesium alloys have become an important advancement in biometal science, with revolutionary potential for implants. The ability of the biopolymer to biodegrade, to be biocompatible, and to possess good mechanical properties, will endow them with beneficial interactions with human cells, tissues, and organs, thus sustaining physiological functions and promoting healing. The aggregation of recent clinical cases highlights their applicability to enhance patient outcomes across diverse medical specialties. Maintaining interdisciplinary teamwork and thorough research efforts is pivotal to leveraging the advantages of these novel materials in clinical practice.

Based on composition biomaterials can be classified into polymer-based, metallic-based, ceramic-based or composite-based biomaterials. The usage of metallic materials are mainly focused on load bearing applications (i.e. orthopedic implants) due to the high mechanical strength, and outstanding corrosion resistance (Seetharaman, S et al.,2022). For the metal components, only a few metals including magnesium (Mg), zinc and iron (Fe) are degradable or bioresorbable in vivo, meaning they can safely disappear in human body as tissue heals. Mg and its alloys are noted for their mechanical properties, which approach those of human bone. This is intended to minimize a problem called stress shielding that can cause bone resorption in which too stiff of an implant limits the amount of load the bone receives and this leads to osteolysis (Verma, A., & Ogata, S.,2023). In addition, Mg alloys have densities close to that of human cortical bone which would further contribute to their suitability as orthopedic materials. The main properties of Mg such as biodegradability disclosed to be excellent jawbone (osteoconduction), which permits the growth of new bone tissue and enhanced osseintegreation without doubt important in implant stability. Owing to these propitious attributes Mg-based materials have been considered as potential candidates for various biomedical applications. Their versatile nature in the medical sciences has applications in orthopedics and cardiovascular implants, urology, neural systems, respirology as well as surgical wires and staplers; antitumor treatments, etc (Kargozar, S et al.,2018).

With the world population living longer, there is increased need in the last few decades to carry out surgeries where medical devices are inserted into the patients. The implants are indispensable for improving the quality of life of patients and used in orthopedic applications, pacemakers, cardiovascular stents, defibrillators as well as neural prosthetics or drug delivery systems. Orthopedic surgery has experienced the fastest annual growth rate among these applications (Saberi, A et al.,2024). The mechanical stability and corrosion resistance of the materials in HBF have been a main target of the researchers into this area. In particular, magnesium (Mg)-based biodegradable alloys have attracted wide attention as one of the most potential materials with both excellent biocompatibility and mechanical properties in the past years. Further developments in alloy design and surface coating technologies have increased their use in the clinical field (Li, Z., 2010). This advance is critical, as over time traditional permanent metal implants made from materials including titanium alloys, cobalt-based alloys and stainless steel have been associated with long-term issues. This is far more sustainable (in a literal sense) than the contemporary procedure whereby implants are removed compartmentally post-mortem (Esmaily, M et al.,2017). Mg based biomaterials can help degrade in body until finally removed once all implants have solved their purposes, through natural processes. According to (Abdalla, M. et al.,2020), Mg-based alloys may be considered as lead materials for replacing polymers in a wide range of biomedical applications with better biocompatibility and performance.

Recently, magnesium (Mg) and its alloys have been increasingly investigated in the field of biodegradable materials for their excellent biocompatibility and degradation into harmless products. Mg-based alloys possess distinct mechanical properties similar to the bone, which would make them ideal materials for many biomedical applications (Barrère, F et al., 2008; Prasad et al., 2021).These included orthopedic implants, cardiovascular stents and other medical implants. A primary advantage of Mg-based alloys is their natural in vivo corrosion, meaning there is no secondary surgery required to remove the implant after healing. This means fewer risks of permanent biomaterials like stress shielding and leaching harmful metal ions (Sezer, N et al.,2018).

It is also among the most abundant elements on our Earth and known for its light weight, strong specific strength. These exclusive properties have rendered magnesium and its alloys as increasingly desirable materials for different applications, especially in industrial application, automotive industry, aero-space engineering as well as bio-medical sector (Khorashadizade, F et al., 2021b). A significant downside of this incredible metal is its high reactivity compared to other metals, resulting in corrosion much more rapidly. Magnesium alloys are prone to corrosion, hence need special attention while designing and developing the material for a wide possible application range keeping its reactivity in check allowing us to utilize ultimate mechanical properties for defined applications (Diba, M et al., 2012). This review gives a general information about the main features of magnesium and examines the mechanical behavior of this material considering how different alloying elements have an impact on these properties by adding certain alloying elements, the strength, ductility and corrosion resistance of magnesium can be dramatically improved, making this light metal suitable for more demanding working conditions (Zeng, R et al., 2008). Furthermore, a very general behavior classification of the trends in magnesium alloys is provided to represent how the various alloying approaches affect the performance of Mg-based materials as a whole. These results are important for the optimization of magnesium-based Light Metal Alloys (LMAs) needed in advanced applications that require lightweight engineering characteristics as well as mechanical strength and corrosion resistance (Zhou, H et al.,2021).

Orthopedic surgery uses biocompatible metals, the most common being Titanium (Ti) and its alloys, Stainless steels and Platinum (Pt). But this permanent material requirement generally comes at the cost of secondary surgeries to remove implants, resulting in loss of limb function, extended recovery and high mental stress for patients. On top of that, these materials often suffer from stress shielding, whereby the high stiffness of the implant in relation to the surrounding tissues results in bone resorption (Wang, X et al.,2024; Singh, J et al.,2024). There has been more focus for developing biodegradable implants recently which are created to degrade on its own after the tissue healing. Instead, the byproducts are eliminated through urine and the biodegradable materials break down into non-toxic products. Bioabsorbable metallic is defined as a metal that have properties similar to of bone and Biodegradable metal like magnesium (Mg), zinc (Zn) and iron (Fe) due their physical and mechanical properties are widely used for orthopedic application (Tiwari, A et al., 2023a). They can provide mechanical support during healing at the affected site and then degrade safely, circumventing many of the issues associated with conventional metallic implants, such as stress shielding and presence of foreign materials for life in the body (Banerjee, M. K et al.,2023b). Comparisons of the properties to bone diagnosed big affected person blessings as a manner to reduce complications and remove secondary procedures that can in any other case be required.

Metallic materials used in conventional devices such as stainless steel, cobalt based alloys and titanium alloys have been well known for their biocompatibility in biomaterial applications. But some of these materials cause problems like stress shielding and yielding metal ions in the body that can be toxic and require removal if deemed maladaptive to the soft tissue (Tiwari, A et al.,2023c). In order to address these concerns, a novel class of biodegradable metals including alloys and metallic glasses have been developed, which has profoundly impacted the field of metallic biomaterials. From bioinert to bioactive multi-functional materials such as anti-bacterial, anti-proliferative, and even anti-cancer properties. As a biodegradable metal, magnesium-based (Mg) biomaterials have been heavily researched and are considered to be the third generation of biodegradable metals (Tiwari, N. A. 2024c). Mg degrades completely in body fluids, with only products of the degradation that are absorbed or excreted when tissue healing is over and no residues will remain. The technique lowers the demand for exposing additional surgeries to take out the implant. Mg has been shown to have good biocompatibility (Erişen, D. E et al., 2021) and its specific mechanical properties closely match those of human bone, making it an ideal material for orthopedic applications. But the uncontroled depravity of Mg can threaten the mechanical stability, as well as result in metal ion release and hydrogen deposition within tissues which bring negative effects.

To address these problems, researchers are concentrating on designing Mg-based biomaterials with specified composition to modulate and adjust the degradation behavior (Chen, S et al.,2018). Numerous synthesis strategies and material formulations have been investigated to improve the mechanical performance, biocompatibility, and resorbability of Mg-based biomaterials. Several in vitro and in vivo studies including in silico characterizations have been performed during the last decade to investigate these materials and consequently developments were made for their biomedical applications (Eren, E. D et al.,2022; Erisen & Uludag, 2024). Magnesium (Mg) and its alloys are more likely to become the degradable implant materials for biomedical applications since there is a wide agreement on their potential benefits by researchers (Brar, H. S et al.,2009). Owing to its biocompatibility and good mechanical and physical properties, magnesium alloys are now recognized as promising biodegradable implant materials that have been widely studied in recent years. In fact, magnesium is so close to human cortical bone in density (~1.75–2.1 g/cm^3) that it has recognized use for orthopedic implants (Mao, L et al., 2017). On the other hand, magnesium is biocompatible and thus likely to promote bone formation, which makes it a good candidate for medical implants (Wang, J et al.,2016).

Nevertheless, its as-cast form presents low strength and deformation limit of Mg-based materials; even when doped with different alloying elements. However, the property profile of magnesium can be remarkably improved by using alloying and processing techniques to surmount these limitations. The formation of intermetallic compounds when a magnesium alloy is combined with other elements significantly influences the microstructure, chemical reactivity and mechanical properties of these materials. These intermetallic phases alter the material's behavior under stress, at elevated temperatures or deformation and corrosion. The mechanical properties of magnesium alloys are usually modified by precipitation hardening, grain-refinement strengthening and solid-solution strengthening mechanisms. The alloying of the magnesium matrix, with elements showing good solubility at a high temperature, can further stabilize this structure. These enhancements are a response to the selection and quantity of specific alloying elements. The optimal properties of the magnesium-based implant for clinical applications are achieved by appropriate combination of all these elements, which enhance overall mechanical strength, and have effects on stability, corrosion resistance or bio-degradability as well.

This chapter covers the development in a particular field of materials, magnesium (Mg)-based biodegradable metals and their uses in biological applications such as orthopedic prosthetics for people whose natural bone tissue has been destroyed. Mg alloys have caught the attention of the class because they can combine biodegradability, biocompatibility, and mechanical properties. It makes them excellent materials for resorbable implants, which can replace permanent implants without any need for later operation. Most importantly, these materials can go through a safe process of degradation inside the body, so that patients don't need to undertake second surgeries after one is finished removing some former device from their lungs or liver a big advantage over traditional permanent materials. The material properties of Mg alloys are also discussed, including what happens when these materials come in contact with the human body. Key to their successful integration into living tissue is bone regeneration science and the ability of implants to "match" with living tissues on a cellular level. This chapter also describes the functional properties of these materials, including mechanical properties and corrosion resistance. The amount of alloying elements used to modify the properties is also presented here, focusing on what kind of performance stage magnesium with calcium alloying (Zn) might best exhibit in biomedicine applications. With particular focus given to the biodegradability of Mg alloys, in order to ensure that the implant will degrade at a safe and predictable rate without causing adverse effects in surrounding tissue.

In the chapter, several categories of Mg-based biomaterials are discussed in detail. With different compositions or processing technologies, for example, these materials have brought improvements in both mechanical and corrosion properties and so are more suitable as medical materials. Finally, the review concludes with a lookout, saying that future research needs to focus upon such issues as how to effectively control corrosion rates and improve mechanical performance. There should be much left in the realm of Mg-based materials if these things can be achieved—important benefits will truly arise for patients. They could offer safer alternatives to present permanent materials, especially in orthopedics as well as other load-bearing biomedical applications.

2. ALLOYING ELEMENTS AND THEIR EFFECT IN MAGNESIUM ALLOY

2.1. Aluminium

This actually has the largest encouraging effect on magnesium. The existence of Aluminum strengthens an alloy by180 Mpa. Therefore it is a widely added Element in Mg (magnesium) Alloys. An alloy is heat treatable if it has more than 6% weight of aluminum. The incorporation of Al in magnesium is exemplified by approximately 12 wt. % and it depends on temperature (Živić, F et al.,2014). However it´s less soluble in alloys containing Al elements due to ambient temperature. More than 3 wt.%. The percentage of Aluminum (% of Al) increases the rate of corrosion by uplifting the cathodic reaction. When Al is dissolved in Mg, the kinetics at anode get decreased below this limit of solubility.

2.2. Arsenic

It can decrease the rate of corrosion by inhibiting the cathodic reaction to take place. Similarly arsenic is called cationic poison as it inhibits recombination of hydrogen, Adding 0.37 wt.% of arsenic would form Mg_3As_2 and this considerably slows corrosion ((Barfield, W. R et al.,2012).

2.3. Bismuth

Through the addition of bismuth, the reactions at both the cathode and the anode are exacerbated, which ultimately results in an increase in corrosion. By prohibiting the sliding of grain boundaries, it enhances the creep and tensile properties of the material ((Bakhsheshi-Rad et al.,2016).

2.4. Beryllium

The grains are coarsened by the addition of small amounts of Beryllium (<30 ppm). In order to the corrosion resistance, too much will have a negative impact of magnesium. It promotes oxidation in the Mg melt during casting process. Beryllium is a slightly soluble in the Mg family alloys ((Yang, K et al.,2015).

2.5. Calcium

Under equilibrium conditions, 1.34% of calcium is soluble in magnesium. It increases the mechanical as well as thermal proper- ties of magnesium alloy. During heat treatment of magnesium alloys, calcium is introduced in casting to suppress the oxygenation process. Calcium increases the rollability of a magnesium sheet to make it more ductile, but the addition should be no more than 0.3. Biomedical applications 1 wt. The violence at acceptable levels of biocompatibility was in terms of calcium to magnesium 28% (Polo, T et al.2020).

2.6. Copper

Because copper is even more incompatible with magnesium than manganese it has an adverse effect on corrosion resistance With Magnesium forms Mg_2Cu The standard tolerance of copper in Mg is 0.1 wt. % but in the case of alloys Al and Mn, the limit falls to 0.01 wt. % 3wt.% (Ahangari, M et al.,2020).

2.7. Iron

Iron to a content of less than 0.005% for corrosion resis- tance in magnesium but over this percentage it is detrimental. It occurs naturally In Mg alloys as an impurity. Fe solubility in magnesium is extraordinarily low (about 0.001 wt.%), permitting it to exist in pure form (Pompa, L et al.,2015).

2.8. Manganese

The solubility of manganese (Mn) in magnesium (Mg) alloys is very low, with it being generally estimated to be at around 2 wt.%. Nevertheless, addition of Mn as a second element in Mg alloys up to 5 wt. % accommodated the power levels used in this plasma beam experiment., does not significantly impact the corrosion kinetics. As a result, it is combined with other elements, such as aluminum typically to improve its performance in the aforementioned applications. Manganese is an essential trace element in human physiology; it has been recognized for its role in osteoporosis and bone fracture avoidance, through its action as a cofactor for several enzymes involved in complex metabolic process associated with development of healthy bone structure). For Mg alloys, the Fe/Mn ratio should be less than or equal to 0.032 to avoid inflicting the worst corrosion resistance; exceeding this value would result in a significant increase of degree in corrosion rate (Cheng, M et al.,2016). New studies have shown that manganese may also entrap iron in aluminium-free Mg alloys, leading to a new strategy to improved corrosion protection of these materials. The encapsulation technique is highly beneficial due to the strong iron adsorption efficiency of manganese, especially in the absence of aluminum, which also enhances wear and durability properties of the alloy.

2.9. Zinc

Zinc (Zn) addition to magnesium (Mg)-based alloys is reported as a promising material to improve the weldability characteristics of Mg alloys. The incorporation of thorium (Th) in Mg-Zn alloys was already utilized in the past to improve their mechanical properties and to increase thermal stability during recrystallization. But now days thorium use have been stoped due to its radioactivity (Zhao, J et al.,2015). The addition of Tin (Sn) showed a potential to reduce the cracking in Mg during forging, however, the solubility of magnesium with Titanium (Ti) is low. Zinc is an essential alloying element in Mg alloys, mainly for mechanical performance improvement and good corrosion resistance especially when co-adding aluminum (Al). Zinc is soluble in magnesium to approximately 2.5 wt%, and although zinc can improve strength, increasing the concentration of zinc beyond this level may increase susceptibility to stress corrosion cracking. The higher Young's modulus of Zn can cause non-uniform load distribution in implants, if used as an orthopedic material, whereas it does not have a major impact on the hydrogen evolution during biocorrosion or in biomedical applications.

2.10. Zirconium

Zirconium (zr) may act as an excellent nucleus to refine magnesium in alloys containing the rare earth metals, zinc (Zn), thorium (Th), or any combination. The stable compounds it forms with aluminum (Al) and manganese (Mn) prevent its use as an alloying element in these systems. Zr improves the filtration of Mg alloys by scavenging Fe to make Fe_2Zr that, as a precipitate feature with high density will segregate at the bottom of molten metal under gravity in order to enhance the purity and cleanness in castings. Besides, due to the low solubility of Zr in Mg, Zr can relatively high hydrogen, carbon and oxygen and Fe as well is inert towards it and does not form intermetallic compounds (Panahi et al., 2020). It has been reported that Zr can deteriorate the surface film by favoring anodic reactions and in turn negatively impact corrosion behavior. Accordingly, the dosage of Zr must be strictly controlled to avoid its impingement on corrosion resistance and meanwhile maintain its advantages in alloy refinement.

Figure 1. Influences of various alloy elements on magnesium alloy

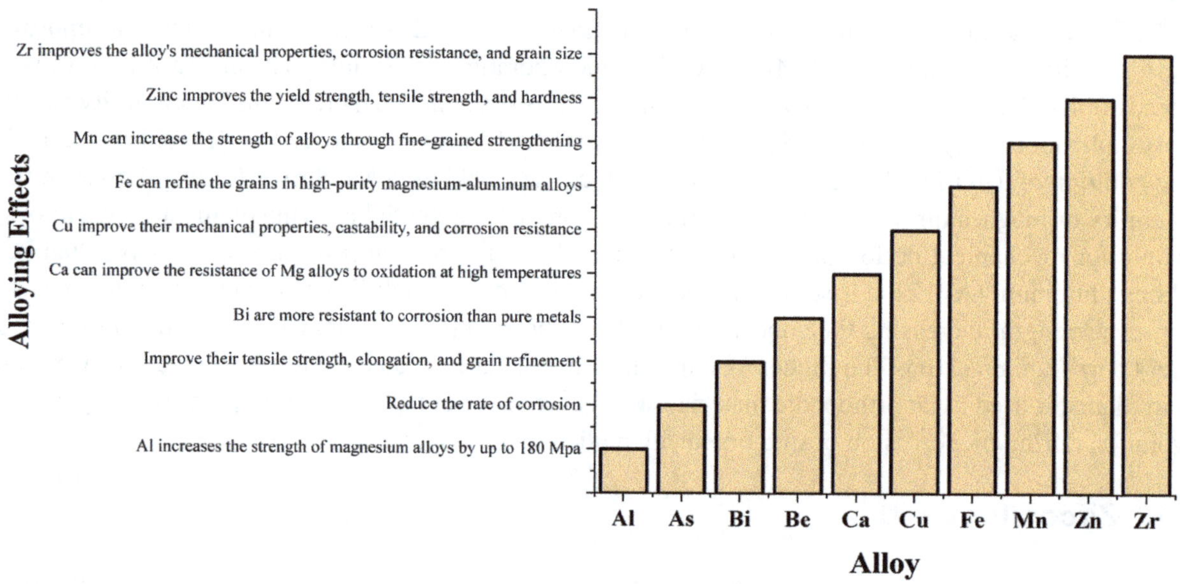

3. CORROSION AND MACHINING OF MG FROM THE PERSPECTIVE OF BIOMEDICAL APPLICATIONS

In magnesium (Mg) alloys, describe how these oxidation reactions are accompanied by a reduction reaction at the cathode, which reduces H^+ ions from water(H_2O), causing release of hydrogen gas. This is a major issue in the context of biomedical applications, owing to possible hazards related to hydrogen gas generation. The problems with hydrogen generation are associated with its productivity per unit time and position as scales of hazardous releases may occur reaches cm/yr or a plot or region may become a TWh-y/H_2-years. Under conditions of lower hydrogen production rates there can be a minimisation of local accumulation, due to the venting away from the generation site (facilitated by diffusion.) Critical factors impacting hydrogen production embrace alloy composition, environmental conditions and surface treatments (Hoop, M et al.,2018). These are the decisive matters at work in relation to parthenogenesis also, with both contributing to how quickly this corrosion happens. In the biomedical applications, controlling the hydrogen release and corrosion is significant to maintain long-term performance and safety of Mg based implants/devices. Some corrosion resistance magnesium alloy based on biomedical applications with their mechanical properties are discussed in Table 1 (Liangjian, C et al.,2015; Pichler, K et al.,2013;Liu, W et al.,2019).

The heterogeneous multi-phase structure of Mg alloys makes uniform corrosion rare for this class of materials. In the alloy the presence of intermetallic compound acted as local cathodes while Mg matrix anode. This galvanic coupling is formed for two reasons: the impurities in the cast Mg metal, and because pure Mg is less noble than its alloying elements. When combined with more noble phases, the dissolution of Mg is also accelerated leading to non-uniform corrosion patterns. The irregular rusting deteriorates the mechanical function of them, which may eventually lead to destruction of structures. In other words, non-uniform dissolution hinders the ability to predict the functional lifetime of these materials, which can be especially important for predictable degradation in applications that require it. Especially relevant in contexts such as biomedical implants that require long-term reliability for patient safety.

Table 1. Corrosion resistance Mg based alloy with their mechanical properties for biomedical application

Alloy	Corrosion Resistance (mm/year)	Tensile Strength (MPa)	Yield Strength (MPa)	Elongation (%)
Mg-Al based	Moderate (0.5 - 2)	150	90	6
Mg-Zn based	Moderate (0.3 - 1.5)	200	110	8
Mg-Zn-Ca	Good (0.2 - 0.8)	150	85	10
Mg-Ca-RE	Very good (0.1 - 0.5)	170	90	7
Mg-Zn-Mn	Good (0.2 - 1)	160	100	8
Mg-RE (Rare Earth)	Excellent (0.05 - 0.2)	240	120	8
Mg-Sr	Good (0.2 - 0.7)	200	110	9
Mg-Ca-Zn-Sr	Good (0.1 - 0.8)	220	130	12
Mg-Li	Fair (0.4 - 1.2)	140	80	16
Mg-Ca	Good (0.2 - 0.9)	160	85	6
Mg-Ca-Zn	Good (0.2 - 0.7)	150	90	8
Mg-Gd-Zn-Zr	Excellent (0.05 - 0.3)	250	140	7
Mg-Y-Nd	Very good (0.1 - 0.4)	230	120	8
Mg-RE-Nd	Excellent (0.05 - 0.2)	240	130	9
Mg-Zn-Cu	Moderate (0.5 - 1.5)	180	90	7

(Liangjian, C et al.,2015; Pichler, K et al.,2013;Liu, W et al.,2019).

3.1. Influence of Proteins and Cells on Mg Corrosion

Proteins are a class of biomolecules that notably influence the corrosion of magnesium (Mg) alloys. However, protein work has instead either seen Mg alloy immunized against corrosion or infected to destroy it (Gopi, D et al.,2014). Protein effect on Mg dissolution is alloy-specific and the use of a different type of protein the activity might be very different towards that class of alloys. Curiously, the effect is time-dependent the proteins initially inhibit corrosion but then they spur it on. In many instances protein-bound metal cations could be a complexed or formed that increases dissolution and therefore corrosion.

Alternatively, cations adsorbed at Mg surface may produce a passivating layer that prevents corrosion. As the corrosion of Mg alloy in proteins can be influenced by differences in metal surface properties andatively, conclusions regarding the mechanism behind protein-mediated Mg corrosion show some discrepancy due to differing protein types. This variability is also one of the complex factors into the study of protein-induced corrosion mechanisms in Mg-based materials.

Magnesium (Mg) surface degradation results in the release of cell-corroded by-products including increased pH, hydrogen gas production and Mg^{2+}. These dissolution by-products highly affect the rate of Mg corrosion. When cells are attached, they form layers on the surface of Mg to serve as a barrier through which corrosion proceeds at a lower rate. Earlier studies have already reported that the pH of cell monolayers on metal surfaces below and accordingly due to cellular metabolism (lactic acid production, reducing pH). Infections, and hydrolysis reactions of the dissolved cations amplifier lower pH values (Mena-Morcillo et al., 2020). An understanding of these findings indicates that cell-to-Mg alloy interactions can be complex and hence influence corrosion behavior. There is some evidence that the cell layers can protect the surface and retard corrosion, but also a possibility of enhanced corrosion stemming from acidic by-products of cellular metabolism. Accordingly, there appear to be stimulatory and inhibitory influences of cells on Mg corrosion.

3.2. Methods to Monitor Mg Corrosion in Biomedical Applications

Given that different applications may require differing levels of corrosion resistance, a large portion of the research on monitoring the biodegradation of Mg and its alloys has concentrated on development of new alloys, along with surface modification techniques (such as coating) to improve their performance (Xin, Y et al.,2007). Currently these alloys are generated by precise alloying magnesium with biocompatible elements that do not affect performance. The use of alloying elements as well as processing methods, heat treatments and microstructure largely determine the properties of the obtained alloy. Single or Multilayer coatings are also employed for increasing the corrosion resistance property of the installed surface by creating different alloy composition. One of the main issues is that these coatings adhere to the metal underneath and this attachment depends on its durability and sustainability just like how it will perform when being attacked. In this way, the thickness of the coating is also determined based on how it is to be used with the Mg alloy where a fine balance needs to be maintained between providing enough corrosion protection without compromising other functional requirements.

3.3. Machining of Biodegradable Implants

It is mandatory to have a percise dimensional accuracy and high surface integrity of bioimplants to provide the optimal cell adhesion. In addition, they required that implants beomer durable and stable, posing little risk of inflammation. The cutting behavior during the machining process is directly related to the surface integrity properties of bimetallic components, which ultimately influences both the feedback response and fatigue life in biomedical implants. While the corrosion rate and composition of the implant were determined by surface characteristics, which play a key role, leading to the importance in selecting also suitable manufacturing technology for biodegradable implants. Excellent machining can improve the mechanical and biological properties of an implant as well as its long-term stability in a biological environment, thus reducing the potential for degradation or adverse reactions (Wasserbauer,

J et al.,2020). Thus, a cautious selection of machining technique is needed to strike a balance between the structural integrity and biocompatibility for bioimplants.

The conventional machining processes are not suitable for applications requiring high surface integrity and burr free surfaces. Applying unsuitable machining parameters in standard machining processes may result in excessive heat generation and thermal ignition of magnesium (Mg). Lower cutting speeds should be used, and parameters optimized to reduce friction and heat generation resulting in improved machinability of Mg alloys. An even more cutting-edge process, cryogenic machining provides the added refinement of grain size and improved surface integrity (Kieke, M et al.,2016). Cryogenic methods provide a cooler material with the basal structure allowing for less deforming and improving quality on the machined surface. This is improved for applications where accuracy and longevity, such as many biomedical or aerospace applications are critical to material performance and safety.

Although certain additive manufacturing modalities such as electron beam melting and selective laser melting are capable of high accuracy and tight tolerances for implant production, their expense has restricted wide commercial adoption. One of the main advantages of non-contact machining processes is that there are fewer chances for material wastage and it can also be used to manufacture complex geometries. The principal non-contact methods in in the laser beam machining (LBM) and Electric discharge machining (EDM). Among its most innovative applications is the ability to create complex geometries and sub-mm-sized details in medical devices, like vascular stents (Witte, F et al.,2008). One major issue associated with LBM is the heat accumulated during processing, which may cause a significant reduction in the corrosion resistance of magnesium (Mg) alloys in vivo. The exothermic plasma species may incriminate the Mg alloy and cause a reduction of its mechanical properties and functionalities in particular. Thus, it is crucial to control the heat generation for the sake of preserving the properties and bio-compatibilities of Mg implants.

3.4. Biomedical Applications

Orthopedic implants with a density similar to human bones and have better biocompatibility that can assist in osteogenesis, are an attractive approach. Due to the intrinsic non-toxicity and full biodegradation in vivo, Mg alloys are potential to be used as bioresorbable biomaterials for orthopaedic applications. Its biocompatibility and low cytotoxicity also confer Mg a high potential as cardiovascular stents materials. Mg, being highly machinable, allows the manufacture of complex and tiny implants or stents. The continued challenge in localized corrosion remains however, with appropriate alloying and coating strategies to minimize the rate of corrosion, done carefully such that biodegradable implants degrade throughout the healing or tissue regeneration (Razavi, M et al.,2020). Biodegradable Implants one of the key benefits of biodegradable implants is that they do not require secondary surgery for removal, given that they decompose into biocompatible by-products, which are excreted naturally from the body.

4. MECHANICAL STRENGTH OF MAGNESIUM (MG)-BASED BIOMEDICAL IMPLANTS

It is essential in assessing and making improvements in the safety and performance of biomedical implants to enable them to have proper mechanical properties for their intended applications throughout their operational life. Implants must be very strong because they have the important job of supporting tissues while it heals. But, does not always mean ideal behavior in biodegradable implants. For instance, an implant with an elastic modulus that differs substantially from that of the bone around it may experience elastic mismatches and stress-shielding. For instance, when the materials in question are metallic biomaterials (used largely for orthopedic applications) stress shielding can occur and lead to a weakening of the surrounding bone undermining implant stability and integration. Moreover, in view of the many challenges facing application of biodegradable implants based on magnesium (Mg) alloys, it is necessary to improve their mechanical performance (Gusieva, K et al.,2014). In Figure 4, discussed about the some mechanical properties of magnesium alloys with processing route for biomedical implant. Two key strategies have been conventionally adopted to enhance the mechanical properties of Mg alloys such as alloying where the composition is modified to improve strength and corrosion resistance, while another strategy is by using advanced manufacturing methods like co-casting or spinning/rolling/manufacturing approach. These approaches are designed to construct implants with adequate mechanical supportive but also degradation in a carefully controlled manner, thereby enhancing tissue regeneration and avoiding stress-related problems.

Mg-based alloys have a combined use in the life science of bone implantable materials and cardiovalvular stents, where both Mg-strength-related challenges stress shielding as well as decoctile/bending labor. When the stiffness of an implant is considerably greater than that of the surrounding bone, stress shielding results and can reduce bone stress with concomitant loss in bone mass. This discrepancy may well affect the performance of the implant in the long time run. Moreover, in case of cardiovascular stents, the Mg-based alloys must possess enough ductility and pliability to endure deformation without fracture during implantation. For this purpose, different materials (e.g., calcium (Ca), aluminum (Al) and zinc (Zn) as well some rare earth elements like as zirconium (Zr)) were alloyed with magnesium during the research of team (Gupta & Wong, 2015). Its hexagonal close-packed (HCP) crystal structure and atomic size (~320 nm) allow it to form solid solutions with various alloying elements (e.g., Al, Zn, Ca), including strontium (Sr), improving mechanical performance. Together, they control significant properties such strength, ductility and corrosion resistance by adjusting the concentration of these elements during alloying. The mechanical strength of the Mg-based alloys could achieve a significant improvement by properly selecting and tuning alloying compositions, so that these are more suitable candidate biomaterials for orthopedic and cardiovascular applications.

(Wong, W. L. E., & Gupta, M. 2006) showed an increase in strength but also a decrease of ductility with increasing yttrium (Y) content between 2.5 wt% and 7.5 wt%, whereas neodymium (Nd) contents ranging from 1 to 4.2 wt% within extruded WE alloys demonstrated the same trend. They have higher strength with most uniform deformation and are therefore better suited to orthopedic implants where high strength is a primary consideration and ductility is less critical, than for cardiovascular applications where enhanced deformability facilitates delivery. On the same note, (Mao, L et al., 2015) have done a research on the effect of aluminum (Al) addition on as-cast mechanical properties of Mg–4Zn–0.2Ca alloys. Introduction of Al and Zn resulted in an enhancement of yield strength (YS) and ultimate tensile strength (UTS) via solid solution strengthening and grain refinement effects. Nevertheless, secondary phases were detected along the grain boundaries when over 3 wt% Al was added; they were identified as $Al_{12}Mg_{17}$. The presence of these phases had a detrimental effect on the elongation and ultimate tensile strength (UTS) of the alloy, resulting in deterioration in its ductility and tensile behavior On the other hand, compressive strength was quite lower than UTS and only a slight increase in this property could be achieved with an increase in Al content. It is observed that the specific Al content must be very precisely controlled in order to achieve this required balance of tensile properties and avoid the unwanted phase formation.

The authors pointed out the cause of this is that a tensile stress and compressive stress on mechanical forms crack propagation sights. In compression, the strength was enhanced by secondary phases formation. For example, a binary Mg–Al alloy foam with 2 wt% Al showed higher compressive strength compared to high Al-containing samples. This was a result of the high secondary phase formation between Mg, Al that deteriorated material mechanical integrity. Similarly, Zhou et al. studied the influence of Neodymium (Nd) on mechanical properties in extruded Mg–1Mn–2Zn alloy, and showed that the addition of Nd refines the microstructure leading to an increase in both ultimate tensile strength (UTS) and ductility (Yuan, B et al.,2022). At the same time, more Nd content did allow for higher UTS and elongation but lower compressive strength due to poor distribution of Mg_7Zn_3 phase in microstructure which resulted in reduced uniformity. Furthermore, Yandong et al. In this study, they investigated the influence of manganese (Mn) addition to as-cast and solution-treated Mg–2Zn–0. 5Ca alloys. It was found that the UTS and elongation of steel were highest when a 1 wt% Mn addition peak was present. Nevertheless, when the Mn content was >1 wt%, the brittle Mg–Zn intermetallic phases formed and deteriorated both in UTS and ductility. Shi et al. (2016) also demonstated that at same Gd content, the presence of Ca content up to 1.2 wt% could improve the mechanical performance of the Mg-Gd-Ca-Zr alloys.

It was evidenced that, as calcium (Ca) is added to Mg alloys, both YS and compressive strength increase with the exception of ductility. This is due to the grain refinement power of Ca, and zirconium (Zr), which may improve mechanical properties of the alloy creating it appropriate for orthopedic used (Qin, Y et al.,2022). Microstructural refinement, enabled by Ca and Zr addition, is another key aspect that can contribute to the improved mechanical properties. Yan et al. published that the YS and ultimate tensile strength (UTS) have excellent when Zn content was up to 14.5 wt%. This improvement was because of the uniform dispersion of Mg–Zn intermetallic phases in the microstructure, introducing strengthening effects. Nevertheless, more excessive Zn content amounts larger grains and YS and UTS were decreased at 14.5 wt% of Zn as the microstructural homogeneity did not occur due to coarser-to-elongated grain formation over this amount. Also, the Mn-Ductility effect was absent in Zn ductility. Under the experimental condition, it can be observed that the elongation of the alloy is increased continuously with increasing Mn content beyond 14.5 wt%, which indicated great potential in high wear ductility, especially for stent applications from higher ductility demand on bending and expansion due to its minimum opportunity under stress phase transformation effect (Bordbar-Khiabani, A et al.,2019). These results underline the need for the control of alloying elements to tailor mechanical properties for biomedical applications.

Mg-Mn-3Sn alloy was developed to improve mechanical properties of the Mg–3Sn alloy for sizing applications, which could form favorable intermetallic phases. (Zhang, Z et al.,2023) reported that the Mg–3Sn. The Sn alloy showed higher ductility and UTS at the same time compared to other widely used magnesium alloys such as WE43, inducting about 12–21% for Al-matrix composite (or lower) by keeping comparable mechanical properties of ultimate tensile strength. The resulting improvement in mechanical properties comes from the presence of Mg_2Sn intermetallic phase; indeed, tin (Sn) is a well-known strengthening element mainly for its effectiveness in intermetallic phase formation. Mg_2Sn was formed as the reinforcement phase in the alloy and improved strength and ductility (Zhao, D et al.,2022). Based on the studies in recent works, along with a variety of depreciation methods such as solid solution strengthening and aging hardening, grain size decrease has also been disclosed to be one of the most significant mechanisms accounting for Mg alloys enhanced mechanical properties. Here, the bond that spot welds magnesium is a solid solution, where alloying elements dissolve into the parent Mg matrix and slow down dislocations. In precipitation strengthening, finely dispersed secondary phases are formed, which pin dislocations, while an increase in grain boundary area due to grain refinement through the Hall-Petch relationship results in decreased slip parameters that strengthen the material. All these mechanisms together lead to improved mechanical performance of biocompatible magnesium alloys.

For the biomedical Mg alloys, two of the most important heat treatment processes for control of the mechanical properties, including solution treating and aging treatments. Solution treatment generally involves the alloying elements dissolving into the matrix which is supposed to increase ductility but normally at a loss in strength. Aging treatment, on the other hand promotes the precipitation of secondary phases which causes an increase in strength but loss in ductility Owing to these conflicting effects, for bone implants such as dental fixtures that require superior strength over high ductility, treatment should contain aging alone; for cardiovascular stents that necessitate excellent balance in ductility and strength, solution treatment or both with aging will be chosen. For instance, a research showed that the aging treatment applied to extruded Mg–3Nd–0.2Zn. While the elongation was kept a 26% even without aging treatment, the ultimate tensile strength (UTS) of 4Zr alloy went up a lot. The superiority of extrusion as a manufacturing route over casting for Mg alloys is due to work-hardening induced by the application of large plastic deformations, results in improved mechanical properties incorporating solid solution strengthening mechanisms within the microstructure (Kumar, A., & Singh, G. 2024). Multiple studies have shown that extrusion, particularly powder extrusion, will lead to improved mechanical properties due to the fine grain structure and homogenous distribution of alloying elements that can be achieved. This makes extrusion a good means to develop both commercial and high-performance Mg alloys in biomedical applications.

(Kanishka, K., & Acherjee, B. 2023) studied the mechanical properties of Mg–4.0wt.% Zn–1. MASConstraintMaker 0Zn–0. Because the mechanical properties of the extruded 2Ca alloy were better than those of melt cast alloys. This enhancement was due to the generation of secondary Ca phases and solution-strengthening impact of Zn. In another study, Wu et al. performed a comprehensive study on as-casted, extruded, and cyclic extrusion–compression processed (CEC) Mg–Zn–Y–Nd alloys which can be used for cardiovascular stent applications. The hot-extruded alloy showed higher yield strength (YS) and ultimate tensile strength (UTS), the CECed alloy exhibited an increased ductility by one- to two-fold. This increased ductility, combined with lower YS and UTS in the CECed material was ascribed to both a finer grain size (<1 μm), precipitation of secondary phases at the grain boundaries and an increased dislocation density due to cyclic plastic strain. More extensive studies on tensile properties and ductility following these three manufacturing techniques, CEC; equal-channel angular pressing (ECAP); hot extrusion for cardio vascular application varied. (Pesode, P., & Barve, S. 2023) findings are somewhat analogous to another study which showed that CECed alloy had the highest ultimate tensile strength (UTS) and yield strength (YS) because intermetallic Mg–Nd phases were formed in this condition and they helped in improving the tensile properties by grain refinement, precipitation strengthening etc. Despite having larger grains in the ECAPed alloy than in the CECed alloy, both alloys had almost similar UTS. It is a thing that Sometimes grain size effect not the only factors, while another important factor could be dislocation at higher strain hardening grains This heavy and complex job of microstructural contribution to mechanical properties impresses the authors that UTS would also be analyzed accordingly.

(Ding, Y et al.,2024) pointed out enhancements of the mechanical properties for the Mg–1.5 Nd 4 wt.-% Sn-CS, as illustrated in Fig. 5Y–1. 2Zn–0. This is mainly due to the formation a long-period stacking ordered (LPSO) structure in the secondary phase ($Mg_{12}ZnY$) in 44Zr alloy after hot extrusion. Grain refinement and LPSO structure significantly increased the hot-extruded alloy strength (strength is here in tensile and compression at high temperature) with plasticity, compared to heat-treatment and as-cast alloys. Uniform distribution of the secondary phase resulted in improved mechanical properties that all hot-extruded alloys present.

Similarly, increased hardness have been reported (Zhang, C et al.,2024) in a ECAPed Mg-2Zn-Ca alloy with 1 wt% of β-tricalcium phosphate (β-TCP). The ECAPed alloy increased greatly in hardness (up to 789 HV) compared with the as-cast counterpart, which possessed hardness of 539 HV. The grain refinement and dispersion of β-TCP particles into the matrix have contributed to this improvement. Gui et al. is consistent with similar results of Yan et al. The hot-extruded Mg–Gd–Zn–Zr–Mn alloy also showed almost twice the yield strength (YS) and ultimate tensile strength (UTS) compared to its as-cast and high temperature-aged conditions. The alloy also showed a larger ductility, with elongation of 18–21.3%.

The mechanical properties of the Mg–Gd–Nd–Zn–Zr alloy were improved greatly by ECAP treatment in the study of (Yang, G et al., 2024) for the UTS, it increased from 161 MPa to 267 MPa and for YS, it rose from 85MPa to 217M Pa. It was improved mainly by formation of the β phase to block dislocation movement and increase strength. Meanwhile, ECAP has enhanced the hardness of alloy, thereby promoting its mechanical properties as well. These studies highlight the relevance of grain size reduction, secondary phase formation and plastic strain to improve the physical/mechanical properties of Mg-alloys that are necessary for their application as biodegradable implant materials.

Figure 2. Mechanical properties of magnesium alloys with processing route for biomedical implant

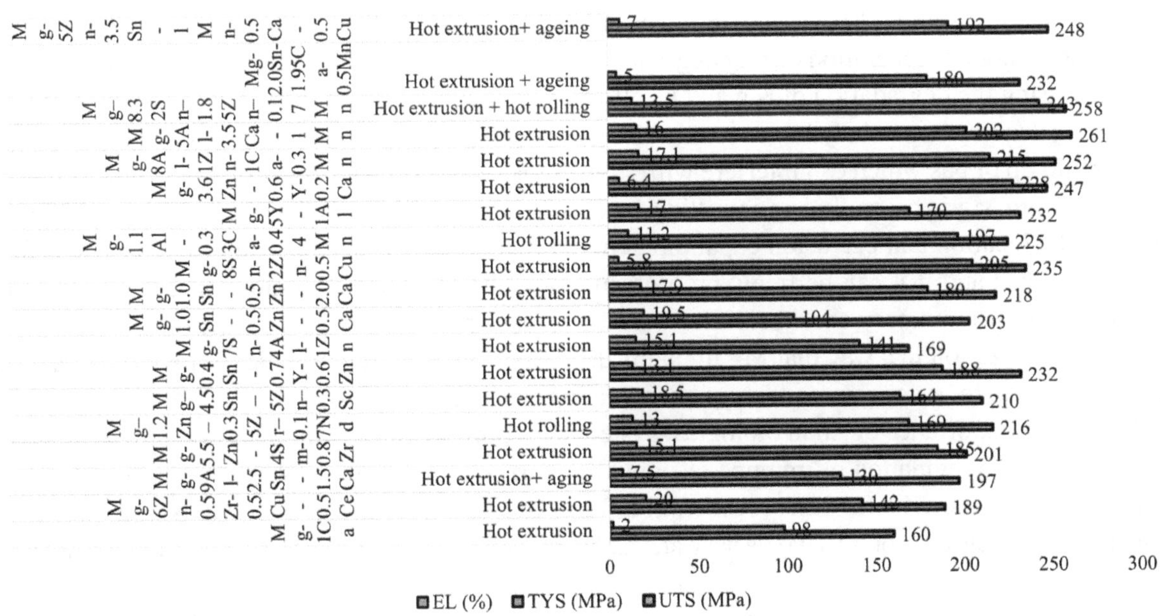

4.1. Biological Properties

Magnesium (Mg) is an essential element in the human body, and it mediates various biological processes such as a whole heap of enzymatic reactions. It takes part in numerous cellular activities such as protein and nucleic acids synthesis, stablization of the plasma membrane, etc. The average human body contains about 21-28 grams of magnesium, with over half in bone tissue. Magnesium makes up 35–40% of the body's magnesium content as soft tissues, while less than 1% is found in serum. Bone magnesium content acts as a depot that participates in the short-term management of large and rapid changes in plasma Mg levels. The property of bone-protecting Mg2+ ions is effective in maintaining the stability and prevent pathological changes which may induce fracture of bones (Bai, J., Wang et al.,2024). The latter is also important for bone formation and repair, since these ions are involved in the mineralization of unmineralized immature bone to fully mineralized mature bone. In fact, Mg^{2+} traces were identified in almost 6 mol % of bone mineral as characterized with a biochemical approach. The higher Mg^{2+} ions present on one side of the cartilage/immature bone in short-term studies in rat growth plate can, however, decrease with time. Bone tissue with more magnesium, it is said to have slightly more elastic bones and therefore has improved biomechanical features. That indicates a double-importance of magnesium for cellular actions as well the structural component and functional adaptation of bone tissue.

A bone formation –enhancing role of magnesium (Mg) has been well reported and demonstrated its determinant for differentiation of osteoblastic cells. The accelerated bone formation observed at and above the degraded Mg implants indicate that Mg degradation can indeed enhance various aspects of bone healing (Jin, Z et al.,2022). Mg^{2+} released into surrounding tissue activates local cellular activity and promotes bone regeneration during degradation of Mg. Study has also shown that the hydrogen gas released during the degradation of Mg and its alloys is typically tolerated in the human body. The issue is when it releases too much hydrogen gas, as this can create problems down the line in terms of healing due to build up of gas which can interfere with tissue regeneration. Thus, the key to reduce possible gas evolution from Mg is basically in controlling its corrosion rate. It may influence the rate of gas clearance, by the location and perfusion capabilities in the blood flow where a certain biomaterial implant has been placed and if it can be freely progressed from a body or not. Examination of the expression profile of proinflammatory cytokines discovered here and previous studies would emphasize on the need for optimization in a way that Mg implants degrade in such a manner which is beneficial, and not counterintuitive, for osteogenesis for any bone healing application.

Due to the importance of Zn in biological systems, for both mechanical and anti-corrosive purposes, as well as in vivo regulation of immune response and enzymatic reactions. Magnesium (Mg) has been proposed an idea material for bone regeneration; however, the direct use of pure Mg is limited because of its too rapid degradation in vivo (Aherwar, A., Singh et al., 2015). Zn as a type of bioactive element would be beneficial for biomedical Mg implants, which apparently increase osteoblastic cell proliferation on their coated surface-responsive capacity to modulation factors or specific ions. Nevertheless, high concentrations of Zn are cytotoxic in vitro. Here Hong et al. explained on their studies such as showed that ZK40–Mg and AZ31–Mg alloys increased cell viability in comparison to pure Mg, but that high concentrations of Zn corrosion products induced cytotoxicity by osmotic shock on surrounding cells.

No cytotoxic or neurotoxic effects were observed for Mg implants with Al concentrations as high up to 9 wt% in another study. In addition, the osteoblastic activity also increased with increasing Al content until this threshold concentration suggesting that low Al ion release may be permissible in the human body. These results indicate that introducing alloy elements, such as Zn and Al into Mg-based implants may improve biocompatibility and even activate cell response for bone healing if the concentrations are well-controlled to avoid toxic ions release. This underlines the need to fine-tune alloy compositions to achieve favorable trade-offs between desired biological properties and biodegradability in the human body. Recently, it has been demonstrated that the incorporation of zirconium (Zr) in magnesium (Mg)-based alloys significantly improves the biocompatibility in vivo. But as found by (Joyce, K et al.,2021) more than 5% Zr will lead to locally forming unalloyed Zr phase, lowering the corrosion resistance. The decrease of corrosion resistance on plated stents, following its damage or loss, for example in crimps performed during balloon expansion might thus result in lower biocompatibility by allowing a larger uncontrolled degradability inducing adverse biological reactions associated with the release of harmful and potentially toxic particles. Consequently, keeping the Zr content in Mg–xZr–ySr alloys below 5% is suggested to prevent the forming of unalloyed Zr phases and promote an acceptable biocompatibility.

In addition, it has been established that the dual doping of Mg-based implants with calcium (Ca) and Zr can increase dramatically the osseointegration ratio and promote implant stabilization in host tissue. These alloying elements assist in the good osseointegration of the implant as well as leads to faster healing and an increased mechanical holding of implants. (Ahirwar, H et al.,2020) suggest that the regulation of Zr content and intentional addition of alloying elements such as Ca are necessary to tailor the mechanical properties, for example UTS,TY,Errand ERl, which could further enhance biomaterial integration into the bone environment with Mg-based metallic glass in orthopedics zone.

4.2. Mg-Based Composites for Potential Biomedical Implant Applications

Biodegradable magnesium (Mg) alloys are promising implant materials in orthopedics due to their desirable characteristic of controlled degradation to help promote bone healing. Many efforts have long been devoted to tuning the degradation rate of Mg alloys accurately in line with bone healing. One approach to overcoming this limitation involves the preparation of new Mg-based composites with other materials. In Figure 3, discussed about various mechanical properties of magnesium based composite with processing route for biomedical implant. Various ceramic and polymer materials have been incorporated in Mg alloys to make composites by the researchers utilizing different manufacturing techniques to present corrosion resistance, mechanical properties, and biocompatible.

Mg and its alloys are used as a matrix, with the addition of ceramic materials for reinforcing—a development that has benefitted both from strengthened mechanical properties resulting from the reinforcement phase and improved corrosion resistance. These reinforcements of ceramics contribute to the prevention of Mg fast degradation and lower support stability and duration during bone healing. By contrast, in polymer-based composites, Mg alloys are often used to reinforce the polymer matrix. Wherein the mechanical toughness and strength of the Mg are utilized by these polymer composites, whereas ferrule the matrix can tune degradation rates and improve biocompatibility (Santos, 2017). These findings provide the possibility of preparing novel magnesium composites having advanced mechanical properties and corrosion resistance while preserving bioactivity and biodegradability essential for effective orthopedic applications. The method provides a customized solution to the problems with Mg alloy implants, and makes them more easily used for clinical applications.

Figure 3. Mechanical properties of magnesium based composite with processing route for biomedical implant

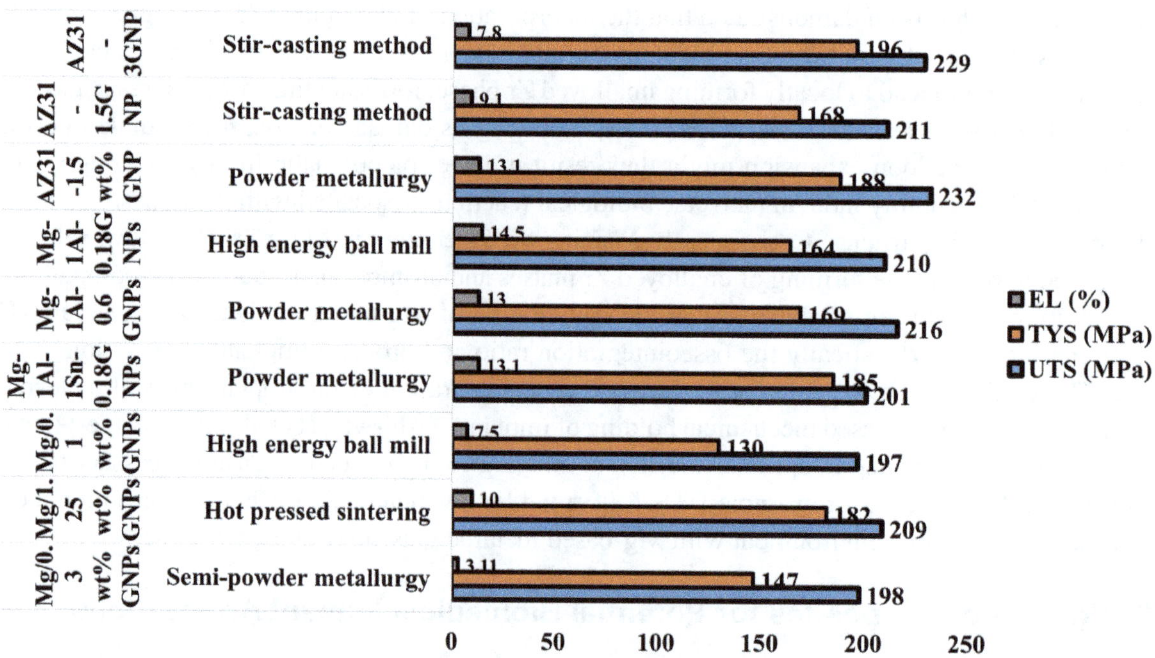

4.3. Mg-Polymer Composites

Additionally, a semicrystalline poly-lactic acid (PLLA)/magnesium (Mg) composite was also studied with the motivation to take an advantage of both materials; higher strength and modulus provided by Mg particles, and better corrosion resistance from protective polymer matrix. The composite was prepared using solvent casting of PLLA with 30 wt% magnesium particles, thereafter compression molding. The mechanical properties of the polymer matrix were greatly improved by incorporating Mg particles with Young's modulus of up to 8 GPa, yield strength (YS) at 100 MPa and hardness at 340 MPa. Hot extrusion was used to evaluate various fabrication methods for PLLA/Mg composites that effectively homogenize Mg particles in the PLLA matrix. The composite displayed increased degradation with the incorporation of Mg particles especially when added up to 5 wt%, and a more pronounced jumps in degradation was perceived at 7 wt%. On the other hand, the mechanical strength of the composite aggravated significantly with increasing Mg contents approaching 7 wt %. (Abbas, Z., et al.,2021) indicates that the addition of Mg reinforcement improves the mechanical properties however, it is necessary to retain an optimum Mg content threshold in PLLA-based composites so as to prevent excessive degradation and rendered mechanical properties. Those results emphasize the need to balance both PLLA/Mg composite compositions for improved mechanical properties and also for controlled biodegradation, which will make them attractive candidate materials for orthopedic implants in the biomedical field.

To improve the mechanical properties of poly-L-lactic acid (PLLA) and decrease its inflammatory reaction in biomedical uses, (Barfield, W. R et al.,2012) a hybrid composite was prepared by introducing oligolactide-grafted magnesium hydroxide (Mg-OLA) into the PLLA matrix. MethodsThe Mg-OLA particles were precipitated by ring-opening polymerization. We have developed a hybrid form, PLLA/Mg80-OLA20, with different Mg-OLA levels (0–20 wt%). The results revealed that the ultimate tensile strength (UTS) and modulus of the PLLA/Mg80-OLA20 composite were significantly higher than those of the PLLA/magnesium hydroxide composite, which showed, improved mechanical. Performance. Particularly, the addition of Mg80-OLA20 (0–20 wt%) not only improved the mechanical properties, but also showed great improvement in biocompatibility. There was 100% rise in the cell viability of the composite which indicated that the presence of Mg80-OLA20 led to less cytotoxicity and a better inflammatory response.

This work illustrates the utility of using oligolactide-grafted magnesium hydroxide with PLLA for developing composites with better mechanical and biocompatibility properties, which can be used in biomedical applications like orthopedic implants, where good mechanical strength along with low inflammatory response are vital for long-term success. This novel hybrid composite system has proven to be a viable solution for biodegradable medical devices, resolving the issue with traditional PLLA/$Mg(OH)_2$ systems of resorbable bone repair materials. (Polo, T. O et al.,2020) reported the reinforcement of AZ31 magnesium alloy fibers onto poly(lactic-co-glycolic acid) (PLGA) matrix that significantly improved the physico-mechanical behavior of the composite AZ31 fibers reinforced the PLGA matrix as evidenced by significant enhancements in both ultimate tensile strength (UTS) and elongation. Moreover, the alkaline corrosion products derived from AZ31 Mg alloy fibers significantly neutralized the acidic degradation products of PLGA to provide a less harsh environment for fiber degradation. The cell direct attachment tests indicated a good biocompatibility of the composite, with recoverable morphology and cellular adhesion and growth.

(Gopi, D et al.,2014) also reported in another study that ADMSCs sensor cells following a chemo-attractant gradient from inflamed cellular tissue areas to the site of these stimulating chemo-attraction solutions or cells; defect. Formulated a new metallic Mg particle/PLGA composite scaffold by solvent casting and salt leaching technique to stimulate bone regeneration after tooth extraction. The compressive strength and modulus of the PLGA scaffold were enhanced by inserting Mg particles into the structure, which had a porous morphous that would be beneficial for cell embedding. The pH buffering effect of the combination of Mg and PLGA, basic-degradation and acidic-degradation materials respectively, could achieve a balanced degradation process with sustained release of Mg during 10 weeks. This is because the degradation products of Mg/PLGA composite scaffolds also had an excellent ability to promote bone marrow stromal cell proliferation, which might be carried out by improving cellular activation in vitro. Moreover, histological analysis supported the good biocompatibility of the composite and indicated its potential for biomedical applications, especially in bone tissue engineering. (Gusieva, K et al.,2014) demonstrates the enhanced mechanical reinforcement and osteogenic properties of Mg-reinforced PLGA composites suitable for both load bearing rehabilitation applications and bone regeneration. Conclusion-Our results highlight the potential for Mg/PLGA composites as a next-generation orthopedic biomaterial to enhance both mechanical and biological activities, promoting tissue regeneration and integration.

The review of literature shows that the reinforcing Mg with biopolymer matrix enhancing mechanical strength of composite mostly outperforms individual attributes. It has also been reported that when Mg is taken up in a polymer matrix, the corrosion resistance of Mg is considerably increased which can provide a barrier between the more aggressive environment and the pure Mg for promoting these alloys. While these advances have been significant, the mechanical strength of Mg-based composites on output is often lower than that of pure Mg by itself, thereby limiting their use in load-bearing biomedical applications, e.g., orthopedic implants . Because of this difference in mechanical performance, further improvement on the composition and structure of Mg-biopolymer composites is required to suit their applications. Future work should aim to locate appropriate biopolymers that can be used and to optimize the reinforcement process so as to achieve mechanical properties at or above those necessary for clinical use. It can address the need to advance the manufacturing techniques, and add more reinforcing agents for improving mechanical properties and degradation rates of Mg-based composites. Apart from being crucial, these efforts are needed to compensate for the shortcomings of strength and stability which are indispensable prerequisites in order to maximize the effectiveness of these composites including applications in biomedicine. The main aim is the fabrication of Mg-based composites having mechanical strength in a permissible range while at the same time being biocompatible and applicable for degradation perspective specifically to medical implants and applications. This balance is key to making Mg-based composites suitable for a wider range of medical applications, especially in load-bearing regions like orthopedic and cardiovascular implants. Moreover, future studies should examine the interplay of reinforcing agents and manufacturing procedures with biological environments to guarantee the best outcome thereof when used in clinical settings.

4.4. Mg-Ceramic Composites

In recent years, composites based on Mg alloys as the matrix and ceramics as reinforcements have been significantly explored with respect to their potential in biomedical applications, particularly in hard tissue engineering. Various mechanical properties of magnesium ceramic composite with processing route for biomedical implant are illustrated in Figure 6. Hydroxyapatite (HAp), beta-tricalcium phosphate (β-TCP), and bicalcium phosphate (BCP, a mixture of HAp and β-TCP) are examples of biomaterials that share many characteristics with the inorganic constituents found in bone, making them particularly attractive alternatives (Yuan, B et al.,2022). Calcium phosphates have been found to adsorb ions and calcium phosphate to provide a surface for the mineralization of bone, and are capable of promoting bone growth on the composite. Calcium phosphate materials, however, have the disadvantage of not being able to achieve the mechanical strength necessary to replace natural bone.

In this context, composites based on calcium phosphates with Mg and Mg alloy matrices (Kanishka, K et al.,2023) have been investigated as an attempt to overcome this limitation. Many research groups have explored the addition of HAp to pure Mg and its alloys (e.g., AZ91D, MgCa) in order to reinforce these materials for improved mechanical properties as well as a solution to their poor corrosion rates. For instance, Witte et al. synthesized a metal matrix composite (MMC) by hot extruding HAp particles in to Mg alloy AZ91D matrix. As a result, we obtained a composite with mechanical characteristics similar to natural bone. The presence of HAp particles was observed to control the corrosion rate, leading to a more homogeneous corrosion trend in artificial seawater and cell culture media (Zhang, C.,et al., 2024). The bone cells adhered, proliferated and survived on the surface of the MMC-HAp composite during immersion tests implying its biocompatibility verifying its suitability for biomedical applications.

(Seetharaman, S et al., 2022) fabricated AZ91/HAp composite through squeeze casting process. Composite compressive strength was significantly higher than pure HAp, reflecting more favourable mechanical behaviour. Nevertheless, the AZ91 alloy exhibits poor natural corrosion resistance compared to HAp alone, which may facilitate homogeneous degradation of the composite and therefore favor quick recurrence of the porous HAp structure capable of supporting bone cell adhesion and proliferation. Consequently, Mg-based composites reinforced with calcium phosphates are potentially a suitable candidate for orthopaedic applications that integrate the bioactivity of calcium phosphates without compromising the mechanical strength and controllable degradation behaviors of Mg alloys. Further work is required in order to tailor the properties of composite materials with sufficient accuracy for their use as hard tissues implants (mechanical properties, biocompatibility and degradation behaviour.

Powder metallurgy (PM) has been considered as a preferred approach to casti ng with regard to the homogeneous dispersion of ceramic particles in Mg ma trix. (Sezer, N et al.,2018) employed this method as well. Mg/hydroxyapatite (HAp) composites with different HAp content (10 wt%, 20 wt%, and 30 wt%) were fabricated using Figure 8 shows the microstructural images of as-cast Mg/10HAp composite and Mg/5WC, WC reinforcing phase were randomly distributed in the entire matrix. Compared to as-extruded pure Mg, the yield strength (YS) of Mg/10HAp composite increased while the ultimate tensile strength (UTS) and elongation decreased. Moreover, the more HAp was incorporated,the higher be the corrosion rate of the composites. Cytotoxicity tests demonstrated that the good biocompatibility of Mg/10HAp composites on L-929 cells.

(Banerjee, M. K et al.,2023b) used friction stir processing (FSP) for fabricating nanohydroxyapatite (n-HAp)-reinforced Mg composites(Mg– nHAp). Bioactivity in simulated body fluid (SBF) is a major indicator which increased together with the refined grain size from 1500 μm to as low as 3.5 μm, in the presence of n-HAp. The Mg–nHAp composite exhibited a slight enhancement in cell viability as observed in cytotoxicity studies. Having also shown that the composite possessed outstanding cell adhesion, this serves to show that it has potential for further use as a biomaterial in the biomedical field especially in orthopedic implant where an improvement on mechanical strength, biocompatibility and integration of cells to bioactive materials are sought.

(Barfield, W. R et al.,2012) first time utilized Microwave-assisted sintering to produce Hydroxyapatite (HAp) reinforced Magnesium (Mg) composites. Microscopic examination indicated that HAp particles were uniformly distributed in the Mg matrix, which resulted in more enhancement on mechanical properties of the composites such as flexural strength and modulus, compressive strength and modulus than those of monolithic Mg. Moreover, the corrosion resistance, cytocompatibility, and bioactivity of HAp/ Mg composite were significantly improved.

The reinforcement of Mg composites by HAp makes it very appealing for biomedical applications, such as orthopedic implants. It was concluded from the study that HAp content is pivotal in controlling mechanical, corrosion resistance and biological response of the composite. This dependence on the Hap content allows the properties of the material to be tuned, and so its optimal behavior with respect to medical implant requirements is concerned regarding strength in conjunction with required degradation control and biocompatibility. Results emphasize the usefulness of microwave-assisted sintering as a solution method for Mg-based composites with designed properties, which presents new opportunities to tailor advanced materials for biomedical engineering.

Figure 4. Mechanical properties of magnesium ceramic composite with processing route for biomedical implant

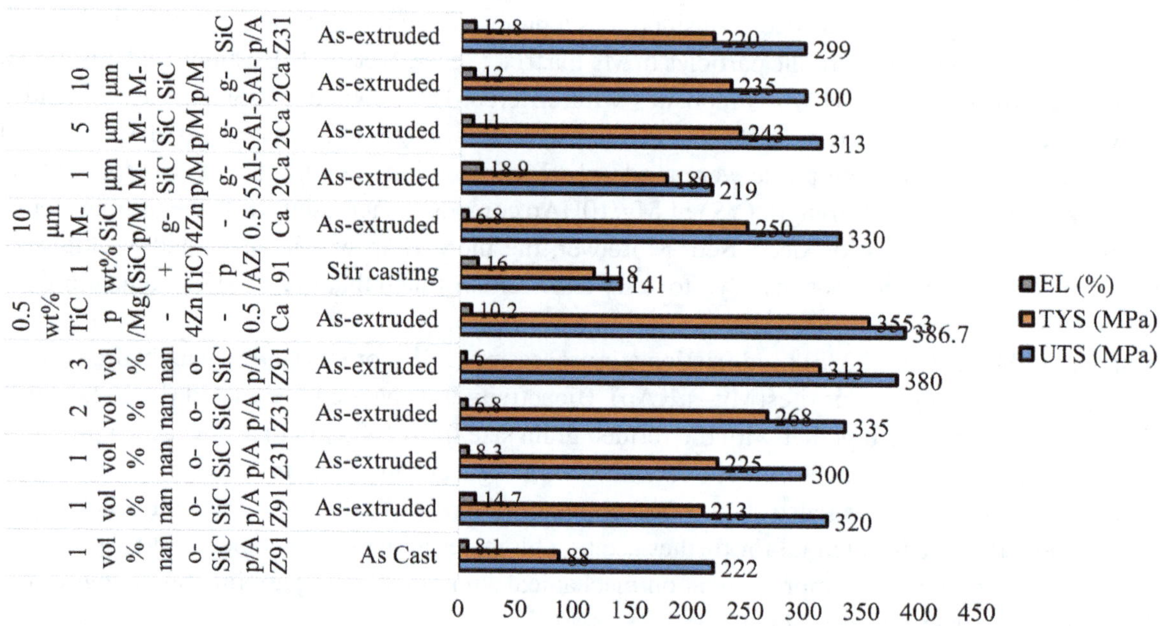

5. CLASSIFICATION OF MAGNESIUM ALLOY

From a biomedical standpoint, the emphasis is now on replacing heavier metal alternatives with lighter and bio-compatible materials to increase energy efficiency in medical equipment and improve patient care. One of the most popular materials in this respect is magnesium (Mg) and its alloys being it low-density, biodegradability which has establish a very strong interest for implants as well other devices. And Mg alloys research as biomaterial is progressing, specially regarding their applications in temporary implants that decompose with the time inside body to avoid secondary surgery for its extraction (Liangjian, C et al.,2015).

The trend in biomedical applications is the increase of usage of nonferrous metals such as titanium, aluminum and magnesium alloys due to their biocompatibility nature among other desirable mechanical properties. Magnesium and its alloys, as the lightest structural metals, are particularly compelling for manufacturing degradable implants to promote regeneration of tissues with no long-term presence of foreign material in the body. Recent developments in Mg-based alloys suitable for biomedical applications include cast and wrought with new alloy systems. These are often used to produce cast magnesium alloys with moderate mechanical properties, as well as salt water applications. It turns out, on the other hand, that wrought magnesium alloys processed by methods such as extrusion (making a shape from an identical cross section of metal), forging and rolling have far stronger mechanical properties because they receive more refined microstructure. Once again this makes wrought Mg alloys more appropriate candidates for load-bearing implants and other medical devices where mechanical strength is paramount. Mg-Zn-Mn and Mg-Ca based alloys like Mg-Al, Mg-Zn,, Z-A,Z-Yb are being studied for their compatibility with bio application conditions

5.1. Mg-Al Based Alloy

As biocompatible metals for medical implants, the magnesium-aluminum (Mg-Al)-based alloys are a promising ones in biomedical applications. Near the eutectic temp of 437 °C, magnesium can dissolve a maximum of 12.7 wt. This results in solid-solution strengthening due to the presence of≈2% aluminum. This process results in formation of the α-Mg and γ-$Mg_{17}Al_{12}$ phase that-effects on its mechanical properties (Ahangari, M et al.,2020). Yet, the mechanical properties of Mg-Al based alloys need improvement for use in biomedical applications as load bearing implants owing to good castability. The Mechanical properties of various Mg-Al alloy for biomedical implant are discussed in Figure 5. Mg-Al based alloys exhibit high corrosion resistance that can be further improved with the percentage of aluminum, an important characteristic in biodegradable implants where controlled degradation rates are mandatory for medical applications. These elements can be additionally augmented with alloying elements that are commonly used in aluminum alloys, such as zinc (Zn) or manganese (Mn), to strengthen the properties further. For example, addition of Mn lowers the iron contamination and enhances anti-corrosion properties in case stronger soluble that affects positive on mechanical performance for these alloys with better Zn maintenance.

The research is almost limited to certain Mg–Al-based alloys, AZ31, AZ61. or other few of the same family (e.g., AM60) have been broadly used for being biodegradable [36]. In these alloys, "A" is for aluminum and "Z" refers to zinc with the numbers being their percent compositions in weight (i.e., a 2% Zn containing AZ61). Due to the absence of zinc-containing eutectic phases, these AM alloys (AM30, AM40, and in certain grades) have greater Al levels along with higher Mn content equate for more extrudability that eventually provides excellent mechanical properties making them beneficial in applications such as medical where you would need both flexibility and strength (Verma, A., & Ogata, S.,2023). Adding calcium (Ca) to Mg-Al alloys has also been found to substantially strengthen these materials and improve creep resistance, both of which are especially important for load-bearing implants. For example, 1.7 wt.% % Ca to an extruded Mg-2. 32Al alloy obviously refined the microstructure and improved ultimate tensile strength (UTS) by 186 MPa, yield strength (YS) at a strength level of 143 Mpa enhanced to YS + elongation up to only and its elongation increased to 8.9%. These improvements in mechanical performance are critical for the development of Mg alloys intended for resorbable medical implants, balancing both strength with controlled bio-degradation within the body.

Figure 5. Mechanical properties of various Mg-Al alloy for biomedical implant

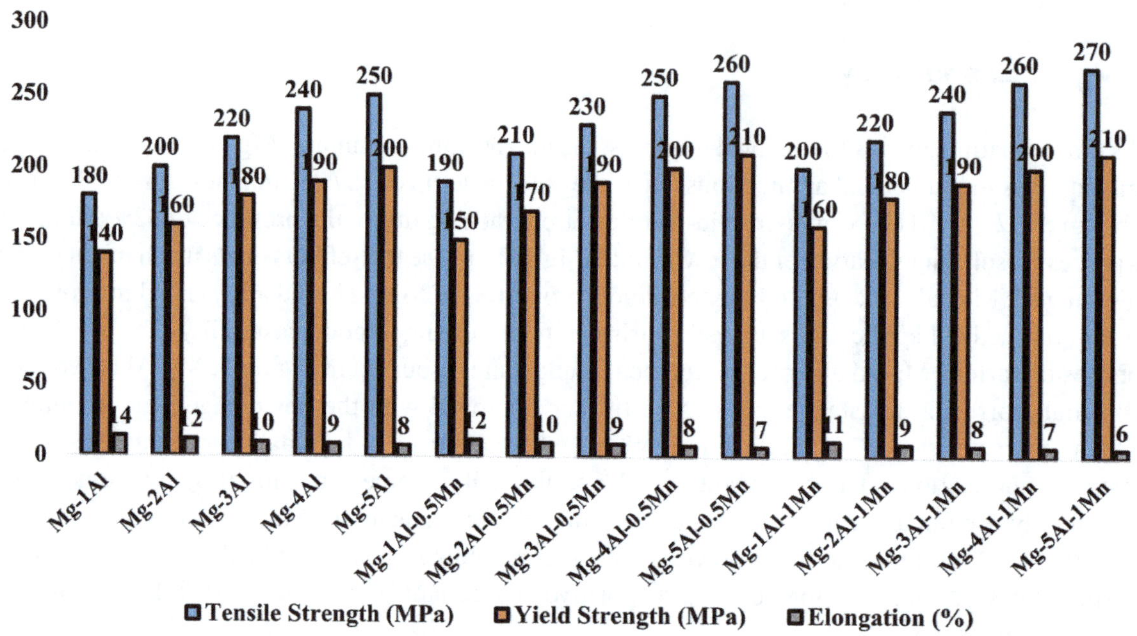

5.2. Mg-Zn Based Alloy

Among the naturally biocompatible elements or alloys, for applications in biomedical goods chemical stabilisation and protection may be provided by zinc (Zn) which is an essential element of human physiology. In contrast to aluminum, the lack of cytotoxicity in addition\- and because \- its biodegradability renders zinc an excellent base for enhanced degradable implant development. On the other hand, Mg-Zn-based alloys are more show promise for applications related to biology rising as a most appealing material candidate in biomedical field owing their biocompatible nature and temporary implantable due safe degradability after degradation inside human body.

Mg-Zn alloys, consisting of an α-Mg matrix with embedded γ-MgZn phases At a eutectic temperature of 340 °C, the maximum solubility in magnesium is demonstrated by zinc (6.2 wt. In fact, most polyoxometalates are water-soluble (from 5 to over >40 wt. %), but it is a prompt decrease in this solubility up to Ca (Li, Z.,2010). Alloys yield strength (YS) will be higher when contain magnesium amount is enhanced and the YS, Mg-Zn alloys also increases with zinc content; for example, 150 °C — % When the content reaches 4 wt.%, P; YS:≤0.007, UTS (Maximum Ultimate Tensile Strength) and (Elongation)% reach maxima of 216.8 MPa and15.9% respectively (Khorashadizade, F., et al.,2021b). Zinc precipitate-phase hardenable magnesium alloy exhibiting both strength and ductility necessary for bone implants applications. The Mechanical properties of various Mg-Zn alloy for biomedical implant are shown in Figure 6. Further improvement in the mechanical properties of Mg-Zn alloys can be achieved by adding some alloying elements such as calcium, zirconium, yttrium and so on. These have led to higher strength, better corrosion resistance and biocompatibility of Mg-Zn based alloys making them as potential candidates for degradable medical devices or implants which can serve the purpose through tissue regeneration without any foreign material over extended intervals.

Figure 6. Mechanical properties of various Mg-Zn alloy for biomedical implant

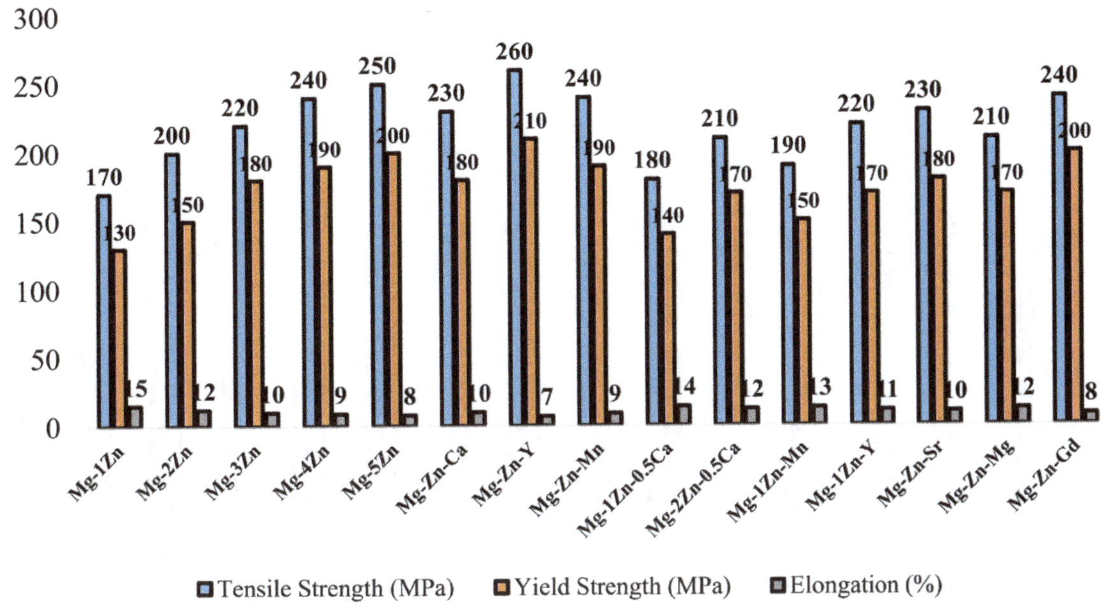

5.3. Mg-Zn-Ca Alloys

Magnesium-zinc-calcium (Mg-Zn-Ca) alloys have attracted great attention in biomedical applications recently, owning to their low cost and excellent mechanical properties as perspective candidates for biodegradable implants are illustrate in Figure 7. Especially calcium has proven to be an excellent grain refiner for magnesium based alloys, leading to better strength and performance. The solubility of Calcium however is limited in Magnesium alloys to 1.34 wt%. %, and the mechanical properties can deteriorate with calcium contained over 1% throughout also, making selection of optimum alloys for biomedical use difficult.

Hot-extruded Mg–Zn–Ca alloys have been demonstrated to possess outstanding mechanical properties under room temperature and are expectedly suitable for medical implant materials that require good mechanical properties at physiological temperatures. This is due to two key factors: processed grains, and a less than basal texture that gives the alloy improved mechanical strength and flexibility. Another case attributed the creation of a magnesium-zinc-calcium alloy to an elongation 44%, Yield tensile strength (YTS) =105 MPa, and Ultimate tensile strength (UTS)=205 Mpa excellent for load bearing in body applications (Tiwari, N. A. et al.,2024c). More interestingly, increasing the extrusion temperature of an alloy increases its ductility but decreases global strength. This trade-off between ductility and strength must be optimally engineered, particularly for biomedical applications where flexibility combined with mechanical stability is paramount to ensure implant safety. To utilize Mg-Zn-Ca alloys for different bio-medical devices and resorbable implants, further research is needed in order to optimize extrusion conditions (atmosphere during process or the feed rate) along with alloy composition.

Figure 7. Mechanical properties of various Mg-Zn-Ca alloy for biomedical implant

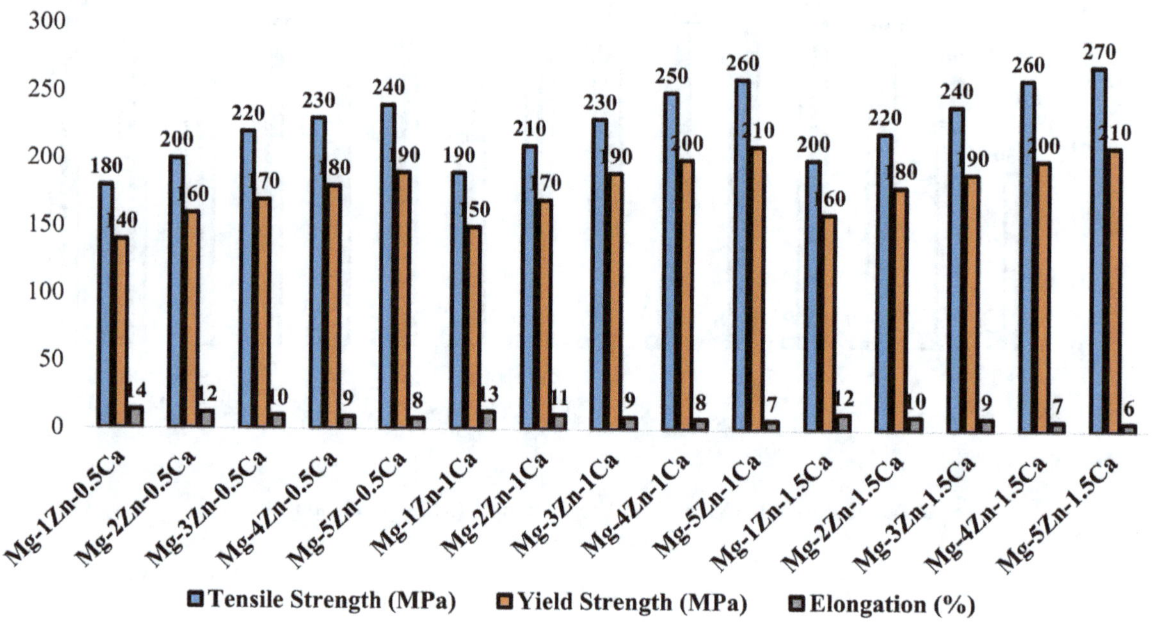

5.4. Mg-Ca-RE Alloys

Calcium in nearly all applications is the most effective alloying element used to refine magnesium granules, and it has an excellent effect on improving tensile properties are detail discussed in Figure 8. Therefore, calcium plays a crucial role in the enhancement of biodegradable implant properties. zinc Magnesium alloy can also add rare earth elements to further improve strength and toughness. This has resulted in the generation of a unique Mg-1Ca-1Y alloy with significantly increased elongation (~15.9%), where yttrium was added to magnesium alloys (Yang, K., & Lin, X.,2015). Although the yttrium addition enhances ductility, it also slightly diminishes corrosion resistance (critical properties in biomedical implants where controlled degradation is necessary). Moreover, the Mg–Ca–Y–Zr alloy would benefit significantly in terms of overall strength when yttrium is associated with other elements. The yield strength of this alloy goes up to 120 MPa, the ultimate tensile strength goes upto 191 MPa and elongation reaching at 8.3% providing medical devices perfect balance between Strength, flexibility with controlled corrosion making them safe for degradation within the body in a predictable way (Pichler, K., et al.,2013). Meanwhile, such enhancements in mechanical and corrosion behavior are crucial for designing resorbable medical implants needing to remain structural sound for an intended period before degrading safely inside the body.

Figure 8. Mechanical properties of various Mg-Ca-Re alloy for biomedical implant

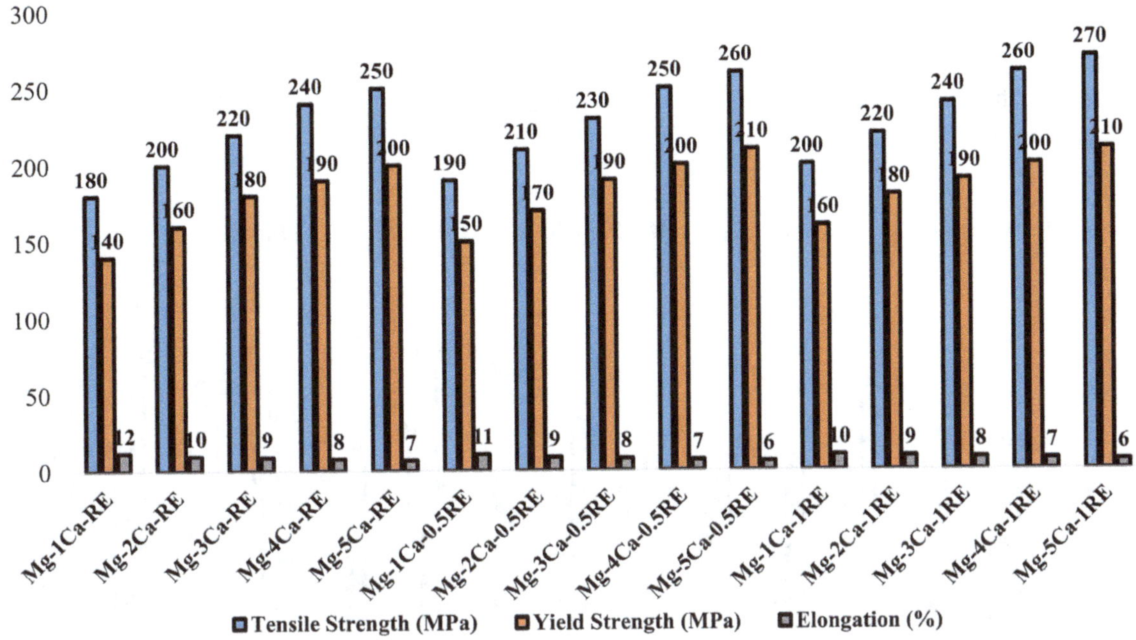

5.5. Mg-Zn-Mn Alloy

Mg–Zn–Mn alloys exhibit a property profile of high strength and ductility, combined with good corrosion resistance which allows them to be reliable candidates in biomedical prospects for use as biodegradable implants. Especially in their extruded form, these alloys offer great mechanical properties. For instance, the Mg-5. 99Zn-1. 76Ca-0.35Mn. With a yield strength specification of 289 MPa and an elongation rate reaching up to 16%, the mechanical property is strong enough for medical devices that need both elasticity and high-strength alloy (Razavi, M et al.,2020).

The mechanical properties of Mg-Zn-Ca alloys were mainly improved by Mn addition. Tensile-yield-strength of Mg-Zn-Ca alloys was increased by 50 MPa when adding 0.3% Mn for instance. Likewise, the addition of ~2 wt. Mn in extruded Mg-2Zn alloy increases both ductility and yield strength. The fine Mn precipitates refine the grains and increase slip resistance during heat deformation process, which could enhance both of strength and ductility. In Figure 9, some of above mechanical properties of various Mg-Zn-Mn alloy for biomedical implant are discussed. The incorporation of Mn into Mg alloys is especially useful in biomedical fields. Part improvements. Recent (Kumar, A., & Singh, G., 2024) have found that Mn-enhanced Mg-6Zn alloys can exhibit increased strength and better age-hardening responses, essential considerations to meet the structural integrity demands of medical implants undergoing long-term corrosion in-vivo consumption. YTS of Mg-Zn-Mn and Mg-Zn-Mg-Si alloys are in the 200–240 MPa range, although they can be further improved by aging treatment. Single step aging boosted the YTS by 80 MPa with a two-stage heat treatment at between 70°C to about 90°C and followed subsequently by around licking metal building materials °C/ lock product into place lifted these values up to more than in some extreme of (338–350) MPa respectively.

Although the as-extruded Mg-Zn-Mn-based alloys have superior mechanical strength at peak aged condition, their poor ductility restricts the application range of these materials. Balancing of the necessary strength and ductility is considered crucial in order to employ these alloys particularly as resorbable implants, which must be mechanically resistant but degrade safely with acceptable rate under physiological environment.

Figure 9. Mechanical properties of various Mg-Zn-Mn alloy for biomedical implant

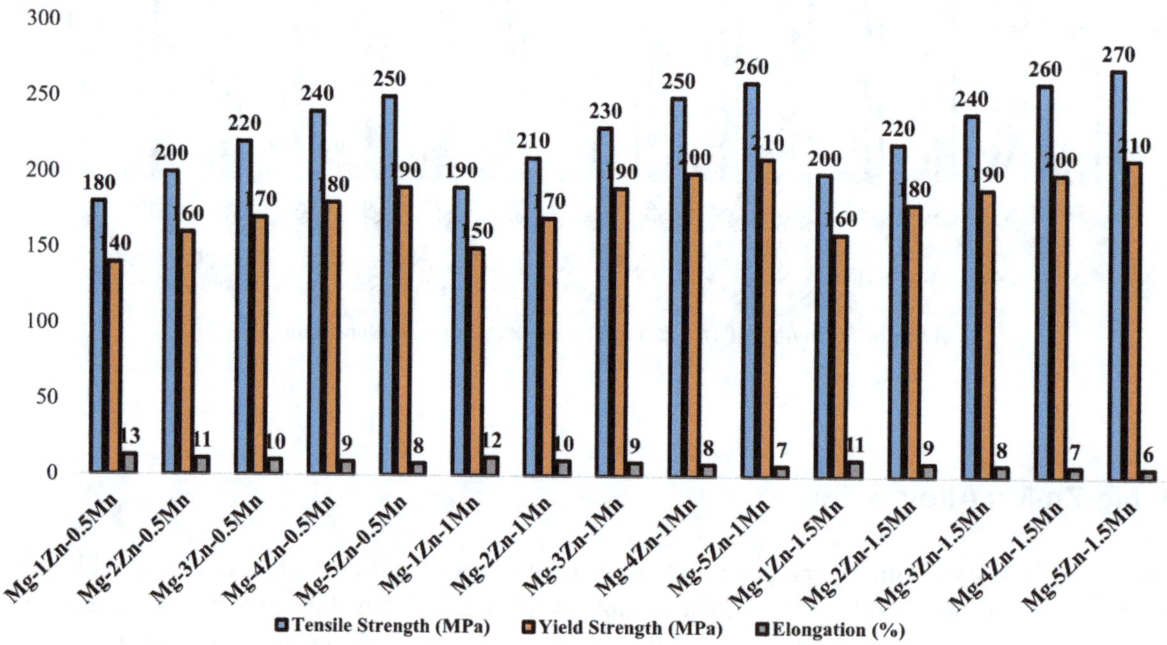

6. THE FUTURE DEVELOPMENTS TRENDS FOR BIOMEDICAL MAGNESIUM ALLOY

The development of biomedical magnesium alloys is a dynamic field, with ongoing research aimed at enhancing their properties for biomedical applications. The following table outlines current advancements and future trends in this area.

Table 2. Future development trends

Development Aspect	Current Strategies	Future Trends
Alloy composition	Incorporation of elements like aluminium (Al), zinc (Zn), and rare earth elements to improve mechanical strength and corrosion resistance.	Exploration of novel alloying elements such as lithium (Li) and calcium (Ca), to achieve better biocompatibility and tailored degradation rates.
Surface Modification	Application of coatings (e.g., hydroxyapatite) and surface treatment (e.g., micro-arc oxidation) to enhance corrosion resistance and biocompatibility.	Development of multifunctional coatings that provide controlled drug release and promote specific cellular responses.
Manufacturing Techniques	Utilization of casting and extrusion method for implant fabrication.	Adoption of additive manufacturing (3D printing) to create patient specific implants with complex geometries and optimized porosity.
Mechanical Properties	Achieving mechanical properties comparable to bone to minimize stress shielding	Designing gradient structures and composite material to further match the mechanical behaviour of various tissues
Degradation control	Use of alloying and surface treatments to control degradation rate.	Implementation of smart material that can adjust degradation behavior in response to physiological conditions
Biocompatibility	Ensuring non toxicity of degradation products and favourable tissue response.	Engineering surfaces that actively promote tissue regeneration and integration.
Clinical Applications	Focus on orthopaedic implants and cardiovascular stents.	Expansion ratio area like dental implants, tissue engineering scaffolds and drug delivery system.

Biomedical engineering is continuously in search of advanced materials, step toward the goal are magnesium (Mg) alloys which provide ultimate attributes of biodegradability, mechanical compatibility with human tissues and inherent biocompatibility. Over the past three decades, scientists have explored the interactions of Mg alloy implants with human cells, tissues, and organs, focusing on protein interactions, cellular responses leading to tissue integration, and systemic effects. Here we summarize recent developments in this space.

Protein adsorption on the surface of Mg implants is a key determinant in regulating cellular behaviors and promoting tissue integration. (Wang, Z.et al., 2023) showed that plasma proteins like fibrinogen and vitronectin adsorb strongly to Mg alloy surfaces and promote osteoblast adhesion and proliferation. By enabling cell adhesion, this integration represents an important initial scaffold for bone cells to promote tissue regeneration. Similarly, (Wei, X et al., 2022) emphasized the dependency of protein adsorption as well as subsequent cellular response on the structure of the surface treatment applied on Mg alloys. Another interest lies in the effect of magnesium ions released during degradation. (Ali et al., 2023) revealed that WE43 Mg alloys after plasma electrolytic oxidation (PEO) treatment had biological abilities which conclude that the release of Mg ions in the PEO region not only contributes to the proliferation and differentiation of osteoblast but also alleviates the inflammatory reaction in surrounding tissues. This bifunctional result demonstrates the significance of controllable ion release in favorable outcomes.

Mg alloys are extensively studied in biomedical applications; however, the interaction between Mg alloys and surrounding tissue is vital to successful implant integration. (Antoniac et al., 2023) conducted a review on the clinical applications of Mg-based implants in orthopedics and highlighted that surface treatments, which can be optimized for specific applications, can promote BFs and accelerate healing. Their investigation emphasized that hydroxyapatite coatings for Mg implants promote nonspecific binding of bone-forming proteins which further enhances osteogenesis. Zhang et al. In vivo studies of Mg alloys were recently performed to study their performance and the results show pronounced stimulation of healing response with significant bone growth around the implantation sites. This finding is consistent with that by (Hu, J et al., 2023), showed that these Mg scaffolds coated with an osteogenic protein leads to greater bone regeneration than when Mg scaffolds were not coated.

Mg alloys are promising for orthopedic applications as they are compatible with bone tissues. (Kuhlmann, J et al., 2013) showed that Mg-based materials deposited onto substrate surfaces serve not only to increase implant biocompatibility, but also to decrease inflammation in surrounding tissues. Furthermore, (Cuartas-Marulanda et al., 2022) has investigated the reaction of soft tissues with Mg alloys and found only slight irritation and rapid tissue recovery. In cardiovascular applications, bioresorbable Mg stents have shown great promise. Wang et al. The controlled biodegradation of Mg-based stents in coronary arteries has been reported positively regarding vessel patency and restenosys-free survival. This study suggests the wider applicability of Mg alloys other than in orthopedics. The biological response of Mg alloys implants depends strongly on the protein adsorption to their surfaces. (Chen, X et al., 2023), investigate the effect of zinc on the biocompatibility of Mg alloys and report zinc-modified surfaces promote selective adsorption of bone-regenerative protein. This selective adsorption is essential for starting and maintaining the healing process. Similarly,development of inorganic biomaterials with controlled chemistries and surface properties have gained significant attention in order to promote their performance and applicability as implant materials. The proteins that are found on the surface of the material can also impact cell–material interactions, as recently established by (Tong, X et al.. 2023), who examined the influence of albumin and serum proteins on cell adhesion and differentiation on Mg alloy surfaces. These findings indicate that protein coated Mg alloys enhance the cellular outcome in comparison to uncoated alloys.

Novel applications of Mg alloys have emerged for instance in pediatric orthopedics and tissue engineering. The use of Mg implants in pediatric fractures proved effective in their study, resulting in prompt healing and minimal complication rates. But controlling how quickly they break down is proving more difficult, one of the main challenges as researchers try to simultaneously accelerate healing by matching the degradation rates to the pace of growing tissue. With the advancements in the field of additive manufacturing, the application of Mg implants has taken a new dimension where the designers are able to develop patient-specific designs. (Tanaka, M et al.,2017), noted (3D) printed Mg scaffolds have been explored for use in complex tissue reconstructions as they can delivery growth factors and proteins to improve regeneration.

The smart materials are designed to achieve these adaptive degradation rates for various biomedical applications which mark the future prospect of Mg alloys. (Costabeber, G et al., 2024), and they showed that these coatings can immobilize proteins, including anti-inflammatory ones, and release them in a controlled fashion. Using this strategy, current drawbacks associated with uncontrolled degradation and localized inflammation may be alleviated. Additionally, the incorporation of bioactive molecules such as growth factors and peptides onto Mg alloy surfaces is an attractive avenue. (Lopez de Armentia et al., 2020) demonstrated the possibility of using biofunctionalized Mg alloys for tissue in addition to substitution, which can also promote regeneration.

The performance of magnesium alloys shows promising application prospects in biomedical engineering. Broad insights into their relationships with human cells, tissues, and organs including protein adsorption and biocompatibility have emerged from three years of research activity. Although still limited in their ability to modulate degradation or to fine-tune systemic effects, continued advancements in surface engineering, alloy design and additive manufacturing will allow for greater clinical adoption. With the incorporation of these smart materials and bioactive molecules, the Mg alloys have the potential to overcome existing limitations of the biomaterials revolutionizing medical implant technology.

CONCLUSION

As a result of their biodegradability, biocompatibility and superior mechanical properties, magnesium (Mg)alloys have gained attention as potential biomedical implant material. Due to its unique biodegradability in vivo, Mg alloys are a favored material for temporary implants including orthopedic devices, cardiovascular stents and other resorbable medical device. Nevertheless, in the medical application it needs to be carefully examined for their mechanical strength and corrosion resistance as well its grain refinement to fulfill required demanding criteria of biomechanical environment. Mg alloys have a wide range of mechanical strengths depending on the composition and manufacture. Pure magnesium is not strong enough for most bio-implant applications but incorporating alloying elements such as aluminum (Al), zinc (Zn), calcium(Ca), manganese(Mn) and rare earth elements(RE) can improve its mechanical strength greatly. As an example, high tensile strength and ductility have been obtained in the case of the magnezisium-alloys Mg-Zn-Mn), edspecially extruded alloys like Mg-5. 99Zn-1. 76Ca-0. 35Mn Yield strengths (289 MPa) Elongation 16% The advantage of these alloys is that they are used for load-bearing implants (such as bone plates, and screws). Moreover, they show even better strengths on account of precipitation strengthening effect if Y and Gd are added to Mg-Zn-RE alloys. Ultimate tensile strengths (UTS) > 300 MPa are possible for these alloys, so may be adequate in more challenging biomedical applications where both strength and bioresorbability is required.

In biomedical aspects, another aspect of Mg alloys is also significant i.e. corrosion resistance and it has to be considered very seriously for the prolonged stability implementations. It is important that the implant helps in healing by providing structural support and then degrades over time without causing a toxic or adverse response. But plain magnesium not only rusted away too fast, it also sheds hydrogen gas fumes that could get in the way of tissue recovery. The addition of alloying elements such as calcium, manganese can effectively enhance the corrosion behavior of magnesium alloys by refining phase microstructure and reducing impurity contents that bring degradation to corroding Mg-matrix. Among the most important corrosion-resistant magnesium-based alloys are some of those based on Mg-Al systems like AZ31 and AZ91, because aluminum oxide (Al_2O_3) is more resistance to aggressive media than that formed by pure magnesium. But aluminium is not biocompatible on long-term, thus other alloys are commonly employed in biomedical implants. However, Mg-Ca and more commonly the new commercialized alloy series of Mg-Zn are recommended for medical applications since all calcium and zinc as a biodegradable elements enhance in vitro corrosion protection effect thus they allow consideration safer to resorbable implants.

Grain refinement is very important in both strength and corrosion resistance improvement of Mg alloys. Fine-grained alloy has the advantages of higher yield strength and better corrosion resistance, because uniform distribution of the alloying elements in nodes can reduce simple closed micro-cracks on its surface. The use of calcium as a grain refiner in magnesium alloys is especially suitable because this element, not only improves mechanical strength but also influences the uniformity of the degradation process. These rare earth elements like yttrium or gadolinium serves to achieve an even more reinforced grain refinement, also aiding in maintaining a fine, stable microstructure precipitation which led it another enhancement; becomes capable of strengthening the material as well while obtaining better ductility.

In conclusion, Mg based alloys combine an uncommon amalgamation of properties which might be suitable for a wide variety of biomedical applications. The Mg alloy could be tailor-made with preferred mechanical strengths for orthopedic application to corrosion resistant properties for cardiovascular stent, by the right elements and processing schemes of choosing. Further investigation of grain refinement, alloy composition, and surface treatments will be needed to maximize the potential applications in Mg alloys so as their degradation properties can translate towards implant materials for medical devices.

ACKNOWLEDGMENT

The authors wish to acknowledge the support received from the Suresh Gyan Vihar University while conducting the research.

REFERENCES

Abbas, Z., Dapporto, M., Tampieri, A., & Sprio, S. (2021). Toughening of bioceramic composites for bone regeneration. *Journal of Composites Science*, *5*(10), 259. DOI: 10.3390/jcs5100259

Abdalla, M., Joplin, A., Elahinia, M., & Ibrahim, H. (2020). Corrosion Modeling of magnesium and its alloys for Biomedical applications [review]. *Corrosion and Materials Degradation*, *1*(2), 219–248. DOI: 10.3390/cmd1020011

Ahangari, M., Johar, M., & Saremi, M. (2020). Hydroxyapatite-carboxymethyl cellulose-graphene composite coating development on AZ31 magnesium alloy: Corrosion behavior and mechanical properties. *Ceramics International*, *47*(3), 3529–3539. DOI: 10.1016/j.ceramint.2020.09.197

Aherwar, A., Singh, A. K., & Patnaik, A. (2015). Current and future biocompatibility aspects of biomaterials for hip prosthesis. *AIMS Bioengineering*, *3*(1), 23–43. DOI: 10.3934/bioeng.2016.1.23

Ahirwar, H., Zhou, Y., Mahapatra, C., Ramakrishna, S., Kumar, P., & Nanda, H. S. (2020). Materials for orthopedic bioimplants: Modulating degradation and surface modification using integrated nanomaterials. *Coatings*, *10*(3), 264. DOI: 10.3390/coatings10030264

Ali, W., Li, M., Tillmann, L., Mayer, T., & González, C., LLorca, J., & Kopp, A. (2023). Bioabsorbable WE43 Mg alloy wires modified by continuous plasma-electrolytic oxidation for implant applications. Part I: Processing, microstructure and mechanical properties. *Biomaterials Advances*, *146*, 213314.

Antoniac, I., Manescu, V., Paltanea, G., Antoniac, A., Nemoianu, I. V., Petrescu, M. I., & Bodog, A. D. (2022). Additive manufactured magnesium-based scaffolds for tissue engineering. *Materials (Basel)*, *15*(23), 8693. PMID: 36500191

Antoniac, I., Manescu, V., Antoniac, A., & Paltanea, G. (2023). Magnesium-based alloys with adapted interfaces for bone implants and tissue engineering. *Regenerative Biomaterials*, *10*, rbad095.

Badkoobeh, F., Mostaan, H., Rafiei, M., & Bakhsheshi-Rad, H. R. Seeram RamaKrishna, Xiongbiao Chen, Additive manufacturing of biodegradable magnesium-based materials: Design strategies, properties, and biomedical applications, Journal of Magnesium and Alloys, Volume 11, Issue 3, 2023, Pages 801-839, ISSN 2213-9567, https://doi.org/DOI: 10.1016/j.jma.2022.12.001

Bai, J., Wang, X., Ye, X., Li, X., Tan, X., Wang, C., Guo, Y., Dong, J., Shao, Y., Chu, C., Xue, F., Tao, L., & Huang, L. (2024). Constructing Mg-phytic acid enhanced hydrogel patch with janus structure for antibacterial and angiogenic urethral repair. *Materials Today. Communications*, *41*, 110621. DOI: 10.1016/j.mtcomm.2024.110621

Bakhsheshi-Rad, H., Hamzah, E., Kasiri-Asgarani, M., Saud, S. N., Yaghoubidoust, F., & Akbari, E. (2016). Structure, corrosion behavior, and antibacterial properties of nano-silica/graphene oxide coating on biodegradable magnesium alloy for biomedical applications. *Vacuum, 131*, 106–110. DOI: 10.1016/j.vacuum.2016.05.021

Banerjee, M. K., Tiwari, A., & Kumar, N. (2023b). Ageing characteristics of stir cast AZ 61 alloy with minor additions. *Recent Patents on Engineering, 18*. Advance online publication. DOI: 10.2174/1872212118666230609122741

Barfield, W. R., Colbath, G., DesJardins, J. D., An, Y. H., & Hartsock, L. A. (2012). The potential of magnesium alloy use in orthopaedic surgery. *Current Orthopaedic Practice, 23*(2), 146–150. DOI: 10.1097/bco.0b013e31824a553b

Barrère, F., Mahmood, T., De Groot, K., & Van Blitterswijk, C. (2008). Advanced biomaterials for skeletal tissue regeneration: Instructive and smart functions. *Materials Science and Engineering R Reports, 59*(1–6), 38–71. DOI: 10.1016/j.mser.2007.12.001

Bordbar-Khiabani, A., Yarmand, B., & Mozafari, M. (2019). Emerging magnesium-based biomaterials for orthopedic implantation. *Emerging Materials Research, 8*(3), 305–319. DOI: 10.1680/jemmr.18.00048

Brar, H. S., Platt, M. O., Sarntinoranont, M., Martin, P. I., & Manuel, M. V. (2009). Magnesium as a biodegradable and bioabsorbable material for medical implants. *JOM, 61*(9), 31–34. DOI: 10.1007/s11837-009-0129-0

Chen, S., Wan, P., Zhang, B., Erişen, D. E., Yang, H., & Yang, K. (2018). A novel polymer critical remelting treatment for improving corrosion resistance of magnesium alloy stent. *Journal of Materials Science and Technology, 35*(1), 19–22. DOI: 10.1016/j.jmst.2018.09.021

Chen, X., Xia, Y., Shen, S., Wang, C., Zan, R., Yu, H., & Zhang, X. (2023). Research on the Current Application Status of Magnesium Metal Stents in Human Luminal Cavities. *Journal of Functional Biomaterials, 14*(9), 462. PMID: 37754876

Cheng, M., Wahafu, T., Jiang, G., Liu, W., Qiao, Y., Peng, X., Cheng, T., Zhang, X., He, G., & Liu, X. (2016). A novel open-porous magnesium scaffold with controllable microstructures and properties for bone regeneration. *Scientific Reports, 6*(1). Advance online publication. DOI: 10.1038/srep24134 PMID: 27071777

Costabeber, G., Guerra, N. B., Brasil, G. S. A. P., da Silva Sasaki, J. C., Scontri, M., Burd, B. S., & Herculano, R. D. (2024). Additive manufacturing of polylactic acid scaffolds dip-coated with polycaprolactone for bone tissue engineering. *Materials Today. Communications, 40*, 109646.

Cuartas-Marulanda, D., Forero Cardozo, L., Restrepo-Osorio, A., & Fernández-Morales, P. (2022). Natural coatings and surface modifications on magnesium alloys for biomedical applications. *Polymers, 14*(23), 5297. PMID: 36501691

Diba, M., Tapia, F., Boccaccini, A. R., & Strobel, L. A. (2012). Magnesium-Containing bioactive glasses for biomedical applications. *International Journal of Applied Glass Science, 3*(3), 221–253. DOI: 10.1111/j.2041-1294.2012.00095.x

Ding, Y., Jiang, H., Dunn, M. L., & Yu, K. (2024). 3D interfacial material-locking structures to promote bonding strength of dissimilar materials. *Journal of Manufacturing Processes*, *112*, 92–102. DOI: 10.1016/j.jmapro.2024.01.031

Eren, E. D., Guisong, G., Mingming, L., Bingchun, Z., Ke, Y., & Shanshan, C. (2022). A novel chitosan and polydopamine interlinked bioactive coating for metallic biomaterials. *Journal of Materials Science. Materials in Medicine*, *33*(10). Advance online publication. DOI: 10.1007/s10856-022-06688-x PMID: 36138240

Erisen, D. E., & Uludag, K. (2024). Estimating biosafety of biodegradable biomedical materials from in vitro ion tolerance parameters and toxicity of nanomaterials in brain. In *Advances in healthcare information systems and administration book series* (pp. 201–221). https://doi.org/DOI: 10.4018/979-8-3693-3661-8.ch010

Erişen, D. E., Zhang, Y., Zhang, B., Yang, K., Chen, S., & Wang, X. (2021). Biosafety and biodegradation studies of AZ31B magnesium alloy carotid artery stent in vitro and in vivo. *Journal of Biomedical Materials Research. Part B, Applied Biomaterials*, *110*(1), 239–248. DOI: 10.1002/jbm.b.34907 PMID: 34236133

Esmaily, M., Svensson, J., Fajardo, S., Birbilis, N., Frankel, G., Virtanen, S., Arrabal, R., Thomas, S., & Johansson, L. (2017). Fundamentals and advances in magnesium alloy corrosion. *Progress in Materials Science*, *89*, 92–193. DOI: 10.1016/j.pmatsci.2017.04.011

Gopi, D., Bhalaji, P., Ramya, S., & Kavitha, L. (2014). Evaluation of biodegradability of surface treated AZ91 magnesium alloy in SBF solution. *Journal of Industrial and Engineering Chemistry*, *23*, 218–227. DOI: 10.1016/j.jiec.2014.08.019

Gupta, M., & Wong, W. (2015). Magnesium-based nanocomposites: Lightweight materials of the future. *Materials Characterization*, *105*, 30–46. DOI: 10.1016/j.matchar.2015.04.015

Gusieva, K., Davies, C. H. J., Scully, J. R., & Birbilis, N. (2014). Corrosion of magnesium alloys: The role of alloying. *International Materials Reviews*, *60*(3), 169–194. DOI: 10.1179/1743280414y.0000000046

Hoop, M., Walde, C. F., Riccò, R., Mushtaq, F., Terzopoulou, A., Chen, X., deMello, A. J., Doonan, C. J., Falcaro, P., Nelson, B. J., Puigmartí-Luis, J., & Pané, S. (2018). Biocompatibility characteristics of the metal organic framework ZIF-8 for therapeutical applications. *Applied Materials Today*, *11*, 13–21. DOI: 10.1016/j.apmt.2017.12.014

Hu, J., Shao, J., Huang, G., Zhang, J., & Pan, S. (2023). In vitro and in vivo applications of magnesium-enriched biomaterials for vascularized osteogenesis in bone tissue engineering: A review of literature. *Journal of Functional Biomaterials*, *14*(6), 326. PMID: 37367290

Hu, J., Shao, J., Huang, G., Zhang, J., & Pan, S. (2023). In vitro and in vivo applications of magnesium-enriched biomaterials for vascularized osteogenesis in bone tissue engineering: A review of literature. *Journal of Functional Biomaterials*, *14*(6), 326. PMID: 37367290

Jin, Z., Zha, M., Wang, S., Wang, S., Wang, C., Jia, H., & Wang, H. (2022). Alloying design and microstructural control strategies towards developing Mg alloys with enhanced ductility. *Journal of Magnesium and Alloys*, *10*(5), 1191–1206. DOI: 10.1016/j.jma.2022.04.002

Joyce, K., Fabra, G. T., Bozkurt, Y., & Pandit, A. (2021). Correction to: Bioactive potential of natural biomaterials: identification, retention and assessment of biological properties. *Signal Transduction and Targeted Therapy*, *6*(1). Advance online publication. DOI: 10.1038/s41392-021-00593-5 PMID: 33953157

Kanishka, K., & Acherjee, B. (2023). Revolutionizing manufacturing: A comprehensive overview of additive manufacturing processes, materials, developments, and challenges. *Journal of Manufacturing Processes*, *107*, 574–619. DOI: 10.1016/j.jmapro.2023.10.024

Kargozar, S., Milan, P. B., Baino, F., & Mozafari, M. (2018). Nanoengineered biomaterials for bone/dental regeneration. In *Elsevier eBooks* (pp. 13–38). https://doi.org/DOI: 10.1016/b978-0-12-813355-2.00002-8

Khorashadizade, F., Abazari, S., Rajabi, M., Bakhsheshi-Rad, H., Ismail, A. F., Sharif, S., Ramakrishna, S., & Berto, F. (2021b). Overview of magnesium-ceramic composites: Mechanical, corrosion and biological properties. *Journal of Materials Research and Technology*, *15*, 6034–6066. DOI: 10.1016/j.jmrt.2021.10.141

Kieke, M., Feyerabend, F., Lemaitre, J., Behrens, P., & Willumeit-Römer, R. (2016). Degradation rates and products of pure magnesium exposed to different aqueous media under physiological conditions. *BioNanoMaterials*, *17*(3–4). Advance online publication. DOI: 10.1515/bnm-2015-0020

Kuhlmann, J., Bartsch, I., Willbold, E., Schuchardt, S., Holz, O., Hort, N., & Witte, F. (2013). Fast escape of hydrogen from gas cavities around corroding magnesium implants. *Acta Biomaterialia*, *9*(10), 8714–8721. PMID: 23069319

Kumar, A., & Singh, G. (2024). Surface modification of Ti6Al4V alloy via advanced coatings: Mechanical, Tribological, Corrosion, Wetting, and Biocompatibility studies. *Journal of Alloys and Compounds*, *989*, 174418. DOI: 10.1016/j.jallcom.2024.174418

Li, J., Zhou, P., Attarilar, S., & Shi, H. (2021). Innovative surface modification procedures to achieve micro/nano-graded Ti-based biomedical alloys and implants. *Coatings*, *11*(6), 647.

Li, Z. (2010). *Mg/hydroxyapatite composites for potential bio-medical applications*. https://bura.brunel.ac.uk/bitstream/2438/4599/1/FulltextThesis.pdf

Liangjian, C., Jun, Z., Kun, Y., Chang, C., Yilong, D., Xueyan, Q., & Zhiming, Y. (2015). Improving of in vitro Biodegradation Resistance in a Chitosan Coated Magnesium Bio-composite. *Rare Metal Materials and Engineering*, *44*(8), 1862–1865. DOI: 10.1016/s1875-5372(15)30114-4

Liu, T., Wang, Y., & Kuang, T. (2024). Oriented porous polymer scaffolds in tissue engineering: A comprehensive review of preparation strategies and applications. *Macromolecular Materials and Engineering*, *309*(1), 2300246.

Liu, W., Yan, Z., Zhang, Z., Zhang, Y., Cai, G., & Li, Z. (2019). Bioactive and anti-corrosive bio-MOF-1 coating on magnesium alloy for bone repair application. *Journal of Alloys and Compounds*, *788*, 705–711. DOI: 10.1016/j.jallcom.2019.02.281

Lopez de Armentia, S., Del Real, J. C., Paz, E., & Dunne, N. (2020). Advances in biodegradable 3D printed scaffolds with carbon-based nanomaterials for bone regeneration. *Materials (Basel)*, *13*(22), 5083. PMID: 33187218

Mamidi, N., García, R. G., Martínez, J. D. H., Briones, C. M., Martinez Ramos, A. M., Tamez, M. F. L., & Segura, F. J. M. (2022). Recent advances in designing fibrous biomaterials for the domain of biomedical, clinical, and environmental applications. *ACS Biomaterials Science & Engineering*, *8*(9), 3690–3716. PMID: 36037103

Mao, L., Shen, L., Chen, J., Wu, Y., Kwak, M., Lu, Y., Xue, Q., Pei, J., Zhang, L., Yuan, G., Fan, R., Ge, J., & Ding, W. (2015). Enhanced bioactivity of MG–ND–ZN–ZR alloy achieved with Nanoscale MGF2 surface for vascular stent application. *ACS Applied Materials & Interfaces*, *7*(9), 5320–5330. DOI: 10.1021/am5086885 PMID: 25705919

Mao, L., Shen, L., Chen, J., Zhang, X., Kwak, M., Wu, Y., Fan, R., Zhang, L., Pei, J., Yuan, G., Song, C., Ge, J., & Ding, W. (2017). A promising biodegradable magnesium alloy suitable for clinical vascular stent application. *Scientific Reports*, *7*(1). Advance online publication. DOI: 10.1038/srep46343 PMID: 28397881

Mena-Morcillo, E., & Veleva, L. (2020). Degradation of AZ31 and AZ91 magnesium alloys in different physiological media: Effect of surface layer stability on electrochemical behaviour. *Journal of Magnesium and Alloys*, *8*(3), 667–675. DOI: 10.1016/j.jma.2020.02.014

Panahi, Z., Tamjid, E., & Rezaei, M. (2020). Surface modification of biodegradable AZ91 magnesium alloy by electrospun polymer nanocomposite: Evaluation of in vitro degradation and cytocompatibility. *Surface and Coatings Technology*, *386*, 125461. DOI: 10.1016/j.surfcoat.2020.125461

Pesode, P., & Barve, S. (2023). Additive manufacturing of metallic biomaterials: Sustainability aspect, opportunity, and challenges. *Journal of Industrial and Production Engineering*, *40*(6), 464–505. DOI: 10.1080/21681015.2023.2229341

Pichler, K., Kraus, T., Martinelli, E., Sadoghi, P., Musumeci, G., Uggowitzer, P. J., & Weinberg, A. M. (2013). Cellular reactions to biodegradable magnesium alloys on human growth plate chondrocytes and osteoblasts. *International Orthopaedics*, *38*(4), 881–889. DOI: 10.1007/s00264-013-2163-3 PMID: 24258151

Polo, T. O. B., Silva, W. P. P., Momesso, G. C., Lima-Neto, T. J., Barbosa, S., Cordeiro, J. M., Hassumi, J. S., Da Cruz, N. C., Okamoto, R., Barão, V. R., & Faverani, L. P. (2020). Plasma Electrolytic Oxidation as a Feasible Surface Treatment for Biomedical Applications: An in vivo study. *Scientific Reports*, *10*(1). Advance online publication. DOI: 10.1038/s41598-020-65289-2 PMID: 32561767

Pompa, L., Rahman, Z. U., Munoz, E., & Haider, W. (2015). Surface characterization and cytotoxicity response of biodegradable magnesium alloys. *Materials Science and Engineering C*, *49*, 761–768. DOI: 10.1016/j.msec.2015.01.017 PMID: 25687006

Prasad, S. V. S., Prasad, S. B., Verma, K., Mishra, R. K., Kumar, V., & Singh, S. (2021). The role and significance of Magnesium in modern day research-A review. *Journal of Magnesium and Alloys*, *10*(1), 1–61. DOI: 10.1016/j.jma.2021.05.012

Qi, L., Zhao, T., Yan, J., Ge, W., Jiang, W., Wang, J., & Zhang, L. (2024). Advances in magnesium-containing bioceramics for bone repair. *Biomaterials Translational*, *5*(1), 3. PMID: 39220661

Qin, Y., Liu, A., Guo, H., Shen, Y., Wen, P., Lin, H., Xia, D., Voshage, M., Tian, Y., & Zheng, Y. (2022). Additive manufacturing of Zn-Mg alloy porous scaffolds with enhanced osseointegration: In vitro and in vivo studies. *Acta Biomaterialia, 145*, 403–415. DOI: 10.1016/j.actbio.2022.03.055 PMID: 35381400

Razavi, M., Fathi, M., Savabi, O., Tayebi, L., & Vashaee, D. (2020). Biodegradable Magnesium Bone Implants Coated with a Novel Bioceramic Nanocomposite. *Materials (Basel), 13*(6), 1315. DOI: 10.3390/ma13061315 PMID: 32183231

Saberi, A., Baltatu, M. S., & Vizureanu, P. (2024). Recent advances in Magnesium–Magnesium oxide nanoparticle composites for biomedical applications. *Bioengineering (Basel, Switzerland), 11*(5), 508. DOI: 10.3390/bioengineering11050508 PMID: 38790374

Santos, G. D. (2017). The importance of metallic materials as biomaterials. *Advances in Tissue Engineering & Regenerative Medicine Open Access, 3*(1). Advance online publication. DOI: 10.15406/atroa.2017.03.00054

Seetharaman, S., Jayalakshmi, S., Singh, R. A., & Gupta, M. (2022). The potential of Magnesium-Based materials for engineering and biomedical applications. *Journal of the Indian Institute of Science, 102*(1), 421–437. DOI: 10.1007/s41745-021-00274-w

Sezer, N., Evis, Z., Kayhan, S. M., Tahmasebifar, A., & Koç, M. (2018). Review of magnesium-based biomaterials and their applications. *Journal of Magnesium and Alloys, 6*(1), 23–43. DOI: 10.1016/j.jma.2018.02.003

Sharma, S. K., Gajević, S., Sharma, L. K., Pradhan, R., Miladinović, S., Ašonja, A., & Stojanović, B. (2024). Magnesium-Titanium Alloys: A Promising Solution for Biodegradable Biomedical Implants. *Materials (Basel), 17*(21), 5157. PMID: 39517433

Shi, Q., Chen, J., Chen, J., Liu, Y., & Wang, H. (2024). Application of additively manufactured bone scaffold: A systematic review. *Biofabrication*. PMID: 38507799

Singh, J., Hashmi, A. W., Ahmad, S., & Tian, Y. (2024). Critical review on biodegradable and biocompatibility magnesium alloys: Progress and prospects in bio-implant applications. *Inorganic Chemistry Communications*, •••, 113111.

Singh, J., Hashmi, A. W., Ahmad, S., & Tian, Y. (2024). Critical review on biodegradable and biocompatibility magnesium alloys: Progress and prospects in bio-implant applications. *Inorganic Chemistry Communications, 169*, 113111. DOI: 10.1016/j.inoche.2024.113111

Tanaka, M., Haniu, H., Kamanaka, T., Takizawa, T., Sobajima, A., Yoshida, K., & Saito, N. (2017). Physico-chemical, in vitro, and in vivo evaluation of a 3D unidirectional porous hydroxyapatite scaffold for bone regeneration. *Materials (Basel), 10*(1), 33. PMID: 28772390

Tipan, N., Pandey, A., & Mishra, P. (2024). Material synthesis and design optimization of biomaterials for biomedical implant applications. *BioImpacts*. PMID: 40161937

Tiwari, A., Kumar, N., & Banerjee, M. (2023a). Applications of genetic algorithm in prediction of the best achievable combination of hardness and tensile strength for graphene reinforced magnesium alloy (AZ61) matrix composite. *Results in Control and Optimization, 14*, 100334. DOI: 10.1016/j.rico.2023.100334

Tiwari, A., Kumar, N., & Banerjee, M. K. (2023c). Effect of Hot Rolling on Microstructure and Mechanical Properties of Stir Cast AZ 61 Alloy with Minor Additions. *Indian Journal of Science and Technology, 16*(24), 1810–1822. DOI: 10.17485/ijst/v16i24.302

Tiwari, N. A. (2024c). Prediction model for hardness and tensile strength of graphene reinforced AZ 61 alloy based composite using Metaheuristic Algorithm. *Global Journal of Engineering and Technology Advances, 20*(3), 042–052. https://doi.org/DOI: 10.30574/gjeta.2024.20.3.0167

Tong, X., Han, Y., Zhou, R., Zeng, J., Wang, C., Yuan, Y., & Lin, J. (2023). Mechanical properties, corrosion and degradation behaviors, and in vitro cytocompatibility of a biodegradable Zn–5La alloy for bone-implant applications. *Acta Biomaterialia, 169*, 641–660. PMID: 37541605

Verma, A., & Ogata, S. (2023). Magnesium based alloys for reinforcing biopolymer composites and coatings: A critical overview on biomedical materials. *Advanced Industrial and Engineering Polymer Research, 6*(4), 341–355. DOI: 10.1016/j.aiepr.2023.01.002

Wang, J., Xu, J., Liu, W., Li, Y., & Qin, L. (2016). Biodegradable Magnesium (Mg) Implantation Does Not Impose Related Metabolic Disorders in Rats with Chronic Renal Failure. *Scientific Reports, 6*(1). Advance online publication. DOI: 10.1038/srep26341 PMID: 27210744

Wang, X., Wang, C., Chu, C., Xue, F., Li, J., & Bai, J. (2024). Structure-function integrated biodegradable Mg/polymer composites: Design, manufacturing, properties, and biomedical applications. *Bioactive Materials, 39*, 74–105. DOI: 10.1016/j.bioactmat.2024.05.024 PMID: 38783927

Wang, Z., Liu, B., Yin, B., Zheng, Y., Tian, Y., & Wen, P. (2022). Comprehensive review of additively manufactured biodegradable magnesium implants for repairing bone defects from biomechanical and biodegradable perspectives. *Frontiers in Chemistry, 10*, 1066103.

Wasserbauer, J., Buchtík, M., Tkacz, J., Fintová, S., Minda, J., & Doskočil, L. (2020). Improvement of AZ91 alloy corrosion properties by duplex NI-P coating deposition. *Materials (Basel), 13*(6), 1357. DOI: 10.3390/ma13061357 PMID: 32192141

Wei, X., Menga, J., Ma, S., Li, Y., Qing, H., Peng, X., . . . Zhao, Q. (2022). MD simulations and experiments of plasma proteins adsorption to the biodegradable magnesium alloys to facilitate cell response. arXiv preprint arXiv:2207.11639.

Witte, F., Hort, N., Vogt, C., Cohen, S., Kainer, K. U., Willumeit, R., & Feyerabend, F. (2008). Degradable biomaterials based on magnesium corrosion. *Current Opinion in Solid State and Materials Science, 12*(5–6), 63–72. DOI: 10.1016/j.cossms.2009.04.001

Wong, W. L. E., & Gupta, M. (2006). Simultaneously Improving Strength and Ductility of Magnesium using Nano-size SiC Particulates and Microwaves. *Advanced Engineering Materials, 8*(8), 735–740. DOI: 10.1002/adem.200500209

Xin, Y., Liu, C., Zhang, X., Tang, G., Tian, X., & Chu, P. K. (2007). Corrosion behavior of biomedical AZ91 magnesium alloy in simulated body fluids. *Journal of Materials Research/Pratt's Guide to Venture Capital Sources, 22*(7), 2004–2011. https://doi.org/DOI: 10.1557/jmr.2007.0233

Yang, G., Liu, L., Xiao, L., Ke, S., Yang, H., & Lu, Q. (2024). Accelerated Scarless Wound Healing by Dynamical Regulation of Angiogenesis and Inflammation with Immobilized Asiaticoside and Magnesium Ions in Silk Nanofiber Hydrogels. *Journal of Materials Chemistry. B*. Advance online publication. DOI: 10.1039/d4tb01584c PMID: 39380345

Yang, K., & Lin, X. (2015). Biocompatibility of surface-modified magnesium and magnesium alloys. In *Elsevier eBooks* (pp. 231–260). https://doi.org/DOI: 10.1016/b978-1-78242-077-4.00007-3

Yuan, B., Chen, H., Zhao, R., Deng, X., Chen, G., Yang, X., Xiao, Z., Aurora, A., Iulia, B. A., Zhang, K., Zhu, X., Iulian, A. V., Hai, S., & Zhang, X. (2022). Construction of a magnesium hydroxide/graphene oxide/hydroxyapatite composite coating on Mg–Ca–Zn–Ag alloy to inhibit bacterial infection and promote bone regeneration. *Bioactive Materials*, *18*, 354–367. DOI: 10.1016/j.bioactmat.2022.02.030 PMID: 35415306

Yuste, I., Luciano, F. C., Anaya, B. J., Sanz-Ruiz, P., Ribed-Sánchez, A., González-Burgos, E., & Serrano, D. R. (2023). Engineering 3D-printed advanced healthcare materials for periprosthetic joint infections. *Antibiotics (Basel, Switzerland)*, *12*(8), 1229. PMID: 37627649

Zeng, R., Dietzel, W., Witte, F., Hort, N., & Blawert, C. (2008). Progress and challenge for magnesium alloys as biomaterials. *Advanced Engineering Materials*, *10*(8). Advance online publication. DOI: 10.1002/adem.200800035

Zhang, C., Lin, J., & Liu, H. (2018). Magnesium-based biodegradable materials for biomedical applications. *MRS Advances*, *3*(40), 2359–2364. DOI: 10.1557/adv.2018.488

Zhang, C., Yang, X., Yu, L., Chen, X., Zhang, J., Zhang, S., & Wu, S. (2024). Electrospun polyasparthydrazide nanofibrous hydrogel loading with in-situ synthesized silver nanoparticles for full-thickness skin wound healing application. *Materials & Design*, *239*, 112818. DOI: 10.1016/j.matdes.2024.112818

Zhang, Y., Liu, W., Liu, Y., Zhang, M., Tian, Y., & Chen, L. (2022). Research Progress on Corrosion Behaviors and Improvement Methods of Medical Degradable Mg– Based Alloys. *Metals*, *13*(1), 71.

Zhang, Z., Liu, A., Fan, J., Wang, M., Dai, J., Jin, X., Deng, H., Wang, X., Liang, Y., Li, H., Zhao, Y., Wen, P., & Li, Y. (2023). A drug-loaded composite coating to improve osteogenic and antibacterial properties of Zn–1Mg porous scaffolds as biodegradable bone implants. *Bioactive Materials*, *27*, 488–504. DOI: 10.1016/j.bioactmat.2023.04.017 PMID: 37180641

Zhao, D., Han, C., Peng, B., Cheng, T., Fan, J., Yang, L., Chen, L., & Wei, Q. (2022). Corrosion fatigue behavior and anti-fatigue mechanisms of an additively manufactured biodegradable zinc-magnesium gyroid scaffold. *Acta Biomaterialia*, *153*, 614–629. DOI: 10.1016/j.actbio.2022.09.047 PMID: 36162767

Zhao, J., Chen, L., Yu, K., Chen, C., Dai, Y., Qiao, X., Yan, Y., & Yu, Z. (2015). Effects of chitosan coating on biocompatibility of Mg–6%Zn–10%Ca3(PO4)2 implant. *Transactions of Nonferrous Metals Society of China*, *25*(3), 824–831. DOI: 10.1016/s1003-6326(15)63669-x

Zhou, H., Liang, B., Jiang, H., Deng, Z., & Yu, K. (2021). Magnesium-based biomaterials as emerging agents for bone repair and regeneration: From mechanism to application. *Journal of Magnesium and Alloys*, *9*(3), 779–804. DOI: 10.1016/j.jma.2021.03.004

Živić, F., Grujović, N., Manivasagam, G., Richard, C., Landoulsi, J., & Petrović, V. (2014). The potential of magnesium alloys as bioabsorbable/ biodegradable implants for biomedical applications. *DOAJ (DOAJ: Directory of Open Access Journals)*. https://doaj.org/article/9afee1a42b824047a615cbf8826996a4

Chapter 14
Biological Interactions of Implants From Biocompatibility to Biodegradability in Modern Medical Applications

Xinyu Fang
https://orcid.org/0000-0001-6337-8009
The First Affiliated Hospital of Fujian Medical University, China

ABSTRACT

This chapter provides a comprehensive exploration of the biological interactions of bone implants, focusing on biocompatibility and biodegradability in modern medical applications. It discusses the classification of biomaterials and medical devices, emphasizing the importance of understanding their interactions with biological systems. The chapter also delves into the challenges of Periprosthetic joint infection (PJI), and the clinical significance of biocompatibility. By integrating multidisciplinary knowledge from materials science, biomechanics, and clinical medicine, this chapter aims to provide readers with comprehensive theoretical foundations and practical guidance, thereby facilitating the development of safer and more durable joint replacement solutions.

1. INTRODUCTION

Biomaterials have become an indispensable component in modern medicine, finding extensive applications in various fields such as cell delivery, drug delivery, and tissue replacement(Eldeeb et al., 2022). Their versatility is evident in their use across a broad spectrum of medical interventions, ranging from simple dental restorations to highly sophisticated drug delivery systems designed to target specific cells or tissues(Hakim et al., 2024; Sheikhpour et al., 2017). Given the critical nature of these applications, it is imperative to thoroughly understand how these materials interact with biological systems. By delving into the nuances of material composition and structural design, we seek to identify strategies that can enhance the overall performance of biomaterials while simultaneously minimizing the risk of adverse immune responses that could compromise their effectiveness and safety (Figure 1).

DOI: 10.4018/979-8-3373-0055-9.ch014

Figure 1. Biological interactions of implants

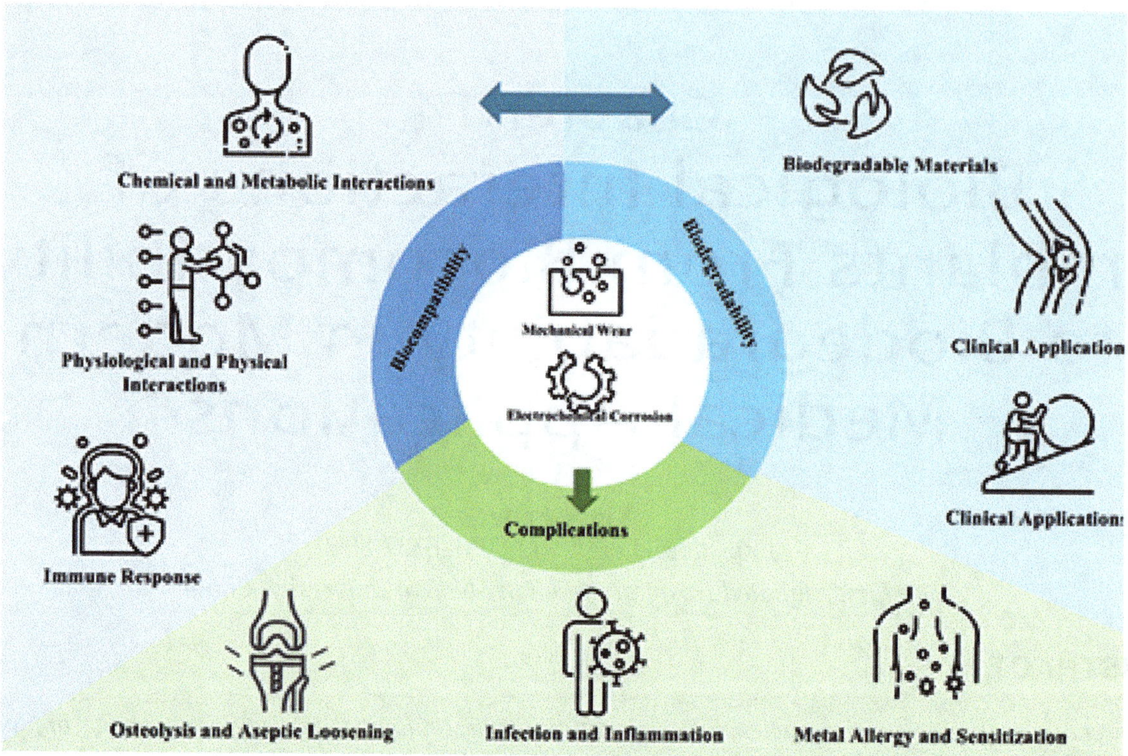

1.1 Applications in Orthopedics

Several key metallic biomaterials, including steel, titanium, cobalt-based alloys, and tantalum, were introduced. These materials are widely used in orthopedic and cardiovascular surgeries due to their unique combination of strength, durability, and biocompatibility. However, the article also explores the challenges associated with these materials, particularly in the context of metal-on-metal (MoM) hip resurfacing arthroplasty (HRA) and large-diameter head total hip replacement (THR). It examines the role of impingement and counter-coup edge loading, which can lead to significant wear and tear of the metal surfaces. Subsequent clinical observations have revealed that nanoparticles and metal ions released from the worn surfaces can trigger adverse reactions to metal debris (ARMD). These reactions include pain, fluid collection, pseudotumors, tissue necrosis, "particle disease," and hypersensitivity reactions such as aseptic lymphocyte-dominated vasculitis-associated lesions (ALVAL or LYDIA). In severe cases, patients may develop systemic toxicity and visceral manifestations, highlighting the critical need for more rigorous pre-market testing.

1.2 Importance of Biocompatibility and Biodegradability

Biocompatibility is a highly complex and multifaceted concept that extends far beyond mere compatibility at the surface level(Singh et al., 2024; Williams, 2008). It encompasses a wide range of interactions between tissues and the body across various systems or conditions, including chemical, metabolic, physiological, and physical aspects(Dee et al., 2003; Rosa et al., 2024). This intricate interplay is crucial for the success of both long-term and short-term implants, as well as for tissue engineering applications(Amini et al., 2012; Amini et al., 2011). For long-term implants, such as those used in orthopedic devices like joint replacements and cardiovascular devices like pacemakers, the materials must be meticulously designed to withstand prolonged use without eliciting adverse reactions that could compromise the patient's health(Oliveira et al., 2022; Škrlová et al., 2019). Tissue engineering scaffolds represent another unique category. These scaffolds are specifically engineered to support cell growth and tissue regeneration while gradually degrading in a controlled manner to facilitate the seamless integration of new tissue into the body(Suhag et al., 2024). The balance between providing structural support and allowing for biological integration is a delicate one, and achieving it is a key challenge in the field of biomaterials.

2. BIOMATERIALS AND MEDICAL DEVICES IN ORTHOPEDICS

Biomaterials and medical devices are integral to modern medicine, particularly in orthopedics, where they are used to repair and replace damaged tissues and organs(Choi et al., 2023; Grimm, 2004). The successful application of these materials hinges on their biocompatibility, mechanical properties, and ability to interact with biological systems without adverse effects(Majeed, 2023; Taraballi et al., 2018). This chapter provides a comprehensive overview of the classification of biomaterials and medical devices based on their composition, surface reactivity, tissue interaction, and risk assessment, integrating insights from recent studies and established standards. Biomaterials can be classified into four main categories based on their composition: metals, ceramics, polymers, and composites. Each category has distinct properties that influence their suitability for specific medical applications.

2.1 Properties and Applications of Key Biomaterials

Stainless steels, cobalt-based alloys, and titanium alloys are commonly used in orthopedic implants(Bai & Gilbert, 2006). These materials are chosen for their excellent corrosion resistance and biocompatibility. For instance, titanium alloys exhibit high strength, low modulus of elasticity, and excellent corrosion resistance, making them ideal for load-bearing implants such as hip and knee replacements(Abd-Elaziem et al., 2024; Geetha et al., 2009). Titanium alloys are particularly favored due to their ability to form a stable oxide layer that enhances their biocompatibility and resistance to wear(Bandyopadhyay et al., 2023).

Ceramics are inorganic, non-metallic materials with high hardness, wear resistance, and excellent biocompatibility(Vaiani et al., 2023). Hydroxyapatite and tricalcium phosphate are examples of bioactive ceramics used in bone replacement and dental applications(Bellucci et al., 2016). These materials can bond with bone tissue, promoting osseointegration and long-term stability. Bioactive ceramics can form a direct bond with living tissues, enhancing their integration with the host. For example, hydroxyapatite is chemically similar with the mineral phase of bone, making it an excellent choice for dental implants and bone grafts(Brochu et al., 2024).

Polymers are versatile materials with a wide range of mechanical properties. They can be tailored for specific applications through chemical modification and processing(AlMaadeed et al., 2020; Fabbri & Messori, 2017). Polymers such as polyethylene, poly(methyl methacrylate), and polylactic acid are used in joint replacements, bone cements, and resorbable implants(Rey-Vinolas et al., 2019). Their biocompatibility and degradation rates are critical factors in their clinical use. For instance, ultra-high molecular weight polyethylene (UHMWPE) is widely used in hip and knee implants due to its excellent wear resistance and biocompatibility(Macuvele et al., 2017). However, polymers are often susceptible to wear debris and may require surface modifications to enhance their durability.

Composites combine two or more materials to achieve enhanced properties. Metal matrix composites (e.g., titanium reinforced with hydroxyapatite) and polymer matrix composites (e.g., high-density polyethylene reinforced with ceramic particles) are examples(Bhong et al., 2023; Macuvele et al., 2017). These materials are used in applications requiring high strength and wear resistance, such as hip and knee implants(Scholz et al., 2011). Composites can be designed to combine the mechanical strength of metals with the bioactivity of ceramics, providing a synergistic effect that enhances their performance in orthopedic applications.

2.2 Surface Characteristics and Mechanical Loading

The surface properties of biomaterials significantly influence their interaction with biological tissues. Surface reactivity includes factors such as wettability, charge, and roughness, which affect protein adsorption, cell attachment, and tissue integration(Dave & Gomes, 2019; Lotfi et al., 2013). Hydrophilic surfaces promote protein adsorption and cell attachment, enhancing biocompatibility. Surface modifications, such as coating with bioactive molecules or creating micro/nanostructures, can further enhance these properties(Cai et al., 2020). Titanium implants coated with hydroxyapatite show improved osseointegration compared to uncoated implants(Neto et al., 2023). Surface roughness also plays a crucial role in promoting cell attachment and bone ingrowth. Studies have shown that micro- and nano-scale surface roughness can significantly enhance the osteoblastic response and bone integration(Hasegawa et al., 2020; Hou et al., 2022).

The interaction between biomaterials and tissues is a critical aspect of their biocompatibility. Biomaterials must elicit minimal inflammatory responses, avoid immune rejection, and promote tissue regeneration. Magnesium alloys, which degrade over time, are being explored for temporary implants due to their ability to support bone healing while gradually being replaced by native tissue(Nasr Azadani et al., 2022; Niranjan et al., 2023). In contrast, permanent implants made from titanium or cobalt alloys must maintain long-term stability without causing adverse tissue reactions. The foreign body response (FBR) is a common reaction to implanted materials, involving the formation of a fibrous capsule around the implant(Capuani et al., 2022; Zhang et al., 2021). This response can be minimized through careful material selection and surface modifications.

3. BIOCOMPATIBILITY: FROM SURFACE TO SYSTEMIC INTERACTIONS

Biocompatibility is a highly complex and multifaceted concept that extends far beyond mere surface compatibility. It encompasses the extensive interactions between tissues and the body under various systems or conditions, including chemical, metabolic, physiological, and physical aspects. In clinical practice, the biocompatibility of implanted materials is crucial for ensuring successful treatment outcomes for patients.

3.1 Chemical and Metabolic Interactions

The chemical properties of implanted materials play a significant role in their biocompatibility. Materials that appear to be ideally chemically inert may still trigger inflammatory reactions due to mechanical stress exerted on surrounding tissues. Silicone elastomers used in breast implants exhibit minimal cytotoxicity in vitro, can still elicit a strong foreign body reaction in vivo. This highlights the importance of understanding the chemical interactions between materials and the biological environment. Additionally, the metabolic fate of degradation products from biodegradable materials must also be considered. For instance, the degradation of Ti implants can lead to the production of acidic by-products, which may cause inflammation and affect local pH levels. These by-products can further disrupt the local tissue environment, leading to chronic inflammation and impaired healing.

3.2 Physiological and Physical Interactions

The physiological and physical interactions between implants and surrounding tissues are equally important. The mechanical properties of implants, such as their Young's modulus, can significantly influence the host response. When the Young's modulus of an implant is higher than that of the surrounding tissue, stress concentration may occur at the interface, leading to inflammation and fibrous capsule formation. This is particularly relevant in soft tissues, which have a wide range of stiffness. Matching the mechanical properties of implants to tissues is a challenge. Stiffness mismatch between metallic implants and soft tissues can lead to tissue damage and poor biocompatibility. The biologic inertness of Ti-Al-V and Co-Cr-Mo alloys ensures their biologic safety. These metals, including cobalt, chromium, molybdenum, and vanadium, play roles in human metabolism by promoting the formation of key enzymes. Although titanium, tantalum, and aluminum are not essential for normal physiological functions, they can be detected in the body through dietary intake. At physiologically acceptable levels, these metals typically cause minimal adverse reactions and can be excreted through multiple organs.

The kidneys play a crucial role in the excretion of these metals, as evidenced by the high concentrations detected in the urine of patients who have undergone total hip arthroplasty (THA) or total knee arthroplasty (TKA). 40% of cobalt (Co) administered intravenously is eliminated within the first 24 hours, 70% within one week, 80% within one month, and 90% within one year(Smith et al., 1972). Mo is rapidly excreted, which accounts for its low blood levels and minimal risk of adverse reactions(Barceloux & Barceloux, 1999a). Al and V are also rapidly excreted, primarily through conversion to hydrophilic forms or binding with hydrophilic substances, although only about 10% of these metals can pass through the glomerular membrane as free ions(Mukherjee et al., 2004). In THA patients, continuous metal release results in an initial rapid phase of excretion, followed by a gradual slowdown that can persist for several years.(Barceloux & Barceloux, 1999b; Catalani et al., 2013; Rehder, 2013)

Defecation serves as another significant pathway for metal excretion. Metals in the gastrointestinal tract, such as Cr, Al, V, and Ti, are excreted in feces due to their low absorption rates(Hong et al., 2017; Tripathi et al., 2018). Additionally, Cr and Ti can be eliminated through the shedding of nails, hair, sweat, and milk(Shi et al., 2013). Titanium particles can also be phagocytosed by cells and expelled as sputum through the respiratory tract.

Surface characteristics, including texture and topography, also affect biocompatibility. Studies have shown that rough or textured surfaces are more effective in reducing foreign body reactions than smooth surfaces. This may be due to the increased surface area and better mechanical interlocking, which reduces relative motion and friction between the implant and tissue. Mechanical loading is another key factor. Proper load transfer is crucial for maintaining the bone-implant interface, as demonstrated by the success of dental and orthopedic implants that are firmly anchored in bone. Conversely, excessive or insufficient loading can lead to bone resorption, as described by Wolff's law of bone remodeling. The mechanical strength of Ti-Al-V and Co-Cr-Mo alloys is vital for treating weight-bearing joints such as the hip and knee(Nortje, 2021; Semlitsch, 1974). The elastic modulus of Ti-Al-V alloy is four to six times that of cortical bone, allowing it to effectively share stress and provide load-bearing capacity(Bidez & Misch, 2005). The elastic modulus of Co-Cr-Mo alloy is about twice that of Ti-Al-V, making it suitable for components that require high wear resistance, such as artificial femoral heads and condyles. However, implants with excessively high elastic modulus may hinder bone integration due to stress shielding, leading to instability and insufficient bone growth.

3.3 Immune Response

The immune response to implanted materials is a key aspect of biocompatibility. The initial inflammatory response, characterized by the recruitment of macrophages and other immune cells, is a natural part of the healing process. However, the long-term presence of these cells, particularly M1-type macrophages, can lead to fibrous capsule formation and foreign body reactions. The shift from an inflammatory phenotype (M1-type macrophages) to a reparative phenotype (M2-type macrophages) is crucial for constructive tissue remodeling and implant integration(Fang et al., 2024; Song et al., 2023). Understanding and modulating this immune response is essential for improving the biocompatibility of implants. The excellent corrosion resistance of these alloys is crucial for long-term stability. After implantation, a thin protective oxide layer forms on the prosthesis surface, providing antioxidant protection and preventing excessive redox reactions. Ti-Al-V forms a stable titanium dioxide layer, while Co-Cr-Mo generates a dense protective oxide layer to resist electrochemical corrosion(Bai & Gilbert, 2006; Kaya, 2011). These layers can self-repair and replenish, ensuring the stability and functionality of the prosthesis over the long term.

4. BIODEGRADABILITY: INNOVATIONS AND CLINICAL IMPLICATIONS

Biodegradable synthetic polymers, including poly(glycolic acid) (PGA), poly(l-lactic acid) (PLLA), and poly(lactic acid-co-glycolic acid) (PLGA), have been developed and approved by the FDA (Makadia & Siegel, 2011).. These degradable implants promote complete tissue healing and functional restoration, eliminating the need for permanent implants. However, their practical application has been limited due to inflammation caused by the acidification of surrounding tissues and early implant fractures resulting from insufficient mechanical strength. Recent advancements in biodegradable metallic materials show potential for treating bone fractures and cardiovascular diseases. Biodegradable metal implants offer greater strength than polymer implants, and their degradation products are fully soluble in body fluids. These particles can be safely eliminated by phagocytic cells such as macrophages. The degradation of biodegradable metals involves anodic and cathodic reactions that form a protective oxide layer. In magnesium alloys, chloride ions in body fluids can compromise this protective layer, accelerating degradation. This ultimately leads to calcium phosphate deposition, enabling cell adhesion and new tissue formation.

The clinical use of biodegradable implants dates back over a century. In 1878, Edward C. Huse successfully used pure magnesium wire for hemostasis, although its biodegradable properties were not recognized at the time (Asgari et al., 2018; Heimann, 2020). Early clinical trials with magnesium-based implants failed due to rapid degradation. However, advancements in metal purification and alloying technologies have renewed interest in biodegradable metals. After nearly two decades of research, these materials now demonstrate significant clinical potential for treating musculoskeletal and cardiovascular conditions. Nonetheless, complications related to implant placement and potential issues in clinical practice remain important concerns.

5. COMPLICATIONS OF PROSTHETIC IMPLANTS

The ideal implant materials should match the elastic modulus of bone, offer high corrosion and wear resistance, and be highly biocompatible. However, commonly used metals in implant designs, including cobalt-chromium-molybdenum (Co-Cr-Mo), Ti and its alloys, and stainless steel (SS), each have distinct advantages and challenges(Mahajan et al., 2023; Zaman et al., 2015). Co-Cr-Mo alloys are valued for their rigidity and long-term corrosion resistance but are costly to manufacture due to their high rigidity. Titanium alloys, while less dense and highly biocompatible, have lower shear and wear resistance and are expensive. Stainless steel, with its high elastic modulus and good short-term corrosion resistance, is less ideal for long-term implants due to potential nickel-related immunogenicity. Despite the excellent corrosion resistance of Ti and Co-Cr-Mo, metal ions, particles, and debris can still be released into the body due to mechanical wear, electrochemical corrosion, and inflammation(Liu & Chen, 2018). Manufacturers use different materials for different implant parts to maximize efficiency and durability. For instance, Co-Cr-Mo is used for femoral heads and components in knee replacements, while Ti alloys are used for femoral shafts and tibial trays. The acetabular cup and tibial surface are often made from ultra-high-molecular-weight polyethylene. The widespread use of Co-Cr-Mo highlights the importance of studying its degradation products for patient safety and implant longevity.

5.1 Mechanical Wear and Electrochemical Corrosion

The clinical implications of mechanical wear and electrochemical corrosion are profound. Aseptic loosening, often attributed to wear debris and corrosion products, is a leading cause of implant failure. This condition typically develops over 15 to 20 years post-implantation and is characterized by chronic inflammation and bone resorption around the implant(Oh et al., 2002). In addition to aseptic loosening, septic loosening caused by microbial infections can also lead to implant failure. Bacterial endotoxins and other pathogen-associated molecular patterns can adhere to wear particles, exacerbating the inflammatory response and accelerating bone loss(Hodges et al., 2021).A clinical case involved a 62-year-old patient with a history of total hip arthroplasty who developed a deep infection. Imaging and laboratory tests revealed septic loosening with significant bone loss. The patient underwent debridement and implant revision, highlighting the complexity of managing such infections.

To address these challenges, clinicians must adopt a multifaceted approach. Regular monitoring of patients with joint prostheses is essential for early detection of wear and corrosion. This includes radiographic imaging, blood tests for metal ion levels, and clinical assessment of pain and mobility. Advanced imaging techniques, such as metal artifact reduction sequence (MARS) MRI, can help in detecting soft tissue masses and early signs of osteolysis(Busch et al., 2022). Furthermore, the development of new materials and surface coatings that enhance wear resistance and reduce corrosion is an active area of research. (Gao & Fu, 2019). Similarly, the use of ceramic-on-ceramic (CoC) and ceramic-on-polyethylene (CoP) bearings has demonstrated lower wear rates and reduced risk of osteolysis(Fang & Shang, 2021; Giuseppe et al., 2021). A clinical example includes a 70-year-old patient who underwent total hip arthroplasty with a CoC bearing(De Smet, 2013). Follow-up imaging and blood tests over five years showed no signs of wear debris or elevated metal ions, demonstrating the effectiveness of advanced materials in reducing wear and corrosion.

5.2 Osteolysis and Aseptic Loosening

Metal debris, often resulting from the combined effects of mechanical wear and electrochemical corrosion, can cause adverse local tissue reactions (ALTR) leading to osteolysis, particularly with titanium alloy implants(Eltit et al., 2019; Lehtovirta, 2020). ALTR can manifest as intra-articular effusion with sterile, watery, yellow or gray, turbid fluid and low cell counts. (Granchi et al., 2018; Lee et al., 2023). In a recent case study, a 65-year-old patient with a titanium alloy hip implant experienced significant intra-articular effusion and pain, which was diagnosed as metallosis due to wear debris(Sundfeldt et al., 2006). Recent estimates suggest that aseptic loosening, often caused by wear debris, leads to early failure in metal-on-metal (MoM) total joint arthroplasty (TJA) at a rate of 4% to 5% within 6 to 7 years post-implantation. Aseptic loosening is a significant complication in total hip (THA) and knee arthroplasty (TKA), often leading to implant failure and the need for revision surgeries(Feng et al., 2022). This condition, characterized by the loss of fixation between the implant and the bone without infection, poses a major challenge in orthopedic surgery. Understanding its clinical presentation, risk factors, and management strategies is crucial for improving patient outcomes.

5.2.1 Clinical Presentation of Aseptic Loosening

In total hip arthroplasty (THA), aseptic loosening is a leading cause of implant failure, with reported rates varying from 10% to over 20% in long-term follow-up studies(Strauss et al., 2017). The condition can manifest as pain, limp, or even frank instability of the hip joint. Clinically, patients may present with progressive discomfort, particularly with weight-bearing activities, and a noticeable decrease in hip function as measured by the Harris Hip Score (HHS)(Haddad et al., 2015). Radiographic findings often include osteolysis, peri-implant radiolucency, and implant migration. Patients with Crowe type IV developmental dysplasia of the hip (DDH) undergoing THA exhibited an 8.5% rate of intraoperative fractures and a 6.1% rate of aseptic loosening over a mean follow-up of 11 years(Kenanidis et al., 2017). One notable case involved a 45-year-old patient with DDH who experienced severe hip pain and instability within 5 years of implantation, requiring revision surgery due to significant osteolysis and loosening(Banaszkiewicz, 2012).

Similarly, in total knee arthroplasty (TKA), aseptic loosening is a common reason for revision surgeries, accounting for up to 32.7% of all revisions(Bonorino & Teves, 2024). Clinical manifestations include persistent pain, stiffness, and a decline in knee function. Patients may also experience instability or a sensation of the knee giving way. Radiographic assessment often reveals signs of loosening such as radiolucency around the implant, component migration, or progressive osteolysis. A study by Abdel et al. found that patients with a BMI $\geq$ 35 kg/m^2 had a higher risk of aseptic tibial loosening, with a revision rate of 9.1% over a mean follow-up of 10 years(Kamal et al., 2024). In one case, a 55-year-old patient with obesity (BMI 38 kg/m^2) underwent TKA and experienced significant pain and instability within 7 years, leading to revision surgery due to aseptic loosening(Tria Jr, 2013).

5.3 Infection and Inflammation

Successful integration of metallic implants requires host cell colonization of the fixation surface. However, bacteria like staphylococci can adhere to the implant surface, forming biofilms that resist clearance by the host's defenses and antibiotics(Lyu et al., 2024). A 70-year-old patient with a history of total knee arthroplasty (TKA) developed a chronic infection due to Staphylococcus aureus, which formed a biofilm on the implant surface, making it extremely difficult to eradicate the infection. Bacteria can remain dormant for months to years before causing infection. Porous implants increase the surface area vulnerable to bacterial colonization. Wear debris, in the form of soluble ions or insoluble particles, can cause localized inflammation, primarily driven by macrophages. The systemic effects of wear debris, particularly metallic and polyethylene (PE) debris, are still debated. Periprosthetic tissue is fibrous and granulomatous, comprising a mix of cell types including histiocytes, fibroblasts, osteoblasts, osteoclasts, and occasionally lymphocytes. Foreign body particles from any part of the prosthesis can stimulate a host cell response, leading to granuloma formation with necrosis, fibrosis, or scar tissue. The study found that patients undergoing total knee arthroplasty with the “medially pedicled infrapatellar fat pad (IPFP) preservation technique” had significantly lower postoperative wound oozing rates compared to the traditional resection group, along with reduced inflammatory markers (CRP and ESR) on postoperative day 5 (Figure 2). However, there was no statistically significant difference in periprosthetic joint infection rates between the two groups. These results suggest that the preservation technique may indirectly mitigate infection risk by reducing wound oozing, but its direct impact on infection prevention requires further validation through larger sample sizes and long-term follow-up.(Zhang et al., 2024)

Figure 2. Animation of surgical procedure. (A) The IPFP is exposed and isolated. The lateral side is carefully separated while the medial side remains pedicled. (B) An intact medial pedicled IPFP flap is formed. The blood and nerve supply are well protected. (C) The IPFP flap is suspended with an Allis clamp to prevent interference in the surgical field, especially the lateral compartment. The prosthesis is implanted smoothly. (D) The IPFP is sutured and fixed in situ

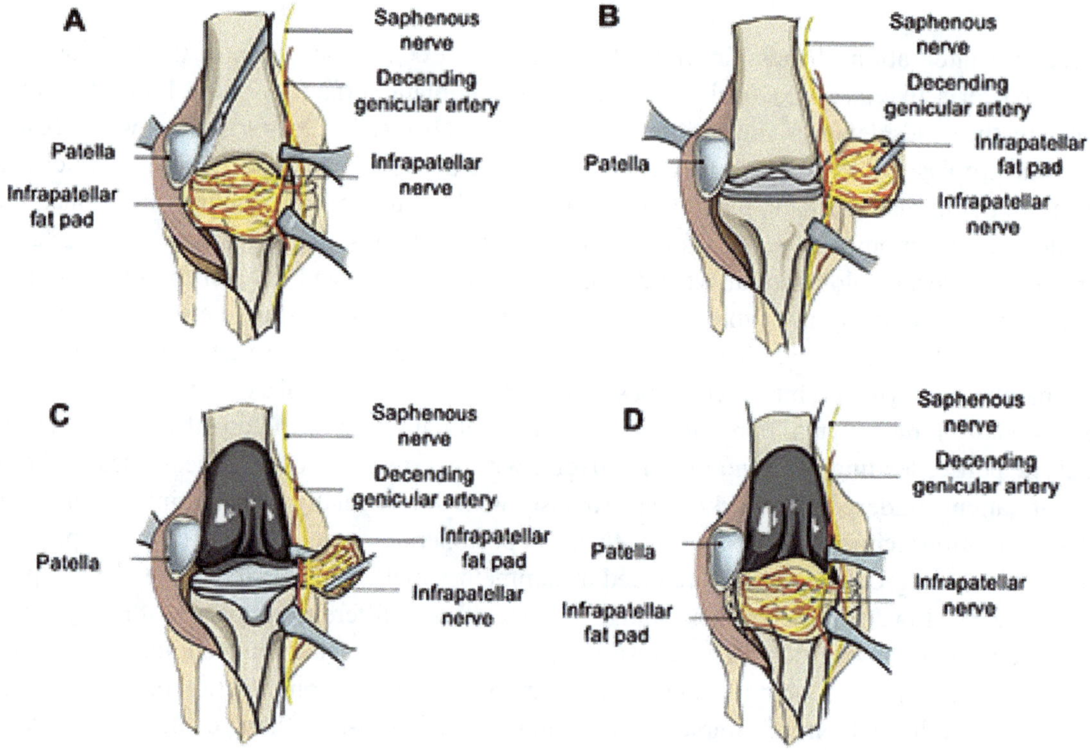

(Zhang et al., 2024)

5.3.1 Periprosthetic joint infection (PJI)

Periprosthetic joint infection (PJI) is a severe complication following artificial joint replacement. Staphylococcus aureus (S. aureus), a Gram-positive coccus, secretes various toxins (such as alpha-hemolysin and phenol-soluble modulinos) that trigger localized and systemic pyogenic infections (Lindsay & Holden, 2004). Outside bone tissue, S. aureus can form abscesses by colonizing skin mucosa or muscle soft tissues, as well as adhere to implant surfaces (e.g., artificial joints) to develop biofilms. The abscess walls and biofilms act as natural barriers, enabling S. aureus to evade host immune attacks and antibiotic eradication(Cheng et al., 2011; Jacqueline & Caillon, 2014), allowing prolonged colonization and leading to persistent or recurrent infections. Karen L. et al.(de Mesy Bentley et al., 2017) observed via electron microscopy that S. aureus can invade the osteocyte lacuna-canalicular network (OLCN) within bone tissue, colonizing, migrating, and proliferating in bone canaliculi. Furthermore, numerous in vitro studies have demonstrated that S. aureus can invade non-professional phagocytes (e.g., osteoblasts, osteoclasts) and osteocytes, establishing intracellular infections through long-term colonization and replication(Kolenda et al., 2020; Krauss et al., 2019).

Previous reports also documented intracellular colonization of S. aureus within osteoblasts in vivo via electron microscopy(Bosse et al., 2005). This bone- and intracellular-colonization strategy not only shields S. aureus from host immunity and antibiotics but also provides a nutrient-rich environment for continued proliferation. This study explored the role of Staphylococcus aureus small colony variants (SCVs) in intraosseous invasion and colonization in periprosthetic joint infection (Figure 3). By collecting surgical samples for microbial culture and transcriptomic analysis, it found that SCVs have stronger invasive and colonization abilities in chronic PJI. The research indicates that SCVs reduce chemokine expression in host bone tissue, inhibiting the recruitment of immune cells like neutrophils. (Cai et al., 2022)

Figure 3. Immunofluorescence and Transmission Electron Microscopy (TEM): Staphylococcus aureus was detected within osteocytes and bone lacunar spaces in the PJI group (indicated by red arrows)

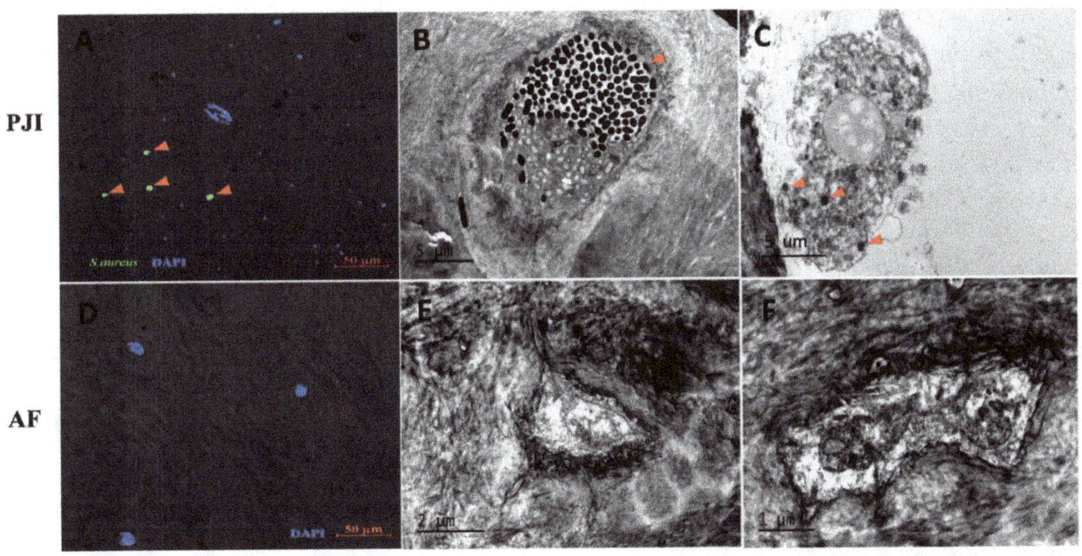

(Cai et al., 2022)

5.3.2 Diagnosis of PJI

Diagnosing PJI is challenging due to its nonspecific clinical presentation and the need for accurate identification of the infecting organism. Clinical signs such as joint pain, swelling, and erythema are common but not definitive. Laboratory tests, including elevated erythrocyte sedimentation rate (ESR) and C-reactive protein (CRP) levels, can indicate inflammation but are not specific to PJI(Zhang et al., 2023). Imaging studies like X-rays, CT scans, and MRI can reveal bone destruction and soft tissue abnormalities but may not always confirm the presence of infection(Cai et al., 2023). Microbiological testing, including joint fluid or tissue cultures and 16S rRNA PCR, is essential for identifying the causative pathogens(Hu et al., 2024). Accurate diagnosis often requires a combination of clinical assessment, laboratory tests, imaging, and microbiological confirmation. For example, a recent study used artificial intelligence (AI) to improve the accuracy of pathological diagnosis of PJI, which showed promising results in identifying infection indicators beyond neutrophil aggregation. A study compared the diagnostic performance of targeted next-generation sequencing (tNGS), metagenomic next-generation sequencing (mNGS), and conventional culture methods for periprosthetic joint infection (PJI), revealing that tNGS demonstrated comparable sensitivity (88.37%) and specificity (95.24%) to mNGS (93.02% and 95.24%), while offering lower costs and shorter turnaround time. (Huang et al., 2023) Additionally, tNGS detected drug resistance genes in 37.5% of cases, suggesting its potential as a cost-effective and efficient tool for multidisciplinary, multi-technology-based PJI diagnosis. A study optimized the metagenomic next-generation sequencing (mNGS) workflow and successfully diagnosed and verified seven cases of mycoplasma periprosthetic joint infection (PJI), as showed as Figure 4. The positive rate was 100%, significantly higher than that of traditional culture (57.14%) and 16S PCR (71.43%). Targeted therapy based on the findings effectively controlled the infections. This study provides a standardized diagnostic strategy for rare bacterial PJIs.(Cai et al., 2023)

Figure 4. The optimized and standardized procedure based on mNGS for Mycoplasma PJI. mNGS, metagenomic next-generation sequencing; PCR, polymerase chain reaction; PJI, periprosthetic joint infection; -, microbial culture negative; +, microbial culture positive

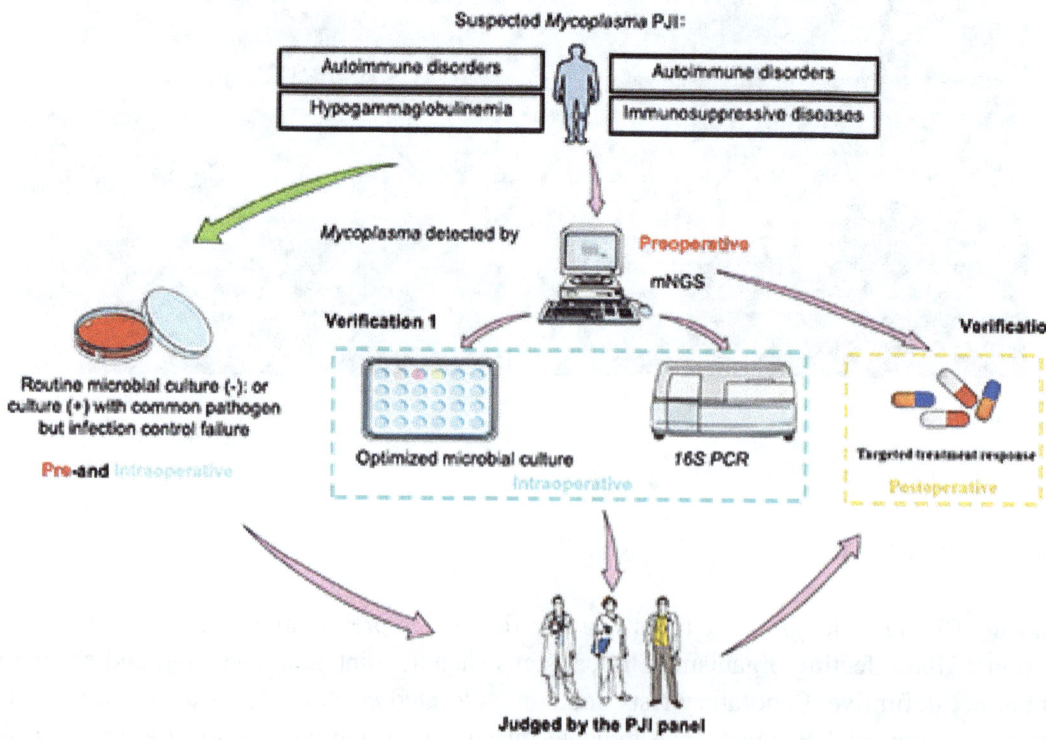

(Cai et al., 2023)

5.3.3 Treatment of PJI

Treatment of PJI is multifaceted and depends on the stage and severity of the infection. Surgical debridement is a cornerstone of treatment, aiming to remove infected tissues and reduce the bacterial load. This often involves thorough cleaning of the joint space and removal of any biofilm-covered implants. Antibiotic therapy is another critical component of PJI management. The choice of antibiotics is guided by the identified pathogen and its susceptibility profile. For instance, vancomycin is commonly used for methicillin-resistant Staphylococcus aureus (MRSA) infections, while fluconazole is preferred for fungal infections. The duration of antibiotic treatment can vary, with some studies suggesting that short-course therapy can be as effective as long-term regimens in certain cases(Chen et al., 2025). After stage II reimplantation surgery of two-stage revision for chronic PJI, extended antibiotic prophylaxis for less than one month can achieve good infection control rate(Fang et al., 2021). In chronic PJI, two-stage revision surgery is often employed. This involves an initial debridement followed by the placement of an antibiotic-loaded cement spacer to maintain local antibiotic concentrations and prevent infection recurrence. altered osteoblast-osteoclast balance (Figure 5). Once the infection is controlled, a second-stage surgery is performed to reimplant a new prosthesis. The use of antibiotic-loaded cement spacers has shown promising results in maintaining joint function and reducing infection recurrence rates(Wang et al., 2024). Prosthetic spacers and cement spacers are both effective at treating chronic PJI because they encourage infection control, and the former improved knee function status between stages(Wu et al., 2024). Both DASR and two-stage arthroplasty achieved a satisfactory infection cure rate and functional recovery for chronic destructive SHA, and DASR demonstrated significantly lower intraoperative blood loss, hospital stay, and hospitalization costs than two-stage arthroplasty(Zhang et al., 2022). A study represents the first application of label-free quantitative proteomics to analyze periprosthetic bone tissue in patients with Staphylococcus aureus periprosthetic joint infection (PJI) (Chen et al., 2022).

Figure 5. Preoperative radiographs, intraoperative surgical field photographs and pathological sections of periprosthetic tissues (HE staining) in the S.aureus PJI group and the aseptic failure group. Preoperative radiographs, surgical field photographs and pathological sections of knee PJI patient (A-D) and knee aseptic failure patient (E-H). Preoperative radiographs, surgical field photographs and pathological sections of hip PJI patient (I-L) and hip aseptic failure patient (M-P). Arrow showed bone destruction (yellow arrows) and neutrephils infiltration (black arrow) in the S.aureus PJI group (A-D and I-L).

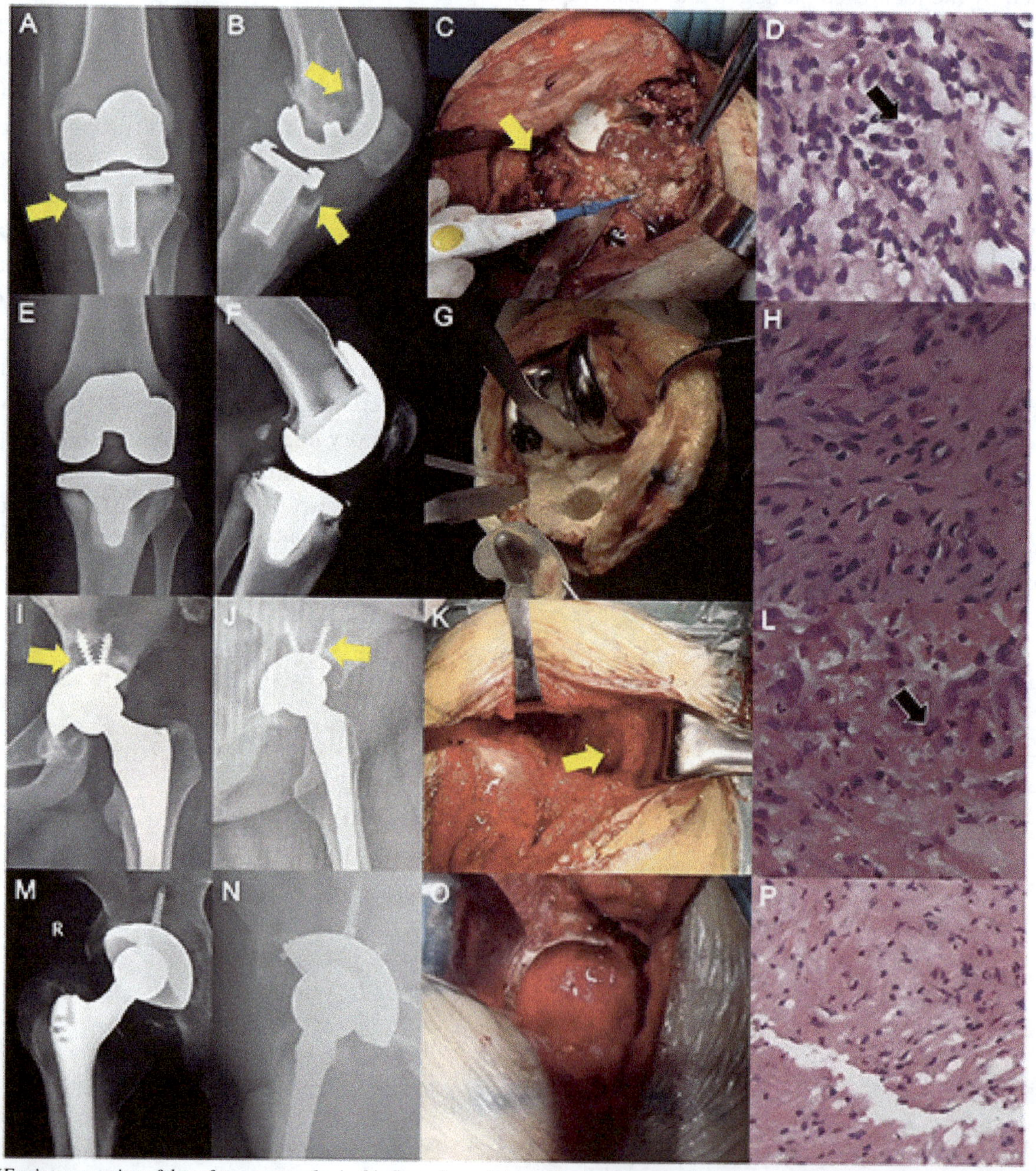

(For interpretation of the references to color in this figure legend, the reader is referred to the web version of this article.)(Chen et al., 2022)

5.4 Metal Allergy and Sensitization

The use of metal implants in orthopedic surgery, particularly metal-on-metal (MoM) prostheses in total hip arthroplasty (THA) and total knee arthroplasty (TKA), has raised significant concerns regarding metal allergies. Certain metals, such as cobalt, titanium, and chromium, can cause allergic reactions due to their ability to be recognized by antigen-presenting cells. Elevated levels of these metals have been observed in patients with MoM implants, prompting regulatory agencies like the UK's Medicines and Healthcare products Regulatory Agency (MHRA) to issue guidance on monitoring metal levels. This review explores the clinical implications of metal allergy in orthopedic implant patients, focusing on prevalence, diagnostic approaches, management strategies, and clinical outcomes.

5.4.1 Prevalence of Metal Allergy

Metal allergy is relatively common, affecting up to 10% of the general population, with nickel being the most frequent allergen (Ahlström et al., 2019). In orthopedic patients, the prevalence of metal allergy varies. Studies have shown that approximately 10%–20% of patients undergoing THA or TKA report metal allergies, with a higher incidence in women(Pacheco, 2019). This gender disparity is attributed to increased exposure to nickel through jewelry and piercings. A study of 94 patients undergoing TKA found that sensitization to metals like cobalt and titanium was significantly higher post-surgery, particularly among those with loosened implant components(Matar et al., 2021). The incidence of metal allergy in patients with well-functioning implants is around 20%–25%, while it rises to approximately 60% in those with loosening prostheses(Basko-Plluska et al., 2011; Rosner & Fonacier, 2017). This suggests that continuous exposure to metal ions from implants can induce allergic reactions.

5.4.2 Clinical Manifestations and Diagnostic Approaches

Clinical manifestations of metal allergy in orthopedic patients can range from mild to severe. Common symptoms include localized or generalized allergic dermatitis, such as eczema, urticaria, and erythema multiforme. Systemic reactions like vasculitis and bullous eruptions have also been reported. In addition to cutaneous symptoms, patients may experience persistent pain, aseptic inflammation, and implant loosening. These symptoms can significantly impact the patient's quality of life and surgical outcomes. For example, a patient with a metal-on-metal hip implant reported persistent pain and swelling around the implant site, along with eczematous skin changes. Imaging revealed signs of implant loosening, which were later confirmed to be associated with a metal allergy.

Diagnosing metal allergy in orthopedic patients can be challenging due to the lack of definitive diagnostic tests. The commonly used tests include skin patch testing, leukocyte migration inhibition test (LMIT), and lymphocyte transformation test (LTT)(Bao et al., 2018; Malahias et al., 2020; Von Blomberg-van der Flier et al., 1989). Skin patch testing is widely used due to its ease and cost-effectiveness but has limitations in specificity(Poon et al., 1998). The LMIT and LTT are more reliable but require specialized laboratories and are more expensive. A positive patch test result does not always correlate with clinical symptoms, and many patients with positive results may never develop metal hypersensitivity(Schalock et al., 2012). Therefore, a combination of clinical symptoms and laboratory tests is essential for accurate diagnosis. For example, a patient with suspected metal allergy underwent skin patch testing, which showed a positive reaction to cobalt. However, the patient had no clinical symptoms of allergy. Further

LTT was performed, which confirmed the presence of sensitization, leading to the recommendation of a metal-free implant for revision surgery.

5.4.3 Differential Diagnosis

Distinguishing metal allergy from infection is crucial, as both can lead to similar symptoms such as pain and implant loosening. Differential diagnosis relies on laboratory analyses, including white blood cell/colloid scintigraphy, ultrasound-guided aspiration, and arthroscopic biopsy. Metal allergy is often associated with lymphocyte infiltration, while infections show polymorphonuclear cells. Fluid aspirated from the joint in metal allergy is typically milky with no bacterial growth, whereas infected joints show grey, rubbery, necrotic tissue. Patch testing and quantification of migration inhibition factor and leukocyte inhibition factor can also aid in differentiating between metal allergy and infection(Hallab et al., 2000). A patient with a loosened knee implant presented with pain and swelling. Initial aspiration showed milky fluid with no bacterial growth, suggesting metal allergy(Browne et al., 2010). Patch testing confirmed sensitization to titanium, and subsequent revision surgery with a metal-free implant resolved the symptoms(Spiewak, 2018).

5.4.4 Clinical Outcomes

Several studies have investigated the impact of patient-reported metal allergy on clinical outcomes. A retrospective study by Nam et al.(Nam et al., 2016) found that patients reporting metal allergy had decreased functional outcomes after TKA and lower mental health scores after THA. This suggests that patient-reported metal allergy may be associated with lower satisfaction and poorer clinical outcomes. However, the study had limitations, including its retrospective nature and the lack of confirmatory testing for metal sensitivity. For instance, a patient with a history of nickel allergy underwent TKA and reported persistent pain and swelling. Despite no confirmatory testing, the patient's functional outcomes were poor, highlighting the need for better diagnostic tools and management strategies(Lionberger et al., 2018; Malahias et al., 2020).

Metal allergy is a clinically relevant issue in orthopedic surgery, particularly with the increasing use of metal implants. While the prevalence of metal allergy is relatively high, especially in women, its clinical manifestations can vary widely. Accurate diagnosis requires a combination of clinical symptoms and laboratory tests. Management strategies range from local treatments to revision surgery with metal-free implants. Further research is needed to better understand the relationship between metal ions and allergic reactions and to develop more effective diagnostic and treatment methods.

6. CONCLUSIONS

Metal prostheses have been a cornerstone of joint replacement surgeries for nearly a century, significantly improving the quality of life for patients with severe joint diseases and injuries. These prostheses, primarily made from materials like cobalt-chromium-molybdenum (Co-Cr-Mo) and titanium-aluminum-vanadium (Ti-Al-V) alloys, have been widely used due to their durability and mechanical strength. However, the long-term use of these materials has raised significant concerns regarding the release of metal ions. Studies have shown that these metal ions can accumulate in various organs and surrounding tissues, potentially leading to toxic and destructive effects on tissues and organs by disrupting inflammatory responses and cellular functions. Elevated metal levels in the blood are often observed in patients undergoing total knee arthroplasty (TKA) and total hip arthroplasty (THA), highlighting the need for safer and more biocompatible materials. Factors such as implant instability or mismatched components can exacerbate metal release, leading to increased wear and tear and accelerated degradation of the metal surface. These issues underscore the importance of precise surgical techniques and the development of more durable materials.

Digital technologies, particularly robotics, offer a promising solution to mitigate these challenges. Robot-assisted arthroplasty enhances surgical precision through 3D preoperative planning, reducing human error and improving prosthesis placement and soft tissue balance. Supported by advanced imaging techniques, 3D printing, and computer-aided design (CAD) tools, joint replacement prostheses are moving toward greater personalization, which is essential for optimizing outcomes, especially in pediatric and young adult patients. Improving prosthetic materials is another key area of research. Key design factors include matching the elastic modulus to human bone (~30 GPa), enhancing biocompatibility and corrosion resistance, using non-toxic alloy elements, and improving the mechanical properties of titanium alloys. Personalized design could further prevent implant damage during removal processes and enhance integration with bone. Despite progress in biomaterials, clinical translation faces several challenges. Ensuring biocompatibility and biodegradability requires thorough testing to prevent adverse reactions and ensure safe degradation. Tailoring material properties for specific applications involves optimizing chemical composition, surface characteristics, and degradation rates. Navigating regulatory and ethical considerations is also crucial to ensure responsible development and use of these materials. Biomaterials have transformed modern medicine, offering innovative solutions for diverse medical needs. Their success hinges on safe and effective interaction with the human body, promoting tissue integration and maintaining functionality within an appropriate timeframe. This chapter provides a comprehensive overview of biomaterials, emphasizing their classification, biocompatibility, biodegradability, and medical applications. Future research will continue to drive innovation, address translation challenges, and ensure the safe and effective use of biomaterials, ultimately improving patient outcomes.

REFERENCES

Abd-Elaziem, W., Darwish, M. A., Hamada, A., & Daoush, W. M. (2024). Titanium-Based alloys and composites for orthopedic implants Applications: A comprehensive review. *Materials & Design*, 112850.

Ahlström, M. G., Thyssen, J. P., Wennervaldt, M., Menné, T., & Johansen, J. D. (2019). Nickel allergy and allergic contact dermatitis: A clinical review of immunology, epidemiology, exposure, and treatment. *Contact Dermatitis*, *81*(4), 227–241. PMID: 31140194

AlMaadeed, M. A. A., Ponnamma, D., & El-Samak, A. A. (2020). Polymers to improve the world and lifestyle: physical, mechanical, and chemical needs. In *Polymer Science and Innovative Applications* (pp. 1–19). Elsevier.

Amini, A. R., Laurencin, C. T., & Nukavarapu, S. P. (2012). Bone tissue engineering: recent advances and challenges. *Critical Reviews™ in Biomedical Engineering, 40*(5).

Amini, A. R., Wallace, J. S., & Nukavarapu, S. P. (2011). Short-term and long-term effects of orthopedic biodegradable implants. *Journal of Long-Term Effects of Medical Implants*, *21*(2). PMID: 22043969

Asgari, M., Hang, R., Wang, C., Yu, Z., Li, Z., & Xiao, Y. (2018). Biodegradable metallic wires in dental and orthopedic applications: A review. *Metals*, *8*(4), 212.

Bai, Z., & Gilbert, J. L. (2006). Corrosion performance of stainless steels, cobalt, and titanium alloys in biomedical applications. In *Corrosion: Environments and Industries* (pp. 837-852). ASM International.

Banaszkiewicz, P. A. (2012). Hip clinical cases. *Postgraduate Orthopaedics: The Candidate's Guide to the FRCS (Tr and Orth) Examination*, 69.

Bandyopadhyay, A., Mitra, I., Goodman, S. B., Kumar, M., & Bose, S. (2023). Improving biocompatibility for next generation of metallic implants. *Progress in Materials Science*, *133*, 101053. PMID: 36686623

Bao, W., He, Y., Fan, Y., & Liao, Y. (2018). Metal allergy in total-joint arthroplasty: Case report and literature review. *Medicine*, *97*(38), e12475. PMID: 30235744

Barceloux, D. G., & Barceloux, D. (1999a). Molybdenum. *Journal of Toxicology. Clinical Toxicology*, *37*(2), 231–237. PMID: 10382558

Barceloux, D. G., & Barceloux, D. (1999b). Vanadium. *Journal of Toxicology. Clinical Toxicology*, *37*(2), 265–278. PMID: 10382561

Basko-Plluska, J. L., Thyssen, J. P., & Schalock, P. C. (2011). Cutaneous and systemic hypersensitivity reactions to metallic implants. *Derm*, *22*(2), 65–79. PMID: 21504692

Bellucci, D., Sola, A., & Cannillo, V. (2016). Hydroxyapatite and tricalcium phosphate composites with bioactive glass as second phase: State of the art and current applications. *Journal of Biomedical Materials Research. Part A*, *104*(4), 1030–1056. PMID: 26646669

Bhong, M., Khan, T. K., Devade, K., Krishna, B. V., Sura, S., Eftikhaar, H., Thethi, H. P., & Gupta, N. (2023). Review of composite materials and applications. *Materials Today: Proceedings*.

Bidez, M., & Misch, C. (2005). Clinical biomechanics in implant dentistry. *Misch's Contemporary. Implant Dentistry*, 140–151.

Bonorino, J. A., & Teves, J. I. (2024). TKA Aseptic Loosening. In *Orthopaedics and Trauma: Current Concepts and Best Practices* (pp. 1585-1591). Springer.

Bosse, M. J., Gruber, H. E., & Ramp, W. K. (2005). Internalization of bacteria by osteoblasts in a patient with recurrent, long-term osteomyelitis: A case report. *JBJS*, *87*(6), 1343–1347. PMID: 15930546

Brochu, B. M., Sturm, S. R., Kawase De Queiroz Goncalves, J. A., Mirsky, N. A., Sandino, A. I., Panthaki, K. Z., Panthaki, K. Z., Nayak, V. V., Daunert, S., & Witek, L. (2024). Advances in Bioceramics for Bone Regeneration: A Narrative Review. *Biomimetics*, *9*(11), 690. PMID: 39590262

Browne, J. A., Bechtold, C. D., Berry, D. J., Hanssen, A. D., & Lewallen, D. G. (2010). Failed metal-on-metal hip arthroplasties: A spectrum of clinical presentations and operative findings. *Clinical Orthopaedics and Related Research*, *468*, 2313–2320. PMID: 20559767

Busch, A., Jäger, M., Beck, S., Wegner, A., Portegys, E., Wassenaar, D., Theysohn, J., & Haubold, J. (2022). Metal Artefact Reduction Sequences (MARS) in Magnetic Resonance Imaging (MRI) after Total Hip Arthroplasty (THA) A non-invasive approach for preoperative differentiation between periprosthetic joint infection (PJI) and aseptic complications? *BMC Musculoskeletal Disorders*, *23*(1), 620. PMID: 35764987

Cai, S., Wu, C., Yang, W., Liang, W., Yu, H., & Liu, L. (2020). Recent advance in surface modification for regulating cell adhesion and behaviors. *Nanotechnology Reviews*, *9*(1), 971–989.

Cai, Y., Ding, H., Chen, X., Chen, Y., Huang, C., Zhang, C., Huang, Z., Huang, Y., Li, W., & Zhang, W. (2023). Optimization and standardization of mNGS-based procedures for the diagnosis of Mycoplasma periprosthetic joint infection: A novel diagnostic strategy for rare bacterial periprosthetic joint infection. *Frontiers in Cellular and Infection Microbiology*, *13*, 1089919. PMID: 36936762

Cai, Y., Huang, C., Chen, X., Chen, Y., Huang, Z., Zhang, C., Zhang, W., & Fang, X. (2022). The role of Staphylococcus aureus small colony variants in intraosseous invasion and colonization in periprosthetic joint infection. *Bone & Joint Research*, *11*(12), 843–853. PMID: 36453022

Capuani, S., Malgir, G., Chua, C. Y. X., & Grattoni, A. (2022). Advanced strategies to thwart foreign body response to implantable devices. *Bioengineering & Translational Medicine*, *7*(3), e10300. PMID: 36176611

Catalani, S., Stea, S., Beraudi, A., Gilberti, M., Bordini, B., Toni, A., & Apostoli, P. (2013). Vanadium release in whole blood, serum and urine of patients implanted with a titanium alloy hip prosthesis. *Clinical Toxicology*, *51*(7), 550–556. PMID: 23845123

Chen, Y., Ding, H., Wang, Q., Huang, Z., Zhang, C., Li, W., Lin, Y., Guo, Y., Fang, X., & Zhang, W. (2025). Can "LITE" Procedure Combined With a Short Course Antibiotic Treatment Be Effective in Treating the Chronic PJI?—A Prospective Randomized Controlled Trial. *Orthopaedic Surgery*.

Chen, Y., Huang, C., Chen, X., Cai, Y., Li, W., Fang, X., & Zhang, W. (2022). Bone protein analysis via label-free quantitative proteomics in patients with periprosthetic joint infection. *Journal of Proteomics, 252*, 104448. PMID: 34883267

Cheng, A. G., DeDent, A. C., Schneewind, O., & Missiakas, D. (2011). A play in four acts: Staphylococcus aureus abscess formation. *Trends in Microbiology, 19*(5), 225–232. PMID: 21353779

Choi, S.-r., Kwon, J.-w., Suk, K.-s., Kim, H.-s., Moon, S.-h., Park, S.-y., & Lee, B. H. (2023). The clinical use of osteobiologic and metallic biomaterials in orthopedic surgery: The present and the future. *Materials (Basel), 16*(10), 3633. PMID: 37241260

Dave, K., & Gomes, V. G. (2019). Interactions at scaffold interfaces: Effect of surface chemistry, structural attributes and bioaffinity. *Materials Science and Engineering C, 105*, 110078. PMID: 31546353

de Mesy Bentley, K. L., Trombetta, R., Nishitani, K., Bello-Irizarry, S. N., Ninomiya, M., Zhang, L., Chung, H. L., McGrath, J. L., Daiss, J. L., & Awad, H. A. (2017). Evidence of Staphylococcus aureus deformation, proliferation, and migration in canaliculi of live cortical bone in murine models of osteomyelitis. In: John Wiley and Sons and The American Society for Bone and Mineral Research….

De Smet, K. (2013). Clinical follow-up of the hip resurfacing patient. In *The Hip Resurfacing Handbook* (pp. 113–127). Elsevier.

Dee, K. C., Puleo, D. A., & Bizios, R. (2003). *An introduction to tissue-biomaterial interactions*. John Wiley & Sons.

Eldeeb, A. E., Salah, S., & Elkasabgy, N. A. (2022). Biomaterials for tissue engineering applications and current updates in the field: A comprehensive review. *AAPS PharmSciTech, 23*(7), 267. PMID: 36163568

Eltit, F., Wang, Q., & Wang, R. (2019). Mechanisms of adverse local tissue reactions to hip implants. *Frontiers in Bioengineering and Biotechnology, 7*, 176. PMID: 31417898

Fabbri, P., & Messori, M. (2017). Surface modification of polymers: chemical, physical, and biological routes. In *Modification of polymer properties* (pp. 109–130). Elsevier.

Fang, X., Sun, D., Li, Y., Han, X., Gan, Y., Jiao, J., Jiang, M., Gong, H., Qi, Y., & Zhao, J. (2024). Macrophages in the process of osseointegration around the implant and their regulatory strategies. *Connective Tissue Research, 65*(1), 1–15. PMID: 38166507

Fang, X., Wang, Q., Yang, X., Zhang, F., Huang, C., Huang, Z., Shen, H., & Zhang, W. (2021). What is the appropriate extended duration of antibiotic prophylaxis after two-stage revision for chronic PJI? *Bone & Joint Research, 10*(12), 790–796. PMID: 34894718

Fang, Y., & Shang, X. (2021). Comparison of Ceramic-on-Ceramic vs. ceramic-on-polyethylene for Primary total hip arthroplasty: A Meta-Analysis of 15 Randomized Trials. *Frontiers in Surgery, 8*, 751121. PMID: 34977138

Feng, X., Gu, J., & Zhou, Y. (2022). Primary total hip arthroplasty failure: Aseptic loosening remains the most common cause of revision. *American Journal of Translational Research, 14*(10), 7080. PMID: 36398241

Gao, G., & Fu, J. (2019). Highly crosslinked UHMWPE for joint implants. *UHMWPE Biomaterials for Joint Implants: Structures, Properties and Clinical Performance*, 21-68.

Geetha, M., Singh, A. K., Asokamani, R., & Gogia, A. K. (2009). Ti based biomaterials, the ultimate choice for orthopaedic implants–a review. *Progress in Materials Science*, *54*(3), 397–425.

Giuseppe, M., Mattia, B., Nadia, B., Raffaele, V., Pasquale, R., Stefano, D. A., Mattia, S., Vincenzo, D. S., & Giulio, M. (2021). Ceramic-on-ceramic versus ceramic-on-polyethylene in total hip arthroplasty: A comparative study at a minimum of 13 years follow-up. *BMC Musculoskeletal Disorders*, *22*(Suppl 2), 1062. PMID: 35039021

Granchi, D., Savarino, L. M., Ciapetti, G., & Baldini, N. (2018). Biological effects of metal degradation in hip arthroplasties. *Critical Reviews in Toxicology*, *48*(2), 170–193. PMID: 29130357

Grimm, M. J. (2004). *Orthopedic biomaterials. Standard handbook for biomedical engineering and design* (2nd ed.). McGraw-Hill.

Haddad, F., Konan, S., & Tahmassebi, J. (2015). A prospective comparative study of cementless total hip arthroplasty and hip resurfacing in patients under the age of 55 years: A ten-year follow-up. *The Bone & Joint Journal*, *97*(5), 617–622. PMID: 25922454

Hakim, L. K., Yari, A., Nikparto, N., Mehraban, S. H., Cheperli, S., Asadi, A., Darehdor, A. A., Nezaminia, S., Dortaj, D., & Nazari, Y. (2024). The current applications of nano and biomaterials in drug delivery of dental implant. *BMC Oral Health*, *24*(1), 126. PMID: 38267933

Hallab, N. J., Mikecz, K., & Jacobs, J. J. (2000). A triple assay technique for the evaluation of metal-induced, delayed-type hypersensitivity responses in patients with or receiving total joint arthroplasty. *Journal of Biomedical Materials Research: An Official Journal of The Society for Biomaterials, The Japanese Society for Biomaterials, and The Australian Society for Biomaterials and the Korean Society for Biomaterials*, *53*(5), 480–489. PMID: 10984695

Hasegawa, M., Saruta, J., Hirota, M., Taniyama, T., Sugita, Y., Kubo, K., Ishijima, M., Ikeda, T., Maeda, H., & Ogawa, T. (2020). A newly created meso-, micro-, and nano-scale rough titanium surface promotes bone-implant integration. *International Journal of Molecular Sciences*, *21*(3), 783. PMID: 31991761

Heimann, R. B. (2020). *Biomaterials–characteristics, history, applications. Materials for Medical Applications* (1st ed.). Walter de Gruyter.

Hodges, N. A., Sussman, E. M., & Stegemann, J. P. (2021). Aseptic and septic prosthetic joint loosening: Impact of biomaterial wear on immune cell function, inflammation, and infection. *Biomaterials*, *278*, 121127. PMID: 34564034

Hong, F., Yu, X., Wu, N., & Zhang, Y.-Q. (2017). Progress of in vivo studies on the systemic toxicities induced by titanium dioxide nanoparticles. *Toxicology Research*, *6*(2), 115–133. PMID: 30090482

Hou, C., An, J., Zhao, D., Ma, X., Zhang, W., Zhao, W., Wu, M., Zhang, Z., & Yuan, F. (2022). Surface modification techniques to produce micro/nano-scale topographies on Ti-based implant surfaces for improved osseointegration. *Frontiers in Bioengineering and Biotechnology*, *10*, 835008. PMID: 35402405

Hu, H., Ding, H., Lyu, J., Chen, Y., Huang, C., Zhang, C., Li, W., Fang, X., & Zhang, W. (2024). Detection of rare microorganisms in bone and joint infections by metagenomic next-generation sequencing: Discrimination, decision-making, and clinical benefits. *Bone & Joint Research, 13*(8), 401. PMID: 39142657

Huang, C., Huang, Y., Wang, Z., Lin, Y., Li, Y., Chen, Y., Chen, X., Zhang, C., Li, W., & Zhang, W. (2023). Multiplex PCR-based next generation sequencing as a novel, targeted and accurate molecular approach for periprosthetic joint infection diagnosis. *Frontiers in Microbiology, 14*, 1181348. PMID: 37275128

Jacqueline, C., & Caillon, J. (2014). Impact of bacterial biofilm on the treatment of prosthetic joint infections. *The Journal of Antimicrobial Chemotherapy, 69*(suppl_1), i37–i40. PMID: 25135088

Kamal, A. S., Safaa, M. M., & Elazab, M. S. (2024). Comparison between Obese and Non-obese Cases after Total Knee Arthroplasty Regarding Their Short Term Functional Outcome. *Fayoum University Medical Journal, 14*(1), 122–130.

Kaya, A. (2011). *Physical and electrochemical characterization of co-deposited TiO2 and MgO layer on Ti6Al4V by micro arc oxidation (MAO)*. Dokuz Eylul Universitesi.

Kenanidis, E., Kakoulidis, P., Tsiridis, E., Atilla, B., Bicanic, G., Sulje, Z., Delimar, D., Bondarenko, S., Filippenko, V., & Akonjom, M. (2017). Acetabular dysplasia. In *The Adult Hip-Master Case Series and Techniques* (pp. 107–213). Springer.

Kolenda, C., Josse, J., Medina, M., Fevre, C., Lustig, S., Ferry, T., & Laurent, F. (2020). Evaluation of the activity of a combination of three bacteriophages alone or in association with antibiotics on Staphylococcus aureus embedded in biofilm or internalized in osteoblasts. *Antimicrobial Agents and Chemotherapy, 64*(3), 02231–02219. DOI: 10.1128/aac PMID: 31871084

Krauss, J., Roper, P., Ballard, A., Shih, C., Fitzpatrick, J., Cassat, J., Ng, P., Pavlos, N., & Veis, D. (2019). Staphylococcus aureus infects osteoclasts and replicates intracellularly. [In.]. *mBio, 10*, e02447–e19. PMID: 31615966

Lee, A., Paiement, G. D., Penenberg, B. L., & Rajaee, S. S. (2023). Metallosis in Total Hip Arthroplasty. *JBJS Reviews, 11*(10), e23. PMID: 37812668

Lehtovirta, L. (2020). Adverse Reaction to Metal Debris in Patients with Metal-on-Metal Hip Replacements: Etiology & Pathogenesis.

Lindsay, J. A., & Holden, M. T. (2004). Staphylococcus aureus: Superbug, super genome? *Trends in Microbiology, 12*(8), 378–385. PMID: 15276614

Lionberger, D. R., Samorajski, J., Wilson, C. D., & Rivera, A. (2018). What role does metal allergy sensitization play in total knee arthroplasty revision? *Journal of Experimental Orthopaedics, 5*, 1–8. PMID: 30109436

Liu, Y., & Chen, B. (2018). In vivo corrosion of CoCrMo alloy and biological responses: A review. *Materials Technology, 33*(2), 127–134.

Lotfi, M., Nejib, M., & Naceur, M. (2013). Cell adhesion to biomaterials: Concept of biocompatibility. *Advances in biomaterials science and biomedical applications, 8*, 208-240.

Lyu, J., Huang, J., Huang, J., Hu, H., Wang, Q., Ding, H., Li, H., Fang, X., & Zhang, W. (2024). Rising challenges in periprosthetic joint infections: A focus on rare pathogens and their clinical implications. *Frontiers in Cellular and Infection Microbiology*, *14*, 1451398. PMID: 39698321

Macuvele, D. L. P., Nones, J., Matsinhe, J. V., Lima, M. M., Soares, C., Fiori, M. A., & Riella, H. G. (2017). Advances in ultra high molecular weight polyethylene/hydroxyapatite composites for biomedical applications: A brief review. *Materials Science and Engineering C*, *76*, 1248–1262. PMID: 28482493

Mahajan, A., Devgan, S., & Kalyanasundaram, D. (2023). Surface alteration of Cobalt-Chromium and duplex stainless steel alloys for biomedical applications: A concise review. *Materials and Manufacturing Processes*, *38*(3), 260–270.

Majeed, T. (2023). Development of Biocompatible Materials for Medical Implants: A Journey Towards Seamless Integration. *The R Journal*, *1*(01), 35–42.

Makadia, H. K., & Siegel, S. J. (2011). Poly lactic-co-glycolic acid (PLGA) as biodegradable controlled drug delivery carrier. *Polymers*, *3*(3), 1377–1397. PMID: 22577513

Malahias, M.-A., Bauer, T. W., Gu, A., Nocon, A. A., De Martino, I., & Sculco, P. K. (2020). Questioning the clinical relevance of positive post-implantation allergy testing for metal hypersensitivity in total knee arthroplasty: A didactic case report. *HSS Journal : the Musculoskeletal Journal of Hospital for Special Surgery*, *16*(2, suppl), 485–489. PMID: 33380984

Matar, H. E., Porter, P. J., & Porter, M. L. (2021). Metal allergy in primary and revision total knee arthroplasty: A scoping review and evidence-based practical approach. *Bone & Joint Open*, *2*(10), 785–795. PMID: 34587776

Mukherjee, B., Patra, B., Mahapatra, S., Banerjee, P., Tiwari, A., & Chatterjee, M. (2004). Vanadium—An element of atypical biological significance. *Toxicology Letters*, *150*(2), 135–143. PMID: 15093669

Nam, D., Li, K., Riegler, V., & Barrack, R. L. (2016). Patient-reported metal allergy: A risk factor for poor outcomes after total joint arthroplasty? *The Journal of Arthroplasty*, *31*(9), 1910–1915. PMID: 26965589

Nasr Azadani, M., Zahedi, A., Bowoto, O. K., & Oladapo, B. I. (2022). A review of current challenges and prospects of magnesium and its alloy for bone implant applications. *Progress in Biomaterials*, *11*(1), 1–26. PMID: 35239157

Neto, J. V. C., Teixeira, A. B. V., & Dos Reis, A. C. (2023). Hydroxyapatite coatings versus osseointegration in dental implants: A systematic review. *The Journal of Prosthetic Dentistry*. PMID: 37845113

Niranjan, C., Raghavendra, T., Rao, M. P., Siddaraju, C., Gupta, M., Jain, V. K. S., & Aishwarya, R. (2023). Magnesium alloys as extremely promising alternatives for temporary orthopedic implants–A review. *Journal of Magnesium and Alloys*.

Nortje, B. D. (2021). *Feasibility of additive manufacturing for patient-specific knee replacements*

Oh, T. J., Yoon, J., Misch, C. E., & Wang, H. L. (2002). The causes of early implant bone loss: Myth or science? *Journal of Periodontology*, *73*(3), 322–333. PMID: 11922263

Oliveira, A., Simões, S., Ascenso, A., & Reis, C. P. (2022). Therapeutic advances in wound healing. *The Journal of Dermatological Treatment*, *33*(1), 2–22. PMID: 32056472

Pacheco, K. A. (2019). Allergy to surgical implants. *Clinical Reviews in Allergy & Immunology*, *56*(1), 72–85. PMID: 30220068

Poon, A. W., Goodman, C. S., & Rubin, R. J. (1998). In vitro and skin testing for allergy: Comparable clinical utility and costs. *The American Journal of Managed Care*, *4*(7), 969–985. PMID: 10181996

Rehder, D. (2013). Vanadium. Its role for humans. *Interrelations between essential metal ions and human diseases*, 139-169.

Rey-Vinolas, S., Engel, E., & Mateos-Timoneda, M. (2019). Polymers for bone repair. In *Bone Repair Biomaterials* (pp. 179–197). Elsevier.

Rosa, V., Silikas, N., Yu, B., Dubey, N., Gopu, S., Zinelis, S., Lima, A. F., Bottino, M. C., Ferreira, J. N., & Schmalz, G. (2024). Guidance on the assessment of biocompatibility of biomaterials: Fundamentals and testing considerations. *Dental Materials*. PMID: 39129079

Rosner, G. A., & Fonacier, L. S. (2017). Hypersensitivity to biomedical implants: Prevention and diagnosis. *Allergy and Asthma Proceedings*.

Schalock, P. C., Menné, T., Johansen, J. D., Taylor, J. S., Maibach, H. I., Lidén, C., Bruze, M., & Thyssen, J. P. (2012). Hypersensitivity reactions to metallic implants–diagnostic algorithm and suggested patch test series for clinical use. *Contact Dermatitis*, *66*(1), 4–19.

Scholz, M.-S., Blanchfield, J. P., Bloom, L., Coburn, B. H., Elkington, M., Fuller, J. D., Gilbert, M. E., Muflahi, S. A., Pernice, M., & Rae, S. I. (2011). The use of composite materials in modern orthopaedic medicine and prosthetic devices: A review. *Composites Science and Technology*, *71*(16), 1791–1803.

Semlitsch, M. (1974). Technical progress in artificial hip joints. *Engineering in Medicine*, *3*(4), 10–19.

Sheikhpour, M., Barani, L., & Kasaeian, A. (2017). Biomimetics in drug delivery systems: A critical review. *Journal of Controlled Release*, *253*, 97–109. PMID: 28322976

Shi, H., Magaye, R., Castranova, V., & Zhao, J. (2013). Titanium dioxide nanoparticles: A review of current toxicological data. *Particle and Fibre Toxicology*, *10*, 1–33. PMID: 23587290

Singh, D., Tirkey, R., Jangde, R., Sahu, P. K., & Bagchi, A. (2024). *Introduction to Biomaterial-Inspired.* Biomaterial-Inspired Nanomedicines for Targeted Therapies.

Škrlová, K., Malachová, K., Muñoz-Bonilla, A., Měřinská, D., Rybková, Z., Fernández-García, M., & Plachá, D. (2019). Biocompatible polymer materials with antimicrobial properties for preparation of stents. *Nanomaterials (Basel, Switzerland)*, *9*(11), 1548. PMID: 31683612

Smith, T., Edmonds, C., & Barnaby, C. (1972). Absorption and retention of cobalt in man by whole-body counting. *Health Physics*, *22*(4), 359–367. PMID: 4625611

Song, Z., Cheng, Y., Chen, M., & Xie, X. (2023). Macrophage polarization in bone implant repair: A review. *Tissue & Cell*, *82*, 102112. PMID: 37257287

Spiewak, R. (2018). Assessment for metal allergy: patch testing. *Metal Allergy: From Dermatitis to Implant and Device Failure*, 107-124.

Strauss, A., Rommelspacher, Y., Nouri, B., Bornemann, R., Wimmer, M., Oldenburg, J., Pennekamp, P., & Schmolders, J. (2017). Long-term outcome of total hip arthroplasty in patients with haemophilia. *Haemophilia, 23*(1), 129–134. PMID: 27480604

Suhag, D., Kaushik, S., & Taxak, V. B. (2024). Scaffold Design in Tissue Engineering. In *Handbook of Biomaterials for Medical Applications, Volume 1: Fundamentals* (pp. 155-184). Springer.

Sundfeldt, M., v Carlsson, L. B., Johansson, C., Thomsen, P., & Gretzer, C. (2006). Aseptic loosening, not only a question of wear: A review of different theories. *Acta Orthopaedica, 77*(2), 177–197. PMID: 16752278

Taraballi, F., Sushnitha, M., Tsao, C., Bauza, G., Liverani, C., Shi, A., & Tasciotti, E. (2018). Biomimetic tissue engineering: Tuning the immune and inflammatory response to implantable biomaterials. *Advanced Healthcare Materials, 7*(17), 1800490. PMID: 29995315

Tria, A. J.Jr. (2013). Bicompartmental knee arthroplasty. *Volume 44, Issue 3. An Issue of Orthopedic Clinics, 44*(3), 17.

Tripathi, D., Mani, V., & Pal, R. P. (2018). Vanadium in biosphere and its role in biological processes. *Biological Trace Element Research, 186*, 52–67. PMID: 29524196

Vaiani, L., Boccaccio, A., Uva, A. E., Palumbo, G., Piccininni, A., Guglielmi, P., Cantore, S., Santacroce, L., Charitos, I. A., & Ballini, A. (2023). Ceramic materials for biomedical applications: An overview on properties and fabrication processes. *Journal of Functional Biomaterials, 14*(3), 146. PMID: 36976070

Von Blomberg-van der Flier, B., Bruynzeel, D., & Scheper, R. (1989). Impact of 25 years of in vitro testing in allergic contact dermatitis. Current topics in contact dermatitis

Wang, Q., Chen, Y., Chen, Y., Lv, J., Ding, H., Huang, J., Huang, J., Huang, Z., Yang, B., & Zhang, W. (2024). Improved cure rate of periprosthetic joint infection through targeted antibiotic therapy based on integrated pathogen diagnosis strategy. *Frontiers in Cellular and Infection Microbiology, 14*, 1388385.

Williams, D. F. (2008). On the mechanisms of biocompatibility. *Biomaterials, 29*(20), 2941–2953. PMID: 18440630

Wu, B., Su, J., Zhang, Z., Zeng, J., Fang, X., Li, W., Zhang, W., & Huang, Z. (2024). Prosthetic spacers in two-stage revision for knee periprosthetic joint infection achieve better function and similar infection control. *Bone & Joint Research, 13*(6), 306. PMID: 38889904

Zaman, H. A., Sharif, S., Idris, M. H., & Kamarudin, A. (2015). Metallic biomaterials for medical implant applications: A review. *Applied Mechanics and Materials, 735*, 19–25.

Zhang, C. f., Fang, X. y., Huang, Z. d., Bai, G. c., Zhang, Z. y., Yang, Y., Zhang, Z. j., Li, W. b., & Zhang, W. m. (2022). Surgical Management for Chronic Destructive Septic Hip Arthritis: Debridement, Antibiotics, and Single-Stage Replacement is as Effective as Two-Stage Arthroplasty. *Orthopaedic Surgery, 14*(6), 1175–1185. PMID: 35583090

Zhang, C., Zhang, Z., Lin, Y., Xin, Y., Li, H., Lin, Y., Huang, Y., Yang, Y., Bai, G., & Huang, Z. (2024). A Novel Infrapatellar Fat Pad Preservation Technique in Total Knee Arthroplasty Reduced Postoperative Pain and Wound Complications. *Orthopaedic Surgery*, *16*(8), 1946–1954. PMID: 38894546

Zhang, D., Chen, Q., Shi, C., Chen, M., Ma, K., Wan, J., & Liu, R. (2021). Dealing with the foreign-body response to implanted biomaterials: Strategies and applications of new materials. *Advanced Functional Materials*, *31*(6), 2007226.

Zhang, Z., Huang, Y., Li, M., Li, W., Fang, X., Huang, Z., & Zhang, W. (2023). Rapid and quantitative detection of neutrophil gelatinase-associated lipocalin in synovial fluid using fluorescence immunochromatographic test strips for diagnosing prosthetic joint infections. *Diagnostic Microbiology and Infectious Disease*, *106*(4), 115929. PMID: 37244008

Chapter 15
Toxicity of Metal and Metal Oxide Nanoparticles:
Impact of Physical State, Mechanisms of Action, and Strategies for Remediation in Biomedical Applications

Elkhatim Hassan Abdelgadir
Naif Arab University of Security Sciences, Saudi Arabia

Fauzia Ashfaq
College of Nursing and Health Sciences, Saudi Arabia

Sakshi Gupta
Amity Institute of Forensic Sciences, India

Nimish Dhakad
https://orcid.org/0009-0004-6987-0335
Amity Institute of Forensic Sciences, India

Sachil Kumar
https://orcid.org/0000-0002-2681-3937
Amity University, Greater Noida, India

Shipra Rohatgi
https://orcid.org/0000-0003-0532-9908
Amity University, India

ABSTRACT

Metal and metal oxide nanoparticles (NPs) show great promise in biomedical applications, but their toxicity varies. Metallic NPs are stable, while metal oxide and alloy NPs dissolve in biological conditions, releasing metal ions that generate reactive oxygen species (ROS), causing oxidative stress and cellular damage. Their rapid distribution and accumulation in tissues raise toxicity concerns. This chapter reviews their toxicological effects, mechanisms, and strategies for risk mitigation, including surface modifications and remediation.

1. INTRODUCTION

With major contributions to biological and pharmaceutical applications, nanotechnology has emerged as a crucial element of many cutting-edge scientific and technical fields. Because of their special physicochemical qualities, such as chemical reactivity, optical traits, and structural plasticity, metal and metal oxide nanoparticles (NPs) have become potential instruments for applications such as drug administration,

DOI: 10.4018/979-8-3373-0055-9.ch015

diagnostics, and sensor technologies. NPs such as copper oxide, iron oxide, zinc oxide, titanium dioxide, gold, and silver are frequently employed. In therapeutic contexts, these nanoparticles offer improved imaging capabilities and new pathways for targeted medication delivery.

Despite their potential, nanoparticles' small size, wide surface area, and special surface characteristics frequently cause them to interact significantly with biological systems, which raises questions regarding their toxicity. For example, oxidative stress, reactive oxygen species (ROS), and consequent cellular damage might result from the dissolution and ion release of metal oxide nanoparticles. Additionally, nanoparticles have the ability to infiltrate the bloodstream, build up in tissues, and disrupt biological functions, all of which may have a negative impact on health. These materials' size, shape, charge, crystallinity, and stability in biological contexts all have a significant impact on their toxicity profiles (Manuja et al., 2021). Understanding how these nanomaterials interact with biological systems and resolving the related toxicity issues are crucial given their growing production and use. The objective of this research is to present a thorough analysis of the physicochemical characteristics that affect the toxicity of nanoparticles, clarify the underlying molecular pathways, and suggest methods for reducing negative effects.

1.1. Overview of Metal and Metal Oxide Nanoparticles

The remarkable chemical, optical, magnetic, and structural characteristics of metal and metal oxide nanoparticles set them apart from their bulk counterparts. Gold (Au), silver (Ag), iron oxide (FeO_4), zinc oxide (ZnO), and titanium dioxide (TiO_2) nanoparticles are a few examples. To accomplish particular functionality, these materials can be synthesised in a variety of shapes, including sheets, rods, and spheres, with customised surface chemistries. Notwithstanding their adaptability, the way nanoparticles behave in biological settings—including how they dissolve, release ions, and aggregate—has a big impact on how reactive and stable they are. Potential cytotoxicity and genotoxicity are raised by the oxidative stress and reactive oxygen species (ROS) that can arise from these interactions (Sengul & Asmatulu, 2020).

1.2. Biomedical Applications of Metal Nanoparticles

The potential of metal nanoparticles to improve therapeutic efficacy and diagnostic accuracy has led to a significant increase in their application in biomedicine. Applications include enhanced imaging methods like magnetic resonance imaging (MRI) and fluorescence imaging, which use nanoparticles as contrast agents, and targeted drug delivery, in which they act as carriers for medicinal medicines (Attarilar et al., 2020). Furthermore, by integrating detection and treatment onto a single platform, metal nanoparticles have shown potential in cancer theranostics. For instance, silver nanoparticles have potent antibacterial qualities, while gold nanoparticles are frequently employed in photothermal therapy. Despite these developments, any toxicity concerns must be addressed before they can be safely incorporated into clinical practice.

1.3. Importance of Understanding Toxicity Profiles

Knowing the toxicity characteristics of nanoparticles is essential as they are used in more biomedical applications. The size, shape, surface charge, and solubility of nanoparticles all affect how they interact with cells and tissues, which frequently results in toxic effects. The degree of systemic exposure and organ accumulation can also be influenced by the administration method, such as intravenous injection, ingestion, or inhalation. According to studies, ion release and ROS generation from specific nanoparticles, like ZnO and CuO, may result in oxidative stress, mitochondrial malfunction, and inflammation. The advancement of nanoparticle-based technology depends on the development of mitigation techniques, such as surface functionalisation or encapsulation with biocompatible polymers.

2. PHYSICOCHEMICAL PROPERTIES OF METAL AND METAL OXIDE NANOPARTICLES

Metal and metal oxide nanoparticles' (NPs') distinct physicochemical characteristics determine how they interact with biological systems, impacting both their toxicity profiles and biomedical uses. Their behaviour and possible biological consequences are greatly influenced by important characteristics such size, surface area, physical condition, surface chemistry, and stability.

2.1. Size and Surface Area Effects

Biological interactions, absorption, and toxicity of nanoparticles are significantly influenced by their size. Because of their larger surface area-to-volume ratio, smaller nanoparticles are more reactive and interact better with biological components. Larger particles (25–50 nm) are usually internalised by phagocytosis or receptor-mediated endocytosis, although nanoparticles smaller than 5 nm might enter cells through non-specific processes. The increased surface area that results from decreasing particle size can improve the particles' ability to aggregate and interact with lipids, proteins, and other biological components. A protein "corona" that alters their biodistribution and cellular absorption may result from these interactions. Furthermore, oxidative stress, inflammation, and disruption of vital cellular functions are some of the ways that smaller nanoparticles may cause toxicity through their increased propensity to aggregate in tissues.

2.2. Physical State: Metallic vs. Metal Oxide Nanoparticles

The stability, solubility, and toxicity of nanoparticles are strongly influenced by their physical condition, whether they are metallic or metal oxide. In biological environments, metallic nanoparticles—like gold and silver—tend to be relatively stable and show little disintegration. They are appropriate for drug delivery and imaging applications due to their stability.

On the other hand, metal oxide nanoparticles—such as copper oxide (CuO) and zinc oxide (ZnO)—are more likely to dissolve and release ions in biological environments. Reactive oxygen species (ROS) and oxidative stress, two major causes of cellular damage, may result from this. Metal oxide nanoparticles' stability and reactivity are further influenced by their physical and chemical characteristics, such as their surface modifications and crystallinity, which makes them both adaptable and potentially dangerous in biomedical settings.

Table 1. Metallic Nanoparticles V/s Metal Oxide Nanoparticles

Aspect	Metallic Nanoparticles	Metal Oxide Nanoparticles
Composition	Pure metals like gold, silver, platinum, etc.	Metal compounds with oxygen (e.g., ZnO, TiO2).
Structure	Metallic bonding, crystalline structure.	Ionic or mixed ionic-covalent bonds.
Optical Properties	Exhibit localized surface plasmon resonance (LSPR).	Absorb/emit light, used in photocatalysis.
Electrical Conductivity	High electrical conductivity.	Typically, semiconductors.
Stability	Prone to oxidation, requires stabilization.	Chemically stable due to oxide layer.
Reactivity	High surface reactivity, influenced by environment.	Reactive in redox processes, photocatalytic.
Biocompatibility	Variable, depending on the metal and coating.	Often higher biocompatibility (e.g., ZnO).
Applications	Catalysis, sensors, electronics, drug delivery.	Photocatalysis, antimicrobial agents, coatings.
Toxicity Concerns	High toxicity due to ion release (e.g., silver).	Toxicity influenced by particle size and dose.

2.3 Surface Chemistry and Functionalization

The way that nanoparticles interact with biological systems and their capacity for functionalisation are determined by their surface chemistry. Chemical alterations, hydrophilicity/hydrophobicity, and surface charge are important factors in toxicity, biodistribution, and cellular absorption. Because of their affinity for negatively charged cell membranes, positively charged nanoparticles frequently show better cellular absorption; however, they may also be more harmful due to lysosomal disruption and ROS generation (Ibrahim et al., 2019). By functionalising the surfaces of nanoparticles with biocompatible coatings like chitosan or polyethylene glycol (PEG), toxicity can be decreased and stability increased. In addition to improving targeting capabilities and preventing aggregation, surface changes can increase the efficacy of nanoparticles in medicinal and diagnostic applications. On the other hand, poor surface functionalisation might have negative consequences, such as the creation of protein corona, which can change the behaviour of nanoparticles in unexpected ways (Jia et al., 2005; Brand et al., 2017).

2.4 Stability and Dissolution Behaviour

The biological effects of nanoparticles are largely determined by their stability and dissolution characteristics. In acidic settings, metal oxide nanoparticles, such ZnO and CuO, dissolve significantly, releasing metal ions that can harm cellular structures and induce oxidative stress. On the other hand, metallic nanoparticles—like those of gold and silver—are safer for some applications because they are more stable and dissolve less readily. Other variables that affect stability include temperature, pH, and the presence of living molecules. When it comes to toxicity and therapeutic efficacy, nanoparticles that assemble or disintegrate randomly in biological settings might present serious problems. Techniques like encapsulation and surface functionalisation can increase the stability of nanoparticles and lessen their negative effects, guaranteeing their safe use in biomedical sectors.

3. MECHANISMS OF TOXICITY IN METAL AND METAL OXIDE NANOPARTICLES

Metal and metal oxide nanoparticles (NPs) have received immense interest in biomedical applications because of their intriguing properties, including high surface-to-volume ratio, reactivity tunability, and size-dependent biological interaction. These features provide them with excellent opportunities in drug delivery, bioimaging, antimicrobial functions, and cancer treatment. Nevertheless, their enhanced reactivity and susceptibility to interacting with biological molecules also trigger anxiety regarding their potential toxicity. The toxicity mechanisms of metal and metal oxide nanoparticles include a mixture of oxidative stress, ion release, protein interactions, inflammation, immune response, and genotoxicity. It is essential to understand these mechanisms in order to design safer nanoparticles with minimal side effects while retaining their therapeutic potential. One of the main mechanisms of nanoparticle toxicity is oxidative stress due to the generation of excess reactive oxygen species (ROS). ROS are free radicals such as superoxide anions, hydroxyl radicals, and hydrogen peroxide that have the potential to inflict substantial damage to cell constituents. Metal and metal oxide nanoparticles, especially the redox-active particles like titanium dioxide (TiO_2), zinc oxide (ZnO), and iron oxide (Fe_2O_3), are able to catalyze electron transfer reactions that result in the ongoing formation of ROS. When ROS levels are higher than cellular antioxidant defense mechanisms, oxidative damage ensues, impairing lipids, proteins, and nucleic acids. Lipid peroxidation destabilizes cell membrane integrity, while protein oxidation modifies enzyme function and interrupts cellular processes. DNA damage caused by oxidative stress may result in mutations, apoptosis, or carcinogenesis. The contribution of oxidative stress to nanoparticle toxicity has been widely investigated, and the necessity for surface modifications and biocompatible coatings has been emphasized as a means of preventing ROS production. Another key contributor to nanoparticle toxicity is dissolution and metal ion release. Certain metal nanoparticles, including silver (AgNPs), zinc oxide (ZnO), and copper oxide (CuO), slowly dissolve in biological fluids, releasing harmful metal ions that interfere with cellular homeostasis. These ions can disrupt crucial biological processes by binding to cellular proteins, enzymes, and DNA, resulting in cytotoxic effects. For example, silver ions (Ag^+) released from AgNPs possess potent antimicrobial activity but also lead to mitochondrial dysfunction and block ATP production in mammalian cells. Likewise, zinc ions (Zn^{2+}) from ZnO nanoparticles can disrupt calcium signaling pathways that result in apoptosis and inflammation. The release rate of ions is influenced by the size of the nanoparticle, surface area, pH of the environment surrounding it, and the presence of biomolecules that aid dissolution. Methods of controlling dissolution rates, including surface passivation and encapsulation, are being developed to minimize the toxic effects of metal ion release. When nanoparticles enter biological systems, they quickly become covered with a protein and other biomolecule "protein corona." The structure and composition of this corona determine nanoparticle interactions with cells, impacting their uptake, distribution, and toxicity. Although the protein corona can decrease nanoparticle-cell interactions and could decrease toxicity, it also has the ability to change protein structure and function and generate unforeseen biological consequences. There are some nanoparticles that induce protein unfolding or aggregation, potentially contributing to conditions like amyloidosis. Even the presence of certain proteins in the corona will enhance immune cell recognition and could enhance nanoparticle clearance or induce immune activation. Elucidating protein corona dynamics is important for nanoparticle behavior prediction in biological systems and nanoparticle design with enhanced biocompatibility (Brand et al., 2017).

3.1. Ion Release and Dissolution in Biological Environments

ZnO and CuO are examples of metal oxide nanoparticles that dissolve significantly in biological settings, releasing metal ions that add to their toxicity. Normal cellular functions may be disrupted as a result of these ions' interactions with cellular constituents. For example, lysosomes' acidic pH speeds up the disintegration of nanoparticles, releasing hazardous metal ions that can build up close to the nucleus and disrupt the functioning of proteins and DNA. The size, content, and surface chemistry of nanoparticles all affect how they dissolve. For instance, ZnO nanoparticles emit Zn2+ ions, which have been linked to harmful consequences like altered enzyme function and decreased cell viability. This method emphasises how crucial it is to regulate the stability of nanoparticles in order to lessen their harmful consequences.

3.2. Reactive Oxygen Species (ROS) Generation and Oxidative Stress

One of the primary mechanisms of nanoparticle toxicity is the generation of reactive oxygen species (ROS). Metal oxide nanoparticles, such as TiO_2 and ZnO, are particularly prone to ROS production due to their interaction with cellular components. ROS can cause oxidative stress, damaging lipids, proteins, and DNA, and leading to mitochondrial dysfunction and apoptosis. ROS generation may result from the intrinsic properties of nanoparticles, such as their ability to catalyze redox reactions or disrupt electron transport chains. For example, the oxidative stress induced by CuO nanoparticles has been shown to deplete cellular antioxidants, such as glutathione, exacerbating their cytotoxic effects.

3.3. Cellular Uptake and Distribution

There are several ways that nanoparticles might enter cells, including passive diffusion, phagocytosis, and endocytosis. The nanoparticles' size, shape, and surface charge all affect the mechanism of uptake. For example, larger particles are internalised by phagocytosis or receptor-mediated endocytosis, but smaller particles are more likely to cross the cell membrane through non-specific translocation. After entering the cell, nanoparticles may gather in particular organelles such the nucleus, mitochondria, and lysosomes. Cell death, ROS generation, and organelle failure can result from this intracellular distribution's disruption of regular cellular processes. Nanoparticles' harmful effects are made worse by their capacity to avoid clearance processes such lysosomal breakdown.

3.4. DNA Damage and Genotoxicity

Genotoxic effects can result from direct or indirect DNA damage caused by nanoparticles. When nanoparticles or the ions they release attach to DNA, direct interactions take place, leading to mutations, cross-linking, or strand breakage. The generation of ROS can oxidise DNA bases, disrupt replication, and trigger DNA repair pathways, all of which can result in indirect genotoxicity. For example, it has been demonstrated that ZnO nanoparticles cause oxidative DNA damage and induce death in human liver cells. Likewise, it has been shown that silver nanoparticles can disrupt the cell cycle, resulting in chromosomal abnormalities and the development of micronuclei. Because these genotoxic effects may have long-term consequences, such as carcinogenesis, they are especially concerning.

3.5. Inflammatory and Immune Responses

Exposure to nanoparticles can activate the immune system, leading to inflammatory responses. This is often mediated by the production of pro-inflammatory cytokines, such as interleukins (IL-1β, IL-6) and tumour necrosis factor-alpha (TNF-α). The activation of macrophages and neutrophils further amplifies the inflammatory response, which can result in tissue damage.

For example, the surface charge and aggregation of nanoparticles can influence their interaction with immune cells. Positively charged nanoparticles are more likely to interact with negatively charged cell membranes, triggering immune activation. Chronic inflammation induced by prolonged exposure to nanoparticles may contribute to the development of diseases, such as fibrosis and cancer.

4. FACTORS INFLUENCING NANOTOXICITY

Various physicochemical and environmental factors determine the toxic effects of metal and metal oxide nanoparticles (NPs). These factors influence how nanoparticles interact with biological systems, their stability, and their potential to cause cellular damage. Understanding these variables is critical for designing safer nanoparticles for biomedical applications.

4.1. Size, Shape, and Surface Modifications

The size and shape of nanoparticles play a central role in determining their interaction with biological systems. Smaller nanoparticles have a higher surface area-to-volume ratio, which enhances their reactivity and cellular uptake. For instance, gold nanoparticles of 6 nm have been shown to penetrate the cell nucleus more effectively than larger particles. However, smaller nanoparticles are also more prone to aggregation, which can exacerbate their toxicity. The shape of nanoparticles also affects their uptake and distribution. Spherical nanoparticles are phagocytosed more easily, while high-aspect-ratio structures, such as rods or fibers, may evade clearance mechanisms and persist longer in tissues. Surface modifications, including functionalization with biocompatible molecules like polyethylene glycol (PEG), can reduce toxicity by enhancing stability, preventing aggregation, and minimizing undesired interactions with biological molecules.

4.2. Interaction with Biological Molecules (Proteins, Lipids, DNA)

Nanoparticles interact dynamically with biological molecules such as proteins, lipids, and DNA upon entering the body. These interactions are often mediated by the formation of a protein "corona," a layer of adsorbed proteins that modifies the nanoparticle's surface properties. The corona composition can influence the biodistribution, cellular uptake, and immunological response to the nanoparticles.

Negatively charged nanoparticles may interact with positively charged regions of proteins or DNA, potentially interfering with their functions. Similarly, negatively charged lipid bilayers can be broken up by cationic nanoparticles, which can cause cytotoxicity and membrane instability. These interactions underscore the need of creating nanoparticles with regulated surface characteristics to prevent unwanted consequences.

4.3. Role of Environmental pH and Biological Fluids

The pH of the surrounding environment and the makeup of biological fluids have a major impact on how nanoparticles behave in biological systems. For example, metal oxide nanoparticles can dissolve more readily in acidic environments, as those found in lysosomes or tumour microenvironments, which can result in the release of hazardous metal ions. Proteins, lipids, and electrolytes included in biological fluids like blood plasma can change how well nanoparticles aggregate and remain stable. It has been demonstrated that ZnO and TiO_2 nanoparticles aggregate in phosphate-buffered saline (PBS) and other common media, increasing their size and decreasing their bioavailability while possibly increasing their local toxicity. The necessity of stability testing under pertinent biological circumstances is highlighted by the dynamic character of these habitats.

4.4. Biodistribution and Accumulation in Tissues

The size, shape, surface charge, and administration method of nanoparticles all influence their biodistribution. The liver, spleen, and kidneys are among the organs where smaller nanoparticles are more prone to concentrate after penetrating tissues. Larger particles, on the other hand, can be restricted to particular areas, such the extracellular matrix or blood vessels. When functionalised with stealth coatings like PEG, nanoparticles can also avoid being cleared by the reticuloendothelial system (RES), increasing their therapeutic efficacy and extending their circulation duration. On the other hand, tissue buildup and extended circulation may raise the chance of chronic toxicity, such as organ failure, fibrosis, or inflammation. Predicting the long-term effects of nanoparticles and improving their safety profiles require an understanding of their pharmacokinetics.

5. SHORT-TERM VS. LONG-TERM TOXICITY

The toxicological effects of metal and metal oxide nanoparticles (NPs) can manifest differently over short-term and long-term exposures. Acute toxicity often results from high doses of nanoparticles introduced into the biological system, whereas chronic toxicity arises from prolonged or repeated exposure to lower doses, leading to cumulative effects on critical organs and systems.

5.1. Acute Toxicity Studies

The rapid consequences of a single or brief exposure to nanoparticles are the main focus of acute toxicity investigations. Rapidly onset toxicities such oxidative stress, inflammation, and cell death are frequently found in these investigations. For example, it has been demonstrated that exposure to ZnO or CuO nanoparticles causes oxidative stress and mitochondrial malfunction in a matter of hours, leading to necrosis or cellular death. Acute toxicity is greatly influenced by the exposure route. Lung irritation and respiratory discomfort can result from inhaling nanoparticles like TiO_2. Similar to this, large local concentrations of nanoparticles administered intravenously may cause fast systemic dispersion and possible harm to organs like the liver and spleen. Finding acceptable exposure limits and possible hazards related to high-dose exposures requires the use of acute toxicity studies.

5.2. Chronic Exposure and Long-Term Risks

Chronic exposure to nanoparticles, even at low doses, poses significant long-term risks due to their persistence and bioaccumulation in tissues. Prolonged exposure can lead to sustained oxidative stress, chronic inflammation, and DNA damage, increasing the risk of carcinogenesis and other long-term health conditions. For example, studies on silver nanoparticles have revealed their tendency to bioaccumulate in organs like the liver and kidneys, leading to tissue damage and impaired organ function over time. Chronic exposure to metal oxide nanoparticles has also been associated with neurotoxicity, as these particles can cross the blood-brain barrier and accumulate in the brain, potentially disrupting neuronal function.

Long-term risks often depend on the physicochemical properties of nanoparticles, including their size, surface charge, and dissolution rate. Understanding these chronic effects is essential for evaluating the safety of nanoparticles in consumer and biomedical products.

5.3. Target Organs and Systems (Liver, Kidney, Brain, etc.)

The liver, kidney, and brain are primary target organs for nanoparticle-induced toxicity due to their roles in metabolism, excretion, and susceptibility to oxidative stress:

- **Liver**: As a key organ for detoxification, the liver is a major site of nanoparticle accumulation. Metallic nanoparticles, such as gold and silver, tend to bioaccumulate in hepatocytes, leading to oxidative damage and disruption of liver enzymes. Prolonged exposure may result in liver fibrosis or failure.
- **Kidney**: The kidney, responsible for nanoparticle excretion, is vulnerable to toxicity due to high local concentrations during filtration. Metal oxide nanoparticles, such as ZnO and CuO, have been shown to induce renal oxidative stress and tubular damage, potentially leading to chronic kidney disease.
- **Brain**: The ability of certain nanoparticles to cross the blood-brain barrier poses significant risks to the central nervous system. For example, silver nanoparticles have been implicated in neurotoxicity, potentially causing inflammation, oxidative damage, and disruption of neuronal pathways.

Understanding the distribution and accumulation of nanoparticles in these target organs provides critical insights into their systemic and long-term toxicological impact.

6. STRATEGIES FOR REMEDIATION AND MITIGATION OF NANOTOXICITY

The toxicity of metal and metal oxide nanoparticles (NPs) presents significant challenges in biomedical applications. However, innovative strategies can mitigate their adverse effects, ensuring safer use in medical and environmental contexts (Manuja et al., 2020; Chopra et al.,2015).

6.1. Surface Functionalization and Coatings

Surface functionalisation is a crucial strategy for lessening nanoparticles' harmful effects. By giving them "stealth" qualities, coating nanoparticles with biocompatible polymers like polyethylene glycol

(PEG) can improve their stability, decrease aggregation, and lessen immunological reactions. Zwitterionic coatings can further reduce cytotoxicity by balancing surface charges and reducing interactions with cellular membranes (Savage et al., 2019).

Additionally, surface-modified nanoparticles show enhanced targeting capabilities. Selective distribution to sick tissues while preserving healthy cells is made possible by functionalisation with particular ligands or antibodies. Hydrophilic coatings, for example, minimise unwanted biological interactions by lowering the possibility of protein corona formation and nanoparticle aggregation (Manuja et al., 2020).

6.2. Biodegradable and Biocompatible Nanomaterials

Another successful tactic is the use of biodegradable and biocompatible nanomaterials. It has been demonstrated that toxicity is decreased by nanoparticles made of substances like chitosan, polylactic acid, and other synthetic or natural biodegradable polymers (Li & Huang, 2010; Cochran et al., 2013). These substances break down into innocuous metabolites that the body can readily eliminate, lowering the possibility of long-term buildup. Concerns regarding the durability of nanoparticles in biological systems are also addressed by including biodegradable elements during manufacturing. For instance, encasing nanoparticles in biodegradable matrices shields them from metal ions, reducing their harmful effects without sacrificing therapeutic effectiveness (Kaur et al., 2015).

6.3. Antioxidant and Detoxification Strategies

Reactive oxygen species (ROS) generation is one of the primary mechanisms of nanotoxicity. Incorporating antioxidants into nanoparticle formulations can significantly reduce oxidative stress. For instance, polyTrolox, a potent antioxidant polymer, has been used to prevent ROS-related damage in cells exposed to metallic nanoparticles. Similarly, detoxification strategies focus on binding and neutralizing free metal ions released by nanoparticle dissolution. Chelating agents and antioxidant-impregnated nanoparticle coatings are examples of strategies designed to mitigate oxidative and inflammatory responses caused by ion release (Sies, 1993).

6.4. Controlled Release Systems and Targeted Nanoparticles

Controlled release systems and targeted delivery techniques ensure that nanoparticles act specifically at the intended site of action. Encapsulation of nanoparticles in stimuli-responsive materials (e.g., pH-sensitive or temperature-sensitive polymers) enables the release of therapeutic agents only under specific conditions, such as in the acidic environment of a tumor Targeted nanoparticles, functionalized with ligands or antibodies, accumulate selectively in diseased tissues, reducing off-target effects. For example, surface-modified nanoparticles designed to escape immune clearance and penetrate the blood-brain barrier (BBB) ensure more effective drug delivery with reduced systemic toxicity. Controlled release also prevents burst release of toxic ions, allowing for gradual therapeutic action while minimizing cellular stress and damage (Kamaly et al., 2016).

6.5. Environmental and Biological Safety Measures

Ensuring the safety of nanoparticles involves considering both their effects on the environment and their medical applications. To reduce their ecological imprint, actions including the creation of eco-friendly nanomaterials, the use of green synthesis techniques, and safer disposal of waste from nanoparticles are essential. Thorough preclinical research is necessary to comprehend the biodistribution, accumulation, and excretion of nanoparticles in biological systems. Safer designs can be made possible by computational methods that offer insights into the toxicity profiles of nanoparticles, such as prediction models based on artificial intelligence. Furthermore, incorporating safety precautions such minimising bio-persistent compounds and steric stabilisation helps lessen the long-term dangers connected to nanoparticle exposure.

7. REGULATORY GUIDELINES AND SAFETY CONSIDERATIONS

The increasing use of metal and metal oxide nanoparticles (NPs) in biomedical applications necessitates a thorough understanding of their toxicity profiles, risk assessments, and compliance with regulatory frameworks. Ensuring the safety of these nanomaterials involves addressing their physicochemical properties, interaction with biological systems, and long-term environmental impact. Various international regulatory agencies have established guidelines for evaluating the safety and biocompatibility of nanomaterials, particularly those intended for medical and pharmaceutical use.

7.1. Regulatory Frameworks for Nanomaterials

Several regulatory bodies oversee the safe use of metal and metal oxide nanoparticles. The U.S. Food and Drug Administration (FDA) regulates nanomaterials across various categories, including pharmaceuticals, medical devices, and cosmetics, emphasizing preclinical toxicity assessments, pharmacokinetics, and long-term effects. Similarly, the European Medicines Agency (EMA) provides specific guidance on nanomedicines, ensuring compliance with safety, efficacy, and quality standards, while risk-based assessments are mandated under the European Chemicals Agency's (ECHA) REACH (Registration, Evaluation, Authorisation, and Restriction of Chemicals) regulation. In the workplace, the Occupational Safety and Health Administration (OSHA) and the National Institute for Occupational Safety and Health (NIOSH) establish exposure limits and safety protocols for handling nanomaterials. Additionally, international organizations such as the International Organization for Standardization (ISO) and the Organisation for Economic Co-operation and Development (OECD) have developed standardized methods for characterizing and assessing nanomaterial toxicity, including ISO 10993 for medical devices.

7.2. Toxicity Evaluation and Risk Assessment

Regulatory agencies stress a tiered process of nanoparticle toxicity assessment, guaranteeing an in-depth examination of their potential hazards. This process starts with physicochemical characterization, where key properties like size, shape, surface charge, aggregation state, and solubility are analyzed. These features have a great impact on nanoparticle interactions with biological systems and their general toxicity. For example, smaller nanoparticles have greater surface-area-to-volume ratios, which enhance their reactivity and capacity to cause cellular damage. Surface charge is also important in cellular uptake, with positively charged nanoparticles tending to be more toxic because they can disrupt negatively charged cell membranes. Likewise, aggregation state and solubility dictate how nanoparticles function in physiological and environmental conditions, influencing their bioavailability and potential toxicity. After physicochemical characterization, in vitro and in vivo toxicity tests are performed to estimate the biological impact of nanoparticles at both the cellular and organismal levels. In vitro investigations entail exposing cultured cells to nanoparticles and measuring prime toxicological endpoints including cytotoxicity, genotoxicity, induction of oxidative stress, and inflammatory reactions. Cytotoxicity assays identify whether nanoparticles impair cell viability by inhibiting metabolic processes or triggering apoptosis. Genotoxicity tests analyze whether nanoparticles induce DNA damage, which may result in mutations or carcinogenesis. Moreover, oxidative stress assays quantify the production of reactive oxygen species (ROS), which may provoke cellular damage and inflammation. Although in vitro investigations are great sources of initial data, in vivo experiments are needed to determine the systemic consequences of nanoparticles within a living organism. Animal models are employed to assess nanoparticle distribution, metabolism, excretion, and possible toxic effects on organs like the liver, kidneys, and lungs. These experiments facilitate the bridging of the gap between laboratory results and actual biological responses. Another essential element of nanoparticle toxicity testing is biodistribution and pharmacokinetics, which investigate how nanoparticles are absorbed, distributed, metabolized, and excreted in the body. Knowledge of nanoparticle accumulation and clearance processes is crucial to long-term health risk prediction. For example, nanoparticles that accumulate in the liver or spleen could potentially have a risk of bioaccumulation, causing chronic toxicity. Pharmacokinetic research informs us about the half-life of nanoparticles in the blood and whether they can pass through biological barriers, like the blood-brain barrier or the placental barrier, which has major implications for human health. Furthermore, scientists examine nanoparticle routes of excretion, such as renal filtration and hepatic metabolism, to determine potential risks due to long-term exposure or inadequate clearance. Apart from human health factors, regulatory agencies also look into environmental and ecotoxicological effects, examining how nanoparticles interact with ecosystems and water bodies. Nanoparticles can find their way into the environment via industrial discharge, biomedical use, and consumer products, having the potential to impact aquatic organisms and soil microorganisms. Research evaluates nanoparticle persistence in various environmental compartments and their ability to bioaccumulate in food webs. For instance, metal oxide nanoparticles employed in sunscreens and cosmetics can be released into water bodies, where they can destabilize aquatic ecosystems by influencing microbial communities or producing toxic byproducts upon UV exposure. Regulatory agencies also investigate how nanoparticles interact with soil and plant systems, affecting agricultural sustainability and biodiversity. Overall, the tiered methodology of nanoparticle toxicity assessment guarantees a comprehensive understanding of their physicochemical characteristics, biological interactions, pharmacokinetics, and environmental impact. Such a methodical assessment structure helps regulatory agencies to set safety guidelines, reduce risks, and encourage responsible development and use of nanotechnology across sectors.

7.3. Safety Considerations in Biomedical Applications

Maintenance of metal-based nanoparticle safety for use in biomedical interventions is vital in reducing potential toxicity while ensuring enhanced therapeutic benefits. Optimization of dosing is an important element in safety evaluation by setting acceptable levels of exposure not to induce toxicity from the presence of nanoparticles. As a consequence of their extremely high surface area-to-volume ratios and unusual physicochemical profiles, metal-based nanoparticles have greater reactivity than their bulk forms, inducing cytotoxicity, oxidative stress, and inflammatory responses at high dose levels. Thus, accurate dosing regimens need to be formulated to achieve efficacy and safety. Scientists establish optimal doses by performing in vitro and in vivo experiments to assess nanoparticle accumulation, clearance rates, and biological interactions. Regulatory agencies also offer guidelines on safe exposure limits based on experimental data to ensure that nanoparticles do not cause unwanted toxic effects when applied in biomedical applications. One key aspect of nanoparticle safety is surface modification in the form of functionalization to improve biocompatibility and diminish unwanted interactions with biological systems. Metal nanoparticles, including silver, gold, and iron oxide nanoparticles, are more likely to bind to proteins, cell membranes, and biomolecules, with attendant risks of aggregation, immune activation, or toxicity. To combat these, techniques of surface engineering like polyethylene glycol coating, ligand conjugation, and polymer encapsulation are adopted. PEGylation, for example, can enhance nanoparticle stability, extend circulation time in blood, and decrease immune recognition. Coating nanoparticles with biocompatible polymers or biomolecules can similarly enhance targeted drug delivery with fewer off-target effects. Such modifications allow for the preservation of functional properties of nanoparticles while reducing possible toxicity in biological systems. In addition to acute toxicity tests, long-term toxicity tests are necessary to comprehend the chronic toxicity of metal-based nanoparticles. Short-term experiments might not accurately reflect the delayed or cumulative effect of long-term nanoparticle exposure, particularly in biomedical applications where repeated administration is the norm, e.g., drug delivery, imaging, and diagnostics. Chronic exposure testing includes long-term animal studies, assessing possible toxic effects on primary organs, immune reactions, and bioaccumulation hazards. For example, nanoparticles that are not effectively removed from the body have the possibility to concentrate in tissues like the liver, spleen, and brain, causing potential inflammation, fibrosis, or neurotoxicity. In addition, scientists study if long-term exposure to metal-based nanoparticles will lead to DNA damage, carcinogenicity, or changes in metabolic pathways. These long-term experiments play an important role in regulatory clearance and the assurance of nanoparticles for safe use before they become generally available for clinical applications. Aside from biological safety, the observance of personal protective equipment (PPE) and handling protocols is important in reducing occupational exposure hazards in research and industrial environments. Protocols for handling nanoparticles entail adequate ventilation, fume hood use, and containment to avoid airborne dispersion of nanoparticles. PPE like gloves, lab coats, respirators, and goggles should be worn by workers and researchers when handling nanoparticle suspensions or dry powders. In addition, training in nanoparticle safety, appropriate disposal, and spill control is a key to creating a safe working environment. By following these protective measures, accidental exposure and health risks associated with nanoparticles can be minimized to a great extent.

7.4. Challenges and Future Directions

Though current regulatory systems present necessary safety guidelines for the biomedical application of metal and metal oxide nanoparticles, a number of challenges still exist. Perhaps the most critical challenge is the absence of standardized testing procedures for assessing nanoparticle toxicity. Unlike traditional chemicals, nanoparticles have varied physicochemical characteristics, such as size, shape, surface charge, and state of aggregation, which can all affect their biological interactions and toxicological profiles. The variability makes it challenging to have universally agreed-upon testing strategies that are effective in determining nanoparticle risks. Existing toxicity assessment methods tend to be based on modified techniques from bulk materials or conventional drugs, which do not necessarily reflect the distinct nanoscale behavior of nanoparticles. Establishing standardized protocols for in vitro and in vivo toxicity testing, such as dose-response, biodistribution, and cellular uptake mechanisms, is essential to ensuring uniform and reproducible safety assessments across various regulatory bodies and industries.

Another major concern is the unknown long-term effect of nanoparticle exposure on human health and the environment. Although short-term studies offer useful information on the acute effects of nanoparticles, the chronic effects of long-term exposure are not well studied. Most metal and metal oxide nanoparticles, including silver, titanium dioxide, and zinc oxide, have been found to have good potential in drug delivery, imaging, and wound healing. Yet, their potential to bioaccumulation in tissues, produce oxidative stress, or interfere with cellular processes over long time periods is still unclear. Furthermore, nanoparticles emitted into the environment via industrial and biomedical uses could remain in ecosystems, impacting microbial communities, aquatic organisms, and food webs. For example, nanoparticles that find their way into water bodies via wastewater outfalls might be transformed so that their toxicity is changed, perhaps creating unsuspected ecological hazards. Epidemiological studies over long periods and environmental monitoring activities are crucial to closing these knowledge gaps and enabling the sustainable utilisation of nanomaterials. In addition, the challenge of regulatory lag is that fast-paced developments in nanotechnology tend to leave behind regulatory policy-making. The evolving nature of nanoscience, with a constant stream of innovations in synthesizing nanoparticles, functionalising them, and exploiting their utility, outclasses prevailing frameworks. Traditional regulatory approval processes may not be well-suited to accommodate the fast-evolving landscape of nanomedicine, leading to delays in bringing novel nanoparticle-based therapies and diagnostics to market. Moreover, the lack of harmonized regulations across different countries further complicates the approval process, creating inconsistencies in safety standards and compliance requirements. To deal with this challenge, the regulatory bodies need to implement adaptive and anticipatory policies that enable timely assessment of new nanotechnologies. This involves incorporating high-throughput screening tools, artificial intelligence-based risk analysis, and monitoring of data in real-time to enable more effective safety assessments. A concerted effort among policymakers, researchers, and industry players will be essential in creating adaptable regulatory frameworks that ensure innovation is balanced with safety. Overcoming these challenges demands interdisciplinary research and regulatory innovation to guarantee the safe and responsible use of metal and metal oxide nanoparticles in biomedical applications. By developing standardized testing protocols, performing thorough long-term studies, and adopting adaptive regulatory approaches, stakeholders can minimize potential risks while realizing the full potential of nanotechnology. It is crucial to promote the safe development of nanoparticle-based innovations for human health and the environment through a concerted effort from researchers, regulators, and industry leaders.

7.5. Standards for Nanomaterial Safety

The research, manufacturing, and use of nanoparticles (NPs) are governed by regulatory frameworks that must be followed to ensure their safe use in biomedical applications. Although there are worldwide standards for nanomaterials, such as those set forth by the OECD and ISO, their application to metal and metal oxide nanoparticles is still developing. These guidelines place a strong emphasis on performing comprehensive preclinical testing, comprehending the interactions of nanoparticles in biological systems, and characterising their physicochemical properties. According to current safety guidelines, nanoparticle stability, size, dissolving behaviour, and biocompatibility are all evaluated. However, the special characteristics of nanoparticles—like their propensity to aggregate or form a protein corona in biological settings—call for further considerations beyond conventional chemical safety evaluations (Soppimath et al., 2001).

7.6. Risk Assessment Models for Nanoparticles

Risk assessment for nanoparticles involves evaluating their exposure pathways, accumulation in tissues, and potential short- and long-term toxic effects. Emerging computational approaches, such as artificial intelligence (AI)-based predictive toxicology, are revolutionizing this field by modeling nanoparticle interactions with biological systems. These models enable high-throughput screening of nanoparticle libraries, predicting toxicity based on physicochemical parameters such as size, shape, and surface charge (Park & Yeo, 2016). Traditional risk assessment models often fail to account for nanoparticles' dynamic behavior in biological environments. As a result, newer frameworks incorporate considerations such as nanoparticle aggregation, biodistribution, and biodegradation to provide more accurate predictions of their safety profiles.

7.7. Challenges in Nanotoxicology Testing and Regulation

One of the major challenges in regulating nanoparticles is the lack of standardized testing protocols that address their unique properties. Current toxicological tests may not accurately capture nanoparticles' dynamic interactions with biological systems, such as their ability to penetrate cellular membranes or accumulate in specific tissues. Another challenge is the difficulty in tracking nanoparticles within biological systems, as they often undergo transformation (e.g., dissolution or protein corona formation) that alters their behavior. Additionally, regulatory agencies face hurdles in developing universal guidelines due 0.to the diversity of nanoparticles and their applications (Bahadar et al., 2016).

8. FUTURE DIRECTIONS IN NANOTOXICOLOGY AND BIOMEDICAL APPLICATIONS

The future of nanotoxicology and biomedical applications lies in refining toxicity assessment methodologies, enhancing the safety of nanoparticles, and ensuring their seamless clinical translation. Advanced techniques such as high-throughput screening, computational toxicology, and multi-omics approaches are expected to play a crucial role in predicting and evaluating nanoparticle-induced toxicity with greater precision. The development of organs-on-a-chip and 3D cell culture models will provide more physiologically relevant testing platforms, reducing reliance on traditional animal studies. Furthermore, long-term and chronic exposure studies are essential for understanding the delayed toxic effects and bioaccumulation risks of metal and metal oxide nanoparticles (Han et al., 2013).

To counter toxicity issues, upcoming research should concentrate on engineering nanoparticles with enhanced biocompatibility and targeted functionality. Biodegradable and stimuli-responsive nanoparticle designs will facilitate controlled drug release and reduce residual toxicity. Surface functionalization approaches, such as PEGylation and biomolecule conjugation, will minimize immunogenicity and off-target effects. Green synthesis strategies based on plant extracts and microbial processes also provide environmentally friendly options that reduce toxic byproducts. The idea of personalized nanomedicine, wherein nanoparticles are customized according to patient-specific biological parameters, will further augment therapeutic efficiency at minimizing side effects. The regulatory framework for nanomedicine needs to keep pace with technological progress in order to provide assurance for safety and efficacy. Standardizing international regulatory guidelines, encouraging open data sharing, and creating open-source databases will make risk assessment more streamlined and facilitate approval of nanoparticles for clinical application. Ethical implications related to nanoparticle exposure, long-term environmental effects, and public acceptance also need to be addressed to ensure safe innovation. Concurrently, nanotechnology will transform personalized and regenerative medicine by facilitating targeted drug delivery, theranostics, and tissue engineering applications. The incorporation of metal oxide nanoparticles in regenerative therapies, including bone and neural tissue scaffolds, will greatly improve cell proliferation and differentiation. Additionally, immunomodulatory nanoparticles will open the door to new treatments in cancer immunotherapy, vaccine delivery, and autoimmune diseases. As metal-based nanoparticles become increasingly used, their occupational and environmental safety concerns need to be actively addressed. Sustainable nanoparticle synthesis, bioremediation approaches, and in-situ workplace exposure monitoring will be essential to reducing ecological and health hazards. Innovation versus safety will be the future of nanotoxicology, demanding a multidisciplinary collaboration among researchers, clinicians, and policymakers. By incorporating sophisticated testing methods, creating more secure nanomaterials, improving regulatory systems, and clarifying ethical and ecological issues, the discipline can realize the full potential of nanotechnology in medical use while minimizing linked risks (Fan et al., 2017).

8.1. Advances in Safer Nanoparticle Design

Recent advances in nanoparticle design focus on enhancing their biocompatibility while minimizing toxicity. Strategies include using biodegradable materials, incorporating targeting ligands for site-specific delivery, and engineering nanoparticles to release therapeutic agents under controlled conditions. Surface modifications, such as PEGylation or zwitterionic coatings, have proven effective in reducing immunogenicity and improving biodistribution. Additionally, research into stimuli-responsive nanoparticles that react to pH or temperature changes offers promising avenues for safer drug delivery systems (Han et al., 2013; Fan et al., 2017).

8.2. Emerging Research Areas in Nanotoxicology

Emerging areas in nanotoxicology research are vital for advancing the safe application of nanotechnology across various fields. One significant focus is on nanoparticle-immune system interactions, aiming to understand how nanoparticles influence immune responses. This includes studying their ability to activate or suppress immune cells, which is crucial for minimizing immunogenicity and unintended inflammatory reactions. Another key area is long-term toxicity studies, which investigate the chronic effects of nanoparticles, such as their potential to accumulate in organs or disrupt cellular processes over extended periods. These studies are essential for assessing the cumulative risks of prolonged exposure, which could lead to diseases like fibrosis or cancer. Finally, computational toxicology is gaining prominence, leveraging AI and machine learning to predict nanoparticle behavior and toxicological outcomes based on their structural and chemical properties (Nel et al., 2013). This approach enables researchers to evaluate risks more efficiently and cost-effectively, reducing the need for extensive laboratory testing while providing detailed insights into nanoparticle safety. Together, these areas address critical gaps in our understanding of nanoparticle toxicity and guide the development of safer nanomaterials.

Table 2. Emerging Areas of Nanotoxicology that can be Researched

Area	Description	Significance
Nanoparticle-Immune System Interactions	Studying how nanoparticles interact with immune cells and modulate immune responses.	Crucial for designing nanoparticles with reduced immunogenicity and inflammation risks.
Long-term Toxicity Studies	Investigating chronic exposure effects, including nanoparticle accumulation in organs and cellular disruptions.	Provides insights into the cumulative risks associated with prolonged nanoparticle exposure.
Computational Toxicology	Using AI and machine learning to predict nanoparticle behavior, toxicity, and interactions with biological systems.	Enables rapid, cost-effective, and precise evaluation of nanoparticle risks without extensive laboratory testing.

(Nel et al., 2013)

8.3. Translating Nanotoxicology Insights into Clinical Practice

Bridging the gap between research and clinical application requires collaboration among toxicologists, material scientists, and clinicians. Translating insights into practice involves incorporating safety measures into the early stages of nanoparticle design, conducting rigorous preclinical and clinical testing, and ensuring compliance with regulatory guidelines (Xuan et al., 2023). Developing safer nanoparticles for clinical use also depends on creating standardized protocols for toxicity testing and addressing concerns about long-term effects through robust post-marketing surveillance (Đorđević et al., 2022).

9. CONCLUSION

The increased use of metal and metal oxide nanoparticles for biomedical purposes is a result of their exclusive physicochemical features, such as high surface area, variable reactivity, and increased biocompatibility. Nanoparticles have been deeply researched to find application in drug delivery, bioimaging, diagnostics, and regenerative medicine, where nanoscale features facilitate enhanced interactions with cells and site-specific therapeutic activity. Yet, the same properties that make them useful also make them potentially toxic, and their safety for human health and the environment is a concern. The toxicity of metal and metal oxide nanoparticles is mainly due to their capacity to produce reactive oxygen species (ROS), cause oxidative stress, release toxic ions, and initiate inflammatory reactions. The extent of these effects depends on parameters like particle size, shape, aggregation state, surface charge, and functionalization. The smaller nanoparticles, because of their larger surface-to-volume ratio, are more reactive and are more prone to interact with biological molecules like proteins, lipids, and nucleic acids. In the same way, the dissolution of some metal nanoparticles, like silver or zinc oxide, can cause the release of toxic metal ions, enhancing their cytotoxic effects. One of the most important features of nanoparticle toxicity is the interaction with cellular and subcellular structures. Metal-based nanoparticles have been shown to interfere with mitochondrial integrity and function and produce excessive levels of ROS, which in turn can result in lipid peroxidation, protein oxidation, and damage to DNA, triggering apoptosis or necrosis in the affected cells. Besides, nanoparticles can also regulate immune responses by inducing inflammatory pathways, enhancing the release of pro-inflammatory cytokines, and recruiting immune cells. Long-term exposure to nanoparticles is associated with long-lasting health impacts, such as genotoxicity, fibrosis, and organ toxicity, which requires a thorough analysis of their biodistribution, accumulation, and elimination from the body. Despite these concerns, metal and metal oxide nanoparticles continue to hold immense promise in nanomedicine. Their potential for targeted drug delivery, enhanced imaging contrast, and bioactive properties make them invaluable tools in modern healthcare. However, ensuring their safe use requires striking a delicate balance between therapeutic efficacy and toxicological risks. One approach to mitigating nanoparticle toxicity involves surface modifications that improve biocompatibility while preserving their functional properties. Strategies such as PEGylation, polymeric coatings, and biomolecule conjugation have been employed to reduce immune recognition and prevent nonspecific interactions with healthy tissues. Additionally, developing controlled release systems that allow for the gradual degradation or dissolution of nanoparticles can minimize toxicity by preventing the excessive accumulation of metal ions or reactive byx`products. Further advancements in nanomedicine must incorporate a multidisciplinary approach, integrating toxicological insights into the rational design of safer and more effective nanoparticles. Computational modeling and artificial intelligence-based predictive toxicology are emerging as valuable tools for assessing the potential risks of nanoparticles before in vivo experimentation. These approaches allow researchers to simulate nanoparticle interactions with biological systems, predicting their toxicity profiles based on physicochemical properties. High-throughput screening methods are also being developed to rapidly evaluate the safety of newly synthesized nanomaterials, accelerating the translation of research findings into clinical applications. Moreover, leveraging biomimetic and bioinspired nanoparticle designs, such as coating nanoparticles with cell membranes or employing naturally occurring nanostructures, can improve their physiological compatibility and reduce undesired side effects.

Regulatory frameworks must evolve in parallel with scientific advancements to ensure the responsible development and commercialization of nanomedicines. Currently, there are inconsistencies in global regulations governing the safety assessment of nanoparticles, leading to challenges in standardizing toxicity testing methodologies. The lack of universally accepted guidelines for nanoparticle characterization, dose-response evaluation, and long-term risk assessment hinders the efficient approval of nanomedicine-based therapies. Addressing these regulatory gaps will require coordinated efforts among government agencies, researchers, and industry stakeholders to establish harmonized safety standards. Additionally, transparent data-sharing initiatives and collaborative research consortia can facilitate the exchange of toxicity data, helping to refine risk assessment models and regulatory policies.

Another critical aspect of nanotoxicology research is the environmental impact of metal-based nanoparticles. The increasing production and utilization of nanoparticles in medical, industrial, and consumer applications raise concerns about their potential release into ecosystems. Once discharged into the environment, nanoparticles can persist in air, water, and soil, where they may interact with living organisms and disrupt ecological balance. Some nanoparticles have been shown to bioaccumulate in aquatic organisms, leading to trophic transfer and potential toxicity in higher-order species. Investigating the environmental fate of nanoparticles, including their degradation pathways, persistence, and potential for bioaccumulation, is essential for assessing their long-term ecological risks. Sustainable nanoparticle synthesis methods, such as green chemistry approaches that utilize plant extracts or microbial processes, offer an environmentally friendly alternative to conventional nanoparticle production, reducing hazardous waste and mitigating ecological impact.

Looking ahead, the future of nanomedicine lies in harnessing the full potential of nanoparticles while minimizing their associated risks. The integration of nanotechnology with emerging fields such as precision medicine, artificial intelligence, and regenerative medicine will open new frontiers for personalized therapies and advanced biomedical applications. Smart nanoparticles capable of responding to physiological cues, such as pH, temperature, or enzyme activity, will enable highly selective and controlled drug release, improving therapeutic outcomes while minimizing systemic toxicity. Nanoparticle-based theranostic platforms, which combine diagnostic and therapeutic functions in a single system, will revolutionize disease management by enabling real-time monitoring of treatment efficacy. In regenerative medicine, metal oxide nanoparticles are being explored for their ability to enhance cell proliferation, differentiation, and tissue repair, offering promising solutions for wound healing, bone regeneration, and neural tissue engineering.

Ensuring the long-term safety and success of nanomedicine will require a commitment to responsible research, ethical considerations, and public engagement. Addressing concerns related to nanoparticle toxicity, environmental sustainability, and regulatory oversight is crucial for fostering public trust and promoting the ethical use of nanotechnology in medicine. Collaborative efforts among scientists, clinicians, regulatory agencies, and industry leaders will be instrumental in overcoming current challenges and shaping the future of safe and effective nanoparticle-based therapeutics. By integrating cutting-edge research with robust safety evaluation frameworks, nanomedicine can continue to advance as a transformative field with profound implications for human health and well-being.

REFERENCES

Attarilar, S., Yang, J., Ebrahimi, M., Wang, Q., Liu, J., Tang, Y., & Yang, J. (2020). The toxicity phenomenon and the related occurrence in metal and metal oxide nanoparticles: A brief review from the biomedical perspective. *Frontiers in Bioengineering and Biotechnology*, *8*, 822. PMID: 32766232

Bahadar, H., Maqbool, F., Niaz, K., & Abdollahi, M. (2016). Toxicity of nanoparticles and an overview of current experimental models. *Iranian Biomedical Journal*. Advance online publication. DOI: 10.7508/ibj.2016.01.001 PMID: 26286636

Brand, W., Noorlander, C. W., Giannakou, C., De Jong, W. H., Kooi, M. W., Park, M. V., & Geertsma, R. E. (2017). Nanomedicinal products: A survey on specific toxicity and side effects. *International Journal of Nanomedicine*, •••, 6107–6129. PMID: 28883724

Chopra, M., Bernela, M., Kaur, P., Manuja, A., Kumar, B., & Thakur, R. (2015). Alginate/gum acacia bipolymeric nanohydrogels-Promising carrier for Zinc oxide nanoparticles. *International Journal of Biological Macromolecules*. Advance online publication. DOI: 10.1016/j.ijbiomac.2014.09.037 PMID: 25304751

Cochran, D. B., Wattamwar, P. P., Wydra, R., Hilt, J. Z., Anderson, K. W., Eitel, R. E., & Dziubla, T. D. (2013). Suppressing iron oxide nanoparticle toxicity by vascular targeted antioxidant polymer nanoparticles. *Biomaterials*. Advance online publication. DOI: 10.1016/j.biomaterials.2013.08.025 PMID: 24016851

Đorđević, S., Gonzalez, M. M., Conejos-Sánchez, I., Carreira, B., Pozzi, S., Acúrcio, R. C., & Vicent, M. J. (2022). Current hurdles to the translation of nanomedicines from bench to the clinic. *Drug Delivery and Translational Research*, •••, 1–26. PMID: 34302274

Fan, W., Yung, B., Huang, P., & Chen, X. (2017). Nanotechnology for multimodal synergistic cancer therapy. *Chemical Reviews*, *117*(22), 13566–13638. PMID: 29048884

Han, S. G., Newsome, B., & Hennig, B. (2013). Titanium dioxide nanoparticles increase inflammatory responses in vascular endothelial cells. *Toxicology*. Advance online publication. DOI: 10.1016/j.tox.2013.01.014 PMID: 23380242

Ibrahim, K., Saeed, K., & Idrees, K. (2019). Nanoparticles: Properties, applications and toxicities. *Arabian Journal of Chemistry*, *12*, 908–931.

Jia, G., Wang, H., Yan, L., Wang, X., Pei, R., Yan, T., & Guo, X. (2005). Cytotoxicity of carbon nanomaterials: Single-wall nanotube, multi-wall nanotube, and fullerene. *Environmental Science & Technology*, *39*(5), 1378–1383. PMID: 15787380

Kamaly, N., Yameen, B., Wu, J., & Farokhzad, O. C. (2016). Degradable controlled-release polymers and polymeric nanoparticles: Mechanisms of controlling drug release. *Chemical Reviews*, *116*(4), 2602–2663. PMID: 26854975

Kaur, P., Thakur, R., Barnela, M., Chopra, M., Manuja, A., & Chaudhury, A. (2015). Synthesis, characterization and in vitro evaluation of cytotoxicity and antimicrobial activity of chitosan-metal nanocomposites. *Journal of Chemical Technology and Biotechnology*. Advance online publication. DOI: 10.1002/jctb.4383

. S.D. Li, L. Huang, Stealth nanoparticles: high density but sheddable PEG is a key for tumor targeting, J. Control. Release (2010), https://doi.org/. jconrel.2010.03.016.DOI: 10.1016/j

Manuja, A., Kumar, B., Kumar, R., Chhabra, D., Ghosh, M., Manuja, M., & Prasad, M. (2021). Metal/ metal oxide nanoparticles: Toxicity concerns associated with their physical state and remediation for biomedical applications. *Toxicology Reports*, *8*, 1970–1978. PMID: 34934635

Manuja, A., Kumar, B., Riyesh, T., Talluri, T. R., & Tripathi, B. N. (2020). Microwave assisted fast fabrication of zinc/iron oxides based polymeric nanocomposites and evaluation on equine fibroblasts. *International Journal of Biological Macromolecules*, *165*, 71–81. PMID: 32987081

Manuja, A., Raguvaran, R., Kumar, B., Kalia, A., & Tripathi, B. N. (2020). Accelerated healing of full thickness excised skin wound in rabbits using single application of alginate/acacia based nanocomposites of ZnO nanoparticles. *International Journal of Biological Macromolecules*, *155*, 823–833. PMID: 32234436

Nel, A., Xia, T., Meng, H., Wang, X., Lin, S., Ji, Z., & Zhang, H. (2013). Nanomaterial toxicity testing in the 21st century: Use of a predictive toxicological approach and high- throughput screening. *Accounts of Chemical Research*. Advance online publication. DOI: 10.1021/ ar300022h PMID: 22676423

Park, H. G., & Yeo, M. K. (2016). Nanomaterial regulatory policy for human health and environment. *Molecular & Cellular Toxicology*, *12*, 223–236.

Savage, D. T., Hilt, J. Z., & Dziubla, T. D. (2019). In vitro methods for assessing nanoparticle toxicity. *Methods in Molecular Biology (Clifton, N.J.)*, •••. Advance online publication. DOI: 10.1007/978-1-4939-8916-4_1 PMID: 30547452

Sengul, A. B., & Asmatulu, E. (2020). Toxicity of metal and metal oxide nanoparticles: A review. *Environmental Chemistry Letters*, *18*(5), 1659–1683.

Sies, H. (1993). Strategies of antioxidant defense. *European Journal of Biochemistry*, *215*(2), 213–219. PMID: 7688300

Soppimath, K. S., Aminabhavi, T. M., Kulkarni, A. R., & Rudzinski, W. E. (2001). Biodegradable polymeric nanoparticles as drug delivery devices. *Journal of Controlled Release*. Advance online publication. DOI: 10.1016/S0168-3659(00)00339-4 PMID: 11166403

Xuan, L., Ju, Z., Skonieczna, M., Zhou, P. K., & Huang, R. (2023). Nanoparticles-induced potential toxicity on human health: Applications, toxicity mechanisms, and evaluation models. *MedComm*, *4*(4), e327. PMID: 37457660

Compilation of References

Aarzoo, N., Naqvi, S., Agarwal, N. B., Singh, M. P., & Samim, M. (2021). Bio-engineered palladium nanoparticles: Model for risk assessment study of automotive particulate pollution on macrophage cell lines. *RSC Advances*, *11*(3), 1850–1861. DOI: 10.1039/d0ra09336j PMID: 35424083

Aarzoo, N., Siddiqui, M. A., Hasan, M., Nidhi, N., Khan, H. A., Rastogi, S., Arora, I., & Samim, M. (2024). Palladium Nanoparticles and Lung Health: Assessing Morphology-Dependent subacute toxicity in rats and toxicity modulation by Naringin – paving the way for cleaner vehicular emissions. *ACS Omega*. Advance online publication. DOI: 10.1021/acsomega.4c02269 PMID: 39100302

Abatangelo, G., Vindigni, V., Avruscio, G., Pandis, L., & Brun, P. (2020). Hyaluronic Acid: Redefining Its Role. Cells 2020, Vol. 9, Page 1743, 9(7), 1743. https://doi.org/DOI: 10.3390/CELLS9071743

Abbasi, R., Shineh, G., Mobaraki, M., Doughty, S., & Tayebi, L. (2023). Structural parameters of nanoparticles affecting their toxicity for biomedical applications: A review. *Journal of Nanoparticle Research*, *25*(3), 43. PMID: 36875184

Abbas, Z., Dapporto, M., Tampieri, A., & Sprio, S. (2021). Toughening of bioceramic composites for bone regeneration. *Journal of Composites Science*, *5*(10), 259. DOI: 10.3390/jcs5100259

Abdalla, M., Joplin, A., Elahinia, M., & Ibrahim, H. (2020). Corrosion Modeling of magnesium and its alloys for Biomedical applications [review]. *Corrosion and Materials Degradation*, *1*(2), 219–248. DOI: 10.3390/cmd1020011

Abd-Elaziem, W., Darwish, M. A., Hamada, A., & Daoush, W. M. (2024). Titanium-Based alloys and composites for orthopedic implants Applications: A comprehensive review. *Materials & Design*, 112850.

Abdel-Ghany, H. M., & Salem, M. E. (2020). Effects of dietary chitosan supplementation on farmed fish; a review. *Reviews in Aquaculture*, *12*(1), 438–452. DOI: 10.1111/raq.12326

Abdel-Salam, F. S., Elkheshen, S. A., Mahmoud, A. A., Basalious, E. B., Amer, M. S., Mostafa, A. A., & Elkasabgy, N. A. (2020). In-situ forming chitosan implant-loaded with raloxifene hydrochloride and bioactive glass nanoparticles for treatment of bone injuries: Formulation and biological evaluation in animal model. *International Journal of Pharmaceutics*, *580*, 119213. DOI: 10.1016/J.IJPHARM.2020.119213 PMID: 32165229

Abouelseoud, I., Salama, S., Mahmoud, B., Dawood, H., Moustafa, M., & Mahmoud, S. (2024). The Effect of Lepidium Sativum L. Extract Loaded on Chitosan Nanoparticles in Experimental Rat Model of Osteoporosis. *Journal of the Medical Research Institute*, *45*(1), 28–39. DOI: 10.21608/jm-alexu.2024.265274.1011

Abrisham, M., Noroozi, M., Panahi-Sarmad, M., Arjmand, M., Goodarzi, V., Shakeri, Y., & Golbaten-Mofrad, H.. (2020). The Role of Polycaprolactone-Triol (PCL-T) in Biomedical Applications: A State-of-the-Art Review. *European Polymer Journal*, *131*(May), 109701. DOI: 10.1016/j.eurpolymj.2020.109701

Acosta, M., Santiago, M. D., & Irvin, J. A. (2022). Electrospun conducting polymers: Approaches and applications. *Materials (Basel)*, *15*(24), 8820. DOI: 10.3390/ma15248820 PMID: 36556626

Adams, C. P., Walker, K. A., Obare, S. O., & Docherty, K. M. (2014). Size-Dependent Antimicrobial Effects of Novel Palladium Nanoparticles. *PLoS One*, *9*(1), e85981. DOI: 10.1371/journal.pone.0085981 PMID: 24465824

Adeli, F., Abbasi, F., Ghandforoushan, P., Külahlı, H. E., Meran, M., Abedi, F., Ghamkhari, A., & Afif, S. (2023). Recent advances in formulation and application of molecular polymer brushes in biomedicine: Therapeutic, diagnostic, and theranostics capabilities. In Nano Today (Vol. 53). https://doi.org/DOI: 10.1016/j.nantod.2023.102010

Adhikari, H. S., & Yadav, P. N. (2018). Anticancer Activity of Chitosan, Chitosan Derivatives, and Their Mechanism of Action. *International Journal of Biomaterials*, *2018*, 1–29. DOI: 10.1155/2018/2952085 PMID: 30693034

Adscheid, A., Selin, M. R.-R., Abdel-Hafez, S. M., Pavone, F. S., Schneider, M., Türeli, A. E., Calamai, M., & Günday-Türeli, N. (2025). Scalable Manufacturing Method for Model Protein-Loaded PLGA Nanoparticles: Biocompatibility, Trafficking and Release Properties. *Pharmaceutics*, *17*(1), 87–87. DOI: 10.3390/pharmaceutics17010087 PMID: 39861735

Afkham, A. A.-M. (2018). Chitosan (CMD)-mediated co-delivery of SN38 and Snail-specific siRNA as a useful anticancer approach against prostate cancer. *Pharmacological Reports*, *70*(3), 418–425. PMID: 29626645

Aflori, M. (2021). Smart nanomaterials for biomedical applications—A review. *Nanomaterials (Basel, Switzerland)*, *11*(2), 396. PMID: 33557177

Afra, S., Koch, M., Żur-Pińska, J., Dolatshahi, M., Bahrami, A. R., Sayed, J. E., Moradi, A., Matin, M. M., & Włodarczyk-Biegun, M. K. (2024). Chitosan/Nanohydroxyapatite/Hydroxyethyl-cellulose-based printable formulations for local alendronate drug delivery in osteoporosis treatment. *Carbohydrate Polymer Technologies and Applications*, *7*, 100418. DOI: 10.1016/j.carpta.2023.100418

Aftab, A. B. (2020). Microfluidic platform for encapsulation of plant extract in chitosan microcarriers embedding silver nanoparticles for breast cancer cells. *Applied Nanoscience*, *10*, 2281–2293.

Agarwal, K. M., Singh, P., Mohan, U., Mandal, S., & Bhatia, D. (2020). Comprehensive Study Related to Advancement in Biomaterials for Medical Applications. *Sensors International*, *1*(January). Advance online publication. DOI: 10.1016/j.sintl.2020.100055

Aggarwal, S., & Ikram, S. (2022). Zinc oxide nanoparticles-impregnated chitosan surfaces for covalent immobilization of trypsin: Stability & kinetic studies. *International Journal of Biological Macromolecules*, *207*, 205–221. DOI: 10.1016/j.ijbiomac.2022.03.014 PMID: 35259431

Aggarwal, V., Tuli, H. S., Varol, A., Thakral, F., Yerer, M. B., Sak, K., Varol, M., Jain, A., Khan, M. A., & Sethi, G. (2019). Role of reactive oxygen species in cancer progression: Molecular mechanisms and recent advancements. *Biomolecules*, *9*(11), 735. DOI: 10.3390/biom9110735 PMID: 31766246

Agnihotri, S. A., Mallikarjuna, N. N., & Aminabhavi, T. M. (2004). Recent advances on chitosan-based micro- and nanoparticles in drug delivery. *Journal of Controlled Release*, *100*(1), 5–28. DOI: 10.1016/j.jconrel.2004.08.010 PMID: 15491807

Agosta, F., Galantucci, S., & Filippi, M. (2017). Advanced magnetic resonance imaging of neurodegenerative diseases. *Neurological Sciences*, *38*(1), 41–51. PMID: 27848119

Agwa, M. M., Soliman, H. S. M., Elmotasem, H., & Sabra, S. (2024). Surface Functionalization for Biomaterials. In *Surface Engineering of Biomaterials Synthesis and Processing Techniques* (pp. 344–361). CRC Press., DOI: 10.1201/9781003429920-20

Ahadiat, S.-A., & Hosseinian, Z. (2023). Astrocytes' innate role in neurodegenerative disorders. *Bulletin of the National Research Center*, *47*(1), 105.

Ahangari, M., Johar, M., & Saremi, M. (2020). Hydroxyapatite-carboxymethyl cellulose-graphene composite coating development on AZ31 magnesium alloy: Corrosion behavior and mechanical properties. *Ceramics International*, *47*(3), 3529–3539. DOI: 10.1016/j.ceramint.2020.09.197

Aherwar, A., Singh, A. K., & Patnaik, A. (2015). Current and future biocompatibility aspects of biomaterials for hip prosthesis. *AIMS Bioengineering*, *3*(1), 23–43. DOI: 10.3934/bioeng.2016.1.23

Ahirwar, H., Zhou, Y., Mahapatra, C., Ramakrishna, S., Kumar, P., & Nanda, H. S. (2020). Materials for orthopedic bioimplants: Modulating degradation and surface modification using integrated nanomaterials. *Coatings*, *10*(3), 264. DOI: 10.3390/coatings10030264

Ahlström, M. G., Thyssen, J. P., Wennervaldt, M., Menné, T., & Johansen, J. D. (2019). Nickel allergy and allergic contact dermatitis: A clinical review of immunology, epidemiology, exposure, and treatment. *Contact Dermatitis*, *81*(4), 227–241. PMID: 31140194

Ahmad, A. (2022). Safety and toxicity implications of multifunctional drug delivery nanocarriers on reproductive systems in vitro and in vivo. *Frontiers in Toxicology*, *4*, 895667. PMID: 35785262

Ahmadi, S., Hivechi, A., Bahrami, S. H., Milan, P. B., & Ashraf, S. S. (2021). Cinnamon extract loaded electrospun chitosan/gelatin membrane with antibacterial activity. *International Journal of Biological Macromolecules*, *173*, 580–590. DOI: 10.1016/j.ijbiomac.2021.01.156 PMID: 33513421

Ahmad, S. I., & Ahmad, R., Khan, Mohd. S., Kant, R., Shahid, S., Gautam, L., Hasan, G. M., & Hassan, Md. I. (2020). Chitin and its derivatives: Structural properties and biomedical applications. *International Journal of Biological Macromolecules*, *164*, 526–539. DOI: 10.1016/j.ijbiomac.2020.07.098 PMID: 32682975

Ahmed, S., Annu, Ali, A., & Sheikh, J. (2018). A review on chitosan centred scaffolds and their applications in tissue engineering. *International Journal of Biological Macromolecules*, *116*, 849–862. DOI: 10.1016/J.IJBIOMAC.2018.04.176 PMID: 29730001

Aibani, N., Rai, R., Patel, P., Cuddihy, G., & Wasan, E. K. (2021). Chitosan nanoparticles at the biological interface: Implications for drug delivery. *Pharmaceutics, 13*(10), 1686. PMID: 34683979

Ai, H., Wang, F., Xia, Y., Chen, X., & Lei, C. (2012). Antioxidant, antifungal and antiviral activities of chitosan from the larvae of housefly, Musca domestica L. *Food Chemistry, 132*(1), 493–498. PMID: 26434321

Ait Hamdan, Y., El Amerany, F., Desbrières, J., Aghrinane, A., Oudadesse, H., & Rhazi, M. (2023). The evolution of the global COVID-19 epidemic in Morocco and understanding the different therapeutic approaches of chitosan in the control of the pandemic. *Polymer Bulletin, 80*(10), 10633–10659. DOI: 10.1007/s00289-022-04579-3 PMID: 36466080

Ajinkya, N., Yu, X., Kaithal, P., Luo, H., Somani, P., & Ramakrishna, S. (2020). Magnetic iron oxide nanoparticle (Ionp) synthesis to applications: Present and future. In Materials (Vol. 13, Issue 20, pp. 1–35). MDPI AG. doi: DOI: 10.3390/ma13204644

Akdeniz, N., Akpolat, V., Kale, A., Erdemoglu, M., Kuyumcuoglu, U., & Celik, Y. (2009). Risk factors for postmenopausal osteoporosis: Anthropometric measurements, age, age at menopause and the time elapsed after menopause onset. *Gynecological Endocrinology, 25*(2), 125–129. DOI: 10.1080/09513590802549817 PMID: 19253109

AKHDAR, R. &. (n.d.). Potentiel biologique des polysaccharides issus de Bifurcaria bifurcata R. Ross: Activité anti complément .

Akinrinmade, O A, CHETTY S, DARAMOLA A K. *et al.* (2017). CD64: An Attractive Immunotherapeutic Target for M1-type Macrophage Mediated Chronic Inflammatory Diseases . *Biomedicines, 5*(3), 56. PMID: 28895912

Aksungur, P., Sungur, A., Ünal, S., Iskit, A. B., Squier, C. A., & Şenel, S. (2004). Chitosan delivery systems for the treatment of oral mucositis: In vitro and in vivo studies. *Journal of Controlled Release, 98*(2), 269–279. DOI: 10.1016/j.jconrel.2004.05.002 PMID: 15262418

Al-Absi, M. Y., Caprifico, A. E., & Calabrese, G. (2023). Chitosan and Its Structural Modifications for siRNA Delivery. *Advanced Pharmaceutical Bulletin, 13*(2), 275–282. DOI: 10.34172/apb.2023.030 PMID: 37342385

Alamdari, S. G., Amini, M., Jalilzadeh, N., Baradaran, B., Mohammadzadeh, R., Mokhtarzadeh, A., & Oroojalian, F. (2022). Recent advances in nanoparticle-based photothermal therapy for breast cancer. *Journal of Controlled Release : Official Journal of the Controlled Release Society, 349*, 269–303. DOI: 10.1016/J.JCONREL.2022.06.050 PMID: 35787915

Alam, M., Rashid, S., Fatima, K., Adnan, M., Shafie, A., Akhtar, M. S., Ganie, A. H., Eldin, S. M., Islam, A., Khan, I., & Hassan, M. I. (2023). Biochemical features and therapeutic potential of α-Mangostin: Mechanism of action, medicinal values, and health benefits. *Biomedicine and Pharmacotherapy, 163*, 114710. DOI: 10.1016/J.BIOPHA.2023.114710 PMID: 37141737

Alaqarbeh, M., Adil, S. F., Ghrear, T., Khan, M., Bouachrine, M., & Al-Warthan, A. (2023). Recent Progress in the Application of Palladium Nanoparticles: A Review. *Catalysts, 13*(10), 1343. DOI: 10.3390/catal13101343

Albright, V., Zhuk, I., & Wang, Y.. (2017). Self-defensive antibiotic-loaded layer-by-layer coatings: Imaging of localized bacterial acidification and pH-triggering of antibiotic release . *Acta Biomaterialia*, *61*, 66–74. PMID: 28803214

Alexeev, V., Donahue, A., & Uitto, J.. (2013). Analysis of chemotactic molecules in bone marrow-derived mesenchymal stem cells and the skin: Ccl27-Ccr10 axis as a basis for targeting to cutaneous tissues . *Cytotherapy*, *15*(2), 171–184.e1. PMID: 23321329

Alfaidi, M., Barker, R. A., & Kuan, W. L. (2024). An update on immune-based alpha-synuclein trials in Parkinson's disease. *Journal of Neurology*, *272*(1), 21. DOI: 10.1007/s00415-024-12770-x PMID: 39666171

Alharbi, H. M. (2024). Exploring the Frontier of Biopolymer-Assisted Drug Delivery: Advancements, Clinical Applications, and Future Perspectives in Cancer Nanomedicine. *Drug Design, Development and Therapy*, *18*, 2063–2087. DOI: 10.2147/DDDT.S441325 PMID: 38882042

Ali, W., Li, M., Tillmann, L., Mayer, T., & González, C., LLorca, J., & Kopp, A. (2023). Bioabsorbable WE43 Mg alloy wires modified by continuous plasma-electrolytic oxidation for implant applications. Part I: Processing, microstructure and mechanical properties. *Biomaterials Advances*, *146*, 213314.

AlMaadeed, M. A. A., Ponnamma, D., & El-Samak, A. A. (2020). Polymers to improve the world and lifestyle: physical, mechanical, and chemical needs. In *Polymer Science and Innovative Applications* (pp. 1–19). Elsevier.

Alqudah, A., Aljabali, A. A., Gammoh, O., & Tambuwala, M. M. (2024). Advancements in neurotherapeutics: Nanoparticles overcoming the blood–brain barrier for precise CNS targeting. *Journal of Nanoparticle Research*, *26*(6), 1–19.

Alrahili, M. R. (2024). Insights into Gold Nanoparticle Optical Behavior: Size-Dependent Absorption and Extinction Peaks. *Journal of Nanoelectronics and Optoelectronics*, *19*(8), 783–790.

Alshubaily, F. A., & Jambi, E. J. (2022). Correlation between Antioxidant and Anti-Osteoporotic Activities of Shilajit Loaded into Chitosan Nanoparticles and Their Effects on Osteoporosis in Rats. *Polymers*, *14*(19), 3972. DOI: 10.3390/polym14193972 PMID: 36235920

Amarzguioui, M. H. (2003). Tolerance for mutations and chemical modifications in a siRNA. *Nucleic Acids Research*, *31*(2), 589–595. PMID: 12527766

Amen, R., Mukhtar, A., Saqib, S., Ullah, S., Al-Sehemi, A. G., Mehdi, S. E. H., . . . Bustam, M. A. (2021). History and development of nanomaterials. In *Nanomaterials: Synthesis, Characterization, Hazards and Safety* (pp. 1-14): Elsevier.

Amini, A. R., Laurencin, C. T., & Nukavarapu, S. P. (2012). Bone tissue engineering: recent advances and challenges. *Critical Reviews™ in Biomedical Engineering, 40*(5).

Amini, A. R., Laurencin, C. T., & Nukavarapu, S. P. (2012). Bone tissue engineering: Recent advances and challenges . *Critical Reviews in Biomedical Engineering*, *40*(5), 363–408. PMID: 23339648

Amini, A. R., Wallace, J. S., & Nukavarapu, S. P. (2011). Short-term and long-term effects of orthopedic biodegradable implants. *Journal of Long-Term Effects of Medical Implants*, *21*(2). PMID: 22043969

Amiryaghoubi, N. F. (2020). Bioactive polymeric scaffolds for osteogenic repair and bone regenerative medicine. *Medicinal Research Reviews*, *40*(5), 1833–1870. PMID: 32301138

Anitha, A., Sowmya, S., Kumar, P. T. S., Deepthi, S., Chennazhi, K. P., Ehrlich, H., Tsurkan, M., & Jayakumar, R. (2014). Chitin and chitosan in selected biomedical applications. *Progress in Polymer Science*, *39*(9), 1644–1667. DOI: 10.1016/J.PROGPOLYMSCI.2014.02.008

Ansari, M. A., Asiri, S. M. M., Alzohairy, M. A., Alomary, M. N., Almatroudi, A., & Khan, F. A. (2021). Biofabricated fatty acids-capped silver nanoparticles as potential antibacterial, antifungal, antibiofilm and anticancer agents. *Pharmaceuticals*, *14*(2), 139. PMID: 33572296

Antoniac, I., Manescu, V., Antoniac, A., & Paltanea, G. (2023). Magnesium-based alloys with adapted interfaces for bone implants and tissue engineering. *Regenerative Biomaterials*, *10*, rbad095.

Antoniac, I., Manescu, V., Paltanea, G., Antoniac, A., Nemoianu, I. V., Petrescu, M. I., & Bodog, A. D. (2022). Additive manufactured magnesium-based scaffolds for tissue engineering. *Materials (Basel)*, *15*(23), 8693. PMID: 36500191

Aranaz, I., Alcántara, A. R., Civera, M. C., Arias, C., Elorza, B., Heras Caballero, A., & Acosta, N. (2021). Chitosan: An overview of its properties and applications. *Polymers*, *13*(19), 3256. PMID: 34641071

Arcos, D. (2023). Nanomaterials in Biomedicine 2022. *International Journal of Molecular Sciences*, *24*(10). Advance online publication. DOI: 10.3390/ijms24109026 PMID: 37240372

Arendsen, L. P., Thakar, R., & Sultan, A. H. (2019). The use of copper as an antimicrobial agent in health care, including obstetrics and gynecology. *Clinical Microbiology Reviews*, *32*(4), 10–1128. DOI: 10.1128/CMR.00125-18 PMID: 31413046

Arif, M., Nawaz, A. F., Mueen, H., Rashid, F., Hemeg, H. A., & Rauf, A. (2023). Nanotechnology-based radiation therapy to cure cancer and the challenges in its clinical applications. *Heliyon*, *9*(6). PMID: 37389057

Arif, U., Haider, S., Haider, A., Khan, N., Alghyamah, A. A., Jamila, N., & Kang, I. K. (2019). Biocompatible polymers and their potential biomedical applications: A review. *Current Pharmaceutical Design*, *25*(34), 3608–3619. DOI: 10.2174/1381612825999191011105148 PMID: 31604409

Arora, R., & Baldi, A. (2024). Revolutionizing Neurological Disorder Treatment: Integrating Innovations in Pharmaceutical Interventions and Advanced Therapeutic Technologies. *Current Pharmaceutical Design*, *30*(19), 1459–1471. PMID: 38616755

Arora, S., Layek, B., & Singh, J. (2020). Design and validation of liposomal ApoE2 gene delivery system to evade blood–brain barrier for effective treatment of Alzheimer's disease. *Molecular Pharmaceutics*, *18*(2), 714–725. PMID: 32787268

Arora, S., Sharma, D., & Singh, J. (2020). GLUT-1: An Effective Target To Deliver Brain-Derived Neurotrophic Factor Gene Across the Blood Brain Barrier. *ACS Chemical Neuroscience*, *11*(11), 1620–1633. DOI: 10.1021/acschemneuro.0c00076 PMID: 32352752

Arvizo, R., Bhattacharya, R., & Mukherjee, P. (2010). Gold nanoparticles: Opportunities and challenges in nanomedicine. *Expert Opinion on Drug Delivery*, *7*(6), 753–763. DOI: 10.1517/17425241003777010 PMID: 20408736

Arya, G. V. (2011). Enhanced antiproliferative activity of Herceptin (HER2)-conjugated gemcitabine-loaded chitosan nanoparticle in pancreatic cancer therapy. *Nanomedicine; Nanotechnology, Biology, and Medicine*, *7*(6), 859–870. PMID: 21550422

Arzaghi, H., Adel, B., Jafari, H., Askarian-Amiri, S., Dezfuli, A. S., Akbarzadeh, A., & Pazoki-Toroudi, H. (2020). Nanomaterial integration into the scaffolding materials for nerve tissue engineering: A review. *Reviews in the Neurosciences*, *31*(8), 843–872. https://doi.org/doi:10.1515/revneuro-2020-0008 PMID: 32776904

Asadi, K., Samiraninezhad, N., Akbarizadeh, A. R., Amini, A., & Gholami, A. (2024). Stimuli-responsive hydrogel based on natural polymers for breast cancer. *Frontiers in Chemistry*, *12*, 1325204. DOI: 10.3389/FCHEM.2024.1325204/BIBTEX PMID: 38304867

Asgari, M., Hang, R., Wang, C., Yu, Z., Li, Z., & Xiao, Y. (2018). Biodegradable metallic wires in dental and orthopedic applications: A review. *Metals*, *8*(4), 212.

Ashrafizadeh, M., Delfi, M., Hashemi, F., Zabolian, A., Saleki, H., Bagherian, M., Azami, N., Farahani, M. V., Sharifzadeh, S. O., Hamzehlou, S., Hushmandi, K., Makvandi, P., Zarrabi, A., Hamblin, M. R., & Varma, R. S. (2021). Biomedical application of chitosan-based nanoscale delivery systems: Potential usefulness in siRNA delivery for cancer therapy. *Carbohydrate Polymers*, *260*(February), 117809. DOI: 10.1016/j.carbpol.2021.117809 PMID: 33712155

Ashrafizadeh, M., Zarrabi, A., Hushmandi, K., Hashemi, F., Rahmani Moghadam, E., Raei, M., Kalantari, M., Tavakol, S., Mohammadinejad, R., Najafi, M., Tay, F. R., & Makvandi, P. (2020). Progress in Natural Compounds/siRNA Co-delivery Employing Nanovehicles for Cancer Therapy. *ACS Combinatorial Science*, *22*(12), 669. DOI: 10.1021/ACSCOMBSCI.0C00099 PMID: 33095554

Ashtari, K., Nazari, H., Ko, H., Tebon, P., Akhshik, M., Akbari, M., Alhosseini, S. N., Mozafari, M., Mehravi, B., Soleimani, M., Ardehali, R., Ebrahimi Warkiani, M., Ahadian, S., & Khademhosseini, A. (2019). Electrically conductive nanomaterials for cardiac tissue engineering. *Advanced Drug Delivery Reviews*, *144*, 162–179. DOI: 10.1016/j.addr.2019.06.001 PMID: 31176755

Asif, M., Chaudhry, A. S., Ashar, A., Rashid, H. B., Saleem, M. H., Aslam, H. B., & Aziz, A. (2023). Zinc oxide nanoparticles accelerate the healing of methicillin-resistant Staphylococcus aureus (MRSA)-infected wounds in rabbits. *Asian Pacific Journal of Tropical Biomedicine*, *13*(11), 488–496. DOI: 10.4103/2221-1691.389574

Astruc, D. (2020). Introduction: Nanoparticles in Catalysis. In Chemical Reviews (Vol. 120, Issue 2, pp. 461–463). American Chemical Society. doi: DOI: 10.1021/acs.chemrev.8b00696

Attallah, O. A., Shetta, A., Elshishiny, F., & Mamdouh, W. (2020). Essential oil loaded pectin/chitosan nanoparticles preparation and optimization via Box–Behnken design against MCF-7 breast cancer cell lines. *RSC Advances*, *10*(15), 8703–8708. DOI: 10.1039/C9RA10204C PMID: 35496538

Attarilar, S., Yang, J., Ebrahimi, M., Wang, Q., Liu, J., Tang, Y., & Yang, J. (2020). The toxicity phenomenon and the related occurrence in metal and metal oxide nanoparticles: A brief review from the biomedical perspective. *Frontiers in Bioengineering and Biotechnology*, *8*, 822. PMID: 32766232

Augustine, R., Gezek, M., & Nikolopoulos, V. K.. (2024). Stem Cells in Bone Tissue Engineering: Progress, Promises and Challenges . *Stem Cell Reviews and Reports*, *20*(7), 1692–1731. PMID: 39028416

Autio, K. A., Dreicer, R., Anderson, J., Garcia, J. A., Alva, A., Hart, L. L., Milowsky, M. I., Posadas, E. M., Ryan, C. J., Graf, R. P., Dittamore, R., Schreiber, N. A., Summa, J. M., Youssoufian, H., Morris, M. J., & Scher, H. I. (2018). Safety and Efficacy of BIND-014, a Docetaxel Nanoparticle Targeting Prostate-Specific Membrane Antigen for Patients with Metastatic Castration-Resistant Prostate Cancer: A Phase 2 Clinical Trial. *JAMA Oncology*, *4*(10), 1344–1351. DOI: 10.1001/jamaoncol.2018.2168 PMID: 29978216

Ayogu, J. I., Elahi, N., & Zeinalipour-Yazdi, C. D. (2025). Emerging Trends in palladium Nanoparticles: Sustainable Approaches for Enhanced Cross-Coupling Catalysis. *Catalysts*, *15*(2), 181. DOI: 10.3390/catal15020181

Azizi-Lalabadi, M., Ehsani, A., Divband, B., & Alizadeh-Sani, M. (2019). Antimicrobial activity of Titanium dioxide and Zinc oxide nanoparticles supported in 4A zeolite and evaluation the morphological characteristic. *Scientific Reports*, *9*(1), 17439. PMID: 31767932

Azmi, N., Tiwari, A., Kaur, N., & Chaskar, A. (2023). Nanotechnology: A Promising Approach in Nerve Regeneration. *Current Nanoscience*, *19*(3), 305–321.

Azuma, K., Osaki, T., Kurozumi, S., Kiyose, M., Tsuka, T., Murahata, Y., Imagawa, T., Itoh, N., Minami, S., Sato, K., & Okamoto, Y. (2015). Anti-inflammatory effects of orally administered glucosamine oligomer in an experimental model of inflammatory bowel disease. *Carbohydrate Polymers*, *115*, 448–456. DOI: 10.1016/j.carbpol.2014.09.012 PMID: 25439918

Azzam, E. I., Jay-Gerin, J. P., & Pain, D. (2012). Ionizing radiation-induced metabolic oxidative stress and prolonged cell injury. *Cancer Letters*, *327*(1-2), 48–60. PMID: 22182453

Babaie, A., Bakhshandeh, B., Abedi, A., Mohammadnejad, J., Shabani, I., Ardeshirylajimi, A., Reza Moosavi, S., Amini, J., & Tayebi, L. (2020). Synergistic effects of conductive PVA/PEDOT electrospun scaffolds and electrical stimulation for more effective neural tissue engineering. *European Polymer Journal*, *140*, 110051. DOI: 10.1016/j.eurpolymj.2020.110051

Babu, A., & Ramesh, R. (2017). Multifaceted Applications of Chitosan in Cancer Drug Delivery and Therapy. *Marine Drugs*, *15*(4), 96. DOI: 10.3390/md15040096 PMID: 28346381

Bach-Toledo, L., Hryniewicz, B. M., Marchesi, L. F., Dall'Antonia, L. H., Vidotti, M., & Wolfart, F. (2020). Conducting polymers and composites nanowires for energy devices: A brief review. *Materials Science for Energy Technologies*, *3*, 78–90. DOI: 10.1016/j.mset.2019.09.006

Badkoobeh, F., Mostaan, H., Rafiei, M., & Bakhsheshi-Rad, H. R. Seeram RamaKrishna, Xiongbiao Chen, Additive manufacturing of biodegradable magnesium-based materials: Design strategies, properties, and biomedical applications, Journal of Magnesium and Alloys, Volume 11, Issue 3, 2023, Pages 801-839, ISSN 2213-9567, https://doi.org/DOI: 10.1016/j.jma.2022.12.001

Bahadar, H., Maqbool, F., Niaz, K., & Abdollahi, M. (2016). Toxicity of nanoparticles and an overview of current experimental models. *Iranian Biomedical Journal*. Advance online publication. DOI: 10.7508/ibj.2016.01.001 PMID: 26286636

Bai, R., & Ge, X. (2024). Blood-brain barrier disruption following brain injury: Implications for clinical practice.

Bai, Z., & Gilbert, J. L. (2006). Corrosion performance of stainless steels, cobalt, and titanium alloys in biomedical applications. In *Corrosion: Environments and Industries* (pp. 837-852). ASM International.

Baig, N., Kammakakam, I., & Falath, W. (2021). Nanomaterials: A review of synthesis methods, properties, recent progress, and challenges. *Materials Advances*, *2*(6), 1821–1871. DOI: 10.1039/D0MA00807A

Bai, J., Wang, X., Ye, X., Li, X., Tan, X., Wang, C., Guo, Y., Dong, J., Shao, Y., Chu, C., Xue, F., Tao, L., & Huang, L. (2024). Constructing Mg-phytic acid enhanced hydrogel patch with janus structure for antibacterial and angiogenic urethral repair. *Materials Today. Communications*, *41*, 110621. DOI: 10.1016/j.mtcomm.2024.110621

Bajpai, A. K., Shukla, S. K., Bhanu, S., & Kankane, S. (2008). Responsive polymers in controlled drug delivery. *Progress in Polymer Science*, *33*(11), 1088–1118.

Bakhsheshi-Rad, H., Hamzah, E., Kasiri-Asgarani, M., Saud, S. N., Yaghoubidoust, F., & Akbari, E. (2016). Structure, corrosion behavior, and antibacterial properties of nano-silica/graphene oxide coating on biodegradable magnesium alloy for biomedical applications. *Vacuum*, *131*, 106–110. DOI: 10.1016/j.vacuum.2016.05.021

Banaszkiewicz, P. A. (2012). Hip clinical cases. *Postgraduate Orthopaedics: The Candidate's Guide to the FRCS (Tr and Orth) Examination*, 69.

Bandyopadhyay, A., Mitra, I., Goodman, S. B., Kumar, M., & Bose, S. (2023). Improving biocompatibility for next generation of metallic implants. *Progress in Materials Science*, *133*, 101053. PMID: 36686623

Banerjee, A. N. (2018). Graphene and its derivatives as biomedical materials: Future prospects and challenges. *Interface Focus*, *8*(3), 20170056. DOI: 10.1098/rsfs.2017.0056 PMID: 29696088

Banerjee, M. K., Tiwari, A., & Kumar, N. (2023b). Ageing characteristics of stir cast AZ 61 alloy with minor additions. *Recent Patents on Engineering*, *18*. Advance online publication. DOI: 10.2174/1872212118666230609122741

Bao, W., He, Y., Fan, Y., & Liao, Y. (2018). Metal allergy in total-joint arthroplasty: Case report and literature review. *Medicine*, *97*(38), e12475. PMID: 30235744

Baranwal, J., Barse, B., Fais, A., Delogu, G. L., & Kumar, A. (2022). Biopolymer: A sustainable material for food and medical applications. *Polymers*, *14*(5), 983. PMID: 35267803

Barbuti, A. (2010). The "hearty" fat: Adipocytes as a source of functional cardiomyocytes . *Cardiovascular Research*, *85*(1), 1–2. PMID: 19887381

Barceloux, D. G., & Barceloux, D. (1999a). Molybdenum. *Journal of Toxicology. Clinical Toxicology*, *37*(2), 231–237. PMID: 10382558

Barceloux, D. G., & Barceloux, D. (1999b). Vanadium. *Journal of Toxicology. Clinical Toxicology, 37*(2), 265–278. PMID: 10382561

Bárcena, C., Sra, A. K., & Gao, J. (2009). Applications of Magnetic Nanoparticles in Biomedicine. In Liu, J. P., Fullerton, E., Gutfleisch, O., & Sellmyer, D. J. (Eds.), *Nanoscale Magnetic Materials and Applications* (pp. 591–626). Springer US., DOI: 10.1007/978-0-387-85600-1_20

Barfield, W. R., Colbath, G., DesJardins, J. D., An, Y. H., & Hartsock, L. A. (2012). The potential of magnesium alloy use in orthopaedic surgery. *Current Orthopaedic Practice, 23*(2), 146–150. DOI: 10.1097/bco.0b013e31824a553b

Baron, R., Ferrari, S., & Russell, R. G. G. (2011). Denosumab and bisphosphonates: Different mechanisms of action and effects. *Bone, 48*(4), 677–692. DOI: 10.1016/J.BONE.2010.11.020 PMID: 21145999

Barrère, F., Mahmood, T., De Groot, K., & Van Blitterswijk, C. (2008). Advanced biomaterials for skeletal tissue regeneration: Instructive and smart functions. *Materials Science and Engineering R Reports, 59*(1–6), 38–71. DOI: 10.1016/j.mser.2007.12.001

Barua, N., & Buragohain, A. K. (2024). Therapeutic Potential of Silver Nanoparticles (AgNPs) as an Antimycobacterial Agent: A Comprehensive Review. *Antibiotics (Basel, Switzerland), 13*(11), 1106. PMID: 39596799

Baryeh, K. (2019). *Development of quantitative lateral flow strip biosensors for the detection of cancer biomarkers.* North Dakota State University, Ch, S., Patel, T. K., Biswas, S., & Ghosh, B. (2024). Micelles-based drug delivery for pancreatic cancer. In *Recent Advances in Nanocarriers for Pancreatic Cancer Therapy* (pp. 115-144): Elsevier.

Bashir, M., Yousuf, I., & Prakash Prasad, C. (2022). Mixed Ni(II) and Co(II) complexes of nalidixic acid drug: Synthesis, characterization, DNA/BSA binding profile and in vitro cytotoxic evaluation against MDA-MB-231 and HepG2 cancer cell lines. *Spectrochimica Acta. Part A: Molecular and Biomolecular Spectroscopy, 271*, 120910. DOI: 10.1016/J.SAA.2022.120910 PMID: 35077983

Basko-Plluska, J. L., Thyssen, J. P., & Schalock, P. C. (2011). Cutaneous and systemic hypersensitivity reactions to metallic implants. *Derm, 22*(2), 65–79. PMID: 21504692

Baus, R. A., Zahir-Jouzdani, F., Dünnhaupt, S., Atyabi, F., & Bernkop-Schnürch, A. (2019). Mucoadhesive hydrogels for buccal drug delivery: In vitro-in vivo correlation study. *European Journal of Pharmaceutics and Biopharmaceutics, 142*, 498–505. DOI: 10.1016/J.EJPB.2019.07.019 PMID: 31330258

Bayda, S., Adeel, M., Tuccinardi, T., Cordani, M., & Rizzolio, F. (2019). The history of Nanoscience and Nanotechnology: From Chemical–Physical applications to Nanomedicine. *Molecules (Basel, Switzerland), 25*(1), 112. DOI: 10.3390/molecules25010112 PMID: 31892180

Begines, B., Ortiz, T., Pérez-Aranda, M., Martínez, G., Merinero, M., Argüelles-Arias, F., & Alcudia, A. (2020). Polymeric Nanoparticles for Drug Delivery: Recent Developments and Future Prospects. *Nanomaterials (Basel, Switzerland), 10*(7). Advance online publication. DOI: 10.3390/nano10071403 PMID: 32707641

Bellucci, D., Sola, A., & Cannillo, V. (2016). Hydroxyapatite and tricalcium phosphate composites with bioactive glass as second phase: State of the art and current applications. *Journal of Biomedical Materials Research. Part A*, *104*(4), 1030–1056. PMID: 26646669

Benhammadi, M., . (2014). Évaluation de l'efficacité des nanoparticules chitosane-siARN à inhiber, in vitro, l'expression du gène MDR1 responsable de la résistance à la chimiothérapie.

Berna, G., De Grazia, A., Antoniazzi, E., Romeo, M., Dell'Antonia, F., Lovero, S., Marchica, P., Rizzetto, C., & Burelli, P. (2022). Novel three-dimensional acellular dermal matrix for prepectoral breast reconstruction: First year in review with BRAXON®Fast. *Frontiers in Surgery*, *9*, 970053. DOI: 10.3389/FSURG.2022.970053/BIBTEX PMID: 36132205

Bhalodi, K., Kothari, C., & Butani, S. (2024). Next-generation cancer nanotherapeutics: Pluronic® F127 based mixed micelles for enhanced drug delivery. *Naunyn-Schmiedeberg's Archives of Pharmacology*, •••, 1–30. PMID: 39527309

Bhardwaj, B., Singh, P., Kumar, A., Kumar, S., & Budhwar, V. (2020). Eco-Friendly greener synthesis of nanoparticles. *Advanced Pharmaceutical Bulletin*, *10*(4), 566–576. PMID: 33072534

Bhatt, P., Joshi, S., Urper Bayram, G. M., Khati, P., & Simsek, H. (2023). Developments and application of chitosan-based adsorbents for wastewater treatments. *Environmental Research, 226*(November 2022), 115530. https://doi.org/DOI: 10.1016/j.envres.2023.115530

Bhattacharjee, R. N. (2023). Nanotheranostics to target antibiotic-resistant bacteria: Strategies and applications. *OpenNano*, *11*, 100138.

Bhattacharya, T., Preetam, S., Ghosh, B., Chakrabarti, T., Chakrabarti, P., Samal, S. K., & Thorat, N. (2023). Advancement in Biopolymer Assisted Cancer Theranostics. *ACS Applied Bio Materials*, *6*(10), 3959–3983. DOI: 10.1021/ACSABM.3C00458/ASSET/IMAGES/LARGE/MT3C00458_0010.JPEG PMID: 37699558

Bhong, M., Khan, T. K., Devade, K., Krishna, B. V., Sura, S., Eftikhaar, H., Thethi, H. P., & Gupta, N. (2023). Review of composite materials and applications. *Materials Today: Proceedings*.

Bianchi, M., Guzzo, S., Lunghi, A., Greco, P., Pisciotta, A., Murgia, M., Carnevale, G., Fadiga, L., & Biscarini, F. (2023). Synergy of nanotopography and electrical conductivity of PEDOT/PSS for enhanced neuronal development. *ACS Applied Materials & Interfaces*, *15*(51), 59224–59235. DOI: 10.1021/acsami.3c15278 PMID: 38091494

Bidez, M., & Misch, C. (2005). Clinical biomechanics in implant dentistry. *Misch's Contemporary. Implant Dentistry*, 140–151.

Biomaterials generations. (n.d.).

Bi, S., Srivastava, R., & Ahmad, T. (2024). The potential antifungal activity of the developed palladium nanoparticles. *Biomedical & Pharmacology Journal*, *17*(4), 2617–2628. DOI: 10.13005/bpj/3053

Bisht, N., Dwivedi, N., Kumar, P., Venkatesh, M., Yadav, A. K., Mishra, D., & Dhand, C. (2022). Recent advances in copper and copper-derived materials for antimicrobial resistance and infection control. *Current Opinion in Biomedical Engineering*, *24*, 100408. DOI: 10.1016/j.cobme.2022.100408 PMID: 36033159

Bivas-Benita, M. L. (2003). Generation of Toxoplasma gondii GRA1 protein and DNA vaccine loaded chitosan particles: Preparation, characterization, and preliminary in vivo studies. *International Journal of Pharmaceutics*, *266*(1-2), 17–27. PMID: 14559390

Bolton, J. L. (2016). Menopausal Hormone Therapy, Age, and Chronic Diseases: Perspectives on Statistical Trends. *Chemical Research in Toxicology*, *29*(10), 1583–1590. DOI: 10.1021/acs.chemrestox.6b00272 PMID: 27636306

Bonilla, J., Fortunati, E., Atarés, L., Chiralt, A., & Kenny, J. M. (2014). Physical, structural and antimicrobial properties of poly vinyl alcohol–chitosan biodegradable films. *Food Hydrocolloids*, *35*, 463–470. DOI: 10.1016/j.foodhyd.2013.07.002

Bonorino, J. A., & Teves, J. I. (2024). TKA Aseptic Loosening. In *Orthopaedics and Trauma: Current Concepts and Best Practices* (pp. 1585-1591). Springer.

Bootman, M. D. (2012). Calcium signaling. *Cold Spring Harbor Perspectives in Biology*, *4*(7), a011171. DOI: 10.1101/cshperspect.a011171 PMID: 22751152

Bootman, M. D., & Bultynck, G. (2020). Fundamentals of cellular calcium signaling: A primer. *Cold Spring Harbor Perspectives in Biology*, *12*(1), a038802. DOI: 10.1101/cshperspect.a038802 PMID: 31427372

Bordbar-Khiabani, A., Yarmand, B., & Mozafari, M. (2019). Emerging magnesium-based biomaterials for orthopedic implantation. *Emerging Materials Research*, *8*(3), 305–319. DOI: 10.1680/jemmr.18.00048

Bosse, M. J., Gruber, H. E., & Ramp, W. K. (2005). Internalization of bacteria by osteoblasts in a patient with recurrent, long-term osteomyelitis: A case report. *JBJS*, *87*(6), 1343–1347. PMID: 15930546

Bouhadjer, R., & Zabar, L. (2023). *Étude bibliographique sur l'activité des flavonoides sur l'organisme humain.*

Boulton, M., & Al-Rubaie, A. (2024). Neuroinflammation and neurodegeneration following traumatic brain injuries. *Anatomical Science International*, •••, 1–12. PMID: 38739360

Boutou, A. K., Pitsiou, G. G., Stanopoulos, I., Kontakiotis, T., Papakosta, D., & Kioumis, I. (2019). Regular Treatment and Disease Outcomes in Patients with Chronic Obstructive Pulmonary Disease Group A. *Respiration*, *99*(5), 458. DOI: 10.1159/000504142 PMID: 31694036

Boyes, W. K., & van Thriel, C. (2020). Neurotoxicology of nanomaterials. *Chemical Research in Toxicology*, *33*(5), 1121–1144. PMID: 32233399

Braak, H., & Braak, E. (1991). Neuropathological stageing of Alzheimer-related changes. *Acta Neuropathologica*, *82*(4), 239–259. PMID: 1759558

Brand, W., Noorlander, C. W., Giannakou, C., De Jong, W. H., Kooi, M. W., Park, M. V., & Geertsma, R. E. (2017). Nanomedicinal products: A survey on specific toxicity and side effects. *International Journal of Nanomedicine*, •••, 6107–6129. PMID: 28883724

Brar, H. S., Platt, M. O., Sarntinoranont, M., Martin, P. I., & Manuel, M. V. (2009). Magnesium as a biodegradable and bioabsorbable material for medical implants. *JOM*, *61*(9), 31–34. DOI: 10.1007/s11837-009-0129-0

Bray, F., Laversanne, M., Sung, H., Ferlay, J., Siegel, R. L., Soerjomataram, I., & Jemal, A. (2024). Global cancer statistics 2022: GLOBOCAN estimates of incidence and mortality worldwide for 36 cancers in 185 countries. *CA: a Cancer Journal for Clinicians*. Advance online publication. DOI: 10.3322/CAAC.21834 PMID: 38572751

Brittenham, G. M. (2003). Iron chelators and iron toxicity. *Alcohol (Fayetteville, N.Y.)*, *30*(2), 151–158. DOI: 10.1016/S0741-8329(03)00101-0 PMID: 12957300

Brochu, B. M., Sturm, S. R., Kawase De Queiroz Goncalves, J. A., Mirsky, N. A., Sandino, A. I., Panthaki, K. Z., Panthaki, K. Z., Nayak, V. V., Daunert, S., & Witek, L. (2024). Advances in Bioceramics for Bone Regeneration: A Narrative Review. *Biomimetics*, *9*(11), 690. PMID: 39590262

Browne, J. A., Bechtold, C. D., Berry, D. J., Hanssen, A. D., & Lewallen, D. G. (2010). Failed metal-on-metal hip arthroplasties: A spectrum of clinical presentations and operative findings. *Clinical Orthopaedics and Related Research*, *468*, 2313–2320. PMID: 20559767

Brown, J. S. (2023). Updating the definition of cancer. *Molecular Cancer Research*, *21*(11), 1142–1147. PMID: 37409952

Brugnerotto, J., Lizardi, J., Goycoolea, F., Argüelles-Monal, W., Desbrières, J., & Rinaudo, M. (2001). An infrared investigation in relation with chitin and chitosan characterization. *Polymer*, *42*(8), 3569–3580. DOI: 10.1016/S0032-3861(00)00713-8

Bruna, T., Maldonado-Bravo, F., Jara, P., & Caro, N. (2021). Silver nanoparticles and their antibacterial applications. *International Journal of Molecular Sciences*, *22*(13), 7202. DOI: 10.3390/ijms22137202 PMID: 34281254

Brusatin, G., Panciera, T., Gandin, A., Citron, A., & Piccolo, S. (2018). Biomaterials and engineered microenvironments to control YAP/TAZ-dependent cell behaviour. *Nature Materials*, *17*(12), 1063–1075. DOI: 10.1038/s41563-018-0180-8 PMID: 30374202

Bukhari, S. N. A., Roswandi, N. L., Waqas, M., Habib, H., Hussain, F., Khan, S., Sohail, M., Ramli, N. A., Thu, H. E., & Hussain, Z. (2018). Hyaluronic Acid, a Promising Skin Rejuvenating Biomedicine: A Review of Recent Updates and Pre-Clinical and Clinical Investigations on Cosmetic and Nutricosmetic Effects. *International Journal of Biological Macromolecules*. Advance online publication. DOI: 10.1016/j.ijbiomac.2018.09.188 PMID: 30287361

Bull, E., Madani, S. Y., Sheth, R., Seifalian, A., Green, M., & Seifalian, A. M. (2014). Stem cell tracking using iron oxide nanoparticles. In International Journal of Nanomedicine (Vol. 9, Issue 1, pp. 1641–1653). Dove Medical Press Ltd. doi: DOI: 10.2147/IJN.S48979

Burgunder, J.-M. (2024). Genetics of Huntington's disease and related disorders: beyond triplet repeats.

Burnstine-Townley, A., Eshel, Y., & Amdursky, N. (2020). Conductive scaffolds for cardiac and neuronal tissue engineering: Governing factors and mechanisms. *Advanced Functional Materials*, *30*(18), 1901369. DOI: 10.1002/adfm.201901369

Busch, A., Jäger, M., Beck, S., Wegner, A., Portegys, E., Wassenaar, D., Theysohn, J., & Haubold, J. (2022). Metal Artefact Reduction Sequences (MARS) in Magnetic Resonance Imaging (MRI) after Total Hip Arthroplasty (THA) A non-invasive approach for preoperative differentiation between periprosthetic joint infection (PJI) and aseptic complications? *BMC Musculoskeletal Disorders*, *23*(1), 620. PMID: 35764987

Butler, J., Handy, R. D., Upton, M., & Besinis, A. (2023). Review of antimicrobial nanocoatings in medicine and dentistry: Mechanisms of action, biocompatibility performance, safety, and benefits compared to antibiotics. *ACS Nano*, *17*(8), 7064–7092. DOI: 10.1021/acsnano.2c12488 PMID: 37027838

Butterfield, D. A., & Halliwell, B. (2019). Oxidative stress, dysfunctional glucose metabolism and Alzheimer disease. *Nature Reviews. Neuroscience*, *20*(3), 148–160. PMID: 30737462

Caballero, D., Abreu, C. M., Lima, A. C., Neves, N. N., Reis, R. L., & Kundu, S. C. (2022). Precision biomaterials in cancer theranostics and modelling. In Biomaterials (Vol. 280). https://doi.org/DOI: 10.1016/j.biomaterials.2021.121299

Cacciatore, I., Ciulla, M., Fornasari, E., Marinelli, L., & Di Stefano, A. (2016). Solid lipid nanoparticles as a drug delivery system for the treatment of neurodegenerative diseases. *Expert Opinion on Drug Delivery*, *13*(8), 1121–1131. DOI: 10.1080/17425247.2016.1178237 PMID: 27073977

Cai, S., Wu, C., Yang, W., Liang, W., Yu, H., & Liu, L. (2020). Recent advance in surface modification for regulating cell adhesion and behaviors. *Nanotechnology Reviews*, *9*(1), 971–989.

Cai, Y., Ding, H., Chen, X., Chen, Y., Huang, C., Zhang, C., Huang, Z., Huang, Y., Li, W., & Zhang, W. (2023). Optimization and standardization of mNGS-based procedures for the diagnosis of Mycoplasma periprosthetic joint infection: A novel diagnostic strategy for rare bacterial periprosthetic joint infection. *Frontiers in Cellular and Infection Microbiology*, *13*, 1089919. PMID: 36936762

Cai, Y., Huang, C., Chen, X., Chen, Y., Huang, Z., Zhang, C., Zhang, W., & Fang, X. (2022). The role of Staphylococcus aureus small colony variants in intraosseous invasion and colonization in periprosthetic joint infection. *Bone & Joint Research*, *11*(12), 843–853. PMID: 36453022

Camorani, S., Tortorella, S., Agnello, L., Spanu, C., d'Argenio, A., Nilo, R., Zannetti, A., Locatelli, E., Fedele, M., Comes Franchini, M., & Cerchia, L. (2022). Aptamer-Functionalized Nanoparticles Mediate PD-L1 siRNA Delivery for Effective Gene Silencing in Triple-Negative Breast Cancer Cells. *Pharmaceutics*, *14*(10), 2225. DOI: 10.3390/PHARMACEUTICS14102225/S1 PMID: 36297659

Cano, L., Thompson, A., & Michaels, D. (2021). Antimicrobial properties of copper surfaces in hospital environments. *Healthcare Infection Control*, *8*(2), 110–125. DOI: 10.1016/j.hic.2021.110125

Cano-Vicent, A., Tuñón-Molina, A., Martí, M., & Serrano-Aroca, Á. (2023). Biocompatible Chitosan Films Containing Acetic Acid Manifested Potent Antiviral Activity against Enveloped and Non-Enveloped Viruses. *International Journal of Molecular Sciences*, *24*(15), 12028. DOI: 10.3390/ijms241512028 PMID: 37569404

Cao, D., & Ding, J. (2022). Recent advances in regenerative biomaterials. *Regenerative Biomaterials, 9*, rbac098. PMID: 36518879

Cao, Y., Tan, Y. F., Wong, Y. S., Liew, M. W. J., & Venkatraman, S. (2019). Recent advances in chitosan-based carriers for gene delivery. *Marine Drugs, 17*(6), 381. PMID: 31242678

Capuani, S., Malgir, G., Chua, C. Y. X., & Grattoni, A. (2022). Advanced strategies to thwart foreign body response to implantable devices. *Bioengineering & Translational Medicine, 7*(3), e10300. PMID: 36176611

Capuani, S., Malgir, G., Corrine, Y. X. C., & Grattoni, A. (2022). *"Advanced Strategies to Thwart Foreign Body Response to Implantable Devices." Bioengineering and Translational Medicine*. John Wiley and Sons Inc., DOI: 10.1002/btm2.10300

Cardozo, B., Karatza, E., & Karalis, V. (2021). Osteoporosis treatment with risedronate: A population pharmacokinetic model for the description of its absorption and low plasma levels. *Osteoporosis International, 32*(11), 2313–2321. DOI: 10.1007/s00198-021-05944-0 PMID: 34002251

Carla, P. D. (2019). *Hydrogels injectables et éponges à base de complexe polyélectrolytes (chitosane/polymère de cyclodextrine) pour une application en ingénierie tissulaire osseuse.*

Carnevale, L., & Lembo, G. (2019). Innovative MRI techniques in neuroimaging approaches for cerebrovascular diseases and vascular cognitive impairment. *International Journal of Molecular Sciences, 20*(11), 2656. PMID: 31151154

Carrow, J. K., & Gaharwar, A. K. (2015). Bioinspired polymeric nanocomposites for regenerative medicine. *Macromolecular Chemistry and Physics, 216*(3), 248–264. DOI: 10.1002/macp.201400427

Casella, A., Panitch, A., & Leach, J. K. (2021). Endogenous electric signaling as a blueprint for conductive materials in tissue engineering. *Bioelectricity, 3*(1), 27–41. DOI: 10.1089/bioe.2020.0027 PMID: 34476376

Catalani, S., Stea, S., Beraudi, A., Gilberti, M., Bordini, B., Toni, A., & Apostoli, P. (2013). Vanadium release in whole blood, serum and urine of patients implanted with a titanium alloy hip prosthesis. *Clinical Toxicology, 51*(7), 550–556. PMID: 23845123

Cavalli, R. (2014). New chitosan nanospheres for the delivery of 5-fluorouracil. *Current Drug Delivery, 11*, 270–278. PMID: 24499357

Chancellor, J. C., Blue, R. S., Cangel, K. A., Aunon-Chancellor, S. M., Rubins, K. H., Katzgraber, H. G., & Kennedy, A. R. (2018). Limitations in predicting the space radiation health risk for exploration astronauts. npj Microgravity 4.

Chang, D. M. (2021). Stimuli-responsive polymeric nanoplatforms for cancer therapy. *Frontiers in Bioengineering and Biotechnology, 9*, 707319. PMID: 34249894

Chang, Y., Jiang, J., Chen, W., Yang, W., Chen, L., Chen, P., Shen, J., Qian, S., Zhou, T., Wu, L., Hong, L., Huang, Y., & Li, F. (2020). Biomimetic metal-organic nanoparticles prepared with a 3D-printed microfluidic device as a novel formulation for disulfiram-based therapy against breast cancer. *Applied Materials Today, 18*. Advance online publication. DOI: 10.1016/J.APMT.2019.100492 PMID: 34746366

Chaudhari, U., Pohjolainen, L., & Ruskoaho, H.. (2023). Genome-wide profiling of miRNA-gene regulatory networks in mouse postnatal heart development-implications for cardiac regeneration . *Frontiers in Cardiovascular Medicine*, *10*, 1148618. PMID: 37283582

Chavda, V. P., Vuppu, S., Bezbaruah, R., Nalla, L. V., Gajula, S. N. R., Balar, P. C., Mishra, T., Sharma, N., Kamaraj, S., Suresh, T., Sairam, A., & Bhattacharjee, B. (2023). Phytochemical Loaded Nanovehicles of Biopolymer for Breast Cancer: A Systemic Review. *Clinical Complementary Medicine and Pharmacology*, *3*(4), 100114. DOI: 10.1016/J.CCMP.2023.100114

Chen, J., Zhang, Z., Li, Y., Zeng, H., Li, Z., Wang, C., Xu, C., Deng, Q., Wang, Q., Yang, X., & Li, Z. (2022). Precise fibrin decomposition and tumor mechanics modulation with hydroxyethyl starch-based smart nanomedicine for enhanced antitumor efficacy.

Chen, A., & Ostrom, C. (2015). Palladium-Based Nanomaterials: Synthesis and Electrochemical Applications. *Chemical Reviews*, *115*(21), 11999–12044. DOI: 10.1021/acs.chemrev.5b00324 PMID: 26402587

Chen, B., Xiang, H., Pan, S., Yu, L., Xu, T., & Chen, Y. (2020). Advanced theragenerative biomaterials with therapeutic and regeneration multifunctionality. *Advanced Functional Materials*, *30*(34), 2002621.

Chen, C., Wu, W., Xu, X., Zhang, L., Liu, Y., & Wang, K. (2014). Chain conformation and anti-tumor activity of derivatives of polysaccharide from Rhizoma Panacis Japonici. *Carbohydrate Polymers*, *105*, 308–316. DOI: 10.1016/j.carbpol.2014.01.089 PMID: 24708985

Chen, G., & Lv, Y. (2018). Decellularized Bone Matrix Scaffold for Bone Regeneration . *Methods in Molecular Biology (Clifton, N.J.)*, *1577*, 239–254. PMID: 28770492

Cheng, A. G., DeDent, A. C., Schneewind, O., & Missiakas, D. (2011). A play in four acts: Staphylococcus aureus abscess formation. *Trends in Microbiology*, *19*(5), 225–232. PMID: 21353779

Cheng, B. G. (2019). Repurposing suramin for the treatment of breast cancer lung metastasis with glycol chitosan-based nanoparticles. *Acta Biomaterialia*, *84*, 378–390. PMID: 30528604

Cheng, M., Wahafu, T., Jiang, G., Liu, W., Qiao, Y., Peng, X., Cheng, T., Zhang, X., He, G., & Liu, X. (2016). A novel open-porous magnesium scaffold with controllable microstructures and properties for bone regeneration. *Scientific Reports*, *6*(1). Advance online publication. DOI: 10.1038/srep24134 PMID: 27071777

Cheng, X., Xie, Q., & Sun, Y. (2023). Advances in nanomaterial-based targeted drug delivery systems. *Frontiers in Bioengineering and Biotechnology*, *11*, 1177151. PMID: 37122851

Chen, H. C. (2015). Grafting chitosan with polyethylenimine in an ionic liquid for efficient gene delivery. *PLoS One*, *10*(4), e0121817. PMID: 25875475

Chen, H., Qin, Z., & Zhao, J.. (2019). Cartilage-targeting and dual MMP-13/pH responsive theranostic nanoprobes for osteoarthritis imaging and precision therapy . *Biomaterials*, *225*, 119520. PMID: 31586865

Chen, J., Dou, Y., Tang, Y., & Zhang, X. (2020). Folate receptor-targeted RNAi nanoparticles for silencing STAT3 in tumor-associated macrophages and tumor cells. *Nanomedicine; Nanotechnology, Biology, and Medicine*, *25*, 102173. DOI: 10.1016/j.nano.2020.102173 PMID: 32084593

Chen, L., Yang, J., Cai, Z., Huang, Y., Xiao, P., Wang, J., Wang, F., Huang, W., Cui, W., & Hu, N. (2024). Electroactive biomaterials regulate the electrophysiological microenvironment to promote bone and cartilage tissue regeneration. *Advanced Functional Materials, 2314079*. Advance online publication. DOI: 10.1002/adfm.202314079

Chen, M., Li, M., & Wei, Y.. (2022). ROS-activatable biomimetic interface mediates in-situ bioenergetic remodeling of osteogenic cells for osteoporotic bone repair . *Biomaterials, 291*, 121878. PMID: 36335716

Chen, P., Zhang, C., & He, P.. (2022). A Biomimetic Smart Nanoplatform as "Inflammation Scavenger" for Regenerative Therapy of Periodontal Tissue . *International Journal of Nanomedicine, 17*, 5165–5186. PMID: 36388874

Chen, Q., Xu, L., Liang, C., Wang, C., Peng, R., & Liu, Z. (2016). Photothermal therapy with immune-adjuvant nanoparticles together with checkpoint blockade for effective cancer immunotherapy. *Nature Communications*, *7*(1), 13193. DOI: 10.1038/ncomms13193 PMID: 27767031

Chen, S., Saeed, A. F. U. H., Liu, Q., Jiang, Q., Xu, H., Xiao, G. G., Rao, L., & Duo, Y. (2023). Macrophages in immunoregulation and therapeutics. *Signal Transduction and Targeted Therapy*, *8*(1), 207. DOI: 10.1038/s41392-023-01452-1 PMID: 37211559

Chen, S., Wan, P., Zhang, B., Erişen, D. E., Yang, H., & Yang, K. (2018). A novel polymer critical remelting treatment for improving corrosion resistance of magnesium alloy stent. *Journal of Materials Science and Technology*, *35*(1), 19–22. DOI: 10.1016/j.jmst.2018.09.021

Chen, W. H., Qi, W. C., Chen, Q., Cui, C., Duan, S., Kang, Y., & Liu, Y.. (2022). *"Biomedical Polymers: Synthesis, Properties, and Applications." Science China Chemistry*. Springer., DOI: 10.1007/s11426-022-1243-5

Chen, X., Xia, Y., Shen, S., Wang, C., Zan, R., Yu, H., & Zhang, X. (2023). Research on the Current Application Status of Magnesium Metal Stents in Human Luminal Cavities. *Journal of Functional Biomaterials*, *14*(9), 462. PMID: 37754876

Chen, Y.-C., Chen, C.-H., Chen, P.-L., Huang, I.-Y., Shen, Y.-S., & Chen, C.-M. (2006). Donor site morbidity after harvesting of proximal tibia bone. *Head & Neck*, *28*(6), 496–500. DOI: 10.1002/hed.20452 PMID: 16673417

Chen, Y., Chen, Z., Yang, D., Zhu, L., Liang, Z., Pang, Y., & Zhou, L. (2022). Novel Microbial Palladium Nanoparticles with a High Photothermal Effect for Antibacterial Applications. *ACS Omega*, *8*(1), 1534–1541. DOI: 10.1021/acsomega.2c07037 PMID: 36643470

Chen, Y., Ding, H., Wang, Q., Huang, Z., Zhang, C., Li, W., Lin, Y., Guo, Y., Fang, X., & Zhang, W. (2025). Can "LITE" Procedure Combined With a Short Course Antibiotic Treatment Be Effective in Treating the Chronic PJI?—A Prospective Randomized Controlled Trial. *Orthopaedic Surgery*.

Chen, Y., Huang, C., Chen, X., Cai, Y., Li, W., Fang, X., & Zhang, W. (2022). Bone protein analysis via label-free quantitative proteomics in patients with periprosthetic joint infection. *Journal of Proteomics*, *252*, 104448. PMID: 34883267

Cheung, R. C. F., Ng, T. B., & Wong, J. H. (2015). Chitosan: An overview of its applications in antiviral treatment. *Applied Biochemistry and Biotechnology, 177*(2), 357–367.

Chi, J. J. (2019). Antitumor evaluation of carboxymethyl chitosan based norcantharidin conjugates against gastric cancer as novel polymer therapeutics. *International Journal of Biological Macromolecules, 136*, 1–12. PMID: 31158420

Chiu, Y. G., & Ritchlin, C. T. (2017). Denosumab: Targeting the RANKL pathway to treat rheumatoid arthritis. *Expert Opinion on Biological Therapy, 17*(1), 119–128. DOI: 10.1080/14712598.2017.1263614 PMID: 27871200

Choi, S.-r., Kwon, J.-w., Suk, K.-s., Kim, H.-s., Moon, S.-h., Park, S.-y., & Lee, B. H. (2023). The clinical use of osteobiologic and metallic biomaterials in orthopedic surgery: The present and the future. *Materials (Basel), 16*(10), 3633. PMID: 37241260

Chopra, M., Bernela, M., Kaur, P., Manuja, A., Kumar, B., & Thakur, R. (2015). Alginate/gum acacia bipolymeric nanohydrogels-Promising carrier for Zinc oxide nanoparticles. *International Journal of Biological Macromolecules*. Advance online publication. DOI: 10.1016/j.ijbiomac.2014.09.037 PMID: 25304751

Chou, W.-C., & Lin, Z. (2024). Impact of protein coronas on nanoparticle interactions with tissues and targeted delivery. *Current Opinion in Biotechnology, 85*, 103046. PMID: 38103519

Chowdhury, S. M. (2017). Ultrasound-guided drug delivery in cancer. *Ultrasonography (Seoul, Korea), 36*(3), 171. PMID: 28607323

Chu, X.-L., Song, X.-Z., Li, Q., Li, Y.-R., He, F., Gu, X.-S., & Ming, D. (2022). Basic mechanisms of peripheral nerve injury and treatment via electrical stimulation. *Neural Regeneration Research, 17*(10), 2185–2193. DOI: 10.4103/1673-5374.335823 PMID: 35259827

Cianciulli, P. (2009). Iron chelation therapy in thalassemia syndromes. *Mediterranean Journal of Hematology and Infectious Diseases, 1*(1), e2009034. DOI: 10.4084/MJHID.2009.034 PMID: 21415999

Citi, L., Carpaneto, J., & Yoshida, K.. (2008). On the use of wavelet denoising and spike sorting techniques to process electroneurographic signals recorded using intraneural electrodes . *Journal of Neuroscience Methods, 172*(2), 294–302. PMID: 18534683

Clarke, B. L., & Khosla, S. (2010). Physiology of Bone Loss. *Radiologic Clinics of North America, 48*(3), 483–495. DOI: 10.1016/j.rcl.2010.02.014 PMID: 20609887

Clunie, G. P., Clark, A., Mortimer, C. J., Stephenson, S., Aitken, J., Smith, C., Sherwin, E., & Archer, T. J. (2009). Evaluating bone health in women with oestrogen receptor positive breast cancer (ERBC) starting aromatase inhibitors. [EJSO]. *European Journal of Surgical Oncology, 35*(5), 475–480. DOI: 10.1016/j.ejso.2008.08.001 PMID: 18950981

Cochran, D. B., Wattamwar, P. P., Wydra, R., Hilt, J. Z., Anderson, K. W., Eitel, R. E., & Dziubla, T. D. (2013). Suppressing iron oxide nanoparticle toxicity by vascular targeted antioxidant polymer nanoparticles. *Biomaterials*. Advance online publication. DOI: 10.1016/j.biomaterials.2013.08.025 PMID: 24016851

Cómitre-Mariano, B., Vellila-Alonso, G., Segura-Collar, B., Mondéjar-Ruescas, L., Sepulveda, J. M., & Gargini, R. (2024). Sentinels of neuroinflammation: The crucial role of myeloid cells in the pathogenesis of gliomas and neurodegenerative diseases. *Journal of Neuroinflammation*, *21*, 304. PMID: 39578808

Cong, X., Zhang, Z., Li, H., Yang, Y. G., Zhang, Y., & Sun, T. (2024). Nanocarriers for targeted drug delivery in the vascular system: Focus on endothelium. *Journal of Nanobiotechnology*, *22*(1), 620. DOI: 10.1186/s12951-024-02892-9 PMID: 39396002

Contera, S., Bernardino de la Serna, J., & Tetley, T. D. (2020). Biotechnology, nanotechnology and medicine. *Emerging Topics in Life Sciences*, *4*(6), 551–554. DOI: 10.1042/etls20200350 PMID: 33295610

Costabeber, G., Guerra, N. B., Brasil, G. S. A. P., da Silva Sasaki, J. C., Scontri, M., Burd, B. S., & Herculano, R. D. (2024). Additive manufacturing of polylactic acid scaffolds dip-coated with polycaprolactone for bone tissue engineering. *Materials Today. Communications*, *40*, 109646.

Costa-Pinto, A. R., Reis, R. L., & Neves, N. M. (2011). Scaffolds Based Bone Tissue Engineering: The Role of Chitosan. *Tissue Engineering. Part B, Reviews*, *17*(5), 331–347. DOI: 10.1089/ten.teb.2010.0704 PMID: 21810029

Crini, G. (2019). Historical Review on Chitin and Chitosan Biopolymers. *Environmental Chemistry Letters*, *17*(4), 1623–1643. DOI: 10.1007/s10311-019-00901-0

Csöbönyeiová, M., Beerová, N., & Klein, M.. (2022). Cell-Based and Selected Cell-Free Therapies for Myocardial Infarction: How Do They Compare to the Current Treatment Options? . *International Journal of Molecular Sciences*, *23*(18), 10314. PMID: 36142245

Cuartas-Marulanda, D., Forero Cardozo, L., Restrepo-Osorio, A., & Fernández-Morales, P. (2022). Natural coatings and surface modifications on magnesium alloys for biomedical applications. *Polymers*, *14*(23), 5297. PMID: 36501691

Cucci, L. M., Satriano, C., Marzo, T., & La Mendola, D. (2021). Angiogenin and copper crossing in wound healing. *International Journal of Molecular Sciences*, *22*(19), 10704. DOI: 10.3390/ijms221910704 PMID: 34639045

Cui, Y., Lv, B., Li, Z., Ma, C., Gui, Z., Geng, Y., Liu, G., Sang, L., Xu, C., Min, Q., Kong, L., Zhang, Z., Liu, Y., Qi, X., & Fu, D. (2024). Bone-Targeted Biomimetic Nanogels Re-Establish Osteoblast/Osteoclast Balance to Treat Postmenopausal Osteoporosis. *Small*, *20*(6). Advance online publication. DOI: 10.1002/smll.202303494 PMID: 37794621

Cui, Z. K., Kim, S., Baljon, J. J., Wu, B. M., Aghaloo, T., & Lee, M. (2019). Microporous methacrylated glycol chitosan-montmorillonite nanocomposite hydrogel for bone tissue engineering. *Nature Communications*, *10*(1). Advance online publication. DOI: 10.1038/s41467-019-11511-3 PMID: 31388014

Culmone, C., Smit, G., & Breedveld, P. (2019). Additive manufacturing of medical instruments: A state-of-the-art review. *Additive Manufacturing*, *27*, 461–473. DOI: 10.1016/j.addma.2019.03.015

Curcio, M., Cirillo, G., Tucci, P., Farfalla, A., Bevacqua, E., Vittorio, O., Iemma, F., & Nicoletta, F. P. (2019). Dextran-Curcumin Nanoparticles as a Methotrexate Delivery Vehicle: A Step Forward in Breast Cancer Combination Therapy. *Pharmaceuticals*, *13*(1), 2. DOI: 10.3390/PH13010002 PMID: 31881645

Curcio, M., Puoci, F., Iemma, F., Parisi, O. I., Cirillo, G., Spizzirri, U. G., & Picci, N. (2009). Covalent insertion of antioxidant molecules on chitosan by a free radical grafting procedure. *Journal of Agricultural and Food Chemistry*, *57*(13), 5933–5938. DOI: 10.1021/jf900778u PMID: 19566085

Dai, Y., Guo, Y., Tang, W., Chen, D., Xue, L., Chen, Y., & Wang, S. (2024). Reactive oxygen species-scavenging nanomaterials for the prevention and treatment of age-related diseases. *Journal of Nanobiotechnology*, *22*(1), 252. PMID: 38750509

Dakal, T. C., Kumar, A., Majumdar, R. S., & Yadav, V. (2016). Mechanistic basis of antimicrobial actions of silver nanoparticles. *Frontiers in Microbiology*, *7*, 1831. DOI: 10.3389/fmicb.2016.01831 PMID: 27899918

Damiri, F. F. (2023). Smart stimuli-responsive polysaccharide nanohydrogels for drug delivery: A review. *Journal of Materials Chemistry. B*. PMID: 37909361

Dang, Y., & Guan, J. (2020). Nanoparticle-based drug delivery systems for cancer therapy. *Smart Materials in Medicine*, *1*, 10–19. PMID: 34553138

Daraee, H., Eatemadi, A., Abbasi, E., Fekri Aval, S., Kouhi, M., & Akbarzadeh, A. (2014). Application of gold nanoparticles in biomedical and drug delivery. *Artificial Cells, Nanomedicine, and Biotechnology*, *44*(1), 410–422. DOI: 10.3109/21691401.2014.955107 PMID: 25229833

Daraghmeh, N. H., Chowdhry, B. Z., Leharne, S. A., Al Omari, M. M., & Badwan, A. A. (2011). Chitin (pp. 35–102). https://doi.org/DOI: 10.1016/B978-0-12-387667-6.00002-6

Dar, M. S., Tabish, T. A., Thorat, N. D., Swati, G., & Sahu, N. K. (2023). Photothermal therapy using graphene quantum dots. *APL Bioengineering*, *7*(3), 031502. DOI: 10.1063/5.0160324 PMID: 37614868

Das, A., Ghosh, S., & Pramanik, N. (2024). Chitosan biopolymer and its composites: Processing, properties and applications-A comprehensive review. *Hybrid Advances*, 100265.

Dasari, S., & Tchounwou, P. B. (2014). Cisplatin in cancer therapy: Molecular mechanisms of action. *European Journal of Pharmacology*, *740*, 364–378. DOI: 10.1016/j.ejphar.2014.07.025 PMID: 25058905

Das, K. K., Basu, B., Maiti, P., & Dubey, A. K. (2024). Interplay of piezoelectricity and electrical stimulation in tissue engineering and regenerative medicine. *Applied Materials Today*, *39*, 102332. DOI: 10.1016/j.apmt.2024.102332

Das, S. S. (2020). Stimuli-responsive polymeric nanocarriers for drug delivery, imaging, and theragnosis. *Polymers*, *12*(6), 1397. PMID: 32580366

Das, S., Carnicer-Lombarte, A., Fawcett, J. W., & Bora, U. (2016). Bio-inspired nano tools for neuroscience. *Progress in Neurobiology*, *142*, 1–22. DOI: 10.1016/j.pneurobio.2016.04.008 PMID: 27107796

Dave, K., & Gomes, V. G. (2019). Interactions at scaffold interfaces: Effect of surface chemistry, structural attributes and bioaffinity. *Materials Science and Engineering C*, *105*, 110078. PMID: 31546353

Davydova, V. N., Kalitnik, A. A., Markov, P. A., Volod'ko, A. V., Popov, S. V., & Ermak, I. M. (2016). Cytokine-inducing and anti-inflammatory activity of chitosan and its low-molecular derivative. *Applied Biochemistry and Microbiology*, *52*(5), 476–482. DOI: 10.1134/S0003683816050070 PMID: 29513410

Dayyoub, T. M. (2022). Shape memory polymers as smart materials: A review. *Polymers*, *14*(17), 3511. PMID: 36080587

de La Torre, J. C. (2012). Cardiovascular risk factors promote brain hypoperfusion leading to cognitive decline and dementia. *Cardiovascular Psychiatry and Neurology*, *2012*(1), 367516. PMID: 23243502

de Mesy Bentley, K. L., Trombetta, R., Nishitani, K., Bello-Irizarry, S. N., Ninomiya, M., Zhang, L., Chung, H. L., McGrath, J. L., Daiss, J. L., & Awad, H. A. (2017). Evidence of Staphylococcus aureus deformation, proliferation, and migration in canaliculi of live cortical bone in murine models of osteomyelitis. In: John Wiley and Sons and The American Society for Bone and Mineral Research….

De Smet, K. (2013). Clinical follow-up of the hip resurfacing patient. In *The Hip Resurfacing Handbook* (pp. 113–127). Elsevier.

Declèves, X. L. (2009). Transport des anticancéreux par les transporteurs ABC et implications en hématologie. La Lettre du cancérologue (Boulogne),, 18(2), 115-125.

Dee, K. C., Puleo, D. A., & Bizios, R. (2003). *An introduction to tissue-biomaterial interactions*. John Wiley & Sons.

Dehghan-Manshadi, M., Azarbayjani, M.-A., Atashak, S., Peeri, M., & Rahmati-Ahmadabad, S. (2022). Effect of resistance training with and without vitamin D calcium chitosan nanoparticle supplements on apoptosis markers in ovariectomized rats: An experimental study. *International Journal of Reproductive Biomedicine*, *20*(7), 549–560. DOI: 10.18502/ijrm.v20i7.11557 PMID: 36187741

De, I., Sharma, P., & Singh, M. (2022). Emerging approaches of neural regeneration using physical stimulations solely or coupled with smart piezoelectric nano-biomaterials. *European Journal of Pharmaceutics and Biopharmaceutics*, *173*, 73–91. PMID: 35227856

Delbianco, M., & Seeberger, P. H. (2020). Materials Science Based on Synthetic Polysaccharides. *Materials Horizons*, *7*(4), 963–969. DOI: 10.1039/C9MH01936G

Demir, D., Bolgen, N., & Vaseashta, A. (2023). Electrospun Nanofibers for Biomedical, Sensing, and Energy Harvesting Functions. Polymers 2023, Vol. 15, Page 4253, 15(21), 4253. https://doi.org/DOI: 10.3390/POLYM15214253

Demir, E., Burgucu, D., Turna, F., Aksakal, S., & Kaya, B. (2013). Determination of TiO2, ZrO2, and Al2O 3 nanoparticles on genotoxic responses in human peripheral blood lymphocytes and cultured embyronic kidney cells. *Journal of Toxicology and Environmental Health - Part A: Current Issues*, *76*(16), 990–1002. DOI: 10.1080/15287394.2013.830584 PMID: 24156722

Deng, C., Fu, H., Xu, J., Shang, J., & Cheng, Y. (2015). Physiochemical and biological properties of phosphorylated polysaccharides from Dictyophora indusiata. *International Journal of Biological Macromolecules*, *72*, 894–899. DOI: 10.1016/j.ijbiomac.2014.09.053 PMID: 25316421

Deng, W., Zhou, Y., Libanori, A., Chen, G., Yang, W., & Chen, J. (2022). Piezoelectric nanogenerators for personalized healthcare. *Chemical Society Reviews*, *51*(9), 3380–3435. DOI: 10.1039/D1CS00858G PMID: 35352069

Deraedt, C., & Astruc, D. (2013). "Homeopathic" Palladium Nanoparticle Catalysis of Cross Carbon–Carbon Coupling Reactions. *Accounts of Chemical Research*, *47*(2), 494–503. DOI: 10.1021/ar400168s PMID: 24215156

Deschênes-Simard, X., Malleshaiah, M., & Ferbeyre, G. (2023). Extracellular Signal-Regulated Kinases: One Pathway, Multiple Fates. *Cancers (Basel)*, *16*(1). Advance online publication. DOI: 10.3390/cancers16010095 PMID: 38201521

Dessay, M., Couture, E., Maaroufi, H., Fournier, F., Gagnon, E., Droit, A., Brown, J. P., & Michou, L. (2022). Attenuated clinical and osteoclastic phenotypes of Paget's disease of bone linked to the p.Pro392Leu/SQSTM1 mutation by a rare variant in the DOCK6 gene. *BMC Medical Genomics*, *15*(1), 41. DOI: 10.1186/s12920-022-01198-9 PMID: 35241069

Detsi, A., Kavetsou, E., Kostopoulou, I., Pitterou, I., Pontillo, A. R. N., Tzani, A., & Zoumpoulakis, P. (2020). Nanosystems for the encapsulation of natural products: The case of chitosan biopolymer as a matrix. *Pharmaceutics*, *12*(7), 669. PMID: 32708823

Dev, A., Binulal, N. S., Anitha, A., Nair, S. V., Furuike, T., Tamura, H., & Jayakumar, R. (2010). Preparation of poly(lactic acid)/chitosan nanoparticles for anti-HIV drug delivery applications. *Carbohydrate Polymers*, *80*(3), 833–838. DOI: 10.1016/j.carbpol.2009.12.040

Dhara, K., & Mahapatra, D. R. (2019). Recent advances in electrochemical nonenzymatic hydrogen peroxide sensors based on nanomaterials: A review. *Journal of Materials Science*, *54*(19), 12319–12357. DOI: 10.1007/s10853-019-03750-y

Díaz-García, D., Díaz-Sánchez, M., Álvarez-Conde, J., & Gómez-Ruiz, S. (2024). Emergence of Quantum Dots as Innovative Tools for Early Diagnosis and Advanced Treatment of Breast Cancer. *ChemMedChem*, *19*(16), e202400172. DOI: 10.1002/CMDC.202400172 PMID: 38724442

Diba, M., Tapia, F., Boccaccini, A. R., & Strobel, L. A. (2012). Magnesium-Containing bioactive glasses for biomedical applications. *International Journal of Applied Glass Science*, *3*(3), 221–253. DOI: 10.1111/j.2041-1294.2012.00095.x

Diez-Pascual, A. M., & Rahdar, A. (2022). Functional Nanomaterials in Biomedicine: Current Uses and Potential Applications. *ChemMedChem*, *17*(16), e202200142. PMID: 35729066

Dilamian, M., Montazer, M., & Masoumi, J. (2013). Antimicrobial electrospun membranes of chitosan/poly(ethylene oxide) incorporating poly(hexamethylene biguanide) hydrochloride. *Carbohydrate Polymers*, *94*(1), 364–371. DOI: 10.1016/j.carbpol.2013.01.059 PMID: 23544550

Ding, L., Agrawal, P., Singh, S. K., Chhonker, Y. S., Sun, J., & Murry, D. J. (2024). Polymer-Based Drug Delivery Systems for Cancer Therapeutics. Polymers 2024, Vol. 16, Page 843, 16(6), 843. https://doi.org/DOI: 10.3390/POLYM16060843

Ding, F., Fu, J., Tao, C., Yu, Y., He, X., Gao, Y., & Zhang, Y. (2020). Recent Advances of Chitosan and its Derivatives in Biomedical Applications. *Current Medicinal Chemistry*, *27*(18), 3023–3045. DOI: 10.2174/0929867326666190405151538 PMID: 30961477

Ding, J., & Guo, Y. (2022). Recent advances in chitosan and its derivatives in cancer treatment. *Frontiers in Pharmacology*, *13*, 888740. PMID: 35694245

Ding, J., & Guo, Y. (2022). Recent Advances in Chitosan and its Derivatives in Cancer Treatment. *Frontiers in Pharmacology*, *13*. Advance online publication. DOI: 10.3389/fphar.2022.888740 PMID: 35694245

Ding, Y. F. (2018). Highly biocompatible chlorin e6-loaded chitosan nanoparticles for improved photodynamic cancer therapy. *ACS Applied Materials & Interfaces*, *10*(12), 9980–9987. PMID: 29498260

Ding, Y., Jiang, H., Dunn, M. L., & Yu, K. (2024). 3D interfacial material-locking structures to promote bonding strength of dissimilar materials. *Journal of Manufacturing Processes*, *112*, 92–102. DOI: 10.1016/j.jmapro.2024.01.031

Dixon, D. T., & Gomillion, C. T. (2023). 3D-Printed conductive polymeric scaffolds with direct current electrical stimulation for enhanced bone regeneration. *Journal of Biomedical Materials Research. Part B, Applied Biomaterials*, *111*(7), 1351–1364. DOI: 10.1002/jbm.b.35239 PMID: 36825765

Do Nascimento, S. da P., de Souza, R. R. M., Sobral, M. V., Xavier-Junior, F. H., da Silva, M. V. S., Viana, M. M., da Silva, F. F., Serpe, M. J., & de Souza, A. L. (2024). Gelatin-Oxidized Alginate and Chitosan-Coated Zein Nanoparticle Hydrogel Composite to Enhance Breast Cancer Cytotoxicity in Dual-Drug Delivery. *ACS Omega*. Advance online publication. DOI: 10.1021/ACSOMEGA.4C06404/ASSET/IMAGES/LARGE/AO4C06404_0011.JPEG PMID: 39554428

Doberenz, F. Z. (2020). Thermoresponsive polymers and their biomedical application in tissue engineering–a review. *Journal of Materials Chemistry. B*, *8*(4), 607–628. PMID: 31939978

Doerksen, R. J., Osborn, H., Socorro, S., Dragojevic, S., Su Ryu, J., Hall, M. E., & Raucher, D. (2022). Targeted Drug Delivery Biopolymers Effectively Inhibit Breast Tumor Growth and Prevent Doxorubicin-Induced Cardiotoxicity. *Molecules (Basel, Switzerland)*, *27*(11), 3371. DOI: 10.3390/MOLECULES27113371 PMID: 35684309

Dong, R., Zhao, X., & Guo, B.. (2016). Self-Healing Conductive Injectable Hydrogels with Antibacterial Activity as Cell Delivery Carrier for Cardiac Cell Therapy . *ACS Applied Materials & Interfaces*, *8*(27), 17138–17150. PMID: 27311127

Dong, X. (2018). Current Strategies for Brain Drug Delivery. *Theranostics*, *8*(6), 1481–1493. DOI: 10.7150/thno.21254 PMID: 29556336

Đorđević, S., Gonzalez, M. M., Conejos-Sánchez, I., Carreira, B., Pozzi, S., Acúrcio, R. C., & Vicent, M. J. (2022). Current hurdles to the translation of nanomedicines from bench to the clinic. *Drug Delivery and Translational Research*, •••, 1–26. PMID: 34302274

Dowlath, M. J. H., Karuppannan, S. K., Sinha, P., Dowlath, N. S., Arunachalam, K. D., Ravindran, B., & Nguyen, D. D. (2021). Effects of radiation and role of plants in radioprotection: A critical review. *The Science of the Total Environment*, *779*, 146431. PMID: 34030282

Dristant, U. M. (2023). An overview of polymeric nanoparticles-based drug delivery system in cancer treatment. *Technology in Cancer Research & Treatment*. PMID: 36718541

Duda, G N, GEISSLER S, CHECA S. *et al.* (2023). The decisive early phase of bone regeneration . *Nature Reviews. Rheumatology*, *19*(2), 78–95. PMID: 36624263

Dukovski, J., Bisera, J. L., Kocbek, P., Bočkor, L., Pepić, I., Hafner, A., & Lovrić, J. (2023). Development of a 3D Corneal Epithelial Model for Early Biocompatibility Screening of Topical Ophthalmic Formulations. *Macedonian Pharmaceutical Bulletin*, *69*(03), 247–248. DOI: 10.33320/maced.pharm.bull.2023.69.03.121

Dumas, A., & Couvreur, P. (2015). Palladium: A future key player in the nanomedical field? *Chemical Science (Cambridge)*, *6*(4), 2153–2157. DOI: 10.1039/c5sc00070j PMID: 28694948

Duncombe, J., Kitamura, A., Hase, Y., Ihara, M., Kalaria, R. N., & Horsburgh, K. (2017). Chronic cerebral hypoperfusion: A key mechanism leading to vascular cognitive impairment and dementia. Closing the translational gap between rodent models and human vascular cognitive impairment and dementia. *Clinical Science*, *131*(19), 2451–2468. PMID: 28963120

Durán, N., & Marcato, P. D. (2013). Nanobiotechnology perspectives. Role of nanotechnology in the food industry: A review. In International Journal of Food Science and Technology (Vol. 48, Issue 6, pp. 1127–1134). doi: DOI: 10.1111/ijfs.12027

Ebrahimi, F., & Dana, H. R. (2022). Poly Lactic Acid (PLA) Polymers: From Properties to Biomedical Applications. *International Journal of Polymeric Materials*, *71*(15), 1117–1130. DOI: 10.1080/00914037.2021.1944140

Ehrlich, H., Krajewska, B., Hanke, T., Born, R., Heinemann, S., Knieb, C., & Worch, H. (2006). Chitosan membrane as a template for hydroxyapatite crystal growth in a model dual membrane diffusion system. *Journal of Membrane Science*, *273*(1–2), 124–128. DOI: 10.1016/J.MEMSCI.2005.11.050

El-Baz, F. K., Salama, A., Ali, S. I., & El-Hashemy, H. A. (2023). Dunaliella salina chitosan nanoparticles as a promising wound healing vehicles: In-vitro and in-vivo study. *OpenNano*, *12*, 100165. DOI: 10.1016/j.onano.2023.100165

Eldeeb, A. E., Salah, S., & Elkasabgy, N. A. (2022). Biomaterials for tissue engineering applications and current updates in the field: A comprehensive review. *AAPS PharmSciTech*, *23*(7), 267. PMID: 36163568

Eleftheriadou, D., Kesidou, D., Moura, F., Felli, E., & Song, W. (2020). Redox-Responsive Nanobiomaterials-Based Therapeutics for Neurodegenerative Diseases. *Small*, *16*(43), e1907308. DOI: 10.1002/smll.201907308 PMID: 32940007

Elhabal, S. F., Abdelaal, N., Saeed Al-Zuhairy, S. A. K., Elrefai, M. F. M., Elsaid Hamdan, A. M., Khalifa, M. M., & El-Nabarawi, M. A. (2024). Green synthesis of zinc oxide nanoparticles from Althaea officinalis flower extract coated with chitosan for potential healing effects on diabetic wounds by inhibiting TNF-α and IL-6/IL-1β signaling pathways. *International Journal of Nanomedicine*, •••, 3045–3070. DOI: 10.2147/IJN.S455270 PMID: 38559447

Elkaliny, N. E., Alzamel, N. M., Moussa, S. H., Elodamy, N. I., Madkor, E. A., Ibrahim, E. M., Elshobary, M. E., & Ismail, G. A. (2024). *"Macroalgae Bioplastics: A Sustainable Shift to Mitigate the Ecological Impact of Petroleum-Based Plastics." Polymers. Multidisciplinary Digital Publishing Institute*. MDPI., DOI: 10.3390/polym16091246

El-Leithy, E. S.-R. (2019). Tamoxifen citrate/coenzyme Q10 as smart nanocarriers bitherapy for breast cancer: Cytotoxicity, genotoxicity, and antioxidant activity. *Journal of Drug Delivery Science and Technology*, *51*, 36–44.

Elsayad, M. K., Mowafy, H. A., Zaky, A. A., & Samy, A. M. (2021). Chitosan caged liposomes for improving oral bioavailability of rivaroxaban: *In vitro* and *in vivo* evaluation. *Pharmaceutical Development and Technology*, *26*(3), 316–327. DOI: 10.1080/10837450.2020.1870237 PMID: 33356742

El-Sherbiny, I. M., El-Sayed, M., & Reda, A. (2020). Superparamagnetic iron oxide nanoparticles (SPIONs) as multifunctional cancer theranostics. *Magnetic nanoheterostructures: diagnostic, imaging and treatment*, 223-241.

Eltit, F., Wang, Q., & Wang, R. (2019). Mechanisms of adverse local tissue reactions to hip implants. *Frontiers in Bioengineering and Biotechnology*, *7*, 176. PMID: 31417898

Elumalai, K. S. (2024). Review of the efficacy of nanoparticle-based drug delivery systems for cancer treatment. *Biomedical Technology*, *5*, 109–122.

Emadi, H., Karevan, M., Masoudi Rad, M., Sadeghzade, S., Pahlevanzadeh, F., Khodaei, M., Khayatzadeh, S., & Lotfian, S. (2023). Bioactive and Biodegradable Polycaprolactone-Based Nanocomposite for Bone Repair Applications. *Polymers*, *15*, 3617. DOI: 10.3390/polym15173617

Emmanuel Jeanpierre. (2014). *L'ostéoporose et son traitement à l'épreuve des faits. Etude rétrospective des pratiques cliniques de la prise en charge médicale proposée à 222 patients, recrutés en 4 ans dans la filière "pour la Qualité de l'Os en Lorraine"*. QOL.

Emmanuelle, N E, MARIE-CÉCILE V, FLORENCE T. *et al.* (2021). Critical Role of Estrogens on Bone Homeostasis in Both Male and Female: From Physiology to Medical Implications . *International Journal of Molecular Sciences*, *22*(4), 1568. PMID: 33557249

Eren, E. D., Guisong, G., Mingming, L., Bingchun, Z., Ke, Y., & Shanshan, C. (2022). A novel chitosan and polydopamine interlinked bioactive coating for metallic biomaterials. *Journal of Materials Science. Materials in Medicine*, *33*(10). Advance online publication. DOI: 10.1007/s10856-022-06688-x PMID: 36138240

Erisen, D. E., & Uludag, K. (2024). Estimating biosafety of biodegradable biomedical materials from in vitro ion tolerance parameters and toxicity of nanomaterials in brain. In *Advances in healthcare information systems and administration book series* (pp. 201–221). https://doi.org/DOI: 10.4018/979-8-3693-3661-8.ch010

Erisen, D. E., & Uludag, K. (2024). Estimating Biosafety of Biodegradable Biomedical Materials from In Vitro Ion Tolerance Parameters and Toxicity of Nanomaterials in Brain. In *Transformative Approaches to Patient Literacy and Healthcare Innovation* (pp. 201–221). IGI Global.

Erişen, D. E., Zhang, Y., Zhang, B., Yang, K., Chen, S., & Wang, X. (2021). Biosafety and biodegradation studies of AZ31B magnesium alloy carotid artery stent in vitro and in vivo. *Journal of Biomedical Materials Research. Part B, Applied Biomaterials*, *110*(1), 239–248. DOI: 10.1002/jbm.b.34907 PMID: 34236133

Ershad-Langroudi, A., Babazadeh, N., Alizadegan, F., Mousaei, S. M., & Moradi, G. (2024). Polymers for Implantable Devices. *Journal of Industrial and Engineering Chemistry*. Advance online publication. DOI: 10.1016/j.jiec.2024.03.030

Eslahi, M., Dana, P. M., Asemi, Z., Hallajzadeh, J., Mansournia, M. A., & Yousefi, B. (2021). The effects of chitosan-based materials on glioma: Recent advances in its applications for diagnosis and treatment. *International Journal of Biological Macromolecules*, *168*, 124–129. DOI: 10.1016/j.ijbiomac.2020.11.180 PMID: 33275978

Esmaeili, H., Patino-Guerrero, A., Hasany, M., Ansari, M. O., Memic, A., Dolatshahi-Pirouz, A., & Nikkhah, M. (2022). Electroconductive biomaterials for cardiac tissue engineering. *Acta Biomaterialia*, *139*, 118–140. DOI: 10.1016/j.actbio.2021.08.031 PMID: 34455109

Esmaily, M., Svensson, J., Fajardo, S., Birbilis, N., Frankel, G., Virtanen, S., Arrabal, R., Thomas, S., & Johansson, L. (2017). Fundamentals and advances in magnesium alloy corrosion. *Progress in Materials Science*, *89*, 92–193. DOI: 10.1016/j.pmatsci.2017.04.011

Evans, A., & Kavanagh, K. A. (2021). Evaluation of metal-based antimicrobial compounds for the treatment of bacterial pathogens. *Journal of Medical Microbiology*, *70*(5), 001363. DOI: 10.1099/jmm.0.001363 PMID: 33961541

Ezoddini-Ardakani, F., Navabazam, A., Fatehi, F., Danesh-Ardekani, M., Khadem, S., & Rouhi, G. (2012). Histologic evaluation of chitosan as an accelerator of bone regeneration in microdrilled rat tibias. *Dental Research Journal (Isfahan)*, *9*(6), 694–699. PMID: 23559943

Fabbri, P., & Messori, M. (2017). Surface modification of polymers: chemical, physical, and biological routes. In *Modification of polymer properties* (pp. 109–130). Elsevier.

Fadilah, N. I. M., Phang, S. J., Kamaruzaman, N., Salleh, A., Zawani, M., Sanyal, A., & Fauzi, M. B. (2023). Antioxidant biomaterials in cutaneous wound healing and tissue regeneration: A critical review. *Antioxidants*, *12*(4), 787. PMID: 37107164

Fahira, A. I., Abdulah, R., Barliana, M. I., Gatera, V. A., & Amalia, R. (2022). Polyethyleneimine (PEI) as a Polymer-Based Co-Delivery System for Breast Cancer Therapy. *Breast Cancer (Dove Medical Press)*, *14*, 71. DOI: 10.2147/BCTT.S350403 PMID: 35422657

Faiz, K., Lam, F. C., Chen, J., Kasper, E. M., & Salehi, F. (2022). The Emerging Applications of Nanotechnology in Neuroimaging: A Comprehensive Review. *Frontiers in Bioengineering and Biotechnology*, *10*, 855195. PMID: 35875504

Fakruddin, M., Hossain, Z., & Afroz, H. (2012). Prospects and applications of nanobiotechnology: A medical perspective. In Journal of Nanobiotechnology (Vol. 10). doi: DOI: 10.1186/1477-3155-10-31

Fan, C., Shi, J., & Zhuang, Y.. (2019). Myocardial-Infarction-Responsive Smart Hydrogels Targeting Matrix Metalloproteinase for On-Demand Growth Factor Delivery . *Advanced Materials (Deerfield Beach, Fla.)*, *31*(40), e1902900. PMID: 31408234

Fan, C., Yang, W., & Zhang, L.. (2022). Restoration of spinal cord biophysical microenvironment for enhancing tissue repair by injury-responsive smart hydrogel . *Biomaterials*, *288*, 121689. PMID: 35931574

Fang, K., Wang, R., & Zhang, H.. (2020). Mechano-Responsive, Tough, and Antibacterial Zwitterionic Hydrogels with Controllable Drug Release for Wound Healing Applications . *ACS Applied Materials & Interfaces*, *12*(47), 52307–52318. PMID: 33183010

Fang, X., Sun, D., Li, Y., Han, X., Gan, Y., Jiao, J., Jiang, M., Gong, H., Qi, Y., & Zhao, J. (2024). Macrophages in the process of osseointegration around the implant and their regulatory strategies. *Connective Tissue Research*, *65*(1), 1–15. PMID: 38166507

Fang, X., Wang, Q., Yang, X., Zhang, F., Huang, C., Huang, Z., Shen, H., & Zhang, W. (2021). What is the appropriate extended duration of antibiotic prophylaxis after two-stage revision for chronic PJI? *Bone & Joint Research*, *10*(12), 790–796. PMID: 34894718

Fang, Y., & Shang, X. (2021). Comparison of Ceramic-on-Ceramic vs. ceramic-on-polyethylene for Primary total hip arthroplasty: A Meta-Analysis of 15 Randomized Trials. *Frontiers in Surgery*, *8*, 751121. PMID: 34977138

Fan, L., Liu, C., Chen, X., Zheng, L., Zou, Y., Wen, H., Guan, P., Lu, F., Luo, Y., Tan, G., Yu, P., Chen, D., Deng, C., Sun, Y., Zhou, L., & Ning, C. (2022). Exosomes-loaded electroconductive hydrogel synergistically promotes tissue repair after spinal cord injury via immunoregulation and enhancement of myelinated axon growth. *Advancement of Science*, *9*(13), e2105586. DOI: 10.1002/advs.202105586 PMID: 35253394

Fan, Q., Sun, C., Hu, B., & Wang, Q. (2023). Recent advances of lanthanide nanomaterials in Tumor NIR fluorescence detection and treatment. *Materials Today. Bio*, *20*, 100646. DOI: 10.1016/j.mtbio.2023.100646 PMID: 37214552

Fan, W., Yung, B., Huang, P., & Chen, X. (2017). Nanotechnology for multimodal synergistic cancer therapy. *Chemical Reviews*, *117*(22), 13566–13638. PMID: 29048884

Fan, Y., Zhai, J., Wang, Z., Yin, Z., Chen, H., Ran, M., Zhu, Z., Ma, Y., Ning, C., Yu, P., & Mao, C. (2025). Piezoelectric heterojunctions as bacteria-killing bone-regenerative implants. *Advanced Materials*, *37*(1), e2413171. DOI: 10.1002/adma.202413171 PMID: 39460412

Farag, M. M. (2023). Recent trends on biomaterials for tissue regeneration applications: review. Journal of Materials Science 2022 58:2, 58(2), 527–558. https://doi.org/DOI: 10.1007/S10853-022-08102-X

Farzin, L., Shamsipur, M., Samandari, L., & Sheibani, S. (2020). HIV biosensors for early diagnosis of infection: The intertwine of nanotechnology with sensing strategies. *Talanta*, *206*, 120201. PMID: 31514868

Fathy, M. M., Mohamed, F. S., Elbialy, N. S., & Elshemey, W. M. (2018). Multifunctional Chitosan-Capped Gold Nanoparticles for Enhanced Cancer Chemo-Radiotherapy: An Invitro Study. *Physica Medica*, *48*(April), 76–83. DOI: 10.1016/j.ejmp.2018.04.002 PMID: 29728233

Fatima, I., Rahdar, A., Sargazi, S., Barani, M., Hassanisaadi, M., & Thakur, V. K. (2021). Quantum Dots: Synthesis, Antibody Conjugation, and HER2-Receptor Targeting for Breast Cancer Therapy. Journal of Functional Biomaterials 2021, Vol. 12, Page 75, 12(4), 75. https://doi.org/DOI: 10.3390/JFB12040075

Fatima, A., Shehzad, A., Shahzad, R., Khan, S., & Al-Suhaimi, E. A. (2024). Impact of nanoparticles on structural elements within the cells. In *Molecular impacts of nanoparticles on plants and algae* (pp. 111–141). Elsevier.

Fatima, M., Sheikh, A., Abourehab, M. A. S., & Kesharwani, P. (2022). Advancements in Polymeric Nanocarriers to Mediate Targeted Therapy against Triple-Negative Breast Cancer. *Pharmaceutics*, *14*(11), 2432. DOI: 10.3390/PHARMACEUTICS14112432 PMID: 36365249

Feki, A. H. (2020). Conception and characterization of a multi-sensitive composite chitosan-red marine alga-polysaccharide hydrogels for insulin controlled-release. *Carbohydrate Polymers*, *236*, 116046. PMID: 32172860

Feng, C., Wang, Y., & Yang, J. (2018). Effects of Reorientation of Graphene Platelets (GPLs) on Young's Modulus of Polymer Composites under Bi-Axial Stretching. *Nanomaterials (Basel, Switzerland)*, *8*(1), 27. DOI: 10.3390/nano8010027 PMID: 29316669

Feng, Q., Zhang, K., Yang, B., & Yu, Y. (2022). Editorial: Biomedical Applications of Natural Polymers. *Frontiers in Bioengineering and Biotechnology*, *10*(November). Advance online publication. DOI: 10.3389/fbioe.2022.1077823 PMID: 36507260

Feng, X., Gu, J., & Zhou, Y. (2022). Primary total hip arthroplasty failure: Aseptic loosening remains the most common cause of revision. *American Journal of Translational Research*, *14*(10), 7080. PMID: 36398241

Fernandes, R. S., Arribada, R. G., Silva, J. O., Silva-Cunha, A., Townsend, D. M., Ferreira, L. A. M., & Barros, A. L. B. (2022). In Vitro and In Vivo Effect of pH-Sensitive PLGA-TPGS-Based Hybrid Nanoparticles Loaded with Doxorubicin for Breast Cancer Therapy. Pharmaceutics 2022, Vol. 14, Page 2394, 14(11), 2394. https://doi.org/DOI: 10.3390/PHARMACEUTICS14112394

Fernandez-Arias, M., Vilas, A. M., Boutinguiza, M., Rodriguez, D., Arias-Gonzalez, F., Pou-Alvarez, P., Riveiro, A., Gil, J., & Pou, J. (2022). Palladium Nanoparticles Synthesized by Laser Ablation in Liquids for Antimicrobial Applications. *Nanomaterials (Basel, Switzerland)*, *12*(15), 2621. DOI: 10.3390/nano12152621 PMID: 35957051

Ferreira, S. S., Passos, C. P., Madureira, P., Vilanova, M., & Coimbra, M. A. (2015). Structure–function relationships of immunostimulatory polysaccharides: A review. *Carbohydrate Polymers*, *132*, 378–396. DOI: 10.1016/J.CARBPOL.2015.05.079 PMID: 26256362

Fogg, K., Tseng, N. H., Peyton, S. R., Holeman, P., Mc Loughlin, S., Fisher, J. P., Sutton, A., Shikanov, A., Gnecco, J. S., Knight, K. M., Slaby, E. M., Weaver, J. D., Hashemi, N. N., Zhang, Y., House, M. D., Vogt, B. J., Aguado, B. A., Bradford, J. C., Robinson, J. L., & Oyen, M. L. (2022). Roadmap on biomaterials for women's health. *JPhys Materials*, *6*(1), 012501. DOI: 10.1088/2515-7639/AC90EE

Fong Yen, W., Basri, M., Ahmad, M., & Ismail, M. (2015). Formulation and evaluation of galantamine gel as drug reservoir in transdermal patch delivery system. *TheScientificWorldJournal*, *495271*. Advance online publication. DOI: 10.1155/2015/495271 PMID: 25853145

Forouzandehdel, S., Meskini, M., & Rami, M. R. (2020). Design and application of (Fe3O4)-GOTfOH based AgNPs doped starch/PEG-poly (acrylic acid) nanocomposite as the magnetic nanocatalyst and the wound dress. *Journal of Molecular Structure*, *1214*, 128142. https://doi.org/https://doi.org/10.1016/j.molstruc.2020.128142

Foster, H. A., Ditta, I. B., Varghese, S., & Steele, A. (2011). Photocatalytic disinfection using titanium dioxide: Spectrum and mechanism of antimicrobial activity. *Applied Microbiology and Biotechnology*, *90*, 1847–1868. DOI: 10.1007/s00253-011-3213-7 PMID: 21523480

Foster, L. J. R., Ho, S., Hook, J., Basuki, M., & Marçal, H. (2015). Chitosan as a Biomaterial: Influence of Degree of Deacetylation on Its Physiochemical, Material and Biological Properties. *PLoS One*, *10*(8), e0135153. DOI: 10.1371/journal.pone.0135153 PMID: 26305690

Foster, V. S., Rash, L. D., & King, G. F.. (2021). Acid-Sensing Ion Channels: Expression and Function in Resident and Infiltrating Immune Cells in the Central Nervous System . *Frontiers in Cellular Neuroscience*, *15*, 738043. PMID: 34602982

Friedman, A. J., Phan, J., Schairer, D. O., Champer, J., Qin, M., Pirouz, A., Blecher-Paz, K., Oren, A., Liu, P. T., Modlin, R. L., & Kim, J. (2013). Antimicrobial and Anti-Inflammatory Activity of Chitosan–Alginate Nanoparticles: A Targeted Therapy for Cutaneous Pathogens. *The Journal of Investigative Dermatology*, *133*(5), 1231–1239. DOI: 10.1038/jid.2012.399 PMID: 23190896

Frolova, M. A., Gryn, S. A., Kovaleva, E. I., Albulov, A. I., Melnik, R. N., Eremets, V. I., ... & Matveeva, I. N. (2020). Study of radioprotective properties of chitosan at gamma-irradiation. *International journal of research in pharmaceutical sciences, 11*(2), 1734-1737.. https://doi.org/.DOI: 10.26452/ijrps.v11i2.2076

Fu, L., Zhang, Y., Farokhzad, R. A., Mendes, B. B., Conde, J., & Shi, J. (2023). 'Passive'nanoparticles for organ-selective systemic delivery: Design, mechanism and perspective. *Chemical Society Reviews*, *52*(21), 7579–7601. PMID: 37817741

Furtado, D., Björnmalm, M., Ayton, S., Bush, A. I., Kempe, K., & Caruso, F. (2018). Overcoming the blood–brain barrier: The role of nanomaterials in treating neurological diseases. *Advanced Materials*, *30*(46), 1801362. PMID: 30066406

Futamura, M., Nagao, Y., Ishihara, K., Takeuchi, M., Nakada, T., Kawaguchi, Y., Asano, M., Kumazawa, I., Shiroko, T., Morimitsu, K., Mori, R., Nawa, M., Shimokawa, T., & Yoshida, K. (2017). Preoperative neoadjuvant chemotherapy using nanoparticle albumin-bound paclitaxel followed by epirubicin and cyclophosphamide for operable breast cancer: A multicenter phase II trial. *Breast Cancer (Tokyo, Japan)*, *24*(4), 615–623. DOI: 10.1007/s12282-016-0748-6 PMID: 28050738

Fu, Y., Ding, Y., Zhang, L., Zhang, Y., Liu, J., & Yu, P. (2021). Poly Ethylene Glycol (PEG)-Related Controllable and Sustainable Antidiabetic Drug Delivery Systems. *European Journal of Medicinal Chemistry*, *217*(May), 113372. DOI: 10.1016/j.ejmech.2021.113372 PMID: 33744689

Galed, G., Miralles, B., Panos, I., Santiago, A., & Heras, A. (2005). -Deacetylation and depolymerization reactions of chitin/chitosan: Influence of the source of chitin. *Carbohydrate Polymers*, *62*(4), 316–320. DOI: 10.1016/j.carbpol.2005.03.019

Gallo, J., Raska, M., Kriegova, E., & Goodman, S. B. (2017). Inflammation and its resolution and the musculoskeletal system. *Journal of Orthopaedic Translation*, *10*, 52–67. DOI: 10.1016/j.jot.2017.05.007 PMID: 28781962

Ganesan, K., Wang, Y., Gao, F., Liu, Q., Zhang, C., Li, P., Zhang, J., & Chen, J. (2021). Targeting Engineered Nanoparticles for Breast Cancer Therapy. *Pharmaceutics*, *13*(11), 1829. DOI: 10.3390/PHARMACEUTICS13111829 PMID: 34834243

Gao, G., & Fu, J. (2019). Highly crosslinked UHMWPE for joint implants. *UHMWPE Biomaterials for Joint Implants: Structures, Properties and Clinical Performance*, 21-68.

Gao, X., Caruso, B. R., & Li, W. (2024). Advanced Hydrogels in Breast Cancer Therapy. Gels 2024, Vol. 10, Page 479, 10(7), 479. https://doi.org/DOI: 10.3390/GELS10070479

Gao, N., Sun, H., Dong, K., Ren, J., & Qu, X. (2015). Gold-nanoparticle-based multifunctional amyloid-β inhibitor against Alzheimer's disease. *Chemistry (Weinheim an der Bergstrasse, Germany)*, *21*(2), 829–835. DOI: 10.1002/chem.201404562 PMID: 25376633

Gao, R., Mao, Y., Ma, C., Wang, Y., Jia, H., Chen, X., & Yu, L. (2022). SERS-based immunoassay of myocardial infarction biomarkers on a microfluidic chip with plasmonic nanostripe microcones. *ACS Applied Materials & Interfaces*, *14*(50), 55414–55422. PMID: 36480247

Gao, S. D.-H. (2009). The effect of chemical modification and nanoparticle formulation on stability and biodistribution of siRNA in mice. *Molecular Therapy : the Journal of the American Society of Gene Therapy*, *17*(7), 1225–1233. PMID: 19401674

Gao, W., Bigham, A., Ghomi, M., Zarrabi, A., Rabiee, N., Saeb, M. R., & Ashrafizadeh, M. (2024). Micelle-engineered nanoplatforms for precision oncology. *Chemical Engineering Journal*, *495*, 153438.

Gao, Y., Lim, J., & Teoh, S. H.. (2015). Emerging translational research on magnetic nanoparticles for regenerative medicine . *Chemical Society Reviews*, *44*(17), 6306–6329. PMID: 26505058

Gao, Z., Mansor, M. H., Winder, N., Demiral, S., Maclnnes, J., Zhao, X., & Muthana, M. (2023). Microfluidic-Assisted ZIF-Silk-Polydopamine Nanoparticles as Promising Drug Carriers for Breast Cancer Therapy. *Pharmaceutics*, *15*(7). Advance online publication. DOI: 10.3390/PHARMACEUTICS15071811 PMID: 37513998

García-Estrada, P., García-Bon, M. A., López-Naranjo, E. J., Basaldúa-Pérez, D. N., Santos, A., & Navarro-Partida, J. (2021). Polymeric implants for the treatment of intraocular eye diseases: Trends in biodegradable and non-biodegradable materials. *Pharmaceutics*, *13*(5), 701. DOI: 10.3390/pharmaceutics13050701 PMID: 34065798

García-González, N., Gonçalves-Sánchez, J., Gómez-Nieto, R., Gonçalves-Estella, J. M., & López, D. E. (2024). Advances and Challenges in Gene Therapy for Neurodegenerative Diseases: A Systematic Review. *International Journal of Molecular Sciences*, *25*(23), 12485. PMID: 39684197

García-Soriano, D., Milán-Rois, P., Lafuente-Gómez, N., Rodríguez-Díaz, C., Navío, C., Somoza, Á., & Salas, G. (2024). Multicore iron oxide nanoparticles for magnetic hyperthermia and combination therapy against cancer cells. *Journal of Colloid and Interface Science*, *670*, 73–85. PMID: 38759270

Gastaldi, L., Battaglia, L., Peira, E., Chirio, D., Muntoni, E., Solazzi, I., Gallarate, M., & Dosio, F. (2014). Solid lipid nanoparticles as vehicles of drugs to the brain: Current state of the art. *European Journal of Pharmaceutics and Biopharmaceutics*, *87*(3), 433–444. DOI: 10.1016/j.ejpb.2014.05.004 PMID: 24833004

Geddes, L., Themistou, E., Burrows, J. F., Buchanan, F. J., & Carson, L. (2021). Evaluation of the in Vitro Cytotoxicity and Modulation of the Inflammatory Response by the Bioresorbable Polymers Poly(D,L-Lactide-Co-Glycolide) and Poly(L-Lactide-Co-Glycolide). *Acta Biomaterialia*, *134*(October), 261–275. DOI: 10.1016/j.actbio.2021.07.049 PMID: 34329786

Geetha, M., Singh, A. K., Asokamani, R., & Gogia, A. K. (2009). Ti based biomaterials, the ultimate choice for orthopaedic implants–a review. *Progress in Materials Science*, *54*(3), 397–425.

Gendler, P. E., Coviak, C. P., Martin, J. T., Kim, K. K., Dankers, J. K., Barclay, J. M., & Sanchez, T. A. (2015). Revision of the Osteoporosis Knowledge Test. *Western Journal of Nursing Research*, *37*(12), 1623–1643. DOI: 10.1177/0193945914537565 PMID: 24923463

Gentile, P, & Nandagiri, V K, PABARI R. *et al.* (2015). Influence of Parathyroid Hormone-Loaded PLGA Nanoparticles in Porous Scaffolds for Bone Regeneration . *International Journal of Molecular Sciences*, *16*(9), 20492–20510. PMID: 26343649

Georgopoulou, A., Papadogiannis, F., Batsali, A., Marakis, J., Alpantaki, K., Eliopoulos, A. G., & Chatzinikolaidou, M. (2018). Chitosan/gelatin scaffolds support bone regeneration. *Journal of Materials Science. Materials in Medicine*, *29*, 1–13. DOI: 10.1007/s10856-018-6064-2 PMID: 29730855

Gerardo Nava, J. L., Rose, J. C., Altinova, H., Dalton, P. D., De Laporte, L., & Brook, G. A. (2021). Nanofibers and Nanostructured Scaffolds for Nervous System Lesions. *Nanomedicines for Brain Drug Delivery*, 61-101.

Ge, Z. (2004). Hydroxyapatite–chitin materials as potential tissue engineered bone substitutes. *Biomaterials*, *25*(6), 1049–1058. DOI: 10.1016/S0142-9612(03)00612-4 PMID: 14615170

Ghaffarlou, M., Rashidzadeh, H., Mohammadi, A., Mousazadeh, N., Barsbay, M., Shara, A., Gharbavi, M., Danafar, H., & Golestan, S. J. (2023). Catechol-Conjugated Alginate Biopolymer for Eco-Friendly Synthesis of Gold Nanoparticles: A Promising Approach for Photothermal and Radiotherapy in Breast Cancer Treatment. https://doi.org/DOI: 10.21203/RS.3.RS-3496758/V1

Ghavimi, S. A. A., Lungren, E. S., Faulkner, T. J., Josselet, M. A., Wu, Y., Sun, Y., Pfeiffer, F. M., Goldstein, C. L., Wan, C., & Ulery, B. D. (2019). Inductive co-crosslinking of cellulose nanocrystal/chitosan hydrogels for the treatment of vertebral compression fractures. *International Journal of Biological Macromolecules*, *130*, 88–98. DOI: 10.1016/J.IJBIOMAC.2019.02.086 PMID: 30779988

Ghaz-Jahanian, M. A., Abbaspour-Aghdam, F., Anarjan, N., Berenjian, A., & Jafarizadeh-Malmiri, H. (2015). Application of Chitosan-Based Nanocarriers in Tumor-Targeted Drug Delivery. *Molecular Biotechnology*, *57*(3), 201–218. DOI: 10.1007/s12033-014-9816-3 PMID: 25385004

Gheorghi ă, D., Moldovan, H., Robu, A., Bi a, A. I., Grosu, E., Antoniac, A., & Băcilă, C. I. (2023). Chitosan-Based Biomaterials for Hemostatic Applications: A Review of Recent Advances. *International Journal of Molecular Sciences*, *24*(13), 10540. PMID: 37445718

Gholap, A. D. (2024). Exploring modified chitosan-based gene delivery technologies for therapeutic advancements. *International Journal of Biological Macromolecules*, *260*, 129581. PMID: 38266848

Ghose, D., & Kumar, D. (2022). Emerging Nanostructures in Dental Applications. In *Emerging Materials: Design, Characterization and Applications* (pp. 265-313): Springer.

Giger, E. V., Castagner, B., & Leroux, J. C. (2013). Biomedical applications of bisphosphonates. *Journal of Controlled Release*, *167*(2), 175–188. DOI: 10.1016/J.JCONREL.2013.01.032 PMID: 23395668

Gim, S., Zhu, Y., Seeberger, P. H., & Delbianco, M. (2019). Carbohydrate-based nanomaterials for biomedical applications. *Wiley Interdisciplinary Reviews. Nanomedicine and Nanobiotechnology*, *11*(5). Advance online publication. DOI: 10.1002/wnan.1558 PMID: 31063240

Giuseppe, M., Mattia, B., Nadia, B., Raffaele, V., Pasquale, R., Stefano, D. A., Mattia, S., Vincenzo, D. S., & Giulio, M. (2021). Ceramic-on-ceramic versus ceramic-on-polyethylene in total hip arthroplasty: A comparative study at a minimum of 13 years follow-up. *BMC Musculoskeletal Disorders*, *22*(Suppl 2), 1062. PMID: 35039021

Godoy-Gallardo, M., Portolés-Gil, N., & López-Periago, A. M.. (2020). Immobilization of BMP-2 and VEGF within Multilayered Polydopamine-Coated Scaffolds and the Resulting Osteogenic and Angiogenic Synergy of Co-Cultured Human Mesenchymal Stem Cells and Human Endothelial Progenitor Cells . *International Journal of Molecular Sciences*, *21*(17), 6418. PMID: 32899269

Goldberg, T., & Fidler, B. (2015). Conjugated Estrogens/Bazedoxifene (Duavee): A Novel Agent for the Treatment of Moderate-to-Severe Vasomotor Symptoms Associated With Menopause And the Prevention of Postmenopausal Osteoporosis. *P&T : a Peer-Reviewed Journal for Formulary Management*, *40*(3), 178–182. PMID: 25798038

Golubovsky, J. L., Ejikeme, T., Winkelman, R., & Steinmetz, M. P. (2021). Osteobiologics. *Operative Neurosurgery (Hagerstown, Md.)*, *21*(Supplement_1), S2–S9. DOI: 10.1093/ons/opaa383 PMID: 34128067

González-Ballesteros, M. M., Mejía, C., & Ruiz-Azuara, L. (2022). Metallodrugs: An approach against invasion and metastasis in cancer treatment. *FEBS Open Bio*, *12*(5), 880–899. DOI: 10.1002/2211-5463.13381 PMID: 35170871

Gopi, D., Bhalaji, P., Ramya, S., & Kavitha, L. (2014). Evaluation of biodegradability of surface treated AZ91 magnesium alloy in SBF solution. *Journal of Industrial and Engineering Chemistry*, *23*, 218–227. DOI: 10.1016/j.jiec.2014.08.019

Granchi, D., Savarino, L. M., Ciapetti, G., & Baldini, N. (2018). Biological effects of metal degradation in hip arthroplasties. *Critical Reviews in Toxicology*, *48*(2), 170–193. PMID: 29130357

Greene, C., & Campbell, M. (2016). Tight junction modulation of the blood brain barrier: CNS delivery of small molecules. *Tissue Barriers*, *4*(1), e1138017. DOI: 10.1080/21688370.2015.1138017 PMID: 27141420

Greenshields, S. (2019). Neurological assessment in children and young people . *British Journal of Nursing (Mark Allen Publishing)*, *28*(16), 1056–1059. PMID: 31518531

Grewal, A. K., & Salar, R. K. (2024). Chitosan nanoparticle delivery systems: an effective approach to enhancing efficacy and safety of anticancer drugs. *Nano TransMed*, 100040.

Grewal, A. K., & Salar, R. K. (2024). Chitosan nanoparticle delivery systems: An effective approach to enhancing efficacy and safety of anticancer drugs. *Nano TransMed*, *3*, 100040. DOI: 10.1016/J.NTM.2024.100040

Grimm, M. J. (2004). *Orthopedic biomaterials. Standard handbook for biomedical engineering and design* (2nd ed.). McGraw-Hill.

Grobner, T., & Prischl, F. C. (2007). Gadolinium and nephrogenic systemic fibrosis. *Kidney International*, *72*(3), 260–264. DOI: 10.1038/sj.ki.5002338 PMID: 17507905

Grosso, R., & de-Paz, M.-V. (2021). Thiolated-Polymer-Based Nanoparticles as an Avant-Garde Approach for Anticancer Therapies—Reviewing Thiomers from Chitosan and Hyaluronic Acid. *Pharmaceutics*, *13*(6), 854. DOI: 10.3390/pharmaceutics13060854 PMID: 34201403

Grünewald, A., Kumar, K. R., & Sue, C. M. (2019). New insights into the complex role of mitochondria in Parkinson's disease. *Progress in Neurobiology*, *177*, 73–93. PMID: 30219247

Guadarrama-Escobar, O. R., Serrano-Castañeda, P., Anguiano-Almazán, E., Vázquez-Durán, A., Peña-Juárez, M. C., Vera-Graziano, R., Morales-Florido, M. I., Rodriguez-Perez, B., Rodriguez-Cruz, I. M., Miranda-Calderón, J. E., & Escobar-Chávez, J. J. (2023). Chitosan Nanoparticles as Oral Drug Carriers. *International Journal of Molecular Sciences*, *24*(5), 4289. DOI: 10.3390/ijms24054289 PMID: 36901719

Guillot, X. (2015). Les différents types d'ostéoporose et leurs stratégies thérapeutiques. *Actualites Pharmaceutiques*, *54*(549), 18–22. DOI: 10.1016/j.actpha.2015.07.004

Gulati, S., Ansari, N., Moriya, Y., Joshi, K., Prasad, D., Sajwan, G., Shukla, S., Kumar, S., & Varma, R. S. (2024). Nanobiopolymers in cancer therapeutics: Advancing targeted drug delivery through sustainable and controlled release mechanisms. *Journal of Materials Chemistry. B*, *12*(46), 11887–11915. DOI: 10.1039/D4TB00599F PMID: 39502076

Gulfam, M., Kim, J. E., Lee, J. M., Ku, B., Chung, B. H., & Chung, B. G. (2012). Anticancer drug-loaded gliadin nanoparticles induce apoptosis in breast cancer cells. *Langmuir*, *28*(21). Advance online publication. DOI: 10.1021/la300691n PMID: 22568862

Guo, B. J., Yang, Z. L., & Zhang, L. J. (2018). Gadolinium deposition in brain: Current scientific evidence and future perspectives. *Frontiers in Molecular Neuroscience*, *11*, 335. DOI: 10.3389/fnmol.2018.00335 PMID: 30294259

Guo, L., Zhang, H., Liu, P., Mi, T., Ha, D., Su, L., Huang, L., Shi, Y., & Zhang, J. (2021). Preclinical Assessment of Paclitaxel- and Trastuzumab-Delivering Magnetic Nanoparticles Fe3O4 for Treatment and Imaging of HER2-Positive Breast Cancer. *Frontiers in Medicine*, *8*, 738775. DOI: 10.3389/FMED.2021.738775/FULL PMID: 34778301

Gupta, P. C., Sharma, N., Mishra, P., Rai, S., & Verma, T. (2024). Role of Gold Nanoparticles for Targeted Drug Delivery. In *Metal and Metal-Oxide Based Nanomaterials: Synthesis, Agricultural, Biomedical and Environmental Interventions* (pp. 243-269): Springer.

Gupta, S. (2022, February 4). Side effects of breast cancer treatment | LBBC. https://www.lbbc.org/about-breast-cancer/side-effects

Gupta, M., & Wong, W. (2015). Magnesium-based nanocomposites: Lightweight materials of the future. *Materials Characterization, 105*, 30–46. DOI: 10.1016/j.matchar.2015.04.015

Gurunathan, S., Kang, M., Jeyaraj, M., & Kim, J. (2021). Palladium Nanoparticle-Induced oxidative stress, endoplasmic reticulum stress, apoptosis, and immunomodulation enhance the biogenesis and release of exosome in human leukemia monocytic cells (THP-1). *International Journal of Nanomedicine, 16*, 2849–2877. DOI: 10.2147/ijn.s305269 PMID: 33883895

Gurunathan, S., Kim, E., Han, J., Park, J., & Kim, J. (2015). Green Chemistry Approach for Synthesis of Effective Anticancer Palladium Nanoparticles. *Molecules (Basel, Switzerland), 20*(12), 22476–22498. DOI: 10.3390/molecules201219860 PMID: 26694334

Gusieva, K., Davies, C. H. J., Scully, J. R., & Birbilis, N. (2014). Corrosion of magnesium alloys: The role of alloying. *International Materials Reviews, 60*(3), 169–194. DOI: 10.1179/1743280414y.0000000046

Gwak, S. J., Nice, J., Zhang, J., Green, B., Macks, C., Bae, S., Webb, K., & Lee, J. S. (2016). Cationic, amphiphilic copolymer micelles as nucleic acid carriers for enhanced transfection in rat spinal cord. *Acta Biomaterialia, 35*, 98–108. DOI: 10.1016/j.actbio.2016.02.013 PMID: 26873365

Haan, N., & Song, B. (2014). Therapeutic application of electric fields in the injured nervous system. *Advances in Wound Care (New Rochelle, N.Y.), 3*(2), 156–165. DOI: 10.1089/wound.2013.0450 PMID: 24761356

Habeeb, M., Vengateswaran, H. T., Tripathi, A. K., Kumbhar, S. T., & You, H. W. (2024). Enhancing biomedical imaging: The role of nanoparticle-based contrast agents. *Biomedical Microdevices, 26*(4), 1–18. PMID: 39441423

Hacker, M. C., Krieghoff, J., & Mikos, A. G. (2019). Synthetic polymers. In *Principles of regenerative medicine* (pp. 559–590). Academic press., DOI: 10.1016/B978-0-12-809880-6.00033-3

Haddad, F., Konan, S., & Tahmassebi, J. (2015). A prospective comparative study of cementless total hip arthroplasty and hip resurfacing in patients under the age of 55 years: A ten-year follow-up. *The Bone & Joint Journal, 97*(5), 617–622. PMID: 25922454

Haikal, C., & Weissert, R. (2019). Editorial for Research Topic: Aging, Peripheral Inflammation, and Neurodegeneration. *Frontiers in Aging Neuroscience, 16*, 1529026. PMID: 39720419

Hakima, el K., Raja, B., Redouan, E. K., Ali, L., A., A., & Ahmed, lahsini. (2017). Physicochemical characterization of chitin and chitosan producted from Parapenaeus Longirostris shrimp shell wastes. Journal of Materials and Environmental Sciences .

Hakim, L. K., Yari, A., Nikparto, N., Mehraban, S. H., Cheperli, S., Asadi, A., Darehdor, A. A., Nezaminia, S., Dortaj, D., & Nazari, Y. (2024). The current applications of nano and biomaterials in drug delivery of dental implant. *BMC Oral Health, 24*(1), 126. PMID: 38267933

Haleem, A., Javaid, M., Singh, R. P., Rab, S., & Suman, R. (2023). Applications of nanotechnology in medical field: A brief review. *Global Health Journal (Amsterdam, Netherlands), 7*(2), 70–77.

Hallab, N. J., Mikecz, K., & Jacobs, J. J. (2000). A triple assay technique for the evaluation of metal-induced, delayed-type hypersensitivity responses in patients with or receiving total joint arthroplasty. *Journal of Biomedical Materials Research: An Official Journal of The Society for Biomaterials, The Japanese Society for Biomaterials, and The Australian Society for Biomaterials and the Korean Society for Biomaterials*, *53*(5), 480–489. PMID: 10984695

Hamdan, Y. A., El-Mansoury, B., Elouali, S., Rachmoune, K., Belbachir, A., Oudadesse, H., & Rhazi, M. (2024). A review of chitosan polysaccharides: Neuropharmacological implications and tissue regeneration. *International Journal of Biological Macromolecules*, •••, 135356. PMID: 39244136

Hamedi, H., Moradi, S., Hudson, S. M., & Tonelli, A. E. (2018). Chitosan based hydrogels and their applications for drug delivery in wound dressings: A review. *Carbohydrate Polymers*, *199*, 445–460. DOI: 10.1016/j.carbpol.2018.06.114 PMID: 30143150

Hanes, J., Dobakova, E., & Majerova, P. (2020). Brain drug delivery: Overcoming the blood-brain barrier to treat tauopathies. *Current Pharmaceutical Design*, *26*(13), 1448–1465. PMID: 32178609

Han, H. D. (2011). Chitosan hydrogel for localized gene silencing. *Cancer Biology & Therapy*, *11*(9), 839–845. PMID: 21358280

Han, P., Frith, J. E., & Gomez, G. A.. (2019). Five Piconewtons: The Difference between Osteogenic and Adipogenic Fate Choice in Human Mesenchymal Stem Cells . *ACS Nano*, *13*(10), 11129–11143. PMID: 31580055

Han, S. G., Newsome, B., & Hennig, B. (2013). Titanium dioxide nanoparticles increase inflammatory responses in vascular endothelial cells. *Toxicology*. Advance online publication. DOI: 10.1016/j.tox.2013.01.014 PMID: 23380242

Han, T., Sun, Y., Jiang, X., Gong, C., Kong, F., Luo, Y., Ge, C., Liu, C., Liu, Y., Mou, Y., Zhang, H., Ju, J., Chen, Y., & Qu, D. (2024). Air bag-embedded MIL-101(Fe) metal-organic frameworks for an amplified tumor microenvironment activation loop through strategic delivery of iron ions and lentinan. *Theranostics*, *14*(15), 5883–5902. DOI: 10.7150/THNO.99303 PMID: 39346539

Han, X., Shen, J., & Chen, S.. (2023). Ultrasonic-controlled "explosive" hydrogels to precisely regulate spatiotemporal osteoimmune disturbance . *Biomaterials*, *295*, 122057. PMID: 36805244

Hardy, J., & Selkoe, D. J. (2002). The amyloid hypothesis of Alzheimer's disease: Progress and problems on the road to therapeutics. *Science*, *297*(5580), 353–356. PMID: 12130773

Harish, V., Tewari, D., Gaur, M., Yadav, A. B., Swaroop, S., Bechelany, M., & Barhoum, A. (2022). Review on Nanoparticles and Nanostructured Materials: Bioimaging, Biosensing, Drug Delivery, Tissue Engineering, Antimicrobial, and Agro-Food Applications. *Nanomaterials (Basel, Switzerland)*, *12*(3), 457. DOI: 10.3390/NANO12030457 PMID: 35159802

Harzand, F. V. (2022). Synthesis of advanced gold nanoparticles for biomedical applications. *Advances in Applied NanoBio-Technologies*, *3*(4), 13–19.

Hasegawa, M., Saruta, J., Hirota, M., Taniyama, T., Sugita, Y., Kubo, K., Ishijima, M., Ikeda, T., Maeda, H., & Ogawa, T. (2020). A newly created meso-, micro-, and nano-scale rough titanium surface promotes bone-implant integration. *International Journal of Molecular Sciences, 21*(3), 783. PMID: 31991761

Hashemi, S. A., Arjmand, M., Lankarani, K. B., Chiang, W.-H., & Ramakrishna, S. (2023). Editorial: Advanced nanotechnological detection and drug delivery configurations [Editorial]. *Frontiers in Pharmacology, 14*. Advance online publication. DOI: 10.3389/fphar.2023.1198100 PMID: 37124201

Hawkins, B. T., & Davis, T. P. (2005). The blood-brain barrier/neurovascular unit in health and disease. *Pharmacological Reviews, 57*(2), 173–185. DOI: 10.1124/pr.57.2.4 PMID: 15914466

Hefnawy, A. K.-S. (2020). Dual-ligand functionalized core-shell chitosan-based nanocarrier for hepatocellular carcinoma-targeted drug delivery. *International Journal of Nanomedicine*, •••, 821–837. PMID: 32103939

Heidari, R. A.-R. (2021). A systematic review of long non-coding RNAs with a potential role in Breast Cancer. *Mutation Research/Reviews in Mutation Research, 787*, 108375. PMID: 34083033

Heilala, M., Lehtonen, A., Arasalo, O., Peura, A., Pokki, J., & Ikkala, O., Nonappa, Klefström, J., & Munne, P. M. (2023). Fibrin Stiffness Regulates Phenotypic Plasticity of Metastatic Breast Cancer Cells. *Advanced Healthcare Materials, 12*(31), 2301137. DOI: 10.1002/ADHM.202301137 PMID: 37671812

Heimann, R. B. (2020). *Biomaterials–characteristics, history, applications. Materials for Medical Applications* (1st ed.). Walter de Gruyter.

He, J., Qiao, L., Li, J., Lu, J., Fu, Z., Chen, J., Zhang, X., & Hu, X. (2024). Advanced strategies for 3D-printed neural scaffolds: Materials, structure, and nerve remodeling. *Bio-Design and Manufacturing, 7*(5), 747–770.

Helmi, O. E. (2021). Targeted doxorubicin delivery and release within breast cancer environment using PEGylated chitosan nanoparticles labeled with monoclonal antibodies. *International Journal of Biological Macromolecules, 184*, 325–338. PMID: 34119547

Heneka, M. T., Golenbock, D. T., & Latz, E. (2015). Innate immunity in Alzheimer's disease. *Nature Immunology, 16*(3), 229–236. PMID: 25689443

Henkel, J., Woodruff, M. A., & Epari, D. R.. (2013). Bone Regeneration Based on Tissue Engineering Conceptions - A 21st Century Perspective . *Bone Research, 1*(3), 216–248. PMID: 26273505

Henna, T. K., & Pramod, K. (2020). Graphene quantum dots redefine nanobiomedicine. *Materials Science and Engineering C, 110*, 110651. DOI: 10.1016/j.msec.2020.110651 PMID: 32204078

Herath, M. E. (2020). Light activated shape memory polymers and composites: A review. *European Polymer Journal, 136*, 109912.

Herawati, A., & Antarnusa, K. Nene, Fe3O4/chitosan nanocomposites for Congo red removal: Adsorption surpasses photodegradation, Materials Chemistry and Physics, Volume 347, 2026, 131431, ISSN 0254-0584, https://doi.org/DOI: 10.1016/j.matchemphys.2025.131431

Herdiana, Y., Husni, P., Nurhasanah, S., Shamsuddin, S., & Wathoni, N. (2023). Chitosan-Based Nano Systems for Natural Antioxidants in Breast Cancer Therapy. In Polymers (Vol. 15, Issue 13). https://doi .org/DOI: 10.3390/polym15132953

Herdiana, Y., Husni, P., Nurhasanah, S., Shamsuddin, S., & Wathoni, N. (2023). Chitosan-based nano systems for natural antioxidants in breast cancer therapy. *Polymers*, *15*(13), 2953. PMID: 37447598

Herdiana, Y., Wathoni, N., Shamsuddin, S., Joni, I. M., & Muchtaridi, M. (2021). Chitosan-based nanoparticles of targeted drug delivery system in breast cancer treatment. *Polymers*, *13*(11), 1717. PMID: 34074020

Herman, H., Rata, D. M., Cadinoiu, A. N., Atanase, L. I., & Hermenean, A. (2024). Colloidal and Biological Characterization of Dual Drug-Loaded Smart Micellar Systems. *Polymers*, *16*(9), 1189. PMID: 38732658

Hersh, A. M., Alomari, S., & Tyler, B. M. (2022). Crossing the blood-brain barrier: Advances in nanoparticle technology for drug delivery in neuro-oncology. *International Journal of Molecular Sciences*, *23*(8), 4153. PMID: 35456971

Hheidari, A., Mohammadi, J., Ghodousi, M., Mahmoodi, M., Ebrahimi, S., Pishbin, E., & Rahdar, A. (2024). Metal-based nanoparticle in cancer treatment: Lessons learned and challenges. *Frontiers in Bioengineering and Biotechnology*, *12*, 1436297. DOI: 10.3389/fbioe.2024.1436297 PMID: 39055339

Hivare, P., Mujmer, K., Swarup, G., Gupta, S., & Bhatia, D. (2023). Endocytic pathways of pathogenic protein aggregates in neurodegenerative diseases. *Traffic (Copenhagen, Denmark)*, *24*(10), 434–452. PMID: 37392160

Hobson, D. W. (2016). *The Commercialization of Medical Nanotechnology for Medical Applications.*, DOI: 10.1007/978-3-319-43525-1_17

Hodges, N. A., Sussman, E. M., & Stegemann, J. P. (2021). Aseptic and septic prosthetic joint loosening: Impact of biomaterial wear on immune cell function, inflammation, and infection. *Biomaterials*, *278*, 121127. PMID: 34564034

Hofmann-Amtenbrink, M., Hofmann, H., Hool, A., & Roubert, F. (2014). Nanotechnology in medicine: European research and its implications. *Swiss Medical Weekly*, *144*, w14044. DOI: 10.4414/smw.2014.14044 PMID: 25473809

Hong, F., Yu, X., Wu, N., & Zhang, Y.-Q. (2017). Progress of in vivo studies on the systemic toxicities induced by titanium dioxide nanoparticles. *Toxicology Research*, *6*(2), 115–133. PMID: 30090482

Hoop, M., Walde, C. F., Riccò, R., Mushtaq, F., Terzopoulou, A., Chen, X., deMello, A. J., Doonan, C. J., Falcaro, P., Nelson, B. J., Puigmartí-Luis, J., & Pané, S. (2018). Biocompatibility characteristics of the metal organic framework ZIF-8 for therapeutical applications. *Applied Materials Today*, *11*, 13–21. DOI: 10.1016/j.apmt.2017.12.014

Hossein, H. H. S., Jabbari, I., Zarepour, A., Zarrabi, A., Ashrafizadeh, M., Taherian, A., & Makvandi, P. (2020). Functionalization of Magnetic Nanoparticles by Folate as Potential MRI Contrast Agent for Breast Cancer Diagnostics. *Molecules (Basel, Switzerland)*, *25*(18), 4053. DOI: 10.3390/MOLECULES25184053 PMID: 32899812

Hou, C., An, J., Zhao, D., Ma, X., Zhang, W., Zhao, W., Wu, M., Zhang, Z., & Yuan, F. (2022). Surface modification techniques to produce micro/nano-scale topographies on Ti-based implant surfaces for improved osseointegration. *Frontiers in Bioengineering and Biotechnology*, *10*, 835008. PMID: 35402405

Hou, D., Youssef, E A S., & Brinton, T. J.. (2005). Radiolabeled cell distribution after intramyocardial, intracoronary, and interstitial retrograde coronary venous delivery: Implications for current clinical trials . *Circulation*, *112*(9, Suppl), I150–I156. PMID: 16159808

Hsu, S.-H., Chen, L.-R., & Chen, K.-H. (2024). Primary Osteoporosis Induced by Androgen and Estrogen Deficiency: The Molecular and Cellular Perspective on Pathophysiological Mechanisms and Treatments. *International Journal of Molecular Sciences*, *25*(22), 12139. DOI: 10.3390/ijms252212139 PMID: 39596206

Hu, Tao, Jie Fang, Yang Shen, Mingyang Li, Bin Wang, Zushun Xu, and Weikang Hu. 2024. "Advances of Naturally Derived Biomedical Polymers in Tissue Engineering." *Frontiers in Chemistry*. Frontiers Media SA. https://doi.org/.DOI: 10.3389/fchem.2024.1469183

Huang, C., Huang, Y., Wang, Z., Lin, Y., Li, Y., Chen, Y., Chen, X., Zhang, C., Li, W., & Zhang, W. (2023). Multiplex PCR-based next generation sequencing as a novel, targeted and accurate molecular approach for periprosthetic joint infection diagnosis. *Frontiers in Microbiology*, *14*, 1181348. PMID: 37275128

Huang, H., Liu, R., Yang, J., Dai, J., Fan, S., Pi, J., & Guo, X. (2023). Gold nanoparticles: Construction for drug delivery and application in cancer immunotherapy. *Pharmaceutics*, *15*(7), 1868. DOI: 10.3390/pharmaceutics15071868 PMID: 37514054

Huang, J., Pan, X., & Yan, N. (2024). Structural biology and molecular pharmacology of voltage-gated ion channels. *Nature Reviews. Molecular Cell Biology*, *25*(11), 904–925. DOI: 10.1038/s41580-024-00763-7 PMID: 39103479

Huang, K., Hu, S., & Cheng, K. (2019). A New Era of Cardiac Cell Therapy: Opportunities and Challenges . *Advanced Healthcare Materials*, *8*(2), e1801011. PMID: 30548836

Huang, Y. zhuo, Ji, Y. rong, Kang, Z. wen, Li, F., Ge, S. fang, Yang, D. P., Ruan, J., & Fan, X. qun. (2020). Integrating eggshell-derived CaCO3/MgO nanocomposites and chitosan into a biomimetic scaffold for bone regeneration. *Chemical Engineering Journal*, *395*. Advance online publication. DOI: 10.1016/j.cej.2020.125098

Huang, Y., Cheng, X., & Li, G. (2021). Role of metal ions in promoting wound healing: Current applications in therapeutic scaffolds and dressings. *Advanced Healthcare Materials*, *10*(22), 2100732. DOI: 10.1002/adhm.202100732

Huang, Y., He, S., Cao, W., Cai, K., & Liang, X.-J. (2012). Biomedical nanomaterials for imaging-guided cancer therapy. *Nanoscale*, *4*(20), 6135–6149. PMID: 22929990

Hubbers, J. (2022). *A health economic evaluation of neurological interventions targeting epilepsy, migraine, dementia and Parkinson's disease in Ethiopia, Malawi, and the United Republic of Tanzania* The University of Bergen].

Hu, H., Ding, H., Lyu, J., Chen, Y., Huang, C., Zhang, C., Li, W., Fang, X., & Zhang, W. (2024). Detection of rare microorganisms in bone and joint infections by metagenomic next-generation sequencing: Discrimination, decision-making, and clinical benefits. *Bone & Joint Research, 13*(8), 401. PMID: 39142657

Hu, H., Qi, Q., Dong, Z., Yu, X., Mo, Y., Luo, J., Wang, Y., Du, S., & Lu, Y. (2020). Albumin coated trimethyl chitosan-based targeting delivery platform for photothermal/chemo-synergistic cancer therapy. *Carbohydrate Polymers, 241*, 116335. DOI: 10.1016/j.carbpol.2020.116335 PMID: 32507205

Hu, J., Shao, J., Huang, G., Zhang, J., & Pan, S. (2023). In vitro and in vivo applications of magnesium-enriched biomaterials for vascularized osteogenesis in bone tissue engineering: A review of literature. *Journal of Functional Biomaterials, 14*(6), 326. PMID: 37367290

Hu, K., & Olsen, B. R. (2016). The roles of vascular endothelial growth factor in bone repair and regeneration . *Bone, 91*, 30–38. PMID: 27353702

Hu, P., & Chen, G. (2024). Extraction, Structural Characterization, and Physicochemical and Biological Properties of Water-Soluble Polysaccharides from Adlay Bran. *Molecules (Basel, Switzerland), 29*(19), 4707. DOI: 10.3390/molecules29194707 PMID: 39407635

Husain, S., Al-Samadani, K. H., Najeeb, S., Zafar, M. S., Khurshid, Z., Zohaib, S., & Qasim, S. B. (2017). Chitosan Biomaterials for Current and Potential Dental Applications. *Materials (Basel), 10*(6), 602. DOI: 10.3390/ma10060602 PMID: 28772963

Hussain, I., Singh, T., & Chittenden, C. (2012). Preparation of chitosan oligomers and characterization: Their antifungal activities and decay resistance. *Holzforschung, 66*(1), 119–125. DOI: 10.1515/HF.2011.130

Hussain, Z., Khan, S., Imran, M., Sohail, M., Syed, W. A. S., & de Matas, M. (2019). *"PEGylation: A Promising Strategy to Overcome Challenges to Cancer-Targeted Nanomedicines: A Review of Challenges to Clinical Transition and Promising Resolution." Drug Delivery and Translational Research*. Springer., DOI: 10.1007/s13346-019-00631-4

Hu, X., Wang, Y., Zhang, L., & Xu, M. (2020). Formation of self-assembled polyelectrolyte complex hydrogel derived from salecan and chitosan for sustained release of Vitamin C. *Carbohydrate Polymers, 234*, 115920. PMID: 32070539

Iacob, A. T., Lupascu, F. G., Apotrosoaei, M., Vasincu, I. M., Tauser, R. G., Lupascu, D., Giusca, S. E., Caruntu, I.-D., & Profire, L. (2021). Recent Biomedical Approaches for Chitosan Based Materials as Drug Delivery Nanocarriers. *Pharmaceutics, 13*(4), 587. DOI: 10.3390/pharmaceutics13040587 PMID: 33924046

Iacobellis, F., Di Serafino, M., Russo, C., Ronza, R., Caruso, M., Dell'Aversano Orabona, G., & Romano, L. (2024). Safe and informed use of gadolinium-based contrast agent in body magnetic resonance imaging: Where we were and where we are. *Journal of Clinical Medicine, 13*(8), 2193. DOI: 10.3390/jcm13082193 PMID: 38673466

Iaconisi, G. N., Lunetti, P., Gallo, N., Cappello, A. R., Fiermonte, G., Dolce, V., & Capobianco, L. (2023). Hyaluronic Acid: A Powerful Biomolecule with Wide-Ranging Applications—A Comprehensive Review. International Journal of Molecular Sciences 2023, Vol. 24, Page 10296, 24(12), 10296. https://doi.org/DOI: 10.3390/IJMS241210296

Ichim, T. E. (2004). RNA interference: A potent tool for gene-specific therapeutics. *American Journal of Transplantation : Official Journal of the American Society of Transplantation and the American Society of Transplant Surgeons*, *4*(8), 1227–1236. PMID: 15268723

Imam, S. S., Alshehri, S., Ghoneim, M. M., Zafar, A., Alsaidan, O. A., Alruwaili, N. K., Gilani, S. J., & Rizwanullah, M. (2021). Recent Advancement in Chitosan-Based Nanoparticles for Improved Oral Bioavailability and Bioactivity of Phytochemicals: Challenges and Perspectives. *Polymers*, *13*(22), 4036. DOI: 10.3390/polym13224036 PMID: 34833334

Iqbal, Y., Amin, F., Fatima, M., Khalid, M., Kanwal, N., Hafez, A. A. A., & Alam, M. (2024). Chitosan-based functional materials combined with plant extract: A promising strategy in the stimulation of wound healing process. *Journal of Drug Delivery Science and Technology*, •••, 106314.

Iravani, S., & Varma, R. S. (2022). Alginate-Based Micro- and Nanosystems for Targeted Cancer Therapy. Marine Drugs 2022, Vol. 20, Page 598, 20(10), 598. https://doi.org/DOI: 10.3390/MD20100598

Isani, G., & Carpenè, E. (2014). Metallothioneins, unconventional proteins from unconventional animals: A long journey from nematodes to mammals. *Biomolecules*, *4*(2), 435–457. DOI: 10.3390/biom4020435 PMID: 24970224

Islam, M. M., Shahruzzaman, M., Biswas, S., Sakib, M. N., & Rashid, T. U. (2020). Chitosan based bioactive materials in tissue engineering applications-A review. *Bioactive Materials*, *5*(1), 164–183. PMID: 32083230

Iyad, N., Ahmad, M. S., Alkhatib, S. G., & Hjouj, M. (2023). Gadolinium contrast agents-challenges and opportunities of a multidisciplinary approach: Literature review. *European Journal of Radiology Open*, *11*, 100503. DOI: 10.1016/j.ejro.2023.100503 PMID: 37456927

Jaber, N., Al-Remawi, M., Al-Akayleh, F., Al-Muhtaseb, N., Al-Adham, I. S. I., & Collier, P. J. (2022). A review of the antiviral activity of Chitosan, including patented applications and its potential use against COVID-19. Journal of Applied Microbiology, 132(1), 41 58. https://doi.org/DOI: 10.1111/jam.15202

Jaber, H. A., Ibrahim, S. A., & Elmansi, M. (2022). The role of chitosan as an antiviral agent. *Antiviral Chemistry & Chemotherapy*, *30*, 204020662210991.

Jacobi, A., & Bareyre, F. M. (2015). Regulation of axonal remodeling following spinal cord injury . *Neural Regeneration Research*, *10*(10), 1555–1557. PMID: 26692837

Jacob, J., More, N., Kalia, K., & Kapusetti, G. (2018). Piezoelectric smart biomaterials for bone and cartilage tissue engineering. *Inflammation and Regeneration*, *38*(1), 2. DOI: 10.1186/s41232-018-0059-8 PMID: 29497465

Jacqueline, C., & Caillon, J. (2014). Impact of bacterial biofilm on the treatment of prosthetic joint infections. *The Journal of Antimicrobial Chemotherapy*, *69*(suppl_1), i37–i40. PMID: 25135088

Jafernik, K., Ładniak, A., Blicharska, E., Czarnek, K., Ekiert, H., Wiącek, A. E., & Szopa, A. (2023). Chitosan-based nanoparticles as effective drug delivery systems—A review. *Molecules (Basel, Switzerland)*, *28*(4), 1963. PMID: 36838951

Jagaran, K., & Singh, M. (2021). Nanomedicine for Neurodegenerative Disorders: Focus on Alzheimer's and Parkinson's Diseases. *International Journal of Molecular Sciences*, *22*(16). Advance online publication. DOI: 10.3390/ijms22169082 PMID: 34445784

Jain, N., Waidi, Y. O., Vannaladsaysy, V., Pathote, D., & Datta, S. (2025). Hydrogels With Nanomaterials Integrated for 3D Bioprinting Technology. In *Design, Fabrication, and Significance of Advanced Nanostructured Materials* (pp. 301-326): IGI Global.

Jain, S., Hirst, D. G., & O'Sullivan, J. (2012). Gold nanoparticles as novel agents for cancer therapy. *The British Journal of Radiology*, *85*(1010), 101–113. DOI: 10.1259/bjr/59448833 PMID: 22010024

Jaisingh, A., & Nebhani, L. (2024). Dendrimers and Hyperbranched Polymers as Promising and Versatile Additives in Polyolefins. *ACS Applied Engineering Materials*, *2*(2), 236–261.

Jarad, N. A., Chami, A., Weitz, J. I., & Didar, T. F. (2024). Advancements in Surface Modification Strategies of Vascular Grafts to Improve Biocompatibility and Tissue Integration. *Exploration of Biomat-X*, *1*(5), 241–265. DOI: 10.37349/EBMX.2024.00018

Jayakumar, R., Prabaharan, M., Nair, S. V., & Tamura, H. (2010). Novel chitin and chitosan nanofibers in biomedical applications. *Biotechnology Advances*, *28*(1), 142–150. DOI: 10.1016/J.BIOTECHADV.2009.11.001 PMID: 19913083

Jayakumar, R., Prabaharan, M., Nair, S. V., Tokura, S., Tamura, H., & Selvamurugan, N. (2010). Novel carboxymethyl derivatives of chitin and chitosan materials and their biomedical applications. *Progress in Materials Science*, *55*(7), 675–709. DOI: 10.1016/j.pmatsci.2010.03.001

Jeandupeux, E. (2018). *Utilisation De l'Acide Poly (Propyl Acrylique) Pour Augmenter l'Efficacité in Vitro Des Nanoparticules à Base De Chitosane Et ARN Messager*. Ecole Polytechnique.

Jeon, H., Kim, M.-S., Park, S. B., Kim, S., Lee, M., Park, S.-A., Hwang, S. Y., Koo, J. M., Oh, D. X., & Park, J. (2023). Improved mechanical properties of biodegradable polycaprolactone nanocomposites prepared using cellulose nanocrystals. *Cellulose (London, England)*, *30*(18), 11561–11574.

Jeon, Y. (2001). Antimicrobial effect of chitooligosaccharides produced by bioreactor. *Carbohydrate Polymers*, *44*(1), 71–76. DOI: 10.1016/S0144-8617(00)00200-9

Jhundoo, H. D., Siefen, T., Liang, A., Schmidt, C., Lokhnauth, J., Béduneau, A., Pellequer, Y., Larsen, C. C., & Lamprecht, A. (2020). Anti-Inflammatory Activity of Chitosan and 5-Amino Salicylic Acid Combinations in Experimental Colitis. *Pharmaceutics*, *12*(11), 1038. DOI: 10.3390/pharmaceutics12111038 PMID: 33138176

Jia, C., Wang, Q., Yao, X., & Yang, J. (2021). The role of DNA damage induced by low/high dose ionizing radiation in cell carcinogenesis. *Exploratory Research and Hypothesis in Medicine*, *6*(4), 177–184.

Jia, G., Wang, H., Yan, L., Wang, X., Pei, R., Yan, T., & Guo, X. (2005). Cytotoxicity of carbon nanomaterials: Single-wall nanotube, multi-wall nanotube, and fullerene. *Environmental Science & Technology*, *39*(5), 1378–1383. PMID: 15787380

Jiang, X., Du, Z., Zhang, X., Zaman, F., Song, Z., Guan, Y., Yu, T., & Huang, Y. (2023). Gelatin-based anticancer drug delivery nanosystems: A mini review. *Frontiers in Bioengineering and Biotechnology*, *11*, 1158749. DOI: 10.3389/FBIOE.2023.1158749/BIBTEX PMID: 37025360

Jiang, X., Yao, Y., Tang, W., Han, D., Zhang, L., Zhao, K., & Meng, Y. (2020). Design of dental implants at materials level: An overview. *Journal of Biomedical Materials Research. Part A*, *108*(8), 1634–1661. DOI: 10.1002/jbm.a.36931 PMID: 32196913

Jia, Y., Chen, S., Wang, C., Sun, T., & Yang, L. (2022). Hyaluronic acid-based nano drug delivery systems for breast cancer treatment: Recent advances. *Frontiers in Bioengineering and Biotechnology*, *10*, 990145. DOI: 10.3389/FBIOE.2022.990145/BIBTEX PMID: 36091467

Jiaying, Y., Bo, S., Xiaolu, W., Yanyan, Z., Hongjie, W., Nan, S., Bo, G., Linna, W., Yan, Z., Wenya, G., Keke, L., Shan, J., Chuan, L., Yu, Z., Qinghe, Z., & Haiyu, Z. (2023). Arenobufagin-loaded PEG-PLA nanoparticles for reducing toxicity and enhancing cancer therapy. *Drug Delivery*, *30*(1), 2177362. DOI: 10.1080/10717544.2023.2177362 PMID: 36772846

Jiménez-Gómez, C. P., & Cecilia, J. A. (2020). Chitosan: A natural biopolymer with a wide and varied range of applications. *Molecules (Basel, Switzerland)*, *25*(17), 3981. DOI: 10.3390/molecules25173981 PMID: 32882899

Jin, G. Z. (2020). Current nanoparticle-based technologies for osteoarthritis therapy. *Nanomaterials (Basel, Switzerland)*, *10*(12), 2368. PMID: 33260493

Jing, C., Li, B., Tan, H., Zhang, C., Liang, H., Na, H., Chen, S., Liu, C., & Zhao, L. (2021). Alendronate-Decorated Nanoparticles as Bone-Targeted Alendronate Carriers for Potential Osteoporosis Treatment. *ACS Applied Bio Materials*, *4*(6), 4907–4916. DOI: 10.1021/acsabm.1c00199 PMID: 35007039

Jin, Z., Zha, M., Wang, S., Wang, S., Wang, C., Jia, H., & Wang, H. (2022). Alloying design and microstructural control strategies towards developing Mg alloys with enhanced ductility. *Journal of Magnesium and Alloys*, *10*(5), 1191–1206. DOI: 10.1016/j.jma.2022.04.002

Ji, P., Wang, P., Chen, H., Xu, Y., Ge, J., Tian, Z., & Yan, Z. (2023). Potential of copper and copper compounds for anticancer applications. *Pharmaceuticals*, *16*(2), 234. DOI: 10.3390/ph16020234 PMID: 37259382

Ji, X., Lu, W., Wu, K., & Cho, W. C. (2017). Influencing Factors of the Pharmacokinetic Characters on Nanopharmaceutics. *Pharmaceutical Nanotechnology*, *5*(1), 24–31. DOI: 10.2174/221173850566616 1214142755 PMID: 28948908

Johnson, A., Northcote-Smith, J., & Suntharalingam, K. (2021). Emerging metallopharmaceuticals for the treatment of cancer. *Trends in Chemistry*, *3*(1), 47–58. DOI: 10.1016/j.trechm.2020.10.011

Jomova, K., Raptova, R., Alomar, S. Y., Alwasel, S. H., Nepovimova, E., Kuca, K., & Valko, M. (2023). Reactive oxygen species, toxicity, oxidative stress, and antioxidants: Chronic diseases and aging. *Archives of Toxicology*, *97*(10), 2499–2574. PMID: 37597078

Joudeh, N., Saragliadis, A., Koster, G., Mikheenko, P., & Linke, D. (2022). Synthesis methods and applications of palladium nanoparticles: A review. *Frontiers in Nanotechnology*, *4*. Advance online publication. DOI: 10.3389/fnano.2022.1062608

Joyce, K., Fabra, G. T., Bozkurt, Y., & Pandit, A. (2021). Correction to: Bioactive potential of natural biomaterials: identification, retention and assessment of biological properties. *Signal Transduction and Targeted Therapy*, *6*(1). Advance online publication. DOI: 10.1038/s41392-021-00593-5 PMID: 33953157

Juncos Bombin, A. D., Dunne, N. J., & McCarthy, H. O. (2020). Electrospinning of natural polymers for the production of nanofibres for wound healing applications. *Materials Science and Engineering C*, *114*, 110994. DOI: 10.1016/J.MSEC.2020.110994 PMID: 32993991

Kagdada, H. L., Bhojani, A. K., & Singh, D. K. (2023). An overview of nanomaterials: history, fundamentals, and applications. *Nanomaterials: Advances and Applications*, 1-26.

Kakkar, A. K., & Dahiya, N. (2015). Management of Parkinson s disease: Current and future pharmacotherapy. *European Journal of Pharmacology*, *750*, 74–81. DOI: 10.1016/j.ejphar.2015.01.030 PMID: 25637088

Kalathil, S., & Chaudhuri, R. (2016). Hollow palladium nanoparticles facilitated biodegradation of an azo dye by electrically active biofilms. *Materials (Basel)*, *9*(8), 653. DOI: 10.3390/ma9080653 PMID: 28773775

Kaltbeitzel, J., & Wich, P. R. (2023). Protein-based Nanoparticles: From Drug Delivery to Imaging, Nanocatalysis and Protein Therapy. *Angewandte Chemie International Edition*, *62*(44), e202216097. PMID: 36917017

Kamal, A. S., Safaa, M. M., & Elazab, M. S. (2024). Comparison between Obese and Non-obese Cases after Total Knee Arthroplasty Regarding Their Short Term Functional Outcome. *Fayoum University Medical Journal*, *14*(1), 122–130.

Kamaly, N., Yameen, B., Wu, J., & Farokhzad, O. C. (2016). Degradable controlled-release polymers and polymeric nanoparticles: Mechanisms of controlling drug release. *Chemical Reviews*, *116*(4), 2602–2663. PMID: 26854975

Kanishka, K., & Acherjee, B. (2023). Revolutionizing manufacturing: A comprehensive overview of additive manufacturing processes, materials, developments, and challenges. *Journal of Manufacturing Processes*, *107*, 574–619. DOI: 10.1016/j.jmapro.2023.10.024

Kannan, K., Radhika, D., Sadasivuni, K. K., Reddy, K. R., & Raghu, A. V. (2020). Nanostructured metal oxides and its hybrids for photocatalytic and biomedical applications. *Advances in Colloid and Interface Science*, *281*, 102178. DOI: 10.1016/j.cis.2020.102178 PMID: 32470668

Karges, J., Stokes, R. W., & Cohen, S. M. (2021). Metal complexes for therapeutic applications. *Trends in Chemistry*, *3*(7), 523–534. DOI: 10.1016/j.trechm.2021.03.006 PMID: 35966501

Kargozar, S., Milan, P. B., Baino, F., & Mozafari, M. (2018). Nanoengineered biomaterials for bone/dental regeneration. In *Elsevier eBooks* (pp. 13–38). https://doi.org/DOI: 10.1016/b978-0-12-813355-2.00002-8

Karimi, M., Bahrami, S., Ravari, S. B., Zangabad, P. S., Mirshekari, H., Bozorgomid, M., Shahreza, S., Sori, M., & Hamblin, M. R. (2016). Albumin nanostructures as advanced drug delivery systems. *Expert Opinion on Drug Delivery*, *13*(11), 1609–1623. DOI: 10.1080/17425247.2016.1193149 PMID: 27216915

Karimzadeh, M., Rashidi, L., & Ganji, F. (2017). Mesoporous silica nanoparticles for efficient rivastigmine hydrogen tartrate delivery into SY5Y cells. *Drug Development and Industrial Pharmacy*, *43*(4), 628–636. PMID: 28043167

Karnwal, A., Jassim, A. Y., Mohammed, A. A., Sharma, V., Al-Tawaha, A. R. M. S., & Sivanesan, I. (2024). Nanotechnology for Healthcare: Plant-Derived nanoparticles in disease treatment and regenerative medicine. *Pharmaceuticals*, *17*(12), 1711. DOI: 10.3390/ph17121711 PMID: 39770553

Karnwal, A., Kumar Sachan, R. S., Devgon, I., Devgon, J., Pant, G., Panchpuri, M., & Kumar, G. (2024). Gold Nanoparticles in Nanobiotechnology: From Synthesis to Biosensing Applications. *ACS Omega*, *9*(28), 29966–29982. PMID: 39035946

Karuppaiah, A. B. (2021). Building and behavior of a pH-stimuli responsive chitosan nanoparticles loaded with folic acid conjugated gemcitabine silver colloids in MDA-MB-453 metastatic breast cancer cell line and pharmacokinetics in rats. *European Journal of Pharmaceutical Sciences*, *165*, 105938. PMID: 34256103

Kashkooli, F. M. (2020). Controlled anti-cancer drug release through advanced nano-drug delivery systems: Static and dynamic targeting strategies. *Biotechnology Reports (Amsterdam, Netherlands)*.

Kashyap, K., & Shukla, R. (2019). Drug Delivery and Targeting to the Brain Through Nasal Route: Mechanisms, Applications and Challenges. *Current Drug Delivery*, *16*(10), 887–901. DOI: 10.2174/1567201816666191029122740 PMID: 31660815

Kathiresan, D. S., Balasubramani, R., Marudhachalam, K., Jaiswal, P., Ramesh, N., Sureshbabu, S. G., Puthamohan, V. M., & Vijayan, M. (2024). Role of Mitochondrial Dysfunctions in Neurodegenerative Disorders: Advances in Mitochondrial Biology. *Molecular Neurobiology*, •••, 1–29. PMID: 39269547

Katsuta, E., Rashid, O. M., & Takabe, K. (2020). Clinical relevance of tumor microenvironment: Immune cells, vessels, and mouse models. *Human Cell*, *33*(4), 930–937. DOI: 10.1007/s13577-020-00380-4 PMID: 32507979

Kaur, P., Thakur, R., Barnela, M., Chopra, M., Manuja, A., & Chaudhury, A. (2015). Synthesis, characterization and in vitro evaluation of cytotoxicity and antimicrobial activity of chitosan-metal nanocomposites. *Journal of Chemical Technology and Biotechnology*. Advance online publication. DOI: 10.1002/jctb.4383

Kaya, A. (2011). *Physical and electrochemical characterization of co-deposited TiO2 and MgO layer on Ti6Al4V by micro arc oxidation (MAO)*. Dokuz Eylul Universitesi.

Kazimir, A., Schwarze, B., Lönnecke, P., Jelača, S., Mijatović, S., Maksimović-Ivanić, D., & Hey-Hawkins, E. (2023). Metallodrugs against Breast Cancer: Combining the Tamoxifen Vector with Platinum(II) and Palladium(II) Complexes. Pharmaceutics 2023, Vol. 15, Page 682, 15(2), 682. https://doi.org/DOI: 10.3390/PHARMACEUTICS15020682

Kean, T., & Thanou, M. (2011). Chitin and Chitosan: Sources, Production and Medical Applications. In Renewable Resources for Functional Polymers and Biomaterials (pp. 292–318). The Royal Society of Chemistry. https://doi.org/DOI: 10.1039/9781849733519 00292

Kean, T., & Thanou, M. (2011). Chitin and Chitosan: Sources, Production and Medical Applications. In *Renewable Resources for Functional Polymers and Biomaterials* (pp. 292–318). The Royal Society of Chemistry. https://doi.org/DOI: 10.1039/9781849733519-00292

Kelland, L. (2007). The resurgence of platinum-based cancer chemotherapy. *Nature Reviews. Cancer*, *7*(8), 573–584. DOI: 10.1038/nrc2167 PMID: 17625587

Kempuraj, D., Dourvetakis, K. D., Cohen, J., Valladares, D. S., Joshi, R. S., Kothuru, S. P., Anderson, T., Chinnappan, B., Cheema, A. K., & Klimas, N. G. (2024). Neurovascular unit, neuroinflammation and neurodegeneration markers in brain disorders. *Frontiers in Cellular Neuroscience*, *18*, 1491952. PMID: 39526043

Kenanidis, E., Kakoulidis, P., Tsiridis, E., Atilla, B., Bicanic, G., Sulje, Z., Delimar, D., Bondarenko, S., Filippenko, V., & Akonjom, M. (2017). Acetabular dysplasia. In *The Adult Hip-Master Case Series and Techniques* (pp. 107–213). Springer.

Kendler, D. L., Cosman, F., Stad, R. K., & Ferrari, S. (2022). Denosumab in the Treatment of Osteoporosis: 10 Years Later: A Narrative Review. *Advances in Therapy*, *39*(1), 58–74. DOI: 10.1007/s12325-021-01936-y PMID: 34762286

Kesharwani, P., Halwai, K., Jha, S. K., Al Mughram, M. H., Almujri, S. S., Almalki, W. H., & Sahebkar, A. (2024). Folate-engineered chitosan nanoparticles: Next-generation anticancer nanocarriers. *Molecular Cancer*, *23*(1), 244. PMID: 39482651

Khadilkar, A., & Mandlik, R. (2015). Epidemiology and treatment of osteoporosis in women: An Indian perspective. *International Journal of Women's Health*, *841*. Advance online publication. DOI: 10.2147/IJWH.S54623 PMID: 26527900

Khan, F., & Ahmad, S. R. (2013). Polysaccharides and Their Derivatives for Versatile Tissue Engineering Application. *Macromolecular Bioscience*, *13*(4), 395–421. DOI: 10.1002/mabi.201200409 PMID: 23512290

Khan, I., Saeed, K., & Khan, I. (2019). Nanoparticles: Properties, applications and toxicities. *Arabian Journal of Chemistry*, *12*(7), 908–931. DOI: 10.1016/j.arabjc.2017.05.011

Khan, M. S., Gowda, B. H. J., Nasir, N., Wahab, S., Pichika, M. R., Sahebkar, A., & Kesharwani, P. (2023). Advancements in dextran-based nanocarriers for treatment and imaging of breast cancer. *International Journal of Pharmaceutics*, *643*, 123276. DOI: 10.1016/J.IJPHARM.2023.123276 PMID: 37516217

Khan, M., Shaik, M. R., Khan, S. T., Adil, S. F., Kuniyil, M., Khan, M., & Nawaz Tahir, M. (2020). Enhanced antimicrobial activity of biofunctionalized zirconia nanoparticles. *ACS Omega*, *5*(4), 1987–1996. PMID: 32039336

Kharaziha, M., Baidya, A., & Annabi, N. (2021). Rational Design of Immunomodulatory Hydrogels for Chronic Wound Healing . *Advanced Materials (Deerfield Beach, Fla.)*, *33*(39), e2100176. PMID: 34251690

Khare, D., Basu, B., & Dubey, A. K. (2020). Electrical stimulation and piezoelectric biomaterials for bone tissue engineering applications. *Biomaterials*, *258*, 120280. DOI: 10.1016/j.biomaterials.2020.120280 PMID: 32810650

Khatri, K. G. (2008). Plasmid DNA loaded chitosan nanoparticles for nasal mucosal immunization against hepatitis B. *International Journal of Pharmaceutics*, *354*(1-2), 235–241. PMID: 18182259

Khodaverdi, K., Naghib, S. M., Mozafari, M. R., & Rahmanian, M. (2024). Chitosan/hydroxyapatite hydrogels for localized drug delivery and tissue engineering: A Review. *Carbohydrate Polymer Technologies and Applications*, •••, 100640.

Khopade, S., Gomte, S. S., Janrao, C., Bavaskar, A., Agnihotri, T. G., Jain, A., & Khatik, R. (2023). Peptide and Protein Delivery through Cellulose, Hyaluronic Acid, and Heparin. In *Peptide and Protein Drug Delivery Using Polysaccharides* (pp. 75–113). Academic Press., DOI: 10.1016/B978-0-443-18925-8.00003-9

Khorashadizade, F., Abazari, S., Rajabi, M., Bakhsheshi-Rad, H., Ismail, A. F., Sharif, S., Ramakrishna, S., & Berto, F. (2021b). Overview of magnesium-ceramic composites: Mechanical, corrosion and biological properties. *Journal of Materials Research and Technology*, *15*, 6034–6066. DOI: 10.1016/j.jmrt.2021.10.141

Khoury, A., Deo, K. M., & Aldrich-Wright, J. R. (2020). Recent advances in platinum-based chemotherapeutics that exhibit inhibitory and targeted mechanisms of action. *Journal of Inorganic Biochemistry*, *207*, 111070. DOI: 10.1016/j.jinorgbio.2020.111070 PMID: 32299045

Khoz, R., Yazdian, F., Pourmadadi, M., Rahdar, A., Fathi-karkan, S., & Pandey, S. (2024). Current Trends in Silica Based Drug Delivery Systems. *European Journal of Medicinal Chemistry Reports*, •••, 100206.

Khursheed, R., Dua, K., Vishwas, S., Gulati, M., Jha, N. K., Aldhafeeri, G. M., Alanazi, F. G., Goh, B. H., Gupta, G., Paudel, K. R., Hansbro, P. M., Chellappan, D. K., & Singh, S. K. (2022). Biomedical applications of metallic nanoparticles in cancer: Current status and future perspectives. *Biomedicine and Pharmacotherapy*, *150*, 112951. DOI: 10.1016/J.BIOPHA.2022.112951 PMID: 35447546

Kieke, M., Feyerabend, F., Lemaitre, J., Behrens, P., & Willumeit-Römer, R. (2016). Degradation rates and products of pure magnesium exposed to different aqueous media under physiological conditions. *BioNanoMaterials*, *17*(3–4). Advance online publication. DOI: 10.1515/bnm-2015-0020

Kim, D., Kim, S., & Na, D. H. (2024). Dendrimer nanoplatforms for oral drug delivery applications. *Journal of Pharmaceutical Investigation*, •••, 1–21.

Kim, H N, Jiao A, Hwang N S. *et al.* (2013). Nanotopography-guided tissue engineering and regenerative medicine . *Advanced Drug Delivery Reviews*, *65*(4), 536–558. PMID: 22921841

Kim, I., Viswanathan, K., Kasi, G., Thanakkasaranee, S., Sadeghi, K., & Seo, J. (2022). ZnO nanostructures in active antibacterial food packaging: Preparation methods, antimicrobial mechanisms, safety issues, future prospects, and challenges. *Food Reviews International*, *38*(4), 537–565.

Kim, J. S., Jang, S. W., Son, M., Kim, B. M., & Kang, M. J. (2016). Enteric-coated tablet of risedronate sodium in combination with phytic acid, a natural chelating agent, for improved oral bioavailability. *European Journal of Pharmaceutical Sciences*, *82*, 45–51. DOI: 10.1016/J.EJPS.2015.11.011 PMID: 26594027

Kim, J., Yu, A. M., Kubelick, K. P., & Emelianov, S. Y. (2021). Gold nanoparticles conjugated with DNA aptamer for photoacoustic detection of human matrix metalloproteinase-9. *Photoacoustics*, *25*, 100307. DOI: 10.1016/J.PACS.2021.100307 PMID: 34703762

Kim, S. (2018). Competitive biological activities of chitosan and its derivatives: Antimicrobial, antioxidant, anticancer, and anti-inflammatory activities. *International Journal of Polymer Science*, *2018*(1), 1708172.

Kim, T H, & Jeon, W Y, JI Y. *et al.* (2021). Electricity auto-generating skin patch promotes wound healing process by activation of mechanosensitive ion channels . *Biomaterials*, *275*, 120948. PMID: 34157562

Kim, Y. S., Ko, M. J., Moon, H., Sim, W., Cho, A. S., Gil, G., & Kim, H. R. (2022). Ultrasound-Responsive Liposomes for Targeted Drug Delivery Combined with Focused Ultrasound. *Pharmaceutics*, *14*(7), 1314. DOI: 10.3390/PHARMACEUTICS14071314/S1 PMID: 35890210

Kim, Y., Zharkinbekov, Z., Raziyeva, K., Tabyldiyeva, L., Berikova, K., Zhumagul, D., & Saparov, A. (2023). Chitosan-based biomaterials for tissue regeneration. *Pharmaceutics*, *15*(3), 807. PMID: 36986668

Kinnel, B., Singh, S. K., Oprea-Ilies, G., & Singh, R. (2023). Targeted Therapy and Mechanisms of Drug Resistance in Breast Cancer. *Cancers (Basel)*, *15*(4), 1320. DOI: 10.3390/CANCERS15041320 PMID: 36831661

Kirtane, A. R., Sadhukha, T., Kim, H., Khanna, V., Koniar, B., & Panyam, J. (2017). Fibrinolytic enzyme co-therapy improves tumor perfusion and therapeutic efficacy of anticancer nanomedicine. *Cancer Research*, *77*(6), 1465. DOI: 10.1158/0008-5472.CAN-16-1646 PMID: 28108516

Koeglsperger, T., Rumpf, S.-L., Schließer, P., Struebing, F. L., Brendel, M., Levin, J., Trenkwalder, C., Höglinger, G. U., & Herms, J. (2023). Neuropathology of incidental Lewy body & prodromal Parkinson's disease. *Molecular Neurodegeneration*, *18*(1), 32. PMID: 37173733

Kolenda, C., Josse, J., Medina, M., Fevre, C., Lustig, S., Ferry, T., & Laurent, F. (2020). Evaluation of the activity of a combination of three bacteriophages alone or in association with antibiotics on Staphylococcus aureus embedded in biofilm or internalized in osteoblasts. *Antimicrobial Agents and Chemotherapy*, *64*(3), 02231–02219. DOI: 10.1128/aac PMID: 31871084

Kołodziejska, M., Jankowska, K., Klak, M., & Wszoła, M. (2021). Chitosan as an underrated polymer in modern tissue engineering. *Nanomaterials (Basel, Switzerland)*, *11*(11), 1–44. DOI: 10.3390/nano11113019 PMID: 34835782

Kolosnjaj-Tabi, J., Gibot, L., Fourquaux, I., Golzio, M., & Rols, M.-P. (2019). Electric field-responsive nanoparticles and electric fields: Physical, chemical, biological mechanisms and therapeutic prospects. *Advanced Drug Delivery Reviews*, *138*, 56–67. DOI: 10.1016/j.addr.2018.10.017 PMID: 30414494

Kong, X., Chen, H., Li, F., Zhang, F., Jiang, Y., Song, J., Sun, Y., Zhao, B., & Shi, J. (2023). Three-dimension chitosan hydrogel loading melanin composite nanoparticles for wound healing by anti-bacteria, immune activation and macrophage autophagy promotion. *International Journal of Biological Macromolecules*, *237*, 124176. DOI: 10.1016/j.ijbiomac.2023.124176 PMID: 37023589

Konnova, S., & Rozhina, E. (2024). *Magnetic Nanoparticles for Biomedical and Imaging Applications* (Vol. 25). MDPI.

Kopeček, J. (2013). Polymer-drug conjugates: Origins, progress to date and future directions. *Advanced Drug Delivery Reviews*, *65*(1), 49–59. DOI: 10.1016/j.addr.2012.10.014 PMID: 23123294

Kordbacheh, H. B. (2024). Co-delivery of Bcl-2 siRNA and doxorubicin using liposome-incorporated poly (ε-caprolactone)/chitosan nanofibers for the treatment of lung cancer. *Journal of Drug Delivery Science and Technology*, *99*, 105994.

Korte, N., Nortley, R., & Attwell, D. (2020). Cerebral blood flow decrease as an early pathological mechanism in Alzheimer's disease. *Acta Neuropathologica*, *140*(6), 793–810. PMID: 32865691

Koszła, O., & Sołek, P. (2024). Misfolding and aggregation in neurodegenerative diseases: Protein quality control machinery as potential therapeutic clearance pathways. *Cell Communication and Signaling : CCS*, *22*(1), 421. PMID: 39215343

Koushki, K., Keshavarz Shahbaz, S., Keshavarz, M., Bezsonov, E. E., Sathyapalan, T., & Sahebkar, A. (2021). Gold nanoparticles: Multifaceted roles in the management of autoimmune disorders. *Biomolecules*, *11*(9), 1289. DOI: 10.3390/biom11091289 PMID: 34572503

Kozielski, K. L.-V.-C. (2019). Cancer-selective nanoparticles for combinatorial siRNA delivery to primary human GBM in vitro and in vivo. *Biomaterials*, *209*, 79–87. PMID: 31026613

Kraegeloh, A., Suarez-Merino, B., Sluijters, T., & Micheletti, C. (2018). Implementation of safe-by-design for nanomaterial development and safe innovation: Why we need a comprehensive approach. *Nanomaterials (Basel, Switzerland)*, *8*(4), 239. PMID: 29661997

Krauss, J., Roper, P., Ballard, A., Shih, C., Fitzpatrick, J., Cassat, J., Ng, P., Pavlos, N., & Veis, D. (2019). Staphylococcus aureus infects osteoclasts and replicates intracellularly. [In.]. *mBio*, *10*, e02447–e19. PMID: 31615966

Kruspe, S. (2017). Aptamer-siRNA Chimeras: Discovery, Progress, and Future Prospects. *Biomedicines*, *209*, 79–87. PMID: 28792479

Kuhlmann, J., Bartsch, I., Willbold, E., Schuchardt, S., Holz, O., Hort, N., & Witte, F. (2013). Fast escape of hydrogen from gas cavities around corroding magnesium implants. *Acta Biomaterialia*, *9*(10), 8714–8721. PMID: 23069319

Kumar, A. (2024). Chitosan-based drug conjugated nanocomposites: Advances and innovation in cancer therapy. *Regenerative Engineering and Translational Medicine*, *10*(1), 1–8.

Kumar, A., & Singh, G. (2024). Surface modification of Ti6Al4V alloy via advanced coatings: Mechanical, Tribological, Corrosion, Wetting, and Biocompatibility studies. *Journal of Alloys and Compounds*, *989*, 174418. DOI: 10.1016/j.jallcom.2024.174418

Kumari, M., Sharma, N., Manchanda, R., Gupta, N., Syed, A., Bahkali, A. H., & Nimesh, S. (2021). PGMD/curcumin nanoparticles for the treatment of breast cancer. *Scientific Reports*, *11*(1), 3824. DOI: 10.1038/S41598-021-81701-X PMID: 33589661

Kumar, P. T. S., Abhilash, S., Manzoor, K., Nair, S. V., Tamura, H., & Jayakumar, R. (2010). Preparation and characterization of novel β-chitin/nanosilver composite scaffolds for wound dressing applications. *Carbohydrate Polymers*, *80*(3), 761–767. DOI: 10.1016/j.carbpol.2009.12.024

Kumar, P., Mangla, B., Javed, S., Ahsan, W., Musyuni, P., Sivadasan, D., Alqahtani, S. S., & Aggarwal, G. (2023). A review of nanomaterials from synthetic and natural molecules for prospective breast cancer nanotherapy. *Frontiers in Pharmacology*, *14*, 1149554. DOI: 10.3389/FPHAR.2023.1149554 PMID: 37274111

Kumar, P., Senthamilselvi, S., Govindaraju, M., & Sankar, R. (2014). Unraveling the caspase-mediated mechanism for phloroglucinol-encapsulated starch biopolymer against the breast cancer cell line MDA-MB-231. *RSC Advances*, *4*(86), 46157–46163. DOI: 10.1039/C4RA06664B

Kumar, R., Aadil, K. R., Ranjan, S., & Kumar, V. B. (2020). Advances in nanotechnology and nanomaterials based strategies for neural tissue engineering. *Journal of Drug Delivery Science and Technology*, *57*, 101617.

Kumbhar, P. R., Kumar, P., Lasure, A., Velayutham, R., & Mandal, D. (2023). An updated landscape on nanotechnology-based drug delivery, immunotherapy, vaccinations, imaging, and biomarker detections for cancers: Recent trends and future directions with clinical success. *Discover Nano*, *18*(1), 156. PMID: 38112935

Kuntamung, K., Jakmunee, J., & Ounnunkad, K. (2021). A label-free multiplex electrochemical biosensor for the detection of three breast cancer biomarker proteins employing dye/metal ion-loaded and antibody-conjugated polyethyleneimine-gold nanoparticles. *Journal of Materials Chemistry. B*, *9*(33), 6576–6585. DOI: 10.1039/D1TB00940K PMID: 34279016

Kurapati, R., Kostarelos, K., Prato, M., & Bianco, A. (2016). Biomedical uses for 2D materials beyond graphene: Current advances and challenges ahead. *Advanced Materials*, *28*(29), 6052–6074. PMID: 27105929

Ku, S. H., Lee, M., & Park, C. B. (2013). Carbon-based nanomaterials for tissue engineering. *Advanced Healthcare Materials*, *2*(2), 244–260. DOI: 10.1002/adhm.201200307 PMID: 23184559

Kwatra, D., Venugopal, A., & Anant, S. (2013). Nanoparticles in radiation therapy: A summary of various approaches to enhance radiosensitization in cancer. *Translational Cancer Research, 2*(4).

Kwon, H. S., & Koh, S. H. (2020). Neuroinflammation in neurodegenerative disorders: The roles of microglia and astrocytes. *Translational Neurodegeneration, 9*(1), 42. DOI: 10.1186/s40035-020-00221-2 PMID: 33239064

La Manna, S., & Marasco, D. (2024). Insights into molecular mechanisms of metallodrugs using metallomic studies. *Inorganica Chimica Acta, 560*, 121816. DOI: 10.1016/j.ica.2023.121816

Lam, M. (2021). Grafting bioactive polymers onto silicone implants to improve biological activity (PhD thesis, Université Paris-Nord-Paris XIII).

Lanoiselée, H. M., Nicolas, G., Wallon, D., Rovelet-Lecrux, A., Lacour, M., Rousseau, S., Richard, A. C., Pasquier, F., Rollin-Sillaire, A., Martinaud, O., Quillard-Muraine, M., de la Sayette, V., Boutoleau-Bretonniere, C., Etcharry-Bouyx, F., Chauviré, V., Sarazin, M., le Ber, I., Epelbaum, S., Jonveaux, T., & Campion, D. (2017). APP, PSEN1, and PSEN2 mutations in early-onset Alzheimer disease: A genetic screening study of familial and sporadic cases. *PLoS Medicine, 14*(3), e1002270. DOI: 10.1371/journal.pmed.1002270 PMID: 28350801

Lan, Q., Lu, R., & Chen, H.. (2020). MMP-13 enzyme and pH responsive theranostic nanoplatform for osteoarthritis . *Journal of Nanobiotechnology, 18*(1), 117. PMID: 32854712

Lara-Velazquez, M. A.-L.-C.-E. (2020). Chitosan-based non-viral gene and drug delivery systems for brain cancer. *Frontiers in Neurology*. PMID: 32849207

Laurent, M. R., Goemaere, S., Verroken, C., Bergmann, P., Body, J.-J., Bruyère, O., Cavalier, E., Rozenberg, S., Lapauw, B., & Gielen, E. (2022). Prevention and Treatment of Glucocorticoid-Induced Osteoporosis in Adults: Consensus Recommendations From the Belgian Bone Club. *Frontiers in Endocrinology, 13*. Advance online publication. DOI: 10.3389/fendo.2022.908727 PMID: 35757436

Law, C. K. Y., Bonin, L., De Gusseme, B., Boon, N., & Kundu, K. (2022). Biogenic synthesis of palladium nanoparticles: New production methods and applications. *Nanotechnology Reviews, 11*(1), 3104–3124. DOI: 10.1515/ntrev-2022-0482

Lee, A. C. (2009). *Global Burden of Neurological Disorders*. Global Health Education Consortium.

Lee, A., Paiement, G. D., Penenberg, B. L., & Rajaee, S. S. (2023). Metallosis in Total Hip Arthroplasty. *JBJS Reviews, 11*(10), e23. PMID: 37812668

Lee, D. S., Woo, J. Y., Ahn, C. B., & Je, J. Y. (2014). Chitosan-hydroxycinnamic acid conjugates: Preparation, antioxidant and antimicrobial activity. *Food Chemistry, 148*, 97–104. DOI: 10.1016/j.foodchem.2013.10.019 PMID: 24262532

Lee, K. Y., & Mooney, D. J. (2012). Alginate: Properties and biomedical applications. *Progress in Polymer Science, 37*(1), 106–126. DOI: 10.1016/j.progpolymsci.2011.06.003 PMID: 22125349

Lee, S. J. (2011). Comparative study of photosensitizer loaded and conjugated glycol chitosan nanoparticles for cancer therapy. *Journal of Controlled Release, 152*(1), 21–29. PMID: 21457740

Lee, S. J., Nah, H., Heo, D. N., Kim, K. H., Seok, J. M., Heo, M., Moon, H. J., Lee, D., Lee, J. S., An, S. Y., Hwang, Y. S., Ko, W. K., Kim, S. J., Sohn, S., Park, S. A., Park, S. Y., & Kwon, I. K. (2020). Induction of osteogenic differentiation in a rat calvarial bone defect model using an In situ forming graphene oxide incorporated glycol chitosan/oxidized hyaluronic acid injectable hydrogel. *Carbon*, *168*, 264–277. DOI: 10.1016/j.carbon.2020.05.022

Lee, Y., Park, Y., Lee, S., Ku, Y., Han, S., Klokkevold, P. R., & Chung, C. (2000). The Bone Regenerative Effect of Platelet-Derived Growth Factor-BB Delivered With a Chitosan/Tricalcium Phosphate Sponge Carrier. *Journal of Periodontology*, *71*(3), 418–424. DOI: 10.1902/jop.2000.71.3.418 PMID: 10776929

Lehmann, S., Dumurgier, J., Ayrignac, X., Marelli, C., Alcolea, D., Ormaechea, J. F., Thouvenot, E., Delaby, C., Hirtz, C., & Vialaret, J. (2020). Cerebrospinal fluid A beta 1–40 peptides increase in Alzheimer's disease and are highly correlated with phospho-tau in control individuals. *Alzheimer's Research & Therapy*, *12*, 1–12. PMID: 33008460

Lehtovirta, L. (2020). Adverse Reaction to Metal Debris in Patients with Metal-on-Metal Hip Replacements: Etiology & Pathogenesis.

Lepage, L., Poivret, D., & Lemétayer, F. (2024). Relationships between prescribed anti-osteoporotic treatment and necessity-concerns assessment: A psychological network analysis. *Current Psychology (New Brunswick, N.J.)*, *43*(5), 4098–4112. DOI: 10.1007/s12144-023-04611-3

Leso, V., & Iavicoli, I. (2018). Palladium Nanoparticles: Toxicological Effects and Potential Implications for Occupational Risk Assessment. *International Journal of Molecular Sciences*, *19*(2), 503. DOI: 10.3390/ijms19020503 PMID: 29414923

Levin, M. (2021). Bioelectric signaling: Reprogrammable circuits underlying embryogenesis, regeneration, and cancer. *Cell*, *184*(8), 1971–1989. DOI: 10.1016/j.cell.2021.02.034 PMID: 33826908

Lewandrowski, K. U. D., Gresser, J., Wise, D. L., & Trantolo, D. J. (2000). Bioresorbable bone graft substitutes of different osteoconductivities: A histologic evaluation of osteointegration of poly(propylene glycol-co-fumaric acid)-based cement implants in rats. *Biomaterials*, *21*(8), 757–764. DOI: 10.1016/S0142-9612(99)00179-9 PMID: 10721744

Li, Z. (2010). *Mg/hydroxyapatite composites for potential bio-medical applications*. https://bura.brunel.ac.uk/bitstream/2438/4599/1/FulltextThesis.pdf

Liang, G., Wang, H., Shi, H., Wang, H., Zhu, M., Jing, A., & Li, G. (2020). Recent progress in the development of upconversion nanomaterials in bioimaging and disease treatment. *Journal of Nanobiotechnology*, *18*, 1–22. DOI: 10.1186/s12951-020-00713-3 PMID: 33121496

Liangjian, C., Jun, Z., Kun, Y., Chang, C., Yilong, D., Xueyan, Q., & Zhiming, Y. (2015). Improving of in vitro Biodegradation Resistance in a Chitosan Coated Magnesium Bio-composite. *Rare Metal Materials and Engineering*, *44*(8), 1862–1865. DOI: 10.1016/s1875-5372(15)30114-4

Liang, X., Li, L., Li, X., He, T., Gong, S., Zhu, S., Zhang, M., Wu, Q., & Gong, C. (2021). A spontaneous multifunctional hydrogel vaccine amplifies the innate immune response to launch a powerful antitumor adaptive immune response. *Theranostics*, *11*(14), 6936–6949. DOI: 10.7150/thno.58173 PMID: 34093863

Liang, Y., & Goh, J. C. (2020). Polypyrrole-incorporated conducting constructs for tissue engineering applications: A review. *Bioelectricity*, *2*(2), 101–119. DOI: 10.1089/bioe.2020.0010 PMID: 34471842

Liao, J. (2020). Review on magnetic natural polymer constructed hydrogels as vehicles for drug delivery. *Biomacromolecules*, *21*(7), 2574–2594. PMID: 32543834

Li, C., Xiao, C., Zhan, L., Zhang, Z., Xing, J., Zhai, J., Zhou, Z., Tan, G., Piao, J., & Zhou, Y. (2022). Wireless electrical stimulation at the nanoscale interface induces tumor vascular normalization. *Bioactive Materials*, *18*, 399–408. DOI: 10.1016/j.bioactmat.2022.03.027 PMID: 35415302

Li, C., Yu, P., Wang, Z., Long, C., Xiao, C., Xing, J., Dong, B., Zhai, J., Zhou, L., & Zhou, Z. (2023). Electro-mechanical coupling directs endothelial activities through intracellular calcium ion deployment. *Materials Horizons*, *10*(11), 4903–4913. DOI: 10.1039/D3MH01049J PMID: 37750251

Li, D. Z. (2020). Redox-responsive self-assembled nanoparticles for cancer therapy. *Advanced Healthcare Materials*, *9*(20), 2000605. PMID: 32893506

Li, D., Li, J., & Wang, S.. (2023). Dually Crosslinked Copper-Poly (tannic acid) Nanoparticles with Microenvironment-Responsiveness for Infected Wound Treatment . *Advanced Healthcare Materials*, *12*(17), e2203063. PMID: 36842067

Li, H. Y., Huang, D. N., Ren, K. F., & Ji, J. (2021). Inorganic-polymer composite coatings for biomedical devices. *Smart Materials in Medicine*, *2*, 1–14. DOI: 10.1016/j.smaim.2020.10.002

Li, H., Xiao, Z., Quarles, L. D., & Li, W. (2021). Osteoporosis: Mechanism, Molecular Target and Current Status on Drug Development. *Current Medicinal Chemistry*, *28*(8), 1489–1507. DOI: 10.2174/0929867327666200330142432 PMID: 32223730

Li, J. J., Tsang, J. Y., & Tse, G. M. (2021). Tumor Microenvironment in Breast Cancer—Updates on Therapeutic Implications and Pathologic Assessment. *Cancers (Basel)*, *13*(16), 4233. DOI: 10.3390/CANCERS13164233 PMID: 34439387

Li, J., & Wang, H. (2023). Selective organ targeting nanoparticles: From design to clinical translation. *Nanoscale Horizons*, *8*(9), 1155–1173. PMID: 37427677

Li, J., Zhou, P., Attarilar, S., & Shi, H. (2021). Innovative surface modification procedures to achieve micro/nano-graded Ti-based biomedical alloys and implants. *Coatings*, *11*(6), 647.

Li, L., Deng, Y., Zeng, Y., Yan, B., Deng, Y., Zheng, Z., & Xiao, X. (2023). The application advances of dendrimers in biomedical field. *VIEW*, *4*(6), 20230023.

Li, M., Tang, Z., Zhang, Y., Lv, S., Yu, H., Zhang, D., Hong, H., & Chen, X. (2014). LHRH-peptide conjugated dextran nanoparticles for targeted delivery of cisplatin to breast cancer. *Journal of Materials Chemistry. B*, *2*(22), 3490–3499. DOI: 10.1039/C4TB00077C PMID: 32261469

Li, M., Yin, H., & Yan, Z.. (2022). The immune microenvironment in cartilage injury and repair . *Acta Biomaterialia*, *140*, 23–42. PMID: 34896634

Li, M., Zhao, Y., Zhang, W., Zhang, S., & Zhang, S. (2021). Multiple-therapy strategies via polysaccharides-based nano-systems in fighting cancer. *Carbohydrate Polymers, 269*, 118323. DOI: 10.1016/j.carbpol.2021.118323 PMID: 34294335

Li, M., Zhao, Z., & Yi, J. (2024). Biomaterials Designed to Modulate Reactive Oxygen Species for Enhanced Bone Regeneration in Diabetic Conditions. *Journal of Functional Biomaterials, 15*(8), 220.

Lima, B. V., Oliveira, M. J., Barbosa, M. A., Gonçalves, R. M., & Castro, F. (2021). Immunomodulatory potential of chitosan-based materials for cancer therapy: A systematic review of *in vitro, in vivo* and clinical studies. *Biomaterials Science, 9*(9), 3209–3227. DOI: 10.1039/D0BM01984D PMID: 33949372

Lim, Z. Z., Li, J. E., & Ng, C. T.. (2011). Gold nanoparticles in cancer therapy. *Acta Pharmacologica Sinica, 32*, 983–990. DOI: 10.1038/aps.2011.82 PMID: 21743485

Lindsay, J. A., & Holden, M. T. (2004). Staphylococcus aureus: Superbug, super genome? *Trends in Microbiology, 12*(8), 378–385. PMID: 15276614

Lingnert, H., Vallentin, K., & Eriksson, C. E. (1979). Measurement of antioxidative effect in model system. *Journal of Food Processing and Preservation.*

Lin, M. T., & Beal, M. F. (2006). Mitochondrial dysfunction and oxidative stress in neurodegenerative diseases. *Nature, 443*(7113), 787–795. PMID: 17051205

Lin, P. H., Sermersheim, M., Li, H., Lee, P. H., Steinberg, S. M., & Ma, J. (2017). Zinc in wound healing modulation. *Nutrients, 10*(1), 16. DOI: 10.3390/nu10010016 PMID: 29295546

Lin, S., Li, J., & Shao, J.. (2022). Anisotropic magneto-mechanical stimulation on collagen coatings to accelerate osteogenesis . *Colloids and Surfaces. B, Biointerfaces, 210*, 112227. PMID: 34838419

Lin, X., & Lai, Y. (2024). Scarring Skin: Mechanisms and Therapies . *International Journal of Molecular Sciences, 25*(3), 1458. PMID: 38338767

Lionberger, D. R., Samorajski, J., Wilson, C. D., & Rivera, A. (2018). What role does metal allergy sensitization play in total knee arthroplasty revision? *Journal of Experimental Orthopaedics, 5*, 1–8. PMID: 30109436

Li, P. Y., Bearoff, F., Zhu, P., Fan, Z., Zhu, Y., Fan, M., Cort, L., Kambayashi, T., Blankenhorn, E. P., & Cheng, H. (2021). PEGylation Enables Subcutaneously Administered Nanoparticles to Induce Antigen-Specific Immune Tolerance. *Journal of Controlled Release, 331*(March), 164–175. DOI: 10.1016/j.jconrel.2021.01.013 PMID: 33450320

Liu, R., Ye, X., & Cui, T. (2020). Recent progress of biomarker detection sensors. *Yosetsu Gakkai Ronbunshu/Quarterly Journal of the Japan Welding Society, 38*(3). doi: DOI: 10.2207/QJJWS.38.193

Liu, Houqin, Liu, Yingke, Tian, Zhipeng, & Li, Jiaxin, Man Li, and Zhihe Zhao. (2024). Coordinating Macrophage Targeting and Antioxidation by Injectable Nanocomposite Hydrogel for Enhanced Rheumatoid Arthritis Treatment. *ACS Applied Materials & Interfaces, 16*(29), 37656–37668. DOI: 10.1021/acsami.4c06840 PMID: 38987704

Liu, J., Xiao, Q., Xiao, J., Niu, C., Li, Y., Zhang, X., Zhou, Z., Shu, G., & Yin, G. (2022). Wnt/β-catenin signalling: Function, biological mechanisms, and therapeutic opportunities. *Signal Transduction and Targeted Therapy*, *7*(1), 3. DOI: 10.1038/s41392-021-00762-6 PMID: 34980884

Liu, J., Yang, C., & Zhou, S. (2016). Polyethylene glycol-modified chitosan nanoparticles for tissue engineering and cancer therapy applications. *Bioengineering & Translational Medicine*, *1*(2), 153–162.

Liu, Q., Wang, W., & Yang, H.. (2024). Microenvironment-Responsive Injectable Conductive Hydrogel for Spinal Cord Injury Repair . *Advanced Functional Materials*, *34*(46), 2406376.

Liu, S., Manshaii, F., Chen, J., Wang, X., Wang, S., Yin, J., Yang, M., Chen, X., Yin, X., & Zhou, Y. (2025). Unleashing the potential of electroactive hybrid biomaterials and self-powered systems for bone therapeutics. *Nano-Micro Letters*, *17*(1), 44. DOI: 10.1007/s40820-024-01536-9 PMID: 39417933

Liu, T., Wang, Y., & Kuang, T. (2024). Oriented porous polymer scaffolds in tissue engineering: A comprehensive review of preparation strategies and applications. *Macromolecular Materials and Engineering*, *309*(1), 2300246.

Liu, W., Yan, Z., Zhang, Z., Zhang, Y., Cai, G., & Li, Z. (2019). Bioactive and anti-corrosive bio-MOF-1 coating on magnesium alloy for bone repair application. *Journal of Alloys and Compounds*, *788*, 705–711. DOI: 10.1016/j.jallcom.2019.02.281

Liu, X., Duan, G., Li, W., Zhou, Z., & Zhou, R. (2017). Membrane destruction-mediated antibacterial activity of tungsten disulfide (WS 2). *RSC Advances*, *7*(60), 37873–37880.

Liu, X., Hao, M., Chen, Z., Zhang, T., Huang, J., Dai, J., & Zhang, Z. (2021). 3D bioprinted neural tissue constructs for spinal cord injury repair. *Biomaterials*, *272*, 120771. DOI: 10.1016/j.biomaterials.2021.120771 PMID: 33798962

Liu, Y., & Chen, B. (2018). In vivo corrosion of CoCrMo alloy and biological responses: A review. *Materials Technology*, *33*(2), 127–134.

Liu, Y., Dzidotor, G., & Le, T. T.. (2022). Exercise-induced piezoelectric stimulation for cartilage regeneration in rabbits . *Science Translational Medicine*, *14*(627), eabi7282. PMID: 35020409

Liu, Y., Li, J., Chen, M., Chen, X., & Zheng, N. (2020). Palladium-based nanomaterials for cancer imaging and therapy. *Theranostics*, *10*(22), 10057–10074. DOI: 10.7150/thno.45990 PMID: 32929334

Liu, Y., & Lu, J. (2024). Research trends of piezoelectric biomaterials in osteochondral tissue engineering. *Materials Today. Communications*, *110264*. Advance online publication. DOI: 10.1016/j.mtcomm.2024.110264

Liu, Y., Ren, Z., Zhang, X., Zhao, Z., Ma, G., Pei, Y., & Pan, J. (2024). Dual-Triggered Peptide-Based Polymeric Micelles Enhance Doxorubicin Delivery for Targeted Cancer Therapy. *ACS Applied Nano Materials*.

Liu, Y., Xu, B., & Lu, M.. (2022). Ultrasmall Fe-doped carbon dots nanozymes for photoenhanced antibacterial therapy and wound healing . *Bioactive Materials*, *12*, 246–256. PMID: 35310377

Liu, Z., Wan, X., Wang, Z. L., & Li, L. (2021). Electroactive biomaterials and systems for cell fate determination and tissue regeneration: Design and applications. *Advanced Materials*, *33*(32), 2007429. DOI: 10.1002/adma.202007429 PMID: 34117803

Li, W., Zhang, M., & Zhang, L. (2020). Chitosan and its derivatives in cancer therapy: Mechanisms of action and potential applications. *Current Drug Targets*, *21*(3), 253–263.

Li, X., Dong, W., Nalin, A. P., Wang, Y., Pan, P., Xu, B., Zhang, Y., Tun, S., Zhang, J., Wang, L.-S., He, X., Caligiuri, M. A., & Yu, J. (2018). The natural product chitosan enhances the anti-tumor activity of natural killer cells by activating dendritic cells. *OncoImmunology*, *7*(6). Advance online publication. DOI: 10.1080/2162402X.2018.1431085 PMID: 29872557

Li, X., Yang, W., & Xie, H.. (2020). CNT/Sericin Conductive Nerve Guidance Conduit Promotes Functional Recovery of Transected Peripheral Nerve Injury in a Rat Model . *ACS Applied Materials & Interfaces*, *12*(33), 36860–36872. PMID: 32649170

Li, Y., Chen, J., Liu, S., Wang, Z., Zhang, S., Mao, C., & Wang, J. (2024). Biodegradable piezoelectric polymer for cartilage remodeling. *Matter*, *7*(4), 1631–1643. DOI: 10.1016/j.matt.2024.01.034

Li, Y., Selvaraj, V., Saravanan, S., Abullais, S. S., & Wankhade, V. (2024). Exploring the osteogenic potential of chitosan-quercetin bio-conjugate: In vitro and in vivo investigations in osteoporosis models. *International Journal of Biological Macromolecules*, *274*, 133492. DOI: 10.1016/J.IJBIOMAC.2024.133492 PMID: 38944072

Li, Y., Wei, Y., Liao, J., Hao, Y., Ning, C., Jiang, L., & Wang, S. (2016). Surface wettability switched cell adhesion and detachment on conducting polymer nanoarray. *Advanced Materials Interfaces*, *3*(19). Advance online publication. DOI: 10.1002/admi.201600598

Li, Y., Yan, X., Liu, J., Li, L., Hu, X., Sun, H., & Tian, J. (2014). Pulsed electromagnetic field enhances brain-derived neurotrophic factor expression through L-type voltage-gated calcium channel-and Erk-dependent signaling pathways in neonatal rat dorsal root ganglion neurons. *Neurochemistry International*, *75*, 96–104. DOI: 10.1016/j.neuint.2014.06.004 PMID: 24937769

Li, Yali, Chen, L., Liu, Y., Zhang, Y., Liang, Y., & Mei, Y. (2018). Anti-inflammatory effects in a mouse osteoarthritis model of a mixture of glucosamine and chitooligosaccharides produced by bi-enzyme single-step hydrolysis. *Scientific Reports*, *8*(1), 5624. DOI: 10.1038/s41598-018-24050-6 PMID: 29618773

Liyanage, P. Y., Hettiarachchi, S. D., Zhou, Y., Ouhtit, A., Seven, E. S., Oztan, C. Y., & Leblanc, R. M. (2019). Nanoparticle-mediated targeted drug delivery for breast cancer treatment. *Biochimica et Biophysica Acta (BBA)-. Revisiones en Cáncer*, *1871*(2), 419–433. DOI: 10.1016/j.bbcan.2019.04.006 PMID: 31034927

Li, Yu., Liu, H., Xu, Q.-S., Du, Y.-G., & Xu, J. (2014). Chitosan oligosaccharides block LPS-induced O-GlcNAcylation of NF-κB and endothelial inflammatory response. *Carbohydrate Polymers*, *99*, 568–578. DOI: 10.1016/j.carbpol.2013.08.082 PMID: 24274545

Li, Z. T. (2010). Chitosan-graft-polyethylenimine with improved properties as a potential gene vector. *Carbohydrate Polymers*, *80*(1), 254–259.

Li, Z., Wang, H., & Zhang, K.. (2022). Bisphosphonate-based hydrogel mediates biomimetic negative feedback regulation of osteoclastic activity to promote bone regeneration . *Bioactive Materials, 13*, 9–22. PMID: 35224288

Li, Z., Zheng, J., & Wan, D.. (2020). Uniaxial Static Strain Promotes Osteoblast Proliferation and Bone Matrix Formation in Distraction Osteogenesis In Vitro . *BioMed Research International, 2020*, 3906426. PMID: 32855965

Logothetidis, S. (2011). Nanotechnology: Principles and applications. In *Nanostructured materials and their applications* (pp. 1-22). Springer.

Loo, C., Lin, A., Hirsch, L., Lee, M.-H., Barton, J., Halas, N., West, J., & Drezek, R. (2004). Nanoshell-enabled photonics-based imaging and therapy of cancer. *Technology in Cancer Research & Treatment, 3*(1), 33–40. PMID: 14750891

Lopez de Armentia, S., Del Real, J. C., Paz, E., & Dunne, N. (2020). Advances in biodegradable 3D printed scaffolds with carbon-based nanomaterials for bone regeneration. *Materials (Basel), 13*(22), 5083. PMID: 33187218

Lotfi, M., Nejib, M., & Naceur, M. (2013). Cell adhesion to biomaterials: Concept of biocompatibility. *Advances in biomaterials science and biomedical applications, 8*, 208-240.

Lo, Y.-L., Chang, C.-H., Wang, C.-S., Yang, M.-H., Lin, A. M.-Y., Hong, C.-J., & Tseng, W.-H. (2020). PEG-coated nanoparticles detachable in acidic microenvironments for the tumor-directed delivery of chemo-and gene therapies for head and neck cancer. *Theranostics, 10*(15), 6695. PMID: 32550898

Luceri, A., Francese, R., Lembo, D., Ferraris, M., & Balagna, C. (2023). Silver nanoparticles: Review of antiviral properties, mechanism of action and applications. *Microorganisms, 11*(3), 629. PMID: 36985203

Lu, H. D. (2014). Chitosan-graft-polyethylenimine/DNA nanoparticles as novel non-viral gene delivery vectors targeting osteoarthritis. *PLoS One*. PMID: 24392152

Luo, R., Liang, Y., Yang, J., Feng, H., Chen, Y., Jiang, X., Zhang, Z., Liu, J., Bai, Y., Xue, J., Chao, S., Xi, Y., Liu, X., Wang, E., Luo, D., Li, Z., & Zhang, J. (2023). Reshaping the endogenous electric field to boost wound repair via electrogenerative dressing. *Advanced Materials, 35*(16). Advance online publication. DOI: 10.1002/adma.202208395 PMID: 36681867

Luo, S., Zhang, C., Xiong, W., Song, Y., Wang, Q., Zhang, H., Guo, S., Yang, S., & Liu, H. (2024). Advances in electroactive biomaterials: Through the lens of electrical stimulation promoting bone regeneration strategy. *Journal of Orthopaedic Translation, 47*, 191–206. DOI: 10.1016/j.jot.2024.06.009 PMID: 39040489

Lu, R., Yoshida, T., Nakashima, H., Premanathan, M., Aragaki, R., Mimura, T., Kaneko, Y., Yamamoto, N., Miyakoshi, T., & Uryu, T. (2000). Specific biological activities of Chinese lacquer polysaccharides. *Carbohydrate Polymers, 43*(1), 47–54. DOI: 10.1016/S0144-8617(99)00198-8

Lu, W., & Wen, J. (2024). Crosstalk Among Glial Cells in the Blood–Brain Barrier Injury After Ischemic Stroke. *Molecular Neurobiology*, •••, 1–14. PMID: 38279077

Lu, Y., Li, M., & Long, Z., Di Yang, Guo, S., Li, J., Liu, D., Gao, P., Chen, G., Lu, X., Lu, J., & Wang, Z. (2019). Collagen/ β -TCP composite as a bone-graft substitute for posterior spinal fusion in rabbit model: A comparison study. *Biomedical Materials (Bristol, England)*, *14*(4), 045009. DOI: 10.1088/1748-605X/ab1caf PMID: 31022713

Lv, W., Xu, J., Wang, X., Li, X., Xu, Q., & Xin, H. (2018). Bioengineered boronic ester modified dextran polymer nanoparticles as reactive oxygen species responsive nanocarrier for ischemic stroke treatment. *ACS Nano*, *12*(6), 5417–5426. PMID: 29869497

Lv, Z., Hu, T., & Bian, Y.. (2023). A MgFe-LDH Nanosheet-Incorporated Smart Thermo-Responsive Hydrogel with Controllable Growth Factor Releasing Capability for Bone Regeneration . *Advanced Materials (Deerfield Beach, Fla.)*, *35*(5), e2206545. PMID: 36426823

Lyu, J., Huang, J., Huang, J., Hu, H., Wang, Q., Ding, H., Li, H., Fang, X., & Zhang, W. (2024). Rising challenges in periprosthetic joint infections: A focus on rare pathogens and their clinical implications. *Frontiers in Cellular and Infection Microbiology*, *14*, 1451398. PMID: 39698321

Mabrouk, M., Das, D. B., Salem, Z. A., & Beherei, H. H. (2021). Nanomaterials for biomedical applications: Production, characterisations, recent trends and difficulties. In Molecules (Vol. 26, Issue 4). MDPI AG. doi: DOI: 10.3390/molecules26041077

Mabsor-Zgandaoui, S., Moussaif, A., El Hessni, A., Mesfioui, A., El Mzibri, M., Iddar, A., & Moutaouakkil, A. (2025a). Ionising and Ultraviolet Radiation Induce the Secretion of Melanin in Tetrahymena Pyriformis. *The International Journal of Environmental Studies*, *82*(1), 1–14. DOI: 10.1080/00207233.2024.2412455

Mabsor-Zgandaoui, S., Rachmoune, K., El Hessni, A., El Mzibri, M., Iddar, A., & Moutaouakkil, A. (2025). Effect of hydrogen peroxide pre-treatment on enhancing the cellular resistance of Tetrahymena pyriformis exposed to ionizing radiation. *Toxicological and Environmental Chemistry*, *107*(2), 294–309.

Mabsor-Zgandaoui, S., Rachmoune, K., El Hessni, A., El Mzibri, M., Iddar, A., & Moutaouakkil, A. (2025b). Effect of Hydrogen Peroxide Pre-Treatment on Enhancing the Cellular Resistance of Tetrahymena Pyriformis Exposed to Ionizing Radiation. *Toxicological and Environmental Chemistry*, *107*(2), 294–309. DOI: 10.1080/02772248.2025.2460804

Machałowski, T., Rusak, A., Wiatrak, B., Haczkiewicz-Leśniak, K., Popiel, A., Jaroszewicz, J., Żak, A., Podhorska-Okołów, M., & Jesionowski, T. (2021). Naturally Formed Chitinous Skeleton Isolated from the Marine Demosponge Aplysina fistularis as a 3D Scaffold for Tissue Engineering. *Materials (Basel)*, *14*(11), 2992. DOI: 10.3390/ma14112992 PMID: 34205950

Machla, F., Bekiari, C., Monou, P. K., Kofidou, E., Theodosaki, A. M., Katsamenis, O. L., & Zisis, V.. (2024). Development of an Oral Epithelial Ex Vivo Organ Culture Model for Biocompatibility and Permeability Assessment of Biomaterials. *Bioengineering (Basel, Switzerland)*, *11*(10), 1035. DOI: 10.3390/bioengineering11101035 PMID: 39451410

Macuvele, D. L. P., Nones, J., Matsinhe, J. V., Lima, M. M., Soares, C., Fiori, M. A., & Riella, H. G. (2017). Advances in ultra high molecular weight polyethylene/hydroxyapatite composites for biomedical applications: A brief review. *Materials Science and Engineering C*, *76*, 1248–1262. PMID: 28482493

Madamsetty, V. S., Tavakol, S., Moghassemi, S., Dadashzadeh, A., Schneible, J. D., Fatemi, I., Shirvani, A., Zarrabi, A., Azedi, F., Dehshahri, A., Aghaei Afshar, A., Aghaabbasi, K., Pardakhty, A., Mohammad-inejad, R., & Kesharwani, P. (2022). Chitosan: A versatile bio-platform for breast cancer theranostics. *Journal of Controlled Release : Official Journal of the Controlled Release Society*, *341*, 733–752. DOI: 10.1016/J.JCONREL.2021.12.012 PMID: 34906606

Mahajan, A., Devgan, S., & Kalyanasundaram, D. (2023). Surface alteration of Cobalt-Chromium and duplex stainless steel alloys for biomedical applications: A concise review. *Materials and Manufacturing Processes*, *38*(3), 260–270.

Mahanty, M., Dutta, B., Ou, W., Zhu, X., Bromberg, J. S., He, X., & Rahaman, S. O. 2024. "Macrophage MicroRNA-146a Is a Central Regulator of the Foreign Body Response to Biomaterial Implants." BioRxiv *: The Preprint Server for Biology*. Elsevier. https://doi.org/DOI: 10.1016/J.BIOMATERIALS.2024.122855

Ma, J.-X., Qian, L., & Zhou, Y. (2015). Stimulation effect of chitosan on the immunity of radiotherapy patients suffered from lung cancer. *International Journal of Biological Macromolecules*, *72*, 195–198. DOI: 10.1016/j.ijbiomac.2014.08.027 PMID: 25175732

Majeed, T. (2023). Development of Biocompatible Materials for Medical Implants: A Journey Towards Seamless Integration. *The R Journal*, *1*(01), 35–42.

Makadia, H. K., & Siegel, S. J. (2011). Poly lactic-co-glycolic acid (PLGA) as biodegradable controlled drug delivery carrier. *Polymers*, *3*(3), 1377–1397. PMID: 22577513

Makino, S., Fukuda, K., & Miyoshi, S.. (1999). Cardiomyocytes can be generated from marrow stromal cells in vitro . *The Journal of Clinical Investigation*, *103*(5), 697–705. PMID: 10074487

Malahias, M.-A., Bauer, T. W., Gu, A., Nocon, A. A., De Martino, I., & Sculco, P. K. (2020). Questioning the clinical relevance of positive post-implantation allergy testing for metal hypersensitivity in total knee arthroplasty: A didactic case report. *HSS Journal : the Musculoskeletal Journal of Hospital for Special Surgery*, *16*(2, suppl), 485–489. PMID: 33380984

Mamidi, N., García, R. G., Martínez, J. D. H., Briones, C. M., Martinez Ramos, A. M., Tamez, M. F. L., & Segura, F. J. M. (2022). Recent advances in designing fibrous biomaterials for the domain of biomedical, clinical, and environmental applications. *ACS Biomaterials Science & Engineering*, *8*(9), 3690–3716. PMID: 36037103

Mamnoon, B., Loganathan, J., Confeld, M. I., De Fonseka, N., Feng, L., Froberg, J., Choi, Y., Tuvin, D. M., Sathish, V., & Mallik, S. (2020). Targeted polymeric nanoparticles for drug delivery to hypoxic, triple-negative breast tumors. *ACS Applied Bio Materials*, *4*(2), 1450. DOI: 10.1021/ACSABM.0C01336 PMID: 33954285

Manuja, A., Kumar, B., Kumar, R., Chhabra, D., Ghosh, M., Manuja, M., & Prasad, M. (2021). Metal/metal oxide nanoparticles: Toxicity concerns associated with their physical state and remediation for biomedical applications. *Toxicology Reports*, *8*, 1970–1978. PMID: 34934635

Manuja, A., Kumar, B., Riyesh, T., Talluri, T. R., & Tripathi, B. N. (2020). Microwave assisted fast fabrication of zinc/iron oxides based polymeric nanocomposites and evaluation on equine fibroblasts. *International Journal of Biological Macromolecules*, *165*, 71–81. PMID: 32987081

Manuja, A., Raguvaran, R., Kumar, B., Kalia, A., & Tripathi, B. N. (2020). Accelerated healing of full thickness excised skin wound in rabbits using single application of alginate/acacia based nanocomposites of ZnO nanoparticles. *International Journal of Biological Macromolecules*, *155*, 823–833. PMID: 32234436

Mao, L., Shen, L., Chen, J., Wu, Y., Kwak, M., Lu, Y., Xue, Q., Pei, J., Zhang, L., Yuan, G., Fan, R., Ge, J., & Ding, W. (2015). Enhanced bioactivity of MG–ND–ZN–ZR alloy achieved with Nanoscale MGF2 surface for vascular stent application. *ACS Applied Materials & Interfaces*, *7*(9), 5320–5330. DOI: 10.1021/am5086885 PMID: 25705919

Mao, L., Shen, L., Chen, J., Zhang, X., Kwak, M., Wu, Y., Fan, R., Zhang, L., Pei, J., Yuan, G., Song, C., Ge, J., & Ding, W. (2017). A promising biodegradable magnesium alloy suitable for clinical vascular stent application. *Scientific Reports*, *7*(1). Advance online publication. DOI: 10.1038/srep46343 PMID: 28397881

Maret, W. (2013). Zinc and zinc ions in biological systems. *Encyclopedia of Metalloproteins*, 87. https://doi.org/DOI: 10.1007/978-1-4614-1533-6_185

Martinelli, C., & Biglietti, M. (2020). Nanotechnological approaches for counteracting multidrug resistance in cancer. *Cancer Drug Resistance (Alhambra, Calif.)*, *3*(4), 1003. DOI: 10.20517/CDR.2020.47 PMID: 35582219

Maryska, L., Jindrichova, B., Siegel, J., Zaruba, K., & Burketova, L. (2023). Impact of palladium nanoparticles on plant and its fungal pathogen. A case study: Brassica napus–Plenodomus lingam. *AoB Plants*, *15*(2). Advance online publication. DOI: 10.1093/aobpla/plad004 PMID: 36970187

Mason, C., & Dunnill, P. (2008). A brief definition of regenerative medicine . *Regenerative Medicine*, *3*(1), 1–5. PMID: 18154457

Matar, H. E., Porter, P. J., & Porter, M. L. (2021). Metal allergy in primary and revision total knee arthroplasty: A scoping review and evidence-based practical approach. *Bone & Joint Open*, *2*(10), 785–795. PMID: 34587776

Mawazi, S. M., Kumar, M., Ahmad, N., Ge, Y., & Mahmood, S. (2024). Recent Applications of Chitosan and Its Derivatives in Antibacterial, Anticancer, Wound Healing, and Tissue Engineering Fields. *Polymers*, *16*(10), 1351. DOI: 10.3390/polym16101351 PMID: 38794545

Mayer, H. F., Coloccini, A., & Viñas, J. F. (2024). Three-Dimensional Printing in Breast Reconstruction: Current and Promising Applications. Journal of Clinical Medicine 2024, Vol. 13, Page 3278, 13(11), 3278. https://doi.org/DOI: 10.3390/JCM13113278

Ma, Z., Hu, X., & Zhang, Y.. (2023). Biomineralized Piezoelectrically Active Scaffolds for Inducing Osteogenic Differentiation . *Chemistry (Weinheim an der Bergstrasse, Germany)*, *29*(15), e202203166. PMID: 36478479

Mbese, Z., & Aderibigbe, B. A. (2021). Bisphosphonate-Based Conjugates and Derivatives as Potential Therapeutic Agents in Osteoporosis, Bone Cancer and Metastatic Bone Cancer. *International Journal of Molecular Sciences*, *22*(13), 6869. DOI: 10.3390/ijms22136869 PMID: 34206757

McCormick, D. A. (2014). Membrane potential and action potential. In *From Molecules to Networks* (pp. 351–376). Elsevier.

McLaughlin, K. A., & Levin, M. (2018). Bioelectric signaling in regeneration: Mechanisms of ionic controls of growth and form. *Developmental Biology*, *433*(2), 177–189. DOI: 10.1016/j.ydbio.2017.08.032 PMID: 29291972

Mekuye, B., & Abera, B. (2023). Nanomaterials: An overview of synthesis, classification, characterization, and applications. *Nano Select*, *4*(8), 486–501. DOI: 10.1002/nano.202300038

Mena-Morcillo, E., & Veleva, L. (2020). Degradation of AZ31 and AZ91 magnesium alloys in different physiological media: Effect of surface layer stability on electrochemical behaviour. *Journal of Magnesium and Alloys*, *8*(3), 667–675. DOI: 10.1016/j.jma.2020.02.014

Meng, H., Mai, W. X., Zhang, H., Xue, M., Xia, T., Lin, S., Wang, X., Zhao, Y., Ji, Z., Zink, J. I., & Nel, A. E. (2013). Co-delivery of an Optimal Drug/siRNA Combination Using Mesoporous Silica Nanoparticle to Overcome Drug Resistance in Breast Cancer In Vitro and In Vivo. *ACS Nano*, *7*(2), 994. DOI: 10.1021/NN3044066 PMID: 23289892

Meyers, J. D., Doane, T., Burda, C., & Basilion, J. P. (2013). Nanoparticles for imaging and treating brain cancer. *Nanomedicine (London)*, *8*(1), 123–143. DOI: 10.2217/nnm.12.185 PMID: 23256496

Mikušová, V., & Mikuš, P. (2021). Advances in Chitosan-Based Nanoparticles for Drug Delivery. *International Journal of Molecular Sciences*, *22*(17), 9652. DOI: 10.3390/ijms22179652 PMID: 34502560

Mikušová, V., & Mikuš, P. (2021). Advances in chitosan-based nanoparticles for drug delivery. *International Journal of Molecular Sciences*, *22*(17), 9652. PMID: 34502560

Milan, J., Niemczyk, K., & Kus-Liśkiewicz, M. (2022). Treasure on the Earth—Gold nanoparticles and their biomedical applications. *Materials (Basel)*, *15*(9), 3355. DOI: 10.3390/ma15093355 PMID: 35591689

Minchin, R. F., & Martin, D. J. (2010). Minireview: Nanoparticles for molecular imaging—an overview. *Endocrinology*, *151*(2), 474–481. PMID: 20016027

Mir, M., Ishtiaq, S., Rabia, S., Khatoon, M., Zeb, A., & Khan, G. M., ur Rehman, A., & ud Din, F. (2017). Nanotechnology: From in vivo imaging system to controlled drug delivery. *Nanoscale Research Letters*, *12*, 1–16. PMID: 28819800

Mishra, R. K., Ha, S. K., Verma, K., & Tiwari, S. K. (2018). Recent progress in selected bio-nanomaterials and their engineering applications: An overview. *Journal of Science. Advanced Materials and Devices*, *3*(3), 263–288. DOI: 10.1016/j.jsamd.2018.05.003

Mishra, S., & Yadav, M. D. (2024). Magnetic nanoparticles: A comprehensive review from synthesis to biomedical frontiers. *Langmuir*, *40*(33), 17239–17269. PMID: 39132737

Mitchell, M. J., Billingsley, M. M., Haley, R. M., Wechsler, M. E., Peppas, N. A., & Langer, R. (2021). Engineering precision nanoparticles for drug delivery. *Nature Reviews. Drug Discovery*, *20*(2), 101–124. DOI: 10.1038/S41573-020-0090-8 PMID: 33277608

Mittal, H., Ray, S. S., Kaith, B. S., Bhatia, J. K., Sharma, J., & Alhassan, S. M. (2018). Recent progress in the structural modification of chitosan for applications in diversified biomedical fields. *European Polymer Journal*, *109*, 402–434.

Mocanu-Dobranici, A.-E., Costache, M., & Dinescu, S. (2023). Insights into the molecular mechanisms regulating cell behavior in response to magnetic materials and magnetic stimulation in stem cell (neurogenic) differentiation. *International Journal of Molecular Sciences*, *24*(3), 2028. DOI: 10.3390/ijms24032028 PMID: 36768351

Mohammed, A. S. A., Naveed, M., & Jost, N. (2021). Polysaccharides; classification, chemical properties, and future perspective applications in fields of pharmacology and biological medicine (a review of current applications and upcoming potentialities). *Journal of Polymers and the Environment*, *29*(8), 2359–2371.

Mohanta, Y. K., Biswas, K., Rauta, P. R., Mishra, A. K., De, D., Hashem, A., Al-Arjani, A. B. F., Alqarawi, A. A., Abd-Allah, E. F., Mahanta, S., & Mohanta, T. K. (2021). Development of Graphene Oxide Nanosheets as Potential Biomaterials in Cancer Therapeutics: An In-Vitro Study Against Breast Cancer Cell Line. *Journal of Inorganic and Organometallic Polymers and Materials*, *31*(11), 4236–4249. DOI: 10.1007/S10904-021-02046-6/METRICS

Mohebali, A., & Abdouss, M. (2020). Layered biocompatible pH-responsive antibacterial composite film based on HNT/PLGA/chitosan for controlled release of minocycline as burn wound dressing . *International Journal of Biological Macromolecules*, *164*, 4193–4204. PMID: 32891643

Mollenhauer, B., Trautmann, E., Otte, B., Ng, J., Spreer, A., Lange, P., Sixel-Döring, F., Hakimi, M., VonSattel, J.-P., & Nussbaum, R. (2012). α-Synuclein in human cerebrospinal fluid is principally derived from neurons of the central nervous system. *Journal of Neural Transmission (Vienna, Austria)*, *119*, 739–746. PMID: 22426833

Mondal, A., Nayak, A. K., Chakraborty, P., Banerjee, S., & Nandy, B. C. (2023). Natural Polymeric Nanobiocomposites for Anti-Cancer Drug Delivery Therapeutics: A Recent Update. *Pharmaceutics*, *15*(8), 2064. DOI: 10.3390/PHARMACEUTICS15082064 PMID: 37631276

Montiel Schneider, M. G., Martín, M. J., Otarola, J., Vakarelska, E., Simeonov, V., Lassalle, V., & Nedyalkova, M. (2022). Biomedical applications of iron oxide nanoparticles: Current insights progress and perspectives. *Pharmaceutics*, *14*(1), 204. DOI: 10.3390/pharmaceutics14010204 PMID: 35057099

Montoya, C., Du, Y., & Gianforcaro, A. L.. (2021). On the road to smart biomaterials for bone research: Definitions, concepts, advances, and outlook . *Bone Research*, *9*(1), 12. PMID: 33574225

MOON, S.-H., JI, S. H., SON, J.-L., SHIN, S.-J., OH, S., KIM, S. H., & BAE, J.-M. (2020). Antibacterial, anti-inflammatory, and anti-osteoclastogenic activities of Potential applications in preventing and treating periodontal diseases. *Dental Materials Journal, 39*(6), 1096–1102. https://doi.org/DOI: 10.4012/dmj.2020-157

Moramkar, N. (2021). Insight into chitosan derived nanotherapeutics for anticancer drug delivery and imaging. *European Polymer Journal*, *154*, 110540.

Morawetz, H. (2014). History of Polymer Science. In *Encyclopedia of Polymer Science and Technology* (pp. 1–30). John Wiley & Sons, Ltd., DOI: 10.1002/0471440264.pst631

Moreno-Alc Antar, G., Casini, A., Moreno-Alc, G., & Casini, A. (2023). Bioinorganic supramolecular coordination complexes and their biomedical applications. *FEBS Letters*, *597*(1), 191–202. DOI: 10.1002/1873-3468.14535 PMID: 36345593

Morin-Crini, N., Lichtfouse, E., Torri, G., & Crini, G. (2019). Applications of chitosan in food, pharmaceuticals, medicine, cosmetics, agriculture, textiles, pulp and paper, biotechnology, and environmental chemistry. *Environmental Chemistry Letters*, *17*(4), 1667–1692. DOI: 10.1007/s10311-019-00904-x

Moro, S. (2023). *Développement d'une nouvelle approche théranostique ciblant la réponse télomérique dans les cellules cancéreuses radio-résistantes du poumon.*

Morozova, O., Isaeva, E., & Klinov, D. (2023). Biodistribution of Fluorescent Albumin Nanoparticles among Organs of Laboratory Animals after Intranasal and Peroral Administration. *Current Issues in Molecular Biology*, *45*(10), 8227–8238. PMID: 37886962

Morris, E., Chavez, M., & Tan, C. (2016). Dynamic biomaterials: Toward engineering autonomous feedback . *Current Opinion in Biotechnology*, *39*, 97–104. PMID: 26974245

Motlani, G., Motlani, V., Acharya, N., Dave, A., Pamnani, S., Somyani, D., & Agrawal, S. (2023). Novel Advances in the Role of Selective Estrogen Receptor Modulators in Hormonal Replacement Therapy: A Paradigm Shift. *Cureus*. Advance online publication. DOI: 10.7759/cureus.49079 PMID: 38125238

Mourdikoudis, S., Pallares, R. M., & Thanh, N. T. (2018). Characterization techniques for nanoparticles: Comparison and complementarity upon studying nanoparticle properties. *Nanoscale*, *10*(27), 12871–12934. PMID: 29926865

Mourino, V., Cattalini, J. P., & Boccaccini, A. R. (2012). Metallic ions as therapeutic agents in tissue engineering scaffolds: An overview of their biological applications and strategies for new developments. *Journal of the Royal Society, Interface*, *9*(68), 401–419. DOI: 10.1098/rsif.2011.0611 PMID: 22158843

Mousazadeh, H. P.-S. (2021). Cyclodextrin based natural nanostructured carbohydrate polymers as effective non-viral siRNA delivery systems for cancer gene therapy. *Journal of Controlled Release*, *330*, 1046–1070. PMID: 33188829

Mozafari, M. (2024). Editorial Note: Biomaterials in Cancer - From Research Breakthroughs to Clinical Implementation. *Translational Oncology*, *48*, 102061. DOI: 10.1016/J.TRANON.2024.102061 PMID: 39134454

Muchtaridi, M., Suryani, A. I., Wathoni, N., Herdiana, Y., Mohammed, A. F. A., Gazzali, A. M., Lesmana, R., & Joni, I. M. (2023). Chitosan/Alginate Polymeric Nanoparticle-Loaded α-Mangostin: Characterization, Cytotoxicity, and In Vivo Evaluation against Breast Cancer Cells. Polymers 2023, Vol. 15, Page 3658, 15(18), 3658. https://doi.org/DOI: 10.3390/POLYM15183658

Mughal, K. A., Khan, A., Sarwar, A., Mahmood, R., & Abbasi, A. A. (2024). Comparative protein structural analyses of α-synuclein linked pathogenic variants reveal the role of N-terminally located critical region in Parkinson's disease pathogenesis. *Biologia*, *•••*, 1–14.

Mukherjee, B., Patra, B., Mahapatra, S., Banerjee, P., Tiwari, A., & Chatterjee, M. (2004). Vanadium—An element of atypical biological significance. *Toxicology Letters*, *150*(2), 135–143. PMID: 15093669

Munhoz, M. A. S., Hirata, H. H., Plepis, A. M. G., Martins, V. C. A., & Cunha, M. R. (2018). Use of collagen/chitosan sponges mineralized with hydroxyapatite for the repair of cranial defects in rats. *Injury, 49*(12), 2154–2160. DOI: 10.1016/j.injury.2018.09.018 PMID: 30268514

Muramatsu, K., Doi, K., Ihara, K., Shigetomi, M., & Kawai, S. (2003). Recalcitrant posttraumatic nonunion of the humerus: 23 patients reconstructed with vascularized bone graft: 23 patients reconstructed with vascularized bone graft. *Acta Orthopaedica Scandinavica, 74*(1), 95–97. DOI: 10.1080/00016470310013734 PMID: 12635801

Mutepfa, A. R., Hardy, J. G., & Adams, C. F. (2022). Electroactive scaffolds to improve neural stem cell therapy for spinal cord injury. *Frontiers in Medical Technology, 4*, 693438. DOI: 10.3389/fmedt.2022.693438 PMID: 35274106

Mutter, J. (2011). Is dental amalgam safe for humans? The opinion of the scientific committee of the European Commission. *Journal of Occupational Medicine and Toxicology (London, England), 6*, 1–17. DOI: 10.1186/1745-6673-6-2 PMID: 21232090

Nadanaka, S., Tamura, J. I., & Kitagawa, H. (2022). Chondroitin Sulfates Control Invasiveness of the Basal-Like Breast Cancer Cell Line MDA-MB-231 Through ROR1. *Frontiers in Oncology, 12*, 914838. DOI: 10.3389/FONC.2022.914838/BIBTEX PMID: 35712490

Naghibi Beidokhti, H. R. (2017). Preparation, characterization, and optimization of folic acid-chitosan-methotrexate core-shell nanoparticles by box-behnken design for tumor-targeted drug delivery. *AAPS PharmSciTech, 18*, 115–129. PMID: 26896317

Nakamoto, Y., Inui, Y., Hotta, M., Wakabayashi, H., & Hanaoka, H. (2024). Recent advancements in new tracers from first-in-human studies. *Annals of Nuclear Medicine*, •••, 1–7. PMID: 39325320

Nam, D., Li, K., Riegler, V., & Barrack, R. L. (2016). Patient-reported metal allergy: A risk factor for poor outcomes after total joint arthroplasty? *The Journal of Arthroplasty, 31*(9), 1910–1915. PMID: 26965589

Nandgude, T., & Pagar, R. (2021). Plausible role of chitosan in drug and gene delivery against resistant breast cancer cells. *Carbohydrate Research, 506*, 108357. DOI: 10.1016/j.carres.2021.108357 PMID: 34146935

Narayanan, K. B., Bhaskar, R., Kim, H., & Han, S. S. (2023). In Vitro Cytocompatibility Assessment of Novel 3D Chitin/Glucan- and Cellulose-Based Decellularized Scaffolds for Skin Tissue Engineering. *Sustainability (Switzerland), 15*(21). Advance online publication. DOI: 10.3390/su152115618

Narkar, A R, Tong Z, Soman P. *et al.* (2022). Smart biomaterial platforms: Controlling and being controlled by cells . *Biomaterials, 283*, 121450. PMID: 35247636

Nasr Azadani, M., Zahedi, A., Bowoto, O. K., & Oladapo, B. I. (2022). A review of current challenges and prospects of magnesium and its alloy for bone implant applications. *Progress in Biomaterials, 11*(1), 1–26. PMID: 35239157

Nasra, S. B. (2022). Nanoscale advances. *Recent advances in nanoparticle-based drug delivery systems for rheumatoid arthritis treatment*, 4(17), 3479-3494.

Nayab, D. E., & Din, F. U., Ali, H., Kausar, W. A., Urooj, S., Zafar, M., Khan, I., Shabbir, K., & Khan, G. M. (2023). Nano biomaterials based strategies for enhanced brain targeting in the treatment of neurodegenerative diseases: An up-to-date perspective. *Journal of Nanobiotechnology*, *21*(1), 477. PMID: 38087359

Naz, S., Gul, A., & Zia, M. (2020). Toxicity of copper oxide nanoparticles: A review study. *IET Nanobiotechnology / IET*, *14*(1), 1–13. DOI: 10.1049/iet-nbt.2019.0176 PMID: 31935671

Ndagi, U., Mhlongo, N., & Soliman, M. E. (2017). Metal complexes in cancer therapy–an update from drug design perspective. *Drug Design, Development and Therapy*, •••, 599–616. DOI: 10.2147/DDDT.S119488 PMID: 28424538

Nel, A., Xia, T., Meng, H., Wang, X., Lin, S., Ji, Z., & Zhang, H. (2013). Nanomaterial toxicity testing in the 21st century: Use of a predictive toxicological approach and high- throughput screening. *Accounts of Chemical Research*. Advance online publication. DOI: 10.1021/ ar300022h PMID: 22676423

Nemati, M. B. (2021). Unraveling the effect of breast cancer patients' plasma on the targeting ability of folic acid-modified chitosan nanoparticles. *Molecular Pharmaceutics*, *18*(12), 4341–4353. PMID: 34779630

Nene, , A., Takahashi, M., & R. Somani, P. (2016). Fe3O4 and Fe Nanoparticles by Chemical Reduction of Fe(acac)3 by Ascorbic Acid: Role of Water. *World Journal of Nano Science and Engineering*, *06*(01), 20–28. DOI: 10.4236/wjnse.2016.61002

Nene, A. G., Takahashi, M., Wakita, K., & Umeno, M. (2016). Size controlled synthesis of Fe3O4 nanoparticles by ascorbic acid mediated reduction of Fe(acac)3 without using capping agent. *Journal of Nano Research*, *40*, 8–19. DOI: 10.4028/www.scientific.net/JNanoR.40.8

Nene, A., Galluzzi, M., Hongrong, L., Somani, P., Ramakrishna, S., & Yu, X. F. (2021). Synthetic preparations and atomic scale engineering of silver nanoparticles for biomedical applications. *Nanoscale*, *13*(33), 13923–13942. DOI: 10.1039/D1NR01851E PMID: 34477675

Nene, A., Phanthong, C., Surareungchai, W., & Somasundrum, M. (2023). Electrochemical detection of creatinine using Au–Ag bimetallic nanoparticles. *Journal of Solid State Electrochemistry*, *27*(10), 2869–2875. DOI: 10.1007/s10008-023-05571-1

Neto, J. V. C., Teixeira, A. B. V., & Dos Reis, A. C. (2023). Hydroxyapatite coatings versus osseointegration in dental implants: A systematic review. *The Journal of Prosthetic Dentistry*. PMID: 37845113

Ng, F., Nicoulin, V., Peloso, C., Curia, S., Richard, J., & Lopez-Noriega, A. (2023). In Vitro and In Vivo Hydrolytic Degradation Behaviors of a Drug-Delivery System Based on the Blend of PEG and PLA Copolymers. *ACS Applied Materials & Interfaces*, *15*(48), 55495–55509. DOI: 10.1021/acsami.2c13141 PMID: 38011651

Ng, S. Y., & Lee, A. Y. W. (2019). Traumatic brain injuries: Pathophysiology and potential therapeutic targets. *Frontiers in Cellular Neuroscience*, *13*, 528. PMID: 31827423

Nguyen, D. D., Luo, L. J., & Lai, J. Y. (2022). Thermogels Containing Sulfated Hyaluronan as Novel Topical Therapeutics for Treatment of Ocular Surface Inflammation. *Materials Today. Bio*, *13*(January). Advance online publication. DOI: 10.1016/j.mtbio.2021.100183 PMID: 34927046

Nguyen, T. P., Nguyen, Q. V., Nguyen, V.-H., Le, T.-H., Huynh, V. Q. N., Vo, D.-V. N., Trinh, Q. T., Kim, S. Y., & Van Le, Q. (2019). Silk Fibroin-Based Biomaterials for Biomedical Applications: A Review. *Polymers*, *11*(12), 1933. DOI: 10.3390/polym11121933 PMID: 31771251

Nief, C., Morhard, R., Chelales, E., Alvarez, D. A., Ioanna Bourla, B. S., Lam, C. T., Sag, A. A., Crouch, B. T., Mueller, J. L., Katz, D., Dewhirst, M. W., Everitt, J. I., & Ramanujam, N. (2021). Polymer-assisted intratumoral delivery of ethanol: Preclinical investigation of safety and efficacy in a murine breast cancer model. *PLoS One*, *16*(1), e0234535. DOI: 10.1371/JOURNAL.PONE.0234535 PMID: 33507942

Nikkhoo, A. R.-N. (2020). Codelivery of STAT3 siRNA and BV6 by carboxymethyl dextran trimethyl chitosan nanoparticles suppresses cancer cell progression. *International Journal of Pharmaceutics*, *581*, 119236. PMID: 32240809

Nikolova, M. P., & Chavali, M. S. (2019). Recent advances in biomaterials for 3D scaffolds: A review. *Bioactive Materials*, *4*, 271–292. DOI: 10.1016/j.bioactmat.2019.10.005 PMID: 31709311

Ning, C., Zhou, Z., Tan, G., Zhu, Y., & Mao, C. (2018). Electroactive polymers for tissue regeneration: Developments and perspectives. *Progress in Polymer Science*, *81*, 144–162. DOI: 10.1016/j.progpolymsci.2018.01.001 PMID: 29983457

Niranjan, C., Raghavendra, T., Rao, M. P., Siddaraju, C., Gupta, M., Jain, V. K. S., & Aishwarya, R. (2023). Magnesium alloys as extremely promising alternatives for temporary orthopedic implants–A review. *Journal of Magnesium and Alloys*.

No, H. K., & Meyers, S. P. (1995). Preparation and Characterization of Chitin and Chitosan—A Review. *Journal of Aquatic Food Product Technology*, *4*(2), 27–52. DOI: 10.1300/J030v04n02_03

No, H. K., Young Park, N., Ho Lee, S., & Meyers, S. P. (2002). Antibacterial activity of chitosans and chitosan oligomers with different molecular weights. *International Journal of Food Microbiology*, *74*(1–2), 65–72. DOI: 10.1016/S0168-1605(01)00717-6 PMID: 11929171

Noh, J.-Y., Yang, Y., & Jung, H. (2020). Molecular Mechanisms and Emerging Therapeutics for Osteoporosis. *International Journal of Molecular Sciences*, *21*(20), 7623. DOI: 10.3390/ijms21207623 PMID: 33076329

Nortje, B. D. (2021). *Feasibility of additive manufacturing for patient-specific knee replacements*

Nowicka-Bauer, K., & Szymczak-Cendlak, M. (2021). Structure and function of ion channels regulating sperm motility—An overview. *International Journal of Molecular Sciences*, *22*(6), 3259. DOI: 10.3390/ijms22063259 PMID: 33806823

O'Dell, D., Schein, P., & Erickson, D. (2016). Simultaneous characterization of nanoparticle size and particle-surface interactions with three-dimensional nanophotonic force microscopy. *Physical Review Applied*, *6*(3), 034010. PMID: 30417030

Odeh, A. B., El-Sayed, B., Knopp, M. M., Rades, T., & Blaabjerg, L. I. (2023). Influence of Polyvinylpyrrolidone Molecular Weight and Concentration on the Precipitation Inhibition of Supersaturated Solutions of Poorly Soluble Drugs. Pharmaceutics 2023, Vol. 15, Page 1601, 15(6), 1601. https://doi.org/DOI: 10.3390/PHARMACEUTICS15061601

Ofridam, F. T. (2021). pH-sensitive polymers: Classification and some fine potential applications. *Polymers for Advanced Technologies*, *32*(4), 1455–1484.

Ogawa, R. (2017). Keloid and Hypertrophic Scars Are the Result of Chronic Inflammation in the Reticular Dermis . *International Journal of Molecular Sciences*, *18*(3), 606. PMID: 28287424

Oh, T. J., Yoon, J., Misch, C. E., & Wang, H. L. (2002). The causes of early implant bone loss: Myth or science? *Journal of Periodontology*, *73*(3), 322–333. PMID: 11922263

Oliveira, A., Simões, S., Ascenso, A., & Reis, C. P. (2022). Therapeutic advances in wound healing. *The Journal of Dermatological Treatment*, *33*(1), 2–22. PMID: 32056472

Omer, H. (2021). Radiobiological effects and medical applications of non-ionizing radiation. *Saudi Journal of Biological Sciences*, *28*(10), 5585–5592. PMID: 34588869

Omidian, H., Wilson, R. L., & Cubeddu, L. X. (2024). Quantum Dot Research in Breast Cancer: Challenges and Prospects. Materials 2024, Vol. 17, Page 2152, 17(9), 2152. https://doi.org/DOI: 10.3390/MA17092152

Omidian, H., Gill, E. J., Dey Chowdhury, S., & Cubeddu, L. X. (2024). Chitosan Nanoparticles for Intranasal Drug Delivery. *Pharmaceutics*, *16*(6), 746. PMID: 38931868

Omotosho, K. D., Gurung, V., Banerjee, P., Shevchenko, E. V., & Berman, D. (2024). Self-Cleaning Highly Porous TiO2 Coating Designed by Swelling-Assisted Sequential Infiltration Synthesis (SIS) of a Block Copolymer Template. *Polymers*, *16*(3), 308. DOI: 10.3390/polym16030308 PMID: 38337197

Onzi, G., Guterres, S. S., Pohlmann, A. R., & Frank, L. A. (2021). Passive targeting and the enhanced permeability and retention (EPR) effect. *the ADME encyclopedia: a comprehensive guide on biopharmacy and pharmacokinetics*, 1-13.

Orive, G., Anitua, E., Pedraz, J. L., & Emerich, D. F. (2009). Biomaterials for promoting brain protection, repair and regeneration. *Nature Reviews. Neuroscience*, *10*(9), 682–692. DOI: 10.1038/nrn2685 PMID: 19654582

Ornaghi, H. L., Monticeli, F. M., & Dall Agnol, L. (2023). *"A Review on Polymers for Biomedical Applications on Hard and Soft Tissues and Prosthetic Limbs." Polymers*. Multidisciplinary Digital Publishing Institute., DOI: 10.3390/polym15194034

Oselusi, S. O., Sibuyi, N. R., Meyer, M., & Madiehe, A. M. (2024). Phytonanotherapeutic Applications of Plant Extract-Synthesized Silver Nanoparticles in Wound Healing—A Prospective Overview. *BioNanoScience*, •••, 1–21.

Osonga, F. J., Kalra, S., Miller, R. M., Isika, D., & Sadik, O. A. (2020). Synthesis, characterization and antifungal activities of eco-friendly palladium nanoparticles. *RSC Advances*, *10*(10), 5894–5904. DOI: 10.1039/c9ra07800b PMID: 35497427

Ossenkoppele, R., van der Kant, R., & Hansson, O. (2022). Tau biomarkers in Alzheimer's disease: Towards implementation in clinical practice and trials. *Lancet Neurology*, *21*(8), 726–734. PMID: 35643092

Ota, D., Akatsuka, S., Nishi, T., Kato, T., Takeuchi, M., Tsuji, M., & Fukuuchi, A. (2019). Phase I study of combination therapy with weekly nanoparticle albumin-bound paclitaxel and cyclophosphamide in metastatic breast cancer patients. *Anticancer Research*, *39*(12), 6903–6907. DOI: 10.21873/anticanres.13910 PMID: 31810960

Ouyang, J., Wen, M., Chen, W., Tan, Y., Liu, Z., Xu, Q., & Liu, Y.-N. (2019). Multifunctional two dimensional Bi 2 Se 3 nanodiscs for combined antibacterial and anti-inflammatory therapy for bacterial infections. *Chemical Communications*, *55*(33), 4877–4880. PMID: 30957120

Öztürk, A. B., Alarcin, E., Yaşayan, G., Avci-Adali, M., Khosravi, A., Zarepour, A., & Zarrabi, A. (2024). Innovative approaches in skin therapy: Bionanocomposites for skin tissue repair and regeneration. *Materials Advances*.

Pacheco, K. A. (2019). Allergy to surgical implants. *Clinical Reviews in Allergy & Immunology*, *56*(1), 72–85. PMID: 30220068

Padmanabhan, P., Kumar, A., Kumar, S., Chaudhary, R. K., & Gulyás, B. (2016). Nanoparticles in practice for molecular-imaging applications: An overview. *Acta Biomaterialia*, *41*, 1–16. PMID: 27265153

Pakzad, S., Taghavi, R., & Hasanzadeh, A.. (2024). A biocompatible cellulose gum based CMC/PVA/SBA-15 film as a colloidal antibacterial agent against MRSA . *RSC Advances*, *14*(49), 36246–36252. PMID: 39539531

Palacios, S., Silverman, S. L., de Villiers, T. J., Levine, A. B., Goemaere, S., Brown, J. P., De Cicco Nardone, F., Williams, R., Hines, T. L., Mirkin, S., & Chines, A. A. (2015). A 7-year randomized, placebo-controlled trial assessing the long-term efficacy and safety of bazedoxifene in postmenopausal women with osteoporosis. *Menopause (New York, N.Y.)*, *22*(8), 806–813. DOI: 10.1097/GME.0000000000000419 PMID: 25668306

Palanisamy, C. P., Pei, J., Alugoju, P., Anthikapalli, N. V. A., Jayaraman, S., Veeraraghavan, V. P., Gopathy, S., Roy, J. R., Janaki, C. S., & Thalamati, D. (2023). New strategies of neurodegenerative disease treatment with extracellular vesicles (EVs) derived from mesenchymal stem cells (MSCs). *Theranostics*, *13*(12), 4138. PMID: 37554286

Panahi, Z., Tamjid, E., & Rezaei, M. (2020). Surface modification of biodegradable AZ91 magnesium alloy by electrospun polymer nanocomposite: Evaluation of in vitro degradation and cytocompatibility. *Surface and Coatings Technology*, *386*, 125461. DOI: 10.1016/j.surfcoat.2020.125461

Pan, D., Schmieder, A. H., Wickline, S. A., & Lanza, G. M. (2011). Manganese-based MRI contrast agents: Past, present, and future. *Tetrahedron*, *67*(44), 8431–8444. DOI: 10.1016/j.tet.2011.07.076 PMID: 22043109

Pandey, N., Singh, K., Kushwaha, S. P., Kumar, A., Khalid, M., Khan, A. A., & Khan, M. (2024). Targeted Drug Delivery to Combat Antibacterial Drug Resistance. In *Contemporary Approaches to Mitigating Antibacterial Drug Resistance* (pp. 250–267). IGI Global.

Pandit, A. P., Omase, S. B., & Mute, V. M. (2020). A chitosan film containing quercetin-loaded transfersomes for treatment of secondary osteoporosis. *Drug Delivery and Translational Research, 10*(5), 1495–1506. DOI: 10.1007/s13346-020-00708-5 PMID: 31942700

Park, J. H., Park, J. U., & Chang, H. (2021). Advances in Biomaterials for Breast Reconstruction. Applied Sciences 2021, Vol. 11, Page 7493, 11(16), 7493. https://doi.org/DOI: 10.3390/APP11167493

Park, H. G., & Yeo, M. K. (2016). Nanomaterial regulatory policy for human health and environment. *Molecular & Cellular Toxicology, 12*, 223–236.

Park, J. K., Chung, M. J., Choi, H. N., & Park, Y. Il. (2011). Effects of the Molecular Weight and the Degree of Deacetylation of Chitosan Oligosaccharides on Antitumor Activity. *International Journal of Molecular Sciences, 12*(1), 266–277. DOI: 10.3390/ijms12010266 PMID: 21339986

Park, J., Akbaba, G. E., Sharma, N., Das, R., Vinikoor, T., Liu, Y., Le, D. Q., Angadi, K., & Nguyen, T. D. (2025). Electrically active biomaterials for stimulation and regeneration in tissue engineering. *Journal of Biomedical Materials Research. Part A, 113*(1), e37871. DOI: 10.1002/jbm.a.37871 PMID: 39806919

Passos, J. S., Lopes, L. B., & Panitch, A. (2023). Collagen-Binding Nanoparticles for Paclitaxel Encapsulation and Breast Cancer Treatment. *ACS Biomaterials Science & Engineering, 9*(12), 6805–6820. DOI: 10.1021/ACSBIOMATERIALS.3C01332/SUPPL_FILE/AB3C01332_SI_001.PDF PMID: 37982792

Patel, J. K., Patel, A., & Bhatia, D. (2021). Introduction to Nanomaterials and Nanotechnology. In Patel, J. K., & Pathak, Y. V. (Eds.), *Emerging Technologies for Nanoparticle Manufacturing* (pp. 3–23). Springer International Publishing., DOI: 10.1007/978-3-030-50703-9_1

Patel, S. G., Patel, M. D., Patel, A. J., Chougule, M. B., & Choudhury, H. (2018). Solid lipid nanoparticles for targeted brain drug delivery. In *Nanotechnology-based targeted drug delivery systems for brain tumors* (pp. 191–244). Elsevier.

Pathak, S., Cao, E., Davidson, M. C., Jin, S., & Silva, G. A. Quantum dot applications to neuroscience: New tools for probing neurons and glia-Supplementary Information. *Journal of Neuroscience*.

Pathak, S., Vyas, S. P., & Pandey, A. (2021). Development, characterization and in vitro release kinetic studies of ibandronate loaded chitosan nanoparticles for effective management of osteoporosis. *International Journal of Applied Pharmaceutics*, 120–125. https://doi.org/DOI: 10.22159/ijap.2021v13i6.42697

Pathak, A., Suthar, T., Nirbhavane, G., & Jain, K. (2024). Polymeric micelles in drug delivery and targeting. In *Molecular Pharmaceutics and Nano Drug Delivery* (pp. 161–182). Elsevier.

Pathak, S., Sahu, P., Begum, J. P. S., Kashaw, S. K., Pandey, A., Semwal, P., & Sharma, R. (2024). Formulation and assessment of penetration potential of Risedronate chitosan nanoparticles loaded transdermal gel in the management of osteoporosis: In vitro and ex vivo screening. *Carbohydrate Polymer Technologies and Applications, 7*, 100440. DOI: 10.1016/J.CARPTA.2024.100440

Patil-Sen, Y. (2021). Advances in nano-biomaterials and their applications in biomedicine. *Emerging Topics in Life Sciences, 5*(1), 169–176. PMID: 33825835

Paula, A., Teixeira, C., Tristão, J. C., Araujo, M. H., Oliveira, L. C. A., Moura, F. C. C., Ardisson, J. D., Amorim, C. C., & Lago, R. M. (2012). Iron: A Versatile Element to Produce Materials for Environmental Applications. *Journal of the Brazilian Chemical Society*, *23*(9), •••.

Pavis, H., Wilcock, A., Edgecombe, J., Carr, D., Manderson, C., Church, A., & Fisher, A. (2002). Pilot study of nasal morphine-chitosan for the relief of breakthrough pain in patients with cancer. *Journal of Pain and Symptom Management*, *24*(6), 598–602. DOI: 10.1016/S0885-3924(02)00522-5 PMID: 12551810

Pawar, K. R., & Babu, R. J. (2010). Polymeric and lipid-based materials for topical nanoparticle delivery systems. *Critical reviews™ in therapeutic drug carrier systems, 27*(5).

Payal, & Pandey, P. (2021). Role of Nanotechnology in Electronics: A Review of Recent Developments and Patents. *Recent Patents on Nanotechnology, 16*(1), 45–66. doi: DOI: 10.2174/1872210515666210120114504

Peña, O. A. & Martin, P. (2024). Cellular and molecular mechanisms of skin wound healing . *Nature Reviews. Molecular Cell Biology*, *25*(8), 599–616. PMID: 38528155

Peng, L., Liang, Y., Zhong, X., Liang, Z., Tian, Y., Li, S., Liang, J., Wang, R., Zhong, Y., & Shi, Y. (2020). Aptamer-conjugated gold nanoparticles targeting epidermal growth factor receptor variant III for the treatment of glioblastoma. *International Journal of Nanomedicine*, •••, 1363–1372. PMID: 32184591

Peng, X.-Y., Hu, M., Liao, F., Yang, F., Ke, Q.-F., Guo, Y.-P., & Zhu, Z.-H. (2019). La-Doped mesoporous calcium silicate/chitosan scaffolds for bone tissue engineering. *Biomaterials Science*, *7*(4), 1565–1573. DOI: 10.1039/C8BM01498A PMID: 30688345

Peng, Y., Jiang, H., & Zuo, H. D. (2023). Factors affecting osteogenesis and chondrogenic differentiation of mesenchymal stem cells in osteoarthritis . *World Journal of Stem Cells, 15*(6), 548–560. PMID: 37424946

Peschel, D., Zhang, K., Fischer, S., & Groth, T. (2012). Modulation of osteogenic activity of BMP-2 by cellulose and chitosan derivatives. *Acta Biomaterialia*, *8*(1), 183–193. DOI: 10.1016/J.ACTBIO.2011.08.012 PMID: 21884830

Pesode, P., & Barve, S. (2023). Additive manufacturing of metallic biomaterials: Sustainability aspect, opportunity, and challenges. *Journal of Industrial and Production Engineering*, *40*(6), 464–505. DOI: 10.1080/21681015.2023.2229341

Petrarca, C., Clemente, E., Di Giampaolo, L., Mariani-Costantini, R., Leopold, K., Schindl, R., Lotti, L. V., Mangifesta, R., Sabbioni, E., Niu, Q., Bernardini, G., & Di Gioacchino, M. (2014). Palladium nanoparticles induce disturbances in cell cycle entry and progression of peripheral blood mononuclear cells: Paramount role of ions. *Journal of Immunology Research*, *2014*, 1–8. DOI: 10.1155/2014/295092 PMID: 25105151

Phan, T. T. V., Huynh, T., Manivasagan, P., Mondal, S., & Oh, J. (2019). An Up-To-Date Review on Biomedical Applications of Palladium Nanoparticles. *Nanomaterials (Basel, Switzerland)*, *10*(1), 66. DOI: 10.3390/nano10010066 PMID: 31892149

Pichler, K., Kraus, T., Martinelli, E., Sadoghi, P., Musumeci, G., Uggowitzer, P. J., & Weinberg, A. M. (2013). Cellular reactions to biodegradable magnesium alloys on human growth plate chondrocytes and osteoblasts. *International Orthopaedics*, *38*(4), 881–889. DOI: 10.1007/s00264-013-2163-3 PMID: 24258151

Piluso, S., Li, Y., & Abinzano, F.. (2019). Mimicking the Articular Joint with In Vitro Models . *Trends in Biotechnology*, *37*(10), 1063–1077. PMID: 31000204

Pimalai, D., Putnin, T., Waiwinya, W., Chotsuwan, C., Aroonyadet, N., & Japrung, D. (2021). Development of electrochemical biosensors for simultaneous multiplex detection of microRNA for breast cancer screening. *Microchimica Acta*, *188*(10), 1–10. DOI: 10.1007/S00604-021-04995-8/METRICS PMID: 34495394

Pinaud, F., King, D., Moore, H. P., & Weiss, S. (2004). Bioactivation and Cell Targeting of Semiconductor CdSe/ZnS Nanocrystals with Phytochelatin-Related Peptides. *Journal of the American Chemical Society*, *126*(19), 6115–6123. DOI: 10.1021/ja031691c PMID: 15137777

Pinon-Castillo, H. A., Martinez-Chamarro, R., Reyes-Martinez, R., Salinas-Vera, Y. M., Manjarrez-Nevarez, L. A., Munoz-Castellanos, L. N., Lopez-Camarillo, C., & Orrantia-Borunda, E. (2021). Palladium Nanoparticles Functionalized with PVP-Quercetin Inhibits Cell Proliferation and Activates Apoptosis in Colorectal Cancer Cells. *Applied Sciences (Basel, Switzerland)*, *11*(5), 1988. DOI: 10.3390/app11051988

Pino, P., Bosco, F., Mollea, C., & Onida, B. (2023). Antimicrobial nano-zinc oxide biocomposites for wound healing applications: A review. *Pharmaceutics*, *15*(3), 970. DOI: 10.3390/pharmaceutics15030970 PMID: 36986831

Pisciotta, A., Lunghi, A., Bertani, G., Di Tinco, R., Bertoni, L., Orlandi, G., Biscarini, F., Bianchi, M., & Carnevale, G. (2022). PEDOT: PSS promotes neurogenic commitment of neural crest-derived stem cells. *Frontiers in Physiology*, *13*, 930804. DOI: 10.3389/fphys.2022.930804 PMID: 36060701

Pitchayyapillai, G., Seenikannan, P., Balasundar, P., & Narayanasamy, P. (2017). Effect of nano-silver on microstructure, mechanical and tribological properties of cast 6061 aluminum alloy. *Transactions of Nonferrous Metals Society of China*, *27*(10), 2137–2145.

Pivovarov, A. S., Calahorro, F., & Walker, R. J. (2018). Na(+)/K(+)-pump and neurotransmitter membrane receptors. *Invertebrate Neuroscience*, *19*(1), 1. DOI: 10.1007/s10158-018-0221-7 PMID: 30488358

Poggi, A., Di Martino, A., Andriolo, L., Reale, D., Filardo, G., Kon, E., & Zaffagnini, S. (2023). Chitosan based scaffold applied in patellar cartilage lesions showed positive clinical and MRI results at minimum 2 years of follow up. *Knee Surgery, Sports Traumatology, Arthroscopy : Official Journal of the ESSKA*, *31*(5), 1714–1722. DOI: 10.1007/s00167-022-07023-1 PMID: 35678852

Polo, T. O. B., Silva, W. P. P., Momesso, G. C., Lima-Neto, T. J., Barbosa, S., Cordeiro, J. M., Hassumi, J. S., Da Cruz, N. C., Okamoto, R., Barão, V. R., & Faverani, L. P. (2020). Plasma Electrolytic Oxidation as a Feasible Surface Treatment for Biomedical Applications: An in vivo study. *Scientific Reports*, *10*(1). Advance online publication. DOI: 10.1038/s41598-020-65289-2 PMID: 32561767

Pompa, L., Rahman, Z. U., Munoz, E., & Haider, W. (2015). Surface characterization and cytotoxicity response of biodegradable magnesium alloys. *Materials Science and Engineering C*, *49*, 761–768. DOI: 10.1016/j.msec.2015.01.017 PMID: 25687006

Poon, A. W., Goodman, C. S., & Rubin, R. J. (1998). In vitro and skin testing for allergy: Comparable clinical utility and costs. *The American Journal of Managed Care*, *4*(7), 969–985. PMID: 10181996

Poongodi, R., Chen, Y. L., & Yang, T. H.. (2021). Bio-Scaffolds as Cell or Exosome Carriers for Nerve Injury Repair . *International Journal of Molecular Sciences*, *22*(24), 13347. PMID: 34948144

Posadas, I., Monteagudo, S., & Ceña, V. (2016). Nanoparticles for brain-specific drug and genetic material delivery, imaging and diagnosis. *Nanomedicine (London)*, *11*(7), 833–849. PMID: 26980585

Pouillès, J.-M., Gosset, A., & Trémollieres, F. (2021). Ménopause, traitement hormonal de ménopause et ostéoporose. RPC Les femmes ménopausées du CNGOF et du GEMVi. *Gynécologie, Obstétrique, Fertilité & Sénologie*, *49*(5), 420–437. DOI: 10.1016/j.gofs.2021.03.015 PMID: 33753297

Pounraj, S., Chen, S., Triscott, H., Lam, A. K., Ma, L., & Rehm, B. H. A. (2024). Engineering neoantigens to form immunogenic biopolymer particles targeting metastatic breast cancer. *Applied Materials Today*, *38*, 102238. DOI: 10.1016/J.APMT.2024.102238

Prabaharan, M., & Mano, J. F. (2004). Chitosan-Based Particles as Controlled Drug Delivery Systems. *Drug Delivery*, *12*(1), 41–57. DOI: 10.1080/10717540590889781 PMID: 15801720

Prajapati, S. K., Malaiya, A., Kesharwani, P., Soni, D., & Jain, A. (2022). Biomedical applications and toxicities of carbon nanotubes. *Drug and Chemical Toxicology*, *45*(1), 435–450. DOI: 10.1080/01480545.2019.1709492 PMID: 31908176

Prasad, S. V. S., Prasad, S. B., Verma, K., Mishra, R. K., Kumar, V., & Singh, S. (2021). The role and significance of Magnesium in modern day research-A review. *Journal of Magnesium and Alloys*, *10*(1), 1–61. DOI: 10.1016/j.jma.2021.05.012

Preetam, S., Ghosh, A., Mishra, R., Pandey, A., Roy, D. S., Rustagi, S., & Malik, S. (2024). Electrical stimulation: A novel therapeutic strategy to heal biological wounds. *Royal Society of Chemistry Advances*, *14*(44), 32142–32173. DOI: 10.1039/d4ra04258a PMID: 39399261

Preserve, L., & Blaney, L. (2007). *Magnetite (Fe3O4): Properties, Synthesis, and Applications*. Retrieved from https://preserve.lehigh.edu/cas-lehighreview-vol-15http://preserve.lehigh.edu/cas-lehighreview-vol-15/5

Prihantono, & Faruk, M. (2021). Breast cancer resistance to chemotherapy: When should we suspect it and how can we prevent it? Annals of Medicine and Surgery, 70, 102793. https://doi.org/DOI: 10.1016/J.AMSU.2021.102793

Provenzale, J. M., & Silva, G. A. (2009). Uses of Nanoparticles for Central Nervous System Imaging and Therapy. *AJNR. American Journal of Neuroradiology*. Advance online publication. DOI: 10.3174/ajnr.A1590 PMID: 19617446

Puertas-Bartolomé, M., Mora-Boza, A., & García-Fernández, L. (2021). Emerging Biofabrication Techniques: A Review on Natural Polymers for Biomedical Applications. *Polymers, 13*(8), 1209. DOI: 10.3390/polym13081209 PMID: 33918049

Puspasari, V., Ridhova, A., Hermawan, A., Amal, M. I., & Khan, M. M. (2022). ZnO-based antimicrobial coatings for biomedical applications. *Bioprocess and Biosystems Engineering, 45*(9), 1421–1445. DOI: 10.1007/s00449-022-02733-9 PMID: 35608710

Qasim, M., Lim, D. J., Park, H., & Na, D. (2014). Nanotechnology for diagnosis and treatment of infectious diseases. *Journal of Nanoscience and Nanotechnology, 14*(10), 7374–7387. DOI: 10.1166/jnn.2014.9578 PMID: 25942798

Qian, L., Jin, F., Wei, Z., Li, T., Sun, Z., Lai, C., Ma, J., Xiong, R., Ma, X., Wang, F., Sun, F., Zheng, W., Dong, W., Sun, K., Wang, T., & Feng, Z. Q. (2023). Wearable, self-powered, drug-loaded electronic microneedles for accelerated tissue repair of inflammatory skin disorders. *Advanced Functional Materials, 33*(27). Advance online publication. DOI: 10.1002/adfm.202209407

Qian, M., Li, S., Xi, K., Tang, J., Shen, X., Liu, Y., & Guo, R.. (2023). ECM-Engineered Electrospun Fibers with an Immune Cascade Effect for Inhibiting Tissue Fibrosis. *Acta Biomaterialia, 171*(November), 308–326. DOI: 10.1016/j.actbio.2023.08.058 PMID: 37673231

Qian, Y., Lin, H., Yan, Z., Shi, J., & Fan, C. (2021). Functional nanomaterials in peripheral nerve regeneration: Scaffold design, chemical principles and microenvironmental remodeling. *Materials Today, 51*, 165–187.

Qiao, X., Bao, L., Liu, G., & Cui, X. (2024). Nanomaterial journey in the gut: From intestinal mucosal interaction to systemic transport. *Nanoscale*. Advance online publication. DOI: 10.1039/d4nr02480j PMID: 39347780

Qi, J., Yu, T., & Hu, B.. (2021). Current Biomaterial-Based Bone Tissue Engineering and Translational Medicine . *International Journal of Molecular Sciences, 22*(19), 10233. PMID: 34638571

Qi, L., Zhao, T., Yan, J., Ge, W., Jiang, W., Wang, J., & Zhang, L. (2024). Advances in magnesium-containing bioceramics for bone repair. *Biomaterials Translational, 5*(1), 3. PMID: 39220661

Qin, N., Han, H., Guan, G., & Han, M.-Y. (2024). Structurally altered size, composition, shape and interface-dependent optical properties of quantized nanomaterials. *Nano Research*, •••, 1–27.

Qin, Y., Liu, A., Guo, H., Shen, Y., Wen, P., Lin, H., Xia, D., Voshage, M., Tian, Y., & Zheng, Y. (2022). Additive manufacturing of Zn-Mg alloy porous scaffolds with enhanced osseointegration: In vitro and in vivo studies. *Acta Biomaterialia, 145*, 403–415. DOI: 10.1016/j.actbio.2022.03.055 PMID: 35381400

Qu, D. J. (2020). Anisamide-functionalized pH-responsive amphiphilic chitosan-based paclitaxel micelles for sigma-1 receptor targeted prostate cancer treatment. *Carbohydrate Polymers, 229*, 115498. PMID: 31826492

Quiñones, J. P., Peniche, H., & Peniche, C. (2018). Chitosan based self-assembled nanoparticles in drug delivery. *Polymers, 10*(3), 1–32. DOI: 10.3390/polym10030235 PMID: 30966270

Rabišková, M., Bautzová, T., Gajdziok, J., Dvořáčková, K., Lamprecht, A., Pellequer, Y., & Spilková, J. (2012). Coated chitosan pellets containing rutin intended for the treatment of inflammatory bowel disease: In vitro characteristics and in vivo evaluation. *International Journal of Pharmaceutics*, *422*(1–2), 151–159. DOI: 10.1016/j.ijpharm.2011.10.045 PMID: 22079717

Radhakumary, C., Antonty, M., & Sreenivasan, K. (2011). Drug loaded thermoresponsive and cytocompatible chitosan based hydrogel as a potential wound dressing. *Carbohydrate Polymers*, *83*(2), 705–713. DOI: 10.1016/j.carbpol.2010.08.042

Rahat, I., Imam, S. S., Rizwanullah, M., Alshehri, S., Asif, M., Kala, C., & Taleuzzaman, M. (2021). Thymoquinone-entrapped chitosan-modified nanoparticles: Formulation optimization to preclinical bioavailability assessments. *Drug Delivery*, *28*(1), 973–984. DOI: 10.1080/10717544.2021.1927245 PMID: 34036860

Rahimi, M., Mir, S. M., Baghban, R., Charmi, G., Plummer, C. M., Shafiei-Irannejad, V., & Pietrasik, J. (2022). Chitosan-based biomaterials for the treatment of bone disorders. *International Journal of Biological Macromolecules*, *215*, 346–367. PMID: 35718150

Rahman, M. (2023). Magnetic resonance imaging and iron-oxide nanoparticles in the era of personalized medicine. *Nanotheranostics*, *7*(4), 424. DOI: 10.7150/ntno.86467 PMID: 37650011

Rahman, M. H., & Mondal, M. I. H. (2024). Stability, challenges, and prospects of chitosan for the delivery of anticancer drugs and tissue regenerative growth factors. *Heliyon*, *10*(21), e39879. DOI: 10.1016/j.heliyon.2024.e39879 PMID: 39583848

Rajitha, A., Kansal, L., Raj, G., Kalra, R., Dhamija, K., & Abdul-Zahra, D. S. 2024. “Biomaterials and Artificial Intelligence: Predictive Modeling and Design.” In *E3S Web of Conferences*. Vol. 505. EDP Sciences. https://doi.org/DOI: 10.1051/e3sconf/202450501003

Rajkumar, D. S. R., Keerthika, K., & Vijayaragavan, V. (2024). Chitosan-Based Biomaterial in Wound Healing: A Review. *Cureus*, *16*(2). PMID: 38562272

Ramalingam, V., Raja, S., & Harshavardhan, M. (2020). In situone-step synthesis of polymer-functionalized palladium nanoparticles: An efficient anticancer agent against breast cancer. *Dalton Transactions (Cambridge, England)*, *49*(11), 3510–3518. DOI: 10.1039/c9dt04576g PMID: 32107506

Rami, M. R., Meskini, M., Qarebaghi, L. M., Salami, M., Forouzandehdel, S., & Cheraghali, M. (2022). Synthesis of magnetic bio-nanocomposites for drug release and adsorption applications. *South African Journal of Chemical Engineering*, *42*, 115–126. https://doi.org/https://doi.org/10.1016/j.sajce.2022.08.003

Ramirez-Rios, S., Michallet, S., Peris, L., Barette, C., Rabat, C., Feng, Y., Fauvarque, M., Andrieux, A., Sadoul, K., & Lafanechère, L. (2020). A new quantitative Cell-Based assay reveals unexpected microtubule stabilizing activity of certain kinase inhibitors, clinically approved or in the process of approval. *Frontiers in Pharmacology*, *11*. Advance online publication. DOI: 10.3389/fphar.2020.00543 PMID: 32425788

Ramzan, F., Ekram, S., & Frazier, T.. (2022). Decellularized Human Umbilical Tissue-Derived Hydrogels Promote Proliferation and Chondrogenic Differentiation of Mesenchymal Stem Cells . *Bioengineering (Basel, Switzerland)*, *9*(6), 239. PMID: 35735483

Ranjbar, M., Pardakhty, A., Amanatfard, A., & Asadipour, A. (2018). Efficient drug delivery of β-estradiol encapsulated in Zn-metal-organic framework nanostructures by microwave-assisted coprecipitation method. *Drug Design, Development and Therapy*, *12*, 2635–2643. DOI: 10.2147/DDDT.S173324 PMID: 30214152

Ransohoff, R. M., & Brown, M. A. (2012). Innate immunity in the central nervous system. *The Journal of Clinical Investigation*, *122*(4), 1164–1171. PMID: 22466658

Rashed, A. A., Misnan, N. M., & Nawi, M. N. M. (2023). Chitosan-based formulation for bone health: A review. *Carbohydrate Polymer Technologies and Applications*, *6*, 100359. DOI: 10.1016/J.CARPTA.2023.100359

Raveendran, J., Resmi, P. E., Ramachandran, T., Nair, B. G., & Satheesh Babu, T. G. (2017). Fabrication of a disposable non-enzymatic electrochemical creatinine sensor. *Sensors and Actuators. B, Chemical*, *243*, 589–595. DOI: 10.1016/j.snb.2016.11.158

Razavi, M., Fathi, M., Savabi, O., Tayebi, L., & Vashaee, D. (2020). Biodegradable Magnesium Bone Implants Coated with a Novel Bioceramic Nanocomposite. *Materials (Basel)*, *13*(6), 1315. DOI: 10.3390/ma13061315 PMID: 32183231

Re, F., Moresco, R., & Masserini, M. (2012). Nanoparticles for neuroimaging. *Journal of Physics. D, Applied Physics*, *45*(7), 073001.

Rehder, D. (2013). Vanadium. Its role for humans. *Interrelations between essential metal ions and human diseases*, 139-169.

Reid, B., & Zhao, M. (2014). The electrical response to injury: Molecular mechanisms and wound healing. *Advances in Wound Care (New Rochelle, N.Y.)*, *3*(2), 184–201. DOI: 10.1089/wound.2013.0442 PMID: 24761358

Ren, W., Lin, Z., & Fan, Y.. (2022). Programmable biological state-switching photoelectric nanosheets for the treatment of infected wounds . *Materials Today. Bio*, *15*, 100292. PMID: 35634172

Rey-Vinolas, S., Engel, E., & Mateos-Timoneda, M. (2019). Polymers for bone repair. In *Bone Repair Biomaterials* (pp. 179–197). Elsevier.

Rezghi Rami, M., Meskini, M., & Ebadi Sharafabad, B. (2024). Fungal-mediated nanoparticles for industrial applications: Synthesis and mechanism of action. *Journal of Infection and Public Health*, *17*(10), 102536. https://doi.org/https://doi.org/10.1016/j.jiph.2024.102536 PMID: 39276432

Rijal, G., Bathula, C., & Li, W. (2017). Application of Synthetic Polymeric Scaffolds in Breast Cancer 3D Tissue Cultures and Animal Tumor Models. *International Journal of Biomaterials*, *8074890*. Advance online publication. DOI: 10.1155/2017/8074890 PMID: 29599800

Robe, A., Pic, E., Lassalle, H. P., Bezdetnaya, L., Guillemin, F., & Marchal, F. (2008). Quantum dots in axillary lymph node mapping: Biodistribution study in healthy mice. *BMC Cancer*, *8*, 111. DOI: 10.1186/1471-2407-8-111 PMID: 18430208

Rodrigues, A. S., Batista, J. G., Rodrigues, M. Á., Thipe, V. C., Minarini, L. A., Lopes, P. S., & Lugão, A. B. (2024). Advances in silver nanoparticles: A comprehensive review on their potential as antimicrobial agents and their mechanisms of action elucidated by proteomics. *Frontiers in Microbiology*, *15*, 1440065. PMID: 39149204

Rodrigues, M. Á. V., Marangon, C. A., & Martins, V. da C. A., & Plepis, A. M. de G. (2021). Chitosan/gelatin films with jatobá resin: Control of properties by vegetal resin inclusion and degree of acetylation modification. *International Journal of Biological Macromolecules*, *182*, 1737–1745. DOI: 10.1016/j.ijbiomac.2021.05.160 PMID: 34051262

Rodríguez, F. C. (2022). Nano-based approved pharmaceuticals for cancer treatment: Present and future challenges. *Biomolecules*, *12*(6), 784. PMID: 35740909

Rokade, S. S., Joshi, K. A., Mahajan, K., Patil, S., Tomar, G., Dubal, D. S., Parihar, V. S., Kitture, R., Bellare, J. R., & Ghosh, S. (2018). Gloriosa superba Mediated Synthesis of Platinum and Palladium Nanoparticles for Induction of Apoptosis in Breast Cancer. *Bioinorganic Chemistry and Applications*, *2018*, 1–9. DOI: 10.1155/2018/4924186 PMID: 30057593

Romero, J. F., Diaz-Barrios, A., & Gonzalez, G. (2021). Biocompatible thermo-responsive N-vinylcaprolactam based hydrogels for controlled drug delivery systems. *Bionatura*, *6*(2), 1712–1719. DOI: 10.21931/RB/2021.06.02.8

Rondon-Cavanzo, E. P. (2022). Biocompatibilité des nanoparticules de chitosane-siRNA in vitro pour la thérapie génique.

Roney, C., Kulkarni, P., Arora, V., Antich, P., Bonte, F., Wu, A., Mallikarjuana, N. N., Manohar, S., Liang, H. F., Kulkarni, A. R., Sung, H. W., Sairam, M., & Aminabhavi, T. M. (2005). Targeted nanoparticles for drug delivery through the blood-brain barrier for Alzheimer's disease. *Journal of Controlled Release*, *108*(2-3), 193–214. DOI: 10.1016/j.jconrel.2005.07.024 PMID: 16246446

Rosa, V., Silikas, N., Yu, B., Dubey, N., Gopu, S., Zinelis, S., Lima, A. F., Bottino, M. C., Ferreira, J. N., & Schmalz, G. (2024). Guidance on the assessment of biocompatibility of biomaterials: Fundamentals and testing considerations. *Dental Materials*. PMID: 39129079

Rosner, G. A., & Fonacier, L. S. (2017). Hypersensitivity to biomedical implants: Prevention and diagnosis. *Allergy and Asthma Proceedings*.

Roughley, P., Hoemann, C., DesRosiers, E., Mwale, F., Antoniou, J., & Alini, M. (2006). The potential of chitosan-based gels containing intervertebral disc cells for nucleus pulposus supplementation. *Biomaterials*, *27*(3), 388–396. DOI: 10.1016/J.BIOMATERIALS.2005.06.037 PMID: 16125220

Roveta, F., Bonino, L., Piella, E. M., Rainero, I., & Rubino, E. (2024). Neuroinflammatory Biomarkers in Alzheimer's Disease: From Pathophysiology to Clinical Implications. *International Journal of Molecular Sciences*, *25*(22), 11941. PMID: 39596011

Ruiz-Alonso, S. L.-M.-B. (2021). Tendon tissue engineering: Cells, growth factors, scaffolds and production techniques. *Journal of Controlled Release*, *333*, 448–486. PMID: 33811983

Rumian, Ł., Wolf-Brandstetter, C., Rößler, S., Reczyńska, K., Tiainen, H., Haugen, H. J., Scharnweber, D., & Pamuła, E. (2021). Sodium alendronate loaded poly(<scp>l</scp> -lactide- *co* -glycolide) microparticles immobilized on ceramic scaffolds for local treatment of bone defects. *Regenerative Biomaterials, 8*(3). Advance online publication. DOI: 10.1093/rb/rbaa012 PMID: 32523731

Rusdin, A., Mohd Gazzali, A., Ain Thomas, N., Megantara, S., Aulifa, D. L., Budiman, A., & Muchtaridi, M. (2024). Advancing Drug Delivery Paradigms: Polyvinyl Pyrolidone (PVP)-Based Amorphous Solid Dispersion for Enhanced Physicochemical Properties and Therapeutic Efficacy. *Polymers, 16*(2), 286. DOI: 10.3390/POLYM16020286 PMID: 38276694

Rusin, A., Seymour, C., & Mothersill, C. (2018). Chronic fatigue and immune deficiency syndrome (CFIDS), cellular metabolism, and ionizing radiation: A review of contemporary scientific literature and suggested directions for future research. *International Journal of Radiation Biology, 94*(3), 212–228. PMID: 29297728

Ryabchikova, E. (2021). *Advances in nanomaterials in biomedicine* (Vol. 11). MDPI.

Saberi, A., Baltatu, M. S., & Vizureanu, P. (2024). Recent advances in Magnesium–Magnesium oxide nanoparticle composites for biomedical applications. *Bioengineering (Basel, Switzerland), 11*(5), 508. DOI: 10.3390/bioengineering11050508 PMID: 38790374

Sachan, R. S. K., Mistry, V., Dholaria, M., Rana, A., Devgon, I., Ali, I., & Bawazeer, S. (2023). Overcoming Mycobacterium tuberculosis drug resistance: Novel medications and repositioning strategies. *ACS Omega, 8*(36), 32244–32257. PMID: 37720746

Safdar, R., Omar, A. A., Arunagiri, A., Regupathi, I., & Thanabalan, M. (2019). Potential of Chitosan and its derivatives for controlled drug release applications – A review. *Journal of Drug Delivery Science and Technology, 49*(July 2018), 642–659. doi: DOI: 10.1016/j.jddst.2018.10.020

Sahani, R. M., & Dixit, A. (2022). A comprehensive review on zinc oxide bulk and nano-structured materials for ionizing radiation detection and measurement applications. *Materials Science in Semiconductor Processing, 151*, 107040.

Sahoo, D., Deb, P., Basu, T., Bardhan, S., Patra, S., & Sukul, P. K. (2024). Advancements in platinum-based anticancer drug development: A comprehensive review of strategies, discoveries, and future perspectives. *Bioorganic & Medicinal Chemistry, 117894*. Advance online publication. DOI: 10.1016/j.bmc.2024.117894 PMID: 39214013

Salehi, F., Behboudi, H., Kavoosi, G., & Ardestani, S. K. (2017). Chitosan promotes ROS-mediated apoptosis and S phase cell cycle arrest in triple-negative breast cancer cells: Evidence for intercalative interaction with genomic DNA. *RSC Advances, 7*(68), 43141–43150. DOI: 10.1039/C7RA06793C

Salehi, M., Khodadadi, A., & Sadeghi, F. (2017). The potential of chitosan nanoparticles in anticancer therapy: A review of biological activities. *Cancer Nanotechnology, 22*(4), 369–382.

Saleh, T. A. (2020). Nanomaterials: Classification, properties, and environmental toxicities. *Environmental Technology & Innovation, 20*, 101067. DOI: 10.1016/j.eti.2020.101067

Salhotra, A, & Shah, H N, Levi B. *et al.* (2020). Mechanisms of bone development and repair . *Nature Reviews. Molecular Cell Biology*, *21*(11), 696–711. PMID: 32901139

Salmani-Zarchi, H., Mousavi-Sagharchi, S. M. A., Sepahdoost, N., Ranjbar-Jamalabadi, M., Gross, J. D., Jooya, H., & Samadi, A. (2024). Antimicrobial Feature of Nanoparticles in the Antibiotic Resistance Era: From Mechanism to Application. *Advanced Biomedical Research*, *13*(1). https://journals.lww.com/adbm/fulltext/2024/11300/antimicrobial_feature_of_nanoparticles_in_the.113.aspx PMID: 39717242

Şalva, E. K. (2012). Local delivery of chitosan/VEGF siRNA nanoplexes reduces angiogenesis and growth of breast cancer in vivo. *Nucleic Acid Therapeutics*, *22*(1), 40–48. PMID: 22217324

Şalva, E. T. (2015). The enhancement of gene silencing efficiency with chitosan-coated liposome formulations of siRNAs targeting HIF-1α and VEGF. *International Journal of Pharmaceutics*, *478*(1), 147–154. PMID: 25445537

Samrot, A. V., Sathiyasree, M., Rahim, S. B. A., Renitta, R. E., Kasipandian, K., Krithika Shree, S., & Chinni, S. V. (2023). Scaffold using chitosan, agarose, cellulose, dextran and protein for tissue engineering—A review. *Polymers*, *15*(6), 1525. PMID: 36987305

San, H. H. M., Alcantara, K. P., Bulatao, B. P. I., Chaichompoo, W., Nalinratana, N., Suksamrarn, A., Vajragupta, O., Rojsitthisak, P., & Rojsitthisak, P. (2022). Development of Turmeric Oil—Loaded Chitosan/Alginate Nanocapsules for Cytotoxicity Enhancement against Breast Cancer. Polymers 2022, Vol. 14, Page 1835, 14(9), 1835. https://doi.org/DOI: 10.3390/POLYM14091835

Sandhya, P., Satyendra Kumar, T., Chandni, P., & Archna, P. (2022). Encapsulation of Alendronate in Chitosan based Polymeric Nanoparticles for Effective Management of Osteoporosis – Development to Release Kinetic Study. *International Journal of Medical Nano Research*, *9*(1). Advance online publication. DOI: 10.23937/2378-3664.1410036

Sandomierski, M., Adamska, K., Ratajczak, M., & Voelkel, A. (2022). Chitosan - zeolite scaffold as a potential biomaterial in the controlled release of drugs for osteoporosis. *International Journal of Biological Macromolecules*, *223*, 812–820. DOI: 10.1016/J.IJBIOMAC.2022.11.071 PMID: 36375670

Sang, S., Cheng, R., Cao, Y., Yan, Y., Shen, Z., Zhao, Y., & Han, Y. (2022). Biocompatible chitosan/polyethylene glycol/multi-walled carbon nanotube composite scaffolds for neural tissue engineering. *Journal of Zhejiang University. Science. B.*, *23*(1), 58–73. PMID: 35029088

Santhanakrishnan, K. R., Koilpillai, J., & Narayanasamy, D. (2024). PEGylation in Pharmaceutical Development: Current Status and Emerging Trends in Macromolecular and Immunotherapeutic Drugs. *Cureus*, (August). Advance online publication. DOI: 10.7759/cureus.66669 PMID: 39262507

Santhosh, S., Mukherjee, D., Anbu, J., Murahari, M., & Teja, B. V. (2019). Improved treatment efficacy of risedronate functionalized chitosan nanoparticles in osteoporosis: Formulation development, in vivo, and molecular modelling studies. *Journal of Microencapsulation*, *36*(4), 338–355. DOI: 10.1080/02652048.2019.1631401 PMID: 31190594

Santin, M., & Phillips, G. J. (2012). *Biomimetic, bioresponsive, and bioactive materials: an introduction to integrating materials with tissues* [M]. John Wiley & Sons.

Santos, G. D. (2017). The importance of metallic materials as biomaterials. *Advances in Tissue Engineering & Regenerative Medicine Open Access, 3*(1). Advance online publication. DOI: 10.15406/atroa.2017.03.00054

Santos, J. D.-F. (2021). Biopolymeric materials used as nonviral vectors: A review. *Polysaccharides, 2*(1), 100–109.

Santos-Lobato, B. L., & Schuh, A. F. S. (2024). Exposure to household pesticides and Parkinson's disease in the Parkinson's Progression Markers Initiative cohort. *Frontiers in Neurology, 15*, 1411468. PMID: 39188706

Saqr, A. Al, Wani, S. U. D., Gangadharappa, H. V., Aldawsari, M. F., Khafagy, E. S., & Abu Lila, A. S. (2021). Enhanced Cytotoxic Activity of Docetaxel-Loaded Silk Fibroin Nanoparticles against Breast Cancer Cells. Polymers 2021, Vol. 13, Page 1416, 13(9), 1416. https://doi.org/DOI: 10.3390/POLYM13091416

Saravanan, S., Vimalraj, S., & Anuradha, D. (2018). Chitosan Based Thermoresponsive Hydrogel Containing Graphene Oxide for Bone Tissue Repair. *Biomedicine and Pharmacotherapy, 107*(November), 908–917. DOI: 10.1016/j.biopha.2018.08.072 PMID: 30257403

Sarmah, M., Neog, A. B., Boruah, P. K., Das, M. R., Bharali, P., & Bora, U. (2019). Effect of Substrates on Catalytic Activity of Biogenic Palladium Nanoparticles in C–C Cross-Coupling Reactions. *ACS Omega, 4*(2), 3329–3340. DOI: 10.1021/acsomega.8b02697 PMID: 31459548

Sartaj, A., Qamar, Z., Qizilbash, F. F., Annu, Md, S., Alhakamy, N. A., Baboota, S., & Ali, J. (2021). Polymeric Nanoparticles: Exploring the Current Drug Development and Therapeutic Insight of Breast Cancer Treatment and Recommendations. Polymers 2021, Vol. 13, Page 4400, 13(24), 4400. https://doi.org/DOI: 10.3390/POLYM13244400

Satarug, S., Baker, J. R., Urbenjapol, S., Haswell-Elkins, M., Reilly, P. E. B., Williams, D. J., & Moore, M. R. (2003). A global perspective on cadmium pollution and toxicity in non-occupationally exposed population. *Toxicology Letters, 137*(1–2), 65–83. DOI: 10.1016/S0378-4274(02)00381-8 PMID: 12505433

Satchanska, G., Davidova, S., & Petrov, P. D. (2024). *"Natural and Synthetic Polymers for Biomedical and Environmental Applications." Polymers*. Multidisciplinary Digital Publishing Institute., DOI: 10.3390/polym16081159

Sau, S., Dey, A., Pal, P., Das, B., Maity, K. K., Dash, S. K., & Das, B. (2024). Immunomodulatory and immune-toxicological role of nanoparticles: Potential therapeutic applications. *International Immunopharmacology, 135*, 112251. PMID: 38781608

Savage, D. T., Hilt, J. Z., & Dziubla, T. D. (2019). In vitro methods for assessing nanoparticle toxicity. *Methods in Molecular Biology (Clifton, N.J.)*, •••. Advance online publication. DOI: 10.1007/978-1-4939-8916-4_1 PMID: 30547452

Schalock, P. C., Menné, T., Johansen, J. D., Taylor, J. S., Maibach, H. I., Lidén, C., Bruze, M., & Thyssen, J. P. (2012). Hypersensitivity reactions to metallic implants–diagnostic algorithm and suggested patch test series for clinical use. *Contact Dermatitis, 66*(1), 4–19.

Schoepf, U., Marecos, E. M., Melder, R. J., Jain, R. K., & Weissleder, R. (1998). Intracellular magnetic labeling of lymphocytes for in vivo trafficking studies. *BioTechniques*, *24*(4), 642–651. DOI: 10.2144/98244rr01 PMID: 9564539

Scholz, M.-S., Blanchfield, J. P., Bloom, L., Coburn, B. H., Elkington, M., Fuller, J. D., Gilbert, M. E., Muflahi, S. A., Pernice, M., & Rae, S. I. (2011). The use of composite materials in modern orthopaedic medicine and prosthetic devices: A review. *Composites Science and Technology*, *71*(16), 1791–1803.

Schreiber, S. B., Bozell, J. J., Hayes, D. G., & Zivanovic, S. (2013). Introduction of primary antioxidant activity to chitosan for application as a multifunctional food packaging material. *Food Hydrocolloids*, *33*(2), 207–214. DOI: 10.1016/j.foodhyd.2013.03.006

Seetharaman, S., Jayalakshmi, S., Singh, R. A., & Gupta, M. (2022). The potential of Magnesium-Based materials for engineering and biomedical applications. *Journal of the Indian Institute of Science*, *102*(1), 421–437. DOI: 10.1007/s41745-021-00274-w

Selkoe, D. J. (2001). Alzheimer's disease: Genes, proteins, and therapy. *Physiological Reviews*. PMID: 11274343

Semlitsch, M. (1974). Technical progress in artificial hip joints. *Engineering in Medicine*, *3*(4), 10–19.

Senapati, S., Mahanta, A. K., Kumar, S., & Maiti, P. (2018). Controlled drug delivery vehicles for cancer treatment and their performance. Signal Transduction and Targeted Therapy 2018 3:1, 3(1), 1–19. https://doi.org/DOI: 10.1038/s41392-017-0004-3

ŞENEL, B. (2019). In vitro preliminary studies of chitooligosaccharide coated nanostructured lipidic nanoparticles for efficient gene delivery. *Journal of Research in Pharmacy*, *23*(4).

Sengul, A. B., & Asmatulu, E. (2020). Toxicity of metal and metal oxide nanoparticles: A review. *Environmental Chemistry Letters*, *18*(5), 1659–1683.

Senjab, R. M., AlSawaftah, N., Abuwatfa, W. H., & Husseini, G. A. (2024). Advances in liposomal nanotechnology: From concept to clinics. *RSC Pharmaceutics*.

Sen, V. D., Balakina, A. A., Stupina, T. S., Mumyatova, V. A., Kulikov, A. V., Tikhonov, I. V., & Pliss, E. M. (2021). Amphiphilic chitosan–polyaminoxyls loaded with daunorubicin: Synthesis, antioxidant activity, and drug delivery capacity. *International Journal of Biological Macromolecules*, *193*, 965–979. PMID: 34751143

Seo, Y., Ghazanfari, L., Master, A., Vishwasrao, H. M., Wan, X., Sokolsky-Papkov, M., & Kabanov, A. V. (2021). Poly(2-oxazoline)-Magnetite NanoFerrogels: Magnetic Field Responsive Theranostic Platform for Cancer Drug Delivery and Imaging. *Nanomedicine; Nanotechnology, Biology, and Medicine*, *39*, 102459. DOI: 10.1016/J.NANO.2021.102459 PMID: 34530163

Sezer, N., Evis, Z., Kayhan, S. M., Tahmasebifar, A., & Koç, M. (2018). Review of magnesium-based biomaterials and their applications. *Journal of Magnesium and Alloys*, *6*(1), 23–43. DOI: 10.1016/j.jma.2018.02.003

Shabani, L., Abbasi, M., Azarnew, Z., Amani, A. M., & Vaez, A. (2023). Neuro-nanotechnology: Diagnostic and therapeutic nano-based strategies in applied neuroscience. *Biomedical Engineering Online*, *22*(1), 1. DOI: 10.1186/s12938-022-01062-y PMID: 36593487

Shah, N. S., Thotathil, V., Zaidi, S. A., Sheikh, H., Mohamed, M., Qureshi, A., & Sadasivuni, K. K. (2022). Picomolar or beyond limit of detection using molecularly imprinted polymer-based electrochemical sensors: A review. *Biosensors (Basel)*, *12*(12), 1107. PMID: 36551073

Shaikh, M. A. J., Alharbi, K. S., Almalki, W. H., Imam, S. S., Albratty, M., Meraya, A. M., Alzarea, S. I., Kazmi, I., Al-Abbasi, F. A., Afzal, O., Altamimi, A. S. A., Singh, Y., Singh, S. K., Dua, K., & Gupta, G. (2022). Sodium alginate based drug delivery in management of breast cancer. *Carbohydrate Polymers*, *292*, 119689. DOI: 10.1016/J.CARBPOL.2022.119689 PMID: 35725179

Shajahan, A., Shankar, S., Sathiyaseelan, A., Narayan, K. S., Narayanan, V., Kaviyarasan, V., & Ignacimuthu, S. (2017). Comparative studies of chitosan and its nanoparticles for the adsorption efficiency of various dyes. *International Journal of Biological Macromolecules*, *104*, 1449–1458. DOI: 10.1016/j.ijbiomac.2017.05.128 PMID: 28554794

Shakeel, A. (2023). Mitigating challenges and expanding the future of vascular tissue engineering—Are we there yet? *Frontiers in Physiology*, *13*, 1079421. PMID: 36685187

Shakeri, S., Ashrafizadeh, M., Zarrabi, A., Roghanian, R., Afshar, E. G., Pardakhty, A., Mohammadinejad, R., Kumar, A., & Thakur, V. K. (2020). Multifunctional polymeric nanoplatforms for brain diseases diagnosis, therapy and theranostics. *Biomedicines*, *8*(1), 13. PMID: 31941057

Shakeri, S., Ashrafizadeh, M., Zarrabi, A., Roghanian, R., Afshar, E. G., Pardakhty, A., Mohammadinejad, R., Kumar, A., & Thakur, V. K. (2020). Multifunctional Polymeric Nanoplatforms for Brain Diseases Diagnosis, Therapy and Theranostics. *Biomedicines*, *8*(1). Advance online publication. DOI: 10.3390/biomedicines8010013 PMID: 31941057

Shakil, M. S. (2021). Using chitosan or chitosan derivatives in cancer therapy. *Polysaccharides*, *2*(4), 795–816.

Shakoor, S., Kibble, E., & El-Jawhari, J. J. (2022). Bioengineering Approaches for Delivering Growth Factors: A Focus on Bone and Cartilage Regeneration . *Bioengineering (Basel, Switzerland)*, *9*(5), 223. PMID: 35621501

Shang, L., Ma, B., & Wang, F.. (2020). Nanotextured silk fibroin/hydroxyapatite biomimetic bilayer tough structure regulated osteogenic/chondrogenic differentiation of mesenchymal stem cells for osteochondral repair . *Cell Proliferation*, *53*(11), e12917. PMID: 33001510

Shan, H., Zhou, X., & Tian, B.. (2022). Gold nanorods modified by endogenous protein with light-irradiation enhance bone repair via multiple osteogenic signal pathways . *Biomaterials*, *284*, 121482. PMID: 35358870

Shanmuganathan, R., & Edison, T. N. J. I., LewisOscar, F., Kumar, P., Shanmugam, S., & Pugazhendhi, A. (2019a). Chitosan nanopolymers: An overview of drug delivery against cancer. *International Journal of Biological Macromolecules*, *130*, 727–736. DOI: 10.1016/j.ijbiomac.2019.02.060 PMID: 30771392

Shan, X. Y. (2020). Effect of chitosan magnetic nanoparticles loaded with Ang2-siRNA plasmids on the growth of melanoma xenografts in nude mice. *Cancer Management and Research*, ●●●, 7475–7485. PMID: 32904466

Shao, X., Lin, S., Peng, Q., ... Tetrahedral DNA Nanostructure: A Potential Promoter for Cartilage Tissue Regeneration via Regulating Chondrocyte Phenotype and Proliferation. Small (Weinheim an Der Bergstrasse, Germany), 2017, 13(12).

Sharma, A., Debik, J., Naume, B., Ohnstad, H. O., Sahlber, K. K., Borgen, E., Børresen-Dale, A. L., Engebråten, O., Fritzman, B., Garred, Ø., Geisler, J., Geitvik, G. A., Hofvind, S., Kristensen, V. N., Kåresen, R., Langerød, A., Lingjærde, O. C., Mælandsmo, G. M., Russnes, H. G., … Giskeødegård, G. F. (2024). Comprehensive multi-omics analysis of breast cancer reveals distinct long-term prognostic subtypes. Oncogenesis 2024 13:1, 13(1), 1–13. https://doi.org/DOI: 10.1038/s41389-024-00521-6

Sharma, A. T. (2019). Commercial application of cellulose nano-composites–A review. *Biotechnology Reports (Amsterdam, Netherlands)*, *21*, e00316. PMID: 30847286

Sharma, L., Khurana, D., Patel, P., Khare, S., & Kurmi, B. D. (2024). Advancements in Micellar Formulation: Drug Delivery Vehicle for Water-Insoluble Drugs. *Recent Advances in Drug Delivery and Formulation: Formerly Recent Patents on Drug Delivery & Formulation*, *18*(3), 188–207.

Sharma, S. K., Gajević, S., Sharma, L. K., Pradhan, R., Miladinović, S., Ašonja, A., & Stojanović, B. (2024). Magnesium-Titanium Alloys: A Promising Solution for Biodegradable Biomedical Implants. *Materials (Basel)*, *17*(21), 5157. PMID: 39517433

Shayesteh, O. H., & Ghavami, R. (2020). A novel label-free colorimetric aptasensor for sensitive determination of PSA biomarker using gold nanoparticles and a cationic polymer in human serum. *Spectrochimica Acta. Part A: Molecular and Biomolecular Spectroscopy*, *226*, 117644. PMID: 31614271

Sheikhpour, M., Barani, L., & Kasaeian, A. (2017). Biomimetics in drug delivery systems: A critical review. *Journal of Controlled Release*, *253*, 97–109. PMID: 28322976

Sheikh, Z., Najeeb, S., Khurshid, Z., Verma, V., Rashid, H., & Glogauer, M. (2015). Biodegradable Materials for Bone Repair and Tissue Engineering Applications. *Materials (Basel)*, *8*(9), 5744–5794. DOI: 10.3390/ma8095273 PMID: 28793533

Shettiwar, A., Gupta, U., Paul, P., Nair, R., Aalhate, M., Mahajan, S., Maji, I., Guru, S. K., Mehra, N. K., & Singh, P. K. (2023). A comprehensive review of the biomaterial-based multifunctional nanocarriers for therapeutic applications in breast cancer. In Journal of Drug Delivery Science and Technology (Vol. 89). https://doi.org/DOI: 10.1016/j.jddst.2023.104990

Sheykhhasan, M., Amini, R., Soleimani Asl, S., Saidijam, M., Hashemi, S. M., & Najafi, R. (2022). Neuroprotective effects of coenzyme Q10-loaded exosomes obtained from adipose-derived stem cells in a rat model of Alzheimer's disease. *Biomedicine and Pharmacotherapy*, *152*, 113224. https://doi.org/https://doi.org/10.1016/j.biopha.2022.113224 PMID: 35679720

Shi, J., Zhou, Z., Du, X., Cavagnaro, M. J., & Cai, J. (2024). New insights and perspectives on traumatic brain injury: integration, translation and multidisciplinary approaches. In (Vol. 15, pp. 1427320): Frontiers Media SA.

Shi, H., Magaye, R., Castranova, V., & Zhao, J. (2013). Titanium dioxide nanoparticles: A review of current toxicological data. *Particle and Fibre Toxicology*, *10*, 1–33. PMID: 23587290

Shi, Q., Chen, J., Chen, J., Liu, Y., & Wang, H. (2024). Application of additively manufactured bone scaffold: A systematic review. *Biofabrication*. PMID: 38507799

Shkotova, L., Bohush, A., Voloshina, I., Smutok, O., & Dzyadevych, S. (2019). Amperometric biosensor modified with platinum and palladium nanoparticles for detection of lactate concentrations in wine. *SN Applied Sciences*, *1*(4). Advance online publication. DOI: 10.1007/s42452-019-0315-9

Siahmansouri, H. S.-N. (2016). Effects of HMGA2 siRNA and doxorubicin dual delivery by chitosan nanoparticles on cytotoxicity and gene expression of HT-29 colorectal cancer cell line. *The Journal of Pharmacy and Pharmacology*, *68*(9), 1119–1130. PMID: 27350211

Siddiqi, K. S., & Husen, A. (2016). Green Synthesis, Characterization and Uses of Palladium/Platinum Nanoparticles. *Nanoscale Research Letters*, *11*(1). Advance online publication. DOI: 10.1186/s11671-016-1695-z PMID: 27807824

Siddiqi, K. S., Husen, A., Sohrab, S. S., & Yassin, M. O. (2018). Recent Status of Nanomaterial Fabrication and Their Potential Applications in Neurological Disease Management. *Nanoscale Research Letters*, *13*(1), 231. DOI: 10.1186/s11671-018-2638-7 PMID: 30097809

Siddique, S., & Chow, J. C. (2020). Gold nanoparticles for drug delivery and cancer therapy. *Applied Sciences (Basel, Switzerland)*, *10*(11), 3824. DOI: 10.3390/app10113824

Sidoryk-Węgrzynowicz, M., Adamiak, K., & Strużyńska, L. (2024). Targeting Protein Misfolding and Aggregation as a Therapeutic Perspective in Neurodegenerative Disorders. *International Journal of Molecular Sciences*, *25*(22), 12448. PMID: 39596513

Sierra-Sánchez, Á., Kim, K. H., Blasco-Morente, G., & Arias-Santiago, S. (2021). Cellular human tissue-engineered skin substitutes investigated for deep and difficult to heal injuries. *NPJ Regenerative Medicine*, *6*(1), 35. PMID: 34140525

Sierri, G., Saenz-de-Santa-Maria, I., Renda, A., Koch, M., Sommi, P., Anselmi-Tamburini, U., Mauri, M., D'Aloia, A., Ceriani, M., & Salerno, D. (2025). Nanoparticle shape is the game-changer for blood–brain barrier crossing and delivery through tunneling nanotubes among glioblastoma cells. *Nanoscale*. PMID: 39588728

Sies, H. (1993). Strategies of antioxidant defense. *European Journal of Biochemistry*, *215*(2), 213–219. PMID: 7688300

Sies, H., Berndt, C., & Jones, D. P. (2017). Oxidative stress. *Annual Review of Biochemistry*, *86*(1), 715–748. PMID: 28441057

Silva Adaya, D., Aguirre-Cruz, L., Guevara, J., & Ortiz-Islas, E. (2017). Nanobiomaterials' applications in neurodegenerative diseases. *Journal of Biomaterials Applications*, *31*(7), 953–984. DOI: 10.1177/0885328216659032 PMID: 28178902

Silva, A. C. (2024). The Role of Inhaled Chitosan-Based Nanoparticles in Lung Cancer Therapy. *Pharmaceutics*, *16*(8), 969. PMID: 39204314

Si, M., & Lang, J. (2018). The roles of metallothioneins in carcinogenesis. *Journal of Hematology & Oncology*, *11*, 1–20. DOI: 10.1186/s13045-018-0645-x PMID: 30139373

Sindhusha, V. B., & Doraiswamy, J. N. (2024). The Role of Chitosan and Gelatin-Based Scaffolds in Bone Regeneration: A Systematic Review. *Cureus*, *16*(9), e69793. PMID: 39435228

Singer, A. J., & Clark, R. A. (1999). Cutaneous wound healing . *The New England Journal of Medicine*, *341*(10), 738–746. PMID: 10471461

Singh Tuli, H., Kaur, P., Nene, A. G., Sharma, D., & Somani, P. R. (2019). Synergistic effect of copper nanoparticles and antibiotics to enhance antibacterial potential Bio-Materials and Technology Synergistic effect of copper nanoparticles and antibiotics to enhance antibacterial potential. In Bio- []). Retrieved from www.bio-mat-tech.com]. *Materials Technology*, *1*(1)

Singh, R. P., Choi, J. W., Tiwari, A., & Pandey, A. C. (2014). Functional nanomaterials for multifarious nanomedicine. *Biosensors nanotechnology*, 141-197.

Singh, D., Tirkey, R., Jangde, R., Sahu, P. K., & Bagchi, A. (2024). *Introduction to Biomaterial-Inspired*. Biomaterial-Inspired Nanomedicines for Targeted Therapies.

Singh, H., Desimone, M. F., Pandya, S., Jasani, S., George, N., Adnan, M., Aldarhami, A., Bazaid, A. S., & Alderhami, S. A. (2023). Revisiting the green synthesis of nanoparticles: Uncovering influences of plant extracts as reducing agents for enhanced synthesis efficiency and its biomedical applications. *International Journal of Nanomedicine*, *18*, 4727–4750. DOI: 10.2147/ijn.s419369 PMID: 37621852

Singh, J., Hashmi, A. W., Ahmad, S., & Tian, Y. (2024). Critical review on biodegradable and biocompatibility magnesium alloys: Progress and prospects in bio-implant applications. *Inorganic Chemistry Communications*, •••, 113111.

Singh, J., & Nayak, P. (2023). pH-responsive polymers for drug delivery: Trends and opportunities. *Journal of Polymer Science*, *61*(22), 2828–2850. DOI: 10.1002/pol.20230403

Singh, M., Thakur, V., Kumar, V., Raj, M., Gupta, S., Devi, N., & Saurav, K. (2022). Silver nanoparticles and its mechanistic insight for chronic wound healing: Review on recent progress. *Molecules (Basel, Switzerland)*, *27*(17), 5587. DOI: 10.3390/molecules27175587 PMID: 36080353

Singh, R., Yadav, D., Ingole, P. G., & Ahn, Y.-H. (2024). Magnetic engineering nanoparticles: Versatile tools revolutionizing biomedical applications. *Biomaterials Advances*, •••, 213948. PMID: 38959651

Singh, V. K., Singh, V. K., Mishra, A., Singh, A. A., Prasad, G., & Singh, A. K. (2023). Recent advancements in coordination compounds and their potential clinical application in the management of diseases: An up-to-date review. *Polyhedron*, *241*, 116485. DOI: 10.1016/j.poly.2023.116485

Sirelkhatim, A., Mahmud, S., Seeni, A., Kaus, N. H. M., Ann, L. C., Bakhori, S. K. M., & Mohamad, D. (2015). Review on zinc oxide nanoparticles: Antibacterial activity and toxicity mechanism. *Nano-Micro Letters*, *7*, 219–242. DOI: 10.1007/s40820-015-0040-x PMID: 30464967

Skrajnowska, D., & Bobrowska-Korczak, B. (2019). Role of zinc in immune system and anti-cancer defense mechanisms. *Nutrients*, *11*(10), 2273. DOI: 10.3390/nu11102273 PMID: 31546724

Škrlová, K., Malachová, K., Muñoz-Bonilla, A., Měřinská, D., Rybková, Z., Fernández-García, M., & Plachá, D. (2019). Biocompatible polymer materials with antimicrobial properties for preparation of stents. *Nanomaterials (Basel, Switzerland)*, *9*(11), 1548. PMID: 31683612

Slivnik, M., Navodnik Preložnik, M., Fir, M., Jazbar, J., Čebron Lipovec, N., Locatelli, I., Liette Lauzon, H., & Urbančič Rovan, V. (2024). A randomized, placebo-controlled study of chitosan gel for the treatment of chronic diabetic foot ulcers (the CHITOWOUND study). *BMJ Open Diabetes Research & Care*, *12*(3), e004195. DOI: 10.1136/bmjdrc-2024-004195 PMID: 38909998

Smith, T., Edmonds, C., & Barnaby, C. (1972). Absorption and retention of cobalt in man by whole-body counting. *Health Physics*, *22*(4), 359–367. PMID: 4625611

Snyder, C. M., Mateo, B., Patel, K., Fahrenholtz, C. D., Rohde, M. M., Carpenter, R., & Singh, R. N. (2024). Enhancement of Triple-Negative Breast Cancer-Specific Induction of Cell Death by Silver Nanoparticles by Combined Treatment with Proteotoxic Stress Response Inhibitors. *Nanomaterials (Basel, Switzerland)*, *14*(19), 1564. DOI: 10.3390/NANO14191564 PMID: 39404291

Sohrabi, H. R., Greenberg, S. M., & Whiley, L. (2024). Cerebral amyloid angiopathy: From bench to bedside. *Frontiers in Neuroscience*, *18*, 1370352. PMID: 38379758

Solangi, N. H., Karri, R. R., Mubarak, N. M., & Mazari, S. A. (2023). *Mechanism of polymer composite-based nanomaterial for biomedical applications*. Advanced Industrial and Engineering Polymer Research.

Soleymani, F., Emadi, R., Sadeghzade, S., & Tavangarian, F. (2020). Bioactivity Behavior Evaluation of PCL-Chitosan-Nanobaghdadite Coating on AZ91 Magnesium Alloy in Simulated Body Fluid. *Coatings*, *10*, 231. DOI: 10.3390/coatings10030231

Song, X., Ding, Q., Wei, W., Zhang, J., Sun, R., Yin, L., Liu, S., & Pu, Y. (2023). Peptide-Functionalized Prussian Blue Nanomaterial for Antioxidant Stress and NIR Photothermal Therapy against Alzheimer's Disease. *Small*, *19*(41), 2206959. PMID: 37322406

Song, Z., Cheng, Y., Chen, M., & Xie, X. (2023). Macrophage polarization in bone implant repair: A review. *Tissue & Cell*, *82*, 102112. PMID: 37257287

Soni, S. K., & Solanki, M. K. (2023). CHEMICAL POTENTIAL AND BIOMEDICAL ACTIVITY OF GOLD NANOPARTICLES IN CANCER AND DRUG DELIVERY SYSTEMS: AN UPDATE. *Vietnam Journal of Science and Technology*.

Soni, S., Ruhela, R. K., & Medhi, B. (2016). Nanomedicine in Central Nervous System (CNS) Disorders: A Present and Future Prospective. *Advanced Pharmaceutical Bulletin*, *6*(3), 319–335. DOI: 10.15171/apb.2016.044 PMID: 27766216

Sood, A., Desseigne, M., Dev, A., Maurizi, L., Kumar, A., Millot, N., & Han, S. S. (2023). A comprehensive review on barium titanate nanoparticles as a persuasive piezoelectric material for biomedical applications: Prospects and challenges. *Small*, *19*(12), 2206401. DOI: 10.1002/smll.202206401 PMID: 36585372

Soppimath, K. S., Aminabhavi, T. M., Kulkarni, A. R., & Rudzinski, W. E. (2001). Biodegradable polymeric nanoparticles as drug delivery devices. *Journal of Controlled Release*. Advance online publication. DOI: 10.1016/S0168-3659(00)00339-4 PMID: 11166403

Souri, M., Soltani, M., Kashkooli, F. M., & Shahvandi, M. K. (2022). Engineered strategies to enhance tumor penetration of drug-loaded nanoparticles. *Journal of Controlled Release*, *341*, 227–246. PMID: 34822909

Spiewak, R. (2018). Assessment for metal allergy: patch testing. *Metal Allergy: From Dermatitis to Implant and Device Failure*, 107-124.

Spillantini, M. G., & Goedert, M. (2000). The α-synucleinopathies: Parkinson's disease, dementia with Lewy bodies, and multiple system atrophy. *Annals of the New York Academy of Sciences*, *920*(1), 16–27. PMID: 11193145

Spuch, C., & Navarro, C. (2011). Liposomes for Targeted Delivery of Active Agents against Neurodegenerative Diseases (Alzheimer's Disease and Parkinson's Disease). *Journal of Drug Delivery*, *469679*. Advance online publication. DOI: 10.1155/2011/469679 PMID: 22203906

Srinageshwar, B., Peruzzaro, S., Andrews, M., Johnson, K., Hietpas, A., Clark, B., McGuire, C., Petersen, E., Kippe, J., Stewart, A., Lossia, O., Al-Gharaibeh, A., Antcliff, A., Culver, R., Swanson, D., Dunbar, G., Sharma, A., & Rossignol, J. (2017). PAMAM Dendrimers Cross the Blood-Brain Barrier When Administered through the Carotid Artery in C57BL/6J Mice. *International Journal of Molecular Sciences*, *18*(3). Advance online publication. DOI: 10.3390/ijms18030628 PMID: 28335421

Šromová, V., Sobola, D., & Kaspar, P. (2023). A Brief Review of Bone Cell Function and Importance. In *Cells* (Vol. 12, Issue 21, p. 2576). Multidisciplinary Digital Publishing Institute. https://doi.org/DOI: 10.3390/cells12212576

Steinmetz, J. D., Seeher, K. M., Schiess, N., Nichols, E., Cao, B., Servili, C., Cavallera, V., Cousin, E., Hagins, H., & Moberg, M. E. (2024). Global, regional, and national burden of disorders affecting the nervous system, 1990–2021: A systematic analysis for the Global Burden of Disease Study 2021. *Lancet Neurology*, *23*(4), 344–381. PMID: 38493795

Stojanovic, M., Rai, V., & Agrawal, D. K. (2024). Effect of electromagnetic field on proliferation and migration of fibroblasts and keratinocytes: Implications in wound healing and regeneration. *Journal of Biotechnology and Biomedicine*, *7*(3), 387. DOI: 10.26502/jbb.2642-91280162 PMID: 39364330

Strauss, A., Rommelspacher, Y., Nouri, B., Bornemann, R., Wimmer, M., Oldenburg, J., Pennekamp, P., & Schmolders, J. (2017). Long-term outcome of total hip arthroplasty in patients with haemophilia. *Haemophilia*, *23*(1), 129–134. PMID: 27480604

Strebhardt, K. (2008). Paul Ehrlich's magic bullet concept: 100 years of progress. *Nature Reviews. Cancer*, *8*(6), 473–480. PMID: 18469827

Subhan, M. A., & Torchilin, V. P. (2023a). Biopolymer-Based Nanosystems for siRNA Drug Delivery to Solid Tumors including Breast Cancer. Pharmaceutics 2023, Vol. 15, Page 153, 15(1), 153. https://doi.org/DOI: 10.3390/PHARMACEUTICS15010153

Subhan, M. A. (2023). Biopolymer-based nanosystems for siRNA drug delivery to solid tumors including breast cancer. *Pharmaceutics, 15*(1), 153. PMID: 36678782

Sudha, P. N., Sangeetha, K., Vijayalakshmi, K., & Barhoum, A. (2018). Nanomaterials history, classification, unique properties, production and market. In *Emerging applications of nanoparticles and architecture nanostructures* (pp. 341–384). Elsevier.

Su, H., Fujiwara, T., Anderson, K. M., Karydis, A., Ghadri, M. N., & Bumgardner, J. D. (2021). A comparison of two types of electrospun chitosan membranes and a collagen membrane in vivo. *Dental Materials, 37*(1), 60–70. DOI: 10.1016/j.dental.2020.10.011 PMID: 33208266

Suhag, D. K. (2024). Scaffold Design in Tissue Engineering. In Handbook of Biomaterials for Medical Applications. *Singapore: Springer Nature Singapore.*, Volume 1: Fundamentals. pp. 155-184.

Suhag, D., Kaushik, S., & Taxak, V. B. (2024). Scaffold Design in Tissue Engineering. In *Handbook of Biomaterials for Medical Applications, Volume 1: Fundamentals* (pp. 155-184). Springer.

Suk, J. S., Xu, Q., Kim, N., Hanes, J., & Ensign, L. M. (2016). "PEGylation as a Strategy for Improving Nanoparticle-Based Drug and Gene Delivery." *Advanced Drug Delivery Reviews. Advanced Drug Delivery Reviews.* Advance online publication. DOI: 10.1016/j.addr.2015.09.012 PMID: 26456916

Sulheim, E., Baghirov, H., von Haartman, E., Bøe, A., Åslund, A. K., Mørch, Y., & Davies Cde, L. (2016). Cellular uptake and intracellular degradation of poly(alkyl cyanoacrylate) nanoparticles. *Journal of Nanobiotechnology, 14*, 1. DOI: 10.1186/s12951-015-0156-7 PMID: 26743777

Summer, M., Ashraf, R., Ali, S., Bach, H., Noor, S., Noor, Q., & Khan, R. R. M. (2024). Inflammatory response of nanoparticles: Mechanisms, Consequences, and Strategies for Mitigation. *Chemosphere*, •••, 142826. PMID: 39002651

Sun, D., Feng, X., Zhu, X., Wang, Y., & Yang, J. (2024). Anticancer agents based on metal organic cages. *Coordination Chemistry Reviews, 500*, 215546. DOI: 10.1016/J.CCR.2023.215546

Sundfeldt, M., v Carlsson, L. B., Johansson, C., Thomsen, P., & Gretzer, C. (2006). Aseptic loosening, not only a question of wear: A review of different theories. *Acta Orthopaedica, 77*(2), 177–197. PMID: 16752278

Sung, H., Ferlay, J., Siegel, R. L., Laversanne, M., Soerjomataram, I., Jemal, A., & Bray, F. (2021a). Global Cancer Statistics 2020: GLOBOCAN Estimates of Incidence and Mortality Worldwide for 36 Cancers in 185 Countries. *CA: a Cancer Journal for Clinicians, 71*(3), 209–249. DOI: 10.3322/CAAC.21660 PMID: 33538338

Sun, P., Jiao, J., & Wang, X.. (2023). Nanomedicine hybrid and catechol functionalized chitosan as pH-responsive multi-function hydrogel to efficiently promote infection wound healing . *International Journal of Biological Macromolecules, 238*, 124106. PMID: 36948329

Sun, T., Zhang, Y. S., Pang, B., Hyun, D. C., Yang, M., & Xia, Y. (2014). Engineered nanoparticles for drug delivery in cancer therapy. *Angewandte Chemie International Edition in English, 53*(46), 12320–12364. DOI: 10.1002/anie.201403036 PMID: 25294565

Sun, T., Zhou, D., Mao, F., & Zhu, Y. (2007). Preparation of low-molecular-weight carboxymethyl chitosan and their superoxide anion scavenging activity. *European Polymer Journal, 43*(2), 652–656. DOI: 10.1016/j.eurpolymj.2006.11.014

Sun, Y., Xiao, Z., Chen, B., Zhao, Y., & Dai, J. (2024). Advances in material-assisted electromagnetic neural stimulation. *Advanced Materials, 36*, 2400346. DOI: 10.1002/adma.202400346 PMID: 38594598

Su, Y., Toftdal, M. S., Le Friec, A., Dong, M., Han, X., & Chen, M. (2021). 3D electrospun synthetic extracellular matrix for tissue regeneration. *Small Science, 1*(7), 2100003. DOI: 10.1002/smsc.202100003 PMID: 40213056

Szczyglewska, P., Feliczak-Guzik, A., & Nowak, I. (2023). Nanotechnology–general aspects: A chemical reduction approach to the synthesis of nanoparticles. *Molecules (Basel, Switzerland), 28*(13), 4932. PMID: 37446593

Talarska, P., Boruczkowski, M., & Zurawski, J. (2021). Current knowledge of Silver and Gold Nanoparticles in Laboratory Research Application, toxicity, Cellular Uptake. *Nanomaterials (Basel, Switzerland), 11*(9), 2454. DOI: 10.3390/nano11092454 PMID: 34578770

Talebian, S., Foroughi, J., Wade, S. J., Vine, K. L., Dolatshahi-Pirouz, A., Mehrali, M., Conde, J., Wallace, G. G., Talebian, S., & Foroughi, J. (n.d.). Article Type: Review Biopolymers for Anti-Tumor Implantable Drug Delivery Systems: Recent Advances and Future Outlook. https://doi.org/DOI: 10.1002/adma.201706665

Tanaka, M., Haniu, H., Kamanaka, T., Takizawa, T., Sobajima, A., Yoshida, K., & Saito, N. (2017). Physico-chemical, in vitro, and in vivo evaluation of a 3D unidirectional porous hydroxyapatite scaffold for bone regeneration. *Materials (Basel), 10*(1), 33. PMID: 28772390

Tang, X., Haddad, P., Mager, N., Geng, X., Reckinger, N., Hermans, S., Debliquy, M., & Raskin, J. (2019). Chemically deposited palladium nanoparticles on graphene for hydrogen sensor applications. *Scientific Reports, 9*(1). Advance online publication. DOI: 10.1038/s41598-019-40257-7 PMID: 30842583

Tang, X., Kurban, M., Hafiz, I., Shen, Q., & Wang, M. (2023). Preparation of hyaluronic acid-loaded Harmine polymeric micelles and in vitro effect anti-breast cancer. *European Journal of Pharmaceutical Sciences, 183*, 106388. DOI: 10.1016/J.EJPS.2023.106388 PMID: 36758771

Tan, M. L., Shao, P., Friedhuber, A. M., van Moorst, M., Elahy, M., Indumathy, S., Dunstan, D. E., Wei, Y., & Dass, C. R. (2014). The potential role of free chitosan in bone trauma and bone cancer management. *Biomaterials, 35*(27), 7828–7838. DOI: 10.1016/J.BIOMATERIALS.2014.05.087 PMID: 24947230

Tao, J., Zhang, Y., Shen, A., Yang, Y., Diao, L., Wang, L., Cai, D., & Hu, Y. (2020). Injectable Chitosan-Based Thermosensitive Hydrogel/Nanoparticle-Loaded System for Local Delivery of Vancomycin in the Treatment of Osteomyelitis. *International Journal of Nanomedicine, 15*, 5855–5871. DOI: 10.2147/IJN.S247088 PMID: 32848394

Taraballi, F., Sushnitha, M., Tsao, C., Bauza, G., Liverani, C., Shi, A., & Tasciotti, E. (2018). Biomimetic tissue engineering: Tuning the immune and inflammatory response to implantable biomaterials. *Advanced Healthcare Materials, 7*(17), 1800490. PMID: 29995315

Tariq, U., Gupta, M., & Pathak, S.. (2022). Role of Biomaterials in Cardiac Repair and Regeneration: Therapeutic Intervention for Myocardial Infarction . *ACS Biomaterials Science & Engineering*, *8*(8), 3271–3298. PMID: 35867701

Teixeira, A. M., & Martins, P. (2023). A review of bioengineering techniques applied to breast tissue: Mechanical properties, tissue engineering and finite element analysis. *Frontiers in Bioengineering and Biotechnology*, *11*. Advance online publication. DOI: 10.3389/FBIOE.2023.1161815 PMID: 37077233

Teleanu, D. M., Chircov, C., Grumezescu, A. M., Volceanov, A., & Teleanu, R. I. (2018). Impact of nanoparticles on brain health: An up to date overview. *Journal of Clinical Medicine*, *7*(12), 490. PMID: 30486404

Teleanu, D. M., Negut, I., Grumezescu, V., Grumezescu, A. M., & Teleanu, R. I. (2019). Nanomaterials for Drug Delivery to the Central Nervous System. *Nanomaterials (Basel, Switzerland)*, *9*(3). Advance online publication. DOI: 10.3390/nano9030371 PMID: 30841578

Teng, Y., Luo, C., Qiu, X., Mu, J., Sriwastva, M. K., Xu, Q., Liu, M., Hu, X., Xu, F., Zhang, L., Park, J. W., Hwang, J. Y., Kong, M., Liu, Z., Zhang, X., Xu, R., Yan, J., Merchant, M. L., McClain, C. J., & Zhang, H. (2025). Plant-nanoparticles enhance anti-PD-L1 efficacy by shaping human commensal microbiota metabolites. *Nature Communications*, *16*(1). Advance online publication. DOI: 10.1038/s41467-025-56498-2 PMID: 39900923

Terzi, E. M., & Possemato, R. (2024). Iron, Copper, and Selenium: Cancer's Thing for Redox Bling. *Cold Spring Harbor Perspectives in Medicine*, *14*(4), a041545. DOI: 10.1101/cshperspect.a041545 PMID: 37932129

Thakor, P. B. (2020). Polymer–drug conjugates: Recent advances and future perspectives. *Drug Discovery Today*, *25*(9), 1718–1726. PMID: 32629170

Thapa, R. K., Soe, Z. C., Ou, W., Poudel, K., Jeong, J., Jin, S. G., Ku, S. K., Choi, H., Lee, Y. M., Yong, C. S., & Kim, J. O. (2018). Palladium nanoparticle-decorated 2-D graphene oxide for effective photodynamic and photothermal therapy of prostate solid tumors. *Colloids and Surfaces. B, Biointerfaces*, *169*, 429–437. DOI: 10.1016/j.colsurfb.2018.05.051 PMID: 29843117

Thirunavukkarasu, G. K. (2018). Magnetic field-inducible drug-eluting nanoparticles for image-guided thermo-chemotherapy. *Biomaterials*, *180*, 240–252. PMID: 30055399

Thomas, A. P., Kasa, V. P., Dubey, B. K., Sen, R., & Sarmah, A. K. (2023). *"Synthesis and Commercialization of Bioplastics: Organic Waste as a Sustainable Feedstock." Science of the Total Environment*. Pergamon., DOI: 10.1016/j.scitotenv.2023.167243

Thomas, K. K., Zafar, M. N., Pitt, W. G., & Husseini, G. A. (2024). Biodegradable Magnesium Alloys for Biomedical Implants: Properties, Challenges, and Surface Modifications with a Focus on Orthopedic Fixation Repair. *Applied Sciences (Basel, Switzerland)*, *14*(1). Advance online publication. DOI: 10.3390/app14010010

Thrivikraman, G., Boda, S. K., & Basu, B. (2018). Unraveling the mechanistic effects of electric field stimulation towards directing stem cell fate and function: A tissue engineering perspective. *Biomaterials*, *150*, 60–86. DOI: 10.1016/j.biomaterials.2017.10.003 PMID: 29032331

Tian, Y., Wu, D., Wu, D., Cui, Y., Ren, G., Wang, Y., & Peng, C. (2022). Chitosan-based biomaterial scaffolds for the repair of infected bone defects. *Frontiers in Bioengineering and Biotechnology, 10,* 899760. DOI: 10.3389/fbioe.2022.899760 PMID: 35600891

Tığlı, R. S., & Gümüşderelioğlu, M. (2009). Evaluation of alginate-chitosan semi IPNs as cartilage scaffolds. *Journal of Materials Science. Materials in Medicine, 20*(3), 699–709. DOI: 10.1007/s10856-008-3624-x PMID: 18987950

Tipan, N., Pandey, A., & Mishra, P. (2024). Material synthesis and design optimization of biomaterials for biomedical implant applications. *BioImpacts*. PMID: 40161937

Tiwari, N. A. (2024c). Prediction model for hardness and tensile strength of graphene reinforced AZ 61 alloy based composite using Metaheuristic Algorithm. *Global Journal of Engineering and Technology Advances, 20*(3), 042–052. https://doi.org/DOI: 10.30574/gjeta.2024.20.3.0167

Tiwari, A., Kumar, N., & Banerjee, M. (2023a). Applications of genetic algorithm in prediction of the best achievable combination of hardness and tensile strength for graphene reinforced magnesium alloy (AZ61) matrix composite. *Results in Control and Optimization, 14,* 100334. DOI: 10.1016/j.rico.2023.100334

Tiwari, A., Kumar, N., & Banerjee, M. K. (2023c). Effect of Hot Rolling on Microstructure and Mechanical Properties of Stir Cast AZ 61 Alloy with Minor Additions. *Indian Journal of Science and Technology, 16*(24), 1810–1822. DOI: 10.17485/ijst/v16i24.302

Toader, C., Dumitru, A. V., Eva, L., Serban, M., Covache-Busuioc, R.-A., & Ciurea, A. V. (2024). Nanoparticle Strategies for Treating CNS Disorders: A Comprehensive Review of Drug Delivery and Theranostic Applications. *International Journal of Molecular Sciences, 25*(24), 13302. PMID: 39769066

Tolaimate, A., Desbrieres, J., Rhazi, M., & Alagui, A. (2003). Contribution to the preparation of chitins and chitosans with controlled physico-chemical properties. *Polymer, 44*(26), 7939–7952. DOI: 10.1016/j.polymer.2003.10.025

Tong, X., Han, Y., Zhou, R., Zeng, J., Wang, C., Yuan, Y., & Lin, J. (2023). Mechanical properties, corrosion and degradation behaviors, and in vitro cytocompatibility of a biodegradable Zn–5La alloy for bone-implant applications. *Acta Biomaterialia, 169,* 641–660. PMID: 37541605

Toosi, S. (2020). Osteogenesis and bone remodeling: A focus on growth factors and bioactive peptides. *BioFactors (Oxford, England), 46*(3), 326–340. PMID: 31854489

Torres Quintas, S., Canha-Borges, A., Oliveira, M. J., Sarmento, B., & Castro, F. (2023). Special Issue: Nanotherapeutics in Women's Health Emerging Nanotechnologies for Triple-Negative Breast Cancer Treatment. In Small. https://doi.org/DOI: 10.1002/smll.202300666

Tottoli, E M, Dorati R, Genta I. *et al.* (2020). Skin Wound Healing Process and New Emerging Technologies for Skin Wound Care and Regeneration . *Pharmaceutics, 12*(8), 735. PMID: 32764269

Tran, D. T., Yadav, A. S., Nguyen, N. K., Singha, P., Ooi, C. H., & Nguyen, N. T. (2023). *"Biodegradable Polymers for Micro Elastofluidics." Small.* John Wiley and Sons Inc., DOI: 10.1002/smll.202303435

Tria, A. J.Jr. (2013). Bicompartmental knee arthroplasty. *Volume 44, Issue 3. An Issue of Orthopedic Clinics, 44*(3), 17.

Tripathi, D., Mani, V., & Pal, R. P. (2018). Vanadium in biosphere and its role in biological processes. *Biological Trace Element Research, 186*, 52–67. PMID: 29524196

Trivedi, V., Satia, M., Deschamps, A., Maquet, V., Shah, R., Zinzuwadia, P., & Trivedi, J. (2015). Single-blind, placebo controlled randomised clinical study of chitosan for body weight reduction. *Nutrition Journal, 15*(1), 3. DOI: 10.1186/s12937-016-0122-8 PMID: 26747458

Trucillo, P. (2024). *"Biomaterials for Drug Delivery and Human Applications." Materials. Multidisciplinary Digital Publishing Institute.* MDPI., DOI: 10.3390/ma17020456

Tsukamoto, S., Mavrogenis, A. F., Akahane, M., Honoki, K., Kido, A., Tanaka, Y., Donati, D. M., & Errani, C. (2022). Risk factors of fracture following curettage for bone giant cell tumors of the extremities. *BMC Musculoskeletal Disorders, 23*(1), 477. DOI: 10.1186/s12891-022-05447-x PMID: 35590280

Tulay, P., Galam, N., & Adali, T. (2018). The Wonders of Silk Fibroin Biomaterials in the Treatment of Breast Cancer. Critical Reviews™ in Eukaryotic Gene Expression, 28(2), 129–134. https://doi.org/DOI: 10.1615/CRITREVEUKARYOTGENEEXPR.2018021331

Twal, S., Jaber, N., Al-Remawi, M., Hamad, I., Al-Akayleh, F., & Alshaer, W. (2024). Dual stimuli-responsive polymeric nanoparticles combining soluplus and chitosan for enhanced breast cancer targeting. *RSC Advances, 14*(5), 3070. DOI: 10.1039/D3RA08074A PMID: 38239437

Ueno, H., Yamada, H., Tanaka, I., Kaba, N., Matsuura, M., Okumura, M., Kadosawa, T., & Fujinaga, T. (1999). Accelerating effects of chitosan for healing at early phase of experimental open wound in dogs. *Biomaterials, 20*(15), 1407–1414. DOI: 10.1016/S0142-9612(99)00046-0 PMID: 10454012

Ulbricht Aus Großenhain. M Sc Juliane. 2017. "Insights into Polymer Biodegradation Investigations on Oxidative, Hydrolytic and Enzymatic Pathways." https://opus.bibliothek.uni-wuerzburg.de/frontdoor/index/index/docId/15868

Umapathy, V. R., Natarajan, P. M., & Swamikannu, B. (2024). Therapeutic Potential of Nanomaterial-based Drug Delivery. *International Journal of Drug Delivery Technology, 14*(1), 556–563.

UN. G. perspective H. stories. (2024, February). Breast cancer cases projected to rise by nearly 40 per cent by 2050, WHO warns | UN News. Webpage. https://news.un.org/en/story/2025/02/1160391

Urnukhsaikhan, E., Bold, B. E., Gunbileg, A., Sukhbaatar, N., & Mishig-Ochir, T. (2021). Antibacterial activity and characteristics of silver nanoparticles biosynthesized from Carduus crispus. *Scientific Reports, 11*(1), 21047. DOI: 10.1038/s41598-021-00520-2 PMID: 34702916

Vaiani, L., Boccaccio, A., Uva, A. E., Palumbo, G., Piccininni, A., Guglielmi, P., Cantore, S., Santacroce, L., Charitos, I. A., & Ballini, A. (2023). Ceramic materials for biomedical applications: An overview on properties and fabrication processes. *Journal of Functional Biomaterials, 14*(3), 146. PMID: 36976070

Van Hees, S., Elbrink, K., De Schryver, M., Delputte, P. L., & Kiekens, F. (2020). Improving Cellular Uptake and Cytotoxicity of Chitosan-Coated Poly(Lactic- *Co* -Glycolic Acid) Nanoparticles in Macrophages. *Nanomedicine (London), 15*(27), 2671–2688. DOI: 10.2217/nnm-2020-0317 PMID: 33112210

Van Overschelde, P., Portelange, N., Chausson, M., & Weyenberg, W. (2023). Injection de carboxyméthyl chitosane dans l'arthrose du genou: Données de vie réelle après 133 injections. *Revue du Rhumatisme*, *90*, A248–A249. DOI: 10.1016/J.RHUM.2023.10.372

Vanivska, K., Dianová, L., Halo, M., Štefunková, N., Lenický, M., Slanina, T., & Massányi, P. (2024). Toxicity of Nanoparticles on Animal and Human Organism: Cell Response. *Journal of Microbiology, Biotechnology and Food Sciences*, *14*, e10844.

Varadharajaperumal, P., Subramanian, B., & Santhanam, A. (2017). Biopolymer mediated nanoparticles synthesized from Adenia hondala for enhanced tamoxifen drug delivery in breast cancer cell line. *Advances in Natural Sciences: Nanoscience and Nanotechnology*, *8*(3), 035011. DOI: 10.1088/2043-6254/AA7253

Vashist, A., Kaushik, A., Vashist, A., Sagar, V., Ghosal, A., Gupta, Y. K., Ahmad, S., & Nair, M. (2018). Advances in carbon nanotubes-hydrogel hybrids in nanomedicine for therapeutics. *Advanced Healthcare Materials*, *7*(9), e1701213. DOI: 10.1002/adhm.201701213 PMID: 29388356

Vashist, A., Manickam, P., Raymond, A. D., Arias, A. Y., Kolishetti, N., Vashist, A., Arias, E., & Nair, M. (2023). Recent Advances in Nanotherapeutics for Neurological Disorders. *ACS Applied Bio Materials*, *6*(7), 2614–2621. DOI: 10.1021/acsabm.3c00254 PMID: 37368486

Velhal, K., Barage, S., Roy, A., Lakkakula, J., Yamgar, R., Alqahtani, M. S., Yadav, K. K., Ahn, Y., & Jeon, B. H. (2022). A Promising Review on Cyclodextrin Conjugated Paclitaxel Nanoparticles for Cancer Treatment. Polymers 2022, Vol. 14, Page 3162, 14(15), 3162. https://doi.org/DOI: 10.3390/POLYM14153162

Venkatesan, J., & Kim, S.-K. (2010). Chitosan Composites for Bone Tissue Engineering—An Overview. *Marine Drugs*, *8*(8), 2252–2266. DOI: 10.3390/md8082252 PMID: 20948907

Verma, J., Kumar, C., Sharma, M., & Saxena, S. (2024). Biotechnological advances in microbial synthesis of gold nanoparticles: Optimizations and applications. *3 Biotech, 14*(11), 263.

Verma, A., & Ogata, S. (2023). Magnesium based alloys for reinforcing biopolymer composites and coatings: A critical overview on biomedical materials. *Advanced Industrial and Engineering Polymer Research*, *6*(4), 341–355. DOI: 10.1016/j.aiepr.2023.01.002

Vindigni, V., Marena, F., Zanettin, C., & Bassetto, F. (2024). Breast Reconstruction: The Oncoplastic Approach. Journal of Clinical Medicine 2024, Vol. 13, Page 4718, 13(16), 4718. https://doi.org/DOI: 10.3390/JCM13164718

Vines, J. B., Lim, D. J., & Park, H. (2018). Contemporary Polymer-Based Nanoparticle Systems for Photothermal Therapy. Polymers 2018, Vol. 10, Page 1357, 10(12), 1357. https://doi.org/DOI: 10.3390/POLYM10121357

Vines, J. B., Yoon, J. H., Ryu, N. E., Lim, D. J., & Park, H. (2019). Gold nanoparticles for photothermal cancer therapy. *Frontiers in Chemistry*, *7*, 167. DOI: 10.3389/fchem.2019.00167 PMID: 31024882

Vinod, C., & Jena, S. (2021). Nano-neurotheranostics: Impact of nanoparticles on neural dysfunctions and strategies to reduce toxicity for improved efficacy. *Frontiers in Pharmacology*, *12*, 612692. PMID: 33841144

Vinodhini, S., Vithiya, B. S. M., & Prasad, T. A. (2022). Green synthesis of palladium nanoparticles using aqueous plant extracts and its biomedical applications. *Journal of King Saud University. Science, 34*(4), 102017. DOI: 10.1016/j.jksus.2022.102017

Virmani, T. K. (2023). Amelioration of cancer employing chitosan, its derivatives, and chitosan-based nanoparticles: Recent updates. *Polymers, 15*(13), 2928. PMID: 37447573

Virmani, T., Kumar, G., Sharma, A., Pathak, K., Akhtar, M. S., Afzal, O., & Altamimi, A. S. A. (2023). Amelioration of Cancer Employing Chitosan, Its Derivatives, and Chitosan-Based Nanoparticles: Recent Updates. *Polymers, 15*(13), 2928. DOI: 10.3390/polym15132928 PMID: 37447573

Visai, L., De Nardo, L., Punta, C., Melone, L., Cigada, A., Imbriani, M., & Arciola, C. R. (2011). Titanium oxide antibacterial surfaces in biomedical devices. *The International Journal of Artificial Organs, 34*(9), 929–946. DOI: 10.5301/ijao.5000050 PMID: 22094576

Viscusi, E. R., & Viscusi, A. R. (2020). Blood–brain barrier: Mechanisms governing permeability and interaction with peripherally acting μ-opioid receptor antagonists. *Regional Anesthesia and Pain Medicine, 45*(9), 688–695. PMID: 32723840

Von Blomberg-van der Flier, B., Bruynzeel, D., & Scheper, R. (1989). Impact of 25 years of in vitro testing in allergic contact dermatitis. Current topics in contact dermatitis

Vyas, J., Shah, I., Singh, S., & Prajapati, B. G. (2023). Biomaterials-based additive manufacturing for customized bioengineering in management of otolaryngology: A comprehensive review. *Frontiers in Bioengineering and Biotechnology, 11*, 1234340. DOI: 10.3389/fbioe.2023.1234340 PMID: 37744247

Wadhawan, A., Singh, J., Sharma, H., Handa, S., Singh, G., Kumar, R., Barnwal, R. P., Pal Kaur, I., & Chatterjee, M. (2022). Anticancer Biosurfactant-Loaded PLA–PEG Nanoparticles Induce Apoptosis in Human MDA-MB-231 Breast Cancer Cells. *ACS Omega, 7*(6), 5231. DOI: 10.1021/ACSOMEGA.1C06338 PMID: 35187338

Waidi, Y. O. (2024a). Carbon-Based Technologies for Water Purification. In *Carbon-Based Materials and Environmental Remediation: Graphene, Biochar, and More* (pp. 113-136): IGI Global Scientific Publishing.

WAIDI, Y. O. (2024b). pH Stimuli Responsive Dextran Gated Mesoporous Silica as a Viable Delivery Vehicle.

Waidi, Y. O., & Jain, N. (2024b). Unlocking New Possibilities: Application of MXenes in 3D Bioprinting for Advanced Therapy. *Nanoscale.*

Waidi, Y. O., Jain, N., Barua, R., Das, S., Vannaladsaysy, V., Prasad, A., . . . Datta, S. (2024). 3 Cellulose: biomedical and engineering applications. *Sustainable Bio-Based Composites: Biomedical and Engineering applications, 20*, 43.

Waidi, Y. O., & Jain, N. (2024a). Unlocking new possibilities: Application of MXenes in 3D bioprinting for advanced therapy. *Nanoscale, 16*(43), 20037–20047. PMID: 39405122

Wanat, K. (2020). Biological barriers, and the influence of protein binding on the passage of drugs across them. *Molecular Biology Reports, 47*(4), 3221–3231. PMID: 32140957

Wang, C., Zhang, R., Wei, X., Lv, M., & Jiang, Z. (2020). Metalloimmunology: The metal ion-controlled immunity. *Advances in Immunology*, *145*, 187–241. DOI: 10.1016/BS.AI.2019.11.007 PMID: 32081198

Wang, G., Li, R., Parseh, B., & Du, G. (2021). Prospects and Challenges of Anticancer Agents' Delivery via Chitosan-Based Drug Carriers to Combat Breast Cancer: A Review. *Carbohydrate Polymers*, *268*(May). Advance online publication. DOI: 10.1016/j.carbpol.2021.118192 PMID: 34127212

Wang, H. (2024). A Review of Nanotechnology in microRNA Detection and Drug Delivery. *Cells*, *13*(15), 1277. DOI: 10.3390/CELLS13151277 PMID: 39120308

Wang, H. L., Wang, Z. G., & Liu, S. L. (2022). Lipid Nanoparticles for mRNA Delivery to Enhance Cancer Immunotherapy. *Molecules (Basel, Switzerland)*, *27*(17), 5607. DOI: 10.3390/MOLECULES27175607 PMID: 36080373

Wang, J., Lin, J., & Chen, L.. (2022). Endogenous Electric-Field-Coupled Electrospun Short Fiber via Collecting Wound Exudation . *Advanced Materials (Deerfield Beach, Fla.)*, *34*(9), e2108325. PMID: 34902192

Wang, J., Xu, J., Liu, W., Li, Y., & Qin, L. (2016). Biodegradable Magnesium (Mg) Implantation Does Not Impose Related Metabolic Disorders in Rats with Chronic Renal Failure. *Scientific Reports*, *6*(1). Advance online publication. DOI: 10.1038/srep26341 PMID: 27210744

Wang, L. W., Peng, C. W., Chen, C., & Li, Y. (2015). Quantum dots-based tissue and in vivo imaging in breast cancer researches: Current status and future perspectives. *Breast Cancer Research and Treatment*, *151*(1), 7. DOI: 10.1007/S10549-015-3363-X PMID: 25833213

Wang, L. Z. (2022). Shape memory polymer fibers: Materials, structures, and applications. *Advanced Fiber Materials*, *4*(1), 5–23. PMID: 35359822

Wang, L., Li, Y., & Ren, M.. (2022). pH and lipase-responsive nanocarrier-mediated dual drug delivery system to treat periodontitis in diabetic rats . *Bioactive Materials*, *18*, 254–266. PMID: 35387157

Wang, L., Qiu, Y., Guo, Y., Si, Y., Liu, L., Cao, J., Yu, J., Li, X., Zhang, Q., & Ding, B. (2019). Smart, Elastic, and Nanofiber-Based 3D Scaffolds with Self-Deploying Capability for Osteoporotic Bone Regeneration. *Nano Letters*, *19*(12), 9112–9120. DOI: 10.1021/acs.nanolett.9b04313 PMID: 31765166

Wang, L., Tang, S., Yu, Y., Lv, Y., Wang, A., Yan, X., Li, N., Sha, C., Sun, K., & Li, Y. (2021). Intranasal delivery of temozolomide-conjugated gold nanoparticles functionalized with anti-EphA3 for glioblastoma targeting. *Molecular Pharmaceutics*, *18*(3), 915–927. PMID: 33417456

Wang, Q., Chen, Y., Chen, Y., Lv, J., Ding, H., Huang, J., Huang, J., Huang, Z., Yang, B., & Zhang, W. (2024). Improved cure rate of periprosthetic joint infection through targeted antibiotic therapy based on integrated pathogen diagnosis strategy. *Frontiers in Cellular and Infection Microbiology*, *14*, 1388385.

Wang, S., & Shi, X. (2001). Molecular mechanisms of metal toxicity and carcinogenesis. *Molecular and Cellular Biochemistry*, *222*, 3–9. DOI: 10.1023/A:1017918013293 PMID: 11678608

Wang, S., Yin, N., Li, Y., Xiang, T., Jiang, W., Zhao, X., Liu, W., Zhang, Z., Shi, J., Zhang, K., Guo, X., Si, P., & Liu, J. (2022). Copper-based metal–organic framework impedes triple-negative breast cancer metastasis via local estrogen deprivation and platelets blockade. *Journal of Nanobiotechnology*, *20*(1). Advance online publication. DOI: 10.1186/s12951-022-01520-8 PMID: 35794596

Wang, W., & Xu, D. (1994). Viscosity and flow properties of concentrated solutions of chitosan with different degrees of deacetylation. *International Journal of Biological Macromolecules*, *16*(3), 149–152. DOI: 10.1016/0141-8130(94)90042-6 PMID: 7981162

Wang, X. J., Gao, Y. P., Lu, N. N., Li, W. S., Xu, J. F., Ying, X. Y., Wu, G., Liao, M. H., Tan, C., Shao, L. X., Lu, Y. M., Zhang, C., Fukunaga, K., Han, F., & Du, Y. Z. (2016). Endogenous Polysialic Acid Based Micelles for Calmodulin Antagonist Delivery against Vascular Dementia. *ACS Applied Materials & Interfaces*, *8*(51), 35045–35058. DOI: 10.1021/acsami.6b13052 PMID: 27750011

Wang, X., Gao, Y., Shi, H., Liu, N., Zhang, W., & Li, H. (2016). Influence of the intensity and loading time of direct current electric field on the directional migration of rat bone marrow mesenchymal stem cells. *Frontiers of Medicine*, *10*, 286–296. DOI: 10.1007/s11684-016-0456-9 PMID: 27324024

Wang, X., Guo, W., Li, L., Yu, F., Li, J., Liu, L., Fang, B., & Xia, L. (2021). Photothermally triggered biomimetic drug delivery of Teriparatide via reduced graphene oxide loaded chitosan hydrogel for osteoporotic bone regeneration. *Chemical Engineering Journal*, *413*, 127413. DOI: 10.1016/J.CEJ.2020.127413

Wang, X., Shan, M., & Zhang, S.. (2022). Stimuli-Responsive Antibacterial Materials: Molecular Structures, Design Principles, and Biomedical Applications . *Advanced Science (Weinheim, Baden-Wurttemberg, Germany)*, *9*(13), e2104843. PMID: 35224893

Wang, X., Teng, Z., Wang, H., Wang, C., Liu, Y., Tang, Y., Wu, J., Sun, J., Wang, H., Wang, J., & Lu, G. (2014). Increasing the cytotoxicity of doxorubicin in breast cancer MCF-7 cells with multidrug resistance using a mesoporous silica nanoparticle drug delivery system. *International Journal of Clinical and Experimental Pathology*, *7*(4), 1337. https://pmc.ncbi.nlm.nih.gov/articles/PMC4014214/ PMID: 24817930

Wang, X., Wang, C., Chu, C., Xue, F., Li, J., & Bai, J. (2024). Structure-function integrated biodegradable Mg/polymer composites: Design, manufacturing, properties, and biomedical applications. *Bioactive Materials*, *39*, 74–105. DOI: 10.1016/j.bioactmat.2024.05.024 PMID: 38783927

Wang, X, & Yu, Y Y, LIEU S. *et al.* (2013). MMP9 regulates the cellular response to inflammation after skeletal injury . *Bone*, *52*(1), 111–119. PMID: 23010105

Wang, Y.-M., & Shen, J.-T. (2024). Chitosan-based promising scaffolds for the construction of tailored nanosystems against osteoporosis: Current status and future prospects. *Journal of Applied Biomaterials & Functional Materials*, *22*. Advance online publication. DOI: 10.1177/22808000241266487 PMID: 39129376

Wang, Y., Yokota, T., & Someya, T. (2021). Electrospun nanofiber-based soft electronics. *NPG Asia Materials*, *13*(1), 22. DOI: 10.1038/s41427-020-00267-8

Wang, Z., Liu, B., Yin, B., Zheng, Y., Tian, Y., & Wen, P. (2022). Comprehensive review of additively manufactured biodegradable magnesium implants for repairing bone defects from biomechanical and biodegradable perspectives. *Frontiers in Chemistry*, *10*, 1066103.

Wang, Z., Yan, Y., Zhang, Z., Li, C., Mei, L., Hou, R., Liu, X., & Jiang, H. (2024). Effect of Chitosan and Its Water-Soluble Derivatives on Antioxidant Activity. *Polymers*, *16*(7). Advance online publication. DOI: 10.3390/polym16070867 PMID: 38611124

Wan, X., Beaudoin, J. J., Vinod, N., Min, Y., Makita, N., Bludau, H., Jordan, R., Wang, A., Sokolsky, M., & Kabanov, A. V. (2019). Co-delivery of paclitaxel and cisplatin in poly(2-oxazoline) polymeric micelles: Implications for drug loading, release, pharmacokinetics and outcome of ovarian and breast cancer treatments. *Biomaterials*, *192*, 1–14. DOI: 10.1016/J.BIOMATERIALS.2018.10.032 PMID: 30415101

Wasserbauer, J., Buchtík, M., Tkacz, J., Fintová, S., Minda, J., & Doskočil, L. (2020). Improvement of AZ91 alloy corrosion properties by duplex NI-P coating deposition. *Materials (Basel)*, *13*(6), 1357. DOI: 10.3390/ma13061357 PMID: 32192141

Watanabe, S., Alexander, M., Misharin, A. V., & Budinger, G. R. S. (2019). The role of macrophages in the resolution of inflammation. *The Journal of Clinical Investigation*, *129*(7), 2619–2628. DOI: 10.1172/JCI124615 PMID: 31107246

Webster, T. J. (2006). Nanomedicine: What's in a definition? *International Journal of Nanomedicine*, *1*(2), 115–116. PMID: 17722527

Wei, X., Menga, J., Ma, S., Li, Y., Qing, H., Peng, X., . . . Zhao, Q. (2022). MD simulations and experiments of plasma proteins adsorption to the biodegradable magnesium alloys to facilitate cell response. arXiv preprint arXiv:2207.11639.

Wei, C., Wang, X., Zheng, M., & Cheng, H. (2012). Calcium gradients underlying cell migration. *Current Opinion in Cell Biology*, *24*(2), 254–261. https://doi.org/https://doi.org/10.1016/j.ceb.2011.12.002 PMID: 22196933

Wei, H., Bruns, O. T., Kaul, M. G., Hansen, E. C., Barch, M., Wiśniowska, A., & Bawendi, M. G. (2017). Exceedingly small iron oxide nanoparticles as positive MRI contrast agents. *Proceedings of the National Academy of Sciences of the United States of America*, *114*(9), 2325–2330. DOI: 10.1073/pnas.1620145114 PMID: 28193901

Wei, H., Cui, J., & Lin, K.. (2022). Recent advances in smart stimuli-responsive biomaterials for bone therapeutics and regeneration . *Bone Research*, *10*(1), 17. PMID: 35197462

Weinberger, F., Breckwoldt, K., & Pecha, S.. (2016). Cardiac repair in guinea pigs with human engineered heart tissue from induced pluripotent stem cells . *Science Translational Medicine*, *8*(363), 363ra148. PMID: 27807283

Wei, T. (2015). *Systèmes innovants de délivrance de médicaments basés sur des nanomicelles pour le traitement du cancer*.

Wei, X., Chen, S., & Xie, T.. (2022). An MMP-degradable and conductive hydrogel to stabilize HIF-1α for recovering cardiac functions . *Theranostics*, *12*(1), 127–142. PMID: 34987638

Wei, X., Chen, X., Ying, M., & Lu, W. (2014). Brain tumor-targeted drug delivery strategies. *Acta Pharmaceutica Sinica. B*, *4*(3), 193–201. PMID: 26579383

Wen, G. L. (2020). Effective phototheranostics of brain tumor assisted by near-infrared-II light-responsive semiconducting polymer nanoparticles. *ACS Applied Materials & Interfaces, 12*(30), 33492–33499. PMID: 32627525

Wihadmadyatami, H., Zulfikar, M. A., Herawati, H., Pratama, D. A. O., Saragih, G. R., Kustiati, U., & Handayani, N. (2023). Chitosan hydrogel nanoparticle enhance therapeutic effect of bovine umbilical mesenchymal stem cell conditioned medium on canine cognitive dysfunction or canine Alzheimer's like mediated by inhibition of neuronal apoptotic. *Open Veterinary Journal, 13*(12), 1504. PMID: 38292722

Williams, D. F. (2008). On the mechanisms of biocompatibility. *Biomaterials, 29*(20), 2941–2953. PMID: 18440630

Witte, F., Hort, N., Vogt, C., Cohen, S., Kainer, K. U., Willumeit, R., & Feyerabend, F. (2008). Degradable biomaterials based on magnesium corrosion. *Current Opinion in Solid State and Materials Science, 12*(5–6), 63–72. DOI: 10.1016/j.cossms.2009.04.001

Wong, W. L. E., & Gupta, M. (2006). Simultaneously Improving Strength and Ductility of Magnesium using Nano-size SiC Particulates and Microwaves. *Advanced Engineering Materials, 8*(8), 735–740. DOI: 10.1002/adem.200500209

Wu, B., Su, J., Zhang, Z., Zeng, J., Fang, X., Li, W., Zhang, W., & Huang, Z. (2024). Prosthetic spacers in two-stage revision for knee periprosthetic joint infection achieve better function and similar infection control. *Bone & Joint Research, 13*(6), 306. PMID: 38889904

Wu, C., Dai, Y., Yuan, G., Su, J., & Liu, X. (2019). Immunomodulatory Effects and Induction of Apoptosis by Different Molecular Weight Chitosan Oligosaccharides in Head Kidney Macrophages From Blunt Snout Bream (Megalobrama amblycephala). *Frontiers in Immunology, 10*. Advance online publication. DOI: 10.3389/fimmu.2019.00869 PMID: 31156612

Wu, S. Y., Parasuraman, V., Arunagiri, V., Gunaseelan, S., Chou, H. Y., Anbazhagan, R., & Prasad, R. (2020). Radioprotective effect of self-assembled low molecular weight Fucoidan–Chitosan nanoparticles. *International Journal of Pharmaceutics, 579*, 119161. PMID: 32081800

Wu, X., Ding, J., & Xu, P.. (2021). A cell-free ROS-responsive hydrogel/oriented poly(lactide-co-glycolide) hybrid scaffold for reducing inflammation and restoring full-thickness cartilage defectsin vivo . *Biomedical Materials (Bristol, England), 16*(6). PMID: 34450597

Wu, X., Xin, Y., Zhang, H., Quan, L., & Ao, Q. (2024). Biopolymer-Based Nanomedicine for Cancer Therapy: Opportunities and Challenges. *International Journal of Nanomedicine, 19*, 7415. DOI: 10.2147/IJN.S460047 PMID: 39071502

Wu, Y., Wang, Y., & Long, L.. (2022). A spatiotemporal release platform based on pH/ROS stimuli-responsive hydrogel in wound repairing . *Journal of Controlled Release: Official Journal of the Controlled Release Society, 341*, 147–165. PMID: 34813880

Xia, H., Li, X., Gao, W., Fu, X., Fang, R. H., Zhang, L., & Zhang, K. (2018). Tissue repair and regeneration with endogenous stem cells. *Nature Reviews. Materials, 3*(7), 174–193. DOI: 10.1038/s41578-018-0027-6

Xiang, J. T. (2018). Near-infrared light-triggered drug release from UV-responsive diblock copolymer-coated upconversion nanoparticles with high monodispersity. *Journal of Materials Chemistry*, *6*(21), 3531–354. PMID: 32254448

Xiao, C., Fan, L., Zhou, S., Kang, X., Guan, P., Fu, R., Li, C., Ren, J., Wang, Z., Yu, P., Wang, Y., Deng, C., Zhou, L., & Ning, C. (2022). One-dimensional ferroelectric nanoarrays with wireless switchable static and dynamic electrical stimulation for selective regulating osteogenesis and antiosteosarcoma. *ACS Nano*, *16*(12), 20770–20785. DOI: 10.1021/acsnano.2c07900 PMID: 36412574

Xia, Y. H. (2021). A review of shape memory polymers and composites: Mechanisms, materials, and applications. *Advanced Materials*, *33*(6), 2000713. PMID: 32969090

Xia, Y., Wang, D., Liu, D., Su, J., Jin, Y., Wang, D., & Liu, B. (2022). Applications of chitosan and its derivatives in skin and soft tissue diseases. *Frontiers in Bioengineering and Biotechnology*, *10*, 894667. PMID: 35586556

Xie, J., Lee, S., & Chen, X. (2010). Nanoparticle-based theranostic agents. *Advanced Drug Delivery Reviews*, *62*(11), 1064–1079. PMID: 20691229

Xie, J., Wang, C., Zhao, F., Gu, Z., & Zhao, Y. (2018). Application of multifunctional nanomaterials in radioprotection of healthy tissues. *Advanced Healthcare Materials*, *7*(20), 1800421. PMID: 30019546

Xie, N. (2024). Synthesis and antibacterial effects of silver nanoparticles (AgNPs) against multi-drug resistant bacteria. [Preprint]. *Bio-Medical Materials and Engineering*, •••, 1–13. PMID: 38995765

Xin, Y., Liu, C., Zhang, X., Tang, G., Tian, X., & Chu, P. K. (2007). Corrosion behavior of biomedical AZ91 magnesium alloy in simulated body fluids. *Journal of Materials Research/Pratt's Guide to Venture Capital Sources, 22*(7), 2004–2011. https://doi.org/DOI: 10.1557/jmr.2007.0233

Xuan, L., Ju, Z., Skonieczna, M., Zhou, P. K., & Huang, R. (2023). Nanoparticles-induced potential toxicity on human health: Applications, toxicity mechanisms, and evaluation models. *MedComm*, *4*(4), e327. PMID: 37457660

Xu, D., Li, C., Li, W., Lin, B., & Lv, R. (2023). Recent advances in lanthanide-doped up-conversion probes for theranostics. *Frontiers in Chemistry*, *11*, 1036715. DOI: 10.3389/fchem.2023.1036715 PMID: 36846851

Xue, Y., Zhu, Z., & Zhang, X.. (2021). Accelerated Bone Regeneration by MOF Modified Multifunctional Membranes through Enhancement of Osteogenic and Angiogenic Performance . *Advanced Healthcare Materials*, *10*(6), e2001369. PMID: 33448103

Xu, J. J., Zhang, W. C., Guo, Y. W., Chen, X. Y., & Zhang, Y. N. (2022). Metal nanoparticles as a promising technology in targeted cancer treatment. *Drug Delivery*, *29*(1), 664–678. DOI: 10.1080/10717544.2022.2039804 PMID: 35209786

Xu, L., Zhang, H., & Wu, Y. (2014). Dendrimer advances for the central nervous system delivery of therapeutics. *ACS Chemical Neuroscience*, *5*(1), 2–13. DOI: 10.1021/cn400182z PMID: 24274162

Xu, M., Qi, Y., Liu, G., Song, Y., Jiang, X., & Du, B. (2023). Size-dependent in vivo transport of nanoparticles: Implications for delivery, targeting, and clearance. *ACS Nano*, *17*(21), 20825–20849. PMID: 37921488

Xu, X. J. (2021). Bioinspired double network hydrogels: From covalent double network hydrogels via hybrid double network hydrogels to physical double network hydrogels. *Materials Horizons*, *8*(4), 1173–1188. PMID: 34821910

Xu, Y. X., Kim, K. M., Hanna, M. A., & Nag, D. (2005). Chitosan–starch composite film: Preparation and characterization. *Industrial Crops and Products*, *21*(2), 185–192. DOI: 10.1016/j.indcrop.2004.03.002

Yadav, H., Malviya, R., & Kaushik, N. (2024). Chitosan in biomedicine: A comprehensive review of recent developments. *Carbohydrate Polymer Technologies and Applications*, •••, 100551.

Yang, J. M. (2004). Evaluation of chitosan/PVA blended hydrogel membranes. journal of Membrane Science,, 236(1-2), 39-51.

Yang, K., & Lin, X. (2015). Biocompatibility of surface-modified magnesium and magnesium alloys. In *Elsevier eBooks* (pp. 231–260). https://doi.org/DOI: 10.1016/b978-1-78242-077-4.00007-3

Yang, C., Ji, J., Lv, Y., Li, Z., & Luo, D. (2022). Application of piezoelectric material and devices in bone regeneration. *Nanomaterials (Basel, Switzerland)*, *12*(24), 4386. DOI: 10.3390/nano12244386 PMID: 36558239

Yang, C.-Y., Huang, W.-Y., Chen, L.-H., Liang, N.-W., Wang, H.-C., Lu, J., Wang, X., & Wang, T.-W. (2021). Neural tissue engineering: The influence of scaffold surface topography and extracellular matrix microenvironment. *Journal of Materials Chemistry. B*, *9*(3), 567–584. PMID: 33289776

Yang, D., Ying, J., Wang, X., Zhao, T., Yoon, S., Fang, Y., Zheng, Q., Liu, X., Yu, W., & Hua, F. (2021). Mitochondrial dynamics: A key role in neurodegeneration and a potential target for neurodegenerative disease. *Frontiers in Neuroscience*, *15*, 654785. PMID: 33912006

Yang, G., Liu, L., Xiao, L., Ke, S., Yang, H., & Lu, Q. (2024). Accelerated Scarless Wound Healing by Dynamical Regulation of Angiogenesis and Inflammation with Immobilized Asiaticoside and Magnesium Ions in Silk Nanofiber Hydrogels. *Journal of Materials Chemistry. B*. Advance online publication. DOI: 10.1039/d4tb01584c PMID: 39380345

Yang, J., Yuan, M., Xu, D., Zhao, H., Zhu, Y., Fan, M., Zhang, F., & Dong, Z. (2018). Highly dispersed ultrafine palladium nanoparticles encapsulated in a triazinyl functionalized porous organic polymer as a highly efficient catalyst for transfer hydrogenation of aldehydes. *Journal of Materials Chemistry. A, Materials for Energy and Sustainability*, *6*(37), 18242–18251. DOI: 10.1039/c8ta07502f

Yang, J., Zeng, H., Luo, Y., Chen, Y., Wang, M., Wu, C., & Hu, P. (2024). *"Recent Applications of PLGA in Drug Delivery Systems." Polymers*. Multidisciplinary Digital Publishing Institute., DOI: 10.3390/polym16182606

Yang, L., Conley, B. M., Cerqueira, S. R., Pongkulapa, T., Wang, S., Lee, J. K., & Lee, K. B. (2020). Effective Modulation of CNS Inhibitory Microenvironment using Bioinspired Hybrid-Nanoscaffold-Based Therapeutic Interventions. *Advanced Materials*, *32*(43), e2002578. DOI: 10.1002/adma.202002578 PMID: 32893402

Yang, L., & Zaharoff, D. A. (2013). Role of chitosan co-formulation in enhancing interleukin-12 delivery and antitumor activity. *Biomaterials*, *34*(15), 3828–3836. DOI: 10.1016/j.biomaterials.2013.02.031 PMID: 23453060

Yang, S., Wang, Y., & Liang, X. (2023). Piezoelectric nanomaterials activated by ultrasound in disease treatment. *Pharmaceutics*, *15*(5), 1338. DOI: 10.3390/pharmaceutics15051338 PMID: 37242580

Yang, Y., Xing, R., Liu, S., Qin, Y., Li, K., Yu, H., & Li, P. (2019a). Hydroxypropyltrimethyl ammonium chloride chitosan activates RAW 264.7 macrophages through the MAPK and JAK-STAT signaling pathways. *Carbohydrate Polymers*, *205*, 401–409. DOI: 10.1016/j.carbpol.2018.10.101 PMID: 30446121

Yang, Y., Xing, R., Liu, S., Qin, Y., Li, K., Yu, H., & Li, P. (2019b). Immunostimulatory Effects of Chitooligosaccharides on RAW 264.7 Mouse Macrophages via Regulation of the MAPK and PI3K/Akt Signaling Pathways. *Marine Drugs*, *17*(1), 36. DOI: 10.3390/md17010036 PMID: 30626153

Yang, Z. Z., Zhang, Y. Q., Wang, Z. Z., Wu, K., Lou, J. N., & Qi, X. R. (2013). Enhanced brain distribution and pharmacodynamics of rivastigmine by liposomes following intranasal administration. *International Journal of Pharmaceutics*, *452*(1-2), 344–354. DOI: 10.1016/j.ijpharm.2013.05.009 PMID: 23680731

Yang, Z., Wu, C., & Shi, H.. (2022). Advances in Barrier Membranes for Guided Bone Regeneration Techniques . *Frontiers in Bioengineering and Biotechnology*, *10*, 921576. PMID: 35814003

Yan, J. Z. (2021). Redox-responsive polyethyleneimine/tetrahedron DNA/doxorubicin nanocomplexes for deep cell/tissue penetration to overcome multidrug resistance. *Journal of Controlled Release*, *329*, 36–49. PMID: 33259850

Yan, J., Xia, D., & Zhou, W.. (2020). pH-responsive silk fibroin-based CuO/Ag micro/nano coating endows polyetheretherketone with synergistic antibacterial ability, osteogenesis, and angiogenesis . *Acta Biomaterialia*, *115*, 220–234. PMID: 32777292

Yan, T. Z. (2020). Chitosan based pH-responsive polymeric prodrug vector for enhanced tumor targeted co-delivery of doxorubicin and siRNA. *Carbohydrate Polymers*, *250*, 116781. PMID: 33049806

Yao, L., & Li, Y. (2016). The role of direct current electric field-guided stem cell migration in neural regeneration. *Stem Cell Reviews and Reports*, *12*, 365–375. DOI: 10.1007/s12015-016-9654-8 PMID: 27108005

Yao, X., Qi, C., Sun, C., Huo, F., & Jiang, X. (2023). Poly(Ethylene Glycol) Alternatives in Biomedical Applications. *Nano Today*, *48*(February), 101738. DOI: 10.1016/j.nantod.2022.101738

Yao, Y., Sangani, C. B., Duan, Y. T., Bhadja, P., & Ameta, R. K. (2021). Molecular modelling, thermal, adsorption and biological studies of conjugate Cu2+-BSA nanoparticles. *Journal of Molecular Liquids*, *331*, 115732. DOI: 10.1016/J.MOLLIQ.2021.115732

Yao, Z., Li, H., & Zhang, L. (2013). Mannosylated chitosan nanoparticles enhance DNA vaccine efficacy in fish vaccination. *Fish & Shellfish Immunology*, *34*(4), 981–988. PMID: 22960101

Yaqoob, S. B., Adnan, R., Khan, R. M. R., & Rashid, M. (2020). Gold, Silver, and Palladium Nanoparticles: A Chemical Tool for Biomedical Applications. *Frontiers in Chemistry*, *8*. Advance online publication. DOI: 10.3389/fchem.2020.00376 PMID: 32582621

Yeh, T.-C., Zhang, W., Ildstad, S. T., & Ho, C. (1993). Intracellular labeling of T-cells with superparamagnetic contrast agents. *Magnetic Resonance in Medicine*, *30*(5), 617–625. DOI: 10.1002/mrm.1910300513 PMID: 8259062

Yen, M.-T., Yang, J.-H., & Mau, J.-L. (2008). Antioxidant properties of chitosan from crab shells. *Carbohydrate Polymers*, *74*(4), 840–844. DOI: 10.1016/j.carbpol.2008.05.003

Yeste, J., Illa, X., Alvarez, M., & Villa, R. (2018). Engineering and monitoring cellular barrier models. *Journal of Biological Engineering*, *12*(1), 18. DOI: 10.1186/s13036-018-0108-5 PMID: 30214484

Yetisgin, A. A., Cetinel, S., Zuvin, M., Kosar, A., & Kutlu, O. (2020). Therapeutic Nanoparticles and Their Targeted Delivery Applications. *Molecules (Basel, Switzerland)*, *25*(9), 2193. DOI: 10.3390/molecules25092193 PMID: 32397080

Yetisgin, A. A., Cetinel, S., Zuvin, M., Kosar, A., & Kutlu, O. (2020). Therapeutic nanoparticles and their targeted delivery applications. *Molecules (Basel, Switzerland)*, *25*(9), 2193. PMID: 32397080

Yingngam, B. 2024. "Sustainable Biomaterials for Pharmaceutical and Medical Applications." In *Advances in Sustainable Biomaterials*, 8:250–74. https://doi.org/DOI: 10.1201/9781003434313-13

Yin, X., Fan, T., Zheng, N., Yang, J., Yan, L., He, S., Ai, F., & Hu, J. (2023). Palladium nanoparticle based smart hydrogels for NIR light-triggered photothermal/photodynamic therapy and drug release with wound healing capability. *Nanoscale Advances*, *5*(6), 1729–1739. DOI: 10.1039/d2na00897a PMID: 36926581

Yi, X., Wu, Y., Tan, G., Yu, P., Zhou, L., Zhou, Z., Chen, J., Wang, Z., Pang, J., & Ning, C. (2017). Palladium nanoparticles entrapped in a self-supporting nanoporous gold wire as sensitive dopamine biosensor. *Scientific Reports*, *7*(1). Advance online publication. DOI: 10.1038/s41598-017-07909-y PMID: 28801614

Yoon, H. (2023). *Polymer Nanomaterials in Biomedicine* (Vol. 24). MDPI.

Yoshida, S., Hagiwara, K., Hasebe, T., & Hotta, A. (2013). Surface Modification of Polymers by Plasma Treatments for the Enhancement of Biocompatibility and Controlled Drug Release. *Surface and Coatings Technology*. Advance online publication. DOI: 10.1016/j.surfcoat.2013.02.042

Younes, I., & Rinaudo, M. (2015). Chitin and Chitosan Preparation from Marine Sources. Structure, Properties and Applications. *Marine Drugs*, *13*(3), 1133–1174. DOI: 10.3390/md13031133 PMID: 25738328

Yousef, M., Pichyangkura, R., Soodvilai, S., Chatsudthipong, V., & Muanprasat, C. (2012). Chitosan oligosaccharide as potential therapy of inflammatory bowel disease: Therapeutic efficacy and possible mechanisms of action. *Pharmacological Research*, *66*(1), 66–79. DOI: 10.1016/j.phrs.2012.03.013 PMID: 22475725

Yuan, B., Chen, H., Zhao, R., Deng, X., Chen, G., Yang, X., Xiao, Z., Aurora, A., Iulia, B. A., Zhang, K., Zhu, X., Iulian, A. V., Hai, S., & Zhang, X. (2022). Construction of a magnesium hydroxide/graphene oxide/hydroxyapatite composite coating on Mg–Ca–Zn–Ag alloy to inhibit bacterial infection and promote bone regeneration. *Bioactive Materials*, *18*, 354–367. DOI: 10.1016/j.bioactmat.2022.02.030 PMID: 35415306

Yuan, X., Zhu, W., Yang, Z., He, N., Chen, F., Han, X., & Zhou, K. (2024). Recent Advances in 3D Printing of Smart Scaffolds for Bone Tissue Engineering and Regeneration. *Advanced Materials*, *36*(34), 2403641. DOI: 10.1002/ADMA.202403641 PMID: 38861754

Yu, C., Morshed, R., Auffinger, B., Tobias, A., & Lesniak, M. (2014). Multifunctional nanoparticles for brain tumor diagnosis and therapy. *Advanced Drug Delivery Reviews*, *66*(1), 42–57. PMID: 24060923

Yu, C., Wang, T., Diao, H., Liu, N., Zhang, Y., Jiang, H., Zhao, P., Shan, Z., Sun, Z., & Wu, T. (2022). Photothermal-triggered structural change of nanofiber scaffold integrating with graded mineralization to promote tendon–bone healing. *Advanced Fiber Materials*, *4*(4), 908–922.

Yu, J., Mu, Q., Fung, M., Xu, X., Zhu, L., & Ho, R. J. Y. (2022). Challenges and Opportunities in Metastatic Breast Cancer Treatments: Nano-Drug Combinations Delivered Preferentially to Metastatic Cells May Enhance Therapeutic Response. *Pharmacology & Therapeutics*, *236*, 108108. DOI: 10.1016/J.PHARMTHERA.2022.108108 PMID: 34999182

Yu, J., Xie, X., Wang, L., Liu, W., Xu, H., Lu, X., Li, X., Ren, J., & Li, W. (2023). Smart Chondroitin Sulfate Micelles for Effective Targeted Delivery of Doxorubicin Against Breast Cancer Metastasis. *International Journal of Nanomedicine*, *18*, 663–677. DOI: 10.2147/IJN.S398802 PMID: 36798532

Yun, U. J. (2019). Anti-cancer effect of doxorubicin is mediated by downregulation of HMG-Co A reductase via inhibition of EGFR/Src pathway. *Laboratory Investigation*, *99*(8), 1157–1172. PMID: 30700846

Yuste, I., Luciano, F. C., Anaya, B. J., Sanz-Ruiz, P., Ribed-Sánchez, A., González-Burgos, E., & Serrano, D. R. (2023). Engineering 3D-printed advanced healthcare materials for periprosthetic joint infections. *Antibiotics (Basel, Switzerland)*, *12*(8), 1229. PMID: 37627649

Yuwen, L., Sun, Y., Tan, G., Xiu, W., Zhang, Y., Weng, L., & Wang, L. (2018). MoS 2@ polydopamine-Ag nanosheets with enhanced antibacterial activity for effective treatment of Staphylococcus aureus biofilms and wound infection. *Nanoscale*, *10*(35), 16711–16720. PMID: 30156245

Zajdel, K., Janowska, J., Frontczak-Baniewicz, M., Sypecka, J., & Sikora, B. (2023). Upconverting nanoparticles as a new bio-imaging strategy—Investigating intracellular trafficking of endogenous processes in neural tissue. *International Journal of Molecular Sciences*, *24*(2), 1122. PMID: 36674638

Zaman, H. A., Sharif, S., Idris, M. H., & Kamarudin, A. (2015). Metallic biomaterials for medical implant applications: A review. *Applied Mechanics and Materials*, *735*, 19–25.

Zare, I., Mirshafiei, M., Kheilnezhad, B., Far, B. F., Hassanpour, M., Pishbin, E., Eftekhar Vaghefi, S. S., Yazdian, F., Rashedi, H., Hasan, A., Wang, X., Adeli, M., & Makvandi, P. (2024). Hydrogel-integrated graphene superstructures for tissue engineering: From periodontal to neural regeneration. *Carbon*, *223*, 118970. DOI: 10.1016/j.carbon.2024.118970

Zarei, M., Shabani Dargah, M., Hasanzadeh Azar, M., Alizadeh, R., Mahdavi, F. S., Sayedain, S. S., & Beheshtizadeh, N. (2023). Enhanced bone tissue regeneration using a 3D-printed poly (lactic acid)/ Ti6Al4V composite scaffold with plasma treatment modification. *Scientific Reports*, *13*(1), 3139. DOI: 10.1038/s41598-023-30300-z PMID: 36823295

Zargar, F., Bhat, H. A., Zargar, M. A., & Malik, S. (2024). Magnetic Nanoparticles in Cancer Thermotherapy: A Mathematical Approach to Optimal Treatment Design.

Zarrintaj, P. M. (2017). Can regenerative medicine and nanotechnology combine to heal wounds? The search for the ideal wound dressing. *Nanomedicine (London)*, *12*(19), 2403–2422. PMID: 28868968

Zarrintaj, P., Seidi, F., Azarfam, M. Y., Yazdi, M. K., Erfani, A., Barani, M., & Mozafari, M. (2023). Biopolymer-based composites for tissue engineering applications: A basis for future opportunities. *Composites. Part B, Engineering*, *258*, 110701.

Zavareh, H. S. (2020). Chitosan/carbon quantum dot/aptamer complex as a potential anticancer drug delivery system towards the release of 5-fluorouracil. *International Journal of Biological Macromolecules*, *165*, 1422–1430. PMID: 32987067

Zeng, R., Dietzel, W., Witte, F., Hort, N., & Blawert, C. (2008). Progress and challenge for magnesium alloys as biomaterials. *Advanced Engineering Materials*, *10*(8). Advance online publication. DOI: 10.1002/adem.200800035

Zeng, W., Luo, Y., Gan, D., Zhang, Y., Deng, H., & Liu, G. (2023). Advances in Doxorubicin-based nano-drug delivery system in triple negative breast cancer. *Frontiers in Bioengineering and Biotechnology*, *11*, 1271420. DOI: 10.3389/FBIOE.2023.1271420 PMID: 38047286

Zeng, X., Wang, G., Liu, Y., & Zhang, X. (2017). Graphene-based antimicrobial nanomaterials: Rational design and applications for water disinfection and microbial control. *Environmental Science. Nano*, *4*(12), 2248–2266.

Zeng, Z., Liu, Y., Wen, Q., Li, Y., Yu, J., Xu, Q., Wan, W., He, Y., Ma, C., Huang, Y., Yang, H., Jiang, O., & Li, F. (2021). Experimental study on preparation and anti-tumor efficiency of nanoparticles targeting M2 macrophages. *Drug Delivery*, *28*(1), 943–956. DOI: 10.1080/10717544.2021.1921076 PMID: 33988472

Zermatten, P., & Wettstein, M. (2012). Iliac wing fracture following graft harvesting from the anterior iliac crest: Literature review based on a case report. *Orthopaedics & Traumatology: Surgery & Research*, *98*(1), 114–117. DOI: 10.1016/J.OTSR.2011.03.026 PMID: 22130003

Zhai, M., Wu, P., Liao, Y., Wu, L., & Zhao, Y. (2024). Polymer Microspheres and Their Application in Cancer Diagnosis and Treatment. International Journal of Molecular Sciences 2024, Vol. 25, Page 6556, 25(12), 6556. https://doi.org/DOI: 10.3390/IJMS25126556

Zhang, B. C. (2020). Efficient CRISPR/Cas9 gene-chemo synergistic cancer therapy via a stimuli-responsive chitosan-based nanocomplex elicits anti-tumorigenic pathway effect. *Chemical Engineering Journal*, *393*, 124688.

Zhang, C. f., Fang, X. y., Huang, Z. d., Bai, G. c., Zhang, Z. y., Yang, Y., Zhang, Z. j., Li, W. b., & Zhang, W. m. (2022). Surgical Management for Chronic Destructive Septic Hip Arthritis: Debridement, Antibiotics, and Single-Stage Replacement is as Effective as Two-Stage Arthroplasty. *Orthopaedic Surgery*, *14*(6), 1175–1185. PMID: 35583090

Zhang, C., Lin, J., & Liu, H. (2018). Magnesium-based biodegradable materials for biomedical applications. *MRS Advances*, *3*(40), 2359–2364. DOI: 10.1557/adv.2018.488

Zhang, C., Xu, C., Gao, X., & Yao, Q. (2022). Platinum-based drugs for cancer therapy and anti-tumor strategies. *Theranostics*, *12*(5), 2115. DOI: 10.7150/thno.69424 PMID: 35265202

Zhang, C., Yang, X., Yu, L., Chen, X., Zhang, J., Zhang, S., & Wu, S. (2024). Electrospun polyaspartydrazide nanofibrous hydrogel loading with in-situ synthesized silver nanoparticles for full-thickness skin wound healing application. *Materials & Design*, *239*, 112818. DOI: 10.1016/j.matdes.2024.112818

Zhang, C., Yu, B., Lin, F., Samanta, S., Yu, H., Zhang, W., Jing, Y., Shang, C., Lin, D., & Si, K. (2023). Deep tissue super-resolution imaging with adaptive optical two-photon multifocal structured illumination microscopy. *PhotoniX*, *4*(1), 38.

Zhang, C., Zhang, Z., Lin, Y., Xin, Y., Li, H., Lin, Y., Huang, Y., Yang, Y., Bai, G., & Huang, Z. (2024). A Novel Infrapatellar Fat Pad Preservation Technique in Total Knee Arthroplasty Reduced Postoperative Pain and Wound Complications. *Orthopaedic Surgery*, *16*(8), 1946–1954. PMID: 38894546

Zhang, D., Chen, Q., Shi, C., Chen, M., Ma, K., Wan, J., & Liu, R. (2021). Dealing with the foreign-body response to implanted biomaterials: Strategies and applications of new materials. *Advanced Functional Materials*, *31*(6), 2007226.

Zhang, J., Zheng, Y., & Lee, J.. (2021). A pulsatile release platform based on photo-induced imine-crosslinking hydrogel promotes scarless wound healing . *Nature Communications*, *12*(1), 1670. PMID: 33723267

Zhang, J., Zhou, R., & Xiang, C.. (2020). Huangbai Liniment Accelerated Wound Healing by Activating Nrf2 Signaling in Diabetes . *Oxidative Medicine and Cellular Longevity*, *2020*, 4951820. PMID: 32566084

Zhang, L., Luo, J., & Zhang, Z. (2017). Chitosan as a promising agent for selective cancer cell targeting and inhibition of tumor growth. *Scientific Reports*, *7*(1), 10262. PMID: 28860532

Zhang, L., Ren, Z., Lü, J., Mo, X., Lin, J., Li, Y., Ma, W., Liu, P., Shen, Y., Zhao, Q., Qian, L., Cheng, X., Yu, Z., & Zhang, B. (2023). Nanoparticles carrying paclitaxel and anti-miR-221 for breast cancer therapy triggered by ultrasound. *Cell Death Discovery*, *9*(1), 1–11. DOI: 10.1038/S41420-023-01594-9;SUBJMETA=1347,152,54,61,631,67;KWRD=BREAST+CANCER,DRUG+DELIVERY PMID: 37582832

Zhang, N. (2023). Promoting the bench-to-bedside translation of nanomedicines. *Medical Review (Berlin, Germany)*, *3*(1), 1–3. PMID: 37724109

Zhang, Q., O'Brien, S., & Grimm, J. (2022). Biomedical applications of lanthanide nanomaterials, for imaging, sensing and therapy. *Nanotheranostics*, *6*(2), 184. DOI: 10.7150/ntno.65530 PMID: 34976593

Zhang, S. J., Lowe, L. A., Anees, P., Krishnan, Y., Fai, T. G., Szostak, J. W., & Wang, A. (2023). Passive endocytosis in model protocells. *Proceedings of the National Academy of Sciences of the United States of America*, *120*(24), e2221064120. PMID: 37276401

Zhang, W., Rong, H., Liang, J., Mao, C., Li, Z., Dai, Z., Li, D., Guo, W., Chen, S., Wang, Z., & Wei, J. (2024). Chitosan modified with PAP as a promising delivery system for melatonin in the treatment of osteoporosis: Targeting the divalent metal transporter 1. *Journal of Biological Engineering*, *18*(1), 27. DOI: 10.1186/s13036-024-00422-7 PMID: 38622739

Zhang, W., Yu, X., & Cheng, S. (2023). Mannosylated chitosan as an effective adjuvant for DNA vaccine development. *Vaccine*, *41*(6), 862–871.

Zhang, X., Guo, M., Guo, Q., Liu, N., Wang, Y., & Wu, T. (2023). Modulating axonal growth and neural stem cell migration with the use of uniaxially aligned nanofiber yarns welded with NGF-loaded microparticles. *Materials Today. Advances*, *17*, 100343.

Zhang, X., Liang, S., Gao, X., Huang, H., Lao, F., & Dai, X. (2021). Protective effect of chitosan oligosaccharide against hydrogen peroxide-mediated oxidative damage and cell apoptosis via activating Nrf2/ARE signaling pathway. *Neurotoxicity Research*, *39*, 1708–1720. PMID: 34622385

Zhang, Y., Chen, S., & Xiao, Z.. (2021). Magnetoelectric Nanoparticles Incorporated Biomimetic Matrix for Wireless Electrical Stimulation and Nerve Regeneration . *Advanced Healthcare Materials*, *10*(16), e2100695. PMID: 34176235

Zhang, Y., Huang, Y., Li, Z., Wu, H., Zou, B., & Xu, Y. (2023). Exploring Natural products as Radioprotective agents for Cancer Therapy: Mechanisms, challenges, and opportunities. *Cancers (Basel)*, *15*(14), 3585. PMID: 37509245

Zhang, Y., Liu, W., Liu, Y., Zhang, M., Tian, Y., & Chen, L. (2022). Research Progress on Corrosion Behaviors and Improvement Methods of Medical Degradable Mg– Based Alloys. *Metals*, *13*(1), 71.

Zhang, Y., Wang, Y., Wu, C., Zhan, L., Wang, A., Cheng, C., Zhao, J., Zhang, W., Chen, J., & Li, P. (2024). Drug–target interaction prediction by integrating heterogeneous information with mutual attention network. *BMC Bioinformatics*, *25*(1), 361. PMID: 39563226

Zhang, Y., Zhang, Y., Li, J., Liang, C., Shi, K., Wang, S., & Liu, X. (2024). Impact of Nanoheater Subcellular Localization on the Antitumor Immune Efficacy of Magnetic Hyperthermia. *Nano Today*, *56*, 102226.

Zhang, Y., Zheng, Z., Zhu, S., Xu, L., Zhang, Q., Gao, J., Ye, M., Shen, S., Xing, J., & Wu, M. (2025). Electroactive electrospun nanofibrous scaffolds: Innovative approaches for improved skin wound healing. *Advancement of Science*, *2416267*. Advance online publication. DOI: 10.1002/advs.202416267 PMID: 40190057

Zhang, Z., Huang, Y., Li, M., Li, W., Fang, X., Huang, Z., & Zhang, W. (2023). Rapid and quantitative detection of neutrophil gelatinase-associated lipocalin in synovial fluid using fluorescence immunochromatographic test strips for diagnosing prosthetic joint infections. *Diagnostic Microbiology and Infectious Disease*, *106*(4), 115929. PMID: 37244008

Zhang, Z., Liu, A., Fan, J., Wang, M., Dai, J., Jin, X., Deng, H., Wang, X., Liang, Y., Li, H., Zhao, Y., Wen, P., & Li, Y. (2023). A drug-loaded composite coating to improve osteogenic and antibacterial properties of Zn–1Mg porous scaffolds as biodegradable bone implants. *Bioactive Materials*, *27*, 488–504. DOI: 10.1016/j.bioactmat.2023.04.017 PMID: 37180641

Zhao, C. H., Zhang, X. P., & Zhang, L. (2022). RGD Peptide Functionalized Graphene Oxide: A Bioactive Surface for Cell-Material Interactions. *Digest Journal of Nanomaterials and Biostructures*, *17*(3), 989–997. DOI: 10.15251/DJNB.2022.173.989

Zhao, C., Tian, S., & Liu, Q.. (2020). Biodegradable nanofibrous temperature-responsive gelling microspheres for heart regeneration . *Advanced Functional Materials*, *30*(21), 2000776. PMID: 33071711

Zhao, C., Zeng, C., Ye, S., Dai, X., He, Q., Yang, B., & Zhu, H. (2020). Yes-associated protein (YAP) and transcriptional coactivator with a PDZ-binding motif (TAZ): A nexus between hypoxia and cancer. *Acta Pharmaceutica Sinica. B*, *10*(6), 947–960. DOI: 10.1016/j.apsb.2019.12.010 PMID: 32642404

Zhao, D., Han, C., Peng, B., Cheng, T., Fan, J., Yang, L., Chen, L., & Wei, Q. (2022). Corrosion fatigue behavior and anti-fatigue mechanisms of an additively manufactured biodegradable zinc-magnesium gyroid scaffold. *Acta Biomaterialia*, *153*, 614–629. DOI: 10.1016/j.actbio.2022.09.047 PMID: 36162767

Zhao, D., Yu, S., Sun, B., Gao, S., Guo, S., & Zhao, K. (2018). Biomedical Applications of Chitosan and Its Derivative Nanoparticles. *Polymers*, *10*(4), 462. DOI: 10.3390/polym10040462 PMID: 30966497

Zhao, J., Chen, L., Yu, K., Chen, C., Dai, Y., Qiao, X., Yan, Y., & Yu, Z. (2015). Effects of chitosan coating on biocompatibility of Mg–6%Zn–10%Ca3(PO4)2 implant. *Transactions of Nonferrous Metals Society of China*, *25*(3), 824–831. DOI: 10.1016/s1003-6326(15)63669-x

Zhao, K., Pu, S., Sun, L., & Zhou, D. (2023). Gentiopicroside-Loaded Chitosan Nanoparticles Inhibit TNF-α-Induced Proliferation and Inflammatory Response in HaCaT Keratinocytes and Ameliorate Imiquimod-Induced Dermatitis Lesions in Mice. *International Journal of Nanomedicine*, *18*, 3781–3800. DOI: 10.2147/IJN.S406649 PMID: 37457802

Zhao, L., Zhou, Y., Zhang, J., Liang, H., Chen, X., & Tan, H. (2023). Natural Polymer-Based Hydrogels: From Polymer to Biomedical Applications. *Pharmaceutics*. Advance online publication. DOI: 10.3390/pharmaceutics15102514 PMID: 37896274

Zhao, M. (2009). Electrical fields in wound healing—An overriding signal that directs cell migration. *Seminars in Cell & Developmental Biology*, *20*(6), 674–682. DOI: 10.1016/j.semcdb.2008.12.009 PMID: 19146969

Zhao, X., Campbell, S., El-Khoury, P. Z., Jia, Y., Wallace, G. Q., Claing, A., Bazuin, C. G., & Masson, J. F. (2021). Surface-Enhanced Raman Scattering Optophysiology Nanofibers for the Detection of Heavy Metals in Single Breast Cancer Cells. *ACS Sensors*, *6*(4), 1649–1662. DOI: 10.1021/ACSSENSORS.1C00332/SUPPL_FILE/SE1C00332_SI_001.PDF PMID: 33847111

Zhao, Y., Liang, Y., & Ding, S.. (2020). Application of conductive PPy/SF composite scaffold and electrical stimulation for neural tissue engineering . *Biomaterials*, *255*, 120164. PMID: 32554132

Zhao, Y., Zheng, J., & Wang, X. (2019). Chitosan nanoparticles for targeted anticancer drug delivery: Design and development. *Journal of Nanobiotechnology*, *17*(1), 10. PMID: 30670021

Zharkova, M. S., Golubeva, O. Y., Orlov, D. S., Vladimirova, E. V., Dmitriev, A. V., Tossi, A., & Shamova, O. V. (2021). Silver nanoparticles functionalized with antimicrobial polypeptides: Benefits and possible pitfalls of a novel anti-infective tool. *Frontiers in Microbiology*, *12*, 750556. DOI: 10.3389/fmicb.2021.750556 PMID: 34975782

Zheng, T., Huang, Y., Zhang, X., Cai, Q., Deng, X., & Yang, X. (2020). Mimicking the electrophysiological microenvironment of bone tissue using electroactive materials to promote its regeneration. *Journal of Materials Chemistry. B*, *8*(45), 10221–10256. DOI: 10.1039/d0tb01601b PMID: 33084727

Zheng, Y., Oz, Y., Gu, Y., Ahamad, N., Shariati, K., Chevalier, J., & Annabi, N. (2024). Rational design of polymeric micelles for targeted therapeutic delivery. *Nano Today*, *55*, 102147.

Zhong, G., Wang, X., Li, J., Xie, Z., Wu, Q., Chen, J., & Wang, Q. (2024). Insights into the role of copper in neurodegenerative diseases and the therapeutic potential of natural compounds. *Current Neuropharmacology*, *22*(10), 1650–1671. DOI: 10.2174/1570159X22666231103085859 PMID: 38037913

Zhong, Y., Peng, Z., Peng, Y., Li, B., Pan, Y., Ouyang, Q., Sakiyama, H., Muddassir, M., & Liu, J. (2023). Construction of Fe-doped ZIF-8/DOX nanocomposites for ferroptosis strategy in the treatment of breast cancer. *Journal of Materials Chemistry. B*, *11*(27), 6335–6345. DOI: 10.1039/D3TB00749A PMID: 37350051

Zhou, B., Lin, W., Long, Y., Yang, Y., Zhang, H., Wu, K., & Chu, Q. (2022). Notch signaling pathway: Architecture, disease, and therapeutics. *Signal Transduction and Targeted Therapy*, *7*(1), 95. DOI: 10.1038/s41392-022-00934-y PMID: 35332121

Zhou, H., Liang, B., Jiang, H., Deng, Z., & Yu, K. (2021). Magnesium-based biomaterials as emerging agents for bone repair and regeneration: From mechanism to application. *Journal of Magnesium and Alloys*, *9*(3), 779–804. DOI: 10.1016/j.jma.2021.03.004

Zhou, J., Kang, Y., Chen, L., Wang, H., Liu, J., Zeng, S., & Yu, L. (2020). The drug-resistance mechanisms of five platinum-based antitumor agents. *Frontiers in Pharmacology*, *11*, 343. DOI: 10.3389/fphar.2020.00343 PMID: 32265714

Zhou, L. D. (2010). The antitumor and antimetastatic effects of N-trimethyl chitosan-encapsulated camptothecin on ovarian cancer with minimal side effects. *Oncology Reports*, *24*(4), 941–948. PMID: 20811674

Zhou, L., Fan, L., Yi, X., Zhou, Z., Liu, C., Fu, R., Dai, C., Wang, Z., Chen, X., Yu, P., Chen, D., Tan, G., Wang, Q., & Ning, C. (2018). Soft conducting polymer hydrogels cross-linked and doped by tannic acid for spinal cord injury repair. *ACS Nano*, *12*(11), 10957–10967. DOI: 10.1021/acsnano.8b04609 PMID: 30285411

Zhou, P., Liang, X., Zhou, C., Qin, J., Hou, C., Zhu, Z., Zhang, W., Wang, S., & Zhong, D. (2019). Glutamine-β-cyclodextrin for targeted doxorubicin delivery to triple-negative breast cancer tumors via the transporter ASCT2. *Journal of Materials Chemistry. B*, *7*(35), 5363–5375. DOI: 10.1039/C9TB01225G PMID: 31403158

Zhou, Q., Li, T., & Wang, K.. (2022). Current status of xenotransplantation research and the strategies for preventing xenograft rejection . *Frontiers in Immunology*, *13*, 928173. PMID: 35967435

Zhou, Y., Hua, S., Yu, J., Dong, P., Liu, F., & Hua, D. (2015). A strategy for effective radioprotection by chitosan-based long-circulating nanocarriers. *Journal of Materials Chemistry. B*, *3*(15), 2931–2934. PMID: 32262491

Zhou, Z., Song, J., Nie, L., & Chen, X. (2016). Reactive Oxygen Species Generating Systems Meeting Challenges of Photodynamic Cancer Therapy. *Chemical Society Reviews*, *45*(23), 6597. DOI: 10.1039/C6CS00271D PMID: 27722328

Zhou, Z., Zheng, J., Meng, X., & Wang, F. (2023). Effects of electrical stimulation on articular cartilage regeneration with a focus on piezoelectric biomaterials for articular cartilage tissue repair and engineering. *International Journal of Molecular Sciences*, *24*(3), 1836. DOI: 10.3390/ijms24031836 PMID: 36768157

Zhu, J., Li, Q., Wu, Z., Xu, Y., & Jiang, R. (2024). Curcumin for Treating Breast Cancer: A Review of Molecular Mechanisms, Combinations with Anticancer Drugs, and Nanosystems. Pharmaceutics 2024, Vol. 16, Page 79, 16(1), 79. https://doi.org/DOI: 10.3390/PHARMACEUTICS16010079

Zhu, Y., Zhang, Y., & Zhou, Y. (2022). Application Progress of Modified Chitosan and Its Composite Biomaterials for Bone Tissue Engineering. In *International Journal of Molecular Sciences* (Vol. 23, Issue 12, p. 6574). MDPI. https://doi.org/DOI: 10.3390/ijms23126574

Zhu, F. D., Hu, Y. J., Yu, L., Zhou, X. G., Wu, J. M., Tang, Y., Qin, D. L., Fan, Q. Z., & Wu, A. G. (2021). Nanoparticles: A Hope for the Treatment of Inflammation in CNS. *Frontiers in Pharmacology*, *12*, 683935. DOI: 10.3389/fphar.2021.683935 PMID: 34122112

Zhu, H., Yang, K., Yao, H., Chen, X., Yan, S., He, Y., Cao, Y., Luo, J., & Wang, D. (2023). Multifunctional Nanoplatform-Mediated Chemo-Photothermal Therapy Combines Immunogenic Cell Death with Checkpoint Blockade to Combat Triple-Negative Breast Cancer and Distant Metastasis. *International Journal of Nanomedicine*, *18*, 3109. DOI: 10.2147/IJN.S408855 PMID: 37323948

Zhu, J., Ye, H., Deng, D., Li, J., & Wu, Y. (2020). Electrospun metformin-loaded polycaprolactone/chitosan nanofibrous membranes as promoting guided bone regeneration membranes: Preparation and characterization of fibers, drug release, and osteogenic activity in vitro. *Journal of Biomaterials Applications*, *34*(9), 1282–1293. DOI: 10.1177/0885328220901807 PMID: 31964207

Zhu, M., Wang, X., Zhou, Y., Li, M., Wang, Y., Wang, Y., Li, T., Song, S. Y., Zhang, B., Hezam, K., & Zhang, J. (2023). Breast tumor-targeted drug delivery via polymer nanocarriers: Endogenous and exogenous strategies. *Journal of Applied Polymer Science*, *140*(31), e54227. DOI: 10.1002/APP.54227

Zhu, S., Chen, W., Masson, A., & Li, Y.-P. (2024). Cell signaling and transcriptional regulation of osteoblast lineage commitment, differentiation, bone formation, and homeostasis. *Cell Discovery*, *10*(1), 71. DOI: 10.1038/s41421-024-00689-6 PMID: 38956429

Zhu, X. Y. (2021). Chitosan-based nanoparticle co-delivery of docetaxel and curcumin ameliorates anti-tumor chemoimmunotherapy in lung cancer. *Carbohydrate Polymers*, *268*, 118237. PMID: 34127219

Zhu, Y., Wang, A., Zhang, S., Kim, J., Xia, J., Zhang, F., Wang, D., Wang, Q., & Wang, J. (2022). Paclitaxel-loaded ginsenoside Rg3 liposomes for drug-resistant cancer therapy by dual targeting of the tumor microenvironment and cancer cells. *Journal of Advanced Research*, *49*, 159. DOI: 10.1016/J.JARE.2022.09.007 PMID: 36167294

Zhu, Z., Wang, X., Li, T., Aime, S., Sadler, P. J., & Guo, Z. (2014). Platinum(II)–Gadolinium(III) Complexes as Potential Single-Molecular Theranostic Agents for Cancer Treatment. *Angewandte Chemie*, *126*(48), 13441–13444. DOI: 10.1002/ANGE.201407406 PMID: 25255766

Zielińska, A., Costa, B., Ferreira, M. V., Miguéis, D., Louros, J. M. S., Durazzo, A., Lucarini, M., Eder, P., Chaud, M. V., Morsink, M., Willemen, N., Severino, P., Santini, A., & Souto, E. B. (2020). Nanotoxicology and Nanosafety: Safety-By-Design and Testing at a Glance. *International Journal of Environmental Research and Public Health*, *17*(13). Advance online publication. DOI: 10.3390/ijerph17134657 PMID: 32605255

Živić, F., Grujović, N., Manivasagam, G., Richard, C., Landoulsi, J., & Petrović, V. (2014). The potential of magnesium alloys as bioabsorbable/ biodegradable implants for biomedical applications. *DOAJ (DOAJ: Directory of Open Access Journals)*. https://doaj.org/article/9afee1a42b824047a615cbf8826996a4

Zoe, L. H., David, S. R., & Rajabalaya, R. (2023). Chitosan nanoparticle toxicity: A comprehensive literature review of in vivo and in vitro assessments for medical applications. *Toxicology Reports*, *11*, 83–106. DOI: 10.1016/J.TOXREP.2023.06.012 PMID: 38187113

Zou, Y., Goei, R., Ong, S.-A., Ong, A. J., Huang, J., & Tok, A. I. Y. (2022). Development of Core-Shell Rh@ Pt and Rh@ Ir Nanoparticle Thin Film Using Atomic Layer Deposition for HER Electrocatalysis Applications. *Processes (Basel, Switzerland)*, *10*(5), 1008.

Zubair, M., Riaz, M., Kiani, M. N., Aslam, H. M., Fatima, A. E. Z., Sultan, H. M., & Rehman, A. U. (2024). Application of Nanotechnology for Targeted Drug Delivery and Nontoxicity. *International Journal of General Practice Nursing*, *2*(2), 57–67.

Zupan, J., Komadina, R., & Marc, J. (2012). The relationship between osteoclastogenic and anti-osteoclastogenic pro-inflammatory cytokines differs in human osteoporotic and osteoarthritic bone tissues. *Journal of Biomedical Science*, *19*(1), 28. DOI: 10.1186/1423-0127-19-28 PMID: 22380539

About the Contributors

Deniz Eren Erişen, originally from Türkiye, is an accomplished researcher in Material Processing Engineering based in China. He earned his PhD in Materials Science and Engineering from the University of Science and Technology of China (USTC) (QS World University Rankings 2024: #63 globally, #3 in China) between 2017 and 2022. Before that, he completed an M.Sc. in Material Science at Nanjing University of Aeronautics and Astronautics (NUAA) (QS 2024: #454 worldwide, #37 in China; SCImago Top 250 in Higher Education, 2023) from 2013 to 2016, and a B.Sc. in Material Science and Engineering at Anadolu University (now Eskisehir Technical University) from 2008 to 2013. His doctoral research was conducted at the Institute of Metal Research (IMR), Chinese Academy of Sciences (ranked #1 globally in the Government sector by SCImago 2023). After completing his doctorate, he continued his postdoctoral research at NUAA's College of Materials Science and Technology and is currently engaged in advanced postdoctoral research at the Nanoencapsulation Laboratory, Institute of Semiconductors, Guangdong Academy of Sciences (GDAS). Dr. Erişen's research focuses on biodegradable metallic biomedical materials, biomaterials, nanoparticles, bioactive coatings for medical devices, and nanomaterial characterization. His notable contributions include studies on the biosafety of AZ31B magnesium alloy stents and chitosan-polydopamine bioactive coatings for implants. He is also the co-editor of the book *Nanomaterials and the Nervous System* (2024), which introduces protocols for neurotoxicity evaluation of industrial nanoparticles and frameworks for neurological safety. His current work combines semiconductor processing techniques—such as vacuum and plasma-based methods—with chemical and biochemical synthesis strategies to advance innovative research in material science and biomedical engineering. He develops functionalized systems by integrating nanomaterials, biomaterials, nanoparticles, and organic materials, incorporating polysaccharides, polymers, and hydrogels, enhanced with noble metals in nanoparticle or ionic form within organic matrices. These combined approaches enable sustainable solutions for medicine and biomedical technologies, as well as related industries like nutrition, agriculture, and livestock production.

Faez Iqbal Khan is currently serving as an Assistant Professor in the Department of Biological Sciences at Xi'an Jiaotong-Liverpool University. He holds a Ph.D. in Computational Chemistry (Bioinformatics) from Durban University of Technology, South Africa, and obtained his Bachelor's and Master's degrees in Biomedical Science and Bioinformatics, respectively. He also earned a Postgraduate Certificate in Teaching and Supporting Learning in Higher Education with merit from the University of Liverpool and holds a Fellowship (FHEA) from Advance HE, a professional recognition scheme for those who teach or support learning in higher education. Throughout his career, Dr. Khan has conducted research and teaching at esteemed institutions such as Rhodes University (South Africa), South China University of Technology, and the University of Electronic Science and Technology of China. His main research areas include protein engineering, protein folding, drug design, and protein dynamics. Dr. Khan has established wide-ranging collaborations with BRICS countries and has mentored several undergraduate and postgraduate students. He has authored over 90 publications in international peer-reviewed journals, which are well cited, and in recognition of his scientific contributions, he was ranked among the top 2% of scientists globally in the single-year rankings for 2022, according to a study conducted by Stanford University. He has received and completed several research grants, including those from the China Postdoctoral Science Foundation, Sichuan Province International Science and Technology Innovation Cooperation, the Hong Kong, Macao and Taiwan Science and Technology Program, and the Research Development Fund (RDF) at XJTLU. He currently holds adjunct faculty appointments at Taylor's University Malaysia and Datta Meghe Institute of Higher Education and Research, as well as an honorary appointment from the University of Liverpool, and is a member of several Suzhou-based key laboratories, including Suzhou Municipal Key Lab for Metabolic Syndrome Drug Research, Suzhou Municipal Key Lab of Biomedical Sciences & Translational Immunology, and AI4Health Suzhou Key Laboratory.

Chang Cheng, currently affiliated with the Guangdong Academy of Sciences (GDAS), Institute of New Materials Research, has been selected for the China Association for Science and Technology's Youth Talent Support Program. He serves on the Medical Metal Materials Committee and Additive Manufacturing Committee of the Nonferrous Metals Society of China. He also holds the positions of Secretary-General and Director of the Additive Manufacturing (3D Printing) Branch of the Guangdong Mechanical Engineering Society. Furthermore, he is a young editorial board member for several SCI journals, including Advanced Powder Materials, Journal of Magnesium and Alloys, and Rare Metals. His publication record includes over 35 high-impact SCI papers, with 20 as the first or corresponding author, amassing more than 1,300 citations. Additionally, he has authored one book and three book chapters, holds two U.S. patents and fifteen Chinese invention patents, and has issued seven group standards. Dr. Chang specializes in laser additive manufactured metal biomedical materials, with a particular focus on magnesium alloys for bioactive materials. His main research interests include the design and post-processing of high-performance alloys for laser additive manufacturing. He has published more than 30 high-level SCI papers, including 19 as the first author or corresponding author, with over 900 citations (according to Google Scholar). He has also published one chapter with Elsevier, applied for two U.S. patents (one authorized), and has 11 authorized Chinese invention patents and two utility models. Dr. Chang has led the development of one national standard, one industry standard, and two group standards in additive manufacturing. As a project leader, he has undertaken a national key research and development sub-project, a Guangdong Province surface project, and a Guangzhou project. He has secured nearly 3 million yuan in funding for additive manufacturing-related enterprises and has participated as a key researcher in nine national key research and development projects. Dr. Chang has received the second prize of Science and Technology from the China Nonferrous Metal Industry and the China Plastic and Aesthetic Association. He also serves as a reviewer for various SCI journals, including J. Mater. Sci. Technol., Addit. Manuf., and Bioact. Mater., with an annual review volume of more than 10. With a strong background in materials science, Dr. Chang is dedicated to advancing research and innovation in the field of new materials. His contributions are highly valued within the scientific community.

Kaicheng Yan is at Nanjing Agricultural University, specializing in fluorescence biosensing and diagnostic systems for pathogenic microorganisms. He is certified as a Senior Public Nutritionist and Psychological Counselor in China. He obtained his Master's degree from East China University of Science and Technology, where he conducted training under distinguished research scientists at the

Feringa Nobel Prize Scientists Joint Research Center. Subsequently, he completed his Ph.D. in Chemistry at the University of Bath, United Kingdom, collaborating with globally renowned specialists in the field of molecular probe development. Dr. Yan has established extensive collaborations with research teams across China and internationally, contributing to interdisciplinary progress in biosensor technology and fluorescence diagnostics.

***

Ilham Aftais holds a PhD in Biochemistry and Nutrition with extensive experience in research, teaching, and publishing. She has taught biochemistry and nutrition, fostering critical thinking and practical skills. Her research on histamine in sardine canning addressed key food safety challenges, resulting in several publications on food safety and nutritional biochemistry.

Navya Aggarwal is a Ph.D. scholar in biotechnology. She is currently specializing in breast cancer research. Her interests lie in drug discovery, hormonal contraception and theragnostics. She has a passion for literature and is an avid reader-writer.

Youssef Ait Hamdan, PhD in Biology, Biochemistry, and Health. His research focuses on exploring the biological and therapeutic effects of chitosans derived from natural resources. He is an associate member of the Center for Regenerative Medicine and Clinic at the University Hospital (CHU), where he contributes to advancing knowledge in the biomedical applications of chitosan.

Adil Aknouch is a Professor of Medical and Nuclear Physics

Imade Armadi is currently pursuing a PhD in biology, biotechnology and biomedicine at the Center for Regenerative Medicine, with research also carried out at the Faculty of Medicine and Pharmacy in Marrakech. His research focuses on the biomedical applications of chitosan, particularly in the context of clinical therapy. His work aims to explore the potential of chitosan in drug delivery systems, tissue engineering and other therapeutic applications, contributing to advances in regenerative medicine and clinical treatments.

Priscilla Albuquerque has a degree in Biomedicine, a master's degree in Biological Sciences/UFPE (2013), a PhD in Biology applied to health from the iLIKA/UFPE (2017), and a specialization in Aesthetic Biomedicine from Nepuga (2018). Between September 2015 and September 2016, she completed a sandwich doctorate (PDSE) and co-tutorship at the University of Minho (Braga/Portugal), with an exchange program at the University of Reading (Reading/England). She is currently an adjunct professor at the Institute of Biological Sciences (ICB) of the University of Pernambuco (UPE), where she teaches biochemistry for the biological sciences, nursing, medicine, and dentistry courses. She is also a researcher at the Laboratory of Bioprospection and Applied Ethnopharmatoxicology (LABEA) of the UPE campus Garanhuns and leader of the line research on the extraction, characterization, and biotechnological application of polysaccharides in the health and food sectors on the same campus, working on the following topics: extraction and physical-chemical characterization of polysaccharides with emphasis on galactomannans; evaluation of the biological activities of polysaccharides; polysaccharide rheology; immobilization of natural products in polysaccharide matrices; and treatment of aesthetic dysfunctions with natural products. He also coordinates the extension project "Belezas do Agreste", whose main objective is to promote the integrality of actions aimed at health and well-being through the digital dissemination of quality content (@belezasdoagreste.upe) and the production of natural soaps and moisturizers for the skin care routine.

Anass Belbachir is a Professor of Pathological Anatomy at the Faculty of Medicine and Pharmacy of Marrakech and the Director of the Center for Regenerative Medicine. His research focuses on cancer, the use of biomaterials, and stem cells for clinical applications, aiming to advance therapeutic strategies in regenerative medicine and cancer treatment.

Hari Shankar Biswas is Assistant Professor in Chemistry Surendranath College, Kolkata. Dr. Biswas is an esteemed Assistant Professor in Chemistry at Surendranath College, Kolkata, with a multifaceted academic background and a strong dedication to research and education. He embarked on his academic journey by obtaining his B.Sc. degree in Chemistry Honours from R.K.M.V.C. College in 2005, followed by his M.Sc. degree from Presidency College, Kolkata, in 2008. Dr. Biswas further pursued his scholarly pursuits by earning a Ph.D. in Material Science from the Saha Institute of Nuclear Physics, University of Calcutta, in 2015. His professional experience spans various roles, including a stint as a School teacher at Keshub Academy, Kolkata, from 2008 to 2010, before transitioning into academia as an Assistant Professor at Surendranath College. Dr. Biswas specializes in the research domain of Carbon thin Film, Graphene, and Graphene Oxide-based nano Composites, focusing on their synthesis, characterization, and application across Environmental, Biological, and Technological fields. Dr. Biswas's scholarly contributions are significant, with authorship of 12 book chapters and over 26 papers published in esteemed International Journals. He is actively involved in academic societies, holding life membership in the Indian Science Congress Association and the Indian Chemical Society, along with an affiliated membership with the Royal Society of Chemistry. In addition to his research and academic endeavors, Dr. Biswas plays a vital role in the dissemination of scientific knowledge as a series book Editor and reviewer for reputable international publishers and journals. He serves as an Editorial board member for the American Journal of Chemical Engineering, Reseapro Journal, Lincoln University College, etc., and is affiliated with Rotary International. Dr. Biswas's editorial contributions extend to editing various books, including titles such as "Emerging Concepts in Chemical and Biological Science" and "Modern Approaches in Chemical and Biological Sciences," among others. Through his comprehensive involvement in research, education, and editorial responsibilities, Dr. Hari Shankar Biswas continues to make valuable contributions to the field of Chemistry, fostering advancements and innovations for the betterment of society.

Shib Shankar Biswas is an Assistant Professor in the Department of Physics at Surendranath College, Kolkata, India. He is a dedicated academician and researcher, known for his contributions to experimental and applied physics. His research interests span across nanomaterials, condensed matter physics, and advanced materials characterization techniques. At Surendranath College, he is actively involved in both undergraduate teaching and academic mentorship, inspiring students through a blend of rigorous teaching methods and innovative research guidance. He has also collaborated on interdisciplinary projects and has contributed to several scientific publications and academic endeavors.

Maryam Bousaid is a PhD in Student Education and studying the mechanical properties of biodegradable polymers.

Kashvi Chaturvedi, is studying B.Tech, specialisation in Artificial Intelligence and Data Science at Ajeenkya D Y Patil University. She started solidifying her programming skills by completing various certification courses like Data Structures in C, Tableau functions and Operating Systems along with doing few projects using languages like Python, C, C++ and creating Web Page using HTML, CSS and Java Script. She had also attended an Arduino-Based Automation course at Dolphin Labs to enhance her software knowledge. There, she learned how to work with sensors, microcontrollers, and circuit design. She has also written multiple review papers on topics such as Geographic Information Systems, Comparative Analysis of BJT Configurations, Swiss Bank vs. European Central Bank: Data Analysis. She is also researching on and learning many topics such as Small Modular Reactors, Bibliometric Analysis, UNSDGs etc. Because of her interest in latest technologies, she also participated in Global Hyperloop Competition hosted by IIT Madras representing her college as team Vac-Vectoor where she learnt more about various subtopics from the mentors.

Susanta Das is a seasoned educator with a proven record of teaching and mentoring diverse students, working at multiple institutions, and providing service to the various initiatives of organizations. He received his Ph.D. and M.A. degrees from Western Michigan University (WMU), USA, and M.Sc. degree from Banaras Hindu University (BHU), India all in Physics. He continued his research as a Marie-Curie post-doctoral fellow at Stockholm University, Stockholm, Sweden on beam diagnostics for the DESIREE (Double ElectroStatic Ion Ring ExpEriment) in the project DITANET (Diagnostic Techniques for particle Accelerators – a Marie-Curie initial training NETwork), at the Indian Institute of Science Education and Research-Kolkata, India on high-pressure physics as a Project Scientist-B, and at the University of Electro-Communications, Tokyo, Japan on ion-surface interactions as a Post-doctoral fellow. Throughout his career, Dr. Das worked and collaborated with many researchers, Ph.D., master, and visiting students from different countries. He further visited many countries to discuss research and an international conference participant (UK, Italy, Germany, Greece, Belgium, Bulgaria, Romania, Brazil etc.). He co-authored several research articles in WoS/Scopus indexed international journals, conference proceedings, and scholarly book chapters published by renowned international publishing houses. Beyond his research endeavors, Dr. Das has a storied history of service to academic and administrative committees, exhibiting his commitment to the growth and development of educational institutions. His experience includes tenure at Central University South Bihar, Sri Sri University, and P.K. University before assuming his current role at Ajeenkya D Y Patil University. He received the Marie-Curie post-doctoral fellowship in Sweden, Science Academies' Summer Research Fellowship in India, Gwen Frostic Doctoral Fellowship, Department Graduate Research and Creative Scholar Award by WMU, and Leo R. Parpart Doctoral Fellowship by Dept. of Physics, WMU, among many others, throughout his academic journey. At present, Dr. Das continues to delve into cutting-edge fields, with a keen interest in nanotechnology, UN SDGs, and data science. His multifaceted contributions, spanning teaching, research, and administrative leadership, showcase a dedicated professional who is instrumental in advancing the vision and mission of the institutions he serves.

Nimish Dhakad is a Master's student in Forensic Science at Amity University, Noida, India. He is pursuing advanced studies in forensic analysis, crime scene investigation, and forensic toxicology. His academic interests focus on the application of scientific principles to criminal justice, with a particular emphasis on forensic biology and chemistry. Nimish is committed to expanding his knowledge in forensic research methodologies and techniques, aiming to contribute to the field's development through practical applications and research initiatives.

Amina El Kaoui ; Master's student at ENCG of Meknes. His field of interest focuses on the socio-economic of natural ressources.

Mohammed El Mzibri is the head of the Life Science Division in the National Centre of Energy, Nuclear Sciences and Techniques (CNESTEN). He has obtained his PhD at the Mediterranée University, Marseilles – France in 1996 and integrated the CNESTEN in 1999 as a researcher to develop medical research program. He has conducted many research projects on oncology, communicable diseases, pharmacology and valorization of medicinal plants. Dr. EL MZIBRI has the opportunity to coordinate many national and international research projects and to participate in many others, leading to the publication of more than 200 international publications, one book and 5 books' chapters. He was the supervisor of many students for the preparation of their PhD and Master Degrees

Fatima Ezzahra Elamrani is currently a PhD student in Medical Biology and Tissue Engineering at Cadi Ayyad University. Her doctoral research focuses on the cell culture of mesenchymal stem cells and their therapeutic and clinical implications, aiming to explore their potential in regenerative medicine and clinical applications.

Samia Elouali is currently a PhD student in Polymer Chemistry at the University of Mons and Cadi Ayyad University. Her research focuses on the extraction and valorization of chitosan-based polymers and their derivatives for applications in metal pollution removal and wastewater treatment, aiming to develop sustainable solutions for environmental remediation.

Xinyu Fang is an orthopedic surgeon whose research focuses on the diagnosis and treatment of osteoarticular diseases. His work aims to elucidate the pathogenesis and improve the diagnostic approaches for bone and joint infections, particularly periprosthetic joint infection, as well as to develop novel therapies for peripheral nerve injury. In addition, he investigates the role of neuromodulation in managing osteoarticular disorders to develop more effective and integrated treatment strategies.

Abdelghani Iddar has been a researcher at the National Center for Nuclear Energy, Science and Technology in Rabat, Morocco. He serves as the research director and head of the Biotechnology and Biomolecule Engineering Unit. Dr. Iddar obtained a Bachelor of Science in Environmental Protection and Management (1996) and a Doctor of Philosophy in Biochemistry and Molecular Biology (2002) from Hassan II University, Casablanca, Morocco. From 1999 to 2000, Mr. Iddar served as a visiting student researcher at the Institute of Plant Biochemistry and Photosynthesis in Seville, Spain, with the support of AECI (Spain). Dr. Iddar is a member of numerous professional societies, has published his works in books and peer-reviewed journals, and has presented his research at conferences around the world. Additionally, he serves as a doctoral advisor and reviewer of scientific journals.

Sachil Kumar is an Associate Professor at Amity Institute of Forensic Sciences, Amity University, India, with a Ph.D. in Forensic Chemistry and Medicine from King George's Medical University, Lucknow. His research focuses on postmortem interval (PMI) estimation using cardiac Troponin-T degradation. Dr. Kumar has extensive teaching experience at various universities, including in India and Saudi Arabia, and has supervised numerous postgraduate dissertations. He has served as Editorial Secretary for the Arab Journal of Forensic Sciences & Forensic Medicine and is actively involved in peer review and editorial roles for international forensic journals.

Kumar Kundu is Assistant Professor & HOD Department of Chemistry Sripat Singh College, Jiaganj, Murshidabad. Dr. Amit Kumar Kundu is an Assistant professor and HOD at the Department of Chemistry, Sripat Singh College, Jiaganj, Murshidabad. Dr. Kundu did B.Sc. from Krishnagar Government College, University of Kalyani in 2007 and completed MSc. Degree from Department of Chemistry, University of Kalyani in 2009. After that he received his PhD in Analytical Chemistry in 2017 from University of Kalyani. His research mainly focuses on Environmental Analytical Chemistry, carbon nano materials, Graphene, Graphene Oxide-basednano composites Synthesis, Characterization and its application. He has huge experience in Groundwater and surface water onsite sampling, sediment sampling from vadose zone and aquifer. He is also proficient in operating and managing various advanced analytical instruments. To date, he has published near about eighteen peer-reviewed journal papers and he is the author of five book chapters. He received World Environment and Livelihood Award form 'International Foundation for Environment and Ecology'.

Loqman Souad is a Professor of Biology at the Faculty of Medicine and Pharmacy, Cadi Ayyad University. His research focuses on microbiology and epidemiology, with an emphasis on understanding microbial behavior and the epidemiological dynamics of infectious diseases.

Saloua Mabsor-Zgandaoui, PhD researcher in the Department of Biology at the Faculty of Sciences, Ibn Tofail University, Kenitra. Her research focuses on the fields of Radiobiology, Biotechnology, Molecular Biology, and Health. As part of her doctoral training, she conducted her research since march 2022 at the National Centre for Nuclear Energy, Sciences, and Technology (CNESTEN), where she explored advanced applications in Radiobiology.

Dilip Kumar Maiti is an eminent chemist with a distinguished academic and professional career. He completed his B.Sc. in Chemistry in 1991 from Ramakrishna Mission Vidyamandir, University of Calcutta, followed by an M.Sc. in 1993 from the University College of Science, Calcutta University. In 1998, he earned his Ph.D. from the Indian Institute of Chemical Biology, Jadavpur University. Dr. Maiti pursued postdoctoral research at Jadavpur University (2000–2002) and later at Wayne State University, Detroit, USA (2002–2004). He began his career as a Scientist in the R&D departments of RPG Life Sciences and Ciba India. Transitioning to academia, he served as a Senior Lecturer at Dumkal College, Kalyani University, before joining the University of Calcutta as a Professor in 2005. Currently, he is the Vice-Chancellor of Biswa Bangla Biswabidyalay, Bolpur. Dr. Maiti's research interests include Organic Synthesis and Material Chemistry. He has been honored with several prestigious awards, including the FRSC (London), UGC Mid-Career Award (2017), and Basudev Banerjee Memorial Award. He is actively involved with numerous scientific societies and editorial boards, contributing significantly to the global scientific community.

Maryem Mekouar is currently a PhD student and holds a Master's degree in Biology and Oral Environment Materials from the Faculty of Medicine, Rabat. Her research focuses on biomaterials and biological sciences, with an emphasis on their applications in oral health and other biomedical fields.

Aicha Moufakkir is a Master's student at Chouaib Doukkali University, El Jadida. His research interests focus on the pathological implications of neurological diseases, such as hepatic encephalopathy, and the therapeutic potential of pharmacological substances derived from natural resources, such as chitosan.

Adnane Moutaouakkil is a researcher at the National Centre for Nuclear Energy, Science, and Technology (CNESTEN, Morocco), within the Biotechnology and Biomolecule Engineering Unit. He holds a PhD in Microbiology and Biotechnology from Hassan II University of Casablanca. His expertise covers enzymology, immunoassays, radiobiology, and microbial interactions, with a particular focus on the effects of ionizing radiation on biomolecules. Dr. Moutaouakkil has developed significant contributions in bioprocess engineering and the development of diagnostic kits. His research integrates nuclear science with biology and environmental sciences, aiming to advance knowledge and provide innovative solutions for health, environmental sustainability, and biotechnology. He is the author and co-author of numerous international publications and is actively involved in mentoring students and promoting the valorization of scientific research.

Ajinkya Girish Nene is a Senior Research Scholar at SUSTech, China, specializing in nanomaterials, mechanobiology, and bio-nano interactions. He earned his Ph.D. in Applied Chemistry from Chubu University, Japan (2017) under the MEXT Fellowship. His work combines nanomaterials synthesis, AFM-based mechanobiology, and advanced microscopy to study immune cell responses to nanoparticles and ionic liquids, including elasticity, adhesion, and autophagy. He also developed agarose–gelatin nanocomposites for cell biomechanics. With experience in both academia and industry (Nanocrystals Technology LP, USA), his research spans biosensing, drug/gene delivery, and protein aggregation, with applications in nanomedicine and regenerative technologies. His publications have 700+ citations (h-index 11, i10-index 12).

Khawla Rachmoune is a PhD student in Biochemistry, Biotechnology, and Radiobiology. Her doctoral research focuses on the radioprotective properties of biopolymers, with the objective of assessing how their composition and structural characteristics influence their performance. Specifically, her work aims to explore the potential of biopolymer-based composite materials as innovative biomedical tools to enhance the therapeutic efficacy of radiation therapy while minimizing its adverse side effects.

Mohammed Rhazi his research has focused on the valorization of terrestrial and marine fishing wastes by the production of chitinous derivatives and their applications in various fields (pharmacology, biomedical, agroalimentary and water treatment). He has been working since 1996 on the complexation of metal ions by chitosan and its derivatives and their interest in wastewater treatment.

Shipra Rohatgi is an Assistant Professor at the Amity Institute of Forensic Sciences, Amity University Noida. With a strong academic background in forensic biology, serology, and cyber forensics, Dr. Rohatgi is dedicated to advancing the field of forensic sciences through both teaching and research. She is particularly interested in the intersection of forensic science and technology, with a focus on innovative methodologies for criminal investigations. Dr. Rohatgi's research interests span various domains of forensic science, including DNA profiling, questioned document examination, and digital forensics. She has co-authored several impactful publications in these areas, contributing to both the theoretical and practical aspects of forensic investigations. Her work on topics such as crime scene management, tampered document analysis, and the use of digital forensics in combating cybercrimes has earned her recognition in the academic community. In addition to her academic pursuits, Dr. Rohatgi has been involved in several forensic projects and collaborations, contributing her expertise to the development of advanced forensic techniques and tools. Her work continues to shape the future of forensic science, particularly in the areas of DNA analysis, and cyber forensics. Dr. Rohatgi is passionate about mentoring students and researchers in the field, and her commitment to the growth of forensic science education is reflected in her active participation in academic conferences, workshops, and seminars.

Sorour Sadeghzade is currently a postdoctoral researcher at the Southern University of Science and Technology (SUSTech), and she was a researcher at Isfahan university of technology, Iran and Pennsylvania State University, USA, before she joined SUSTech. She received her PhD from Isfahan university of technology in 2021 and worked as a postdoctoral researcher at that university, too. She received several scholarship and awards for her research in bioinspired structure and biomaterial. Most of her research experience is in the fields of fracture mechanics, bioinspired structure, biomaterials, additive manufacturing and material characterization under various forming procedures. Energy harvester based on hydrogel, and organ on a chip, OCA and material characterization are her current research interest.

Sunil Sankathala is pursuing B. Tech Computer Science and Engineering at Ajeenkya D Y Patil University, with a diverse range of interests that bridge technology, sustainability, and innovation. He academic focus lies in computer science fundamentals like programming, data structures, algorithms, and software development. He actively supplements this with hands-on experience in web development and robotics. Beyond the core curriculum, He has a strong research inclination in emerging and impactful fields. He current research interests include Artificial Intelligence in Healthcare, bioplastics, nano bioplastics, e-waste management, and solid waste management—areas where he believes technology can make a meaningful difference in solving real-world problems. He has been publishing research articles and book chapters in reputed national and international journals and books. He is also part of an ambitious Hyperloop project, which involves interdisciplinary collaboration and cutting-edge engineering aimed at revolutionizing transportation. Working on this project has enhanced his skills in teamwork, innovation, and systems thinking. In the tech domain, he has developed proficiency in web development using technologies such as HTML, CSS, JavaScript, and React. He interests in robotics further reflects building smart, responsive systems that interact with the physical world. Throughout his course, he has helped strengthen problem-solving and collaborative skills. He proficient in languages such as Python, C, C++, MySQL, Ubuntu, Machine Learning and familiar with frameworks and tools like React, Git. He actively participates in hackathons, coding competitions, and tech events to challenge himself and learn from peers. He is also a quick learner and highly adaptable, with an eagerness to stay updated with the latest trends in the tech world. He aspires to become a skilled software engineer and contribute to building innovative solutions that have a positive impact on society. The long-term goal is to work in a dynamic environment where knowledge, grow professionally, and make meaningful contributions to the field of technology. Driven by a spirit of exploration and sustainability, he aims to contribute to technological solutions that address both human and environmental needs. The long-term goal is to become a versatile engineer and researcher, working at the intersection of computer science, environmental innovation, and human well-being.

Aditya Shrivastav is pursuing a B.Tech in Computer Engineering at the School of Engineering, Ajeenkya D Y Patil University. His research interests include the biosynthesis of nanoparticles, the application of ionic liquids, and the integration of Artificial Intelligence across various domains. Aditya is particularly passionate about leveraging these technologies in synergy to address real-world challenges. He aspires to advance research in these fields, contributing to innovative solutions and practical problem-solving. With a keen interest in interdisciplinary approaches, Aditya is committed to pushing the boundaries of science and technology for meaningful and impactful advancements.

Prakash Somani is Founder and Director Applied Science Innovations Pvt. Ltd., India Research Fields Green Chemistry Advanced Materials Nanotechnology Qualifications 2002, Ph. D., National Chemical Laboratory and Pune University, Pune, India 1996, M. Sc., Department of Physics, Pune University, Pune, India 1994, B. Sc., Agasti Arts, Commerce and D. R. Science College, Pune University, Pune, India

Amit Tiwari is currently working as an Assistant Professor in the Department of Mechanical Engineering at Suresh Gyan Vihar University, Jaipur Rajasthan India. He received his PhD degree from Suresh Gyan Vihar University, Jaipur, Rajasthan, India Accredited with "A+" Grade by NACC. He has 5 Indian Design Patent, 1 Australian Patent and 3 UK Design patent. He has serving as the Editor Board Member of various International Journal. He has published 50 research papers in reputed International and National Journals. He has also associated with a life member of Technical societies like The Society of Mechanical and Automotive Engineers [SMAE], LM NO – 02112, Research Scholar Hub [RSH], India, Life Membership No. - RSH/IN/RJ/LM-09/2017, International Association of Engineers [IAENG], Hongkong, Membership No.-193169, Society for Environmental Resources and Biotechnology Development, Annual Membership, Membership No. AN: 119321and Scholars Academics & Scientific Society, Membership No: SAS/RMSASS/352. Dr. Tiwari research area includes Metal Casting, Heat Treatment, Composite Material, and Optimization Techniques.

Yang Wenjie study the impact of nano/microplastic to immunity and immune cells, which provide a new vision on this area. He is also focusing on the nanotoxicity, innate memory and tumour immunity. During his research, he explored the biomechanical proprieties of monocytes and macrophages, using innate memory mechanism can help us to understand and design innate memory-reprogramming immunotherapies.

Wenming Zhang is a seasoned orthopedic surgeon with extensive experience in complex joint disorders. His clinical practice is deeply involved in primary and revision total joint arthroplasty, minimally invasive surgery, and the surgical management of bone and joint infections. Complementing his surgical expertise, Dr. Zhang also pursues clinical research in nerve injury and regeneration, striving to integrate innovative regenerative therapies into orthopedic practice to enhance patient recovery.

Index

C

D

E

F

H

I

M

N

O

P

R

S

T

www.ingramcontent.com/pod-product-compliance
Lightning Source LLC
LaVergne TN
LVHW080530300926
849162LV00007B/1579